Y0-BSD-919

ACCELERATION AND TRANSPORT OF ENERGETIC PARTICLES OBSERVED IN THE HELIOSPHERE

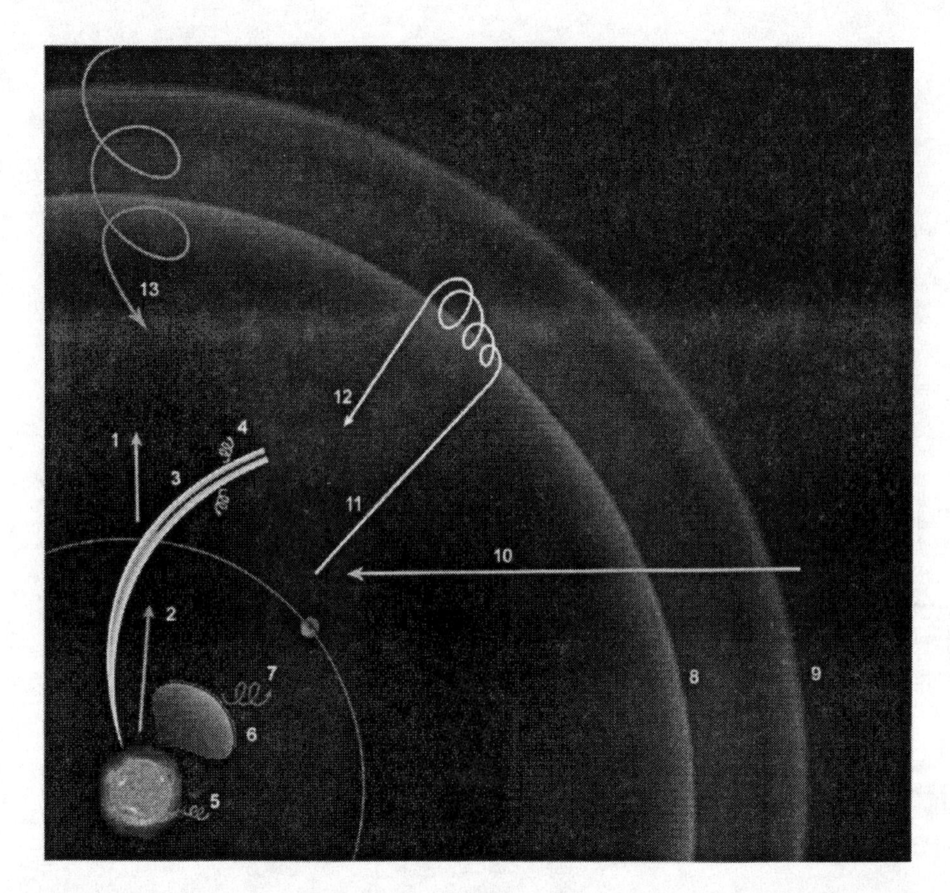

Cover Illustration: The cover drawing illustrates schematically many of the energetic particle populations observed within the heliosphere and discussed in this book. The expanding solar wind and solar rotation results in a spiraling interplanetary magnetic field. In the ecliptic plane the average velocity of the solar wind [1] is ~400 km/sec, but occasional long-lived high-speed streams [2] travel at 500 to 700 km/sec. When high-speed wind overtakes slower wind the magnetic field is compressed, forming a co-rotating interaction region [3]. Beyond Earth's orbit, typically at 2 to 4 AU, energetic particles [4] are accelerated to energies of ~10 MeV by the forward and reverse shocks formed at the interface of fast and slow solar wind streams.

Solar flares that erupt from active regions on the Sun can accelerate particles [5] to energies >100 MeV in minutes. Active regions also sometimes eject giant blobs of plasma called coronal mass ejections (CMEs) [6]. Interplanetary shocks created by the fastest CMEs are responsible for the largest solar particle events [7] which can fill much of the inner heliosphere and last for days.

As the supersonic solar wind approaches the interstellar medium a roughly spherical termination shock [8] forms, most likely between 80 and 100 AU. Somewhat beyond this lies the heliopause [9] – the interface between the solar wind and interstellar medium. As the Sun moves through the Galaxy interstellar ions are deflected around the heliopause, but neutral atoms [10] flow through the heliosphere. Some neutrals are ionized by solar UV radiation or by charge-exchange processes, becoming "pickup ions" [11] that are convected into the outer heliosphere. At the termination shock some pickup ions are accelerated to much higher energies to become "anomalous cosmic rays" [12].

The highest energy particles observed in the heliosphere are the galactic cosmic rays [13], with energies ranging from ~10 to ~10^{14} MeV. Most galactic cosmic rays are believed to be accelerated by shock waves from supernova explosions as they pass through the interstellar medium. (Cover illustration by Eric Christian of Goddard Space Flight Center).

ACCELERATION AND TRANSPORT OF ENERGETIC PARTICLES OBSERVED IN THE HELIOSPHERE

ACE-2000 Symposium

Indian Wells, California 5–8 January 2000

EDITORS

Richard A. Mewaldt
California Institute of Technology

J. R. Jokipii
University of Arizona

Martin A. Lee
Eberhard Möbius
University of New Hampshire

Thomas H. Zurbuchen
University of Michigan

Melville, New York, 2000

AIP CONFERENCE PROCEEDINGS ■ VOLUME 528

Sep/Ae
PHYS

Editors:

Richard A. Mewaldt
California Institute of Technology
220-47 Downs Laboratory
Pasadena, CA 91125
E-mail: RMewaldt@srl.caltech.edu

J. R. Jokipii
University of Arizona
Department of Planetary Sciences
Tucson, AZ 85721
E-mail: jokipii@lpl.arizona.edu

Martin A. Lee
Eberhard Möbius
University of New Hampshire
Space Science Center/Morse Hall
Durham, NH 03824
E-mail: Marty.Lee@unh.edu
Eberhard.Moebius@unh.edu

Thomas H. Zurbuchen
University of Michigan
Space Research Building
Ann Arbor, MI 48109
E-mail: thomasz@umich.edu

The articles on pp. 21–31, 79–86, and 181–184 were authored by U. S. Government employees and are not covered by the below mentioned copyright.

Authorization to photocopy items for internal or personal use, beyond the free copying permitted under the 1978 U.S. Copyright Law (see statement below), is granted by the American Institute of Physics for users registered with the Copyright Clearance Center (CCC) Transactional Reporting Service, provided that the base fee of $17.00 per copy is paid directly to CCC, 222 Rosewood Drive, Danvers, MA 01923. For those organizations that have been granted a photocopy license by CCC, a separate system of payment has been arranged. The fee code for users of the Transactional Reporting Service is: 1-56396-951-3/00/$17.00.

© 2000 American Institute of Physics

Individual readers of this volume and nonprofit libraries, acting for them, are permitted to make fair use of the material in it, such as copying an article for use in teaching or research. Permission is granted to quote from this volume in scientific work with the customary acknowledgment of the source. To reprint a figure, table, or other excerpt requires the consent of one of the original authors and notification to AIP. Republication or systematic or multiple reproduction of any material in this volume is permitted only under license from AIP. Address inquiries to Office of Rights and Permissions, Suite 1NO1, 2 Huntington Quadrangle, Melville, NY 11747-4502; phone: 516-576-2268; fax: 516-576-2450; e-mail: rights@aip.org.

L.C. Catalog Card No. 00-105240
ISSN 0094-243X
ISBN 1-56396-951-3

Printed in the United States of America

QB
464
.15
A24
2000
PHYS

CONTENTS

OVERVIEW

SOLAR ENERGETIC PARTICLES

INTERPLANETARY ACCELERATION

ANOMALOUS AND GALACTIC COSMIC RAYS IN THE HELIOSPHERE

COSMIC RAYS IN THE GALAXY

PREFACE

Particle acceleration is ubiquitous in the Universe. The process of shock acceleration, in particular, is observed to occur on scales that range from $\sim 10^5$ km to $\sim 10^{15}$ km, including the Earth's bow shock, shocks driven by coronal mass ejections from the Sun, the solar wind termination shock that surrounds our solar system, and shocks driven by supernovae blast waves. Historically, most of our understanding of particle acceleration has come through observations and theoretical modeling of the characteristics of energetic particle populations that can be observed in local interplanetary space.

The ACE-2000 Symposium, held on January 5 to 8, 2000 in Indian Wells, California, had as its focus understanding the acceleration and transport of four energetic particle components that can be observed in local interplanetary space:

- Solar energetic particles
- Particles accelerated in interplanetary space
- Interstellar pickup ions and anomalous cosmic rays
- Galactic cosmic rays

In particular, the Symposium focused on new observations, now becoming available from the Advanced Composition Explorer (ACE) and other missions that include Wind, Ulysses, Voyager, SAMPEX, and SOHO, and on recent theoretical advances in modeling particle acceleration and transport.

ACE-2000 was sponsored by the ACE Science Working Team. ACE was launched in August, 1997 carrying six high-resolution instruments designed to measure the composition of energetic nuclei from solar wind to galactic cosmic ray energies, and three instruments to provide the interplanetary context for these studies. Data from ACE and other spacecraft are now providing an opportunity to study the acceleration and transport of solar, interplanetary, and galactic particles with unprecedented precision. In particular, the combination of detailed elemental, isotopic, and ionic charge state information with excellent time resolution now makes it possible to provide sensitive tests of particle acceleration and transport models over a broad energy range, and to identify the seed populations that get accelerated.

ACE-2000 was attended by 94 scientists, representing a broad cross-section of the energetic-particle community, and including representatives from the various space missions as well as ground-based and balloon-borne experiments. The scientific program, which included a total of 102 talks and poster papers, was organized into four sessions: 1) Solar Energetic Particles 2) Interplanetary Acceleration 3) Anomalous and Galactic Cosmic Rays in the Heliosphere, and 4) Cosmic Rays in the Galaxy. Prior to the several poster sessions, short, entertaining reviews of the posters were presented by the six Poster Reporters. On the last day of the symposium, four teams of Discussion Organizers led a wide-ranging discussion of issues that had been raised.

The Scientific Organizing Committee for ACE-2000, consisting of the five editors, was responsible for selecting the focus of the symposium and the invited speakers, and for arranging the scientific program. The organization of the papers in this volume reflects the four sessions of the Symposium. All papers were refereed by one referee and one of the editors. We are grateful to Marjorie L. Miller of Caltech for providing critical editorial assistance in managing the distribution of papers to the referees and collecting and assembling this volume on a tight schedule. We also thank Debbie K. Eddy of the University of Michigan for editorial assistance and Charles Doering of AIP for valuable advice on transforming a symposium into published proceedings.

The Local Organizing Committee was headed by Alan Cummings, who located the site of the symposium, and who managed to negotiate the myriad of details necessary to see that the participants were registered, badged, housed, fed, equipped with a laser pointer and microphone, and (usually) supplied with their morning paper. Alan also presided over the after-dinner awards for dubious achievements in categories that included: best answer to an intimidating question; most plots on a graph; best display of chutzpah after the bell; and most ironhanded chair. Other members of the Local Organizing Committee that warrant special thanks include Andrew Davis, Cherylinn Rangel, and Steve Sears, as well as Ed Stone, whose wise advice guided planning throughout this effort.

The venue for the ACE-2000 Symposium was the beautiful Miramonte Resort in Indian Wells, CA, which provided excellent facilities and very comfortable desert surroundings in the shadow of the Santa Rosa Mountains. Special thanks are due Mr. Terry Sullivan, the National Sales Manager, and Ms. Linda Boyd, Catering Manager, for the fine service and hospitality.

This Symposium occurred exactly three years after a workshop held at Caltech during January 1997, that is summarized in the book "The Advanced Composition Explorer Mission". That workshop, attended by many of the same scientists, was designed to assess the state of energetic particle studies in the context of other on-going missions and to identify key scientific goals prior to the launch of ACE. A look back indicates a number of cases where key questions have now been answered, but even more where new puzzles and challenges have emerged.

Richard Mewaldt
Randy Jokipii
Marty Lee
Eberhard Möbius
Thomas Zurbuchen

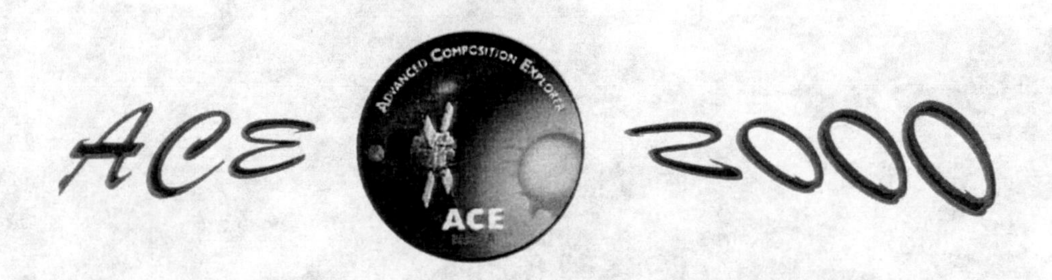

The ACE-2000 Symposium was sponsored by the Science Working Team of NASA's Advanced Composition Explorer (ACE) Mission, led by Principal Investigator Ed Stone. Listed below are the members of the Scientific and Local Organizing Committees. The Discussion Organizers were responsible for the sessions on the final day of the Symposium. In addition, six Poster Reporters provided overviews of the poster sessions.

Scientific Organizing Committee
Richard Mewaldt (Chair)
Randy Jokipii
Marty Lee
Eberhard Möbius
Thomas Zurbuchen

Local Organizing Committee
Alan Cummings (Chair)
Christina Cohen
Andrew Davis
Rick Leske
Richard Mewaldt
Cherylinn Rangel
Steve Sears
Mark Wiedenbeck

Discussion Organizers
Solar Energetic Particles:
Ilan Roth+, Gerry Share+ and Allan Tylka+
Interplanetary Acceleration:
Miriam Forman+ and Tom Krimigis
Cosmic Rays in the Heliosphere:
Horst Fichtner, Berndt Klecker+ and Frank McDonald
Galactic Cosmic Rays:
Bernd Heber, Marty Israel+ and Frank Jones

+ Poster Reporters

1. Tycho von Rosenvinge
2. Lev Dorman
3. Donald Reames
4. Alan Cummings
5. George Ho
6. Paulett Liewer
7. Penny Slocum
8. Steve Sears
9. Karim Meziane
10. George Gloeckler
11. Mark Wiedenbeck
12. Jim Ryan
13. Miriam Forman
14. Toni Galvin
15. Bob Lin
16. Richard Lingenfelter
17. Jim Higdon
18. Georgia de Nolfo
19. Christina Cohen
20. Marty Lee
21. Richard Mewaldt
22. Abe Falcone
23. Rob Gold
24. Carol Maclennan
25. Conrad Steenberg
26. Mihir Desai
27. Ed Cliver
28. Andrew Davis
29. Bob Binns
30. Gary Zank
31. Tohru Hada
32. Jack Gosling
33. Masato Yoshimori
34. Roger Williamson
35. Jake Waddington
36. Ed Hawkins
37. Susan Mahan Niebur
38. Eric Christian
39. Len Fisk
40. Andrew Lukasiak
41. Ke Chiang Hsieh
42. Dennis Haggerty
43. Devrie Intriligator
44. Jim Ling
45. Ron Zwickl
46. Richard Leske
47. Erhard Keppler
48. Bernd Heber
49. Danny Summers
50. Dalmiro Maia
51. Marty Israel
52. Joe Giacalone
53. Daniel Morris
54. Manfred Scholer
55. Joe Mazur
56. Luke Sollitt
57. Sam Krucker
58. Jakobus le Roux
59. Frank Jones
60. Jon Ormes
61. Berndt Klecker
62. Ted Freeman
63. Randy Jokipii
64. Vladimir Ptuskin
65. Frank McDonald
66. Jeff George
67. Nathan Yanasak
68. Horst Fichtner
69. Thomas Zurbuchen
70. Stamatios Krimigis
71. Rob Decker
72. Eberhard Moebius
73. Don Ellison
74. Joe Dwyer
75. Gerry Share
76. Ed Roelof
77. Ilan Roth
78. Paul Boberg
79. Paul Hink
80. Mike Lijowski
81. Glenn Mason
82. Mark Popecki
83. Peter Bochsler
84. Allan Tylka
85. Jim Miller
86. Nasser Barghouty
87. Yuri Litvinenko

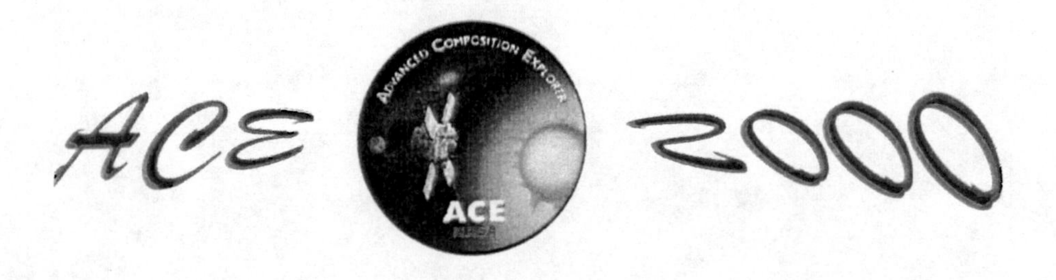

Overview

Acceleration of Energetic Particles on the Sun, in the Heliosphere, and in the Galaxy

Martin A. Lee

Space Science Center and Institute for the Study of Earth, Oceans and Space, Morse Hall, University of New Hampshire, Durham, NH 03824-3525

Abstract. This review describes the energetic particle populations observed in the heliosphere including gradual and impulsive solar energetic particle (SEP) events, interplanetary energetic particle enhancements, the anomalous cosmic ray (ACR) component, and galactic cosmic rays (GCRs). Also described briefly are the interstellar pickup ions and the solar wind itself, which are not energetic in the usual sense but are substantially heated and/or accelerated and provide the seed particles for the heliospheric populations. Firstly several overarching themes are presented which are important for all the energetic particles: acceleration mechanisms, particle transport theory, species-dependent transport and acceleration, injection mechanisms, and energetic particles as probes of heliospheric structure and dynamics. Secondly each population of energetic particles is reviewed including our current understanding of the source of the particles, the acceleration mechanism, and the transport, and any puzzles or challenges for theory. Shock acceleration is the dominant acceleration mechanism. Gradual SEP events are accelerated at coronal/interplanetary shocks, the corotating ion events are accelerated at the shocks bounding corotating interaction regions (CIRs) in the solar wind, diffuse ions are accelerated at planetary bow shocks, the ACRs are accelerated at the solar wind termination shock, and GCRs are accelerated at supernova remnant shocks. Only impulsive SEP events do not originate at shocks; they are accelerated by direct electric fields or stochastic acceleration at sites of magnetic reconnection in solar flares. For GCRs in particular, transport in the heliosphere and interstellar space affects their observed energy spectra and composition.

1. INTRODUCTION

Energetic particles are an essential component of the plasma Universe. Because of the unimportance of binary collisions between particles throughout most of the Universe (away from compact objects), plasma processes can be very undemocratic; a large fraction of the energy available from mass motion or rotation can be bestowed on very few high-energy particles. Thus, these particles play an important dynamical role in the evolution of supernovae, and in the structure and dynamics of galaxies, the outer portions of stellar cavities, and more exotic objects such as pulsars and jets in radio galaxies. Furthermore, through synchrotron emission, bremsstrahlung, nuclear interactions producing γ-rays, and even plasma interactions producing radio waves, energetic particles produce the photons essential to high-energy astronomy.

The heliosphere is a microcosm in which most of the high-energy processes of the plasma Universe take place, but at lower energy. Here various populations of energetic particles can be, or have been, observed *in situ* and in detail by an extensive array of past and present spacecraft covering the heliosphere from heliocentric radial distances of 0.3 AU to 77 AU, and out of the ecliptic to ~80° latitude. Knowledge gained here may be applied to distant sites in the plasma Universe.

Actually, a platform near 1 AU is not a bad vantage point from the point of view of energetic particles. It may be about the best vantage point in the heliosphere, unless your speciality is the anomalous cosmic ray (ACR) component. But even then Earth collects ACRs in a special radiation belt easily accessible to observation from near-Earth polar orbit! Since August 1997 near vantage point L1 we have had the Advanced Composition Explorer (ACE) with advanced instrumentation, which has been providing a fresh view of our energetic particle environment. The space physics community already has an impressive new collection of data on the elemental, isotopic and charge-state composition of these particles. It is the purpose of this Symposium to provide a forum for new ideas and data from ACE and other spacecraft such as Ulysses, Wind, SAMPEX and Voyager, and an opportunity to test old ideas and have new insights. Particularly for the study of solar energetic particles (SEPs) this is an exciting time as we anticipate the peak of solar activity in a year or so and hope for some impressive events in the declining phase of Cycle 23.

In this introductory talk I hope to present an overview of heliospheric energetic particles, in NASA jargon to provide a "roadmap" of specific populations, acceleration mechanisms, modes of transport, and current understanding. As I soon realized after Christmas, this is a vast task. In particular the many

CP528, *Acceleration and Transport of Energetic Particles Observed in the Heliosphere: ACE 2000 Symposium,* edited by Richard A. Mewaldt, et al.
© 2000 American Institute of Physics 1-56396-951-3/00/$17.00

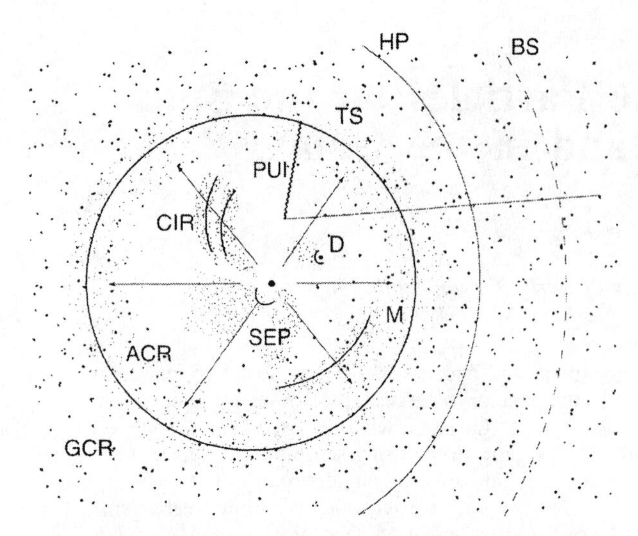

FIGURE 1. Schematic diagram of the heliosphere showing the Sun, the solar wind (arrows directed away from the Sun), the termination shock (TS), the heliopause (HP), and a possible heliospheric bow shock (BS). Also shown are gradual and impulsive solar energetic particle (SEP) events, a corotating ion event associated with a CIR, diffuse (D) ions associated with a planetary bow shock, an event associated with a merged interaction region in the outer heliosphere (M), anomalous cosmic rays (ACR), galactic cosmic rays (GCR), and a pickup ion trajectory (PUI) following ionization of an inflowing interstellar atom.

abundance variations and fractionation processes are overwhelming. I began to wish we were embarking on these studies nearer the beginning of the Universe when matter consisted of hydrogen and helium only. Let alone the fact that we require the heavy elements for our existence! However, I have been able to rely on some excellent reviews including those by Miller et al. (1) and Reames (2). And indeed the detailed data on composition and its variation present a major challenge for theory, which must account for the behavior of each species.

The scope of this keynote address precludes a comprehensive list of references. These should be included in the reviews of specific energetic particle populations which follow in this volume.

2. HELIOSPHERIC ENERGETIC PARTICLES

Since a picture is worth a thousand words, Figure 1 introduces the major heliospheric energetic particle populations. With some artistic license Figure 1 depicts the heliosphere, formed by the solar wind which is shown by radial vectors pointing outwards from the Sun. Since the solar wind is supersonic (actually super-fast-magnetosonic) forced deflection or acceleration/deceleration of the wind creates shock waves. Shown by solid lines are the termination shock marking the termination of the solar wind, a coronal/interplanetary shock driven from the Sun by a coronal mass ejection (CME), a planetary bow shock, a forward/reverse shock pair bounding a corotating interaction region (CIR), and a large shock in the outer heliosphere resulting from a coalescence of smaller shocks at maximum solar activity. The heliopause separating solar and interstellar plasma is denoted by a light solid line. A possible heliospheric bow shock in the interstellar medium is denoted by a dashed line.

Following the Archimedean spiral pattern of the interplanetary magnetic field which channels their transport are the solar energetic particles (SEPs). They occur in "gradual" events accelerated by coronal/interplanetary shocks and in "impulsive" events accelerated by solar flares in the lower corona. Gradual events last for days and occur at a rate of about 10/year during solar maximum, whereas impulsive events last for hours and occur at a rate of about 1000/year during solar maximum. The interplanetary particle populations consist of "diffuse" ions at Earth's and other planetary bow shocks; the corotating ion events peaking at the forward and reverse shocks in CIRs ($\gtrsim$ 3 AU) and extending inwards to the orbit of Earth; "energetic storm particle" (ESP) events at interplanetary traveling shocks, which are actually a phase of gradual SEP events; and large events in the outer heliosphere which are a coalescence of many SEP events and are associated with "merged interaction" regions often with a large embedded shock (3). Presumably peaking at the termination shock but extending into the inner heliosphere, particularly during solar minimum when solar modulation is reduced, is the anomalous cosmic ray (ACR) component . These ions result from a remarkable sequence of events: neutral interstellar gas flows into the heliosphere and is ionized by photons from the Sun or charge exchange with solar wind ions; these newborn ions are picked up by the solar wind to form pickup ions which are advected to the outer heliosphere; the pickup ions are accelerated at the solar wind termination shock to form the ACRs; because they are predominantly singly ionized, the ACR minor ions have high rigidity and can propagate back into the inner heliosphere. With "thermal" speeds on the order of the solar wind speed, the pickup ions are legitimate suprathermal or energetic particles in their own right; they affect solar wind dynamics and shock propagation in the outer heliosphere. Galactic cosmic rays (GCRs) pervade space. This population of energetic particles is the "old timer" of heliospheric physics. They were discovered by V.F. Hess during balloon flights in 1912 with instruments which measured increasing electrical conductivity with increasing height in the atmosphere. The conductivity was ascribed to ionization by an extra-terrestrial and extra-solar radiation. The GCR are accelerated in the Galaxy, roam around the interstellar medium and galactic halo for ~ 10^7 years, and penetrate

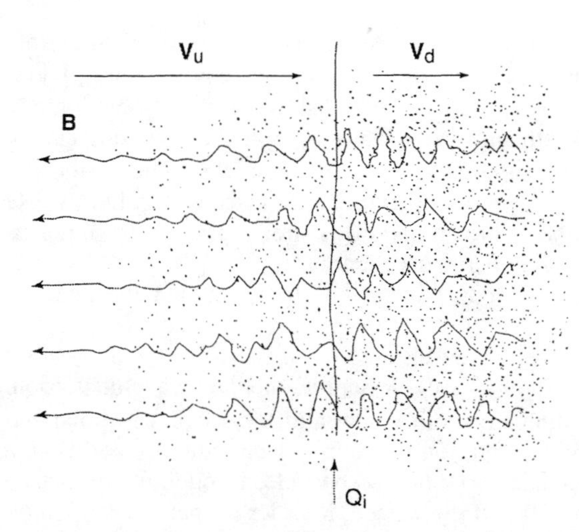

FIGURE 2. Schematic diagram of a parallel shock showing upstream ($\mathbf{V}_u$) and downstream ($\mathbf{V}_d$) flows relative to the shock, magnetic field lines with fluctuations, and energetic particles accelerated by the shock (dots). Q_i describes the injection rate of species i from the thermal plasma into the acceleration process.

the heliosphere with reduced intensity in the inner heliosphere due to the process of solar modulation. They provided the first view of a new world of high-energy astrophysics, revealed new subatomic particles, and attracted high-energy physicists like John Simpson, Frank McDonald and James Van Allen to forge the new field of space physics. We should also not forget the solar wind itself. The solar wind is heated and accelerated in the low corona enough to escape the solar gravitational field by processes not yet well understood, and it continues to be energized in interplanetary space.

3. OVERARCHING THEMES

Let me first discuss five important themes relevant to all these energetic particle populations.

3.1. Acceleration Mechanisms

If you will excuse the vernacular, you don't have to be a rocket scientist to note the close association of heliospheric energetic particle populations with shock waves! The correspondence is nearly one-to-one. The mechanism of shock acceleration is illustrated in Figure 2 for a parallel shock, for which the average magnetic field **B** is aligned with the shock normal. In the normal-incidence frame the upstream flow in the frame of the shock is $\mathbf{V}_u$, while the downstream flow is $\mathbf{V}_d$. Particles of species i are extracted from the plasma at the shock front with rate Q_i and injected into the mechanism of shock acceleration. The particles are accelerated by traversing the shock front many times by pitch-angle scattering in the magnetic turbulence adjacent to the shock as shown. As they scatter back and forth across the shock the particles are scattered between converging flows due to the compression at the shock, and are accelerated as is a ping-pong ball bouncing between ping-pong paddles which approach each other. The magnetic turbulence need not be ambient turbulence. The streaming of the energetic particles (mostly protons) relative to the upstream flow excites the required hydromagnetic waves. Thus, the ions, described by omnidirectional distribution functions $f_i\,(v,z)$ for each species i, and the waves, described by wave intensity $I\,(k,z)$, form a coupled system (4,5). Here we have tacitly assumed a stationary planar configuration dependent only on coordinate z measuring distance upstream of the shock; v is ion speed and k is wavenumber (waves are mostly aligned with **B**). A good recent review of shock acceleration is the review by Jones and Ellison (6).

If the shock is oblique, in addition to gaining energy by shock compression, particles drift in the inhomogeneous average magnetic field at the shock front parallel (for ions) to the motional electric field ($\mathbf{E} = -c^{-1}\,\mathbf{V} \times \mathbf{B}$) and gain energy by the process of shock drift. These apparently separate mechanisms of energy gain are actually features of the same mechanism, shock acceleration; indeed their relative contributions are frame-dependent (7). At an oblique shock particles can be accelerated in the absence of pitch-angle scattering; the energy gain is then entirely due to shock drift acceleration. If scattering is effective and the resulting particle distributions are nearly isotropic, then the process is known as diffusive shock acceleration.

At planar stationary shocks $f_i\,(v, z \leq 0) \propto v^{-\beta}$ with index $\beta = 3X/(X-1)$, where X ($= V_u/V_d$) is the shock compression ratio. Technically V_u and V_d refer to the average wave frames rather than the plasma rest frames. Shock-associated particle distributions observed in the heliosphere, however, are not generally power laws. Finite shock lifetime, particle escape from the shock, and competing adiabatic deceleration in the solar wind can all limit acceleration and result in energy spectra with high-energy cutoffs.

Shock acceleration at a parallel shock is a type of Fermi acceleration. The essence of Fermi acceleration is elastic scattering of a particle in the frame of a wave or "magnetic cloud". Viewed in a different frame the particle may gain (head-on collision) or lose (overtaking collision) energy during the interaction. At a rate depending on the configuration of waves or "magnetic clouds", some particles achieve high energies by many sequential interactions. Three configurations relevant to Fermi acceleration are shown schematically in Figure 3.

In Figure 3(a) elastic scattering of protons by Alfvén waves with velocity $V_A\,\mathbf{e}_z$ (neglecting dispersion) is shown by a cut through the proton velocity distribution

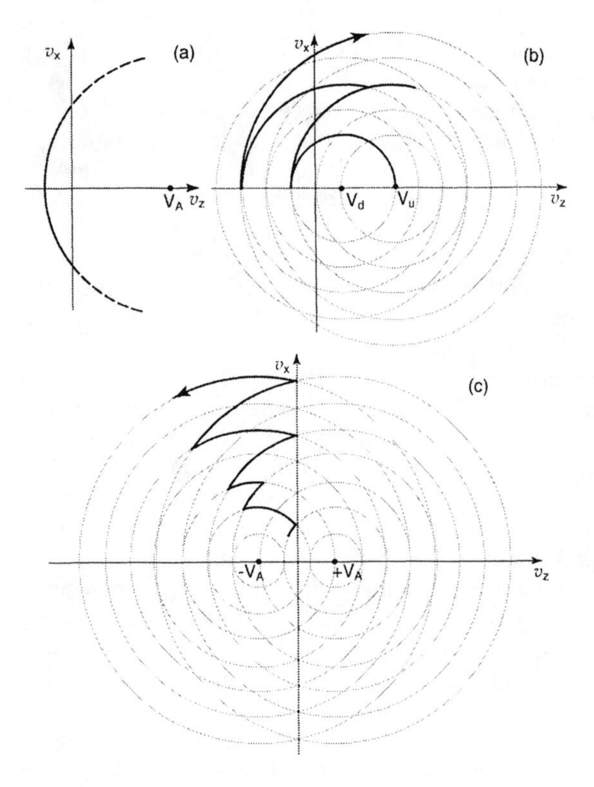

FIGURE 3. Schematic diagram of protons accelerated by Fermi acceleration showing sample trajectories in the v_x (>0) - v_z plane. (a) shows scattering by Alfvén waves propagating with phase velocity $V_A\,\mathbf{e}_z$, (b) shows first-order Fermi acceleration at a parallel shock, where V_u and V_d are the upstream and downstream plasma velocities relative to the shock. (c) shows second-order Fermi acceleration due to Alfvén waves propagating in both directions along **B**.

in the plasma frame. Particles with $v_x \sim 0$, $v_z < 0$ are scattered on a shell centered on $V_A\,\mathbf{e}_z$ and gain energy in the plasma frame. For $v \ll V_A$ as shown this energization is limited by a substantial reduction in the rate of pitch-angle scattering for $v_z \gtrsim 0$, as shown by dashed lines. This reduction is due to a "resonance gap" for protons in the range $0 \lesssim v_z \lesssim 2\,V_A$, which cannot resonate with the scattering turbulence. In any case, for this configuration the particle speed is limited to $v \sim V_A$.

In Figure 3(b) shock acceleration is illustrated as a Fermi process by showing the trajectory of a particle in v_x (>0) – v_z space, where only in this figure $z > 0$ measures distance downstream of the shock. Neglecting wave phase speeds, the particle moves on shells centered on $V_d\,\mathbf{e}_z$ if it is downstream of the shock and on shells centered on $V_u\,\mathbf{e}_z$ if it is upstream. The transition from an upstream to a downstream shell occurs when the particle crosses the shock, which requires $v_z > 0$. Conversely, the transition from a downstream to an upstream shell occurs when the particle crosses the shock in the opposite direction, which requires $v_z < 0$. These requirements yield a trajectory, such as the one shown for a particle initially downstream with velocity $V_u\,\mathbf{e}_z$, with increasing energy in the frame of the shock as shown. The systematic energy gain by sequential elastic scattering, in this case off the upstream and downstream plasma, is known as first-order Fermi acceleration.

In Figure 3(c) a configuration is shown consisting of Alfvén waves propagating in both directions along **B**. Particles scatter elastically on shells centered either on $+V_A\,\mathbf{e}_z$ or $-V_A\,\mathbf{e}_z$, where we neglect wave dispersion. (Technically both right- and left-circularly polarized wave components are required, and in addition a mechanism to transport particles through the resonance gap at $v_z \sim 0$ is required, in order for particles to scatter over the full shell.) Now particles can transfer between the two families of shells at any point in velocity space. As a result particles can either gain energy (for the sample trajectory shown) or lose energy (for the reverse of the trajectory shown). This process is known as second-order Fermi acceleration or stochastic acceleration, and is often described by diffusion in v space.

In all of these processes the pitch-angle scattering is generally due to scattering by wave magnetic fluctuations. If the wave amplitudes are small, the scattering only has a secular effect on the particle velocity if the wave and particle satisfy the resonance condition

$$\omega - k_z v_z + n\Omega \cong 0, \tag{1}$$

where ω is the wave frequency, k_z is the z-component of the wavevector, $\Omega = eB/mc$ is the cyclotron frequency (assumed to be nonrelativistic), e and m are the particle charge and mass, and n is an integer. Therefore, the scattering essential to Fermi acceleration can only occur if resonant waves are present. The resonance gap mentioned previously is evident for $v_z \to 0$ since $k_z \to \infty$, where there is vanishing wave power. For large amplitude fluctuations there is no such resonance condition.

In addition to Fermi acceleration, particles may also be accelerated directly by an electric field, as in shock drift acceleration. In most astrophysical plasmas the ambient electric field vanishes in the plasma rest frame since the very mobile electrons can short-out the field. There are exceptions to this "rule" if the electron density is low; for example, within Earth's magnetosphere in auroral regions electrons are probably accelerated into the upper atmosphere to create the aurorae by induction electric fields parallel to **B**. However, the exceptions appear to be limited to cases in which a strong magnetic field is produced by external currents, e.g. magnetospheres. Generally the electric field is limited to the motional electric field ($\mathbf{E} = -c^{-1}\,\mathbf{V} \times \mathbf{B}$) due to plasma motion with velocity $\mathbf{V}$.

However, in this case the potential difference is across **B**, and particle energization can occur only if a mechanism exists, such as shock drift, to transport particles normal to **B**. An important application of direct electric field acceleration is at sites of magnetic reconnection where $\mathbf{B} \cong 0$ so that **B** does not inhibit direct particle transport along **E**, or where $\mathbf{E} \neq -c^{-1}\,\mathbf{V} \times \mathbf{B}$ (8).

Other acceleration mechanisms, less important for the energetic particle populations described above, include acceleration by electric field fluctuations parallel to **B** and/or associated with electrostatic waves (Landau damping of these waves is accompanied by plasma energization) and the betatron effect or magnetic pumping. A special mechanism involving the Landau resonance [$n = 0$ in equation (1)] is "transit-time" damping (9), which may accelerate the high-energy tails of the solar wind ion distributions (see § 5). Magnetic pumping occurs in planetary magnetospheres and is not central to the theme of this Symposium.

3.2. Particle Transport

In order to describe quantitatively particle acceleration and the transport of particles to an observer from remote sites of acceleration, we require an appropriate transport equation. Two transport equations are commonly used to describe heliospheric energetic particle populations, depending on the effectiveness of pitch-angle scattering. If scattering is strong so that the particle distribution function is nearly isotropic in the plasma frame, and in addition $v >> |\mathbf{V}|$ so that the distribution function is nearly isotropic in the frame of the Sun, then the appropriate equation, due to Gene Parker (10) [see also (11,12)], is

$$\frac{\partial f_i}{\partial t} + (\mathbf{V} + \mathbf{V}_{Di}) \cdot \nabla f_i - \nabla \cdot (\mathbf{K}_i \cdot \nabla f_i)$$

$$-\frac{1}{3} \nabla \cdot \mathbf{V}\, p \frac{\partial f_i}{\partial p} - p^{-2} \frac{\partial}{\partial p} (p^2 D_{pp} \frac{\partial f_i}{\partial p}) = Q_i \qquad (2)$$

Here $f_i\,(p, \mathbf{r}, t)$ is the omnidirectional distribution function of energetic particle species i, $\mathbf{r}$ is the spatial coordinate, p is momentum magnitude, $\mathbf{V}_{Di}$ [$= pvc\,(3e_i)^{-1}\, \nabla \times (B^{-2}\,\mathbf{B})$] is the drift velocity, $\mathbf{K}_i$ is the symmetric spatial diffusion tensor, D_{pp} is the momentum-magnitude diffusion coefficient, and Q_i is a source term. D_{pp} describes stochastic acceleration. $\mathbf{K}_i$ describes spatial diffusion parallel to **B** due to pitch-angle scattering. It also describes spatial diffusion perpendicular to **B** due to magnetic field-line wandering, scattering of particles onto neighboring field lines, stochastic drift of particles due to field irregularities, and the inherent stochasticity of particle transport due to finite gyroradius. Although there exist estimates of $\mathbf{K}_\perp$, (e.g. 13), no comprehensive theory exists, and applications usually neglect it or choose it *ad hoc*.

Equation (2) is generally applicable to ACRs, GCRs, most interplanetary particle populations, and the later phases of gradual SEP events for which particle distribution functions are nearly isotropic. However, caution must be exercised, for example, in applying equation (2) to "diffuse" ions escaping Earth's bow shock, which can exhibit large anisotropies. In all these applications D_{pp} is negligible. However, equation (2) with $D_{pp} \neq 0$ is also appropriate to describe the acceleration of impulsive SEP events at the Sun if stochastic acceleration by isotropizing turbulence is the relevant acceleration mechanism.

Equation (2) is not in general appropriate to describe the transport of SEPs in the inner heliosphere since particle scattering mean free paths parallel to **B** in the ambient interplanetary turbulence are comparable to heliocentric radial distance r and anisotropies in impulsive events, and early in gradual events, are observed to be large. Particle gyration about **B** does insure gyrotropy. An appropriate transport equation for these events is the focused transport equation, here specialized for radial magnetic field, e.g. (14,15,16,17,18)

$$\frac{\partial f_i}{\partial t} + (V + v\mu) \frac{\partial f_i}{\partial r} - \frac{(1-\mu^2)}{r} V v \frac{\partial f_i}{\partial v}$$

$$+ \frac{1-\mu^2}{r} (v + \mu V) \frac{\partial f_i}{\partial \mu} = \frac{\partial}{\partial \mu} [(1-\mu^2)\, D_{\mu\mu} \frac{\partial f_i}{\partial \mu}] + Q_i \qquad (3)$$

Here $f_i\,(v, \mu, r, t)$ is the gyro-phase-averaged distribution function of species i, $\mu = \cos\theta$ where θ is the particle pitch angle in the wind frame, v is particle speed in the wind frame, and $D_{\mu\mu}$ is the pitch-angle diffusion coefficient. Clearly terms 2, 3 and 5 describe particle motion along the field (2), adiabatic deceleration (3) and pitch-angle diffusion (5). Term 4 describes magnetic focusing of particles toward smaller pitch angles. The obvious disadvantage in using this equation is that it contains an extra independent variable μ.

3.3. Species-dependent Transport and Acceleration

The transport equations (2) and (3) depend on v explicitly and on v and rigidity $R = pc/e$ through $\mathbf{V}_D$, $\mathbf{K}$, D_{pp}, $D_{\mu\mu}$ and Q_i. If $m_i = A_i m_p$ and $e_i = Q_i e_p$, then these quantities depend on v and A_i/Q_i. Thus each element and ion charge state satisfies a different equation and has its own story to tell. Each ion becomes a unique tool in probing transport and acceleration processes on the Sun, in the heliosphere, and in interstellar space.

For example, in impulsive SEP events the more massive ions are generally enhanced relative to hydrogen by factors of up to 10 (e.g. 19), presumably because they possess larger A/Q. A remarkable exception is $^3He^{++}$ which can be enhanced by factors of up to 10^5, presumably because it is the only ion with $1 < A/Q < 2$. Gradual SEP events exhibit dramatic abundance variations through the event (e.g. 20). Since ACRs are mostly singly-charged, the different elements span a wide range of rigidity for a fixed speed. Those few which are doubly-charged or triply-charged can be accelerated to higher energies by shock drift acceleration in the heliospheric motional electric field (21,22).

For galactic cosmic ray ions the source term Q_i in equation (2) must include the production, or loss, of a species through spallation in the interstellar medium. The production of a specific secondary ion species depends on the cross section for its production from heavier nuclei. The radioactive secondaries such as ^{10}Be provide measures of cosmic ray lifetime in the Galaxy. Similarly if radioactive nuclei are present in the source material for GCRs (produced, for example, in a supernova explosion), then the electron-capture radioisotopes can provide a measure of the time delay between production and acceleration.

Electrons are unique, but are generally a minor component of these heliospheric particle populations. Therefore, I shall not discuss them thoroughly. At shocks their acceleration is less efficient because they do not excite the required scattering turbulence. However, in impulsive SEP events they are accelerated by the electric field in the reconnection current sheet of the flare and may excite the turbulence responsible for the stochastic acceleration of the ions.

3.4. Injection Mechanisms

Since equation (2) is restricted to $v >> V$, it does not generally describe the acceleration of the ions out of the background thermal distribution. Acceleration processes operating in this low-energy domain may be modified versions of those that operate at higher energies or they may be unique to lower energies. In either case, since they cannot be described quantitatively by equation (2), their effect must be incorporated into Q_i as an injection rate of species i into the transport and acceleration processes operating at energies which satisfy $v >> V$. For example, in the description of diffusive shock acceleration based on equation (2), the injection rate of species i is often taken to be the same small fraction of the upstream plasma flux of that species in the shock frame. However, the composition of the corotating ion events is not identical to that of the slow or fast solar wind (23), indicating that ion fractionation does occur in the shock injection process. A quantitative description of injection processes is challenging since wave dispersion, large-amplitude waves and, in the case of shocks, shock structure are all

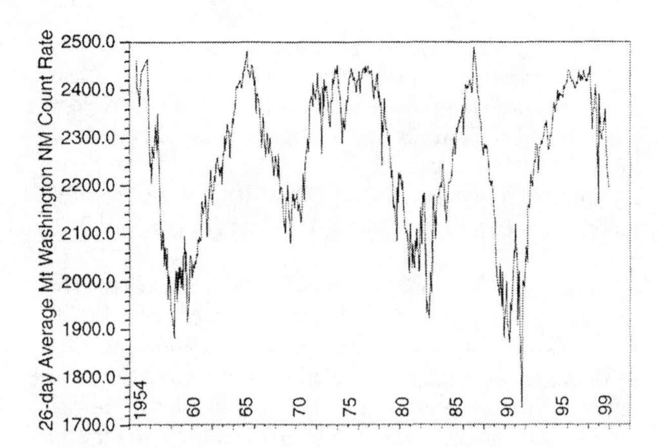

FIGURE 4. 26-day averages of the Mt. Washington neutron monitor count rate from 1954-2000. The count rate is proportional to the intensity of protons with a mean energy of ~2 GeV. Two solar magnetic cycles are evident. (Courtesy J.A. Lockwood)

important for low-energy ions. Numerical simulations using hybrid codes or Monte Carlo codes have been very useful in investigating these complexities at shocks for a range of Mach number and obliquity, and in estimating injection rates (e.g. 24,25,26,27,28). A promising analytical theory for injection at parallel shocks has been proposed by Malkov (29), in which incoming ions are first trapped in the large-amplitude magnetosonic wave of the shock and then may leak back upstream. At perpendicular shocks a unique injection mechanism appears to be required. Lee et al. (30) and Zank et al. (31) have suggested that shock surfing is the required injection mechanism (see § 7).

3.5. Energetic Particles as Probes of the Heliosphere

As a result of their high mobility, energetic particles can act as effective probes of heliospheric structure or conditions in solar active regions. Impulsive SEP events provide information on solar flares and sites of magnetic reconnection; gradual SEP events are precursors of large CME-driven shocks (32); the corotating ion events reveal the structure of the CIR further from the Sun; the energy spectra of ACRs provide the best estimate of the distance to the solar wind termination shock (33).

Figure 4 shows the 26-day-average count rate of the Mt. Washington neutron monitor during the period 1954-2000. The rate is proportional to the intensity of GCR protons with a mean energy of ~ 2 GeV which impinge on Earth's atmosphere at Mt. Washington, New Hampshire. Since the interstellar intensity of these particles is expected to be constant over such short cosmic timescales, every bump and wiggle in Figure 4 may be attributed to heliospheric processes. The 11-year solar activity cycle is evident in the broad

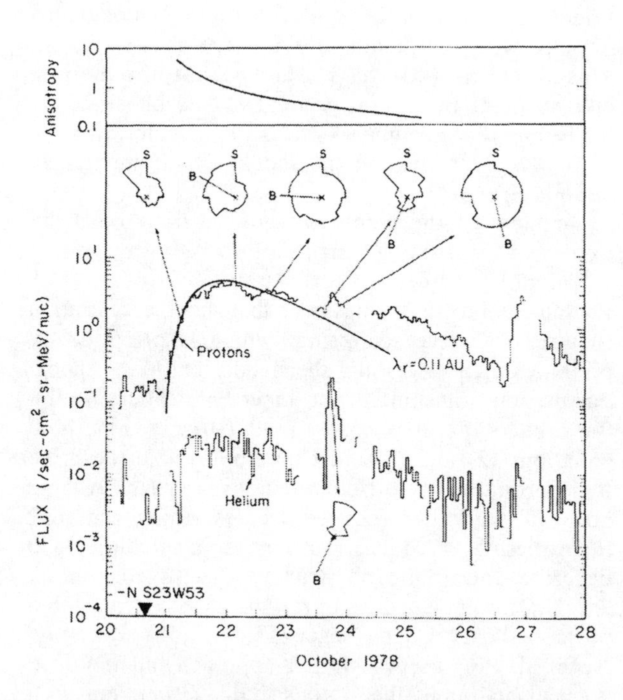

FIGURE 5. A gradual and an impulsive SEP event observed by ISEE-3 in October 1978. The gradual event presumably originates at the indicated flare on 20 October 1978, and exhibits decreasing anisotropy through the event. The large anisotropy of the He-rich impulsive event indicates a radial scattering mean free path λ_r ~1AU. The slow decay of the gradual event is not due to diffusion from an impulsive source, which would imply λ_r ~ 0.11 AU, but rather to continual acceleration of ions at a coronal/interplanetary shock. (Fig. 1 from (36))

minima in GCR intensity corresponding to enhanced solar modulation during maximum solar activity. The impulsive minima are Forbush decreases due to the temporary expulsion of GCRs from the inner heliosphere by large shock waves and merged interaction regions (3). The broad minima during maximum solar activity are most likely the result of an accumulation of Forbush decreases. The two largest Forbush decreases in 1983 and 1993 were caused by particularly large merged interaction regions, which excited traveling shocks beyond the solar wind termination shock, which in turn generated the radio bursts observed by Voyagers 1 and 2 when they reached the heliopause (34,35). The alternating "sharp" and "more rounded" peaks in GCR intensity are evidence of the 22-year solar magnetic cycle and are due to the alternating sense of drift circulation of GCRs in alternating solar cycles.

4. SOLAR ENERGETIC PARTICLES

Figure 5 shows two SEP events in October 1978 evident in both energetic protons and helium observed

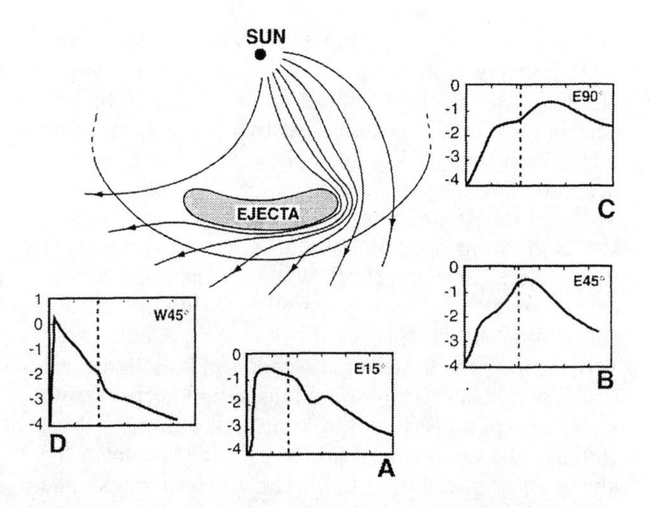

FIGURE 6. A schematic diagram of the CME-ejecta and the CME-driven shock showing sample SEP time profiles observed at different longitudes relative to that of the CME. (modified Fig. 15 from (37))

by ISEE-3 (36). The first event is a gradual event lasting a few days and presumably associated with the indicated flare on 20 October 1978, which was magnetically well connected to the spacecraft. The second event is an impulsive event enhanced in helium and lasting a few hours. The gradual event is characterized by a nearly isotropic distribution function after an initial anisotropic phase, whereas the impulsive event is highly anisotropic. If gradual and impulsive events are fit using equations (2) or (3) assuming impulsive injection of the ions at the Sun, the fits are often respectable but yield much smaller radial scattering mean free paths λ for gradual events than for impulsive events. The fit for the gradual event shown in Figure 5 yields λ = 0.11 AU, in contrast to λ ~ 1 AU obtained for most impulsive events. Since the two events shown in Figure 5 occur at about the same time, their transport must involve the value of λ characteristic of impulsive events. The extended evolution of the gradual event is not due to the slow diffusive escape of ions from the Sun, but rather to the relatively slow propagation of a CME-driven shock wave which accelerates ions continually. In contrast impulsive events are released impulsively in flares, propagate nearly scatter-free to the orbit of Earth, and show a clear signature of velocity dispersion. Thus these two classes of events are very different in their origins, transport and composition, and they must be considered separately.

4.1. Gradual SEP Events

The morphology of gradual events arises from a fast CME, with an extent in solid angle of about 1 steradian, which erupts into the solar wind and drives a shock ahead of it. The shock accelerates the particles by the mechanism of diffusive shock acceleration. The resulting geometry of the CME ejecta, shock and interplanetary field is shown in Figure 6 from Cane et al. (37). The origin of the CME and shock in the low corona may often be identified by a solar flare. The different panels in Figure 6 show how the geometry accounts for the SEP intensity profiles observed at different longitudes relative to the flare site. Panel D shows an event for which the observer is magnetically well connected to the shock when it is near the Sun so that particles arrive promptly following shock formation. At the time of shock arrival, denoted by a vertical line, the observer is on the weak eastern flank of the shock and sees reduced intensity. Panel A shows an event observed at the longitude of the CME. Initially the observer is either disconnected from the shock or connected to the weak western flank, and observes reduced intensity. When the shock arrives the observer is at the strong "nose" of the shock and sees a high intensity. Inside the magnetically closed CME ejecta the particle intensity is locally reduced. Panels B and C show events for which the CME origin is east of the observer. In these events the initial magnetic connection is very weak and maximum particle intensity occurs at, or even after, shock arrival.

Although the basic shock acceleration mechanism of the particles is straightforward, as described in § 3.1, the geometry shown in Figure 6 complicates the theoretical description considerably. The shock evolves in time and space. Initially the shock is weak due to the high Alfvén speed V_A in the low corona. Further from the Sun the shock increases in strength as V_A decreases. Finally the shock weakens as V_A levels off and the shock, no longer driven by the CME, propagates freely into the expanding solar wind. The magnetic field geometry is also complicated, and affects particle transport through the anisotropic diffusion tensor. Presumably these features account for the paradoxical fact that the shock can accelerate ions to ~100 MeV/nucleon within its short lifetime for $r \lesssim 20$ solar radii, whereas at Earth orbit the shock normally only accelerates ions as ESP events up to ~1 MeV/nucleon within its much longer lifetime (38).

The injection and acceleration of ions in gradual events produces elemental and charge-state fractionation of the energetic ions with respect to their presumed solar wind origin. Breneman and Stone (39) showed that the event-integrated elemental composition of gradual events exhibits a systematic fractionation according to Q/A. Recent observations by Tylka et al. (20) for the 20 April 1998 gradual event show complex patterns of elemental fractionation during the event. For example, Fe/O first increases and then decreases prior to shock passage, while He/O shows the reverse behavior. These variations appear to be representative of gradual events. ACE is providing a wealth of elemental and charge-state composition data for gradual events which are a challenge for theory. However, it must be noted that not all fractionation is due to the injection, acceleration and transport mechanisms. Shocks will accelerate any ion which is present. Mason et al. (40) have shown that the helium enhancement in several gradual events observed by ACE can be accounted for by reacceleration of impulsive event ions at the shock, which are rich in helium.

In view of the complexity of acceleration at an evolving coronal/interplanetary shock, early calculations were mostly illustrative (41,42,43). A fundamental difficulty has been that shock acceleration requires effective scattering, which is provided by proton-excited waves in a sheath adjacent to the shock, but particle transport in the inner heliosphere within the orbit of Earth is nearly scatter-free. Thus, both equations (2) and (3) must be employed to describe transport in the two different regions of space. A large body of work has focused on describing the SEP intensities observed near Earth by using equation (3) to describe interplanetary transport with the shock acceleration included in an *ad hoc* source term with a prescribed energy dependence (15,44,17,45). Recently, Ng et al. (46) used a similar approach but included wave excitation by the protons so that $D_{\mu\mu}$ in equation (3) is determined self-consistently. Intensity profiles were then obtained for several ion species and the remarkable trends observed by Tylka et al. (20) in the elemental variations were recovered.

I was inspired by the work of Ng et al. (46) to develop a theory which treats both shock acceleration and interplanetary transport by matching the solution of equation (2) in the turbulent sheath adjacent to the shock to the solution of a simplified version of equation (3) in interplanetary space (47). The shock acceleration is assumed to be planar and includes wave excitation by protons, while the interplanetary transport includes the mirror force arising from the spherical geometry. The mirror force extracts ions from the outer edge of the turbulent sheath and "squirts" them into interplanetary space. This ion loss produces an exponential energy cutoff. Preliminary results yield an expression for the upstream omnidirectional distribution function f_i of the form

$$f_i\,(r,\upsilon,t) = [G\,N_i\,(r_s/r)^2\,(\upsilon_o/\upsilon)^{\beta}][\exp(-B(\upsilon))] \cdot [C(\upsilon)]\,[1 + D(\upsilon,r,t)] \qquad (4)$$

The first factor in brackets is the ion distribution at a planar shock reduced by the r^{-2} dependence of the spherical geometry. Here N_i is the injection rate, G is a constant, and υ_o is the injection speed. The second factor describes the exponential energy cutoff. The third factor describes the fraction of the ions at the shock which escape the shock upstream. Finally, $D(\upsilon,r,t)$ describes the ESP enhancement associated withshock passage. Expressions B, C and D all depend

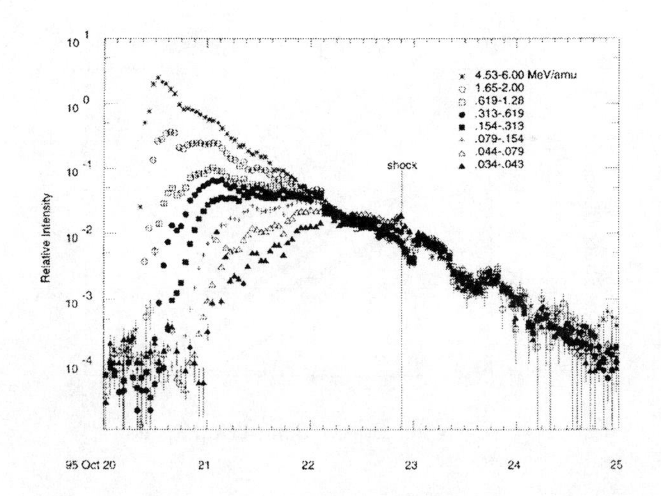

FIGURE 7. Proton intensities observed by Wind in the 8 indicated energy channels for the event of October 1995. The energy channels are normalized in the period following shock passage. The "invariant" energy spectrum during 22-25 October 1995 is evident. (Fig. 6 from (48))

on A/Q. Expression (4) should be able to reproduce the time-intensity profiles and compositional variations observed by Tylka et al. (20).

Figure 7 from Reames et al. (48) shows a simple feature of gradual events. Even though the details of shock geometry and particle transport cause the energy spectrum early in the event to be variable, later in the event the spectrum is an invariant (approximate) power law which decays exponentially. This feature is best portrayed by normalizing the energy channels to be equal late in the event as in Figure 7. This behavior can be explained by noting that late in the event the ions are distributed over a very large volume and have small gradients. Neglecting ∇f_i, V_{Di}, D_{pp} and Q_i in equation (2), only $\partial f_i/\partial t$ and the term describing adiabatic deceleration remain. If furthermore $f_i\,(p,t_o) = f_{oi}\,p^{-\Gamma}$, then the solution is

$$f_i = f_{oi}\,p^{-\Gamma} \exp\left[-\frac{2V}{3r}\,\Gamma\,(t-t_o)\right] \qquad (5)$$

which exhibits exponential decay with time. Fitting equation (5) to the decay for the event of 20 October 1995 shown in Figure 7 yields $\Gamma \cong 12$ assuming $V = 400$ km s^{-1}, which agrees reasonably well with the value $\Gamma \cong 8$ inferred from the invariant power law energy spectrum.

4.2. Impulsive SEP Events

The biggest mystery of heliospheric energetic particles is the origin of the impulsive SEP events. On the basis of their high charge state (indicative of 10^7 ^{o}K plasma (49)), their short temporal and spatial extent, and their association with γ-rays produced by particle interactions near the photosphere, they appear to originate in solar flares low in the corona. Flare electromagnetic emissions indicate that most of the energetic particles remain in the low corona; only a small fraction find their way to magnetic field lines open to interplanetary space. The impulsive events tend to be electron-rich, heavy-ion-rich, and greatly enhanced in ^{3}He (they are often called "^{3}He-rich" events). Furthermore, the variations in relative ion abundances and ^{3}He from event to event are very large (19). In view of these diverse signatures impulsive events appear to require more than one acceleration mechanism.

Virtually all possible acceleration mechanisms have been suggested: stochastic acceleration in the turbulence generated by the flare (50,51); direct electric field acceleration at the site of magnetic reconnection (52,8); direct wave acceleration (53); shock acceleration by the shock formed as the reconnection plasma jet impinges on the denser plasma below (54). Since the energetic electron energy content is larger than the energetic ion content in the flare, an important possibility is that streaming accelerated electrons excite waves which in turn provide the turbulence for stochastic ion acceleration (55,56,53). The acceleration of ^{3}He by the hydrogen-cyclotron waves excited in this process is favored since it is the only ion with Q/A in the range $0.5 < Q/A < 1$.

Again in the spirit of a picture being worth 1000 words, and with the freedom afforded by no *in situ* measurements and firm theoretical models, Figure 8 presents a schematic diagram for the flare origin of an impulsive event. A sheared arcade of magnetic loops at the solar surface is shown below a site of magnetic reconnection. The reconnected plasma above it is ejected into space, dragging the twisted magnetic field with it to form a CME. The current sheet reconnects through the "tearing" mode (shown by the "bubbles") and creates an electric field. The lower reconnection jet shown forms a shock at the top of the loop arcade. The shock-heated plasma at the top of the loop creates soft X-ray emission as observed by Yohkoh. Characteristic values for the magnetic field strength, electric field strength, and plasma flow velocity into the reconnecting current sheet are ~ 100 G, ~5 V cm^{-1}, and ~10 km s^{-1}. The reconnection electric field accelerates electrons and ions directly to energies of ~30 keV and ~1 MeV/nucleon, respectively (8). Direct acceleration ceases when the particles escape the current sheet along the magnetic field component transverse to the sheet either into interplanetary space or down the legs of the arcade to their photospheric footpoints. At the footpoints the electrons collide with the photosphere and create hard X-rays by bremsstrahlung, while the higher-energy ions produce γ-rays and neutrons through nuclear interactions. The directly accelerated electrons (and possibly ions) in the current sheet are unstable tothe excitation of hydrogen-cyclotron waves and other

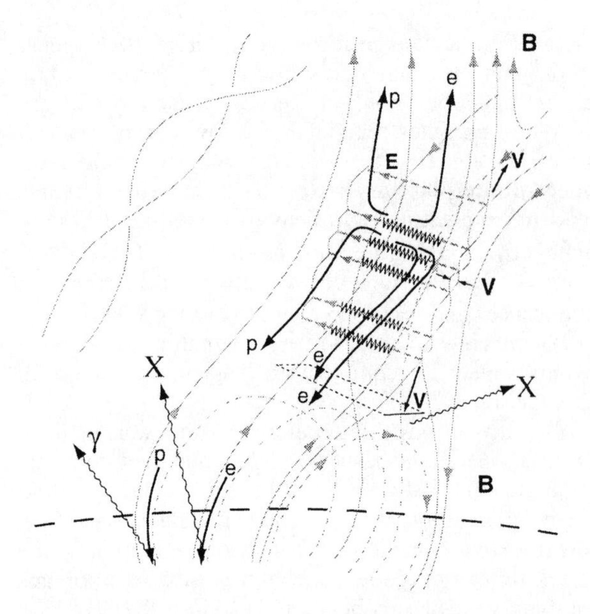

FIGURE 8. Schematic diagram of magnetic reconnection above an arcade of magnetic loops releasing a CME with twisted magnetic field into space. The reconnecting current sheet with electric field **E**, plasma flow **V**, and tearing-mode bubbles is shown. A shock is generated at the loop-top by the downward flow. Electrons and protons accelerated by the electric field are shown escaping downwards or into interplanetary space. The electrons excite turbulence in the current sheet, while both electrons and protons produce *X*-rays and γ-rays when they interact near the loop footpoints.

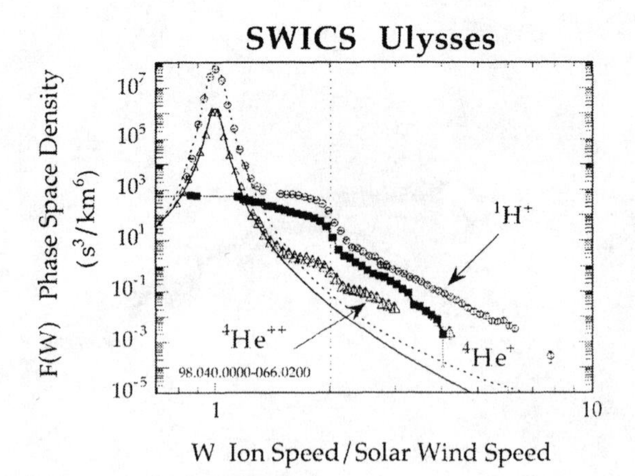

FIGURE 9. Phase-space densities of H^+, $^4He^+$ and $^4He^{++}$ measured by Ulysses in slow in-ecliptic solar wind plotted versus ion speed relative to 375 km/s, the average wind speed during the 26-day period. The solar wind peaks of H^+ and $^4He^{++}$ centered on $W = 1$ are evident with suprathermal tails fit by κ-functions. The pickup ion components of all three species are seen for $W \lesssim 2$ with extended high-energy tails for $W \gtrsim 2$. There were no shocks observed during this period. (Fig. 7.7 from (58))

modes, which permeate the region of space above the arcade. These are responsible for the stochastic acceleration of ^{3}He. They might also preferentially scatter the lighter ions from the current sheet with lower energy, which could account for the enhancement of the heavier ions in impulsive events. This entire scenario has the advantage that it does not depend on a postulated turbulent wave field or shock. It relies only on magnetic reconnection, which causes the flare energy release. The turbulence arises naturally from the streaming particles accelerated by the reconnection electric field.

With enhanced sensitivity and temporal and energy resolution, and excellent ability to measure elemental, charge-state and isotopic abundances, I anticipate that ACE will contribute greatly to our understanding of the origin of impulsive events.

5. SOLAR WIND AND INTERSTELLAR PICKUP IONS

The solar wind creates the heliosphere and is the medium in which most of the processes I have discussed take place. Together with the interstellar pickup ions (57) it provides the seed population for the acceleration processes which take place in the heliosphere. With energies of ~1 keV/nucleon these ions are not usually considered to be energetic. Yet the acceleration and heating of the solar wind core distribution is a subject of current research. And both the solar wind and the interstellar pickup ions exhibit high-energy tails which are intriguing and not well understood.

Figure 9 from Gloeckler et al. (58) shows Ulysses observations of the phase space densities (distribution functions) of $^1H^+$, $^4He^+$ and $^4He^{++}$ accumulated during a ~26 day period near the ecliptic in slow solar wind. Normalized to the average solar wind speed (375 km s^{-1}) for the interval, these ion speed spectra span nearly 12 orders of magnitude! $^1H^+$ and $^4He^{++}$ show a solar wind peak at W ~1 and a solar wind high-energy tail (fit in the figure by a κ-function). All three species show a pickup-ion hump extending to W ~2 and a high-energy tail extending beyond the pickup-ion hump. The solar wind itself is probably accelerated and heated by Alfvén/ion-cyclotron waves generated at the base of the corona (59,60,61,62). Although the origin of these waves at the high frequencies required to resonate with the ions in the low corona is unknown, they can in principle heat the ions perpendicular to the magnetic field as shown in Figure 3(a). The heating is very effective because the Alfvén speed is high (~2000 km s^{-1}). The mirror force in the approximately radial magnetic field then accelerates the wind (62).

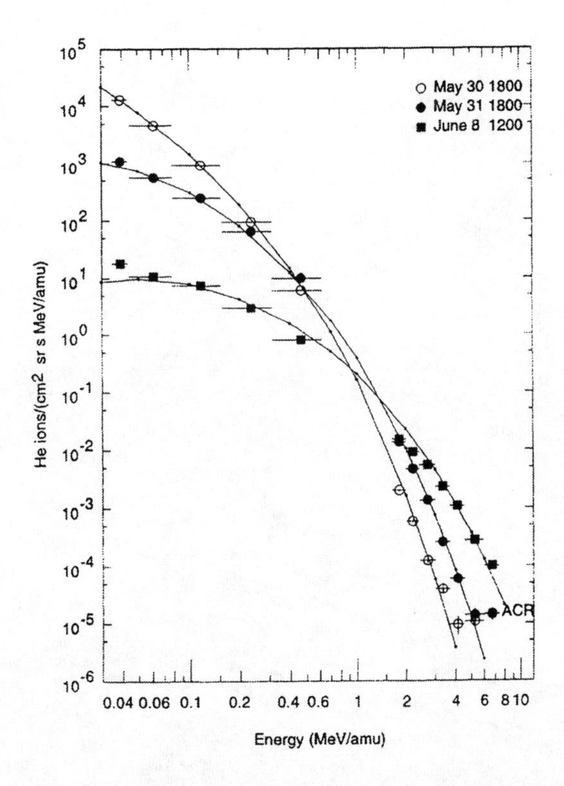

FIGURE 10. Energy spectra of energetic helium measured by Wind for three periods during the corotating ion event of May/June 1995. The spectra harden with time during the event as the magnetically-connected portion of the shock is more distant and stronger. The observations are fit using the theory of Fisk and Lee (64). (Fig. 2 from (65))

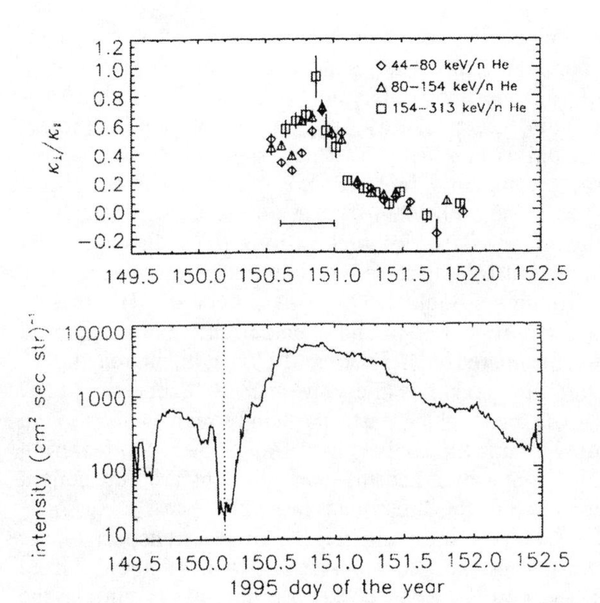

FIGURE 11. Wind observations of the corotating ion event of May/June 1995, the same event shown in Fig. 10. The lower panel shows the intensity of 44-620 keV/nucleon He with the dashed vertical line denoting the stream interface. The upper panel shows calculated values of $K_{\perp}/K_{\parallel}$ in the fast stream for the three energy channels indicated. (Fig. 4 from (68))

The origin of the high-energy tails is unclear. There were apparently no shocks present during the period shown in Figure 9. Since the Alfvén speed is small ($V_A << V$) and pitch-angle scattering rates are also small, stochastic acceleration is unlikely. A possible mechanism is "transit-time" damping of compressive magnetosonic waves (9,63). These waves damp by accelerating ions in Landau resonance ($\omega - k_z v_z \cong 0$) toward the velocity ω/k_z by mirroring the ions in the "magnetic bottles" associated with wave compression. If the waves propagate obliquely to **B** so that $\omega/k_z >> V_A$, this acceleration can be substantial.

6. INTERPLANETARY ENERGETIC PARTICLE EVENTS

The dominant population of interplanetary energetic particles is associated with corotating interaction regions (CIRs) in the solar wind. These corotating ion events consist of ions accelerated at the forward and reverse shocks bounding the CIR. The shocks generally form beyond ~ 3AU. The ion intensities peak at the two shocks but extend inwards to the orbit of Earth from the forward shock in the slow solar wind and from the reverse shock in the fast wind. The theory of diffusive shock acceleration was first applied to these events by Fisk and Lee (64), who were able to account for most of their observed features. The acceleration is best considered in the corotating frame where the ion distribution is stationary; the ion spectrum results from a balance between shock acceleration and adiabatic deceleration in the diverging solar wind. Interestingly the energy-asymptotic expression derived by Fisk and Lee (64) for $f_i\,(p,r)$ was shown by Scholer et al. (13) to be valid at low energies as well, as long as $r \gtrsim 1$ AU. Thus, at least under their assumptions that perpendicular diffusion is negligible and $K_{rr} \propto vr$, the asymptotic expression should be quite robust. Figure 10 from Reames et al. (65) shows fits of that expression to the energy spectra observed by Wind at three times during a large event in 1995. The fits are clearly very good. The spectra later in the event are harder since the path along the magnetic field through the spacecraft to the shock is longer, which favors the access of higher-energy ions, and the shock at the point of connection is stronger, which results in more effective shock acceleration. Nonetheless, it is puzzling that a decrease in the ion differential intensity at very low energies is expected but not observed.

The composition of the corotating ion events is generally consistent with their origin in the fast and slow solar wind. However, there are some puzzles which require resolution (23). Helium and carbon are enhanced by about a factor of two compared with abundances in the fast solar wind, which supplies most of the energetic CIR ions at 1AU. The additional helium ions probably originate as interstellar pickup helium, which has a "thermal" speed equal to V and is more efficiently injected and accelerated at the shock (66). The additional carbon ions may originate from the inner source of pickup ions (67), although that source also includes oxygen and does not appear to be sufficient to yield C/O ~ 1 as observed. The pickup ion origin of the He and C enhancements is supported by the increase of He/O and C/O with increasing V. Another puzzle is the constancy of energetic Mg/O across the slow and fast solar wind streams even though the streams themselves have values of Mg/O differing by a factor of 2 and the energetic ions are not thought to penetrate the stream interface (23).

A very interesting feature of the transport of the corotating energetic ions is shown in Figure 11 from Dwyer et al. (68). The second panel shows the intensity of 44-620 keV He in the 30 May 1995 event including a small (large) enhancement in the slow(fast) stream and a precipitous decrease at the stream interface. The ion flux streaming away from the shock in the frame of the fast wind is observed to have the form of a diffusive flux with a fixed spatial gradient and different values of $K_\perp$ and $K_{||}$. Fitting the flux data for different orientations of **B** yields the values of $K_\perp/K_{||}$ shown in the top panel. These large values of $K_\perp/K_{||}$ are remarkable. Turbulence levels in the fast wind at 1 AU are low and impulsive events at similar energies in the fast wind appear to have small values of $K_\perp/K_{||}$. A satisfactory explanation of this observation does not exist. Large values of $K_\perp/K_{||}$ would invalidate the theory of Fisk and Lee (64).

The diffuse ions at Earth's and other planetary bow shocks are understood in principle as a result of diffusive shock acceleration (69,70,71,72). However, the finite size and the curvature of a bow shock are not easy to include with rigor in a theory. Near the foreshock boundary the acceleration process has just begun and ion distributions are not nearly isotropic. Rather they are observed as "field-aligned beams", or as "intermediate" ion distributions in association with large-amplitude turbulence excited by the streaming ions. Also it is not entirely clear whether the soft energy spectrum of the diffuse ions, which is exponential in energy per charge, is due to finite field-line connection time to the shock, ion drift along the shock front to the weak flanks of the shock, diffusive escape of ions across field lines to the weak flanks, or upstream escape at an effective "free-escape" boundary.

7. THE ANOMALOUS COSMIC RAY COMPONENT

The anomalous cosmic rays originate predominantly as interstellar pickup ions accelerated at the solar wind

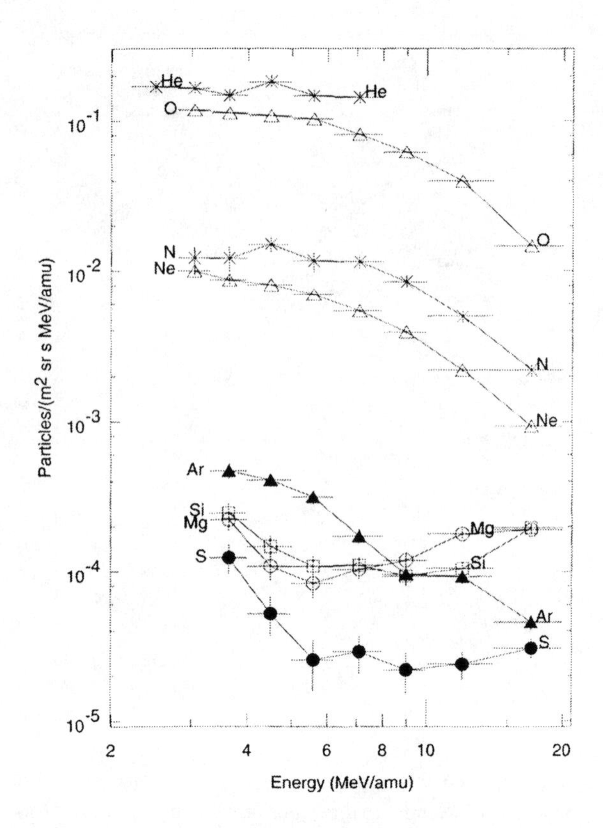

FIGURE 12. Wind observations of ion energy spectra measured in quiet times during 1994-1998 for elements with a clear (He, O, N, Ne, Ar) or a possible (Si, Mg, S) ACR enhancement at low energy. (Fig. 4 from (75))

termination shock (73,74). In principle, the termination shock will also accelerate ambient solar wind or energetic particles from, for example, CIRs. However, pickup ions should be enhanced over ambient solar wind because of their much larger "thermal" speed of order V, which should favor their injection rate. Also pickup ions are mostly singly-ionized and have a larger rigidity at the same speed, so that they have easier access to the inner heliosphere as ACRs. The exception is hydrogen. Since ACR hydrogen is underabundant even assuming all ACRs originate as interstellar pickup ions (33), it is very unlikely that ions are extracted from the solar wind and accelerated at the termination shock.

Figure 12 from Reames (75) shows quiet-time spectra measured by Wind in the energy range 3-20 MeV/nucleon during 1994-1998. Apart from H, which is difficult to disentangle from GCR hydrogen, the elements He, O, N, Ne and Ar are those expected from interstellar pickup ions. The spectra peak at energies below ~ 3 MeV/nucleon by virtue of the solar modulation of the rather soft spectra at the shock. The origin of the low-energy enhancements in Si, Mg and S is unclear. They could arise from a background population of SEPs at quiet times, or they could be

ACRs originating from inner source pickup ions or Jovian ions (at least for sulphur).

The ACR also provides us with our best estimate of the location of the termination shock. As the Voyager spacecraft travel further from the Sun they observe the demodulation of the ACR energy spectra toward a spectrum at the shock which is expected to be a power law at low energies with an exponential cutoff above ~250 MeV/charge. The cutoff is due to the potential difference between the pole and the equator of the heliosphere. Careful fits to the data of solutions of the spherically-symmetric transport equation (2) neglecting $\mathbf{V}_{Di}$ and D_{pp}, with the required form at the shock, yield an estimate of the shock location. Based on 1994 data, Cummings and Stone (33) find a shock location of $r_s = 81 \pm 3$ AU in the general direction of Voyagers 1 and 2. Noting that the shock "breathes" in response to variations in solar wind ram pressure and can move in or out by a few AU, we can expect Voyager 1, currently at about 77 AU, to encounter the shock in a few years. Although drift transport of cosmic rays is crucial to understanding their global distribution throughout the heliosphere, its neglect in the solutions used to fit the data relatively close to the shock should be approximately valid, since the ACR gradient should be nearly radial and therefore normal to $\mathbf{V}_{Di}$.

The preferential injection of pickup ions at the shock to form the ACR component is at first sight puzzling. Although the "thermal" speed of pickup ions is far greater than that of solar wind ions, the upstream speeds of both species relative to the plasma downstream of the shock are similar and should determine scattering rates downstream and the probability of upstream escape. However, the termination shock on average is a perpendicular shock since the Archimedean spiral magnetic field is wrapped into a very tight spiral in the outer heliosphere. At perpendicular shocks ions which approach the shock front slowly may be trapped between the electrostatic shock potential and the upstream Lorentz force. While oscillating in the trap, these ions bounce or surf along the shock front gaining energy in the motional electric field (30,31). When they finally overcome the shock potential and move downstream, their energy, primarily due to motion along the shock surface, can exceed the threshold for diffusive shock acceleration. Since no solar wind ions approach the shock slowly, whereas many pickup ions do, this injection mechanism can in principle account for the pickup ion origin of ACRs. However, it appears to be in conflict with the inference that the interstellar pickup ion injection rate increases with ion mass (33); shock surfing should favor the lighter ions, more of which can be reflected by the shock potential.

Another interesting feature of the termination shock is that the ACR hydrogen may have sufficient pressure to modify the structure of the shock (e.g., 76). In a modified shock the velocity jump occurs in a gradual precursor decrease followed by a weaker subshock. At a modified shock, massive pickup ions like N, O, and Ne sample more of the full shock compression within a gyroradius than do H and He during the acceleration at low energies and acquire harder spectra. If there is sufficient turbulence to obviate the injection problem, then this mechanism could also result in enrichment of the heavy ions (77). Clearly there are still many puzzles concerning the structure of the termination shock. We await with excitement Voyager's traversal of the shock within the next few years.

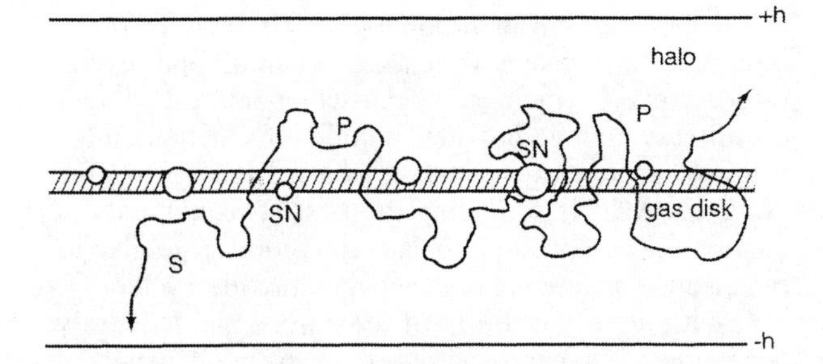

FIGURE 13. Schematic diagram of a cross-section of the Galaxy showing the gas disk and the halo extending a distance h above and below the disk. Supernova remnants in the disk bounded by shocks are the origin of primary (P) cosmic rays, which diffuse through the Galaxy and escape into intergalactic space. Primary cosmic rays may collide with gas in the disk to create secondary (S) cosmic rays.

8. GALACTIC COSMIC RAYS

Figure 13 is a schematic diagram of the life-cycle of galactic cosmic rays with energies up to ~ 10^{17} eV. It shows a cross section of the Galaxy with a gas disk of width ~100 parsec and a halo of width $2h$ ~1 kiloparsec. The halo has low gas density and consists of magnetic loops blown out of the gas disk by the cosmic ray pressure. The primary (P) cosmic rays originate at the shock waves surrounding supernovae bubbles resulting from explosions of massive stars born in the gas disk. These cosmic rays are channeled throughout the Galaxy by the galactic magnetic field and scattered by magnetic irregularities generated by turbulent gas motion. They remain in the Galaxy for about 10^7 yr before escaping into intergalactic space. The schematic trajectories shown are "guiding center" trajectories averaged over each gyroperiod; the irregular shape results from diffusion.

While traveling through the gas disk some cosmic rays suffer nuclear collisions with ambient gas, which may destroy the original primary ion and create a new secondary (S) ion as shown. The composition of the primary and secondary nuclei are very different. Primaries are accelerated by the shock wave out of the interstellar medium with a possible small admixture of debris from the supernova explosion. Secondaries consist of spallation products which include many otherwise rare species such as Li, Be and B and several radioisotopes, which are of particular interest. Every cosmic ray nuclide provides unique information on the origin of cosmic rays and their propagation through the Galaxy. ACE is providing the best measurements of cosmic ray composition to date and promises to give us the clearest picture yet of the cosmic-ray life-cycle.

Although the bulk of cosmic rays is surely accelerated at supernovae shock waves, many details of the process are unknown. The *X*-ray emission from supernovae remnants is due to electron bremsstrahlung and provides direct information on the electron component but no information on the energetically dominant ions. Since supernovae shock waves are very strong (Mach number >> 1) and the galactic cosmic ray pressure is comparable to the interstellar gas pressure, the cosmic ray pressure strongly modifies the structure of supernovae shocks. These modified shocks are very efficient at transforming the kinetic energy of the supernova explosion into cosmic ray energy. If relativistic particles dominate the cosmic ray pressure they decrease the adiabatic index of the gas from 5/3 towards 4/3, which increases the shock compression ratio and hardens the energy spectrum of the accelerated cosmic rays. Also important is adiabatic deceleration of the cosmic rays behind the shock and escape of the accelerated particles into interstellar space. Supernova shock waves cannot accelerate cosmic rays to energies $\gtrsim 10^{16}$ - 10^{17} eV due to their limited lifetime as strong shocks; such cosmic rays are also not easily confined to the shock. This high-energy cutoff must correspond to the "knee" in the cosmic ray energy spectrum observed at $\sim 10^{15}$ eV with a softer spectrum at higher energies (78). The highest energy cosmic rays are either reaccelerated in the Galaxy, are extragalactic, or originate at objects like young pulsars.

The galactic propagation of cosmic rays can be investigated using equation (2) with an appropriate choice of source Q and a spallation loss term. A simplified set of equations for a primary species P and a secondary species S for the geometry of Figure 13 is, for example, given by

$$-K\frac{\partial^2 P}{\partial z^2}+\frac{\partial}{\partial E}(\dot{E}P)=Q(E)\,\delta(z)-\lambda\delta(z)P \qquad (6)$$

$$-K\frac{\partial^2 S}{\partial z^2}+\frac{\partial}{\partial E}(\dot{E}S)=-\Gamma S+\sigma\delta(z)P \qquad (7)$$

where $P(z,E)$ and $S(z,E)$ are the differential densities of the primary and secondary species ($\int dE\,P$ is the primary number density, for example), E is kinetic energy, K is the spatial diffusion coefficient, the cosmic ray configuration is assumed to be planar and stationary, and free escape at the halo boundaries requires $P(\pm h, E) = S(\pm h, E) = 0$. The term $Q(E)\,\delta(z)$ describes the source of the primary species in the gas disk, $-\lambda\delta(z)\,P$ describes the spallation loss of that species in the gas disk, $\sigma\,\delta(z)\,P$ describes the production of the secondary species S due to the spallation of species P, and $-\Gamma S$ describes the decay of a radioisotope. The second term on the left side of both equations describes possible reacceleration of cosmic rays by shock waves. Equations (6) and (7) can be solved and provide a somewhat more detailed description of cosmic ray propagation than the traditional "leaky box" model. The important point is that each nuclide is described by an equation such as (6) or (7) and its abundance gives information about galactic structure. The quantities λ and σ depend on spallation cross sections, which must be measured separately, and the line-of-sight integrated density (grammage) of the gas disk. Using the propagation equations, measured abundances determine the (energy-dependent) grammage traversed by the average particle during its lifetime, and, in the case of radioactive nuclides, the cosmic ray lifetime. The most important cosmic ray "clocks" (radioactive nuclides) are ^{10}Be with a half-life at rest of 2.3 x 10^6 yr and ^{26}Al with a half-life of 1.3 x 10^6 yr (79). Also important as diagnostics of the cosmic ray life-cycle are the radionuclides which decay by electron capture. These only decay if orbital electrons are present. As high-energy cosmic rays these nuclides are stripped of their electrons and cannot decay. Thus, if they are produced in supernovae explosions with known abundance, their abundance in cosmic rays can provide a measure of the time interval between supernova explosion and acceleration. Similarly, secondary electron-capture nuclides may provide information on cosmic ray reacceleration since decay would imply that they spent some of their lifetime at lower energy where the probability of electron capture is much larger.

The overall abundance of primary cosmic rays is remarkably similar to solar system abundances. This supports the idea that cosmic rays are accelerated by a supernova blast wave out of an interstellar medium not dissimilar from that which collapsed to form the solar system, with little admixture of exotic material from the supernova explosion. However, the primary cosmic rays are systematically biased with respect to solar system abundances with enhancements by about a factor of 5 for the refractory elements, those with low first ionization potential (FIP) (e.g. 80). This must indicate that the seed particles for cosmic ray acceleration include not only interstellar gas, but also an admixture of refractory elements from interstellar dust (81), dust grains formed in the supernova (82), or stellar energetic particles with the same enhancement in low-FIP elements evident in SEPs (80).

CONCLUSIONS

The variety of energetic particles in the heliosphere is very rich. In some cases, such as ACRs and gradual

SEP events, the basic acceleration mechanism is known but a quantitative evaluation of the consequences remains challenging. In other cases, such as impulsive SEP events, the basic mechanism is a subject of controversy. In either case the observations of ACE and other spacecraft provide the detailed composition and the temporal and energy resolution necessary to discover new features, stimulate new ideas and challenge theory. This interplay between observations, ideas and theory is the essence of good science. Let the ACE-2000 Symposium begin!

ACKNOWLEDGMENTS

I wish to thank the Caltech Local Organizing Committee for an extremely well organized Symposium and a well chosen venue! This work was supported, in part, by NASA Sun-Earth-Connections Theory Program Grant NAG5-1479, by NASA Grant NAG5-7792, and by NSF Grant ATM - 9633366.

REFERENCES

1. Miller, J.A., et al., *J. Geophys. Res.* **102**, 14,631 (1997).
2. Reames, D.V., *Space Sci. Rev.* **90**, 413 (1999).
3. Burlaga, L.F., McDonald, F.B., Ness, N.F., Schwenn, R., Lazarus, A.J., and Mariani, F., *J. Geophys. Res.* **89**, 6579 (1984).
4. Lee, M.A., *J. Geophys. Res.* **88**, 6109 (1983).
5. Gordon, B.E., Lee, M.A., Möbius, E., and Trattner, K.J., *J. Geophys. Res.* **104**, 28,263 (1999).
6. Jones, F.C., and Ellison, D.C., *Space Sci. Rev.* **58**, 259 (1991).
7. Jokipii, J.R., *Astrophys. J.* **255**, 716 (1982).
8. Litvinenko, Y.E., *Astrophys. J.* **462**, 997 (1996).
9. Fisk, L.A., *J. Geophys. Res.* **81**, 4633 (1976).
10. Parker, E.N., *Planet. Space Sci.* **13**, 9 (1965).
11. Gleeson, L.J., and Axford, W.I., *Astrophys. J.* **149**, L115 (1967).
12. Jokipii, J.R., Levy, E.H., and Hubbard, W.B., *Astrophys. J.* **213**, 861 (1977).
13. Scholer, M., et al., *Space Sci. Rev.* **89**, 369 (1999).
14. Roelof, E.C., "Propagation of solar cosmic rays in the interplanetary magnetic field", in *Lectures in High Energy Astrophysics*, edited by H. Ogelmann and J.R. Wayland, *NASA Spec. Publ. SP-199*, 1969, p.111.
15. Kallenrode, M.B., *J. Geophys. Res.* **98**, 19,037 (1993).
16. Ng, C.K., and Reames, D.V., *Astrophys. J.* **424**, 1032 (1994).
17. Heras, A.M., Sanahuja, B., Lario, D., Smith, Z.K., Detman, T., and Dryer, M., *Astrophys. J.* **445**, 497 (1995).
18. Isenberg, P.A., *J. Geophys. Res.* **102**, 4719 (1997).
19. Mason, G.M., Reames, D.V., Klecker, B., Hovestadt, D., and von Rosenvinge, T.T., *Astrophys. J.* **303**, 849 (1986).
20. Tylka, A.J., Reames, D.V., and. Ng, C.K., *Geophys. Res. Lett.* **26**, 2141 (1999).
21. Mewaldt, R.A., Selesnick, R.S., Cummings, J.R., Stone, E.C., and von Rosenvinge, T.T., *Astrophys. J.* **466**, L43, (1996).
22. Jokipii, J.R., *Astrophys. J.* **466**, L47 (1996).
23. Mason, G.M., Mazur, J.E., Dwyer, J.R., Reames, D.V., and von Rosenvinge, T.T., *Astrophys. J.* **486**, L149 (1997).
24. Quest, K.B., *J. Geophys. Res.* **93**, 9649 (1988).
25. Ellison, D.C., Möbius, E., and Paschmann, G., *Astrophys. J.* **352**, 376 (1990).
26. Giacalone, J., Burgess, D., Schwartz, S.J., and Ellison, D.C., *Geophys. Res. Lett.* **19**, 433 (1992).
27. Liewer, P.C., Goldstein, B.E., and Omidi, N., *J. Geophys. Res.* **98**, 15,211 (1993).
28. Kucharek, H., and Scholer, M., *J Geophys. Res.* **100**, 1745 (1995).
29. Malkov, M.A., *Phys. Rev. E* **58**, 4911 (1998).
30. Lee, M.A., Shapiro, V.D., and Sagdeev, R.Z., *J. Geophys. Res.* **101**, 4777 (1996).
31. Zank, G.P., Pauls, H.I., Cairns, I.H., and Webb, G.M., *J. Geophys. Res.* **101**, 457 (1996).
32. Bieber, J.W., and Evenson, P., *Geophys. Res. Lett.* **25**, 2955 (1998).
33. Cummings, A.C., and Stone, E.C., *Space Sci. Rev.* **78**, 117 (1996).
34. Kurth, W.S., Gurnett, D.A., Scarf, F.L., and Poynter, R.L., *Nature* **312**, 27-31 (1984).
35. Gurnett, D.A., Kurth, W.S., Allendorf, S.C., and Poynter, R.L., *Science* **262,** 199-203 (1993).
36. Mason, G.M., Ng, C.K., Klecker, B., and Green, G., *Astrophys. J.* **339**, 529 (1989).
37. Cane, H.V., Reames, D.V. and von Rosenvinge, T.T., *J. Geophys. Res.* **93**, 9555 (1988).
38. Lee, M.A., "Particle acceleration and transpsort at CME-driven shocks", in *Coronal Mass Ejections*, edited by N. Crooker, J.A. Joselyn, and J. Feynman, AGU, Washington, DC, 1997, p. 227.
39. Breneman, H.H., and Stone, E.C., *Astrophys. J.*, **299**, L57 (1985).
40. Mason, G.M., Mazur, J.E., and Dwyer, J.R., *Astrophys. J.* **525**, L133 (1999).
41. Achterberg, A., and Norman, C.A., *Astron. Astrophys.* **89**, 353 (1980).
42. Lee, M.A., and Fisk, L.A., *Space Sci. Rev.* **32**, 205, (1982).
43. Lee, M.A., and Ryan, J.M., *Astrophys. J.* **303**, 829 (1986).
44. Ruffolo, D., *Astrophys. J.* **442,** 861 (1995).
45. Lario, D., Sanahuja, B., and Heras, A.M., *Astrophys. J.* **509**, 415 (1998).
46. Ng, C.K., Reames, D.V., and Tylka, A.J., *Geophys. Res. Lett.* **26**, 2145 (1999).
47. Lee, M.A., *EOS*, *Spring Meeting Suppl.*, 1999, S255(Abstract).
48. Reames, D.V., Barbier, L.M., von Rosenvinge, T.T., Mason, G.M., Mazur, J.E., and Dwyer, J.R., *Astrophys. J.* **483**, 515 (1997).
49. Luhn, A., et al., *Adv. Space Res.* **4** (2-3), 161 (1984).
50. Miller, J.A., Guessoum, N., and Ramaty, R., *Astrophys. J.* **361**, 701, (1990).
51. Ryan, J.M., and Lee, M.A., *Astrophys. J.* **368**, 316, (1991).
52. Holman, G.D., *Astrophys. J.* **452**, 451 (1995).
53. Miller, J.A., and Viñas, A.F., *Astrophys. J.* **412**, 386 (1993).

54. Tsuneta, S., and Naito, T., *Astrophys. J.* **495**, L67 (1998).
55. Fisk, L.A., *Astrophys. J.* **224**, 1048 (1978).
56. Temerin, M., and Roth, I., *Astrophys. J.* **391**, L105, (1992).
57. Möbius, E., Hovestadt, D., Klecker, B., Scholer, M., Gloeckler, G., and Ipavich, F.M., *Nature* **318**, 426 (1985).
58. Gloeckler, G., Geiss, J., and Fisk, L.A., "Heliospheric and interstellar phenomena revealed from observations of pickup ions", in *The Heliosphere near Solar Minimum: the Ulysses Perspectives*, edited by A. Balogh, E.J. Smith and R.G. Marsden, Springer-Praxis, Berlin, in press, 2000.
59. McKenzie, J.F., Banaszkiewicz, M., and Axford, W.I., *Astron. Astrophys.* **303**, 45 (1995).
60. Hollweg, J.V., *J. Geophys. Res.* **104**, 24,781 (1999).
61. Tam, S.W.Y., and Chang, T., *Geophys. Res. Lett.* **26,** 3189 (1999).
62. Isenberg, P.A., Lee, M.A., and Hollweg, J.V., *Sol. Phys.*, in press, 2000.
63. Fisk, L.A., Gloeckler, G., Zurbuchen, T.H., and Schwadron, N.A., this volume, 2000.
64. Fisk, L.A., and Lee, M.A., *Astrophys. J.* **237**, 620 (1980).
65. Reames, D.V., Ng, C.K., Mason, G.M., Dwyer, J.R., Mazur, J.E., and von Rosenvinge, T.T. , *Geophys. Res. Lett.* **24**, 2917 (1997).
66. Gloeckler, G., et al., *J. Geophys. Res.* **99**, 17,637 (1994).
67. Geiss, J., Gloeckler, G., Fisk, L.A., and von Steiger, R., *J. Geophys. Res.* **100**, 23,373 (1995).
68. Dwyer, J.R., Mason, G.M., Mazur, J.E., Jokipii, J.R., von Rosenvinge, T.T., and Lepping, R.P., *Astrophys. J.* **490**, L115 (1997).
69. Eichler, D., *Astrophys. J.* **244**, 711 (1981).
70. Ellison, D.C., *Geophys. Res. Lett.* **8**, 991 (1981).
71. Lee, M.A., *J. Geophys. Res.* **87**, 5063 (1982).
72. Smith, C.W., and Lee, M.A., *J. Geophys. Res.* **91**, 81 (1986).
73. Pesses, M.E., Jokipii, J.R., and Eichler, D., *Astrophys. J.* **246**, L85 (1981).
74. Jokipii, J.R., *J. Geophys. Res.* **91**, 2929 (1986).
75. Reames, D.V., *Astrophys. J.* **518**, 473 (1999).
76. leRoux, J.A., and Fichtner, H., *J. Geophys. Res.* **102**, 17,365 (1997).
77. Ellison, D.C., Jones, F.C., and Baring, M.G., *Astrophys. J.* **512**, 403 (1999).
78. Axford, W.I., *Proc. Internat. Conf. Cosmic Rays 17th* **12,** 155 (1981).
79. Ptuskin, V.S., and Soutoul, A., *Space Sci. Rev.* **86**, 225 (1998).
80. Meyer, J.P., *Astrophys. J. Suppl.* **57**, 151 (1985).
81. Meyer, J.P., Drury, L.O'C., and Ellison, D.C., *Astrophys. J.* **487**, 182 (1997).
82. Lingenfelter, R.E., Ramaty, R., and Kozlovsky, B., *Astrophys. J.* **500**, L153 (1998).

Solar Energetic Particles

Solar Flare Photons and Energetic Particles in Space

E. W. Cliver

Air Force Research Laboratory, Hanscom AFB, MA 01731-3010

Abstract. I review the evolution of research on solar energetic particle events, beginning with Forbush's report of the ground level event of 1946, through the most recent observations of the Advanced Composition Explorer (ACE). The emphasis is on research that attempted to link solar flare electromagnetic emissions with the solar energetic particles (SEPs) observed in space following flares. The evolution of thought on this topic is traced from the initial paradigm in which SEPs were accelerated at the flare site (a δ-function in space and time) to the current two-class picture accommodating both impulsive acceleration at the flare site (small ^{3}He-rich events) and prolonged acceleration at extended shocks driven by coronal mass ejections (large proton events). I conclude with some open questions; the most prominent of these concerns the relative contributions of the flare and shock acceleration processes to "mixed" or hybrid SEP events in which the distinguishing characteristics of the impulsive and gradual classes are blended.

INTRODUCTION

Within the last 15 years our views of particle acceleration in association with solar flares have undergone a remarkable revision and codification (1). In this review, I trace the steps that led to the current picture. The emphasis will be on observations of solar flare electromagnetic emissions and the solar energetic particles (SEPs) that are observed in space following flares and coronal mass ejections (CMEs). This emphasis on observations reflects my bias but, more importantly, it indicates that ours is a field driven by data. New instruments, such as those on the Advanced Composition Explorer (ACE), inevitably lead to revisions of earlier ideas based on less sophisticated measurements. ACE has already challenged and refined the existing picture. Knowing how our present understanding developed can serve as a guideline to accommodate additional surprises that are certain to come as more events are recorded.

In this paper, I first review the origins of this field beginning with the 25 July 1946 ground level event and continuing through the association of coronal mass ejections (CMEs) and SEPs based on Skylab coronagraph data. The next section of the paper addresses correlations/associations between the intensities of various flare emissions and the sizes of SEP events and the section following that recounts the remarkable flurry of research during the decade from 1984-1993 that established the two-class paradigm. I conclude with a list of open questions for further investigation.

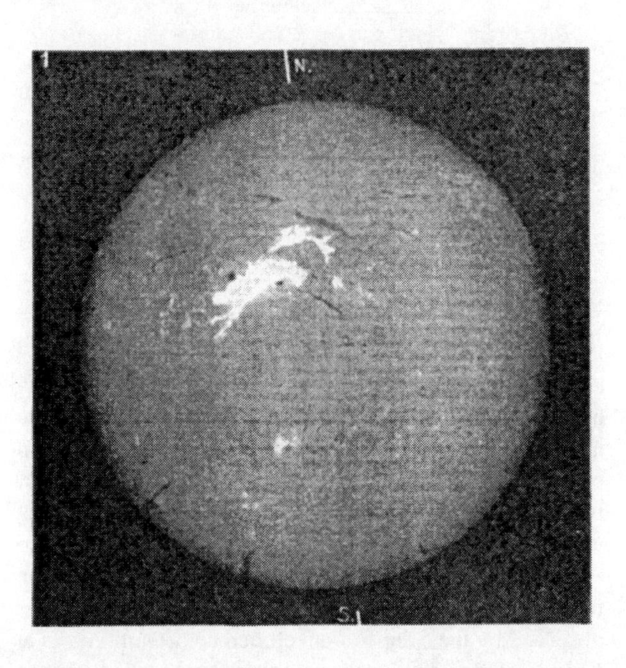

FIGURE 1. Meudon Observatory Hα image of the solar flare associated with the ground level event of 25 July 1946. From (3), with kind permission of Société Astronomique de France.

ORIGINS

The history of this field began with the report by Forbush (2) of a sudden transient increase in the counting rate of his ionization chamber on 25 July 1946 in association with a large solar flare. The flare, observed in Hα

CP528, *Acceleration and Transport of Energetic Particles Observed in the Heliosphere: ACE 2000 Symposium,*
edited by Richard A. Mewaldt, et al.
2000 American Institute of Physics 1-56396-951-3

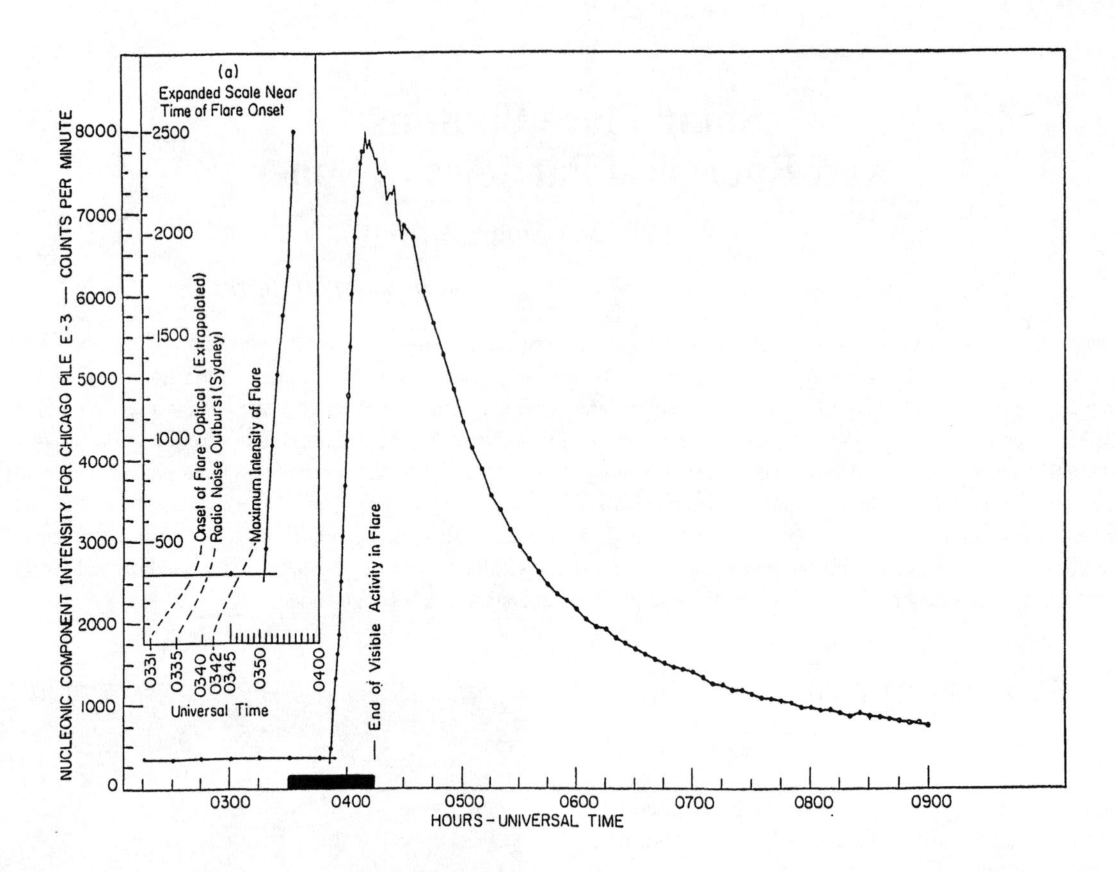

FIGURE 2. Chicago neutron monitor record of the ground level event of 23 February 1956 (adapted from 5).

at Meudon, is shown in Figure 1 (3). Lange and Forbush (4) had earlier observed two similar counting rate increases, occurring in association with solar activity in early 1942, but Forbush did not draw the connection between the earlier increases and the simultaneous flaring until he observed the 1946 event.

The first SEP event to be analyzed in detail was the famous ground level event of 23 February 1956 that was observed by the Simpson-developed neutron monitors at Chicago and five other sites. Meyer, Parker, and Simpson (5) drew attention to the short duration ($\sim$45 minutes) of the flare, marked with a heavy line in Figure 2, and the long duration ($\sim$15 hours) of the particle event. They concluded that a short-lived acceleration was followed by diffusive propagation. Thus the initial paradigm for particle acceleration in association with solar flares was that all particles were rapidly accelerated at the flare site. SEP acceleration could be modeled as a δ-function injection in space and time (Figure 3), given the relatively small angular extent and short time span of the flare. Actually Figure 3 is anachronistic because the Parker spiral had not yet been hypothesized in 1956.

As a result of radio observations of flares, various authors suggested that proton events (observed early on as polar cap absorption events with riometers) might be associated with only certain flares, specifically those with long-duration type IV emission. For example, Kundu and Haddock (6) suggested that intense microwave type IV bursts were a signature of proton acceleration. The most systematic expression of the emerging view that "proton flares" were different from other flares was given by J.P. Wild, S.F. Smerd, and A.A. Weiss in an Annual Reviews paper in 1963 (7). By this time, these Australian radio astronomers had already been observing the Sun at meter wavelengths for over a decade. They separated flares into two characteristic types (or phases) as shown in Figure 4 (8). The vast majority of flares exhibited only the phenomena to the left of the dashed line in the figure. Type III radio bursts, attributed to escaping beams of $\sim$100 keV electrons, were the characteristic emission of impulsive flares (or of the impulsive phase of fully-developed flares). Certain flares - generally the largest or most intense events - were accompanied by a second phase of radio emission characterized by slow-drifting type II bursts,

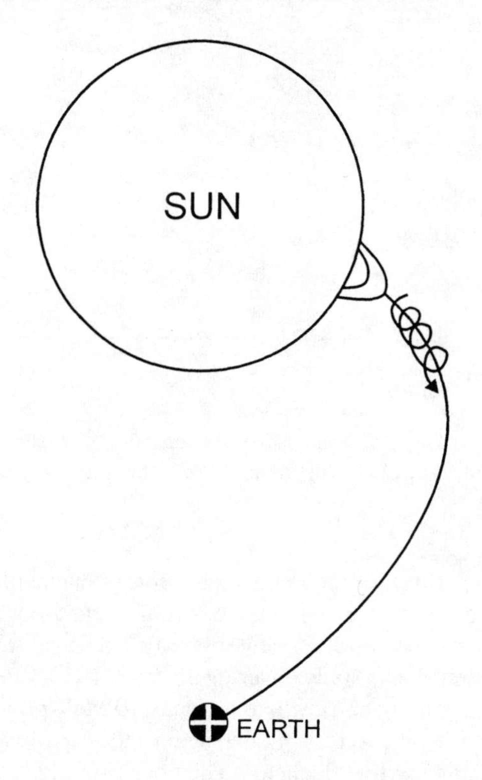

FIGURE 3. The initial paradigm for particle acceleration at the Sun: it was thought that all SEPs were rapidly accelerated at the flare site.

attributed to magnetohydrodynamic shocks moving outward in the corona with speeds $\sim$1000 km s^{-1}, as well as by various kinds of type IV emission. Wild, Smerd, and Weiss suggested that protons and high-energy (relativistic) electrons were accelerated in this second phase.

Their suggestion was so close to our current picture that it is worth quoting directly, "Studies of radio emission ... give striking evidence that two separate phases [of particle acceleration] are involved. The first (which is sometimes preceded by a gradual preliminary heating of the flare region) is a succession of bursts of electrons ($\sim$100 keV), the acceleration of each being accomplished in a very short time ($\sim$1 sec); this requires a catastrophic event, probably involving the conversion of magnetic into kinetic energy ... The acceleration of protons to high energies need not be involved in this phase. The second phase, occurring only in large flares, is initiated directly by the first: the sudden release of energy sets up a magnetohydrodynamic shock wave which travels out through the coronal plasma and creates conditions suitable for Fermi acceleration of protons and electrons to very high energies ($\sim$1 BeV)." The main differences between this picture and the current paradigm are: (a) the role ascribed to CMEs for shock formation in the modern view (9, 10); and (b) the recognition that protons and high-energy electrons are also accelerated in the impulsive phase of energetic flares (11).

Evidence for the "two-phase" picture, as it came to be called, was provided by Lin (12) who reported on the basis of early SEP measurements from space that "pure" electron events were preferentially associated with flares that only exhibited metric type III emission while "mixed" events with protons and relativistic electrons tended to follow flares with type II/IV radio events. Thus, the work of Wild, Smerd, and Weiss and Lin laid the foundation for the current two-class picture of SEP events championed by Reames and colleagues (1, and references therein), although the general acceptance of this view - with a host of new details - would take nearly a quarter century following Lin's work.

The last key element of the modern picture was recognized in 1978 when Kahler, Hildner, and Van Hollebeke pointed out an association between CMEs observed by the coronagraph on Skylab and proton events. There was still considerable support for the flare δ-function paradigm (Figure 3) at the time, however, so Kahler et al. hedged their bets on the interpretation. In their conclusion, they suggested that the CME might facilitate observation of protons either by opening field lines so particles accelerated by the flare could escape or, alternatively - the current view - that "there may exist a proton acceleration region above or around the outward moving ejecta far above the flare site."

CORRELATIONS AND ASSOCIATIONS

Lin and Hudson (13) presented evidence that a large hard X-ray burst was a necessary condition for a major $\sim$10 MeV proton event at Earth. They viewed the intense X-ray emission as evidence for explosive heating of the solar atmosphere, producing a blast wave that drove material out of the Sun's gravitational field and also accelerated the energetic particles. Thus the flare was primary and the ejected mass secondary to the particle acceleration. Kahler (14) injected a note of caution regarding findings such as that of Lin and Hudson that drew associations between phenomena observed in large flares. He posited the existence of a "big flare syndrome" which said, in effect, that big flares have more of everything and that correlations and associations observed in samples of big flares should not be taken as evidence of causality.

How to beat the big flare syndrome? Cliver, Kahler, and McIntosh (15) identified several large proton flares

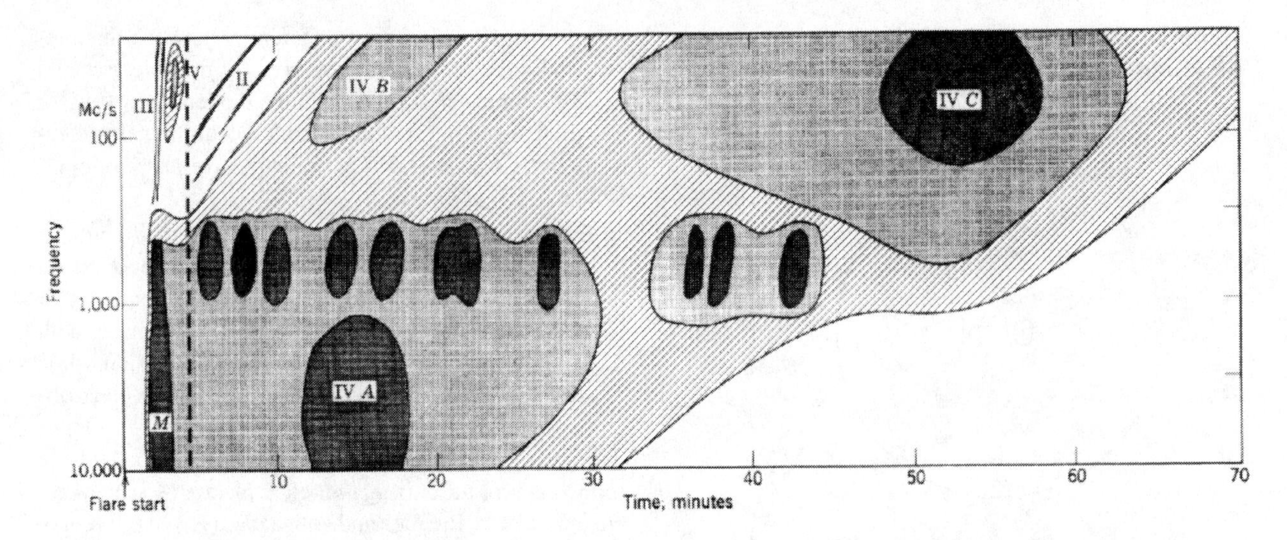

FIGURE 4. Schematic of radio emission for a large "fully-developed" solar flare (8). The dashed line separates the first and second phases. From M. R. Kundu and S. F. Smerd, *Inf. Bull. Europ. Solar Radio Obs.*, **11**, 4, Copyright © 1962. Reprinted by permission of John Wiley & Sons, Inc.

that lacked the strong impulsive phases seemingly required by the analysis of Lin and Hudson (13). They determined that these "weak impulsive phase" proton flares (having peak ~9 GHz flux densities $\leq$ 100 solar flux units) exhibited clear evidence for mass ejection and shock formation. Cliver et al. concluded their paper as follows: "...the fact that significant proton events can originate in flares without prominent impulsive phases leads us to question the importance of the flash [impulsive] phase in the production of protons observed at 1 AU in even the classic "big flares". For the "big flares" as well as for the weak impulsive phase proton flares, we suspect that the key element leading to shock formation and the subsequent acceleration of the protons observed at Earth is a *magnetically driven* mass ejection [italics in the original]." This stood in contrast to the picture of Lin and Hudson in which the impulsive phase energy release was a requirement for CME formation. Kahler et al. (16) took the study of Cliver et al. (15) one step further when they reported a SEP event that lacked an associated flare. Actually the double-ribbon brightening that accompanied a filament disappearance in this event (Figure 5) can be thought of as a "soft" version of a flare but the absence of the impulsive phase fireworks (e.g., big hard X-ray and microwave bursts) in this event underscored the notion that these emissions are not indicative of the process whereby the protons observed at Earth are accelerated.

Before the launch of the Solar Maximum Mission (SMM) satellite with its Gamma-Ray Spectrometer (GRS) in 1980 (17), proton event intensities at 1 A.U. could only be compared with flare radio and X-ray emissions from electrons back at the Sun. The GRS enabled comparisons of the numbers of protons interacting in the solar atmosphere as deduced from gamma-ray line observations with SEP event intensities in space (e.g., 18). The result of a study covering the years 1980-1985 (19) is shown in Figure 6, where the peak 10 MeV proton event intensity at 1 A.U. is plotted against the 4-8 MeV prompt gamma-ray line fluence. The poor correlation between

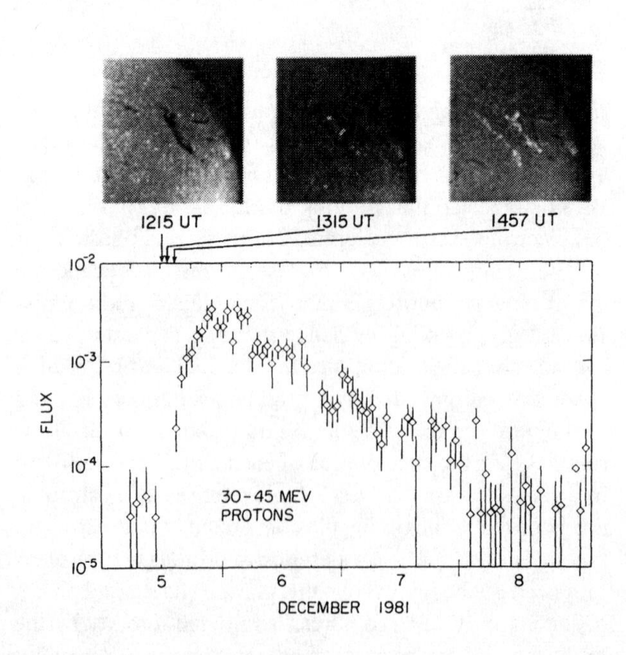

FIGURE 5. The Hα erupting filament and SEP event of 5 December 1981 (adapted from 16).

these parameters results primarily from a class of large SEP events that lacked detectable gamma-ray line emission. These events are in the same class as the weak impulsive phase events of Cliver et al. (15) and include the SEP event associated with a disappearing solar filament reported by Kahler et al. (16); in each case they exhibited evidence for mass ejection and/or shock formation. The poor correlation is consistent with the idea that the protons observed in space are not accelerated in the solar flare.

As a result of these various studies, it has gradually come to be accepted that the largest SEP events originate at a CME-driven coronal/interplanetary shock (Figure 7; 20). As will be discussed later, however, recent ACE observations have raised the possibility that flare accelerated particles may regularly contribute to well-connected SEP events.

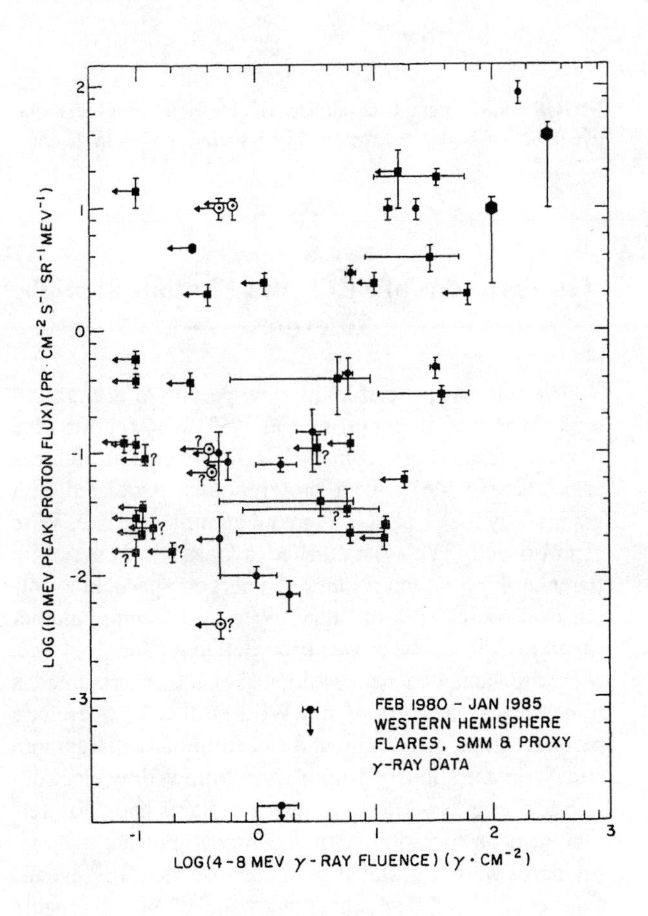

FIGURE 6. Peak ~10 MeV proton flux vs. 4-8 MeV gamma-ray line fluence for well-connected flares for the first five years of the SMM mission (19).

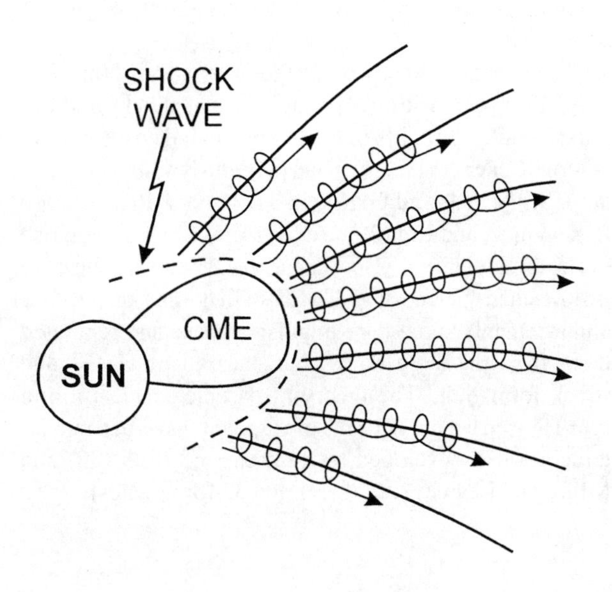

FIGURE 7. Schematic showing SEP acceleration at a CME-driven coronal/interplanetary shock wave (20).

THE CURRENT PARADIGM

The current paradigm is the outgrowth of a remarkable period of progress spanning the decade from ~1985 to ~1995. The research followed low- and high-energy tracks simultaneously and, in the end, these paths combined to reveal a remarkable synthesis and broad advance in understanding.

Low-Energy SEP Events

The progress in this area was driven by new instrumentation on the International Sun Earth Explorer (ISEE-3) satellite, specifically the ULEZEQ (21) and ULEWAT (21) / VLET (22) instruments, that measured charge states and composition, respectively, of low-energy (~1 MeV/nucleon) ions. The measured charge states were crucial to the adoption of the new paradigm. Klecker et al. (23) were the first to show that ^{3}He-rich SEP events (24) and large "normal" SEP events had quite different charge states (Figure 8; see also 25). Hurford et al. (26) found that the ^{3}He-rich flares were also enhanced in elements with $Z \geq 6$. Mason et al. (27) subsequently showed that the composition of the high-Z elements in the large sample of ^{3}He-rich SEP events observed by ISEE-3 exhibited a characteristic pattern of enhancement with respect to the elemental abundances observed in large proton events (Figure 9). Thus the charge states and composition measurements gave clear evidence for two types of SEP events.

Kahler et al. (28) examined the association of the ^{3}He-rich SEP events with CMEs and coronal shocks and obtained a null result, in contrast to the high degree of association Kahler et al. (9, 10) had previously found between large SEP events and CMEs. In a key study, Reames, von Rosenvinge, and Lin (29) used velocity dispersion in one event to associate a ^{3}He enhancement with an increase in low-energy electrons. These studies thus linked ^{3}He enhancements to the flare impulsive phase and separated them from the large flares characterized by CMEs and shock formation. The theory of particle acceleration in the ^{3}He-rich events in terms of resonant wave-particle interaction was developed by Temerin and Roth (30) and Miller and Reames (31) (see 1 and 32 for updates).

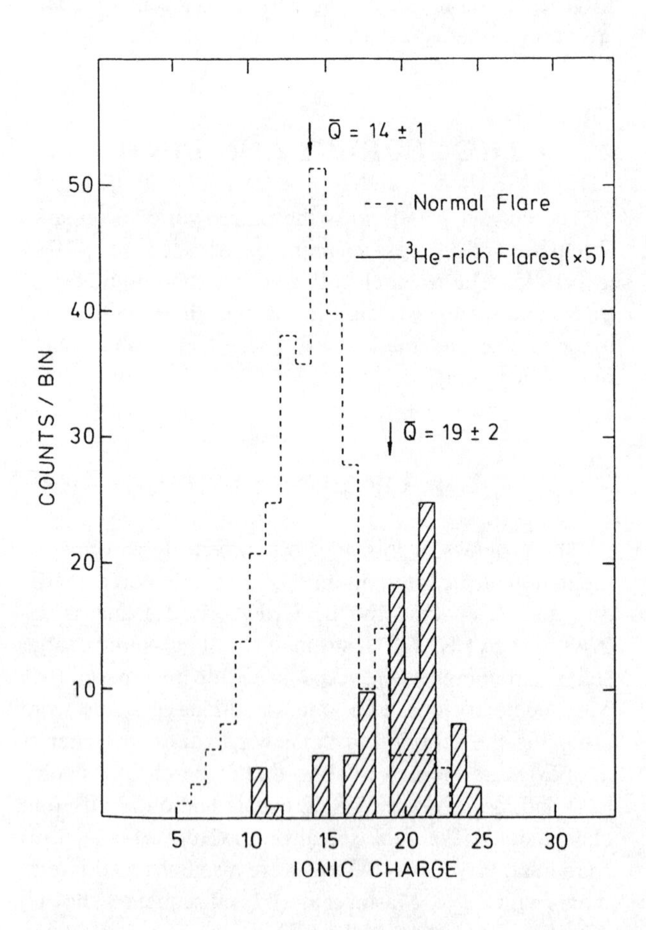

FIGURE 8. Histograms showing the difference in charge state distributions for a large normal SEP event and a sample of ^{3}He-rich SEP events (23).

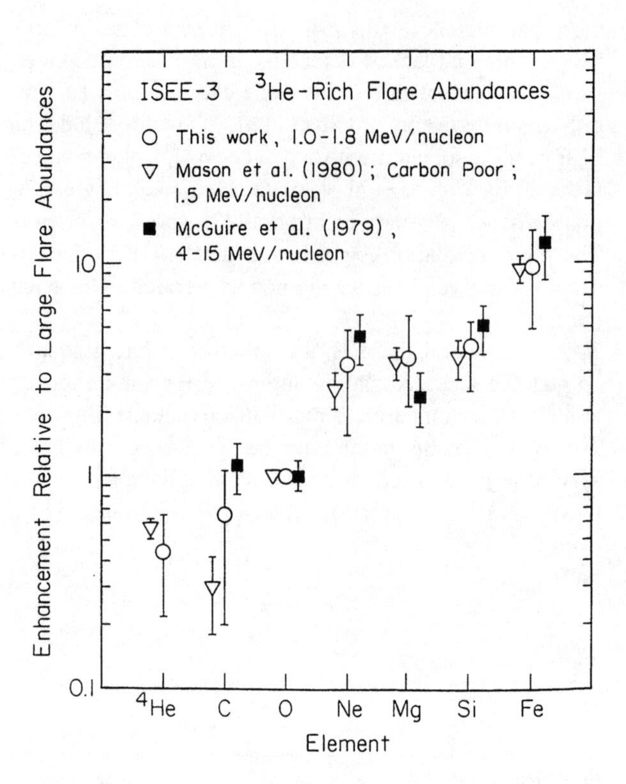

FIGURE 9. Elemental abundances of ^{3}He-rich SEP events relative to SEP abundances observed in association with large flares (27).

Observations of SEPs and Photons at High Energies

The seminal paper for the new paradigm at high energies was that of Evenson et al. (33) who reported the puzzling result that electron-rich SEP events at high energies (25-45 MeV) were preferentially associated with gamma-ray flares, although not all gamma-ray flares were electron-rich. This established a link between certain flares at the Sun and a class of energetic particle events but it also created a mystery. Why only some gamma-ray flares? The answer was provided in a paper by Cane, McGuire, and von Rosenvinge (34) that represented a reincarnation of sorts of the Wild et al. (7) two-phase picture. Cane et al. showed that impulsive flares with soft X-ray durations < 1 hour were followed by electron-rich SEP events while long-duration flares had "normal" low electron to proton ratios. Only impulsive gamma-ray flares were associated with electron-rich SEP events. Cane et al. also showed that the gradual flares had broader SEP "cones of emission" than their impulsive counterparts (Figure 10). The analysis of Cane et al. received strong confirmation from a paper by Moses et al. (35) in which it was shown that electron events associated

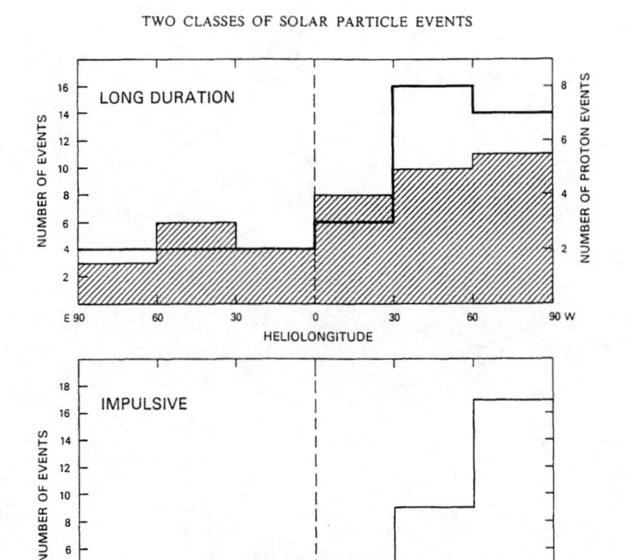

FIGURE 10. Histograms showing the locations of long-duration (top) and impulsive (bottom) SEP-associated flares (34). The shaded and unshaded distributions in the upper panel refer to two different samples of large SEP events.

with gradual solar flares had unbroken power-law rigidity spectra while electron events following impulsive flares exhibited a spectral break, with flattening above ~2 MV (Figure 11; cf., 36). At much lower energies, Lin (37) showed that electron spectra for large gradual flares continued to increase without flattening down to the lowest energy measured (~2 keV), indicating a small amount of mass traversal and acceleration high in the corona, while one of two impulsive events from the Cane et al. list that he had data for exhibited a spectral rollover at ~20 keV, implying acceleration at heights ~10^4 km.

In a pioneering paper in 1991, Murphy et al. (38) modeled the gamma-ray measurements of a large gradual flare on 27 April 1981 in terms of an accelerated beam of particles with a heavy element composition similar to that observed for ^{3}He-rich SEP events associated with impulsive flares (Figure 12). The basic result of Murphy et al. was confirmed by Share and Murphy (39) by deconvolving the composite spectrum of the 19 largest gamma-ray flares observed by SMM. In addition, Mandzhavidze et al. (40) showed that interacting particles were likely ^{3}He-enriched, as expected. These results suggested that while the particles in space (SEPs) came in two basic flavors, the interacting particles in flares came in only one. Ramaty et al. (41) presented additional evidence for this point of view (cf., 42). For example, Ramaty et al. found that interacting particles had enhanced electron to proton ratios, albeit somewhat higher than those observed for the impulsive events of Cane et al. (34) (cf., 43).

The Two-Class Paradigm

The new two-class paradigm was articulated in a series of review papers by Reames (1, 44, 45, 46). His synthesis is shown in Table 1, a summation that has become an icon in this field. The cartoons that have been added below

TABLE 1. PROPERTIES OF IMPULSIVE AND GRADUAL EVENTS (45)

	IMPULSIVE	GRADUAL
PARTICLES:	ELECTRON-RICH	PROTON-RICH
^{3}He/^{4}He	~1	~0.0005
Fe/O	~1	~0.1
H/He	~10	~100
Q_{Fe}	~20	~14
DURATION	HOURS	DAYS
LONGITUDE CONE	<30°	~180°
RADIO TYPE	III, V(II)	II, IV
X-RAYS	IMPULSIVE	GRADUAL
CORONAGRAPH	---	CME
SOLAR WIND	---	IP SHOCK
EVENTS/YEAR	~1000	~10

SEPs
CME
SEPs
γ-RAYS
γ-RAYS
IMPULSIVE
GRADUAL

the Table columns show that the new paradigm encompasses the initial δ-function paradigm of particle acceleration in flares (diagram on the left) shown in Figure 3 and the notion that prolonged acceleration could occur at CME-driven shocks. Reames used the Cane et al. (34) two-class picture and the results of the low energy charge state and composition observations to fashion the current picture. In the Reames' synthesis, the "pure" low-energy (~100 keV) electron events of Lin (12) are equated with the more energetic impulsive events of Cane et al. while the results on electron spectra and interacting particles are omitted. Given two basic types of events, one would expect that on occasion hybrids might be observed in which characteristics of both classes are mixed. This can be seen in the diagram on the right in Table 1 where both a flare and a CME-shock can accelerate particles. In fact, for certain well-connected events, there is evidence for

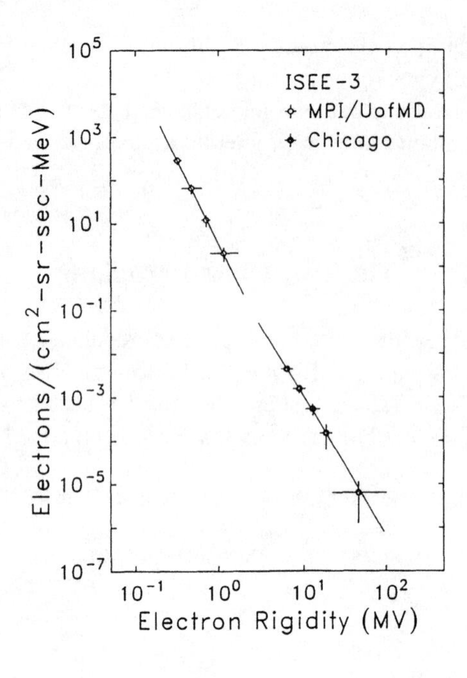

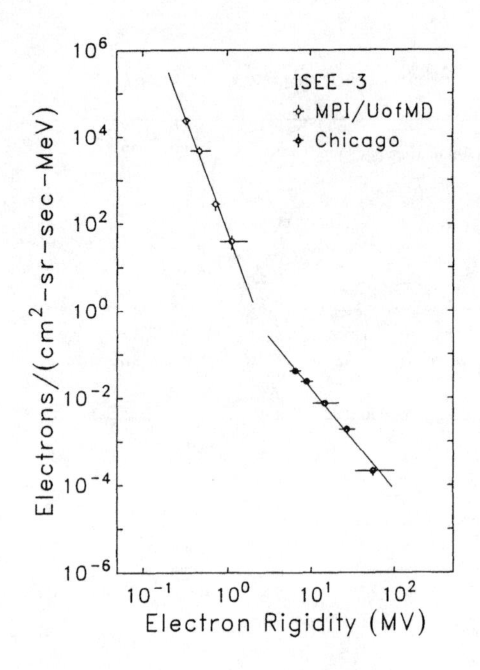

FIGURE 11. Electron rigidity spectra of SEP events associated with gradual (left) and impulsive (right) flares (35).

the evolution of composition from an "impulsive-like" composition early on to a "normal" composition for later stages of the event (e.g., 47, 48). To address the omissions in the standard picture in Table 1, Cliver (42) introduced an expanded table (Table 2) which adds: (1) two additional columns to account for hybrid events (mixed-impulsive and mixed-gradual); (2) four rows at the bottom for interacting particles (common characteristics for both gradual and impulsive flares, i.e., a single flavor); and (3) four additional rows for SEPs to account for electron to proton ratios, the ratio of interacting to interplanetary protons, and high- and low-energy electron spectra.

OPEN QUESTIONS

It came as somewhat of a surprise that four of the nine largest SEP events observed by ACE through 1998 had compositions and inferred charge states more characteristic of impulsive than gradual events (49; cf., 50). On the basis of the soft X-ray profiles, at least two of the four associated flares would have been classified as impulsive events; a third originated in an active region about two days beyond the west limb, so no time-scale determination is possible. LASCO observations were available for three of the four events and in each case a high speed (> 1000 km s^{-1}) CME was observed. The ^{3}He/^{4}He ratios in these events were enriched by factors of 10-100 above nominal solar wind values (49, 51), vs. the nominal factor of 2500 enhancement required for a "pure" impulsive event (Table 1). In the classification scheme of Table 2, these SEP events would be classified as "mixed-impulsive" events based on their soft X-ray durations, compositions, charge states, and CME associations. Mason, Mazur, and Dwyer (51) suggest that the ^{3}He enhancements in events like these may be caused by shock acceleration of ^{3}He particles previously accelerated in pure impulsive flares. Reames and his colleagues explain the high Fe charge states and enhancements of high-Z particles in certain large (and therefore presumably gradual) ACE (and Wind) events in terms of electron stripping (52) and a model of particle acceleration by shocks that takes self-generated waves into account (53), respectively. While they present a compelling case, particularly for the 20 April 1998 SEP event (54), acceptance of their picture implies: (1) that two quite different acceleration/transport processes can produce identical SEP composition; (2) and that the stripping process results in an Fe charge state similar to that observed in impulsive flares.

The need for such coincidences gives one pause and suggests an alternative picture: simply that flare accelerated SEPs account for the observed composition and

Table 2. Expanded SEP Classification System (42)

	Pure Impulsive	Mixed-Impulsive	Mixed-Gradual	Pure Gradual
Sun				
Radio Type	III(km III?)	II,III,V	II,III,IV	IV(I?)
SXR Duration	<1 hr(weak)	< 1 hr	> 1 hr	> 1 hr (weak)
CME Width	---	~20°	>45°	>45°
Solar Wind				
IP Shock	---	---	yes	yes
SEPs				
0.5 MeV e^-/ 10 MeV pr	---	>100	<100	<100
H/He	~10	>10?	~100	~100
$^3He/^4He$	~1	<0.1	~ 0.0005	~0.0005
Fe/O	~1	<1?	~1 --> ~0.1	~0.1
Q_{Fe}	~20	~20	(~20 ---> ~14)	~14
Electron Spectra				
High Energy	---	Flatten	No Break	No Break
Low Energy	No Roll-over	Roll-over	No Roll-over	No Roll-over
Solar/IP pr (~10 MeV)	---	~1-100	~0.1-10	<0.1
Longitude Cone	<30°	~100°	~200°	~ 200°
Interacting Beam				
Fe/O	(~1)	(~1)	~1	?
H/He	(<10)	(<10)	<10	?
0.5 MeV e^-/ 10 MeV pr	---	>100	>100	?
Bremsstrahlung Index (~0.5 MeV)	---	~3	~3	?

high Fe charge states in these events. We know from the gamma-ray observations of large flares such as those associated with the large SEP events analyzed by Cohen et al. (49) that particles of impulsive phase composition are produced at the Sun (38, 39, 40). Moreover, we know that such particles escape from impulsive flares and that only a relatively small fraction of interacting particles needs to escape to produce a detectable SEP event (e.g., 18, 19, 41). It is hard to argue that none of the flare-accelerated particles escape in large gradual flares. The question is

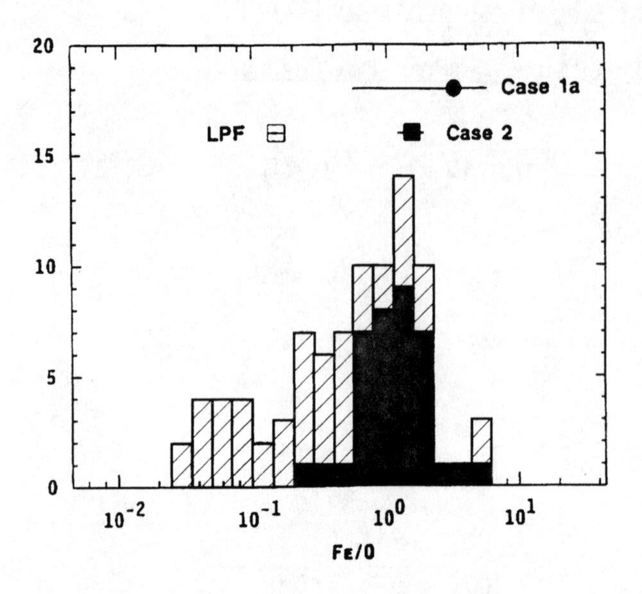

FIGURE 12. Histogram of the distribution of Fe/O in SEP events with the ^{3}He-rich subset blackened (38). The modeled composition of the accelerated beam based on gamma-ray observations for two separate cases (1a and 2) are shown at the top of the plot. The typical large proton flare (LPF) Fe/O ratio of 0.1 is shown by an open square.

- in what proportion in relation to the shock-accelerated particles and at what energies? As pointed out by Cohen et al. (49), the four large ACE events they analyzed do not show the strong temporal or spatial variabilities in their compositions that might be expected for the hybrid picture suggested by Reames (48) and Cliver (42) in which flare particles early in an event are followed by shock-accelerated SEPs. As an alternative, Cohen et al. (55) suggest that an energy-dependent superposition of particle populations may be involved.

Other open questions include: (1) the cause of the remarkable separation in classes of electron spectra based on soft X-ray event duration (Figure 11) found by Moses et al. (35); (2) the nature of spatially- (56) and temporally-extended (57) gamma-ray events; and (3) the possible influence of separated coronal and interplanetary shocks on SEP production (58, 59, cf., 60, 61).

CONCLUSION

From roughly 1985-1995, the research field of solar energetic particle events experienced a paradigm shift from the intial δ-function injection picture (Figure 3) to the current two-class viewpoint (Table 1). Given the general lack of development of the δ-function paradigm, this change can arguably be characterized as a transition from a "pre-paradigmatic state" to one with a well-accepted and detailed framework (1, and references therein). The value of such a framework is clearly revealed in the focus that is brought to the new ACE observations. The first question that is asked is: Do these data fit the paradigm (Table 1)? And, if not, why not? Clearly the new ACE observations have raised interesting challenges to the current paradigm. Ongoing investigations of ACE data, particularly in combination with data from the High Energy Solar Spectroscopic Imager (HESSI; 62) scheduled to fly in July 2000, promise to provide further insight to a field that has witnessed rapid progress in the last 15 years.

ACKNOWLEDGMENTS

I thank Dick Mewaldt and Alan Cummings for organizing a stimulating and productive conference and Christina Cohen and Allan Tylka for useful comments.

REFERENCES

1. Reames, D.V., *Space Science Rev.*, **90**, 413 (1999).
2. Forbush, S.E., *Phys. Rev.*, **70**, 771 (1946).
3. Servajean, R., and Olivieri, G., *Astronomie*, **60**, 215 (1946).
4. Lange, I., and Forbush, S.E., *Terr. Mag.*, **47**, 331 (1942).
5. Meyer, P., Parker, E.N., and Simpson, J.A., *Phys. Rev.*, **104**, 768 (1956).
6. Kundu, M.R., and Haddock, F.T., *Nature*, **186**, 610 (1960).
7. Wild, J.P., Smerd, S.F., and Weiss, A.A., *Ann. Rev. Astron. Astrophys.*, **1**, 291 (1963).
8. Kundu, M.R., and Smerd, S.F., *Inf. Bull. Europ. Solar Radio Obs.*, **11**, 4 (1962).
9. Kahler, S.W., Hildner, E., and Van Hollebeke, M.A.I., *Solar Phys.*, **57**, 429 (1978).
10. Kahler, S.W., et al., *J. Geophys. Res.*, **89**, 9683 (1984).
11. Forrest, D.J., and Chupp, E.L., *Nature*, **305**, 291 (1983).
12. Lin, R.P., *Solar Phys.*, **12**, 266 (1970).
13. Lin, R.P., and Hudson, H.S., *Solar Phys.*, **153**, (1976).
14. Kahler, S.W., *J. Geophys. Res.*, **87**, 3439 (1982).
15. Cliver, E.W., Kahler, S.W., and McIntosh, P.S., *Astrophys. J.*, **264**, 699 (1983).
16. Kahler, S.W., et al., *Astrophys. J.*, **302**, 504 (1986).
17. Forrest, D.J., et al., *Solar Phys.*, **65**, 15 (1980).
18. von Rosenvinge, T.T., Ramaty, R., and Reames, D.V., *Proc. 17th Int. Cosmic Ray Conf.*, **3**, 28 (1981).
19. Cliver, E.W., et al., *Astrophys. J.*, **343**, 953 (1989).

20. Kahler, S.W., Reames, D.V., and Sheeley, N.R., Jr., *Proc. 21st Int. Cosmic Ray Conf.*, **5**, 183 (1990).

21. Hovestadt, D., et al., *IEEE Trans. Geos. Electr.*, **GE-16**, 166 (1978).

22. von Rosenvinge, T.T., McDonald, F.B., Trainor, J.H., Van Hollebeke, M.A.,and Fisk, L.A., *IEEE Trans. Geos. Electr.*, **GE-16**, 208 (1978).

23. Klecker, B., et al., *Astrophys. J.*, **281**, 458 (1984).

24. Hsieh, K.C., and Simpson, J.A., *Astrophys. J. (Lett.)*, **162**, L191 (1970).

25. Luhn, A., Klecker, B., Hovestadt, D., and Möbius, E., *Astrophys. J.*, **317**, 951 (1987).

26. Hurford, G.J., Mewaldt, R.A., Stone, E.C., and Vogt, R.E., *Astrophys. J. (Lett.)*, **201**, L95 (1975).

27. Mason, G.M., Reames, D.V., Klecker, B., Hovestadt, D., and von Rosenvinge,T.T., *Astrophys. J.*, **303**, 849 (1986).

28. Kahler, S.W., Reames, D.V., Sheeley, N.R., Jr., Howard, R.A., Koomen, M.J.,and Michels, D.J., *Astrophys. J.*, **290**, 742 (1985).

29. Reames, D.V., von Rosenvinge, T.T., and Lin, R.P., *Astrophys. J.*, **292**, 716 (1985).

30. Temerin, M., and Roth, I., *Astrophys J. (Lett.)*, **391**, L105 (1992).

31. Miller, J.A., and Reames, D.V., *High Energy Solar Physics*, eds., R. Ramaty, N.Mandzhavidze, and X.-M. Hua, (AIP Press: Woodbury, NY) AIP Conf. Proc. #374, p. 450 (1996).

32. Miller, J.A., et al., *J. Geophys. Res.*, **102**, 14631 (1997).

33. Evenson, P., Meyer, P., Yanagita, S., and Forrest, D., *Astrophys. J.*, **283**, 43 (1984).

34. Cane, H.V., McGuire, R.E., and von Rosenvinge, T.T., *Astrophys. J.*, **301**, 448 (1986).

35. Moses, D., Dröge, W., Meyer, P., and Evenson, P., *Astrophys. J.*, **346**, 523 (1989).

36. Evenson, P., Hovestadt, D., Meyer, P., and Forrest, D.J., *Proc. 19th Int. Cosmic Ray Conf.*, **4**, 74 (1985).

37. Lin, R.P., *Proc. 21st Int. Cosmic Ray Conf.*, **5**, 88 (1990).

38. Murphy, R.J., Ramaty, R., Kozlovsky, B., and Reames, D.V., *Astrophys. J.*, **371**, 793 (1991).

39. Share, G.H., and Murphy, R.J., *Astrophys. J. (Lett.)* (in press) (2000).

40. Mandzhavidze, N., Ramaty, R., Kozlovsky, B., *Astrophys. J.*, **518**, 918 (1999).

41. Ramaty, R., Mandzhavidze, N., Kozlovsky, B., and Skibo, J.G., *Adv. Space Res.*, **13**(9), 273 (1993).

42. Cliver, E.W., in *High Energy Solar Physics*, eds., R. Ramaty, N.Mandzhavidze, and X.-M. Hua, (AIP Press: Woodbury, NY) AIP Conf. Proc. #374, p. 45 (1996).

43. Kallenrode, M.-B., Cliver, E.W., and Wibberenz, G., *Astrophys. J.*, **391**, 370 (1992).

44. Reames, D.V., *Astrophys. J. Suppl.*, **73**, 235 (1990).

45. Reames, D.V., *Adv. Space Res.*, **13**(9), 331 (1993).

46. Reames, D.V., *Rev. Geophys. (Suppl.)*, **33**, (U.S. National Report to the IUGG) p. 585 (1995).

47. Mason, G.M., Gloeckler, G., and Hovestadt, *D., Astrophys. J.*, **267**, 844 (1983).

48. Reames, D.V., *Astrophys. J. (Lett.)*, **358**, L63 (1990).

49. Cohen, C.M.S., et al., *Geophys. Res. Lett.*, **26**, 2697 (1999).

50. Mazur, J.E., Mason, G.M., Looper, M.D., Leske, R.A., and Mewaldt, R.A., *Geophys. Res. Lett.*, **26**, 173 (1999).

51. Mason, G.M., Mazur, J.E., and Dwyer, J.R., *Astrophys. J. (Lett.)*, **525**, L133 (1999).

52. Reames, D.V., Ng, C.K., and Tylka, A.J., *Geophys. Res. Lett.*, **26**, 3585 (1999).

53. Ng, C.K., Reames, D.V., and Tylka, A.J., *Geophys. Res. Lett.*, **26**, 2145 (1999).

54. Tylka, A.J., Reames, D.V., and Ng, C.K., *Geophys. Res. Lett.*, **26**, 2141 (1999).

55. Cohen, C.M.S., et al., *Geophys. Res. Lett.*, **26**, 149 (1999).

56. Vestrand, W.T., and Forrest, D.J., *Astrophys. J. (Lett.)*, **409**, L69 (1993).

57. Kanbach, G.O., et al., *Astron. Astrophys. Suppl.*, **97**, 349 (1993).

58. Gopalswamy, N., et al., *J. Geophys. Res.*, **103**, 307 (1998).

59. Cliver, E.W., Webb, D.F., and Howard, R.A., *Solar Phys.*, **187**, 89 (1999).

60. Torsti, J., Kocharov, L.G., Teittinen, M., and Thompson, B.J., *Astrophys. J.*, **510**, 460 (1999).

61. Krucker, S., Larson, D.E., Lin, R.P., and Thompson, B.J., *Astrophys. J.*, **519**, 864 (1999).

62. Lin, R.P., and the HESSI Team, *SPIE*, **3442**, 2 (1998).

Energetic Electrons Accelerated in Solar Particle Events

R. P. Lin

Space Sciences Laboratory, University of Berkeley, CA 94720

Abstract. New measurements of energetic solar electrons from the WIND spacecraft are reviewed, and the implications for particle acceleration mechanisms discussed. In non-relativistic electron/^{3}He-rich (so-called impulsive) events the electron energy spectrum is often found to extend below ~1 keV, indicating that acceleration occurs high in the corona. Comparison of the escaping electrons with the electrons producing the associated hard X-ray burst suggests that acceleration is occurring over a wide range of altitudes. For Large Solar Energetic Particle (LSEP, or so-called gradual) events, WIND observations show the low energy ~1-10 keV electron component is sometimes missing. In many LSEP events the electrons are released from the Sun up to ~0.5 hour later than the onset of the solar type III radio burst, and coronal transient waves are detected traveling across the Sun by the SOHO EIT instrument. Onset timing analyses show two types of LSEPs; in some events the first arriving ~0.1-6 MeV protons are released ~0.5-2 hours after the electrons and travel a path length of ~1.2 AU (essentially scatter-free), while in other events the protons are released at the same time as the electrons but appear to travel ~2 AU. If we assume the observed energetic particles are accelerated by a shock in front of an outward propagating fast CME, the electrons are accelerated earlier and lower in the corona (~0.5 R_{Sun}) and the protons later and higher, ~4 R_{Sun} for the first type of event, and from ~4 to > ~10 R_{Sun}, with the more energetic protons accelerated lower for the second type. In mid-2000 the High Energy Solar Spectroscopic Imager (HESSI) mission will be launched to provide detailed X-ray and gamma-ray imaging and spectroscopy observations to study particle acceleration and energy release processes at the Sun. Comparisons between HESSI and ACE/WIND should provide new insights into the origins of energetic solar particles.

INTRODUCTION

Observations of impulsive non-relativistic electron events at energies above ~40 keV [1, 2] provided the first interplanetary evidence for electrons accelerated in solar particle events. No energetic solar ions were detected above background in these events, indicating the electron to proton (*e/p*) ratio is very large. Soon thereafter energetic electrons up to relativistic energies were detected accompanying the "classical" Large Solar Energetic Particle (LSEP) events [3]. These events are dominated by energetic, >10MeV protons, with small *e/p* ratios at a given energy.

Later, high sensitivity measurements from the ISEE-3 spacecraft of electrons down to 2 keV showed that impulsive acceleration of non-relativistic electrons occurred, on average, several times a day or more during solar maximum making these the most common type of particle acceleration by the Sun [4]. Many of these non-relativistic electron events are unaccompanied by reported Hα flares, and many are observed only at energies below ~15 keV. With more sensitive measurements accelerated ions were detected above background for some of these non-relativistic electron events. The associated ion emission is primarily at low energies (~MeV/nucleon and below) and ^{3}He-rich [4, 5]; that is, the ions have ^{3}He/^{4}He ratios of order unity while the typical ratios for the solar atmosphere or solar wind are a few times 10^{-4}. The soft x-ray (SXR) bursts accompanying these events are typically impulsive, with duration ~<10 minutes, so these non-relativistic electron-^{3}He-rich events are sometimes called *impulsive* solar energetic particle events. As these electrons escape they produce solar and interplanetary type III radio bursts through beam-plasma interactions (see [6] for review).

On the other hand, LSEP events usually occur after a large solar flare. Tens of LSEP events are detected per year near solar maximum. The observed ionization states of LSEP particles are typical of a 1-2 × 10^6 K plasma, suggesting that these come from the quiescent solar corona and/or solar wind. It is believed that shock waves driven by fast Coronal Mass Ejections (CMEs), propagating over a wide longitude range of the solar corona, are the

CP528, *Acceleration and Transport of Energetic Particles Observed in the Heliosphere: ACE 2000 Symposium,* edited by Richard A. Mewaldt, et al.
© 2000 American Institute of Physics 1-56396-951-3/00/$17.00

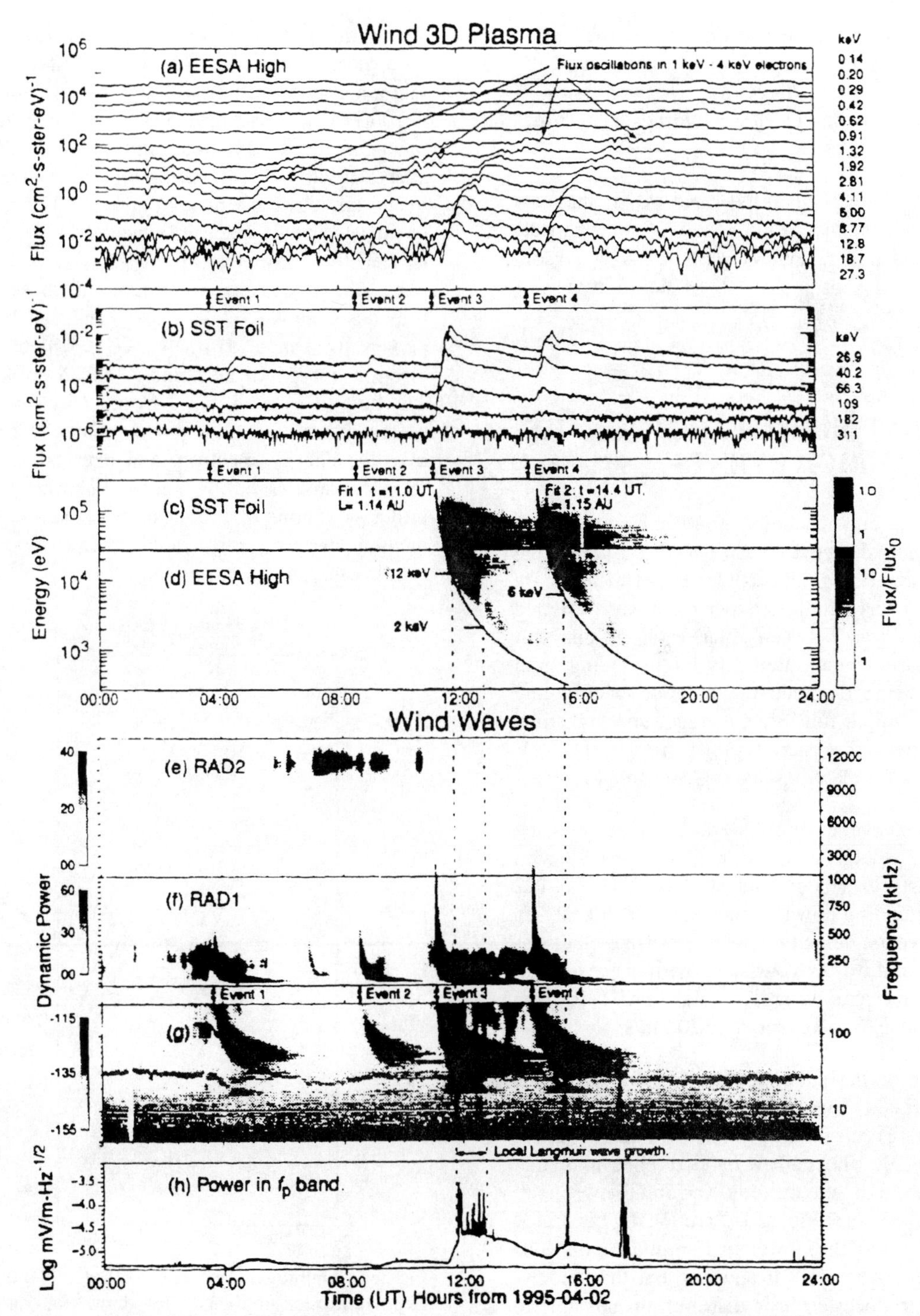

Figure 1. (a) The omnidirectional electron fluxes at 96s resolution as measured by EESA-H on WIND. The center energies are listed on the right. (b) The omnidirectional electron fluxes as measured by the SST-Foil detector. (c) and (d) Spectrograms of the ratio (F/F0) of electron fluxes (F) to a reference flux (F0) prior to the first event. Superimposed on the plot are linear fits of the arrival time of the electrons versus their inverse velocity. The event start times are marked on the plot. (e), (f), and (g) Solar type III radio bursts as observed by the WIND WAVES instrument. The RAD2 panel displays the radio emissions from 1 MHz to 14 MHz (linear frequency axis). The TNR (Thermal Noise Receiver) panel displays the frequency range from 4 kHz to 250 kHz. (h) The electric field wave power in the frequency band (19 kHz to 41.5 kHz) encompassing the local Langmuir frequency From [7], © 1998. The American Astronomical Society.

accelerating agent. LSEP events are also called "*gradual*" events since they are typically accompanied by flare soft X-ray (SXR) emission of relatively long duration, with e-folding decay times of >~ tens of minutes.

Since the launch of WIND and ACE, the new observations have changed our view of solar particle events. This paper reviews particle, X-ray, radio observations relevant to electron acceleration in solar particle events.

NONRELATIVISTIC ELECTRON-^{3}HE-RICH EVENTS

Figure 1a and 1b show four solar impulsive electron events detected on April 2, 1995 [7] by the 3D Plasma and Energetic Particles Experiment [8] on the WIND spacecraft. This experiment was designed to bridge the gap between solar wind plasma and energetic particle measurements by providing high sensitivity, wide dynamic range, good energy and angular resolution, full 3-D coverage, and high time resolution over the energy range from a few eV to~>300 keV for electrons and to ~>6 MeV for ions.

The impulsive events can be clearly identified through the velocity dispersion of the onsets, with the strongest event beginning at ~ 1110 UT at 182 keV and extending down in energy to ~ 0.6 keV. In all of the events, velocity dispersion of the electrons (Figure 1c and 1d) is consistent with the expected Archimedean spiral length (~ 1.15 AU) for the measured solar wind velocity (~ 370 km/s^{-1}).

All four of the solar impulsive electron events were associated with solar type III radio bursts (Figure 1e-1g) observed by the WIND WAVES instrument [9]. The narrow-banded emissions that persisted through out the day varying between 20 kHz and 40 kHz (see Figure 1g) are locally generated Langmuir waves; the power in Langmuir emissions is shown in Figure 1h. It appears that the velocity dispersion produces electron distributions unstable to the generation of Langmuir waves. The Langmuir waves then produce radio emission through wave scattering. These observations confirm that the non-relativistic electrons are the source of the emission, and therefore the radio bursts can be used to track the escaping electrons from the Sun into the interplanetary medium.

Figure 2 shows electron energy spectra for a typical impulsive event. The left-most spectrum shows the pre-event solar wind electron core and halo (dotted lines give Maxwellian fits), and shows that a highly non-thermal "superhalo" extends out to ~ 100 keV at this time. Due to velocity dispersion the peak in the event spectra progresses to lower energies with time; contributions from other smaller events are also evident, e.g., at 11:45-11:55 UT. The event-integrated spectrum displays a peak at ~<1 keV, with significant flux at ~0.5 keV. Above the peak, the spectrum is similar to those reported previously above ~2 keV [10], and can be fit to a power-law shape $dJ/dE = AE^{-\delta}$, where E is the electron energy in keV and A and δ are constants. The best fit gives $\delta = 3.0$ from ~1 keV to 40 keV, steepening to $\delta = 4.4$ above ~40 keV. Integrating over the energy spectrum and over the duration of the event, and assuming that the electrons are emitted into a ~40° cone of propagation in the interplanetary medium, we estimate a total energy of ~>3 × 10^{26} ergs in escaping electrons.

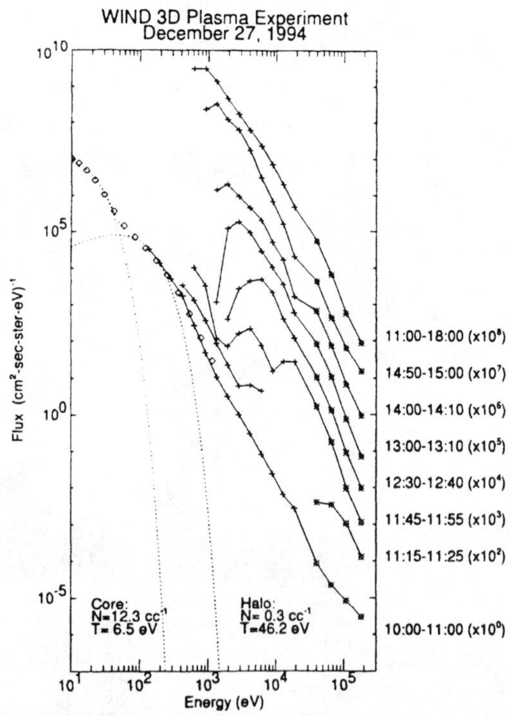

Figure 2. Omnidirectional electron spectra (with pre-event electron fluxes subtracted) are shown for various times during the 27 December 1994 event, and averaged over the entire event. The pre-event electron spectrum shows the solar wind electron core and halo components, as well as a 'super halo' extending to >~100 keV [11].

Because at coronal temperatures electrons are not gravitationally bound while protons are, an ambipolar electric field (the Pannekoek-Rosseland field [12, 13]) is set up with a total potential drop of about 1 kV from the base of the corona to 1 AU.

This potential varies inversely with distance from the center of the Sun. It accelerates protons outward and decelerates electrons. Thus, the peak in the spectrum of the electrons just as they escape the corona could be up to ~1 keV more than measured at 1 AU, e.g., ranging from ~<1 up to ~2 keV for the event of Figure 2, depending on the height of the acceleration.

The fact that the event spectrum extends down to such low energies indicates that at least some of the electron acceleration must occur high in the corona, since the range of ~ keV energy electrons in ionized hydrogen, due to Coulomb collisions, is short compared to the column depth through the corona. Assuming that the initial accelerated electron spectrum is represented by a power-law with the same exponent as seen at energies above the peak, the maximum overlying column density can be calculated [14]. For a peak at ~1.5 keV, the column density must be less than ~9 × 10^{17} cm^{-2}. This value implies that the lowest energy electrons must have been accelerated at altitudes of ~1 R_{Sun}, for the typical active coronal density models [15], or ~0.2 R_{Sun}. for the quiet equatorial corona at sunspot minimum [16].

A ^{3}He enhancement was detected in solar energetic ions during this electron event (J. Mazur, private communication, 1995). If the ^{3}He originated at the same altitude in the corona as these low energy electrons, the magnetic field would likely be much too weak for acceleration by electromagnetic hydrogen cyclotron waves [17].

ELECTRON – HARD X-RAY COMPARISON

The escaping electron population can be compared to those electrons at the Sun, inferred from the bremsstrahlung X-ray emission they produced through collisions. These comparisons show the number of escaping electrons is generally 10^{-2} to 10^{-3} the number of electrons at the Sun [4]. The typical numbers of electrons in an interplanetary event, however, would produce too few X-rays at the Sun for the present hard X-ray instrumentation to detect. On the other hand, even the largest electron events observed at 1 AU have far too few electrons to produce the observed hard X-ray emission in a large flare.

Direct comparison of the X-ray producing electrons and escaping electrons from a solar flare on 8 November, 1978 [18], showed that the energy spectrum of the escaping electrons observed from 2 to 100 keV differs significantly from the spectra of the X-ray producing electrons (Figure 3) and of the accelerated electrons (computed assuming that energy loss by Coulomb collisions is the dominant electron loss process). Even taking into account the material traversed by the escaping electrons, we find that they could not have come directly from the acceleration region. The accelerated electron spectrum at >25 keV energies, where energy losses are negligible, is much steeper than the escaping electron spectrum. If we match the accelerated and escaping spectra at 25 keV, a discrepancy of several orders of magnitude results at low energies. If the amount of material traversed is increased to reduce the low energy flux, the escaping spectrum would bend over in the energy range 2-10 keV.

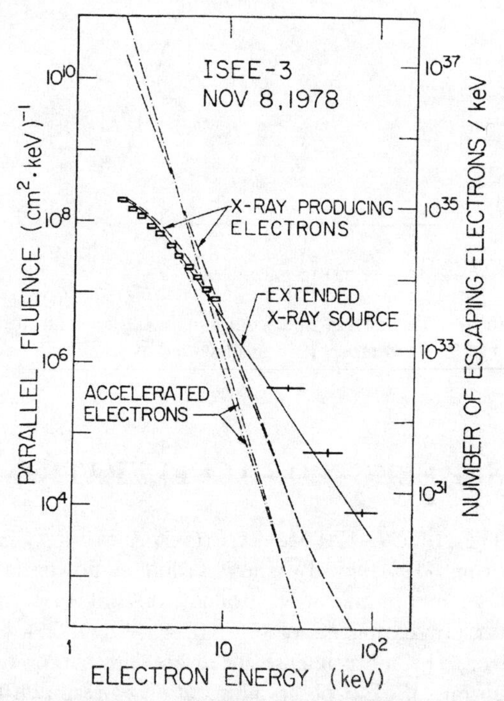

Figure 3. The spectra of the X-ray producing electrons (dashed lines) and the accelerated electrons (dash-dot lines) are shown here for comparison with the escaping electron spectrum (boxes and crosses) observed at 1 AU. The lower curved spectra of each pair of lines show how energy loss from the passage of the escaping electrons through the overlying solar atmosphere might modify these spectra. The solid curve is computed for electrons escaping from and X-ray source which extends from the chromosphere to high in the corona. From [18], with kind permission from Kluwer Academic Publishers.

The observations suggest a model where the escaping electrons come from an extended X-ray producing region which ranges from the chromosphere to high in the corona. In this model the low energy escaping electrons (2-10 keV) come from the higher part of the extended X-ray source where the overlying column density is low, while the high energy electrons (20-100 keV) come from the entire X-ray source.

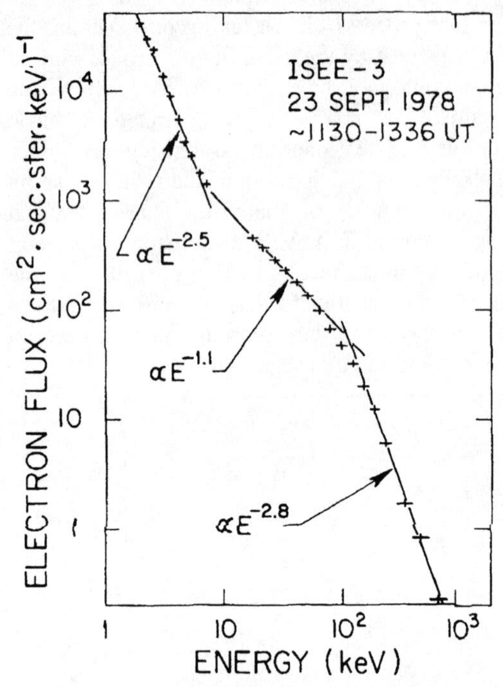

Figure 4. The energy spectrum for the large solar flare event of 23 September, 1978 as observed by ISEE-3.

LSEP ELECTRON ACCELERATION

The 20 keV – 20 MeV energy spectra for LSEP electrons [19] typically show a double power law with a smooth transition around 100-200 keV and power law exponents of 0.6-2.0 below and 2.4-4.3 above. The more intense the event, the harder the spectrum. ISEE-3 observations at energies down to ~2 keV showed that most LSEP events near solar maximum had a low-energy component with a steep power law spectrum (Fig. 4) extending from ~10keV down to < ~2keV [4], similar to the spectra of non-relativistic electron events. This was taken as evidence that in LSEP events the acceleration accurred high in the corona, as would be expected for CMEs. The three events found without this low energy component were those with impulsive SXR bursts [20].

Figure 5 shows the 22 April 1995 event observed by WIND. The presence of significant fluxes of >10 MeV protons and the gradual SXR burst confirm this is an LSEP event. Electrons are detected down to ~8 keV in this event (Fig. 5) with no increase below ~6 keV. The electron spectrum for the event (Figure 6) shows a peak at ~12 keV, and a rapid falloff below ~8 keV. Above ~12 keV the spectrum fits a double power law with a break around 50 keV, and power law exponents of δ = 1.1 below and δ = 3.1 above the break.

If the peak is due to the electrons traversing overlying material, the source region must be at a column depth of ~1 × 10^{19} cm^{-2}. For quiet coronal density models the electrons would have been accelerated in the chromosphere, while for active corona models the height would be ~0.15 R_{Sun}. Thus, this event is inconsistent with the simple picture for LSEP events of particle acceleration by CME shocks high (>~ few R_{Sun}) in the corona. In surveying the WIND data, a number of such LSEP events, with no low energy (<~10keV) electron emission, have been found [21].

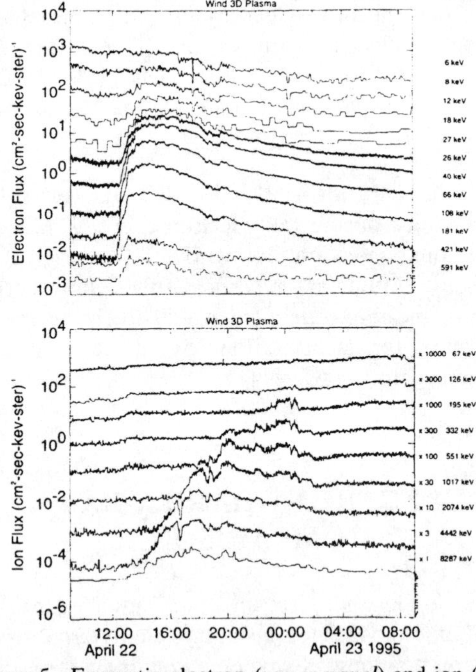

Figure 5. Energetic electron (*upper panel*) and ion (*lower panel*) fluxes for 22-23 April 1995. The energies (and flux factors) are indicated on the right. The gradual (LSEP) event is evident at energies above ~12 keV in electrons.

ONSET TIMING STUDIES

Krucker *et al.* [21] surveyed the solar electron events observed by Wind. Timing analysis of the velocity dispersion of the onsets reveals two different kinds of electron events: (1) events released from the Sun at the onset of a radio type III burst, which suggest that these electrons are part of the population producing the type III radio emission; and (2) events in which the electrons are released up to half an hour later than the onset of the type III burst. The latter events are proton rich, i.e. LSEP events.

For about ¾ of the delayed LSEP events, large-scale coronal transient waves, also called EIT waves or coronal Moreton waves, are observed by the Extreme Ultraviolet Imaging Telescope (EIT) on board *SOHO*. These waves originate at the flare site and often traversed a substantial fraction of the solar disk to reach the footpoint of the spiral interplanetary field line connected to the Wind spacecraft. It's not clear whether these waves are the chromospheric/lower coronal signatures of CME shocks or a separate flare driven shock.

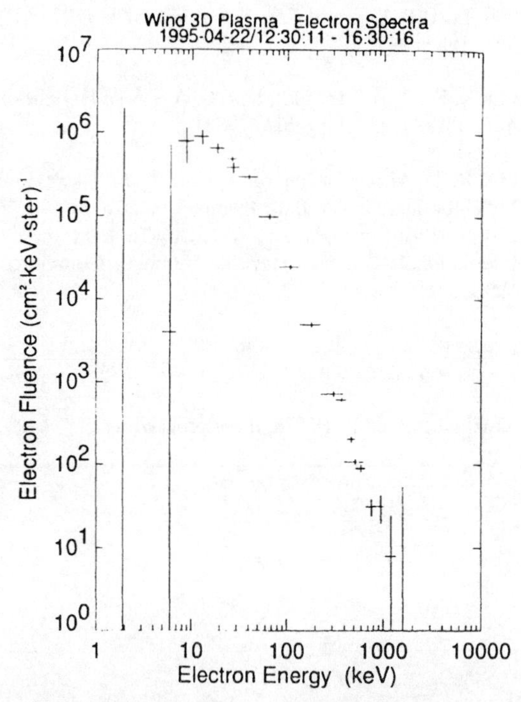

Figure 6. Omnidirectional electron energy spectrum (with pre-event background subtracted) averaged over the first four hours of the event.

Recently, Krucker and Lin [22] analyzed the velocity dispersion of ~30 keV – 6 MeV proton and ~1-300 keV electrons in LSEP type events. Two classes of events were found. For one class (70% of the events), the first arriving protons and electrons travel essentially scatterfree, with derived path lengths between 1.1 and 1.3 AU, but the protons are released ~0.5 to 2 hours later. For events of the second class the protons show significantly larger path lengths, around 2 AU, than the electrons (still ~1.2 AU) but the protons and electrons are to be released simultaneously, within an uncertainty of 20 minutes.

For both classes of events the ions in the onset phase are observed to be streaming outward, aligned within ~45° pitch angle of the magnetic field, with no obvious difference between classes. Thus, it is unlikely that pitch angle scattering can explain the difference in path lengths. The alternative explanation is that the acceleration/ injection of protons varies monotonically with energy, with the highest energy protons accelerated/ injected earliest.

SUMMARY

It appears that non-relativistic (impulsive SXR) electron events are generally the result of an acceleration that occurs high in the corona. The electrons then produce type III radio emission as they escape from the Sun. Whether this acceleration is part of the acceleration of the electrons which produce the hard X-ray burst, or a simultaneous but separate phenomenon, is not clear. How these are related to the acceleration of ^{3}He is also unclear.

The LSEP electron acceleration process sometimes produces only higher energy electrons, from ~8-10 keV to relativistic energies. The simplest interpretation of the electrons from <~ 1 keV to ~10-20 keV sometimes seen in LSEP events (and sometimes not) is that they are due to a mixed event, where the non-relativistic (impulsive SXR) electron acceleration occurs in the LSEP event.

The LSEP electron acceleration appears related to the EIT wave which may or may not be related to the CME shock. If the CME shock accelerates LSEP electrons and ions, the ions appear to be accelerated later, at a higher altitude than the electrons. For one class of events, ions of all energies are accelerated at ~4 R_{Sun} altitude. For the second class the ions are accelerated at altitudes from ~4 to >~10 R_{Sun}, with the lower energy ions accelerated higher. The reasons for the two classes are not presently understood.

Clearly, spatially resolved observations of X-ray and γ-ray continuum and γ-ray line signatures of energetic electrons and ions, respectively, at the Sun, will help to identify the acceleration mechanisms. The High Energy Solar Spectroscopic Imager (HESSI) planned for launch in July 2000 [21], will provide such observations for comparison with solar particle measurements from ACE, WIND, and other spacecraft.

ACKNOWLEDGMENTS

This research was supported in part by NASA grants NAG5-6928 and NAG5-2815.

REFERENCES

1. Van Allen, J. A., and S. M. Krimigis, *J. Geophys. Res.* **70**, 5737- (1965).

2. Anderson, K.A., and R.P. Lin, *Phys. Rev. Lett.* **16**, 1121-1124 (1966).

3. Cline, T. L., and F. B. McDonald, *Solar Phys.* **5**, 507-530 (1968).

4. Lin, R. P., *Solar Phys.* **100**, 537-561 (1985).

5. Reames, D.V., and T.T. von Rosenvinge, and R.P. Lin, *Astrophys. J.* **292**, 716-724 (1985).

6. Lin, R.P., 'Electron beams and Langmuir turbulence in solar Type III radio bursts observed in the interplanetary medium', in *Basic Plasma Processes on the Sun*, edited by E. Priest and V. Krishan, International Astronomical Union, The Netherlands, 1990, pp.467-481.

7. Ergun, R.E., D. Larson, R.P. Lin, J.P. McFadden, C.W. Carlson, K.A. Anderson, L. Muschietti, M. McCarthy, G. Parks, H. Reme, J. M. Bosqued, C. d'Uston, T.R. Sanderson, K.P. Wenzel, M. Kaiser, R.P. Lepping, S.D. Bale, P. Kellogg, and J.-L. Bougeret, *Astrophys. J.* **503**, 435-445 (1998).

8. Lin, R. P., K. A. Anderson, S. Ashford, C. Carlson, D. Curtis, R. Ergun, D. Larson, J. McFadden, M. McCarthy, G. K. Parks, H. Rème, J.M. Bosqued, J. Coutelier, F. Cotin, C. d'Uston, K.-P. Wenzel, T.R. Sanderson, J. Henrion, and J.C. Ronnet, *Space Sci. Rev.* **71**, 125-153 (1995).

9. Bougeret, J.-L., M.L. Kaiser, P.J. Kellogg, R. Manning, K. Goetz, S.J. Monson, N. Monge, L. Friel, C.A. Meetre, C. Perche, L. Sitruk, and S. Hoang, *Space Sci. Rev.* **71**, 231-263 (1995).

10. Potter, D. W., R. P. Lin and K. A. Anderson, *Astrophys. J.* **236**, L97-L100 (1980).

11. Lin, R. P., D. Larson, J. McFadden, C.W. Carlson, R.E. Ergun, S. Ashford, K.A. Anderson, M. McCarthy, G.K. Parks, H. Rème, J.M. Bosqued, C. d'Uston, T.R. Sanderson, and K.-P. Wenzel, *Geophys. Res. Lett.* **23**, 1211-1214, 1996.

12. Pannekoek, A., 'Ionization in stellar atmospheres', *Bull. Astron. Inst. Neth.*, **1**, 107- (1922).

13. Rosseland, S., *Monthly Notices Royal Astron. Soc.* **84**, 720-, (1924).

14. Lin, R.P., *Space Sci. Rev.* **16**, 189-256 (1974).

15. Dulk, G.A., and D.J. McLean, *Solar Physics* **57**, 279-295 (1978).

16. Saito, K., A. I. Poland, and R. H. Munro, *Solar Phys.* **55**, 121-134 (1977).

17. Temerin, M., and I. Roth, *Astrophys. J.* **391**, L105-L108 (1992).

18. Pan, L., R.P. Lin, and S.R. Kane, *Solar Phys.* **91**, 345-357 (1983).

19. Lin, R. P., R. A. Mewaldt, and M. A. I. Van Hollebeke, *Astrophys. J.* **253**, 949-962 (1982).

20. Lin, R. P., "Electron energy spectra from 2 keV to 1 MeV for large solar flare events," in *21st Internat'l. Cosmic Ray Conf. Papers*, edited by R. J. Protheroe, International Cosmic Ray Conference Committee, 1990, pp. 88-91.

21. Krucker, S., D. E. Larson, R. P. Lin, and B. J. Thompson,. *Astrophys. J.* **519**, 864-875 (1999).

22. Krucker, S., and R. P. Lin, these proceedings.

23. Lin, R. P., these proceedings.

Particle Acceleration at Sites of Magnetic Reconnection

Yuri E. Litvinenko

Institute for the Study of Earth, Oceans, and Space, University of New Hampshire, Durham, NH 03824-3525, USA

Abstract. Electric fields induced by the changing magnetic field at sites of magnetic reconnection can efficiently accelerate charged particles in the solar corona. This review begins with estimates for the electric field magnitude in flare models and presents some of the theoretical results for the electron and proton acceleration in reconnecting current sheets in solar flares. Particular emphasis is placed on models for collisionless acceleration in a large-scale reconnecting current sheet with a nonzero magnetic field and a highly super-Dreicer electric field of order a few V cm^{-1}. Particle orbits in model current sheets are discussed using an approximate analytical approach that allows one to identify the effects of both the electric and magnetic field components on the particle motion. Formulae for the particle energy gains and acceleration times are presented. Given a super-Dreicer electric field in the sheet, it is the magnetic field structure in the sheet that determines both the electron to proton ratio for the accelerated particles and their typical energies and spectra. The analytical results form the basis for the electric field acceleration models in solar flares. In particular, physical conditions can be identified that lead to either flares in which electrons primarily generate hard X-rays in the energy range of tens of keV or flares with unusually large electron fluxes at gamma-ray energies extending up to a few tens of MeV.

INTRODUCTION

A large body of research has been devoted to the question of charged particle orbits in reconnecting current sheets in the context of particle acceleration on the Sun and in the geotail (e.g., Speiser, 1965; Martens, 1988; Zhu and Parks, 1993; Litvinenko and Somov, 1993; Litvinenko, 1996, 1997). DC electric field acceleration in current sheets has been demonstrated to be a natural mechanism for the generation of hard X-ray producing electrons with typical energies of tens of keV in solar flares (Litvinenko, 1996, and references therein).

The reconnection electric field in the current sheet is determined by the plasma inflow speed to the sheet and the local magnetic field. For inflow speeds of order a few km s^{-1} and the magnetic fields of a few hundred Gauss, the electric field can be as strong as a few V cm^{-1} and hence can lead to relativistic energies of charged particles, provided their acceleration length—the particle displacement along the electric field in the current sheet—is large enough. The acceleration length itself is controlled by the magnetic field inside the sheet. The principal point is that although the magnetic field cannot change the particle energy, it can change the orbit, determining the displacement along the electric field and hence the energy gain. It is conceptually important that particle acceleration in this approach is considered as an inherent part of the flare energy release in a large-scale current sheet (or several current sheets) in the solar corona. It appears possible, in particular, to relate the electron to proton ratio of accelerated particles to the magnetic field geometry in the reconnection region (Fletcher and Martens, 1998).

"The particle acceleration problem of electron-rich γ-ray flares is perhaps the most challenging" (Hudson and Ryan, 1995). These flares are defined by an unusually intense gamma-ray continuum above 1 MeV, which can be fit well by a hard power law (for a review of observations, see Hudson and Ryan, 1995; Rieger, Gan, and Marschhäuser, 1998). Nuclear gamma-line radiation in electron-rich events is dominated by electron bremsstrahlung. This implies the presence of large fluxes of relativistic electrons with energies up to a few tens of MeV and a high electron to proton ratio in this energy range.

The goal of this paper is to review the theory of charged particle acceleration by the direct electric field associated with magnetic reconnection and to show that this process can be responsible for both X-ray generating electrons in "normal" flares and electrons with energies exceeding 10 MeV in the electron-rich flares. Section 2 presents arguments for highly super-Dreicer electric fields in solar flares and delineates some general aspects of particle acceleration by the electric field. Section 3 gives a review of results on charged particle orbits in the current sheet far from a singular line of magnetic field in the sheet. The results are used to explain the acceleration of hard X-ray generating electrons in solar flares. Section 4 discusses the energy gains for electrons near the singular line. It is shown that impulsive electron acceleration to tens of MeV

CP528, *Acceleration and Transport of Energetic Particles Observed in the Heliosphere: ACE 2000 Symposium*, edited by Richard A. Mewaldt, et al.
© 2000 American Institute of Physics 1-56396-951-3/00/$17.00

is possible in the flare reconnecting current sheet undergoing the tearing instability.

ELECTRIC FIELDS AND PARTICLE ACCELERATION IN SOLAR FLARES

The particle acceleration model discussed in this paper is based on the idea that magnetic reconnection in solar flares corresponds to electric fields E of order 10 V cm^{-1} in the corona. This is a few orders of magnitude larger than the Dreicer field, which means that the acceleration process is essentially collisionless. Wave-particle interactions can typically be ignored as well. Thus the usual Ohm's law is not applicable in the current sheet, and the particle escape itself provides an effective plasma resistivity (Lyons and Speiser, 1985; Litvinenko, 1997).

Before performing the analysis of charged particle orbits in current sheets with a large (super-Dreicer) E, it is instructive to review the estimates for direct electric fields in solar flares. The simplest argument is based on the flare energy requirements. The electrodynamic power dissipated in a flare, which is up to $10^{28} - 10^{29}$ erg s^{-1}, is determined by the free magnetic energy in the corona and the corresponding electric current I. The rate of work of an electromagnetic field on a system of electric currents and charges is $\mathbf{j} \cdot \mathbf{E}$ per unit volume, where $\mathbf{j}$ is the electric current density. Hence the flare energy release rate

$$P = IU = IEl, \quad (1)$$

where from observations $I \lesssim 3 \times 10^{21}$ cgs (10^{12} A) and the active region length scale is $l \lesssim 10^{10}$ cm. This leads to $E \gtrsim 3 \times 10^{-3}$ cgs $= 1$ V cm^{-1} (cf., Melrose, 1990).

Other approaches lead to essentially the same value of the electric field in the reconnection region $E = 1 - 10$ V cm^{-1}. For example, the analysis of the current sheet structure, which uses balance equations based on the conservation laws and the Maxwell equations (Somov, 1992), shows that for the flare energy requirements to be satisfied, reconnection inflow has to be fast: $v_{\text{in}} \gtrsim 10$ km s^{-1}. The corresponding motional electric field is $v_{\text{in}} B_0 / c = 1 - 10$ V cm^{-1}, where the coronal magnetic field $B_0 = 10^2 - 10^3$ G, and continuity dictates that the same electric field be present in the reconnection region itself. Numerical simulations also show that filament eruptions in the corona give rise to magnetic reconnection that indeed proceeds at a rate corresponding to the electric field of about 10 V cm^{-1} at the reconnection site (e.g., Forbes, 1992). Finally, this theoretical estimate is confirmed by the measurements based upon the Stark effect. In particular, Foukal and Behr (1995) reported $E \approx 35$ V cm^{-1} in a flare surge (see also the review by Foukal and Hinata, 1991).

It is immediately clear from the extremely large value of the total potential $U = El$ that the particle acceleration length l_{acc} in current sheets in solar flares has to be much less than the total length of the sheet l (Martens, 1988; Litvinenko, 1996; see also Equations (10) and (14) below). The relation $l_{\text{acc}} \ll l$ is a salient feature of particle acceleration in current sheets with a nonzero magnetic field, which prevents electrons from gaining unreasonably large energies of order eU. It is also important that a small acceleration length limits the total electric current I through the sheet:

$$I = e\dot{N}\frac{l_{\text{acc}}}{l}, \quad (2)$$

where $\dot{N}$ is the number of particles flowing into and out of the sheet per unit of time. The factor $l_{\text{acc}}/l \ll 1$ appears because the particles that had left the sheet cannot contribute to the current inside it.

A simpler particle runaway model that ignores the magnetic field in the sheet altogether provides an alternative approach to the particle acceleration problem, in particular in application to the hard X-ray generating electrons in solar flares. This model envisions the formation of an electron beam by postulating that $l_{\text{acc}} = l$ for all particles since they are presumed to move from one end of the sheet to the other. The correspondingly small value of E also has to be postulated. This model encounters a difficulty, however. In order to be consistent with observational estimates of the number of energetic electrons ($10^{34} - 10^{36}$ s^{-1}), the electric current associated with the beam would have to be so large that its magnetic field would exceed typical coronal values by several orders of magnitude (Holman, 1985). To avoid this contradiction, one would have to postulate the existence of at least 10^5 acceleration regions producing oppositely directed electron beams (see Miller *et al.*, 1997, for a review).

I tend to favor the former model, in which $l_{\text{acc}} \ll l$ and the charged particles in the current sheet are accelerated locally all along its length. This approach not only drastically decreases the electric current through any cross-section of the sheet but also has the advantage of treating both particle acceleration and global flare energy release as parts of a single physical process—collisionless magnetic reconnection in a large-scale current sheet. One should not forget, however, that both the current sheet model and the particle runaway model are likely to require unreasonably large electric currents outside of the current sheet. Hence the return current of thermal electrons appears to be inevitable in the solar atmosphere (Knight and Sturrock, 1977). The return current has to neutralize the direct current of accelerated electrons, thus avoiding the problems of the huge magnetic field of the electron beam and the enormous charge displacement (van den Oord, 1990).

PARTICLE ORBITS IN THE RECONNECTING CURRENT SHEET

Given the complicated three-dimensional nature of the magnetic and velocity fields in the solar corona, not only the reconnecting component of the magnetic field, say $B_x = B_x(y)$, but also the longitudinal (along the electric field) and the transverse (perpendicular to the plane of the sheet) magnetic field components are likely to be present in the current sheet. The usual approach in the study of the charged particle trajectories in the sheet is to approximate the field by the first nonzero terms in the Taylor expansion inside the sheet located at $y = 0$:

$$\mathbf{B} = -(y/a)B_0\hat{\mathbf{x}} - B_\perp\hat{\mathbf{y}} + B_\parallel\hat{\mathbf{z}}. \qquad (3)$$

Here the minus signs correspond to the electric current in the positive z-direction, and a is the current sheet half-thickness. The reconnection electric field inside the sheet is

$$\mathbf{E} = E\hat{\mathbf{z}}. \qquad (4)$$

Both E and the nonreconnecting component $B_\parallel$ may be assumed constant. The transverse field $B_\perp = B_\perp(x)$ changes sign at the center of the sheet and reaches a maximum at its edges $x = \pm b$. When the half-width of the sheet $b \gg a$, this component is a very slowly varying function of x. Hence $B_\perp$ is often also assumed constant on a given particle trajectory.

It should be stressed that the magnetic field in the sheet is neither uniform nor static.The reconnecting field lines move into the sheet with speed v_{in} and out of the sheet with the characteristic speed of order the Alfvén speed

$$v_{\text{A}} = \frac{B_0}{\sqrt{4\pi m_{\text{p}} n}} \qquad (5)$$

and carry the magnetized particles with them. Here n is the particle density and m_{p} is the proton mass. This familiar "sling-shot effect" causes the reconnected field lines to straighten out so that $B_\perp$ increases from zero at $x = 0$ to the maximum value $\pm B_{\perp,\text{max}}$ at $x = \pm b$ for each reconnected field line, leading to a dependence $B_\perp = B_\perp(x)$ in a steady state and to the corresponding temporal evolution of each reconnected field line. It is only when the length scale of particle acceleration in the sheet is small enough that the spatial dependence of the field lines can be ignored with the exception of the variation of $B_x \sim y/a$ across the current sheet thickness $2a$. This simplification appears to be justified for the hard X-ray generating electrons in flares but will have to be relaxed below when considering MeV electrons in the electron-rich flares.

Even under the simplifying assumption $B_\perp = \text{const}$, the character of the charged particle motion for various relative values of the magnetic field components in the current sheet is nontrivial (Litvinenko, 1996). In the limit $B_\perp = 0$, whether $B_\parallel = 0$ or not, the motion consists of the acceleration along the electric field $\mathbf{E} = E\hat{\mathbf{z}}$ and finite oscillations along the y-axis caused by the Lorentz force $\sim v_z B_x$ (Speiser, 1965; Zhu and Parks, 1993). This idealized, highly symmetric situation, however, is unlikely to occur. In fact any sheet model requires a nonzero $B_\perp$ as a result of reconnection itself (cf., Martens, 1988).

Particle orbits in current sheets with $B_\perp \neq 0$ are very complex in general, but the situation is simpler in two limiting cases.

If the longitudinal field $B_\parallel$ is small enough, then the maximum displacement along the electric field and the energy gain are determined by the particle gyroradius in the transverse field $B_\perp$:

$$\mathcal{E} = 2mc^2 \left(\frac{E}{B_\perp}\right)^2, \qquad (6)$$

where m is the particle mass and c is the speed of light (Speiser, 1965). Litvinenko and Somov (1993) showed that the corresponding energy gain is too small to explain the electron acceleration to energies above 1 keV in solar flares.

Since the magnetic field in the solar corona is known to have a significant axial component along the coronal loops, the other limit of a strong longitudinal field $B_\parallel$ on the order of the main reconnecting field B_0 could be appropriate for flaring current sheets. The strong longitudinal field $B_\parallel > B_{\parallel,\text{c}}$ magnetizes the particles and makes them follow the field lines. The critical field $B_{\parallel,\text{c}}$, leading to the transition to this new type of motion is given by

$$B_{\parallel,\text{c}} = \left(\frac{mc^2 E B_0}{eaB_\perp}\right)^{1/2} \qquad (7)$$

(Litvinenko, 1996), where m and e are the particle mass and electric charge. For electrons in solar flares, Litvinenko and Somov (1993) obtained $B_{\parallel,\text{c}} \leq 0.1B_0$ for typical parameters of current sheets in the solar corona (see also below). Therefore electrons are magnetized efficiently by the longitudinal field $B_\parallel$ in the sheet. The effect is absent, however, for much heavier protons that will still follow the Speiser-type orbits.

A potential complication in the problem is the charge separation electric field that arises in the sheet because electrons and protons follow different orbits. This effect is present in particle simulations of collisionless reconnection (Horiuchi and Sato, 1997). Litvinenko (1997) showed, however, that the simulations are in a surprisingly good agreement with the results of the test particle approach.

Thus the magnetized electrons will mainly move along the magnetic field lines in the current sheet. The adiabatic particle motion in principle can be described by drift theory. The main effect though is acceleration along the field lines that will cease when the particles leave the sheet. Integrating the magnetic field line equations

$$-\frac{a}{y}\frac{dx}{B_0} = -\frac{dy}{B_\perp} = \frac{dz}{B_\parallel} \tag{8}$$

defines the acceleration length l_{acc} as the displacement δz along the electric field, which corresponds to $|\delta y| = a$ when the magnetized electrons initially inside the sheet at $y = 0$ leave the sheet along the field lines:

$$l_{acc} = \frac{B_\parallel}{B_\perp} a \tag{9}$$

(see Litvinenko, 1996, for a detailed discussion of particle orbits in this case). The displacement perpendicular to the electric field is given by a similar formula $\delta x = \frac{1}{2} a B_0 / B_\perp$. The typical energy gain for the magnetized particles

$$\mathcal{E} = \frac{B_\parallel}{B_\perp} eEa \tag{10}$$

allows to explain the electron acceleration in flares.

In what follows the reconnecting and nonreconnecting components of the field are assumed to be of the same order, $B_\parallel = B_0 = 100$ G. This choice ensures that $B_\parallel \gg B_{\parallel,c}$ for electrons. Using $B_0 = 100$ G and the numbers below leads to $B_{\parallel,c} \approx 4$ G for electrons whereas $B_{\parallel,c} \approx 180$ G for protons.

Models for fast reconnection in solar flares suggest that the average transverse field is of order $\langle B_\perp \rangle = 1$ G (Somov, 1992). The particle density in the current sheet is $n = 10^{10}$ cm^{-3}. The sheet dimensions are as follows: $a = 10^2$ cm, $b = 10^8$ cm, $l = 10^9$ cm. These are typical estimates used in studies of collisionless acceleration processes in flares (see Martens, 1988, for discussion of stability of a current sheet with $a/b \ll 1$). In accordance with the arguments of the previous section, the reconnection electric field is taken to be $E = 10^{-2}$ cgs $= 3$ V cm^{-1}, implying fast reconnection as conventionally measured by the Alfvén Mach number $M = v_{in}/v_A$. The inflow speed is determined by the plasma electric drift speed into the sheet:

$$v_{in} = \frac{E}{B_0} c. \tag{11}$$

The numbers adopted above imply that the reconnection regime considered is fast since $v_{in} \approx 3 \times 10^6$ cm s^{-1} and $v_A \approx 2 \times 10^8$ cm s^{-1}, leading to $M \approx 10^{-2}$. Fast reconnection is necessary to ensure the large magnetic energy release rate P implied by observations of the flare impulsive phase. The power output is determined by the Poynting flux into the sheet:

$$P = \frac{1}{\pi} v_{in} B_0^2 bl. \tag{12}$$

The choice of parameters above leads to a reasonable value of 10^{27} erg s^{-1}. A larger power output would be possible in a larger current sheet.

Using the formulae and numbers above, it is straightforward to see that the hard X-ray generating electrons in solar flares can indeed be accelerated in a large-scale current sheet. Equation (10) gives the electron energy gain of about 30 keV, which would lead to X-rays in the same energy range. The acceleration time, which is simply the time spent by a particle inside the sheet, can be as small as 10^{-6} s (Litvinenko, 1996). The particle influx to the sheet is defined as

$$\dot{N} = 4lbv_{in}n, \tag{13}$$

which is about 10^{34} s^{-1} for the parameters above. This is clearly enough to cover the needed supply of electrons in small impulsive flares (Dennis, 1985). The influx can be much larger in gradual flares that presumably correspond to large-size current sheets (Martens, 1988). For example, assuming the length scale of the current sheet to be of order the size of an active region, $l \approx b \approx 10^{10}$ cm, leads to the electron flux $\dot{N} \approx 10^{37}$ s^{-1}. The flux can be even higher when reconnection is faster and v_{in} is larger.

Recall that the magnetizing longitudinal field $B_{\parallel,c}$ in Equation (7) is proportional to the square root of the particle mass. This is why even $B_\parallel = B_0$ is not high enough to make the proton motion adiabatic. The typical proton energy is found from Equation (6) to be about 200 keV. Martens and Young (1990) presented several arguments in support of proton beams in solar flares.

Allowing for a nonzero magnetic field inside the sheet has important consequences for electron orbits. First of all, the particle escape is much more efficient across the sheet than along it. If the electrons could simply move along the electric field direction through the total current sheet length, their typical energy would be determined by the total potential drop

$$eU = eEl \approx 3\,\text{GeV}. \tag{14}$$

Electrons with GeV energies are hardly ever observed in solar flares. The effect of the magnetic field is to decrease the average acceleration length by about five orders of magnitude, resulting in $\mathcal{E} \ll eU$. This makes the electric field acceleration a local acceleration mechanism that can occur throughout the reconnection region. In contrast to electron runaway models, the particles leave the current sheet sideways—perpendicular to the electric field—rather than at its top or bottom.

Equation (10) with $B_\perp = \langle B_\perp \rangle$ leads to electron energies that are rather modest in the context of the electron-rich flares. The next section investigates the possibility that the electric field acceleration in the current sheet can lead to MeV energies in those parts of the sheet where $B_\perp$ is much less than its average value.

ACCELERATION AT SINGULAR LINES

This is a long section, but what are you going to do, read other papers in this book?

Because the transverse field $B_\perp$ varies along the current sheet, going through zero at least at one point (its center), the magnetic field projection onto the xy-plane has the geometry of a standard magnetic X-point. In other words, this point is a projection of the singular magnetic field line with $B_x = B_y = 0$ onto the xy-plane. This is where the field lines are "cut" and "reconnected." More complicated geometries with multiple singular lines are also possible due to the tearing instability in the sheet, for example. Since the particle acceleration length scale is typically much less than the length scale of $B_\perp$ variation that can be of order b, Equations (9) and (10) derived for $B_\perp = \text{const}$ remain valid for $B_\perp = B_\perp(x)$, unless $B_\perp \to 0$. Now, however, the energy gain depends on the location in the sheet as a parameter with $B_\perp = B_\perp(x)$.

It is important that the particles should be accelerated to very high energies in the vicinity of the singular lines where the electric and magnetic field lines are coaligned. Following Syrovatskii (1975), I assume that in some flares the tearing instability leads to multiple singular lines with **E**||**B** inside a current sheet. Efficient electron acceleration in the vicinity of such singular lines may be responsible for the electron-rich impulsive flares.

According to Syrovatskii (1973, 1975), the impulsive phase of a flare corresponds to instability of a current sheet formed previously in the solar atmosphere. The tearing instability causes the formation of multiple singular lines inside the sheet. The lines are characterized by the electric field directed along the magnetic field line, which causes very efficient charged particle acceleration. Fast particles themselves, however, create the electric current that corresponds to a new current sheet growing at the original singular line of the magnetic field. This eventually leads to a nonlinear stabilization of the instability The process repeats itself at the newly formed singular lines, leading to a nonuniform internal structure of the global current sheet that contains numerous singular lines. Under these conditions the acceleration of particles is the main mechanism for the release of magnetic energy accumulated in the preflare current sheet.

The electron motion remains adiabatic even at the singular line itself because of a strong longitudinal field $B_\parallel$. In spite of this, integrating the equations of particle motion, whether directly or in the guiding center approximation, is a very complicated task (cf., Bruhwiler and Zweibel, 1992). Moreover, the simplified representation of the electric and magnetic fields in the sheet by the first nonzero terms in the Taylor expansion may not be sufficient to determine rigorously the time spent by the particles near the singular line and their energy gain. Therefore I use the following simple approach based on the picture of the magnetic line motion (see Litvinenko, 2000, for a detailed description).

The electrons are assumed to follow the reconnected field lines that move out of the current sheet with the speed of order v_A, much like beads sliding on a moving wire. The effect of the reconnection electric field is to accelerate the electrons along the magnetic field lines until the particles leave the singular line vicinity. Goldstein, Matthaeus, and Ambrosiano (1986) were evidently the first to use this approximation to estimate the particle energy gains in two-dimensional current sheets ($B_\parallel = 0$) in various astrophysical environments. Taking the magnetic field dynamics into account is what makes this approach different from the studies of charged particle acceleration at singular lines with a static magnetic field (Bulanov and Sasorov, 1976; Bulanov, 1980; Bruhwiler and Zweibel, 1992).

For simplicity, I assume that the reconnected field lines move out of the sheet with a constant speed v_A along the x-axis:

$$B_\perp(t) \approx \frac{2}{b}\langle B_\perp \rangle v_A t. \tag{15}$$

Here $\langle B_\perp \rangle = B_{\perp,\max}/2$, and the scale of the field variation from zero to $B_{\perp,\max}$ is half of the current sheet width, b, where the effect of multiple singular lines on the motion of a given particle is ignored also for simplicity. The approximation of a constant speed v_A corresponds to a linear dependence $B_\perp \sim x$ in the steady state when $x = v_A t$.

It is clear from Equation (10) that higher particle energies can be reached close to the singular line where $B_\perp \to 0$. One might think that arbitrarily large energy gains (up to those given by Equation (14)) could be possible near the singular line. This is not the case though because the acceleration time t_{acc} is finite. Eventually the magnetized electrons are carried with the reconnected field lines away from the singular line and acceleration ceases. This effect limits the electron energy.

It is easy to show that the temporal variation of the transverse magnetic field can be ignored for the deka-keV electrons responsible for the flare hard X-rays. The relative error in the energy gain from Equation (10) due to ignoring the variation of the field is

$$\frac{\delta \mathcal{E}_e}{\mathcal{E}_e} = \frac{\delta B_\perp}{B_\perp}. \tag{16}$$

The change in $B_\perp$ during the acceleration time is determined from Equation (15) as

$$\delta B_\perp = \frac{2}{b}\langle B_\perp \rangle v_A t_{acc}. \tag{17}$$

The acceleration time $t_{acc} = l_{acc}/c$ is evaluated for the acceleration length l_{acc} given by Equation (9). Here the electron speed v_e is replaced by the speed of light c without significant error because $v_e \approx 0.3c$ already for a 30-keV electron. Combining these equations leads to the estimate

$$\frac{\delta \mathcal{E}_e}{\mathcal{E}_e} = 2\frac{\langle B_\perp \rangle}{B_\perp}\frac{B_\parallel}{B_\perp}\frac{a}{b}\frac{v_A}{c} \tag{18}$$

with $B_\perp = B_\perp(x)$. This is very small in most of the current sheet. For instance, $\delta\mathcal{E}_e/\mathcal{E}_e \approx 10^{-6} \ll 1$ for $B_\perp = \langle B_\perp \rangle = 10^{-2}B_\parallel$. Nevertheless the approximation obviously breakes down at singular lines where $B_\perp \to 0$ and this fact is used below to determine the maximum energy of the accelerated electrons $\mathcal{E}_{e,max}$.

The maximum electron energy can be estimated as follows (see Goldstein, Matthaeus, and Ambrosiano (1986) for the two-dimensional case $B_\parallel = 0$). The magnetized electrons move almost along **B** inside the sheet, and their relativistic kinetic energy increases with time as

$$\mathcal{E}_e(t) \approx ceEt. \tag{19}$$

The maximum energy is defined by Equation (10). The maximum energy itself, though, is a function of time because the particles move out of the sheet together with the magnetic field lines. Therefore the time-dependent $B_\perp$ given by Equation (15) should be substituted into Equation (10):

$$\mathcal{E} = \frac{B_\parallel}{\langle B_\perp \rangle}\frac{eEab}{2v_A t}. \tag{20}$$

Now, since the actual electron energy is given by Equation (19) and the maximum energy is given by Equation (20), equating the two expressions gives an equation for the electron acceleration time, which is solved to give

$$t_{e,acc} \approx \left(\frac{B_\parallel}{\langle B_\perp \rangle}\frac{ab}{2cv_A}\right)^{1/2} \approx 3\times 10^{-4}\ \text{s}. \tag{21}$$

Substituting this result back into Equation (19) gives the sought-after maximum electron energy

$$\mathcal{E}_{e,max} \approx ceE\left(\frac{B_\parallel}{\langle B_\perp \rangle}\frac{ab}{2cv_A}\right)^{1/2} \approx 30\ \text{MeV}. \tag{22}$$

The same numerical values as before have been employed in these estimates.

It is gratifying to note that the same value for $\mathcal{E}_{e,max}$ is obtained from Equations (10) and (18) under condition that $\delta\mathcal{E}_e \approx \mathcal{E}_e$. Hence the derived $\mathcal{E}_{e,max}$ is actually the energy when the effects of the reconnected field line motion become noticeable on a given electron orbit. The derived maximum energy of the accelerated electrons in the current sheet is compatible with the observations that imply electron energies of a few tens of MeV in impulsive electron-rich solar flares. These electrons create the strong bremsstrahlung that dominates the nuclear gamma-line radiation.

Recall that it was possible to ignore the variation of $B_\perp$ in Equations (9) and (10) because the acceleration time is much less than the time scale of the field variation. It is the increasing acceleration time near the singular line that limits the particle energy gain. Physically, the energy of a magnetized electron increases with time but the maximum possible energy decreases as the particles move out of the sheet and the transverse magnetic field "felt" by the particles becomes larger, which makes it easier for them to escape the current sheet.

Clearly the maximum acceleration length is still much less than the total length of the current sheet:

$$l_{e,acc} = \frac{\mathcal{E}_{e,max}}{eE} \approx 10^7\ \text{cm} \ll l. \tag{23}$$

Thus even for the highest energies, the strong DC electric field acceleration remains a local acceleration mechanism. In other words the maximum energy is still much less than the total potential drop given by Equation (14). This confirms the previous results (Martens, 1988; Litvinenko, 1996) according to which the acceleration by strong electric fields in the reconnection region is a local mechanism acting all along the length of the sheet. The acceleration length remains small enough to ignore the Coulomb losses as well. Thus the particle energization process in impulsive flares is essentially collisionless as assumed throughout this paper. It is also interesting to note that $t_{e,acc}$ in Equation (21) does not depend on the reconnection electric field E.

It is worth repeating that for time intervals and energy gains smaller than those given by Equations (21) and (22) the particle motion with the magnetic field lines can be ignored, so that the electron acceleration far from the magnetic singular line can be studied assuming a constant instantaneous value of $B_\perp$ that depends on $x = v_A t$ as a parameter. This leads to a continuous electron spectrum extending to high energies (cf., Martens, 1988). Calculation of the detailed spectrum and the total number of accelerated electrons is a complicated problem that requires numerical simulations including the effects of nonuniform electric and magnetic fields, particle escape from the sheet, and possibly the charge separation electric fields. Nevertheless, it can be demonstrated that a power-law spectrum may result (Litvinenko, 1996). As before, consider acceleration in the case of a linear magnetic X-point in the xy-plane: $B_\perp \sim x$. The energy spectrum $f(\mathcal{E}_e)$ below $\mathcal{E}_{e,max}$ follows from the continuity equation $f(\mathcal{E}_e)\mathrm{d}\mathcal{E}_e = f(x)\mathrm{d}x$ with $\mathcal{E}_e \sim B_\perp^{-1} \sim x^{-1}$ from Equation (10). Assuming a spatially uniform inflowing distribution $f(x) = \text{const}$ leads to

$$f(\mathcal{E}_e) \sim \frac{\mathrm{d}x}{\mathrm{d}\mathcal{E}_e} \sim \mathcal{E}_e^{-2}. \tag{24}$$

Mori, Sakai, and Zhao (1998) considered a similar problem of charged particle acceleration in the vicinity of a singular line in the solar corona and demonstrated numerically the formation of a power-law spectrum with the index of about $2-2.2$ for a wide range of parameters. It appears, however, that a somewhat steeper spectrum would be necessary to interpret observations of the continuous gamma-ray radiation in electron-rich solar flares (Yoshimori *et al.*, 1992). The discrepancy is not suprising given the simplifying assumptions used in the estimate above. A more complicated geometry, for example, could make the actual spectrum steeper. Equation (24) is valid for a linear magnetic X-point. This result is easily generalized for a singular line of any order. Assuming $B_\perp \sim x^\alpha$ leads to the energy spectrum

$$f(\mathcal{E}_e) \sim \mathcal{E}_e^{-(1+\alpha)/\alpha}, \tag{25}$$

which gives $f(\mathcal{E}_e) \sim \mathcal{E}_e^{-3}$ for $\alpha = 1/2$. Yet another possibility is to relax the assumption of the electric field homogeneity. Schopper, Birk, and Lesch (1999) argue that particle acceleration by inhomogeneous electric fields in reconnection regions can result in observed energy spectra.

Other properties of the electron-rich events can be addressed in the context of the model. The event duration, in particular, should be determined by the Alfvén transit time along the current sheet width:

$$t_A = \frac{b}{v_A} \approx 0.5 \text{ s} \tag{26}$$

(cf., Syrovatskii, 1975), which is in agreement with the typically observed electron-rich event durations of a few seconds. As far as the number of energetic electrons produced in an "elementary" acceleration pulse is concerned, it can be estimated as the number of particles accelerated at one singular line:

$$\delta N = 4lav_A t_{e,acc} n \approx 2 \times 10^{26}, \tag{27}$$

where the acceleration time $t_{e,acc}$ is given by Equation (21). This estimate appears to be consistent with the electron fluence data for the energy range of a few MeV in the electron-rich events (e.g., Rieger, Gan, and Marschhäuser, 1998). It is of interest that a typical length scale δb associated with each acceleration site is given by the Alfvén transit time:

$$\delta b = v_A t_{e,acc} \approx 6 \times 10^4 \text{ cm}. \tag{28}$$

Because $\delta b \ll b$, each singular line can indeed be treated independently, justifying the notion of multiple acceleration sites in a single current sheet.

The question remains whether or not the considered mechanism of particle acceleration at singular lines of three-dimensional magnetic field is relatively inefficient for proton acceleration. This indeed appears to be so (Litvinenko, 2000), explaining a high electron to proton ratio in the energy range of a few tens of MeV, which is the essential property of the electron-rich flares. Substituting the numerical values given in the previous section into Equation (7) gives the magnetizing field $B_{\parallel,c} \approx 180$ G for protons. Because $B_\parallel < B_{\parallel,c}$, the protons are not magnetized by the longitudinal field in the current sheet. Hence the protons move rapidly out of the singular line vicinity, and their energy gains are smaller than those for electrons.

CONCLUSION

The electric field acceleration model is in nice agreement with studies of hard X-ray impulsive flares (e.g., Aschwanden *et al.*, 1996). These studies strongly suggest that electron acceleration in impulsive flares occurs in the cusp region above the flare loop, leading to the formation of the coronal hard X-ray sources discovered by Masuda *et al.* (1994). The inferred geometry is that of a reconnecting current sheet outside the flare loop. The present paper envisions electron acceleration in such a current sheet in the corona. The particles are accelerated at the reconnection site and ejected from the current sheet sideways into the flare loop where they produce the observed hard X-rays.

Super-Dreicer electric fields should be expected in nearly collisionless space plasmas, even when the turbulence is taken into account. Solar flares appear to provide an example of highly super-Dreicer electric fields associated with magnetic reconnection.

Studies of the charged particle motion and acceleration in reconnecting current sheets have demonstrated very interesting effects associated with the three-dimensional magnetic field structure in the sheet. Values of the field components determine the character of particle orbits and acceleration efficiency. Direct electric field acceleration leads to electron energies and fluxes required to explain the flare hard X-ray radiation. The acceleration time is very short, perhaps only a few milliseconds, which may correspond to the observed strong variability of the hard X-rays. The deka-keV electron beams produced in the reconnecting current sheet can also be responsible for the generation of various waves that later interact with ions and create heavy ion anomalous abundancies in flares (for a review see, e.g., Miller *et al.*, 1997).

Of particular interest is the acceleration of electrons in the vicinity of a singular line of magnetic field in the current sheet with a strong longitudinal component of the field. A physical mechanism that may underlie the electron-rich impulsive solar flares is the tearing instability of a pre-flare current sheet in the corona, which leads to effective electron acceleration at the singular magnetic field lines where the electric and magnetic fields are coaligned (Syrovatskii, 1975). There are several arguments in favor of this model. First, typical electron energy gains of a few tens of MeV explain the strong gamma-ray continuum in the electron-rich events. Second, proton energy gains at the singular lines are noticeably smaller (at most a few MeV), thus explaining the fact that the particle flux at energies of a few tens of MeV is primarily comprised of electrons that generate the intense gamma-ray bremsstrahlung. Third, the electron acceleration times are small enough to be consistent with the observed temporal evolution of the flare radiation. The total event duration is predicted to be on the order of the Alfvén time of a few seconds, which is also consistent with observations. Fourth, the model has the advantage of being able to interpret flares with either dominantly hard X-ray or gamma-ray emission. Different regimes are determined by the magnetic field structure in the reconnection region.

Finally, the results described in this paper may be applicable not only to solar flares but also to other phenomena in space, characterized by efficient particle acceleration. Litvinenko (1999) demonstrated that electron acceleration in a current sheet can lead to the observed synchrotron radiation by relativistic electrons in extragalactic jets.

ACKNOWLEDGEMENTS

This work was supported by NSF grant ATM-9813933 and NASA grant NAG5-7792.

REFERENCES

1. Aschwanden, M. J., Kosugi, T., Hudson, H. S., Wills, M. J., and Schwartz, R. A., *Astrophys. J.* **470**, 1198–1217 (1996).
2. Bruhwiler, D. L. and Zweibel, E. G., *J. Geophys. Res.* **97A**, 10825–10830 (1992).
3. Bulanov, S. V., *Soviet Astron. Lett.* **6**, 206–208 (1980).
4. Bulanov, S. V. and Sasorov, P. V., *Soviet Astron.* **19**, 464–468 (1976).
5. Dennis, B. R., *Solar Phys.* **100**, 465–490 (1985).
6. Fletcher, L. and Martens, P. C. H., *Astrophys. J.* **505**, 418–431 (1998).
7. Forbes, T. G., in *Eruptive Solar Flares*, edited by Z. Švestka et al., Springer-Verlag, Berlin, 1992, pp. 79–88.
8. Foukal, P. V. and Behr, B. B., *Solar Phys.* **156**, 293–314 (1995).
9. Foukal, P. V. and Hinata, S., *Solar Phys.* **132**, 307–334 (1991).
10. Goldstein, M. L., Matthaeus, W. H., and Ambrosiano, J. J., *Geophys. Res. Lett.* **13**, 205–208 (1986).
11. Holman, G. D., *Astrophys. J.* **293**, 584–594 (1985).
12. Horiuchi, R. and Sato, T., *Phys. Plasmas* **4**, 277–289 (1997).
13. Hudson, H. S. and Ryan, J. M., *Annual Rev. Astron. Astrophys.* **33**, 239–282 (1995).
14. Knight, J. W. and Sturrock, P. A., *Astrophys. J.* **218**, 306–310 (1977).
15. Litvinenko, Y. E., *Astrophys. J.* **462**, 997–1004 (1996).
16. Litvinenko, Y. E., *Phys. Plasmas* **4**, 3439–3441 (1997).
17. Litvinenko, Y. E., *Astron. Astrophys.* **349**, 685–690 (1999).
18. Litvinenko, Y. E., *Solar Phys.*, in press (2000).
19. Litvinenko, Y. E. and Somov, B. V., *Solar Phys.* **146**, 127–133 (1993).
20. Lyons, L. R. and Speiser, T. W., *J. Geophys. Res.* **90A**, 8543–8546 (1985).
21. Martens, P. C. H., *Astrophys. J.* **330**, L131–L133 (1988).
22. Martens, P. C. H. and Young, A., *Astrophys. J. Suppl.* **73**, 333–342 (1990).
23. Masuda, S., Kosugi, T., Hara, H., Tsuneta, S., and Ogawara, Y., *Nature* **371**, 495–497 (1994).
24. Melrose, D. B., *Aust. J. Phys.* **43**, 703–752 (1990).
25. Miller, J. A., Cargill, P. J., Emslie, A. G., *et al.*, *J. Geophys. Res.* **102A**, 14631–14659 (1997).
26. Mori, K.-I., Sakai, J.-I., and Zhao, J., *Astrophys. J.* **494**, 430–437 (1998).
27. Rieger, E., Gan, W. Q., and Marschhäuser, H., *Solar Phys.* **183**, 123–132 (1998).
28. Schopper, R., Birk, G. T., and Lesch, H., *Phys. Plasmas* **6**, 4318–4327 (1999).
29. Somov, B. V., *Physical Processes in Solar Flares*, Kluwer, Dordrecht, 1992, Ch. 3.
30. Speiser, T. W., *J. Geophys. Res.* **70**, 4219–4226 (1965).
31. Syrovatskii, S. I., *Comments on Astrophys. and Space Phys.* **4**, 65–70 (1973).
32. Syrovatskii, S. I., *Bulletin Acad. Sci. USSR* **39**, 359–373 (1975).
33. van den Oord, G. H. J., *Astron. Astrophys.* **234**, 496–518 (1990).
34. Yoshimori, M., Takai, Y., Morimoto, K., Suga, K., and Ohki, K., *Publ. Astron. Soc. Japan* **44**, L107–L110 (1992).
35. Zhu, Z. and Parks, G., *J. Geophys. Res.* **98A**, 7603–7608 (1993).

The Mixing Of Interplanetary Magnetic Field Lines: A Significant Transport Effect In Studies Of The Energy Spectra Of Impulsive Flares

J. E. Mazur[1], G. M. Mason[2,3], J. R. Dwyer[2], J. Giacalone[4], J. R. Jokipii[4], and E. C. Stone[5]

[1]*The Aerospace Corporation, 2350 E. El Segundo Blvd., El Segundo, CA 90245-4691 USA*
[2]*Department of Physics, University of Maryland, College Park, MD 20742 USA*
[3]*Institute for Physical Science & Technology, University of Maryland, College Park, MD 20742 USA*
[4]*Lunar and Planetary Laboratory, University of Arizona, Tucson, AZ 85721*
[5]*California Institute of Technology, Pasadena, CA 91125*

Abstract. Using instrumentation on board the *ACE* spacecraft we describe short-time scale (~3 hour) variations observed in the arrival profiles of ~20 keV $nucleon^{-1}$ to ~2 MeV $nucleon^{-1}$ ions from impulsive solar flares. These variations occurred simultaneously across all energies and were generally not in coincidence with any local magnetic field or plasma signature. These features appear to be caused by the convection of magnetic flux tubes past the observer that are alternately filled and devoid of flare ions even though they had a common flare source at the Sun. In these particle events we therefore have a means to observe and measure the mixing of the interplanetary magnetic field due to random walk. In a survey of 25 impulsive flares observed at ACE between 1997 November and 1999 July these features had an average time scale of 3.2 hours, corresponding to a length of ~0.03 AU. The changing magnetic connection to the flare site sometimes lead to an incomplete observation of a flare at 1 AU; thus the field-line mixing is an important effect in studies of impulsive flare energy spectra.

INTRODUCTION

Interplanetary particle events from impulsive solar flares have several characteristics that distinguish them from more intense particle events associated with traveling interplanetary shocks. One key characteristic is the pattern of abundance enhancements in the flare events compared to the solar wind composition: ^3He is ~10-1000 more abundant, and Ne-Si and Fe are enhanced by factors of ~3 to 5 and ~10, respectively (e.g. Reames [1] and references therein). An observer at 1 AU detects the electrons and ions from flares that are only at western solar longitudes near the intersection of the nominal interplanetary magnetic field line and the solar corona (e.g. Kahler et al. [2]; Cane, McGuire, & von Rosenvinge [3]). The unique abundance signatures that possibly result from wave-particle resonances (e.g. Temerin & Roth [4]; Miller & Viñas [5]), the high ionization states that imply temperatures above 10MK (Luhn et al. [6]), and the requirement of favorable magnetic connection to a flare site all indicate that the energetic particle source is close to the Sun, within a solar radius of the photosphere.

The propagation of flare electrons and ions from the flare site to 1 AU yields another characteristic of these particle events. The

CP528, *Acceleration and Transport of Energetic Particles Observed in the Heliosphere: ACE 2000 Symposium,* edited by Richard A. Mewaldt, et al.
© 2000 American Institute of Physics 1-56396-951-3/00/$17.00

acceleration process may take only tens of seconds to fully accelerate ions and electrons (Miller & Viñas [5]). This is much shorter than the minutes, hours, or even nearly day-long time scales associated with the particle propagation along magnetic field lines to 1 AU. The distribution of particle speeds therefore produces a measurable velocity dispersion wherein propagation effects dominate. Larson et al. [7] found that the effect of velocity dispersion revealed details of the propagation of ~0.14 to 100 keV electrons from impulsive flares within a magnetic cloud. They interpreted the numerous abrupt decreases of the electron fluxes as signatures of the disconnection of one end of the cloud's magnetic field from the solar corona. The electron event profiles in the study of Larson et al. [7] lasted as long as ~ 6 hours.

Here we show observations of the temporal structures in the arrival profiles of ~20 keV nucleon^{-1} to 5 MeV nucleon^{-1} impulsive solar flare ions made between 1997 November and 1999 July with instrumentation on board the *ACE* spacecraft (Mazur et al. [8]). At the lowest energies the ions have minimum times-of-flight from the flare to 1 AU of ~24 hours, with some variation due to the solar wind speed and the length of the resulting magnetic field line. These low energy ions propagate to 1 AU faster than the solar wind but slower than the energetic electrons that Larson et al. [7] discussed. We therefore use these ions as probes of pre-existing interplanetary magnetic field structures on the scale of ~0.03 AU, allowing us to study the magnetic connection to a flare with ions over a much longer time scale than previously possible.

OBSERVATIONS

The ion observations presented here were made with the Ultra-Low Energy Isotope Spectrometer (*ULEIS*) sensor on board the Advanced Composition Explorer (*ACE*) spacecraft, which was launched into orbit about the L1 Lagrangian point in late 1997 (Stone et al. [9]). *ULEIS* is a time-of-flight mass spectrometer that measures composition and energy spectra of H-Ni in the energy range of ~0.02 – 10 MeV nucleon^{-1} (Mason et al. [10]). We have observed numerous interplanetary particle events whose abundances and occasional associations with Type-III radio bursts and streaming ~100 keV electrons indicate that the ions were accelerated at an impulsive flare site (e.g. Mason et al. [11]; Reames [1]).

Event Profiles: Impulsive Versus Shock-Associated Energetic Particles

Figure 1a-d shows an example of impulsive flares observed with *ACE* during 1999 January 9 – 10. Figure 1a shows the energy (MeV nucleon^{-1}) of H-Fe ions as a function of their arrival time. The histogram in Figure 1b plots the smoothed ion count rate binned in ~14-minute intervals in order to better show the intensity variations. At least 2 particle injections occurred within this 2-day interval: one event began at ~1400UT on 9 January 1999, and particles at 1 MeV nucleon^{-1} from another injection arrived at 1AU on 1999 January 10 at ~0900UT

We focus on the more intense event beginning on 1999 January 9 at 1300UT. The velocity dispersion tells us that the ions in this event are from the same source at the Sun; we could not conclude this from the event counting rates alone shown in Figure 1b. Notice the occasional interruptions of the event profile on the time scale of ~1 hour wherein the particle counting rate decreased by a factor of 5-10 (e.g. at 1920 UT on 1999 January 9). The modulation of the event profile occurred simultaneously across all energies, and did *not* correlate with large changes in the direction of the local interplanetary magnetic field (Figure 1 c-d). The transitions from the relatively high intensity event rates to pre-existing particle rates were quite sharp, lasting <2.5 minutes.

In order to quantify these short-term intensity variations, we analyzed the event histogram of Figure 1b in the following way: beginning with the first arrival of ions from the flare at 1407UT on 1999 January 9, we used the velocity dispersion profile in Figure 1a to distinguish the particles of interest from the low fluxes of particles from previous events. From this time, we marked the end of a sub-interval (labeled 1 in Figure 1b) and the beginning of another (labeled 2) when the counting rate increased by a factor of ~1.3. This procedure was continued through the entire event, measuring the start and end times with a peak-to-valley ratio criterion of ~1.3. This particular threshold for deciding when to label a new sub-interval was sufficiently low to pick out the features clearly seen with the eye in the velocity profiles. The velocity dispersion profile clearly shows when the next event began on 1999 January 10 at 0910UT, where we measured a new set of intervals using the same technique. We return to the statistics of the sub-intervals measured in these and other events in the next section.

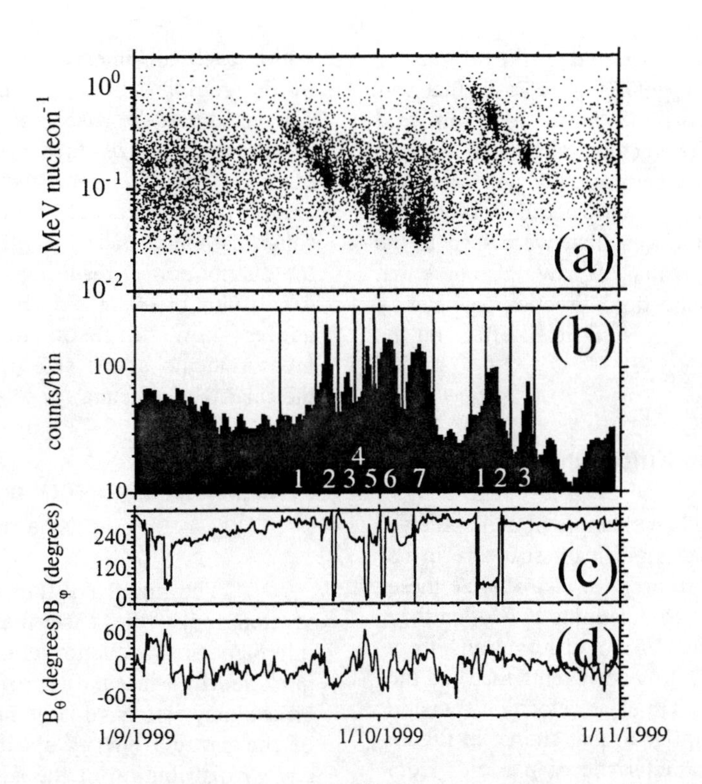

FIGURE 1. Energetic particle velocity dispersion and interplanetary magnetic field direction during 1999 January 9-10. (a) H-Fe ion energy versus time; (b) histogram of ion count rate; (c) magnetic field azimuth direction in the geocentric-solar-ecliptic (GSE frame); (d) magnetic field altitude direction (GSE frame).

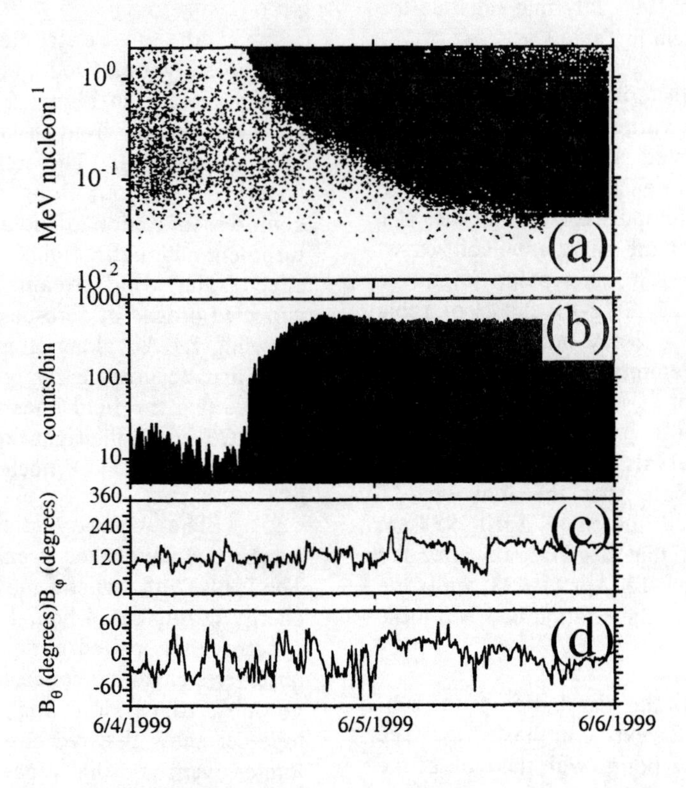

FIGURE 2. Oxygen velocity dispersion and interplanetary magnetic field direction during the 1999 June 4 event.

We next contrast the impulsive time profile of Figure 1 with an event on 4 June 1999 (Figure 2a-d) that was associated with a coronal mass ejection and interplanetary shock. This event also had velocity dispersion for ions and electrons with an onset near 0700UT on 1999 June 4, but we did not observe intensity variations of the kind that were seen in the impulsive flare of 1999 January 9. We also note that the interplanetary magnetic field direction changed by 10's of degrees without any significant effect on the event's profile (Figures 2c & d).

Size Of The Fine Structure

Figure 1 showed an example of an impulsive solar flare event that had significant structure in its time-of-arrival profile. In order to characterize these structures more fully and to determine how often they occur, we surveyed the *ULEIS* observations from 1997 November to 1999 July for events that had the following characteristics: (1) clear velocity dispersion in heavy ions (Z>2), similar to that shown in Figure 1; (2) Fe/O~1 as is characteristic of particles from impulsive flares; (3) low-intensity events, similar to Figure 1, in order to eliminate the heavy-ion rich onsets commonly seen in shock-associated particle events. We found 25 impulsive flare particle events from 1997 November to 1999 July that satisfied the survey criteria and list them in Table 1.

In order to characterize the time scale of the flare particle intensity variations, we applied the same technique as presented in Figure 1 above (peak-to-valley ratio of ~1.3) to each event of Table 1. To show the wide variety of impulsive event profiles at 1 AU, we plot in Figure 3 the energy/nucleon versus time of heavy ions (C-Fe) in four separate impulsive event periods (events #5, 7-11,12-13, and 21 of Table 1). Each panel covers a 2-day interval; the color scale is logarithmic in counts/sec with an arbitrary normalization. The event of 1998 April 12 had a ~6 hour-long interruption in its profile, in contrast to the many shorter-lived intervals observed in the 1999 January 9 events. Note that the ions near 1 MeV/nucleon that arrived late on 11 April 1998 are part of the same event that began again after the drop-out near ~0200 on 12 April 1998; only the velocity dispersion shows the relation between these two flux increases.

Figure 4a shows the distribution of the sub-interval durations of all 25 events in the survey. The average duration was 3.2 hours, with the bulk of the sub-intervals shorter than ~3 hours. We next took the average solar wind speed, also measured on *ACE*, within each sub-interval to calculate the spatial size of the region filled with energetic particles that convected past the spacecraft. Note that this estimate of the spatial size does not include any possible effects of the flux tube geometry or its angle with respect to the radial solar wind flow and is therefore a rough estimate. The histogram in Figure 4b shows the distribution of resulting sizes, with an average of 4.7×10^6 km or ~0.03 AU. In the discussion below we address how this estimate compares to previous measurements of the size of structures observed in the solar wind plasma.

Magnetic Connection Effects On Energy Spectra

The impulsive flares shown in the previous section make the case that the structure of the interplanetary magnetic field has a large effect on the particle time-intensity profiles. The connection effects may also have large impacts on measurements of the particle energy spectra. As records of the energy distribution of the particles that escape from the flare, it is critical to know whether features such as peaked spectra represent a source effect or a transport effect (e.g. Reames, Richardson, & Wentzel [12]).

As an extreme example how an incompletely observed event distorts the particle energy spectra, in Figure 5 we show the dispersion profile of Fe ions from an impulsive flare on 16 May 1998 (event #6). The ions arrived at ACE only within a ~4-hour long interval, just after a compressed region of solar wind and just before turbulent magnetic fields associated with a high-speed solar wind stream. The curve shows the expected profile of zero-degree pitch angle particles traveling 1.1 AU along an archimedian spiral from a flare that occurred early on 16 May 1998 at W44. We see that the field lines from this flare corotated past ACE just at the right time for ions between ~100 and a few hundred keV/nucleon to arrive at 1 AU.

The velocity dispersion yields energy spectra that are peaked in energy as show in Figure 6. The "pulse" of particles moved from higher to lower energy during the 4-hour long interval, resulting in spectra that rolled-over arose because of the observer's changing connection. In this example, the velocity dispersion and interplanetary context together show that we observed only a subset of a longer event and that processes near the source did not lead to the peaked energy spectra.

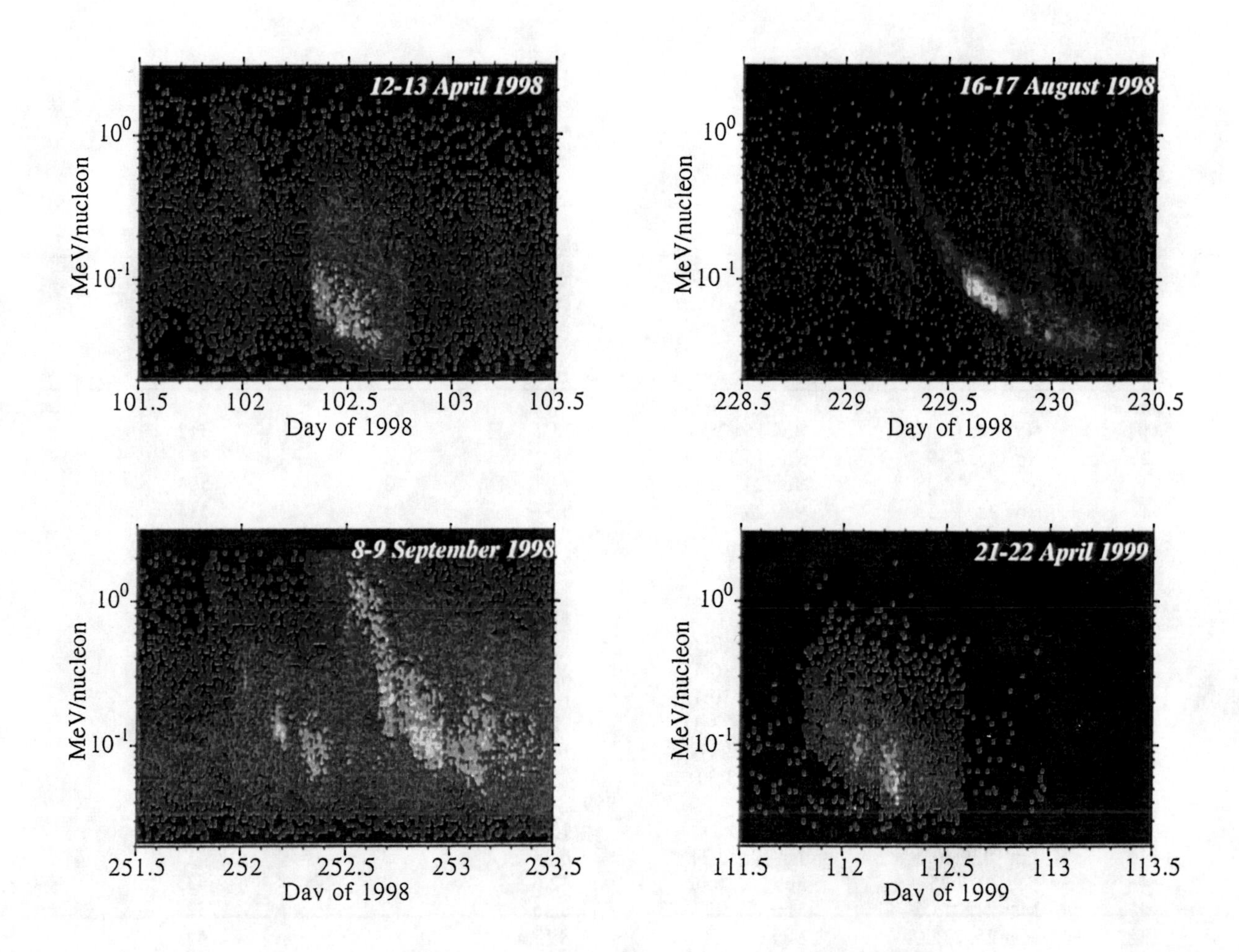

FIGURE 3. Velocity dispersion profiles of heavy ions in several impulsive flare events at 1 AU.

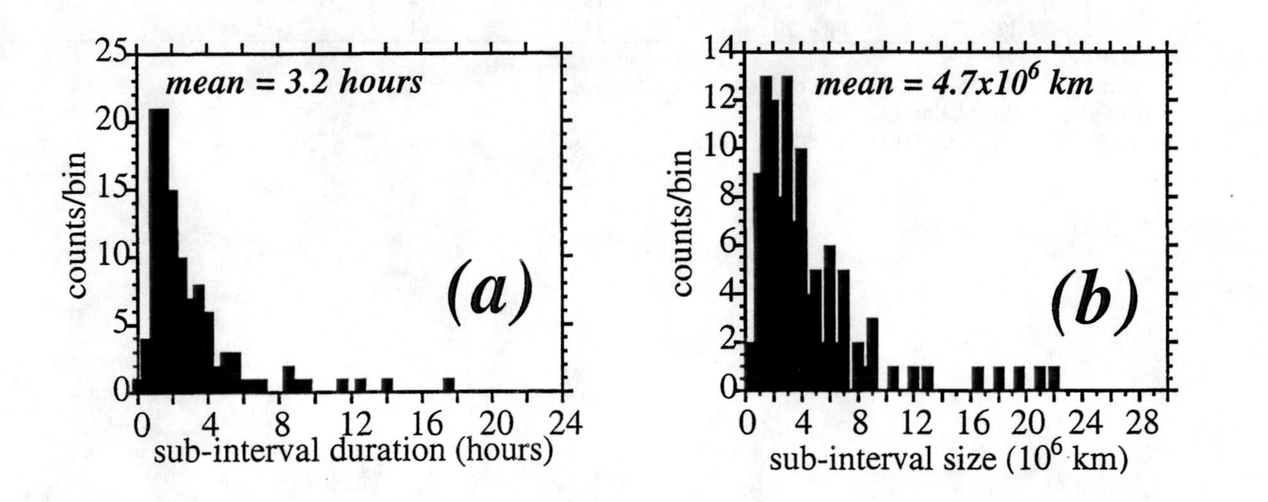

FIGURE 4. Time and size distributions of sub-intervals in 25 impulsive flares.

TABLE 1. Impulsive event time periods

Event	Start [a]	Stop [a]	Duration [b] (hr)	Number of sub-intervals [c]	Average solar wind speed [d] (km sec^{-1})
			1997		
1	Nov 14, 13:34	Nov 15,12:58	23.4	3	342
2	Nov 15, 4:19	Nov 15, 22:34	18.3	4	347
3	Nov 22, 18:28	Nov 23, 2:24	7.93	3	504
4	Nov 22, 19:55	Nov 23, 2:25	6.51	5	514
			1998		
5	Apr 11, 19:18	Apr 12, 19:35	24.3	5	382
6	May 16, 8:09	May 16, 11:52	3.73	2	507
7	Aug 15, 12:53	Aug 16, 6:33	17.6	7	369
8	Aug 17, 2:35	Aug 17, 17:30	14.9	2	336
9	Aug 17, 6:26	Aug 18, 10:45	28.3	11	310
10	Aug 17, 21:11	Aug 18, 10:12	13.0	2	298
11	Aug 18, 10:33	Aug 18, 23:31	12.9	6	330
12	Sep 8, 19:58	Sep 9, 12:17	16.3	8	359
13	Sep 9, 7:37	Sep 9, 23:00	15.4	4	373
			1999		
14	Jan 9, 14:07	Jan 10, 5:31	15.4	7	412
15	Jan 10, 9:10	Jan 10, 15:31	6.35	3	411
16	Feb 20, 7:25	Feb 20, 20:02	12.6	1	433
17	Mar 21, 21:06	Mar 22, 3:27	6.35	2	327
18	Mar 25, 23:00	Mar 26, 7:35	8.57	3	441
19	Apr 2, 16:49	Apr 3, 9:29	16.7	4	420
20	Apr 5, 9:31	Apr 6, 21:23	35.9	6	524
21	Apr 21, 19:12	Apr 22, 14:00	18.8	7	467
22	May 12, 9:38	May 12, 17:15	7.6	6	425
23	Jun 6, 2:32	Jun 9, 17:18	14.8	6	600
24	Jul 3, 20:31	Jul 4, 12:02	15.5	4	456
25	Jul 4, 6:12	Jul 4, 20:26	14.2	1	435

[a] All times UT, 1997-1999.

[b] Duration of heavy ion fluxes, ~0.02 – 2 MeV $nucleon^{-1}$.

[c] Sub-intervals as defined in section 2.1.

[d] 1-hour averaged solar wind speeds from ACE/SWEPAM.

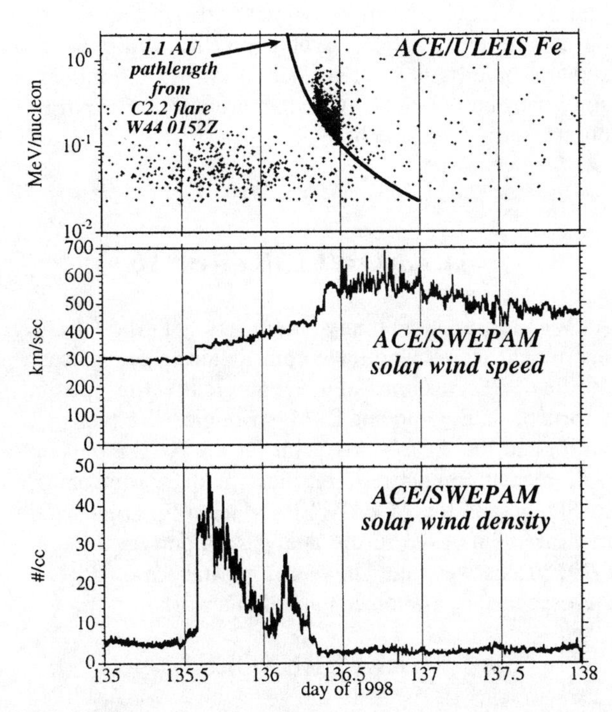

FIGURE 5. Fe dispersion profile and solar wind context for the 16 May 1998 impulsive flare.

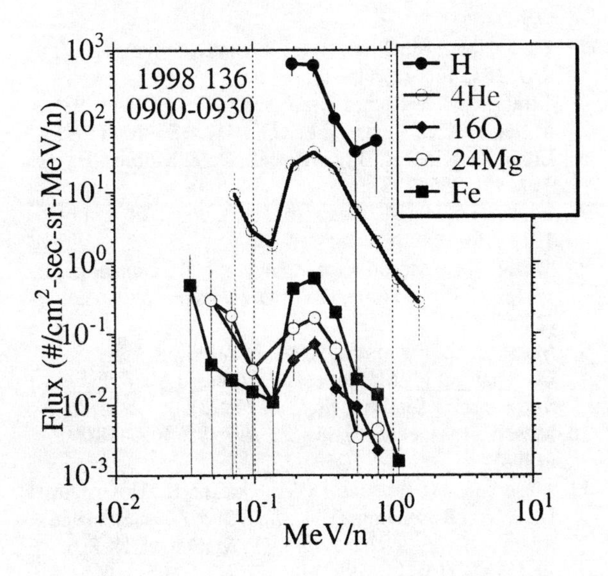

FIGURE 6. Average energy spectra in a 30-minute period during the 16 May 1998 flare.

DISCUSSION

Using the *ACE/ULEIS* instrument we have surveyed energetic particles from 25 impulsive solar flares and have found that the low energy ion intensity profiles exhibited sharp (<2.5 minutes) boundaries and occasional drop outs. The intensity changes occurred simultaneously across all energies and generally were not coincident with signatures of the local interplanetary magnetic field. Since the spacecraft must be magnetically connected to the flare sites in order to see the escaping ions, and since impulsive flare particles originate at compact sites, the observed waxing and waning of the particle intensities is puzzling.

These effects may be caused as follows (Mazur et al. [8]; Giacalone, Jokipii, & Mazur [13]). The solar wind continuously convects magnetic flux tubes that are connected to a solar active region to 1 AU and beyond. During the convection to 1 AU, which takes about 3-4 days, the footpoints of these flux tubes become mixed with the footpoints of other flux tubes that are not magnetically connected to the active region. The footpoint motion has the effect that adjacent flux tubes at 1 AU can be significantly separated at the Sun, and, conversely, that adjacent flux tubes at the Sun may be separated at 1 AU on scales of ~0.01 AU. The flare occurs and populates those flux tubes that are connected to the acceleration site with energetic particles; the other unconnected flux tubes remain empty. The particles within each connected flux tube arrive at 1 AU with velocity dispersion. As the mixed flux tubes move past *ACE*, we see either filled flux tubes and hence the on-going particle velocity dispersion, or empty flux tubes and hence the particle drop-outs. In this picture, the local interplanetary magnetic field does not necessarily correlate with the dramatic intensity changes we have seen. Variations of the particle acceleration process at the Sun are ruled out as a cause of the intensity changes because the drop-outs occur simultaneously at all energies.

The effects of field-line mixing are not as obvious in the more intense particle events associated with coronal mass ejections and interplanetary shocks. We can understand this as a result of the relative sizes of the acceleration sites: in shock-related events, the temporal structure observed at 1 AU reflects the ever-changing magnetic connection of an observer to the expanding shock that covers a wide extent in heliolongitude (e.g. Cane, Reames, & von Rosenvinge [14]). Consequently, any flux tube that *ACE* intersects will be populated with energetic ions as the shock propagates outward. This is not the case for impulsive flares because the acceleration site of an impulsive flare is much smaller and therefore it is more likely that an observer's magnetic connection will wax and wane depending on the detailed magnetic topology near the flare.

These effects have only been glimpsed in prior studies of *impulsive* flare heavy ions (e.g. Reames, Richardson, & Wenzel [12]) mainly because these earlier studies covered energies above ~1 MeV/nucleon. At these higher energies, the particle events do not last as long (c.f. Figure 1a) making it less likely to see structure on the time scale of an hour. The falling energy spectra also make the statistical accuracy poorer at the higher energies, putting the detection of such structures reported here outside the capability of prior studies of impulsive flares.

Anderson and Dougherty [15] observed numerous instances of modulation of electron and ion intensities during long-lasting solar or interplanetary events. Also, Buttighofer [16] noted similar abrupt modulations of low-energy electrons observed on Ulysses that had no obvious coincident plasma signature. We believe that the particle channels observed by Anderson and Dougherty [15] and Buttighofer [16] are the same kind of interplanetary structures we have examined in this work. We also note that the structures discussed here may have led to the electron disconnection events Larson et al. [7] observed within a magnetic cloud.

The interplanetary magnetic field structure that we have deduced from these observations should be related to measurements of the solar wind and the interplanetary magnetic field. Matthaeus et al. [15] studied the characteristic time of correlations between the interplanetary magnetic field and solar wind speed measured at two different times from a single spacecraft. The average correlation time (3.8 hours) and length (4.9×10^6 km) measured in the solar wind by Matthaeus et al. [17] are indeed similar to the average flare sub-interval duration (3.2 hours) and inferred length (4.7×10^6 km) reported here.

We have seen cases with ACE where the changing magnetic connection to the flare causes an incomplete observation of the particle event. The resulting energy spectra may then have features such as low-energy roll-overs that arise not because of processes such as energy loss, but because of the mixing of the interplanetary magnetic field. Measurements well below ~1 MeV/nucleon where the fluxes are highest are key to observing the intermittent magnetic connection. In future studies of impulsive flare energy spectra we can thus account for a significant transport effect.

Giacalone, Jokipii, & Mazur [13] & [18] discuss a model that relates these observations to a random walk of the magnetic field lines that are line-tied in the supergranulation network of the sun's photosphere. Particle events on the scale of a supergranule (tens of thousands of km) in the model show structures at 1 AU that are similar to the observations discussed here.

ACKNOWLEDGMENTS

We thank the many members of the *ACE* instrument teams who have contributed to the success of the ACE mission, and acknowledge the special efforts of R. E. Gold and S. M. Krimigis. We thank C. Smith and the ACE/MAG team for the ACE magnetic field measurements reported here, and D. McComas, R. Skoug, and the ACE/SWEPAM team for solar wind measurements used in the analysis. Contract number Q295801 between the University of Maryland and The Aerospace Corporation supported this work.

REFERENCES

1. Reames, D. V., *Space Sci. Rev.*, **90**, 413-489 (1999).
2. Kahler, S. W., Lin, R. P., Reames, D. V., Stone, R. G., & Liggett, M., *Solar Phys.*, **107**, 385-394 (1987).
3. Cane, H. V., McGuire, R. E., & von Rosenvinge, T. T., *ApJ*, **301**, 448-459 (1986).
4. Temerin, M. & Roth, I., *ApJ*, **391**, L105-L108 (1992).
5. Miller, J. A. & Viñas, A. F., *ApJ*, **412**, 386-400 (1993).
6. Luhn, A., Klecker, B., Hovestadt, D., & Möbius, E., *ApJ*, **317**, 951-955 (1987).
7. Larson, D. E. et al., *Geophys. Res. Letters*, **24**, 1911-1914 (1997).
8. Mazur, J. E., Mason, G. M., Dwyer, J. R., Giacalone, J., Jokipii, J. R., & Stone, E. C., *ApJ. Letters, L79-L83* (2000).
9. Stone, E. C., Frandsen, A., M., Mewaldt, R. A., Christian, E. R., Margolies, D., Ormes, J. F., & F. Snow, *Space Sci. Rev.*, **86**, 1-22 (1998).
10. Mason, G. M. et al., *Space Sci. Rev.*, **86**, 409-448 (1998)
11. Mason, G. M., Reames, D. V., Klecker, B., Hovestadt, D., & von Rosenvinge T. T., *ApJ*, **303**, 849-860 (1986).
12. Reames, D. V., Richardson, I. G., & Wenzel, K. P., *ApJ*, **387**, 715-721 (1992).
13. Giacalone, J., Jokipii, J. R., & Mazur, J. E., *ApJ. Letters, L75-L78,* (2000).
14. Cane, H. V., Reames., D. V., & von Rosenvinge, T. T., *J. Geophys. Res.*, **93**, 9555-9567 (1988).
15. Anderson, K. A., & Dougherty, W. M., *Sol. Phys.*, **103**, 165-175 (1986).
16. Buttighoffer, A., *A&A.*, **335**, 295-302 (1998).
17. Matthaeus, W. H., Goldstein, M. L., & King, J. H., *J. Geophys. Res.*, **91**, 59-69 (1986).
18. Giacalone, J., Jokipii, J. R., & Mazur, J. E., "Solar-Energetic Particles vs. Global Cosmic-Ray Diffusion", *these proceedings*, (2000).

The Isotopic Composition of Solar Energetic Particles

C. M. S. Cohen[1], R. A. Leske[1], E. R. Christian[2], A. C. Cummings[1], R. A. Mewaldt[1], P. L. Slocum[3], E. C. Stone[1], T. T. von Rosenvinge[2], M. E. Wiedenbeck[3]

[1]*California Institute of Technology, Pasadena, CA 91125*
[2]*NASA/Goddard Space Flight Center, Greenbelt, MD 20771*
[3]*Jet Propulsion Laboratory, Pasadena, CA 91109*

Abstract. Since the launch of ACE in August 1997, the Solar Isotope Spectrometer (SIS) has observed 11 large solar particle events in which elemental and isotopic composition was determined over a large energy range. The composition of these events has raised many issues and challenged generally accepted characterizations of solar energetic particle (SEP) events. In particular, $^{3}He/^{4}He$ enhancements have been observed in several large events as well as enhancements of heavy ions typically associated with smaller impulsive events. The isotopic composition varies substantially from event to event (a factor of 3 for $^{22}Ne/^{20}Ne$) with enhancements and depletions that are generally correlated with elemental composition. This correlation suggests that the isotopic enhancements may be related to the Q/M fractionation typically evident in the elemental composition of SEP events. However, there are also significant deviations from this pattern, which may imply that wave-particle resonances or other mass fractionation processes may be involved. We review the recent isotopic observations made with ACE and discuss their implications for particle acceleration and transport.

INTRODUCTION

One of the windows into the composition of the Sun's atmosphere is solar energetic particle (SEP) events. Measuring the composition of these events provides information regarding the particle source as well as acceleration and transport processes which can alter the initial composition. Understanding these processes furthers our knowledge of solar dynamics and allows SEP event composition to be used as a probe of solar composition.

Although there are event-to-event variations, SEP events generally can be classified into one of two categories, gradual or impulsive (1). Gradual events are large events that usually last for days. On average the composition is very similar to that of the solar wind and solar corona. This and the observed correlation between gradual events and coronal mass ejections (CMEs) has led to the conclusion that these particles are accelerated out of the outer corona and interplanetary medium by the interplanetary shocks that are sometimes formed by outwardly moving CMEs. Since typically the shocks are spatially large structures, the resulting SEPs can be observed over a broad longitudinal range.

Impulsive events are short in duration (usually hours) and have composition that differs substantially from that of the solar wind. Dramatic enhancements of ^{3}He have been observed, sometimes as great as 10^{4} times that of the solar wind. The heavy ion composition is quite different from that of the gradual events; Ne, Mg, and Si are often enhanced by a factor of ~3 (normalized to C, N, and O) and Fe is typically enhanced by a factor of ~10 over average gradual event values. The correlation with optical and x-ray solar flares and the limited solar longitude range that these events originate from suggests that the sources of these particles are the solar flare sites themselves.

Within the two classes, the event-to-event variation can be substantial and is most likely a result of acceleration and/or transport processes. When compared to a multi-event average, both Fe-poor and Fe-rich gradual events have been observed. Often the

CP528, *Acceleration and Transport of Energetic Particles Observed in the Heliosphere: ACE 2000 Symposium*, edited by Richard A. Mewaldt, et al.
© 2000 American Institute of Physics 1-56396-951-3/00/$17.00

deviation in composition of a single gradual event from the long-term average composition can be well represented by a power law in charge/mass (Q/M) (2).

While the heavy ion composition of impulsive events varies by less than a factor of 2 (3), the $^3He/^4He$ ratio can vary by orders of magnitude from one event to another. The large enhancements of ^3He have been attributed to a resonance between the particles and ion cyclotron waves (4,5). This same mechanism has been suggested as a cause of the enhanced $^{22}Ne/^{20}Ne$ values observed in impulsive events (6). In this case, the particles would be resonating with the waves at the second harmonic of the ion cyclotron frequency (7).

Although there have been previous reports of events with both impulsive and gradual characteristics (e.g. 8), such events were considered anomalous. However, data from a new generation of energetic particle detectors have found a large fraction of long duration events that appear to have a mixture of impulsive and gradual characteristics (9,10) indicating that such events are not rare and need to be understood. Several large events have been observed which have significant enhancements of $^3He/^4He$ and Fe/O ~ 1 at energies ≥ 1 MeV/nucleon (9,10). Coronal mass ejections were often observed in conjunction with these events, typical of gradual events, yet the measured elemental and isotopic composition is more similar to that of impulsive events than gradual events (9,10) and the observed charge state composition includes charge states characteristic of impulsive events as well as those from gradual events (11).

Recently, observations of finite enhancements of ^3He over the solar wind value during time periods of low to moderate solar activity have prompted the suggestion that the enhancements of ^3He measured in these large SEP events may result from a re-acceleration of remnant impulsive material that is present in the interplanetary medium (10). It is further suggested that this may also account for the 'mixed' charge state composition of these events.

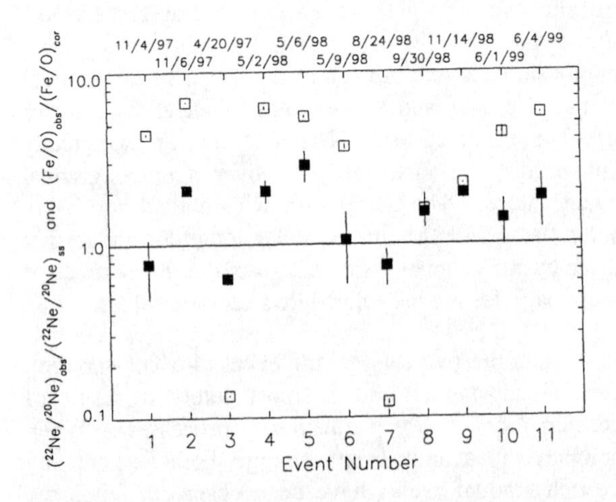

FIGURE 1. Enhancements of $^{22}Ne/^{20}Ne$ (filled symbols) and Fe/O (open symbols), relative to solar system (14) and gradual SEP values (1), respectively, for the 11 events measured by SIS.

With new data from instruments with improved resolution and sensitivity it is becoming clear that the composition of SEP events deserves more examination. The event-to-event variation in elemental and isotopic composition provides clues to the acceleration and transport processes involved. In this study we review the recent composition measurements taken at ~10 – 60 MeV/nucleon in 11 large SEP events and the observed deviations from classic gradual event composition. These variations are examined in relation to the ideas of Q/M fractionation, wave-particle resonance, and the presence of remnant impulsive material.

OBSERVATIONS AND THEORIES

The data presented here were obtained from the Solar Isotope Spectrometer (SIS) (12) on the Advanced Composition Explorer (ACE). The instrument contains two telescopes, each composed of two position-sensing matrix detectors, followed by a stack of large-area silicon solid-state detectors. Using the standard ΔE versus residual energy technique, the nuclear charge, mass, and kinetic energy of the particles can be determined. The mass resolution of SIS allows the isotopes of elements of $2 \le Z \le 28$ to be resolved, while the large geometry factor permits this to be accomplished with excellent statistical accuracy.

Since the launch of ACE, 11 large SEP events have been observed and studied with SIS. The compositional variation between these events is dramatic, with $^{22}Ne/^{20}Ne$ ranging from 0.056 ± 0.014 to 0.212 ± 0.044 (13) and Fe/O ranging from 0.018 ± 0.001 to 0.900 ± 0.006 (9). As can be seen from Figure 1, the $^{22}Ne/^{20}Ne$ and Fe/O ratios, relative to solar system (14) and gradual SEP event values (1) respectively, vary substantially from event to event and are significantly enhanced in many of the events. Other elemental and isotopic ratios also show comparably large fractionation for these events (see 15,9,16).

In many of these events a significant overabundance of ^3He was measured (9,10). Mass histograms for each event are presented in Figure 2

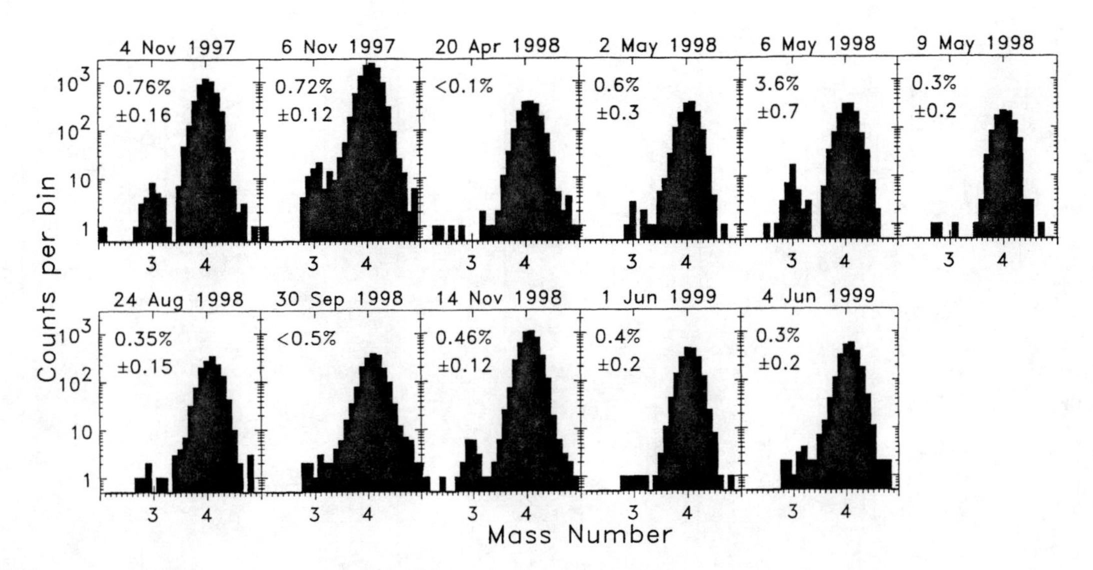

FIGURE 2. Mass histograms of 7.3 to 9.7 MeV/nucleon He for the 11 SEP events (16, copyright Astronomical Society of the Pacific 2000). Event average ^{3}He/^{4}He ratios are indicated for each period.

along with the ^{3}He/^{4}He ratio. In at least 4 events, the ^{3}He/^{4}He ratio is clearly enhanced by more than an order of magnitude above the solar wind value of 0.04% (17), and in most of the remaining events significant enhancements are also possible. Neither this nor the large heavy ion enhancements are expected for gradual events, suggesting either a mixture of impulsive and gradual material or significant fractionation during acceleration and/or transport.

Fractionation Theories

The Q/M fractionation process identified in (2) can produce elemental and isotopic enhancements. Assuming the fractionation can be expressed as a power law in Q/M, the measured abundance can be related to the source abundance as follows:

$$N_{SEP} = N_{Source}\left(\frac{Q}{M}\right)^{\gamma} \tag{1}$$

If two isotopes have the same charge state then equation (1) reduces to a mass fractionation in the measured isotopic ratio:

$$\left(\frac{N_{M1}}{N_{M2}}\right)_{SEP} = \left(\frac{N_{M1}}{N_{M2}}\right)_{Source}\left(\frac{M_2}{M_1}\right)^{\gamma} \tag{2}$$

If the same process is responsible for both the observed elemental and isotopic fractionations then they should be correlated. This is the case for species heavier than He and it has been reported that time periods with large ^{22}Ne/^{20}Ne enhancements also generally have large Fe/O enhancements (13), which is also evident in Figure 1. Since the Fe/O ratio also depends on the first-ionization-potential (FIP)-related fractionation, and the degree of FIP fractionation varies from event to event (18,19,20), the correlation is not as tight as might be expected. In addition, the power law behavior of the Q/M fractionation may not continue over the large Q/M range from Ne (Q/M ~ 0.4) to Fe (Q/M ~ 0.25).

A better correlation is found between the ^{22}Ne/^{20}Ne and the Na/Mg abundance ratios (21). Here the Q/M values of Na and Mg are similar to that of Ne and are relatively insensitive to temperature in the 2-5 x 10^6 K range due to the He-like electron configuration of the ions (22,23). Additionally, Na and Mg are both low-FIP elements, so the variation of the FIP fractionation is not an issue. Under the assumption that a power law Q/M fractionation is responsible for both the elemental and isotopic fractionation, we have investigated the degree to which this fractionation process represents the composition of the 11 events using the following three simple models.

A source population is created with elemental and isotopic composition as given in (1) and (14) and a charge state distribution determined from thermal equilibrium at a specified temperature as given in (22) and (23). The abundance of each ion is multiplied by an enhancement factor of $A(Q/M)^{\gamma}$. The resulting population is compared to the measured composition of an event and a χ^2 value is determined from

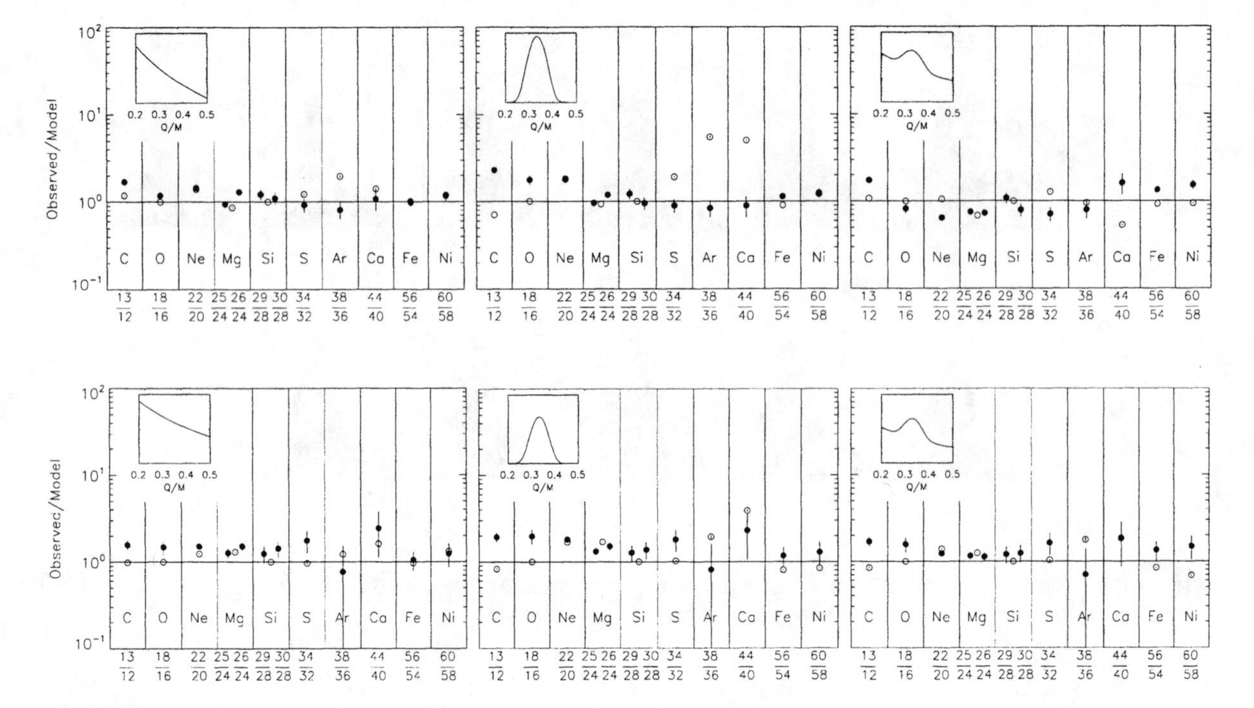

FIGURE 3. Results of the 3 models divided by measured ratios. Solid points are isotope ratios; open points are elemental ratios; inset graph shows the form of the enhancement function in arbitrary units. The top row shows data from the 11/6/97 event, the bottom row contains data from the 11/14/98 event. A perfect fit would result in points forming a horizontal line at 1.

$$\chi^2 = \sum \left(\frac{\frac{R_{Obs}}{R_{Mod}} - 1}{\frac{\sigma_{Obs}}{R_{Mod}}} \right)^2 \tag{3}$$

where, R_{Obs} is the observed composition relative to that of average gradual events, R_{Mod} is the ratio resulting from the model, and σ_{Obs} is the uncertainty in the measurements. The model parameters, A, γ and the source temperature, are varied until χ^2 is minimized.

The fact that $^3He/^4He$ ratio is significantly enriched in some of these events relative to the solar wind suggests a wave-particle resonance may be operating during the acceleration of these ions. This can result in the enhancement of heavy ions with Q/M values near 0.333, those resonating with the second harmonic of the ion cyclotron frequency (7). This effect is investigated using a similar model to the one described above, but instead of a power law the enhancement factor is taken to be a Gaussian in Q/M with the peak at 0.333 and a standard deviation of 0.03. The choice of this standard deviation was based on a discussion of Q/M values expected to be affected by the resonance (7). The amplitude of the Gaussian and the source temperature are the fit parameters in this model. Since it is possible that both the Q/M and resonance processes are operating, a final version of the model uses the product of the Q/M power law and the Gaussian as the enhancement factor. In this case, the parameters are γ, an overall amplitude of the enhancement factor at Q/M = 0.333, and the source temperature.

The results of the models for the 3 different enhancement functions are given in Figure 3 for two events which have the best statistical accuracy, 6 November 1997 and 14 November 1998. It is evident that the resonance enhancement function alone (middle panel of both rows of Figure 3) does not yield results consistent with the SIS data. The best fit result still has extremely large inconsistencies in both the elemental (e.g., Ar/Si and Ca/Si) and isotopic (e.g., $^{13}C/^{12}C$, $^{18}O/^{16}O$, $^{22}Ne/^{20}Ne$) ratios. When the parameters are altered to best reproduce the Ne isotopic data (the impetus for the model), the resonance model results (not shown) show significant deviations from the observations for $^{13}C/^{12}C$ and Ne/O. Deviations, both isotopic and elemental, are also present for heavier ions such as Ca, Fe and Ni.

While the power law function and the combined power law and resonance function yield results of similar quality, there are significant outliers where this simple model does not replicate the observed data, specifically $^{13}C/^{12}C$ for both events and Ar/Si for the

14 November 1998 event. As the inclusion of resonance-related fractionation does not significantly improve the results over those of a simple power law fractionation, resonant enhancements of heavy ions must be a minor effect in these events. Overall, it is clear that the process taking place is more complicated than is modeled here.

The assumption of thermal equilibrium was adopted for the sake of keeping the models simple. In fact, thermal equilibrium for all species is not realistic. It is often observed in the solar wind that charge state distributions for heavy elements such as Fe are not well represented by thermal equilibrium distributions (24,25). Modeling has also shown that different elements freeze-in (stop altering their charge state distributions) at different heights in the corona, corresponding to different source temperatures (26). Additionally, stripping of the ions may be occurring before and/or during acceleration (e.g., 27), altering the initial charge state distribution. Such deviations from thermal equilibrium would result in different enhancement patterns than our models produce. We conclude that more sophisticated models than those presented here are required to understand the abundance patterns in these SEP events.

Remnant Material Theory

Significant amounts of ^{3}He have been detected at energies of ~1 MeV/nucleon during moderately active and quiet time periods, suggesting that there is a substantial volume of the interplanetary medium that contains remnant material from impulsive SEP events (10). It is thought that this material, which is already suprathermal, could be preferentially accelerated by a passing CME-related shock and accelerated to the observed SEP energies. Thus a gradual event (i.e, one in which the particles were accelerated by an interplanetary shock rather than at a flare site) could

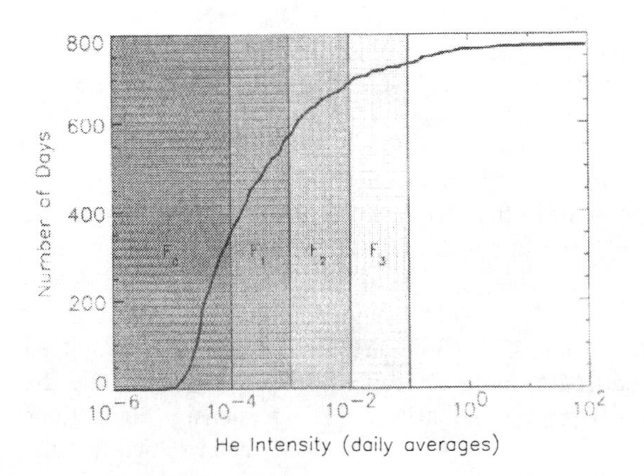

FIGURE 4. Time period selection based on He intensity.

exhibit some characteristics of an impulsive event, the degree of which would depend on the relative amounts of remnant and coronal (or solar wind) material accelerated by the shock.

In such a scenario, one would expect other impulsive characteristics to be evident. In addition to being ^{3}He-rich, impulsive events typically have enhanced Fe/O ratios and often have increased ^{22}Ne/^{20}Ne ratios (6,28). Periods of low to moderate solar activity have been examined in search of such signatures in the SIS data in an effort to test the remnant material theory. We note, however, that the vast majority of this suprathermal material is expected to be below the energy range of SIS.

Daily averages of the SIS He intensity (I_{He}) at 3.8 – 4.9 MeV/nucleon were sorted by increasing value and the integral number of days with He intensity below a given value was calculated (Figure 4). The data were divided into 4 sets according to different levels of activity: F_0: $I_{He} \leq 10^{-4}$, F_1: $10^{-4} < I_{He} \leq 10^{-3}$, F_2: $10^{-3} < I_{He} \leq 10^{-2}$, and F_3: $10^{-2} < I_{He} \leq 10^{-1}$, where I_{He} is in (cm^2 sec sr MeV/nuc)$^{-1}$. In addition, days

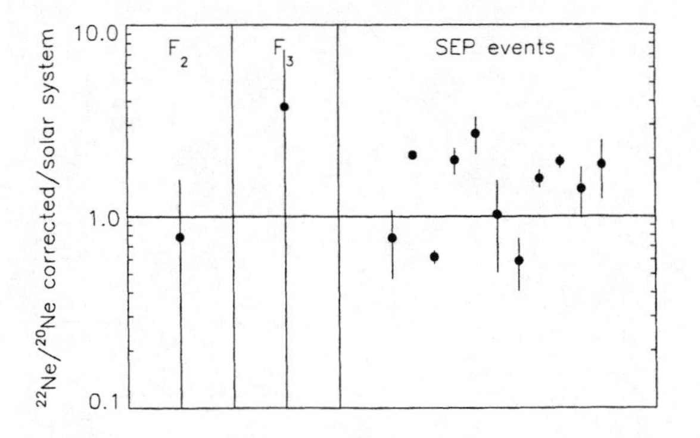

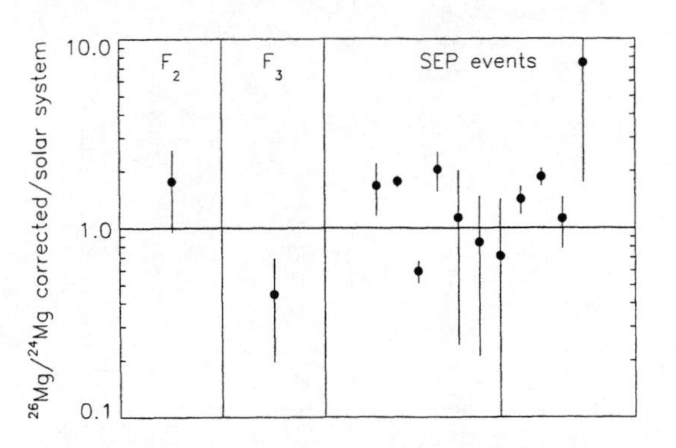

FIGURE 5. Ne and Mg enhancements for different time selections and the 11 SEP events.

containing the tails of the 11 large events were removed from the 4 data sets (the majority of the events themselves were excluded due to the He intensity restrictions).

In order to obtain meaningful Ne and Mg isotope ratios from the different activity levels, the anomalous and galactic cosmic ray (ACR and GCR, respectively) contributions have to be subtracted. From 9/1997 to 11/1999, the ACR component has decreased by a factor of ~20, while the GCR component has been reduced by only a factor of ~2 (29). Additionally the ACR level exhibits factors of ~5 intensity fluctuations on short time scales. Since it has been shown that the ACR and GCR intensity levels scale approximately as powers of the Climax Neutron Monitor data (e.g., 29, 30), we use daily averaged Climax data as a proxy for the overall levels of ACR and GCR contributions. An exponential and a power law energy spectrum are used to represent the ACR and the GCR components respectively. Daily composite spectra are formed from the independently scaled components and summed to produce an overall 'background' spectrum for the days falling within each activity level.

Good agreement is found between the calculated F_0 background and measured Ne spectrum indicating that little solar material is contributing to the F_0 data and that this method of estimating the GCR and ACR background is adequate. The background-corrected Ne and Mg isotopic ratios (as compared to solar system values (14)) are plotted in Figure 5 for the F_2 and F_3 data, and the 11 large SEP events. The F_1 data are consistent with the background spectra and are not shown.

No excess of $^{22}Ne/^{20}Ne$ is apparent in the F_2 or the F_3 data as would be expected if substantial remnant material were present, although the statistical uncertainties are so large that enhancements cannot be ruled out. The uncertainties in the $^{26}Mg/^{24}Mg$ ratios are smaller and the data are also consistent with no enhancement. It should be noted that the energy range of these data is ~20 - ~55 MeV/nucleon and that the solar contribution at these energies is small except in large events. Currently we have not yet resolved the Ne and Mg isotopes at lower energies in the SIS data with sufficient accuracy to present ratios at energies where the solar component is stronger.

Elemental composition measurements in SIS extend down to ~10 MeV/nucleon and a test of the presence of remnant impulsive material is iron, which is typically a factor of ~3 higher in impulsive events (as compared to silicon) than in gradual events. Iron spectra obtained for the 4 activity levels are presented in Figure 6. At high energies, which are dominated by GCRs, all levels exhibit similar intensities. At low energies the solar contribution increases with increasing activity level as expected. Although they are not shown, the silicon spectra show similar trends.

Also shown in Figure 6 is the Fe/Si ratio as a function of energy. An enhancement of ~2 over F_0 levels is apparent at low energies for the F_2 and F_3 data. The sharp decrease in this enhancement near 40 MeV/nucleon is a result of the decrease in the solar contribution to the Fe and Si spectra. From these data we conclude that significant amounts of Fe-enhanced material (possibly from impulsive events) is present during periods of moderate solar activity.

Related to this, a search for ^3He-rich time periods in the SIS data resulted in 92 identified periods greater than 5 hours in length, many of which were not clearly associated with an impulsive SEP event (31). Since the hourly averaged $^3He/^4He$ ratios were required to be consistently greater than 0.10 during these time periods, a lower $^3He/^4He$ enhancement criterion would certainly result in many more time periods, indicating enhancements of ^3He are present during a considerable

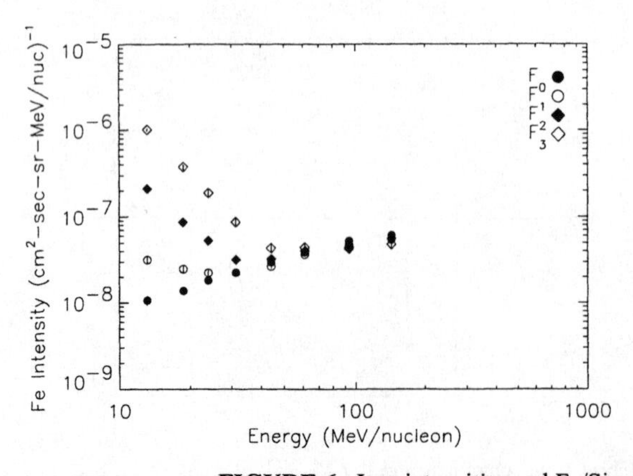

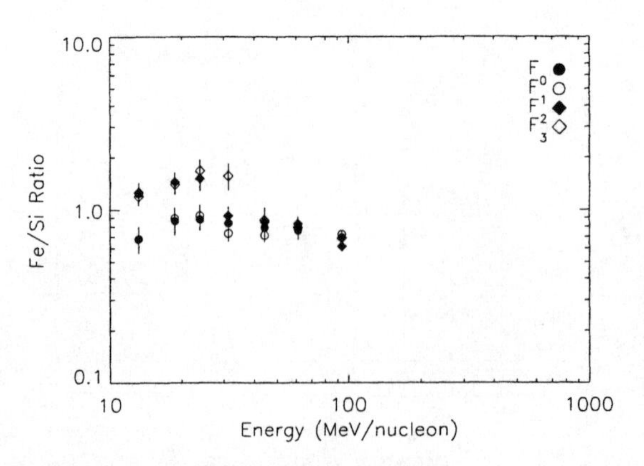

FIGURE 6. Iron intensities and Fe/Si ratios versus energy for different time selections.

portion of the SIS data.

The energy range examined here in searching for evidence of remnant material is much higher than that used by (10), so the vast majority of this suprathermal material is not available for study with the SIS data. However, both the Fe and ^{3}He data suggest that some remnant material might be present even at SIS energies. The absence of any signature in the Ne and Mg isotopic data is not inconsistent with the Fe and He data in that both the energy range and statistical accuracy are much more limited.

While remnant material appears to be present in the interplanetary medium, it is difficult for its presence alone to explain the composition of several of the 11 large events observed by SIS. For example, assuming Fe/O ratios of 1 and 0.134 for impulsive and gradual material respectively (1), to obtain an Fe/O ratio of ~0.8 (typical of these events), a mix of 77% impulsive and 23% gradual material is required. This would result in observed ^{3}He/^{4}He ratios of 8-38% (assuming a ^{3}He/^{4}He ratio of ~10-50% for impulsive and 0.04% for gradual material), inconsistent with the SIS observations of ~0.5% ^{3}He/^{4}He (equivalent to ~1-3% impulsive material).

The high Fe/O (and ^{22}Ne/^{20}Ne) ratio could be a result of Q/M fractionation of the source material which is then subsequently mixed with a small amount residual impulsive material, causing an increase in the ^{3}He/^{4}He ratio but not significantly affecting the heavier ion composition. However, this cannot explain the observed charge states in these Fe-rich events. In addition, Q/M-fractionation effects that favor Fe over O would favor low Fe charge states. However, the average Fe charge state in the 6 November 1997 event is ~15 at ~1 MeV/nucleon (11,32) and ~19 at $\geq$ 10 MeV/nucleon (32,33), several units higher than the value of ~11 that is typical of 'pure' gradual events (11). Again, this would require essentially all of the material to be impulsive (with an average Fe charge state of ~20 (11)), contrary to the observed ^{3}He abundances.

CONCLUSIONS

The elemental and isotopic data from 11 large SEP events observed by SIS show significant deviations from the composition of the solar system and solar corona. In an effort to understand the fractionation processes that may be occurring we have examined the data in light of three models. While general trends in the data can be reproduced with a Q/M fractionation model, there are particular elemental and isotopic ratios that are not well fit. We find that a simple resonance between the particles and waves at the second harmonic of the ion cyclotron frequency does not adequately represent the observations and when combined with Q/M fractionation does little to improve the fit over that from Q/M fractionation alone.

The idea of remnant impulsive-flare material being swept up by a passing shock may provide an explanation for the observed ^{3}He/^{4}He ratios in many of the large events observed by SIS. However, this theory alone can not consistently explain the high Fe/O enhancements, the high average Fe charge states, and the moderate ^{3}He/^{4}He enhancements found in these events. Combining Q/M-fractionated material with remnant material might explain the elemental and isotopic abundances, but the observed high average charge states would remain a problem.

The improved resolution and statistical accuracy of recent SEP data provide a wealth of information regarding the composition of the source plasma as well as properties of transport and acceleration processes. None of the current fractionation theories adequately account for the data, indicating that there are processes occurring that are not well understood. Clearly Q/M (or a related parameter, such as rigidity) is an important parameter. A complex source or multiple sources (with differing compositions) may also be required (34). In light of these new results the development of a detailed acceleration/transport theory that can be used to model all the observed elemental, isotopic, and ionic charge state fractionations is strongly needed.

ACKNOWLEDGMENTS

This work was supported by NASA at the California Institute of Technology (under grant NAS5-6912), the Jet Propulsion Laboratory, and the Goddard Space Flight Center. Climax neutron monitor data were obtained from the University of Chicago, National Science Foundation grant ATM-9613963.

REFERENCES

1. Reames, D. V., *Space Sci. Rev.* **90**, 413-491 (1999).

2. Breneman, H. and Stone, E. C., *Astrophys. J. Letters* **299**, L57-L61 (1985).

3. Mason, G. M., et al., *Astrophys. J.* **303,** 849-860 (1986).

4. Fisk, L. A., *Astrophys. J.* **224,** 1048-1055 (1978).

5. Temerin, M. and Roth, I., *Astrophys. J. Letters* **391,** L105-L108 (1992).

6. Mason, G. M., et al., *Astrophys. J.* **425,** 843-848 (1994).

7. Bochsler, P. and Kallenbach, R., *Meteoritics* **29**, 653-656 (1994).

8. van Hollebeke, M. A. I., et al., *Astrophys. J. Suppl.* **73,** 285-296 (1990).

9. Cohen, C. M. S., et al., *Geophys. Res. Letters* **26,** 2697-2700 (1999).

10. Mason, G. M., et al., *Astrophys. J. Letters* **525,** L133-L136 (1999).

11. Möbius, E., et al., *Proc. 26th Intern. Cosmic Ray Conf.(Salt Lake City)* **6,** 87-90 (1999).

12. Stone, E. C., et al., *Space Sci. Rev.* **86,** 357-408 (1998).

13. Leske, R. A., et al., *Geophys. Res. Letters* **26**, 2693-2696 (1999).

14. Anders, E., and Grevesse, N., *Geochim. Cosmochim.Acta* **53**, 197-214 (1989).

15. Leske, R. A., et al., *Geophys. Res. Letters* **26**, 153-156 (1999).

16. Leske, R. A., et al., "Measurements of the Heavy-Ion Elemental and Isotopic Composition in Large Solar Particle Events from ACE" in *High Energy Solar Physics: Anticipating HESSI*, edited by R. Ramaty and N. Mandzhavidze, ASP Conference Series in press, 2000.

17. Gloeckler, G., and Geiss, J., *Space Sci. Rev.* **84**, 275-284 (1998).

18. Garrard, T. L. and Stone, E. C., *Adv. Space Res.* **14,** 589-598 (1994).

19. Williams, D. L., et al. *Space Sci. Rev.* **85,** 379-386 (1998).

20. Mewaldt, R. A., et al. "The FIP-Factor in Solar Energetic Particles: Variations from Event to Event" in *ACE 2000 Symposium*, edited by R. A. Mewaldt et al., AIP Conference Proceedings this volume, New York: American Institute of Physics, 2000.

21. Mewaldt, R. A., et al., *Proc. 26th Intern. Cosmic Ray Conf.(Salt Lake City)* **6,** 127-130 (1999).

22. Arnaud, M., and Rothenflug, R., *Astron. Astrophys. Suppl. Ser.* **60,** 425-457 (1985).

23. Arnaud, M., and Raymond, J., *Astrophys. J.* **398,** 394-406 (1992).

24. Ipavich, F. M., et al., "Solar Wind Iron and Oxygen Charge States and Relative Abundances Measured by SWICS on Ulysses" in *Solar Wind 7*, edited by E. Marsch and R. Schwenn, COSPAR Colloquia Series 3, Pergamon Press, 1992, pp. 369-373.

25. Galvin, A. B., et al., "Silicon and Oxygen Charge State Distributions and Relative Abundances in the Solar Wind Measured by SWICS on Ulysses" in *Solar Wind 7*, edited by E. Marsch and R. Schwenn, COSPAR Colloquia Series 3, Pergamon Press, 1992, pp. 337-340.

26. Ko, Y-K., et al. *J Geophys Res,* **104,** 17005-17020 (1999).

27. Barghouty, N. and Mewaldt, R. A., "Simulation of Charge State Equilibrium and Acceleration of Solar Energetic Ions" in *ACE 2000 Symposium*, edited by R. A. Mewaldt et al., AIP Conference Proceedings this volume, New York: American Institute of Physics, 2000.

28. Dwyer, J. R., et al., *Proc. 26th Intern. Cosmic Ray Conf.(Salt Lake City)* **6,** 147-150 (1999).

29. Leske, R. A., et al., "Observations of Anomalous Cosmic Rays at 1 AU" in *ACE 2000 Symposium*, edited by R. A. Mewaldt et al., AIP Conference Proceedings this volume, New York: American Institute of Physics, 2000.

30. Wiedenbeck, M. E., et al., *Proc. 26th Intern. Cosmic Ray Conf.(Salt Lake City)* **7,** 508-511 (1999).

31. Cohen, C. M. S., et al., "Observations of ^{3}He-rich Solar Energetic Particle Events with the Solar Isotope Spectrometer" in *High Energy Solar Physics: Anticipating HESSI*, edited by R. Ramaty and N. Mandzhavidze, ASP Conference Series in press, 2000.

32. Mazur, J. E., et al., *Geophys. Res. Letters* **26,** 173-176 (1999).

33. Cohen, C. M. S., et al., *Geophys. Res. Letters* **26,** 149-152 (1999).

34. Cliver, E., "Proton and Photon Signatures from Solar Flares" in *ACE 2000 Symposium*, edited by R. A. Mewaldt et al., AIP Conference Proceedings this volume, New York: American Institute of Physics, 2000.

Ionic Charge State Measurements in Solar Energetic Particle Events

Mark Popecki[1], E. Möbius[1], B. Klecker[2], A. B. Galvin[1], L. M. Kistler[1], A. T. Bogdanov[2]

(1) Space Science Center and Department of Physics, University of New Hampshire, Durham, NH, USA; (2) Max-Planck-Institut für extraterrestrische Physik, Garching, Germany

Abstract. With the launch of the Advanced Composition Explorer, it has become possible through the SEPICA instrument to make direct ionic charge state measurements for individual Solar Energetic Particle events. In large events, the charge state may even be measured as a function of time, revealing changes that may be created by phenomena such as injections from different acceleration mechanisms, or confinement by magnetic field structures. The charge state can be a sensitive indicator of separate SEP populations. Several examples of SEP events will be presented. One of these, the November, 1997 event, displayed a trend in which the mean charge state for several ions increased with energy. These measurements may be the result of several processes, including a mixture of plasma with different source and acceleration histories, and abundance formation and possibly additional charge state modification by collisional or other means in the corona. A wide range of iron charge states have been measured for a variety of SEP events, ranging from $\langle Q \rangle = 10+$ to 20+. The mean charge states of C, O, Ne, Mg and Si all increased as the iron charge state increased. In events with the highest iron charge states, there were abundance enhancements in Ne with respect to oxygen in those cases, even though the mass/charge of the O and Ne were similar. In events with the lowest iron charge states, all these ions except Mg showed mean charge states generally consistent with coronal material of an equilibrium temperature of 1.3-1.6 million degrees K.

INTRODUCTION

Ionic charge state measurements in Solar Energetic Particle (SEP) events provide information about the SEP source, acceleration and propagation history. Charge state distributions are the combined result of several processes. First, the ionization and recombination processes at the source create the signature of a source temperature. This signature may be modified by passage of the SEPs through dense matter during and after acceleration, and by a rigidity-dependent of the acceleration mechanism. Direct charge state measurements are difficult, however, because of the high energy per charge and low fluxes of SEPs.

The first SEP charge state masurements were made with the ULEZEQ instrument on ISEE3 (Hovestadt et al. (17), (18); Klecker, et al. (21)). Using the direct deflection method, the mean charge states were determined for several ions from C to Fe. These measurements provided additional evidence that SEPs are accelerated principally in two environments: deep in the corona, in association with flares, and in the outer corona or solar wind, by shocks. Charge state measurements have also been made by other means, for example by using the Earth's magnetic field as a rigidity filter (e.g. Leske et al. (25); Klecker et al (22) and Mason et al. (30)) by inference from mass composition (e.g. Cohen et al. (11, 12)), or by the assumption that all ions of a given rigidity have the same propagation time (Dietrich and Lopate (13)). All these methods have provided only mean charge states so far, and temporal and/or energy variations have become visible only in the largest events (e.g. Oetliker, et al. (38); Mazur et al. (31)).

The SEPICA instrument on ACE has a sufficiently large geometric factor to provide charge state distributions, and measure temporal and energy variations for a variety of events. It also allows a charge analysis even for individual small impulsive events. Since they provide insight into the source temperature, passage through dense matter and

CP528, *Acceleration and Transport of Energetic Particles Observed in the Heliosphere: ACE 2000 Symposium*, edited by Richard A. Mewaldt, et al.
© 2000 American Institute of Physics 1-56396-951-3/00/$17.00

acceleration, charge state measurements are a valuable complement to mass composition measurements.

A brief summary of charge state measurement methods will be presented next. Following that, examples of charge state measurements for individual SEP events will be shown. Some events had temporal variations. Another example exhibited increasing charge states with increasing energy.

Studies involving multiple SEP events will then be presented. One of these studies focused on events associated with CME-driven shocks. The charge state measurements were used to infer the temperature of the corona. A second study compiled the mean charge states and mass composition of several species in the first year of ACE operation.

METHODS OF CHARGE STATE DETERMINATION

Several techniques of measuring or estimating SEP charge states have been developed. Most require large events or high fluxes. The electrostatic deflection method can provide a charge state distribution, while the others solely provide the mean charge state.

A method that works over a wide energy range uses the Earth's magnetic field as a rigidity (pc/Q) filter. This method has been applied by the SAMPEX group (Mazur et al. (31); Leske et al. (25); Mason et al. (29); Oetliker et al. (38)), with several sensors covering the energy range of 0.5-50 MeV/nuc. It requires large events or steady state populations that can be integrated over a long time (e.g. the anomalous component of cosmic rays, Klecker et al. (22)). The rigidity is mapped as a function of latitude using ions with known charge states such as protons or He^{++}. Heavy ion species are identified, and their momentum derived from energy measurements at every latitude. The ionic charge state is then calculated. Corrections are made for temporal changes in the magnetosphere from the proton cutoff rigidities on each orbit.

An alternate method operates with the assumption that the time to maximum flux of an SEP event will be the same for all $Z \geq 2$ ions with the same rigidity (Dietrich and Lopate (13)). Helium is assumed to be fully stripped and is used as the standard for rigidity. The charge states of the other ions are chosen to replicate the time to maximum flux of He. This method has been used with the IMP8 instrument at 10-500 MeV/nuc. It requires SEP events with sufficient fluxes at these energies, and it applies to the early ($\leq$20 hours) of an event. In addition, it is assumed that the charge state of an ion is constant over all observed energies.

Inferral of charge states has been used at high energies, 12-60 MeV/nuc, (e.g. Cohen et al (11,12)) with the ACE/SIS instrument. Breneman and Stone (7) used the Luhn et al. (27) charge state measurements, together with abundances in SEP events, and found that abundances were organized by $(Q/M)^{\gamma}$, where γ could be different for each event, and even change sign. This was attributed to acceleration and propagation processes. This relationship is being considered again for recent SEP events, with contemporary abundances and charge states from ACE/SEPICA (Galvin et al., 2000 (15)).

The charge inferral method makes use of the relationship between $(Q/M)^{\gamma}$ and abundance. γ is determined by measuring the relative abundance of the isotopes of the same element. This value is expected to apply to all the elements. An element, such as C, whose charge state does not vary much with temperature and hence can be estimated, provides a single value of Q, M and abundance. This is used together with γ to form the relation between abundances of all elements and $(Q/M)^{\gamma}$, and to calculate the charge states. Solar source charge states are derived by an iterative process, assuming that the original charge state distribution of an element has been altered in abundance via acceleration and transport processes by the measured γ, to the composition that is observed at the Earth. The $(Q/M)^{\gamma}$ abundance factor is applied as a correction to each charge state of an element.

In order to directly measure the charge state of each ion, an electrostatic deflection method is employed. This method is limited to energies below a few MeV/Q because of practical limitations on instrument size and maximum deflection voltage. This technique was used in the ISEE3/ULEZEQ (Hovestadt, et al. (17) and ACE/SEPICA (Möbius et al. (32,33)) instruments for SEPs, and in SOHO/STOF at lower energies suprathermal ions (Hovestadt et al. (19)). First, an electrostatic deflection is applied to incoming ions, which is proportional to the charge per energy (Q/E) of the ion. In the STOF instrument, a time-of-flight and solid state detector measurement are added to provide mass, energy and ionic charge.

In the ULEZEQ and SEPICA instruments, the energy and ion species are determined by the measurement of energy loss and residual energy in a proportional counter/solid state detector telescope. This total energy is used together with the deflection (Q/E) to derive the charge state Q. The energy limits are set at the upper end by a minimum deflection (Q/E) requirement, and at the lower end, where the energy loss and residual energy measurements become indistinguishable for two or more ion species. A typical energy for Fe for which charge state measurements can be obtained is about 0.5 MeV/nuc. This method yields charge state distributions, which can carry information about whether thermal equilibrium conditions prevailed at the source. They may also indicate the nature of an SEP event, i.e. gradual or impulsive, or even a combination, as discussed in the next section.

SEP EVENT CLASSIFICATION

SEP observations have become classified in two categories, whose names have been derived from the associated X-ray observations: impulsive and gradual events. A detailed discussion may be found in Reames (44), Cane et al. (9), Reames, (1997) and Cane (10). A summary of observational characteristics is presented in Table 1.

TABLE 1. Observational Characteristics of Gradual and Impulsive Events

Gradual (CME Related)	Impulsive
• High ion fluxes	• Low ion fluxes
• Low e/i ratio	• High e/i ratio
	• Short e profile (min to hr)
• ≈ Coronal abundance with some bias in Q/M due to acceleration	• Strong enrichment of heavy ions, Fe/O>1, often ^{3}He enrichment
• Type II radio burst	• Type III radio burst
• CME shock	

Previous Charge State Results	
• No significant difference in Q for C, O, Ne and Mg	
• Si ≈ 11+	• Si ≈ 14+
•Fe ≈ 14-15 (ISEE/ULEZEQ) Fe ≈ 14-15 (high E; SAMPEX)	• Fe ≈ 20

Gradual energetic particle events have been associated with coronal mass ejections (CMEs). They typically last up to several days, with large ion fluxes and SEP abundances that are similar to coronal values (Reames et al. (43, 44) and references therein). SEP ions have higher fluxes than electrons (Reames (44), Kahler et al. (20)). They also have a wide longitudinal source distribution across the solar disk (Reames (44), Cane et al. (8)), which is consistent with the broad longitudinal extent of a CME-driven shock. They are associated with Type II radio bursts, which are diagnostic indications of shocks, and can be used to infer the densities and hence the altitude of the shocks. Metric type II emission has been associated with shocks at solar altitudes of less than 3 solar radii, and hectometric Type II with shocks in the interplanetary medium (Cane 10).

Impulsive events, on the other hand, have a shorter duration. They show substantial enhancements in certain ions, including Ne through Fe, and especially in ^{3}He, relative to coronal values (Reames, et al. (42)). Energetic ion fluxes are generally weaker than in gradual events. However, electron fluxes are relatively high in these events and show a rapid rise and relatively brief (minutes to hours) duration (Reames (44)). Impulsive events are associated with Type III radio bursts, produced by electron beams that propagate from a location near a flare site (Wild, et al. (49) and Reames (44)) out to interplanetary space. Impulsive events are also associated with flares at western solar longitudes, particularly near W60, from where the interplanetary magnetic field is optimally connected to the Earth. This localization suggests that most of the particle acceleration takes place at the footpoints of these field lines.

The first SEP charge state measurements were made by Hovestadt et al. (17,18), Klecker et al. (21) and Luhn et al. (26), using the ULEZEQ instrument on ISEE3. Ions in gradual events had lower charge states than those in impulsive events. For example, Fe in gradual events had a mean charge state of 14, while the average for impulsive events was approximately 20. Silicon also showed a noticeable difference, with charge states of 11 and 14, respectively (Luhn, et al. (26)).

Charge state measurements provide essential information for understanding the processes in both classes of event. For example, if SEPs are accelerated out of a region of thermal equilibrium with no significant modification of the charge state distribution, the temperature can be inferred using ionization and recombination between the various ionization states of an element. The charge distribution for a given temperature has been modeled (e.g., Arnaud & Rothenflug, (3) and Arnaud & Raymond, (2)). If the individual charge states can be resolved, the abundance ratio of adjacent charge states can be used to derive the source temperature. This method has been used successfully in the solar wind, for which sufficient charge state resolution is available (von Steiger (48), Aellig et al. (1), Galvin (14)). Often, the mean of the charge state distribution is used, with the assumption that it represents a source in thermal equilibrium. In thermal equilibrium, usually only one or two adjacent charge states contribute significantly to the distribution. Distributions that are substantially wider suggest the presence of other physical processes. For

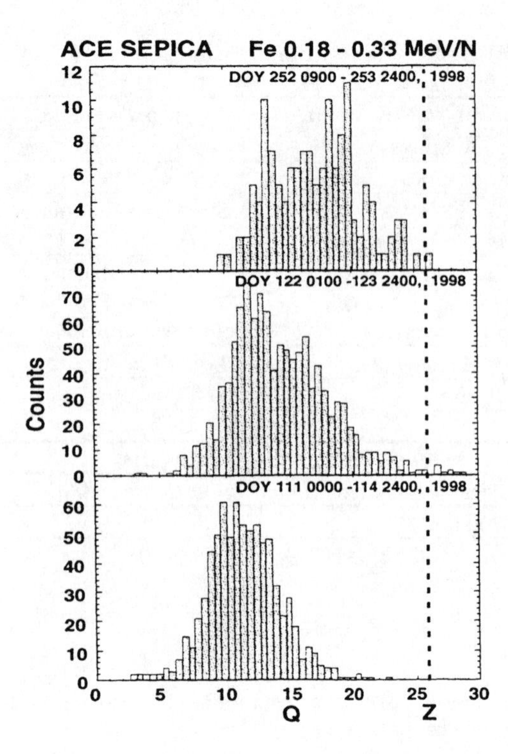

FIGURE 1. Three examples of iron charge state histograms are shown. The top panel is typical of an impulsive event, the bottom is a gradual event, and the middle may be a combination.

example, SEPs may come from several sources at different temperatures, or the initial state may have been altered by stripping in a dense medium. Rigidity-dependent processes, such as wave-particle interactions or possibly shock acceleration, may also modify the initial charge state distribution.

CHARGE STATE OBSERVATIONS IN INDIVIDUAL EVENTS AND TEMPORAL EVOLUTION

Iron charge state histograms for three different SEP events are shown in Figure 1. The bottom panel has a low iron charge indicative of a CME related event. It has a peak near Q_{Fe}=11, which is similar to iron in the solar wind (Gloeckler et al. (16); Aellig, et al. (1)) and consistent with earlier observations by Mason et al. (28) in the same energy range. The observed charge state distribution is consistent with one or two dominant charge states. The upper panel shows high charge states from an impulsive SEP event. It has a broad distribution that extends from Q_{Fe}=10+ to 25+. This is substantially wider than expected from a source in thermal equilibrium.

The middle panel shows an event with intermediate iron charge states. It contains a peak near Q_{Fe}=11+ to 12+ and a tail extends to fully ionized iron. This event is discussed in more detail in a companion paper in this volume (Popecki et al. (40)). Temporal variations were observed in charge states and mass composition. They suggest that three separate iron populations were observed, in three separate flux tubes, within and just after a magnetic cloud.

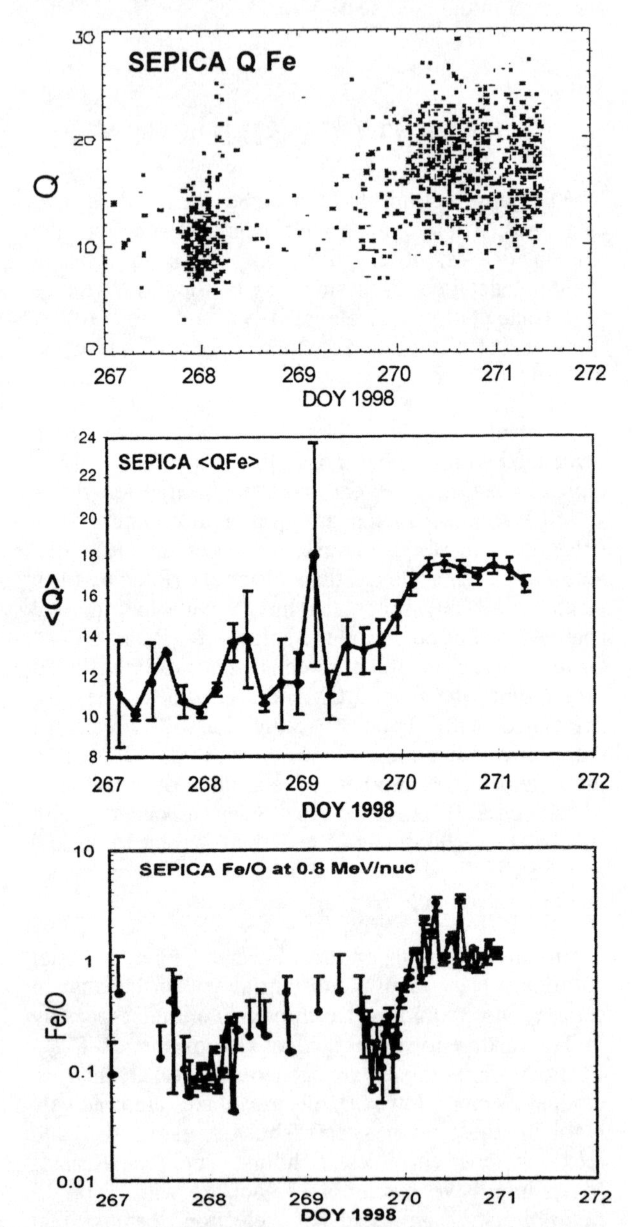

FIGURE 2. A sequence of a gradual and an impulsive event is shown. The top panel shows the charge state of individual iron ions. The middle panel shows the average iron charge state, and the Fe/O ratio is at the bottom.

Another example of individual event variations is shown in Figure 2 (Popecki, et al. (39)). From top to bottom are presented the charge state of individual iron ions, the average iron charge state and the Fe/O ratio. A gradual event occurred on day 268, with a low Fe/O ratio. At the end of this period the injection of a small quantity of high charge iron occurred. X-ray flares with optical counterparts were observed on the solar disk during this event, yet they apparently did not contribute substantially to this event. Later, on days 269-271, a broad distribution of high charge iron was observed. The mean charge state increased from Q_{Fe}=15+ to 18+. During this time period, when the charge state of iron was high, X-ray flares occurred at western longitudes, where the magnetic connection from the Sun to ACE should be optimum. The Fe/O ratio increased during this high charge event, consistent with the conventional classification of gradual and impulsive events. It is important to note that the charge state and Fe/O ratio rose during the impulsive phase, instead of adopting a single state, characteristic of an impulsive event.

VARIATION OF CHARGE STATES WITH ENERGY: NOVEMBER, 1997

The November 7-9, 1997 SEP event exhibited a substantial increase of the charge state with energy for C through Fe, below 1 MeV/nuc (Möbius et al. (33)). The same trend was observed by the SAMPEX team, using the rigidity cutoff method (Mazur et al. (31)). This period included strong interplanetary shocks and X-ray flares (Mason et al., 1999). In an earlier SAMPEX observation, Oetliker et al. (38) found similar behavior only for iron. For the November, 1997 SEP event, and at much higher energies, Cohen et al. (11) inferred an iron charge state of 18.4 ±0.3 for 12-60 MeV/nuc, thus extending this trend to higher energies. Dietrich and Lopate (13) used the time to maximum flux analysis to calculate a mean iron charge state of 13±3, for 10-500 Mev/nuc. The difference between this result and Cohen et al. (11) could be related to two aspects of the time to maximum flux method. The method is sensitive to the initial phase of the event, with rising fluxes, and it is assumed that the charge state is the same at all energies.

Direct charge state measurements from SEPICA for this event were presented in Möbius et al. (33) at four energies. Iron charge state histograms at these energies show the following features: at the lowest energy (0.18-0.26 MeV/nuc), there was a peak at approximately Q_{Fe}=10-11+, with a tail extending up to Q_{Fe}=19+. At the highest energy (0.44-0.54 MeV/nuc), there was a wide peak centered at Q_{Fe}=14-15+. The peaks at low and high energies are similar to those observed in gradual and impulsive events (e.g., Figure 3), which may suggest that the peak and tail are the superposition of CME accelerated particles with a flare associated population at higher charge states. However, the location of the peak in the November event also appears to vary with energy, which may not be readily explained by a superposition of these two sources.

It has been suggested recently (Barghouty and Mewaldt (4); Barghouty and Mewaldt (5); Ostryakov and Stovpyuk (47) and Reames et al. (45)) that a variation of charge states with energy could be created by shock acceleration deep in the corona at densities of $10^9 cm^{-3}$, through the presence of ionizing collisions. Interestingly, Type II metric radio bursts occur at solar altitudes of three solar radii or less and are often associated with flares. These bursts are also observed for SEP events that cannot be clearly associated with CMEs (Cane (10)). This observation would put an associated shock, and thus shock acceleration, into an altitude regime with the right densities for stripping to occur during acceleration. This results quite naturally in an energy dependent charge state.

Furthermore, Barghouty has suggested that the peak and tail distribution could be related to the varying probability for ionization and recombination for different charge states (private communication). Starting from a thermal equilibrium distribution, it is easier to ionize iron further than it is to recombine to a lower charge state. Therefore, variations toward high charge states would tend to produce an asymmetric distribution. However, the timescales for coronal shocks to traverse the dense regions are 1000 sec or less. This may be difficult to reconcile with the time profile of the November event. It was a typical gradual event, with the passage of an interplanetary shock observed at the spacecraft. If acceleration in the lower corona was adopted as an explanation of the observed relation of charge with energy, the temporal characteristics of an implusive event would be expected. In addition, the visibility of such a shock might be restricted to western solar longitudes with optimum magnetic connection to Earth, since all the acceleration takes place at the footpoints of the field lines. This condition was satisfied in the November 7-9 event, with an X-ray flare at W63 (Mason et al. (29)). Thus, a shock low in the corona could have contributed to the observed SEP population.

So far, the SEPICA instrument has only observed such a pronounced Q(E) relation during the November 1997 activity, although there have been indications at other times. Unfortunately, the high charge resolution fan of the instrument, with the widest energy range, stopped operating due to a control valve failure before any other large events occurred. Higher energy information can be extracted from the low resolution fans with modeling efforts of the detailed instrument response that are currently underway.

The energy range has recently been extended to lower energies through joint observations with ACE/SEPICA and SOHO /STOF (Bogdanov et al. (6)). They find a relatively strong variation of the charge state with energy for an event in June, 1999 (DOY 177-178). The result appears to be consistent with additional stripping during acceleration in the high density plasma of the lower corona, as shown in a comparison with a relation suggested by Reames et al. (44). Another example, from May 1, 1998, shows a much weaker dependence on energy, which could also be explained by rigidity dependent acceleration of solar wind material at the interplanetary shock. This point was also made by Klecker, et al. (23) in their comparison of charge states in CME related SEPs with simultaneously observed solar wind. They find a systematic increase of the mean charge state of iron over that in the solar wind by $\Delta Q \approx 1$-2. The observed abundance appears to be consistent with the reported variation in the charge states, assuming a Q/M dependent modification of the abundance ratios in these events over that of the solar wind, according to Breneman and Stone (7). In summary, there are three possible contributors to energy variations in charge state distributions: the mixture of impulsive and gradual acceleration, stripping, and rigidity dependent acceleration. All three may be present. Only a careful analysis of charge distributions over the entire energy range from the solar wind to SEPs can reveal their respective contributions.

MULTI-EVENT STUDIES

In addition to these examples, a variety of events in 1998 were examined in a more systematic way. Two studies were carried out with these events.

In the first study, the cleanest examples of gradual events were selected. These events were chosen for minimum contamination by impulsive activity: they had low Fe charge states of $Q_{Fe} \approx 11+$, low solar wind speeds and Fe/O ≤ 0.3 (Klecker et al. (24)). There were no significant temporal changes in charge state during the events. Except for Mg, which was consistently one charge state lower than what is found in the slow solar wind, the other ion species had mean charge states that were consistent with a source temperature of 1.3-1.6 10^6 K, which is consistent with coronal values.

In another study, charge states of several ions (C, O, Ne, Mg and Si) were measured for events in 1998, without regard to impulsive or gradual character. As the mean charge of iron increased, the others did as well. Some, including C, Ne and Mg, reached a fully stripped condition (Möbius et al. (34, 35)). The charge states of O, Ne and Mg were compared to what would be expected from the mean charge of iron, assuming the iron came from a region of thermal equilibrium. There were deviations from equilibrium for O and Mg (Möbius et al. 2000) in the events with the highest iron charge states. In those events, O had lower charge states than expected, while Mg had higher charge states. In the low iron charge events, i.e., typical of gradual events, the agreement between the oxygen and iron was good, but the charge of Mg was low, as described by Klecker, et al. (1999).

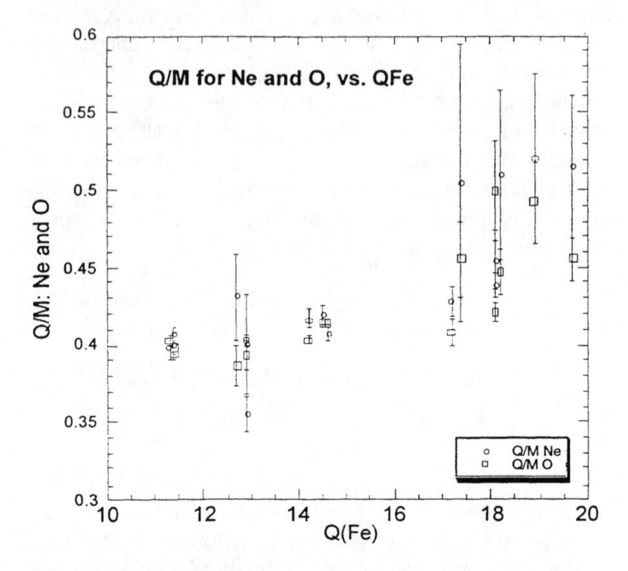

FIGURE 3. The Q/M ratio for Ne (circles) and O (squares) is shown for events during 1998.

The charge states measured in the high iron charge events seem to be inconsistent with a source at thermal equilibrium. If there was equilibrium at the source, the charge state signature may have been lost, possibly by collisions in the corona, during or after acceleration. Alternatively, there may not have been a state of thermal equilibrium at the source at all.

When SEP events are organized by Fe charge state, abundance enhancements of Ne and Fe with respect to O fall into the high charge state events. These enhancements have been associated with impulsive events (Reames, et al. (42)). Abundance ratios from SEPICA are given in Möbius et al. (35), vs. the mean iron charge state. The actual Q/M ratios for Ne and O are shown in Figure 3. The Q/M ratio for Ne is similar to that for O over a wide range of iron charge states. When the iron charge state is high, there is even some indication that Q/M for O may be lower than for Ne, although the uncertainties are large. If the neon abundance is determined by the observed Q/M ratio, as suggested by Breneman and Stone (7), it should be similar to that of O, but in fact it is substantially enhanced when the charge state of iron is high. This is another indication that the charge state observed at ACE may have been altered from the one that organized the abundances.

Mullan and Levine (36) and Mullan (37) suggested for heavy ions in impulsive events that competition between energy gain through stochastic wave heating and energy loss via Coulomb collisions can lead to a preferred acceleration for low charge states. If thermal equilibrium initially exists under these conditions, Ne and Fe enhancements may occur because there are larger quantities of Fe and Ne at lower charge states. This model has recently been considered by Litvinenko (private communication) in connection with the observations discussed above. However, the observed abundance ratios would require temperatures of less than 10^6 K.

SUMMARY

In the first year of ACE/SEPICA measurements, a wide variety of SEP events have been observed. Mean iron charge states varied from 10+ to 20+. This is a wider range than reported by Luhn et al. (26). Charge state data complement the more plentiful composition measurements in an ideal way, because they can provide additional insight into the physical mechanisms involved, such as the source region temperature, the possibility of stripping during or after acceleration, shock acceleration in the corona or interplanetary space and rigidity dependent acceleration. The lowest charge state events were consistent with interplanetary shock acceleration of outer coronal material or the solar wind, with an inferred source temperature and SEP mass composition similar to coronal values.

The highest charge state events were consistent with a heated source region, with abundance enhancements of ^{3}He, Ne and Fe. In these events, the width of the iron charge state distributions suggests that they take place in a source region that is either not in thermal equilibrium, or else the charge state signature of thermal equilibrium has been modified by ionizing collisions. Furthermore, the abundance enhancements of Ne/O in these events cannot be easily explained by rigidity dependent acceleration at the *observed* charge states, since the Q/M ratio is similar for the two ions in those cases. This suggests that the charge states that established the abundances before or during acceleration have been modified to those we observe at 1 AU.

Events with intermediate mean charge states were sometimes found to be superpositions of high and low charge state populations. Temporal variations in iron charge state and mass composition revealed SEP populations with either gradual or impulsive characteristics within a single event. However, iron charge state distributions peaked at $Q \approx 15+$ have been observed, which cannot be represented by a simple superposition of two types of SEP events. They may be a form of impulsive event with a mean charge state reduced by collisional modification, or by rigidity dependent acceleration.

Some events with low or intermediate charge states show a dependence of the charge state on energy, such that higher energies correspond to higher charge states. One possibility to explain this observation would be a superposition of gradual and impulsive events, with different energy spectra. However, in at least one example even the peak of the charge state distribution moved with energy, which requires an alternative explanation. It has been shown that the relation could be produced by shock acceleration of SEPs in the low corona, in the presence of ionizing collisions. A comparison of the solar wind charge states to those in CME shock-related SEPs also shows this trend, thus suggesting that it may result from rigidity dependent acceleration at the interplanetary shock.

ACKNOWLEDGMENTS

The authors are grateful to the many unnamed individuals at the University of New Hampshire and the Max-Planck-Institut fur extraterrestrische Physik for their enthusiastic contributions to the completion of the ACE SEPICA instrument. They thank F. Gliem, K. -U. Reiche, K. Stockner and W. Wiewesieck for the implementation of the S3DPU. The work on the SEPICA instrument was supported by NASA under Contract NAG5-6912.

REFERENCES

1. Aellig, M.R., et al., *JGR*, **103**, A8, 17215-17222 (1998).

2. Arnaud, M. and J. Raymond, *ApJ*, **398**, 394-406 (1992).

3. Arnaud, M. and R. Rothenflug, *Astronomy and Astrophysics Supplement Series*, **60**, 425-457 (1985).

4. Barghouty A.F. and R.A. Mewaldt, *ApJ Letters*, **520**, L127-L130 (1999).

5. Barghouty, A.F, and R.A. Mewaldt, *ACE2000 Symposium Proceedings*, 2000.

6. Bogdanov, A., B. et al., *ACE2000 Symposium Proceedings*, 2000.

7. Breneman, H.H. and E.C. Stone, *ApJ*, **299**, L57-L61 (1985).

8. Cane, H.V., D.V. Reames and T.T. von Rosenvinge, *JGR*, **93**, 9555-9567 (1988).

9. Cane, H.V., R.E. McGuire and T.T. von Rosenvinge, *ApJ*, **301**, 448-459 (1986).

10. Cane, H.V., *Coronal Mass Ejections*, edited by N. Crooker, J.A. Joselyn and Joan Feynman, Washington, D.C., AGU, 1997, pp.205-215.

11. Cohen, C.M.S., et al., *GRL*, **26**, 149-152 (1999a).

12. Cohen, C.M.S., et al., *GRL*, **26**, 2697-2700 (1999b).

13. Dietrich, W. and C. Lopate, *Proceedings of the 26th International Cosmic Ray Conf., Salt Lake City, Utah*, eds. D. Kieda, M. Salamon and B. Dingus, 1999, 6, pp. 91-94.

14. Galvin, A.B., F.M. Ipavich, C.M.S. Cohen, G. Gloeckler, *Sp. Sci. Rev.*, **72**, 65-70 (1995).

15. Galvin, A.B., et al., *ACE2000 Symposium Proceedings*, 2000.

16. Gloeckler, G., et al., *GRL*, **26**, 157-160 (1999).

17. Hovestadt, D. et al., *IEEE Trans. Geos. Electr.*, **GE-16**, 166 (1978).

18. Hovestadt, D., et al., *Adv. Sp. Res.*, **1**, 61 (1981).

19. Hovestadt, D. et al., *Solar Physics*, **162**, 441-481 (1995).

20. Kahler, S.W., et al., ApJ, **302**, 504 (1986).

21. Klecker, B., et al., *ApJ*, **281**, 458-462 (1984).

22. Klecker, B., et al., *ApJ*, **442**, L69-L72 (1995).

23. Klecker, B., et al., *ACE2000 Symposium Proceedings*, 2000.

24. Klecker,B., et al., *Proceedings of the 26th International Cosmic Ray Conf., Salt Lake City, Utah*, eds. D. Kieda, M. Salamon and B. Dingus, 1999, 6, pp. 83-86.

25. Leske, R.A., et al., *ApJ (Letters)*, **442**, L149-L152 (1995).

26. Luhn, A et al., *ApJ*, **317**, 951-955 (1987).

27. Luhn, A., et al., *Adv. Space Res.*, **4**, 161-164 (1984).

28. Mason, G.M., J.E. Mazur, M.D. Looper and R.A. Mewaldt, *ApJ*, **452**, 901-911 (1995).

29. Mason, G.M.,et al., *GRL*, **26**, 141-144 (1999).

30. Mason, G.M., J.E. Mazur, M.D. Looper, and R.A. Mewaldt, *ApJ*, **452**, 901-911 (1995).

31. Mazur, J.E., G.M. Mason, M.D. Looper, R.A. Leske, and R.A. Mewaldt, *GRL*, **26**, 173-176 (1999).

32. Mobius, E., et a., *Space Sci. Rev.*, **86**, 449-495 (1998).

33. Möbius, E., et al., *GRL*, **26**, 145-148 (1999).

34. Möbius, E., et al., *ACE2000 Symposium Proceedings*, 2000.

35. Moebius, E., et al., *Proceedings of the 26th International Cosmic Ray Conf., Salt Lake City, Utah*, eds. D. Kieda, M. Salamon and B. Dingus, 1999, 6, pp. 87-90.

36. Mullan D.J. and R.H. Levine, *ApJ Suppl. Ser.*, **47**, 87-102 (1981).

37. Mullan, D.J., *ApJ*, **268**, 385 (1983).

38. Oetliker, M.B., et al., *ApJ*, **477**, 495-501 (1997).

39. Popecki, M.A., et al., *Proceedings of the 26th International Cosmic Ray Conf., Salt Lake City, Utah*, eds. D. Kieda, M. Salamon and B. Dingus, 1999, 6, pp. 187-190.

40. Popecki, M.A., et al., *ACE2000 Symposium*, 2000b

41. Reames, D.V., *Coronal Mass Ejections*, edited by N. Crooker, J.A. Joselyn and Joan Feynman Washington, D.C., AGU, 1997, pp. 217-226.

42. Reames, D.V., J.P. Meyer and T.T. von Rosenvinge, *ApJ Suppl.*, **90**, 649-667 (1994).

43. Reames, D.V, *Sp. Sci. Rev.*, **85**, 327-340 (1998).

44. Reames, D.V., *Sp Sci Rev*, **90 (3/4),** 413-491 (1999).

45. Reames, D.V., C.K. Ng, A. J. Tylka, *GRL*, **26**, 3585-3588 (1999a).

46. Skoug, R.M., et al., *GRL*, **26**, 161-164 (1999).

47. Ostryakov , V.M. and M.F. Stovpyuk, *Solar Physics*, **189**, 357-372 (1999).

48. von Steiger, R. et al., 1995, *Space Sci Rev.*, **72**, 71 (1995).

49. Wild, J.P, S.F. Smerd, and A.A. Weiss, *Ann.Rev. of Astronomy and Astrophysics*, **1**, 291 (1963).

Simulation of Charge-Equilibration and Acceleration of Solar Energetic Ions

A.F. Barghouty and R.A. Mewaldt

California Institute of Technology, Pasadena, CA 91125, USA

Abstract.
Recent measurements of the mean ionic charge states of solar energetic iron and silicon by SAMPEX and ACE during the large solar events of 1992 November 1 and 1997 November 6 show a mean ionic charge that increases with energy. This feature has implications for the use of the observed charge state as a probe of the coronal electron temperature and density, as well as for models of ion acceleration and transport in the coronal plasma. In this paper, we show results of a nonequilibrium model for the mean ionic charge that includes shock-induced acceleration in addition to charge-changing processes. The model is able to reproduce the general features observed without, however, specifying uniquely the acceleration time and the plasma electron density. Based on our simulations for iron and silicon for the 1992 and 1997 events, and assuming a characteristic shock-acceleration time of $\sim$ 10 sec, our model suggests an equilibration-acceleration site at heights $\sim$ 1 solar radius above the solar surface, a density $\sim 10^9$ cm^{-3}, and an electron temperature $\sim 1 - 1.33$ MK. For ions with kinetic energy $\gtrsim$ 30 MeV/nucleon we estimate the amount of coronal material the ions traverse to be $\sim$ 100 μg/cm^2.

INTRODUCTION

Solar energetic particles (SEPs) observed near 1 AU are known to belong to one of two classes, one associated with impulsive flare events and the other, gradual events, with interplanetary shocks associated with coronal mass ejections (CMEs). Although the distinction between the two classes of events is sometimes blurred, there are, however, population-discriminating features, including the measured charge states of these ions. For example, iron ions with a mean charge of around 20 are found to be associated with impulsive events (1), whereas those with a mean charge of around 14 (similar to the coronal and solar-wind iron seed population) are typically associated with gradual events (2). These charge states can be characterized by temperatures $\sim 10^7$ MK for impulsive events while the gradual ones indicate a temperature $\sim 1 - 3$ MK, typical of the ambient corona. Thus, the measured charge state of SEP ions can be used to help discriminate between gradual and impulsive solar events.

The use of equilibrium ionization temperatures to infer the local plasma temperature from the measured mean charge suggests somewhat different temperatures for different ion species (3, 4, 5, 6). An inferred average temperature $\sim 1 - 3$ MK seems, nonetheless, to be consistent with most measured species. Effects like plasma heating, non-Maxwellian velocity distribution for the electrons, and high relative velocities between the energetic ions and ambient plasma electrons were explored and found unable to account for the variation in the inferred temperature (3, 6).

Recent measurements of the charge states of solar energetic ions associated with the large solar events of 1992 November 1 and 1997 November 6 by SAMPEX (5, 7) and ACE (8) show an energy dependence of the measured mean charge over the energy range $\sim 0.5 - 50$ MeV/nucleon. Higher energy ions, especially for iron and silicon, have a higher mean charge. This feature was not reported in earlier observations by ISEE-3 (2) for the 1978-1979 events which covered a much smaller energy interval near $\sim$ 1 MeV/nucleon. Clearly an energy-dependent mean charge will affect the interpretation of inferred plasma temperatures.

To appreciate the significance of this energy dependence of the mean charge, one has to realize that the observed charge states of SEPs are expected to be sensitive indicators of the local density and temperature of the charge-equilibration and acceleration site in the coronal plasma. In the case of SEP ions associated with large gradual events, the acceleration is thought to be by large-scale shocks in the outer corona driven by CMEs (e.g., (9)). The charge state is expected to be established via ion-electron and ion-proton collisions in the ambient dense and hot coronal plasma. The rates that establish the equilibrated charge, or the mean charge, depend sen-

CP528, *Acceleration and Transport of Energetic Particles Observed in the Heliosphere: ACE 2000 Symposium,* edited by Richard A. Mewaldt, et al.
© 2000 American Institute of Physics 1-56396-951-3/00/$17.00

sitively on both the plasma parameters, its density and temperature, as well as on the energy of the ion.

According to Kahler (10), shock acceleration is most effective at heights ~ 2 solar radii above the solar surface, above which the electron density begins to drop appreciably. Hence, the observed energy-dependent mean charge is also expected to reflect conditions of the site. Transport effects, from the local equilibration-acceleration site to near 1 AU, due to the much lower temperature and density of the solar wind, are not expected to have any appreciable effect on the energy or the charge state of the SEP ions (11, 12). Therefore, the observed charge states of SEP ions near 1 AU can, in principle, be used as indicators of the coronal plasma parameters as well as measures of the characteristics of the acceleration mechanism.

Understanding the physical reasons behind the energy dependence of the mean charge is crucial, then, to our understanding of the acceleration processes in SEP events as well as to our ability to infer the plasma parameters characterizing those events. In this paper, we present simulations for the charge distributions of iron and silicon at different energies and for the mean charge as a function of energy, for the 1992 November 1 and 1997 November 6 events. The simulations are based on a nonequilibrium model that couples the charge-changing and acceleration processes in a dynamic fashion. This coupling can produce an energy-dependent mean charge when, e.g., the timescales for acceleration and ionic equilibration are comparable (13).

This model, as well as that of Ostryakov and Stovpyuk (14), is derived assuming that both plasma and shock parameters are stationary in space and time. This assumption is justified when the charge-changing and acceleration processes act over timescales much shorter than those characterizing changes in the plasma and shock parameters. For example, assuming a charge-equilibration and acceleration timescale of ~ 10 sec, an interplanetery shock propagating at $\sim 10^3$ km/s will sample a change in the coronal plasma density of only $\sim 25\%$. An important characteristic of both models, explored in some detail here, is that under this assumption the local density and the characteristic acceleration time cannot be uniquely specified; only their product can be specified. However, by using other means of characterizing the acceleration time, this feature can be quite useful in placing limits on the plasma density for different events.

In addition to these two models (which we term homogeneous models), two recent suggestions addressing the observed energy-dependence of the mean charge have been put forward. Reames et al. (15) argue that under ionic equilibrium conditions the energy dependence of the mean charge arises from electron stripping in "moderately" dense coronal plasma during acceleration. In our model, ionic equilibrium conditions are treated as a limiting case. Mason et al. (16), although with a different focus from ours, suggest that the wide range of ionization states is a result of a mixed seed population; one characteristic of a higher-temperature, thus higher charge state, impulsive solar-flare population with the other being of a lower-temperature, gradual-event population.

DESCRIPTION OF THE MODEL

The model is developed to follow the evolution of the ions' phase-space distribution function in time and momentum subject to the processes of charge-changing ionization and recombination reactions, and shock-induced acceleration. Due to the assumption of homogeneity, the ions are assumed to leak out of the local acceleration region at a rate parameterized by an escape term. This assumption also allows us to ignore propagation effects from the acceleration region to near 1 AU. Upon leakage, the ions' steady-state distribution function is frozen-in, both in charge and momentum.

In a spatially homogeneous acceleration region, the shock is idealized as a static and finite plane shock with its normal parallel to the flow of the coronal plasma. In order to avoid introducing another free parameter in the model for the injection energy of the suprathermal ions, we assume that the thermal ions can be sufficiently scattered in momentum space so as to attain (via the second-order Fermi process) the Alfv́en mometum, interact with the shock, and be accelerated to tens of MeV/nucleon via the first-order Fermi process. These ions are also assumed to be energetic enough to be able to leak out of the acceleration region at an escape rate that depends on their energy as well as the strength of the scattering.

Energy losses are assumed to be due only to the passage of the energetic ions through the plasma. In the homogeneous limit, adiabatic energy loss due to the diverging solar wind plasma is ignored.

The ions' balance equation then takes the form:

$$\frac{\partial f^q}{\partial t} = \frac{1}{p^2}\frac{\partial}{\partial p}\left[p^2 D_{pp}\frac{\partial f^q}{\partial p}\right] - \frac{1}{p^2}\frac{\partial}{\partial p}\left[p^2\frac{dp}{dt}f^q\right] - \frac{f^q}{\tau_{esc}} + Q^q_{ir}(p,t)\,, \qquad (1)$$

where $f^q(p,t)$ is the phase-space density function of ion with charge q as a function of momentum p at time t.

The source and sink function $Q^q_{ir}(p,t)$, due to ionization and recombination, is given by:

$$Q^q_{ir}(p,t) = n_e\Big[R_i^{q-1}(p)f^{q-1} + R_r^{q+1}(p)f^{q+1} - (R_i^q(p)f^q + R_r^q(p)f^q)\Big]\,, \qquad (2)$$

where n_e is the mean electron density and $R_i(p)$ and $R_r(p)$ are the momentum-dependent ionization and recombination rates. For energetic ions, charge-changing processes are mainly due to ion-electron and ion-proton collisions. Ion-electron ionization cross sections are estimated using the formalism and parameters of (17, 18). For ion-proton collisions the cross sections are estimated using the recently developed formalism of (19). The rates are then calculated from the cross sections by multiplying the cross sections by the relative velocities of the energetic ion and ambient thermal electrons and protons, respectively, and by the ambient electron density of the plasma. Calculated this way, the rates depend on the charge and momentum of the ion in addition to the temperature and density of the plasma.

In Eq. (1) the term $f^q/\tau_{\rm esc}$ describes the rate of escape from the shock acceleration region, where

$$\mathcal{P}_{\rm esc} = \frac{\tau_s}{\tau_{\rm esc}} = \frac{4v_s}{\eta v} , \qquad (3)$$

is the escape probability per shock crossing (e.g., (20)) with v_s being the shock propagation speed, v the ion speed, and η is the shock compression ratio. In the rest frame of the plasma,

$$v_s = \left(\frac{3k_B T_e}{m_0 c^2}\right)^{1/2} c , \qquad (4)$$

with T_e the electron temperature, k_B Boltzmann's constant, m_0 the nucleon's rest mass, and c the speed of light. The cycle time τ_s, or time to complete a crossing of the shock front, is given by

$$\tau_s = 4(1+\eta^{-2})\frac{D_\parallel}{v_s v} . \qquad (5)$$

The diffusion coefficient $D_\parallel$ along the magnetic field line can be expressed as (21)

$$D_\parallel(p;q) = D_0 P^{2-\alpha}\beta\frac{B_0}{B_s} , \qquad (6)$$

with D_0 being a constant expressing the strength of the diffusion, P the ion's rigidity, α the spectral index of the magnetic turbulence, and $\beta = v/c$. The ratio B_0/B_s is the strength of the magnetic field at 1 AU to that at the acceleration site, typically included to ensure that the particle remains tied to the field line (22). In the hard-sphere approximation (23), the diffusion coefficient in momentum space D_{pp} is related to $D_\parallel$ as

$$D_{pp}(p;q) = \frac{v_{sw}^2}{9D_\parallel} p^2 , \qquad (7)$$

where v_{sw} is the solar wind speed.

The rate of momentum gain per shock crossing is (20)

$$\left.\frac{dp}{dt}\right|_+ = \frac{\zeta v_{sw}^2}{4D_\parallel} p , \qquad (8)$$

where $\zeta = (\eta-1)/6\eta$. Momentum losses due to passage of the energetic ions through the plasma are assumed to be of the standard form:

$$\left.\frac{dp}{dt}\right|_- = -{\rm const}\,\frac{m_0}{p} q^2 n_e \left(4\ln\frac{p}{m_0 c} - \ln n_e + 74.1\right) . \qquad (9)$$

For the numerical solution of the system of equations (1), the initial $f^q(p,t=0) = f_0^q(p)$ is assumed Maxwellian in both p and q characterized by the electron temperature T_e. The boundary conditions in p are taken to be

$$f^q(p=p_0,t) = f_0^q(p=p_0) , \qquad (10)$$

$$\left.\frac{\partial f^q(p,t)}{\partial p}\right|_{p=p_0} = \left.\frac{\partial f_0^q(p)}{\partial p}\right|_{p=p_0} , \qquad (11)$$

where p_0 is the lower limit for p ($p_0 < p_{\rm thermal}$). Other physically reasonable boundary conditions can be used, e.g., requiring $f^q(t)$ to remain finite for all p (14). However, we find our simulations to be insensitive to the exact forms of the boundary conditions in p. Shock acceleration is assumed to affect only those ions pre-accelerated by diffusing in momentum space and attaining a momentum $p > p_A$, where p_A is the Alfvén momentum given by $p_A = m_0 B_s/\sqrt{\mu n_e}$, μ being the magnetic permeability constant.

SIMULATIONS

In Figs. 1 and 2 we show the simulated steady-state solutions of Eq. (1) for the Fe and Si charge distributions and the mean charge as a function of energy during the 1992 November event. For these simulations, based on a best fit for the mean charge observations, the electron temperature is taken to be 1.33 MK, the density is 3×10^8 cm^{-3}, and the spectral index α is assumed to be 5/3. Expressed in units of GV for rigidity, the strength of the diffusion coefficient $D_0 = 10^{19}$ cm^2/s. The strength of the magnetic field at the acceleration site is taken to be 100 Gauss, for which the Alfvén energy is ≈ 0.25 MeV/nucleon.

In Figs. 1 and 2 are plotted the normalized charge distributions at energies of 0.32, 3.2, and 32 MeV/nucleon. This choice of energies spans the range available from present measurements. The smooth dotted curves represent simulated charge distributions using only the first and second moments of the phase-space distribution, i.e., assuming a pure Gaussian. The simulated mean charge is

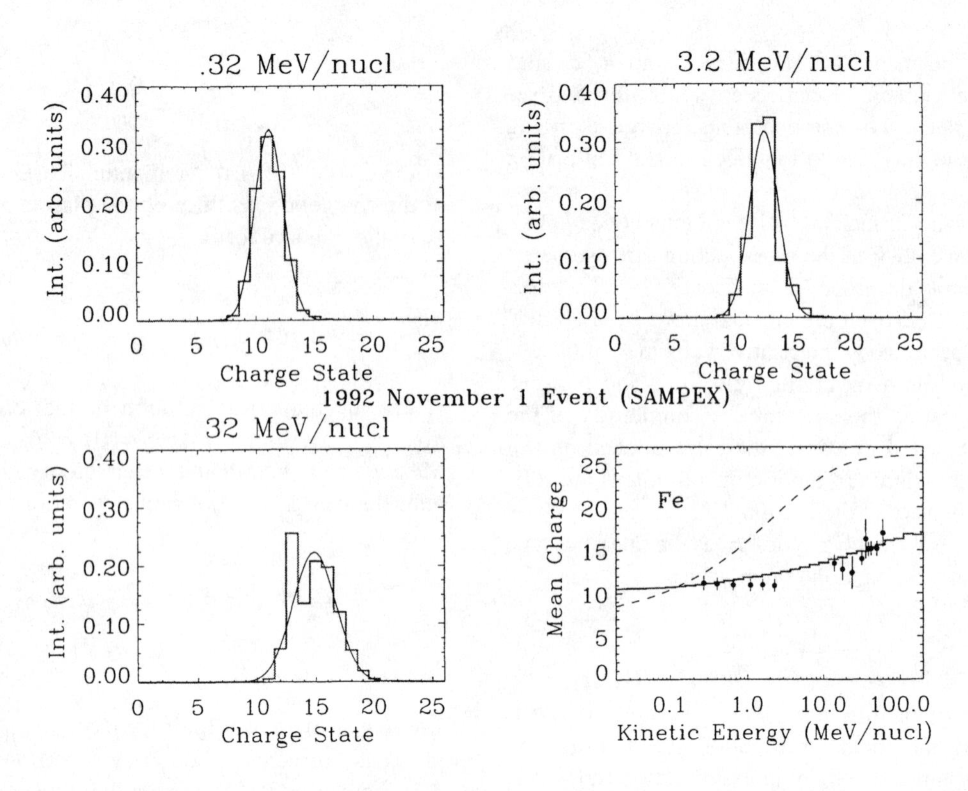

FIGURE 1. Simulated charge distributions (histograms) and mean charge (jagged curve) of iron for the 1992 November event. Solid smooth curves are simulated charge distributions assuming pure Gaussians. The smooth dashed curve is from a semi-empirical formula for the equilibrium mean charge dependence on energy (15). Data points are SAMPEX observations (4,5,24).

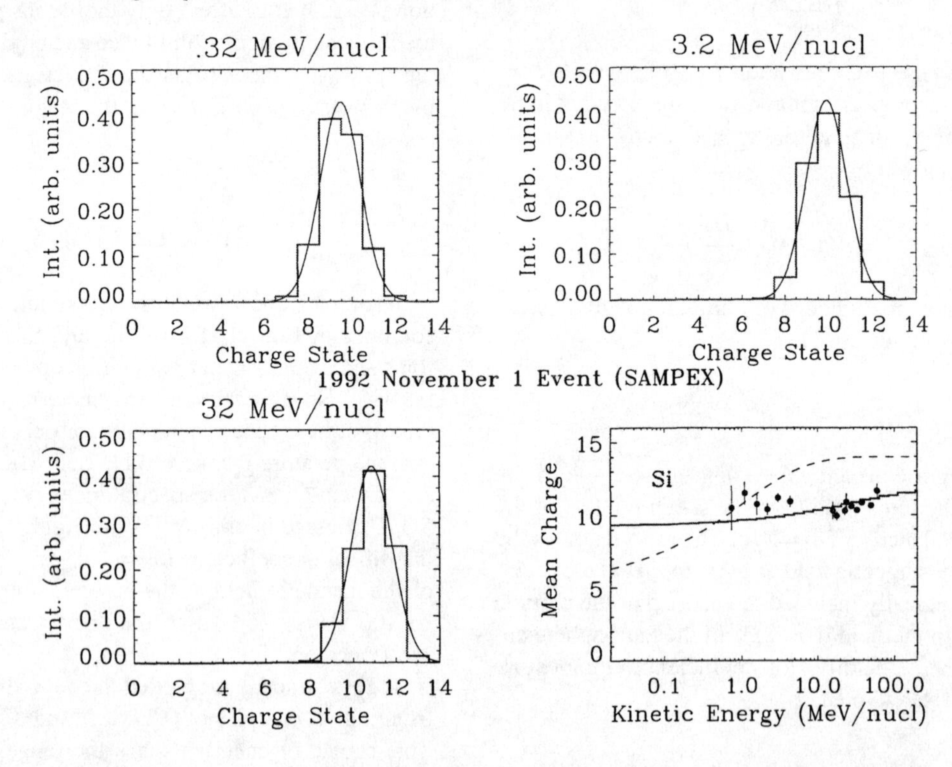

FIGURE 2. Simulated charge distributions and mean charge of silicon for the 1992 November event. [See Fig. 1 and text for further explanation.]

plotted on the lower right panels (jagged curves). Data points for the mean charge are SAMPEX observations (4, 5, 24). The dashed smooth curves are the results of a semi-empirical formula for the equilibrium mean charge (15).

The simulations are repeated for the 1997 November event in Figs. 3 and 4. A good fit to the observed mean charge as a function of energy is found for a temperature of 1 MK, density = 2×10^9 cm^{-3}, spectral index = 5/3, and $D_0 = 2 \times 10^{19}$ cm^2/s. The open-symbol data points for the mean charge are ACE observations (8), while the solid-symbol ones are from SAMPEX (7). (The apparent discrepancy between the ACE and SAMPEX data for the mean charges for the 1997 events is not resolved as of yet.)

Since observations of SEP charge distributions are presently available only for low energy ions (< 1 MeV/nucleon)(8), we are deferring detailed discussion of these, as well as energy spectra, to a future paper.

DISCUSSION

We find that the simulated charge distributions are well approximated by Gaussian distributions with some important qualifications. At low energies, the distribution can depart from a Gaussian due to the lower limit set by the temperature, i.e., the distribution will be skewed toward higher charges. Figs. 3 and 4 at the low energy of 0.32 MeV/nucleon depict this behavior (note that the simulation temperature for this event is lower than that for the 1992 event). The observed charge distributions for iron in the 1997 event by Möbius et al. (8) at comparable energies appear to be similarly skewed. Conversely, at high energies (where no observations of charge state distributions are available yet), the distribution is expected to be skewed toward lower charges due to the upper limit set by the fully stripped ion.

Only for energies well between the two limits do the simulations give distributions closely resembling a pure Gaussian. The third moment (the asymmetry or skewness parameter) of the phase-space distribution is found in our simulations and in (14) to be positive at low energies, cross the zero-line at moderate energies, and be negative at high energies. Deviation from a pure Gaussian at moderate energies can also arise from the structure of the ionization cross sections (e.g., Fig. 3).

The simulated mean charge as a function of energy seen in Figs. 1-4 appears to closely resemble the measured one, but less so for silicon for the 1997 event. It is also clear that both the simulated and measured mean charges in both events do not represent the equilibrium charge state at a given energy (15), even though silicon for the 1997 event appears to be approaching this limit.

For the same Alfvén energy, the simulations shown in Figs. 1-4 are reproduced with different densities and diffusion strengths. We find the results of the simulations unchanged as long as the product $D_0 \times \sqrt{n_e}$ is kept constant. For example, for the 1992 event increasing the density by a factor of 10 to 3×10^9 cm^{-3} while decreasing D_0 to $10^{19}/\sqrt{10}$ cm^2/s produces identical results to those shown on Figs. 1 and 2. Or, for the 1997 event decreasing the density by a factor of 10 to 2×10^8 cm^{-3} while increasing D_0 to $2\sqrt{10} \times 10^{19}$ cm^2/s gives identical results to those shown on Figs. 3 and 4.

In this model, the overriding parameters for the simulated mean charge are the magnitude of the diffusion coefficient and the plasma density. Both directly affect the acceleration and equilibration timescales. For both simulated events, even though the two parameters are allowed to vary, the timescale for acceleration remains comparable to the equilibration timescale in those simulations that fit the observed energy dependence of the mean charge. (Other parameters in this model also affect the timescales but less sensitively, e.g., the Alfvén momentum.) Altering the acceleration timescale relative to equilibration appreciably, on the other hand, produces simulated mean charges that do not resemble the measured ones.

Comparable timescales in this homogeneous model translate into charge-equilibration and acceleration processes taking place concurrently. This is equivalent to stating that at a given energy comparable timescales suggest comparable plasma grammage inferred from either process. Since this grammage depends on the ion energy, ions with higher energy traverse more grammage, and vice versa, giving rise to an energy-dependent mean charge. However, the grammage also depends on the product of the plasma density × the characteristic residence time (for equilibration and acceleration). Below, we explore this further.

The acceleration time τ_{acc} is proportional to D_0 via

$$\tau_{acc} \sim \int_{p_i}^{p_c} \left(\frac{dp}{dt}\right)^{-1} dp, \qquad (12)$$

where $p_i \sim p_A$ and $p_c > p_i$, and with $dp/dt \propto 1/D_0$ from Eqs. (6) and (8). The residence or escape time τ_{esc} and equilibration time (time to reach ionic equilibrium) τ_{eql} are related to the amount of coronal material the ions encounter as

$$\lambda_{esc} \propto \tau_{esc} v n_e \ , \ \lambda_{eql} \propto \tau_{eql} v n_e \ . \qquad (13)$$

Now, from Eq. (1) in steady-state, and for a fixed Alfvén momentum, a sufficient condition for the mean charge to be an increasing (but bounded) function of momentum (i.e., $d\langle q\rangle/dp > 0$) is

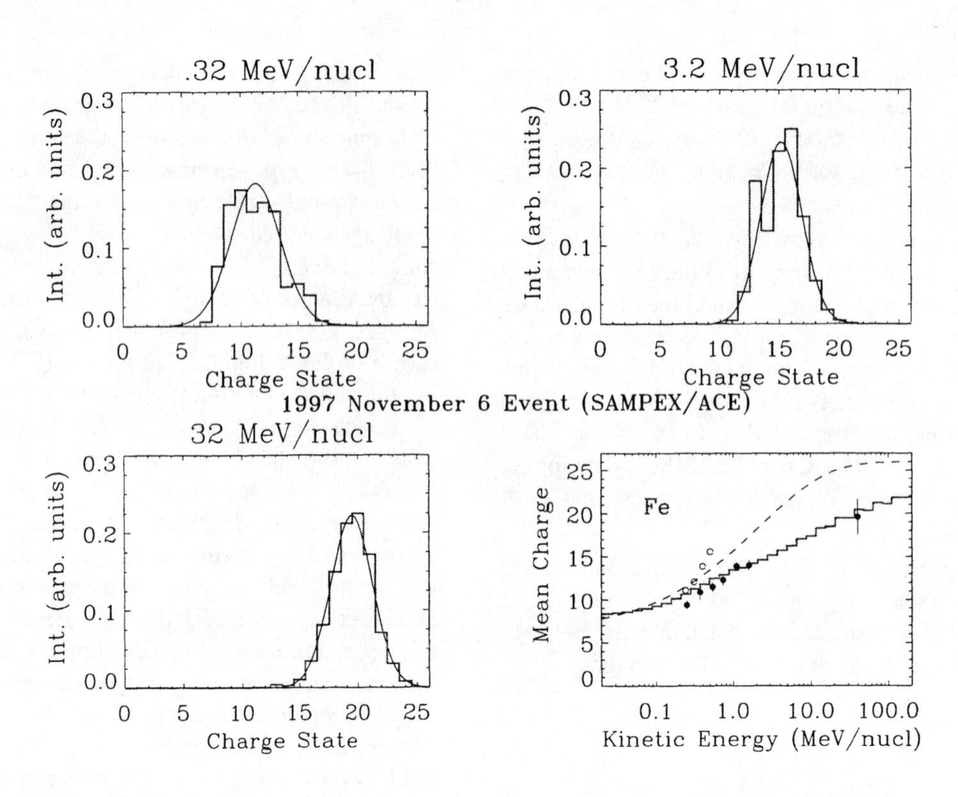

FIGURE 3. Simulated charge distributions and mean charge of iron for the 1997 November event. Solid data points are SAMPEX observations (7) while open ones are ACE observations (8). [See also Fig. 1.]

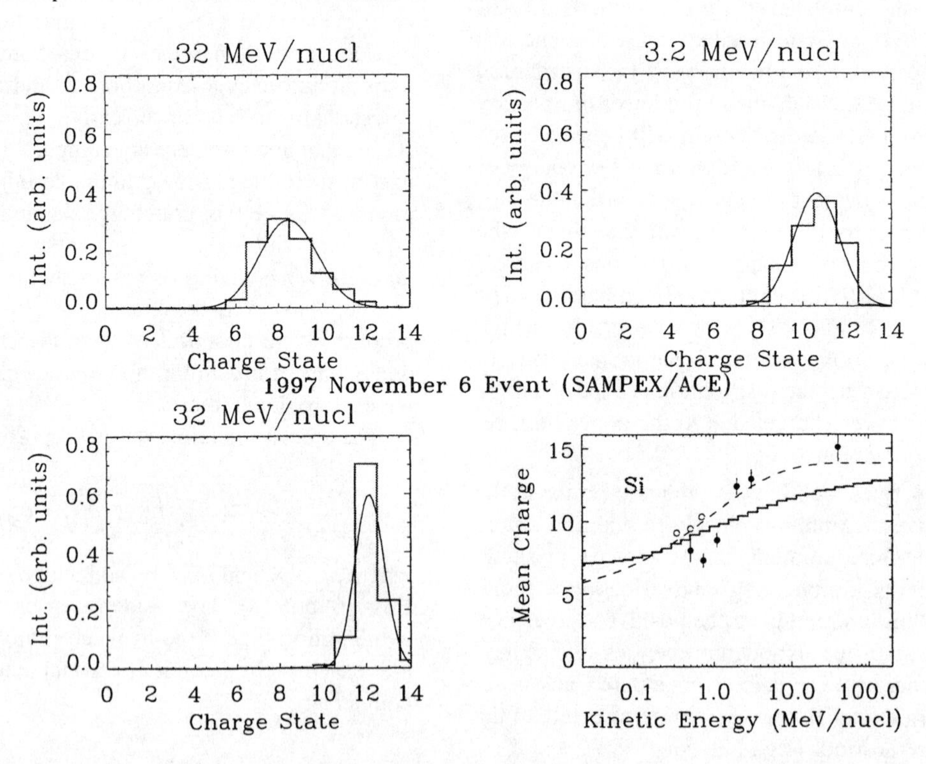

FIGURE 4. Simulated charge distributions and mean charge of silicon for the 1992 November event. [See Fig. 3 and text for further explanation.]

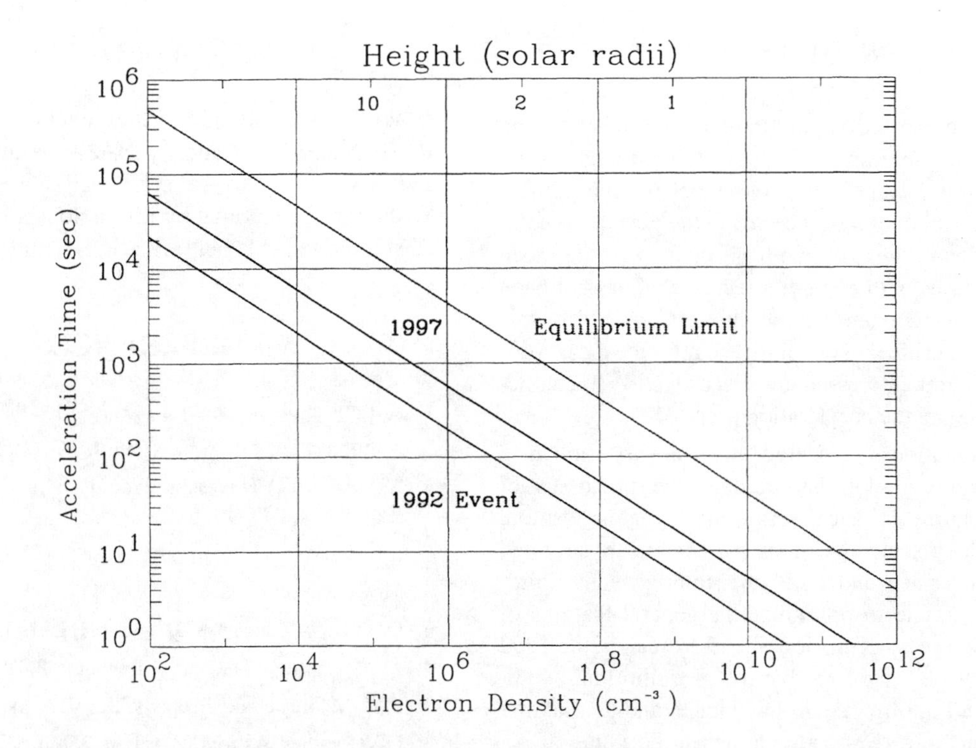

FIGURE 5. Calculated loci of acceleration time vs. electron density (or height above the solar surface) for the 1992- and 1997-events simulations. These loci are for ions with kinetic energy $\gtrsim$ 30 MeV/nucleon characterized by a constant acceleration time $\times$ $\sqrt{}$plasma electron density. The "equilibrium limit" depicts charge-state equilibrium (15). All points on the 1992-event locus, for example, give the same results shown on Figs. 1 and 2.

$$\lambda_{esc}^{-1} - \lambda_{eql}^{-1} > \frac{\text{const}}{\tau_{acc}\sqrt{n_e}} . \tag{14}$$

Fig. 5 depicts the loci of the acceleration time vs. electron density for iron and silicon ions with kinetic energy $\gtrsim$ 30 MeV/nucleon for the 1992 and 1997 events, as well as for equilibrium conditions (15) in which, at these energies, both ions are fully stripped of their electrons. All three loci on Fig. 5 are characterized by a constant $\tau_{acc} \times \sqrt{n_e}$, e.g., all points on the 1992-event locus give the same results shown on Figs. 1 and 2. In Fig. 5 height above the solar surface is calculated assuming hydrostatic equilibrium with $T_e = 1.2$ MK and with density varying with height h as (25)

$$n_e = 4.2 \times 10^{4+4.32/h} . \tag{15}$$

For the 1997 event an upper limit for the acceleration time can be estimated from the time interval from when SOHO first observed the CME (12:10 UT) to when ACE first observed $\gtrsim$ 30 MeV/nucleon ions at L1 (13:00 UT), corrected for the transit time. This gives an estimate for the acceleration time of $\sim$ 10 min. From Fig. 5, the corrsponding lower limit for density for the 1997 event is then $\sim 10^6$ cm^{-3}, and the corresponding upper limit for height $\sim$ 3 solar radii.

Although we cannot specify a more precise acceleration time from these results alone, we can use measurements of hard X-rays and γ-rays from other SEP events to infer a timescale. Time-profile measurements of hard X-ray and γ-ray emission are known to depend sensitively on electron and ion acceleration time (e.g., (26, 27)). For protons and ions accelerated to energies $\gtrsim$ 10 MeV/nucleon, a characteristic acceleration time of $\lesssim$ 10 sec can be inferred from the measured time-profile of the X-ray intensity (27, 28). Although this timescale is expected to vary from one event to another, below we assume an acceleration timescale of $\sim$ 10 sec to characterize the 1992 and 1997 events.

With a characteristic acceleration time $\sim$ 10 sec for ions with energy $\gtrsim$ 30 MeV/nucleon, our model (see Fig. 5) is consistent with an equilibration-acceleration site at heights $\sim$ 1 solar radii above the solar surface, where the density is $\sim 10^9$ cm^{-3} and the typical temperature is $\sim 1.0 - 1.33$ MK. Based on our simulations for the 1992 and 1997 events, we estimate that for these ions the amount of coronal material traversed during equilibration and acceleration is $\sim$ 100 μg/cm^2. On the other hand, equilibrium conditions (15) suggest a grammage of $\sim$ 10 mg/cm^2 and densities of $\sim 10^{11}$ cm^{-3}, which is typical of the lower corona at heights of $\sim$ 0.1 solar radius.

SUMMARY

We have presented simulations for the mean charge and charge distributions of iron and silicon ions for the 1992 November 1 and 1997 November 6 large particle events. Measurements of the mean charge state of solar energetic ions in these events have shown that the mean charge increases with energy; a feature that has not been seen before. Understanding the physical reasons for this new feature is critical to our ability to infer the local coronal plasma temperature and density, as well as to characterize and model the acceleration process.

The simulations presented here are based on a nonequilibrium model that couples the processes of charge-changing and acceleration in a dynamic fashion. In this model plasma and shock parameters are assumed stationary in space and time (the homogeneous limit). The model is able to reproduce the general features of the mean charge as a function of energy seen in the 1992 and 1997 events. We find that ionic equilibrium conditions to be a limiting case in this model, and that neither event appears to correspond to ionic equilibrium.

The simulations suggest that the energy dependence of the mean charge is due to the charge-changing and acceleration processes taking place concurrently. In the homogeneous limit in which this model is derived, this is equivalent to stating that at a given energy the amount of coronal material that the ions encounter inferred from either process is the same. Since this grammage depends on the ion energy, ions with higher energy traverse more grammage, and vice versa, giving rise to an energy-dependent mean charge.

The simulations have also shown that a characteristic of models derived in this limit is the ability to specify the plasma grammage without being able to specify uniquely the acceleration time or the electron density of the plasma; lacking additional information only the product $\tau_{acc} \times \sqrt{n_e}$ can be specified. However, we do find the range of this product consistent with a wide range of solar-wind conditions.

In order to infer the location of the equilibration-acceleration region, we need to specify the acceleration timescale. This timescale can be inferred from hard X-rays and γ-rays from SEP events. Assuming a characteristic shock-acceleration time of $\sim$ 10 sec, and based on our simulations for the 1992 and 1997 events, this model is consistent with an equilibration-acceleration site at heights $\sim$ 1 solar radius above the solar surface, electron temperatures $\sim 1 - 1.33$ MK, and plasma electron densities $\sim 10^9$ cm^{-3}. Our calculations also suggest that for ions with kinetic energy $\gtrsim$ 30 MeV/nucleon the amount of coronal material the ions traverse during equilibration and acceleration is $\sim 100\ \mu$g/cm^2.

ACKNOWLEDGMENTS

Work is supported by NSF grant 9810653 and NASA-JOVE NAG8-1208 and by NASA grants NAS5-30704 and NAG5-6912 at Caltech. A.F.B. thanks Prof. M. Yoshimori for stimulating discussions and Dr. Christina Cohen for valuable help with SOHO and ACE data.

REFERENCES

1. A. Luhn et al., Astrophys. J. **317**, 951 (1987).
2. A. Luhn et al., Adv. Space Res. **4**, 161 (1984).
3. A. Luhn and D. Hovestadt, Proc. 19th Int. Cosmic-Ray Conf. (La Jolla) **4**, 245 (1985).
4. R.A. Leske et al., Astrophys. J. Lett. **452**, L149 (1995).
5. M. Oetliker et al., Astrophys. J. **477**, 495 (1997).
6. D. Ruffolo, Astrophys. J. Lett. **481**, L119 (1997).
7. J.E. Mazur et al., Geophys. Res. Lett. **26**, 173 (1999).
8. E.M. Möbius et al., Geophys. Res. Lett. **26**, 145 (1999).
9. D.V. Reames, Astrophys. J. Lett. **358**, L63 (1990).
10. S.W. Kahler, Astrophys. J. **428**, 837 (1994).
11. D. Hovestadt et al., Astrophys. J. **281**, 463 (1984).
12. A. Luhn and D. Hovestadt, Astrophys. J. **317**, 852 (1987).
13. A.F. Barghouty and R.A. Mewaldt, Astrophys. J. Lett. **520**, L127 (1999).
14. V.M. Ostryakov and M.F. Stovpyuk, Solar Phys. **189**, 357 (1999).
15. D.V. Reames, C.K. Ng, and A.J. Tylka, Geophys. Res. Lett. **26**, 3585 (1999).
16. G.M. Mason, J.E. Mazur, and J.R. Dwyer, Astrophys. J. Lett. **525**, L133 (1999).
17. M. Arnaud and J. Raymond, Astrophys. J. **398**, 394 (1992).
18. M. Arnaud and R. Rothenflug, Astron. and Astrophys. **60**, 425 (1985).
19. A.F. Barghouty, Phys. Rev. A **61**, 052702 (2000).
20. M.A. Forman and G.M. Webb, in *Collisionless Shocks in the Heliosphere: A Tutorial Review*, eds. R.G. Stone and B. Tsurutani (Geophys. Monog. 34), (AGU: Washington, DC, 1985), 91.
21. J.R. Jokipii, Rev. Geophys. Space Phys. **9**, 27 (1971).
22. J.R. Jokipii and G. Morfill, Astrophys. J. **312**, 170 (1987).
23. J. Skilling, Mon. Not. R. Astron. Soc. **172**, 557 (1975).
24. G.M. Mason et al., Astrophys. J. **452**, 901 (1995).
25. H. Zirin, *Astrophysics of the Sun*, (Cambridge U. Press: Cambridge, England, 1989), 226.
26. S.R. Kane et al., Astrophys. J. Lett. **300**, L95 (1986).
27. E.L. Chupp, Ann. Rev. Astron. Astrophys. **22**, 359 (1994).
28. M. Yoshimori et al., Adv. Space Res. **25**, 1801 (2000).

The Observational Consequences Of Proton-Generated Waves At Shocks

Donald V. Reames

NASA Goddard Space Flight Center
Greenbelt, MD 20771

Abstract. In the largest solar energetic particle (SEP) events, acceleration takes place at shock waves driven out from the Sun by fast coronal mass ejections. Protons streaming away from strong shocks generate Alfvén waves that trap particles in the acceleration region, limiting outflowing intensities but increasing the efficiency of acceleration to higher energies. Early in the events, with the shock still near the Sun, intensities at 1 AU are bounded and spectra are flattened at low energies. Elements with different charge-to-mass ratios, Q/A, differentially probe the wave spectra near shocks, producing abundance ratios that vary in space and time. An initial rise in He/H, while Fe/O declines, is a typical symptom of the non-Kolmogorov wave spectra in the largest events. Strong wave generation can cause cross-field scattering near the shock and unusually rapid reduction in anisotropies even far from the shock. At the highest energies, shock spectra steepen to form a "knee." For protons, this spectral knee can vary from ~10 MeV to ~1 GeV depending on shock conditions for wave growth. In one case, the location of the knee scales approximately as Q/A in the energy/nucleon spectra of other species.

INTRODUCTION

There is now a general understanding that the largest and most energetic of the solar energetic particle (SEP) events are associated with shock waves driven out from the Sun by coronal mass ejections (CMEs) (*e.g.* 5, 6, 7, 18). Differences between these large 'gradual' SEP events and the smaller, but more numerous, events from impulsive flares have been recently reviewed (18) and will not be discussed here.

The idea of shock acceleration of SEPs is by no means new. The first suggestion that SEP events can have either shocks or flares as progenitors was made in 1963 by Wild, Smerd, and Weiss (26) from radio observations of type II and type III bursts, respectively. However, this evidence was lost for 20 years while purveyors of the "flare myth" (5), that all particles come from flares, held sway. Difficulties in transporting SEPs across magnetic field lines over 180° in solar longitude were surmounted by the mathematically contrived "coronal diffusion," which had scant basis in physics. The demise of the flare myth began when SEP ionization states were measured that were much too low for the temperatures in a solar flare (12, 24). At about the same time, the relationship between CMEs, shocks, and SEPs began to emerge (8).

It is well known that particles streaming out along magnetic field lines generate Alfvén waves (22) and that those waves then scatter the particles that come behind. Self-generated waves were first used as the basis of an equilibrium shock-acceleration theory for galactic cosmic rays (GCRs) by Bell (1). This theory was adapted to SEP acceleration by Lee (9). This is a seminal theory that has helped us understand many features of shock acceleration as we discuss below. However, we must remember that it is an equilibrium theory, not a dynamic theory. It does not give us realistic asymptotic behavior far from the shock or initial behavior of the first particles accelerated. In addition, Lee theory assumes a planar shock of fixed characteristics that does not describe the CME shock evolution in a nearly spherical geometry where there are substantial changes in the plasma parameters with time. In fact, Lee theory was originally applied only to locally accelerated particles; it was not well recognized that the shock, now seen at 1 AU, had been accelerating particles all the way out from the Sun.

CP528, *Acceleration and Transport of Energetic Particles Observed in the Heliosphere: ACE 2000 Symposium*,
edited by Richard A. Mewaldt, et al.
2000 American Institute of Physics 1-56396-951-3

To understand time-dependent processes in a curved geometry we must use a numerical model such as that of Ng (14, 15). This model follows the coupled transport of particles and waves through space and time along a magnetic flux tube. At each point, the pitch-angle distribution of the particles is determined by scattering on waves, and the proton distribution function defines the growth rate of those waves.

When we discuss particle transport and scattering by waves, it is often noted that the wave turbulence, measured as field lines are convected across a magnetometer, differs from that we deduce from particle transport. This problem has been known for over 20 years (4). Energetic particles are scattered by only a small fraction of the measured magnetic turbulence. While this is often cast in dire terms as a "failure of quasi-linear theory," it is better described as a poor understanding of those wave modes that do not scatter the particles (18). Bieber *et al.* (2) discuss these modes in terms of two-dimensional turbulence moving across the magnetic field. Meanwhile, it is clear that *energetic particles themselves are the best possible probes of that part of the turbulence that affects energetic particles.*

WAVES AND SHOCKS

The Streaming Limit

It was observed 10 years ago (17) that intensities of MeV protons arriving early in large SEP events did not exceed a limiting value of a few hundred $(cm^2\ sr\ s\ MeV)^{-1}$, as seen in Figure 1. This was understood as defining the intensity point, near the shock, where production of self-generated resonant waves decreases until the wave intensity is no longer adequate to constrain the outward streaming of the particles. From this point, particles stream outward with intensities decreasing with distance in the diverging magnetic field (14, 20).

The streaming limit applies only to particle transport from the source to the observer. Higher intensities can certainly exist near the source where streaming is just adequate to produce a wave-particle equilibrium. At equilibrium the intensities of both particles and waves declines in a simple way with distance from the shock (9). When shocks are sufficiently strong, these intensity peaks, called ESP events or shock spikes, survive out to 1 AU. The intensity in these peaks can rise above the streaming-

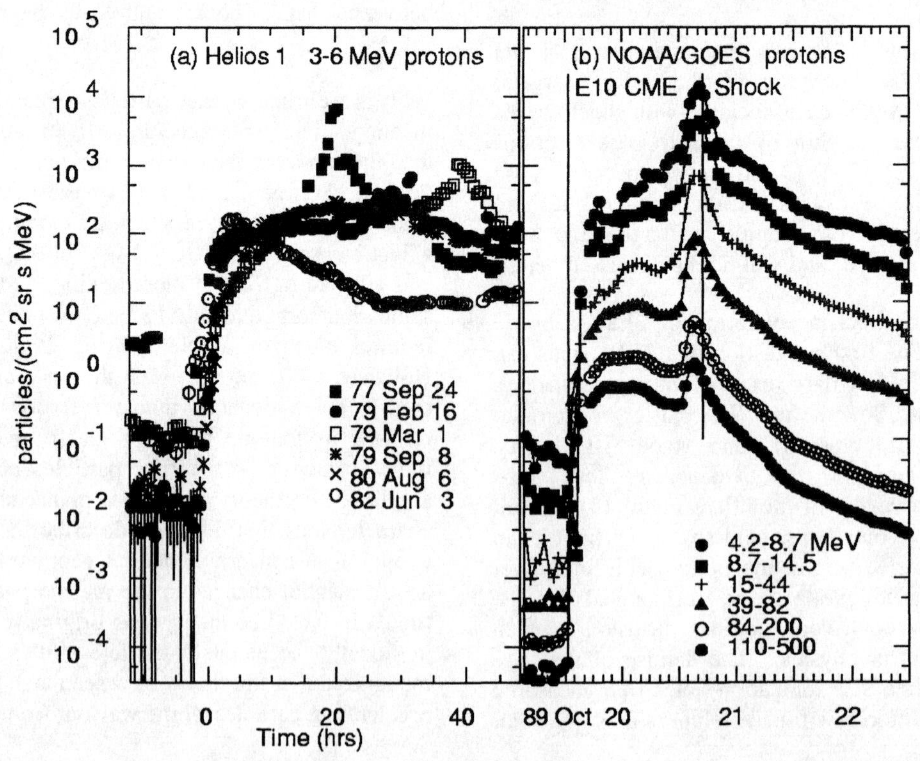

FIGURE 1. Panel (a) shows superposed intensity-time profiles of 3-6 MeV protons in several events with streaming-limited intensities early in the events. Panel (b) shows similar limits as a function of energy in the large 1989 October 19 event. Intensities often peak at the time of shock passage at values that are 10-100 times the streaming limit.

limited value by factors of 10-100, as seen in Figure 1. The intensity value of the streaming limit decreases with increasing energy, as seen in the right panel, Figure 1(b).

Shock Acceleration

Lee (9) theory describes the region near the shock where the particle and wave intensities are in equilibrium, *i.e.* it describes a planar shock peak of infinite extent after an infinite time. The theory has no temporal 'streaming limit' since the latter implies a time scale that is too short for resonant waves to grow and for equilibrium to be established. At the shock, the energy spectrum is a power law with a spectral index that depends upon the shock compression ratio. If we move a distance x away from the shock, the spectrum we observe is flattened at low energies, as shown in Figure 2. It is not surprising that more resonant waves have grown at low energies where intensities are higher.

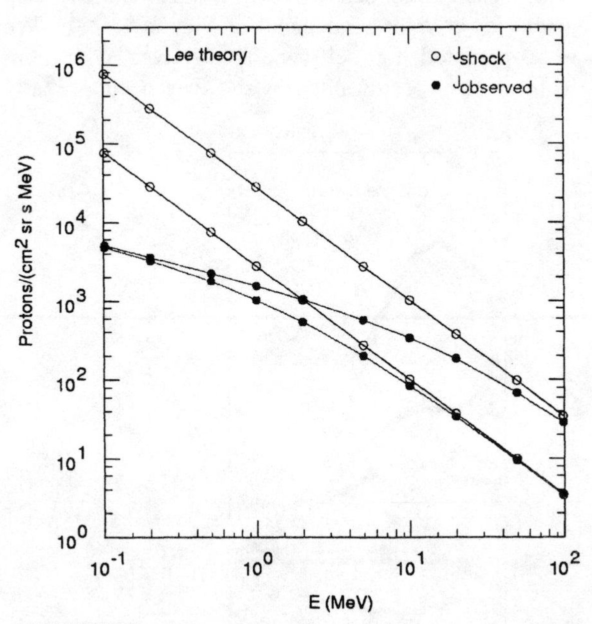

FIGURE 2. Proton intensity is shown at the shock and a flattened spectrum is seen a fixed distance away, according to Lee (9) theory. Increasing the source *does not* increase the intensity observed at low energy.

However, if we simply increase the normalization (injection) at the shock, Figure 2 shows that the low-energy intensities observed at a distance x are unchanged, but the flattening has extended to higher energy. This behavior is the Lee-theory's spatial equivalent of the streaming limit. *The effect of the increase at the shock is only seen as an extension of the observed spectrum to higher energy.* We will return to this idea again later.

It is instructive to take the simple pedestrian view of a shock shown in Figure 3. Each time particles scatter back and forth across the shock they gain an increment of velocity. Initially, just above their injection energy, they must scatter on ambient turbulence. As the particles begin to gain energy, some stream away and generate resonant Alfvén waves of wave number, $k_{res}=B/\mu P$, where P is the particles rigidity and μ the cosine of its pitch angle. As later particles arrive at that rigidity, they are scattered and trapped by the resonant waves and are more likely to be accelerated further. As these more-energetic particles stream away, they generate waves that resonate at their higher rigidity, and so forth.

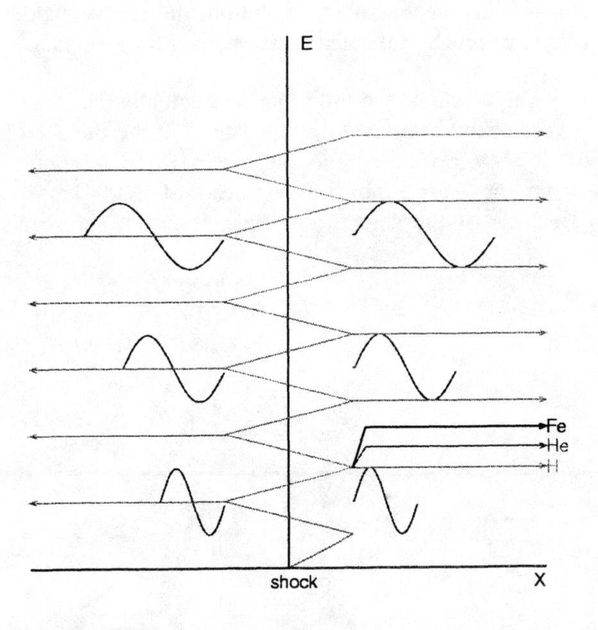

FIGURE 3. Cartoon showing trapping of particles near the shock by self-generated waves.

This process leads to the growth of a 'wall of waves' that resonates with particles of higher and higher rigidity. As equilibrium is established at each level, the intensity of escaping ions becomes fixed. Hence, as we inject more particles at the bottom, the wall must grow higher. The increase in acceleration efficiency caused by wave growth is essential for acceleration to high energies. At sufficiently high energy, where the upward flow of particles becomes inadequate to sustain the waves, a spectral 'knee' develops, and intensities plummet as particles leak easily from the shock. At the energy where intensities return to the power law in Figure 2, the real-world spectrum would actually develop a knee if there were not time to scatter and accelerate those particles.

ABUNDANCE VARIATIONS

Most of the waves at the shock are generated by protons, the most abundant species. Other ions, accelerated to a given velocity or energy/nucleon, resonate with different wave numbers, k, depending inversely upon their rigidity, hence upon Q/A. Thus, these ions probe the shape of the wave spectrum, often producing dramatic abundance variations with time as the shock and its wave spectra evolve. To follow this evolution theoretically, we must use a numerical model such as that of Ng *et al.* (15). A comparison of observations and theory for the 1998 April 20 event is shown in Figure 4. The calculated abundances depend strongly upon the assumed values of Q/A and on time variation of the shock strength, which is assumed to decrease linearly in this calculation. Considering the uncertainty in the shock evolution, the simple model captures much of the qualitative behavior of the data.

The event shown in Figure 4 is actually the largest event of the solar cycle, as measured in the fluence of >10 MeV protons. The associated CME is emitted from the west limb with a speed of 1600 km s^{-1}. Because of the source longitude, it is not surprising that the shock weakens considerably with time.

The abundances at any specific time result from the cumulative effects of transport through the spatially and temporally varying wave spectra from the shock to the observer. Sometimes this complexity makes it difficult to understand the physics, even when experiment and theory agree. For example, the study of large SEP events revealed a surprising feature of the initial behavior of He/H when compared with Fe/O. Early in an event, one would expect the first particles that arrive to propagate through a pre-existing wave spectrum that might have a Kolmogorov, $k^{-5/3}$, form. With this form, high-rigidity ions will scatter less than those of low rigidity, at the same velocity. For rising time profiles, an abundance ratio such as Fe/O or He/H should begin at high values and then decline with time since scattering delays ions in the denominator initially, relative to those in the numerator.

When we began to study this behavior in large SEP events, both experiment and our new theory showed the opposite initial behavior for Fe/O and He/H as seen in the lower right-hand panel of Figure 5 (21). We were prepared for self-generated waves with non-Kolmogorov spectra, but why did they occur so early

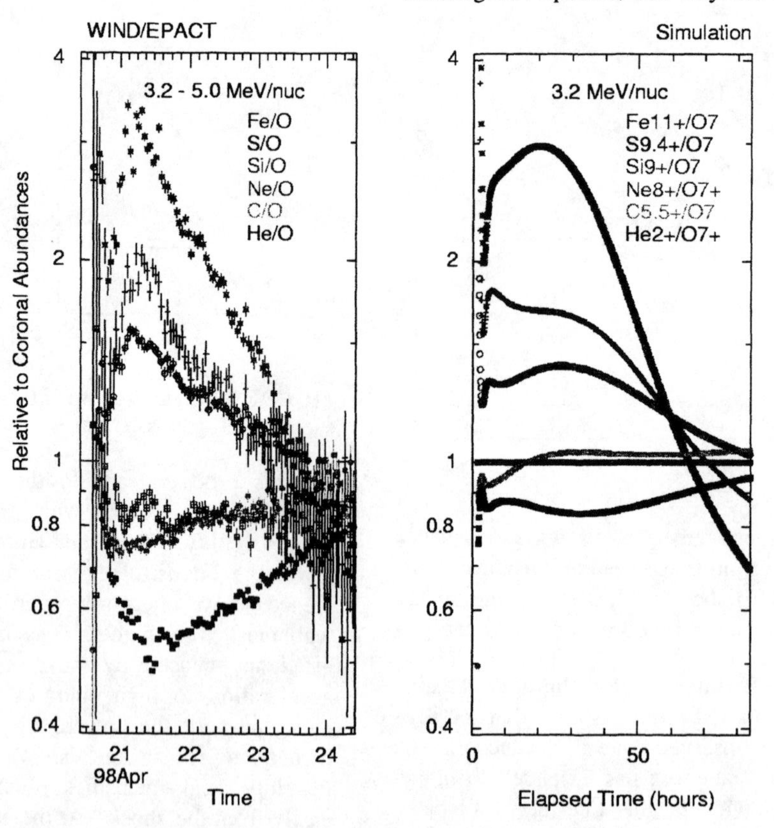

FIGURE 4. A comparison of element abundances, normalized to coronal values, in the 1998 April 20 event (25) with those calculated using the theory of Ng *et al.* (15) for specific values of the ionization states shown.

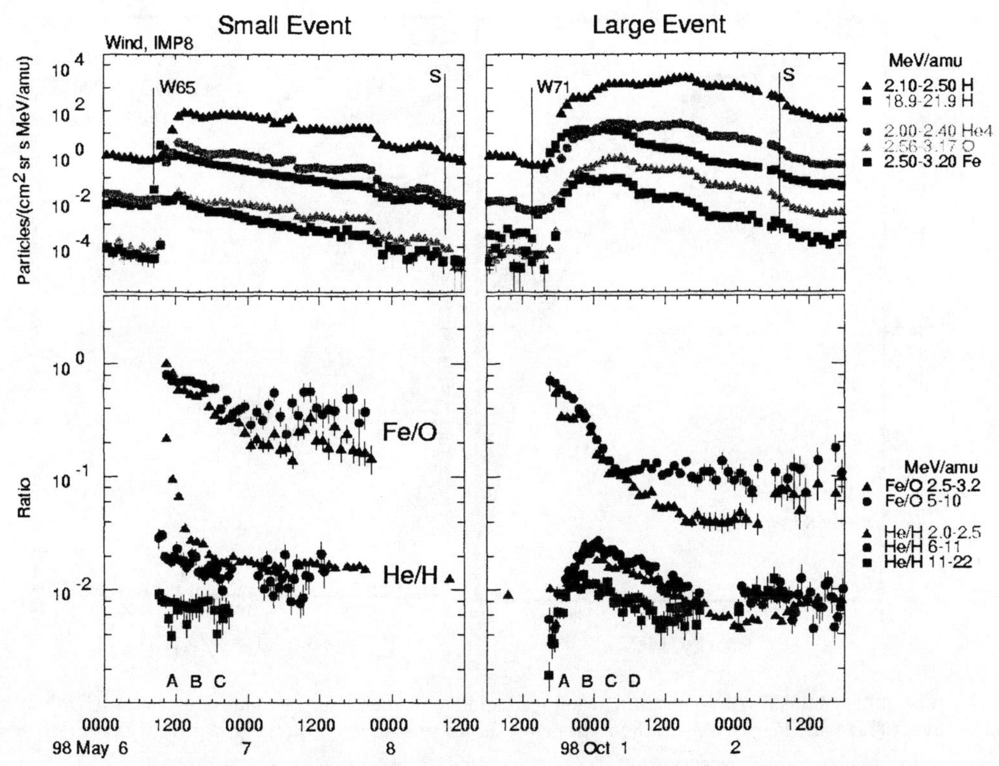

FIGURE 5. Intensities and abundances of ion species *vs.* time are compared for the small 1998 May 6 SEP event and the large 1998 September 30 event (21). The initial rise in He/H and fall in Fe/O are clear in the September 30 event.

in the event? In smaller events, with less wave growth, both Fe/O and He/H initially decline, as seen in the lower-left panel of Figure 5.

The explanation lies in the velocity of the protons generating the resonant waves (15, 21). The 2 MeV protons, for example, have just arrived and are just beginning to generate resonant waves. However, the 2 MeV amu^{-1} He resonates with waves generated by protons of twice the velocity, ~8 MeV, that arrived much earlier. Thus the 2 MeV amu^{-1} He can be strongly scattered when intensity of 8 MeV protons is sufficiently high.

Figure 6 compares the proton spectral evolution for the 2 events shown in Figure 5. Spectra are taken at the times labeled by letters along the abscissa in Figure 5. Spectra in the September event are flattened at low energies at intensities we would expect from the streaming limit but they have a strong high-energy component. Spectra for the smaller May event rapidly attain power-law shapes with no excess of high-energy protons. Wave generation occurs in a shell near the shocks in both events; however, copious 2-20 MeV protons can rapidly generate significant intensities of resonant waves all the way out to 1 AU in the September event but not in the May event, as the spectra show.

Hints of this inverted behavior in He/H were noticed 20 years ago (27, 13), but no satisfactory explanation was offered until recently (15, 21).

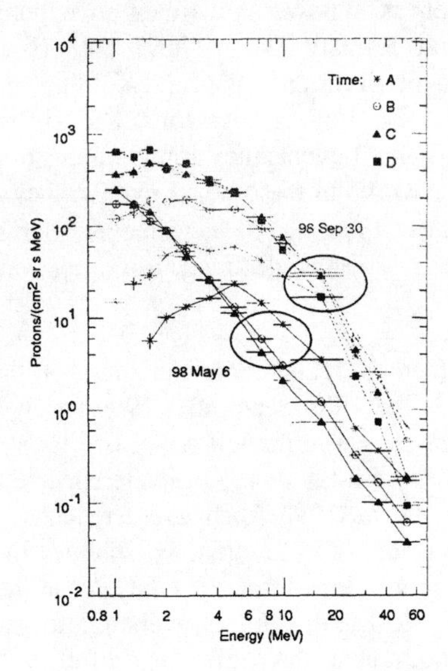

FIGURE 6. Proton spectral evolution for the two events shown in Figure 5 (21).

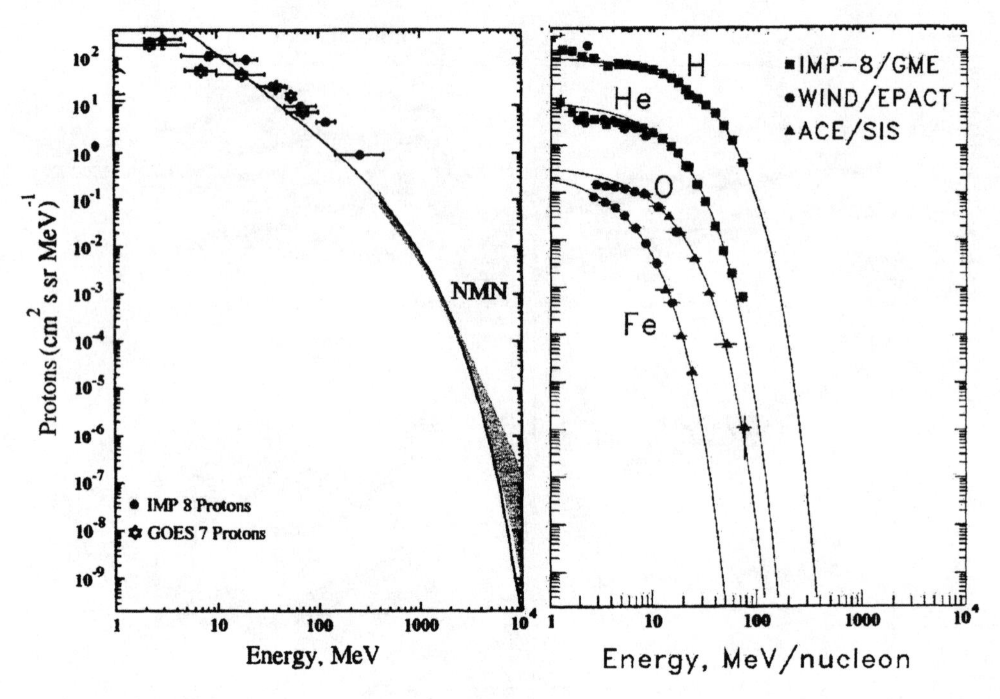

FIGURE 7. The left panel shows a spectrum from spacecraft and the neutron monitor network (NMN) in the 1989 September 30 event (11) with *Eo* = 1 GeV. The right panel shows spectra from the 1998 April 20 event (23) with *Eo* = 15

THE SPECTRAL KNEE

We suggested earlier that acceleration would begin to fail at an energy where proton intensities could no longer sustain the growth of resonant waves. For this reason, and perhaps others, high-energy particles begin to leak from the acceleration region to form a spectral knee. The shape of the energy spectrum in this region was written as a power-law times an exponential by Ellison and Ramaty (3). Those authors examined spacecraft observations of the e-folding or "knee" energy, E_o, for protons, electrons, and He in 8 large SEP events. In 3 events they found values of E_o of 20, 25, and 30 MeV; in the other 5 events, they found E_o =∞, *i.e.* above the range of instruments. For 2 ground-level events, spectra deduced from neutron-monitor data gave E_o = 5 GeV.

Data from a recent determination of the proton spectrum in the 1989 September 29 event by Lovell *et al.* (11) are shown in the left panel of Figure 7. This is contrasted with the more complete spectra of H, He, O and Fe, in the 1998 April event recently compiled by Tylka *et al.* (23) for this workshop. In the first event the proton knee is at E_o = 1 GeV, in the second, it is at E_o = 15 MeV, a dramatic change for two events that are both near the west solar limb with similar CME speeds of ~1800 km s^{-1} and ~1600 km s^{-1}, respectively.

The two events in Figure 7 actually have similar proton intensities below ~50 MeV. However, differences in the knee energies cause vastly different behavior above ~100 MeV. This striking difference can have a profound application to the safety of astronauts on deep space missions, as shown in Figure 8. Soft radiation, with E ~40 MeV, begins to penetrate spacecraft walls, while hard radiation, with E >130 MeV, can penetrates 5 cm of Al and is difficult to shield. Behind 10 g cm^{-2} of material astronauts would receive a dose ~4 rem hr^{-1} at intensities in the 1989

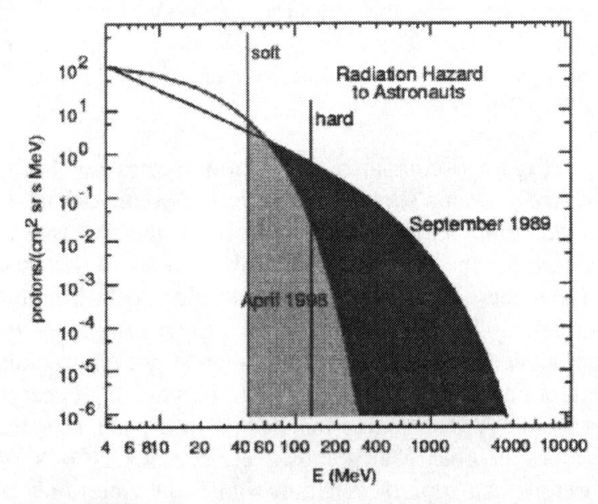

FIGURE 8. The proton spectra from Figure 7 present a drastically different radiation hazard to astronauts.

September event, accumulating their annual dose limit of 50 rem in relatively few hours. Differences in the knee energy alone can turn a benign event into a significant radiation hazard.

The new observations of spectra for all species in the 1998 April event show knee energies that scale like *Q/A* throughout this event. However, knee energies seen late in the 1998 August event scale as a higher power of *Q/A*, and E_o decreases with time (23). Unfortunately, most of the knee energies in other events of this solar cycle are above the observation range of the available instruments.

DISCUSSION AND CONCLUSIONS

Our focus on SEP events has shifted drastically in recent years. Where we once viewed event-averaged element abundances, we now look at detailed time variations in composition and spectra within an event. We see abundant evidence of wave-particle interactions, but only about half of the examples of that evidence have been included here. Other features are not presented here, such as the way SEP abundances average to coronal values, or the rate at which particle angular distributions attain isotropy in large and small events. These features were presented in an earlier review (19).

The new theory that describes the effects of wave-particle interactions (15, 16) does not yet fit the observations precisely (*e.g.* Figure 4). However, it describes the qualitative behavior of the newly observed abundance variations and it has surprised us with explanations of long-standing problems. These include the initial behavior of He/H, and differences in the angular distributions and scattering mean free paths deduced from large and small SEP events. Observational consequences of self-generated waves that have been identified (19) are the following:

1) Streaming-limited intensities early in large events (14, 17, 18, 20).
2) Flattened low-energy spectra in large events (8, 16, 18, 25).
3) Systematic time variations in abundances (16, 18, 25).
4) Abundance variations that average spatially to coronal (FIP-dependent) values (18, 19). (In contrast to acceleration in flares where abundances such as Fe/O can be enhanced everywhere in space).
5) He/H uncorrelated with Fe/O; breakdown of the power law of abundances *vs.* *Q/A* resulting from non-Komolgorov wave spectra (15, 18, 19, 21).
6) Initial *rise* in He/H in large events (15, 19, 21, 25).
7) Rapid onset of isotropy in large events, even at 1 AU (19).
8) Large variations in the energy of the spectral "knee" (18, 19).

A somewhat surprising aspect of these results is the importance of injection. More particles produce more resonant waves, but also, the resulting increase in efficiency causes particles to be accelerated to higher energy. When a 2000 km s^{-1} shock moves out from the Sun through a 500 km s^{-1} solar wind, ions of the bulk solar wind are injected into the shock at ~12 keV amu^{-1}. All species are injected at the same velocity or energy/nucleon, *not* the same momentum, rigidity, or energy/charge. Perhaps ~1% of the ions will be successfully scattered and subsequently accelerated. As the shock speed increases, the incident flow rate and the injection velocity both increase. However, this does not fully explain why peak particle intensities are observed to increase at such a high power of the shock speed, roughly as the fourth power (19). Although, injection is clearly a nonlinear process, as we discussed in connection with Figure 3.

In large SEP events, it is no longer possible to treat the accelerated ions as test particles in the interplanetary plasma. They profoundly modify that plasma, not only near the source, but also throughout the inner heliosphere. One of the more difficult theoretical questions under investigation is the manner in which proton-generated waves are dissipated in the plasma. Intensity-dependent cascading or absorption of waves could certainly modify the particle spectra and abundances, although these effects have not yet been included in the calculations.

There will always be competing models. It is relatively easy to achieve time-varying abundances, for example, by mixing two sources with arbitrary abundances and time scales. However, the new shock models attempt to simultaneously account for intensity-time behavior, energy spectra, abundance variations and angular distributions using well-known physical processes in a single source. Where possible, parameters of the model are taken from plasma observations. The initial results are very promising.

ACKNOWLEDGMENTS

I gratefully acknowledge the contribution made by Chee Ng, not only to this paper, but also to my personal education during the last few years. I am also deeply indebted to Allan Tylka who has generously contributed many of his results, prior to publication, for inclusion herein. I thank Marty Lee for his comments on the manuscript.

REFERENCES

1. Bell, A. R.: 1978, *Mon. Not. Roy. Astron. Soc.*, **182**, 147.

2. Bieber, J. W., Wanner, W., and Matthaeus, W. H., *J. Geophys. Res.* **101**, 2511 (1996).

3. Ellison, D., and Ramaty, R., *Astrophys. J.* **298**, 400 (1995).

4. Fisk, L. A., in *Solar System Plasma Physics*, Vol. **1**, edited by E. N. Parker, C. F. Kennel, L. J. Lanzerotti (Amsterdan: North Holland), p. 177 (1979).

5. Gosling, J. T., *J. Geophys. Res.* **98**, 18949 (1993).

6. Kahler, S. W., *Ann. Rev. Astron. Astrophys.* **30**, 113 (1992).

7. Kahler, S. W., *Astrophys. J.* **428**, 837 (1994).

8. Kahler, S. W., *et al.*, *J. Geophys. Res.* **89**, 9683 (1984).

9. Lee, M. A., *J. Geophys. Res.* **88**, 6109 (1983).

10. Lee, M. A., in: *Coronal Mass Ejections*, edited by N. Crooker, J. A. Jocelyn, J. Feynman, Geophys. Monograph **99**, (AGU press) p. 227 (1997).

11. Lovell, J. L., Duldig, M. L., Humble, J. E., *J. Geophys. Res.* **103**, 23,733 (1998).

12. Luhn, A., Klecker, B., Hovestadt, D., and Möbius, E., *Astrophys. J.* **317**, 951 (1987).

13. Mason, G. M., Gloeckler, G. and Hovestadt, D., *Astrophys. J.*, **267**, 844 (1983).

14. Ng, C. K., and Reames, D. V. *Astrophys. J.* **424**, 1032 (1994).

15. Ng, C. K., Reames, D. V., and Tylka, A. J., *Geophys. Res. Lett.* **26**, 2145 (1999).

16. Ng, C. K., Reames, D. V., and Tylka, A. J., *Proc. 26th ICRC* (Salt Lake City) **6**, 151 (1999).

17. Reames, D. V., *Astrophys. J.* (*Letters*), **358**, L63 (1990).

18. Reames, D. V., *Space Science Revs.* **90**, 413 (1999).

19. Reames, D. V. *Proc. 26th ICRC* (Salt Lake City) Highlight paper, AIP Press, in press (2000).

20. Reames, D. V., and Ng, C. K., *Astrophys. J.* **504**, 1002 (1998).

21. Reames, D. V., Ng, C. K., and Tylka, A. J., *Astrophys. J. (Letters)* **531**, L83 (2000).

22. Stix, T. H., *The Theory of Plasma Waves* (New York: McGraw-Hill) (1962).

23. Tylka, A. J., Boberg, P. R., McGuire, R. E., Ng, C. K., and Reames, D. V., *ACE 2000 Workshop*, this volume.

24. Tylka, A. J., Boberg, P. R., Adams, J. H., Jr., Beahm, L. P., Dietrich, W. F., and Kleis, T., *Astrophys. J.* (*Letters*) **444**, L109 (1995).

25. Tylka, A. J., Reames, D. V., and Ng, C. K., *Geophys. Res. Lett.* **26**, 145 (1999).

26. Wild, J.P., Smerd, S. F., and Weiss, A. A., *Ann. Rev. Astron. Astrophys.* **1**, 291 (1963).

27. Witte, M., Wibberenz, G., Kunow, H., and Muller-Mellin, R., *Proc. 16th ICRC*, (Kyoto) **5**, 79, (1979).

On the Solar Release of Energetic Particles detected at 1 AU

Säm Krucker & Robert P. Lin

Space Sciences Lab, University of California, Berkeley, CA 94720-7450

Abstract. The 3-D Plasma and Energetic Particles experiment on the WIND spacecraft was designed to provide high sensitivity measurements of both suprathermal ions and electrons down to solar wind energies. A statistical survey of 26 solar proton events has been investigated. For all these proton events, a temporally related electron event is observed. The presented results focus on the properties of protons released near the Sun which show a velocity dispersion when detected at 1 AU. The particle flux onset times observed at 1 AU in the energy range between 30keV and 6 MeV suggest that there are two classes of proton events: (1) For one class (70% of the events), the first arriving protons are traveling almost scatterfree as indicated by the derived path lengths between 1.1 and 1.3 AU, (2) whereas the events of the second class show significantly larger path lengths of around 2 AU. Relative to the electron release time at the Sun, the almost scatterfree traveling protons of the first class of events are release delayed by 0.5 to 2 hours. For the events of the second class, protons and electrons seemed to be released simultaneously within the accuracy of 20 minutes.

INTRODUCTION

The Wind 3-D Plasma and Energetic Particles experiment (5) has a good temporal resolution for analyzing onset times of solar events at 1 AU. Onset time analysis allow to approximate the coronal release times of the particles observed at 1 AU and it is then possible to relate the in-situ observations to events occurring at the Sun. This helps to understand the acceleration mechanisms of solar energetic particles. Timing analysis of electron events observed by WIND/3DP have been published by Ergun et al. (1) and Krucker et al. (4). ACE/EPAM observations of electron events are discussed in this proceeding by Hawkins et al. (2). In the present work, the timing of proton events is investigated and compared to the electron timing.

ONSET TIME ANALYSIS

The particle flux onset time at 1 AU, $t_{1AU}(\varepsilon)$, of a particle with energy ε is given by the particle release time at the Sun, $t_{Sun}(\varepsilon)$ plus the travel time:

$$t_{1AU}(\varepsilon) = t_{Sun}(\varepsilon) + L(\varepsilon)\, v_{rel}^{-1}(\varepsilon) \qquad (1)$$

where $v_{rel}(\varepsilon)$ is the component of the relativistic velocity parallel to the magnetic field, and L is the path length. Assuming a simultaneous particle release and a constant path length for all energies ($t_{sun}(\varepsilon) = t_{sun} = const$ & $L(\varepsilon) = L = const$), the observed onset time $t_{1AU}(\varepsilon)$ is a linear function of $v_{rel}^{-1}(\varepsilon)$ with a slope equal to the path length, L, and an intersection at the time of the particles release at the Sun, t_{Sun}:

$$t_{1AU}(v_{rel}^{-1}(\varepsilon)) = t_{Sun} + L \cdot v_{rel}^{-1}(\varepsilon). \qquad (2)$$

The onset times at different energies, as well as a lower and upper limits of these values, are determined by eye. The limits are used to describe the uncertainties of the onset times. The error bars shown in the following plots are therefore to be understood as maximum uncertainties.

OBSERVATIONS

An example of a solar ion event observed by WIND/3DP is presented in Figure 1. The observed velocity dispersion - high energy ions arrive earlier than lower energy ones - shows the solar origin of these particles. The ion flux is found to travel along the magnetic field lines as seen in the pitch angle distribution presented in the bottom two panels of Figure 1.

All ion events presented in this work are proton dominated. This can be proven by the existence of penetrating ions in the 500 to 850 keV electron channels, which can only be explained by penetrating protons. Heavier ions at these energies do not penetrate the foil in front of the electron detectors. Therefore, it can be excluded that the observed ion time profiles are produced by heavier ions than protons, at least for energies between 500 to 850 keV. ACE/ULEIS observations show that the events presented in this work are dominated by protons also at lower energies (J.E. Mazur, private communication).

CP528, *Acceleration and Transport of Energetic Particles Observed in the Heliosphere: ACE 2000 Symposium,* edited by Richard A. Mewaldt, et al.
© 2000 American Institute of Physics 1-56396-951-3/00/$17.00

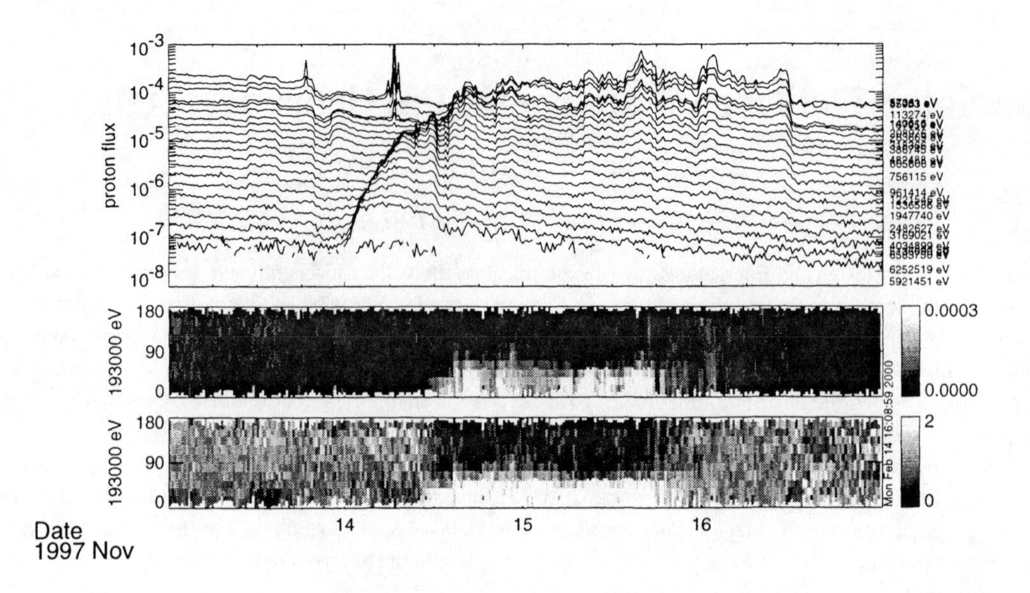

FIGURE 1. Overview plot of the proton event on November 13, 1997. Top: The omnidirectional proton flux at different energies is show. Bottom: Pitch angle distribution for 115-279 keV protons. Enhanced flux is shown white. In the bottom panel, the flux values are normalized by the pitch angle averaged flux, whereas the flux in the panel shown above is not normalized.

ANALYSIS OF TWO TYPICAL EVENTS

Onset time analysis allow to investigate the particle release times at the Sun and the distance traveled until the particles reach the spacecraft (cf. Eq.2). The investigation of these two parameters reveal that there are two classes of events. One event of each class is discussed here in detail followed by a presentation of some statistical results. An example of the more common class of event (18 out of a total of 26 events) is shown in Figure 2 (left). The linear fits to the onset times as a function of inverse velocity show the same slope for protons (dash-dotted line) and electrons (dashed line) indicating that the first arriving protons and electrons are traveling about the same distance until they reach the spacecraft. The derived path length are $L_p = 1.20 \pm 0.05$ AU for protons and $L_e = 1.19 \pm 0.04$ AU for electrons, respectively. Furthermore, the approximated path lengths are comparable to the Parker spiral length of 1.16 AU calculated from the observed daily averaged solar wind speed. The intersections of the fitted lines give information about the solar release time of protons and electrons. A particle release during the onset of the radio type III burst would give an intersection around zero in the shown plot as indicated by the dotted line in Figure 2. The intersections of the lines fitted to the observed onset times are at later times: The first electrons are found to be released at 14:19±3UT, i.e. 26 ± 5 minutes after the type III onset at the Sun (cf. Krucker et al. 1999), and the first protons seem to be released an additional 54 ± 12 minutes later than the first electrons.

An example of the second class of event is shown on the right in Figure 2. The electron onset times give again a similar path length ($L_e = 1.29 \pm 0.13$ AU) as the Parker spiral length (1.26 AU), but in this event, protons seem to travel a much longer distance until they arrive at the spacecraft. The proton path length ($L_p = 2.02 \pm 0.07$ AU) is 60 % longer than the Parker spiral length. Contrary to the previously presented event, the solar release times of protons (21:36±12 UT) and the electrons (21:26±3 UT) seem to be simultaneous within the uncertainties. Compared to the type III radio burst onset, both, protons and electrons are again released delayed by 10 minutes or more.

STATISTICAL RESULTS

Statistical results of the 26 analyzed events are presented in Figure 3. On the left, the derived path length for protons and electrons are compared. There are clearly two classes of events: For one class (18 out of a total of 26 events), electrons and protons have a similar path length, and for the second class (8 out of 26), protons have a significantly larger derived path length than electrons. For events of the first class, there is a linear correlation between L_p and L_e, $L_p \propto 0.7 \pm 0.3 L_e$, with L_p general slightly larger than L_e. The two classes additionally show

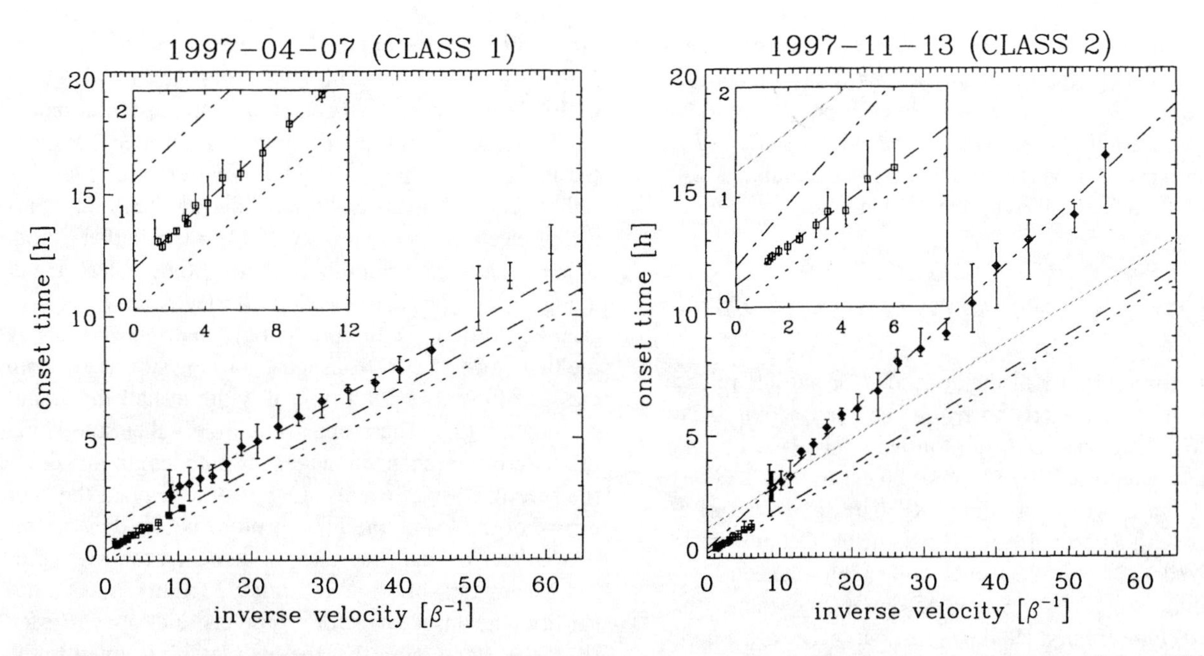

FIGURE 2. Comparison of proton (diamonds) and electron (squares) onset times at 1 AU. The onset time after the radio type III burst onset at the Sun as a function of the inverse velocity is shown for two typical events of each class. The dashed-dotted and dashed lines are linear fits to the proton and electron onset times, respectively. The gray solid line in the plot on the right is a linear fit to the >3 MeV proton data only. The dotted lines show the expected onset times for particles traveling scatterfree along the Parker spiral assuming they are released at the time of the radio type III burst onset. The inserts show a zoom-in of the same plots providing a better overview of the electron onset times.

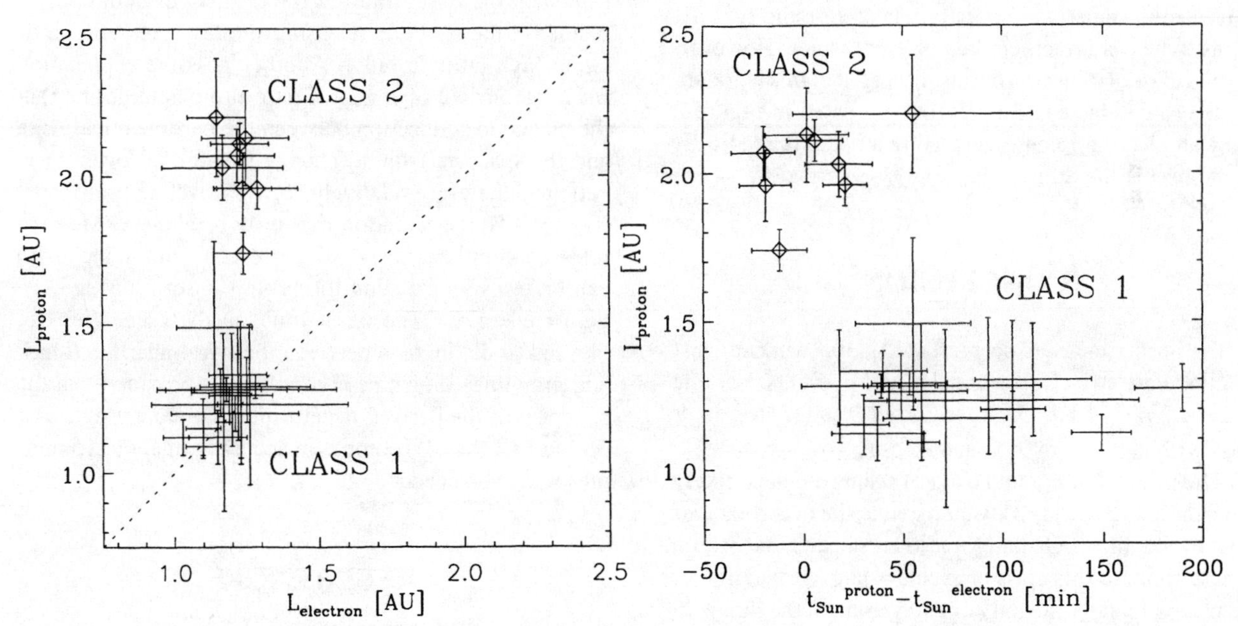

FIGURE 3. Statistical results of the onset time analysis. On the left, the path lengths derived from the proton onset, L_{proton}, are compared with the path lengths derived from the electron onset, $L_{electron}$. Events with $L_{proton} > L_{electron}$ (class 2 events) are marked with diamonds. On the right, the delay between the proton and electron release times at the Sun, $t_{Sun}^{proton} - t_{Sun}^{electron}$, are plotted against the path lengths derived from the proton onset, L_{proton}. Again, events with $L_{proton} > L_{electron}$ (class 2 events) are marked with diamonds.

a different relative particle release for protons and electrons: Events of the first class ($L_p \approx L_e$) show a delayed release time for protons relative to electrons of about one hour, whereas for events of the second class ($L_p > L_e$), protons and electrons seem to be released simultaneously within the uncertainties of about 20 minutes.

CORRELATION WITH SOLAR FLARES

Within ±2 hours of the derived solar particle release times, it is very likely to find some solar activity. For the 26 events analyzed, the following numbers of solar events are reported: (i) 21 out of 26 events with a GOES SXR flare, 3 with a possible SXR flare (ii) 17 out of 26 events with a coronal type III burst (Sol. Geophy. Data, SGD) (iii) 25 out of 26 events with an interplanetary type III burst (Wind WAVES) (iv) 15 out of 26 events with a coronal type II burst (SGD) (v) 14 out of 16 with a coronal mass ejection (SoHO/LASCO). A detailed temporal comparison with the derived particle release times, however, does not show a clear correlation with a particular solar event, nor does it show any convincing correlation outlining again the two classes of proton events presented above. In particular, the two classes of proton events do not correspond to the two classes of electron events reported by Krucker et al. (4). They found that one class of electron events is temporally related to radio type III bursts, whereas the other class of event is not. However, the two classes of proton events discussed in this work are not the same classes: Both events presented in Figure 2 show an electron release delayed relative to the radio III burst onset at the Sun.

CONCLUSION

The onset time analysis presented in this work suggest that there are two classes of solar proton events: (1) one class with $t_{Sun}^{proton} > t_{Sun}^{electron}$ and $L_p \approx L_e$, and (2) the other class with $t_{Sun}^{proton} \approx t_{Sun}^{electron}$ and $L_p > L_e$.
(1) The protons of the first class of events are most likely shock accelerated (cf. 3). Compared to the occurrence of solar events like SXR flares, radio bursts, etc., the proton release time for this class of events is late. At the time of the proton release, the only ongoing event at the Sun is in most of the cases the coronal mass ejection (CME) moving away from the Sun. Therefore, the protons are most likely shock accelerated. Assuming the proton release is related to a shock front of a coronal mass ejection, the late proton release can be interpreted as a release at high altitude. A fast CME shock with a speed of 1000 km/s travels about 5 solar radii within an hour. The observed late proton release might therefore suggest a proton release at several solar radii away from the Sun. LASCO CME observations could corroborate this speculation.
(2) The derived proton path lengths of around 1.5 times the Parker spiral length in the second class of events is rather surprising, especially considering that the first arriving electrons seem to travel along the Parker spiral. It is therefore rather unlikely that protons indeed travel along a much larger distance until they reach the spacecraft. Additionally, the pitch angle distributions during the first hours after the onset are looking similar for both classes of events with a typical width at half maximum of roughly 45°. Therefore it is unlikely that some kind of scattering mechanism makes the path length longer for the second class of events. One way to explain the later arrival of the lower energetic protons is that they are released later or that they escape later than the MeV protons. The gray solid line in Figure 2 (right) shows a linear fit to the onset times for proton energies above 3 Mev ($\beta^{-1} < 12$) assuming that the path length is equal to the Parker spiral length. This fit suggests that the $>$ 3MeV protons are released delayed relative to the electron release time as seen in the events of the first class. The later onset time of the lower energetic protons might then be explained by a later release or escape. For 0.3 MeV ($\beta^{-1} \approx 40$) protons, it would be a later release or escape by about 5 hours. Assuming the protons are shock accelerated, the onset times at lower energies could be explained if the shock is releasing the low energetic particle at high altitude only. Another possible explanation for a late arrival of low energetic protons might be that the magnetic connection between the particle release site and the spacecraft might change before the lower energetic and therefore relatively slowly traveling protons arrive (6). The speculation that protons above 3 Mev are released simultaneously (or can escape simultaneously) can be test by analyzing the proton onset times at even higher energies. The onset times at these energies can be used to distinguish between the two linear fits (black and gray line) shown in Figure 2 (left) provided that the accuracy of the derived onset times is good enough. For 50 MeV ($\beta^{-1} \approx 3$) protons, an accuracy of about 15 minutes would be needed.

REFERENCES

1. Ergun, R. E. et al., 1998, ApJ, 503, 435
2. Hawkins, S. E. et al., 2000, in press, these proceedings
3. Kahler, S. W., 1996, AIP Conf. Pro., 374, 61
4. Krucker, S. et al.,1999, ApJ, 519, 864
5. Lin, R. P. et al., 1995, Space Sci. Rev., 71, 125
6. Mazur, J. E., 2000, in press, these proceedings

Coronal Origin of Particle Events Detected by EPAM: Multi-instrument Observations

Dalmiro Maia[a,b] and Monique Pick[b]
S. Eduard Hawkins III[c] and Edmond C. Roelof[c]

[a]*Faculdade de Ciências da Universidade de Lisboa, Lisbon, Portugal*
[b]*Département d'Astronomie Solaire de l'observatoire de Paris, UMR 8645 CNRS, Meudon, France*
[c]*The Johns Hopkins University Applied Physics Laboratory, Laurel, MD 20723, USA*

Abstract. We investigate the solar origin and propagation of a series of well collimated energetic electron events measured in situ by the EPAM experiment on the ACE spacecraft. EPAM measures electrons in the energy of range of 40 to 300 keV over a wide range look directions and with better than 1 minute time resolution. During the events in our study, these particles are strongly collimated along the magnetic field. As such, these near-relativistic (beta=0.4-0.7) particles tend to be scatter free and their observed arrival at ACE provides a good estimate of the release time back to the Sun. We combine these observations with fast imaging of the solar corona in the meter wave domain provided by the Nançay radioheliograph and dynamic spectral information from the WAVES experiment on the WIND spacecraft. Together, this complement of observations of solar energetic particles events provides insight into the onset times and sites of particle acceleration in the low corona.

INTRODUCTION

Coronal and interplanetary type III bursts are tracers of electron beams traveling along open magnetic field lines with velocities about one third the speed of light. At each level in the corona or in the heliosphere the electrons excite plasma waves at the local plasma frequency. These waves are partially converted into electromagnetic radiation at the fundamental or at the harmonic of the local plasma frequency. Some of the electrons can propagate along field lines connected to spacecraft experiments in the interplanetary medium capable of in situ particle measurements, giving rise to the so called impulsive electron events. Electrons of different energies released at roughly the same time at the sun will take different amounts of time to cover the same trajectory. The resulting dispersion in arrival times versus electron energies is one of the distinguishing features of electron impulsive events. Another indicative feature is the well collimated nature of many of these events (Lin et al. , 1981).

The onsets of in situ impulsive particle events are almost always associated with radio type III bursts. The values given in the literature are that about 90% of impulsive electrons events onsets are associated with type III radio bursts (Lin , 1985).

We present here data on a group of 3 consecutive impulsive electron events measured by the Electron Proton and Alpha Monitor (EPAM) experiment on the Advanced Composition Explorer (ACE) spacecraft (Gold et al. , 1998). We compare the onset times from the energetic electron impulsive events, measured at ACE/EPAM, with remote multifrequency imaging of solar corona at metric wavelengths, provided by the Nançay radioheliograph (Kerdraon & Delouis , 1997), from which we are able to obtain the exact times and positions of type III sources.

RADIO OBSERVATIONS

The multifrequency imaging observations of the Nançay radioheliograph (NRH) on November 28, 1997 show a series of short duration, relatively weak bursts, originating from regions widely separated in the east-

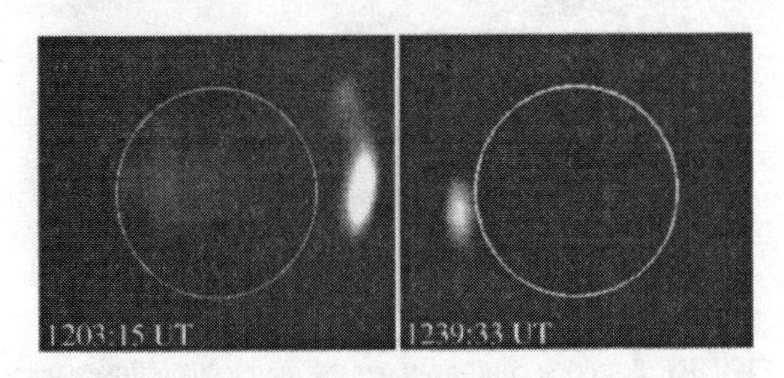

FIGURE 1. Radio images at 164 MHz showing the positions of the eastern and western sites of emission.

CP528, *Acceleration and Transport of Energetic Particles Observed in the Heliosphere: ACE 2000 Symposium,* edited by Richard A. Mewaldt, et al.
© 2000 American Institute of Physics 1-56396-951-3/00/$17.00

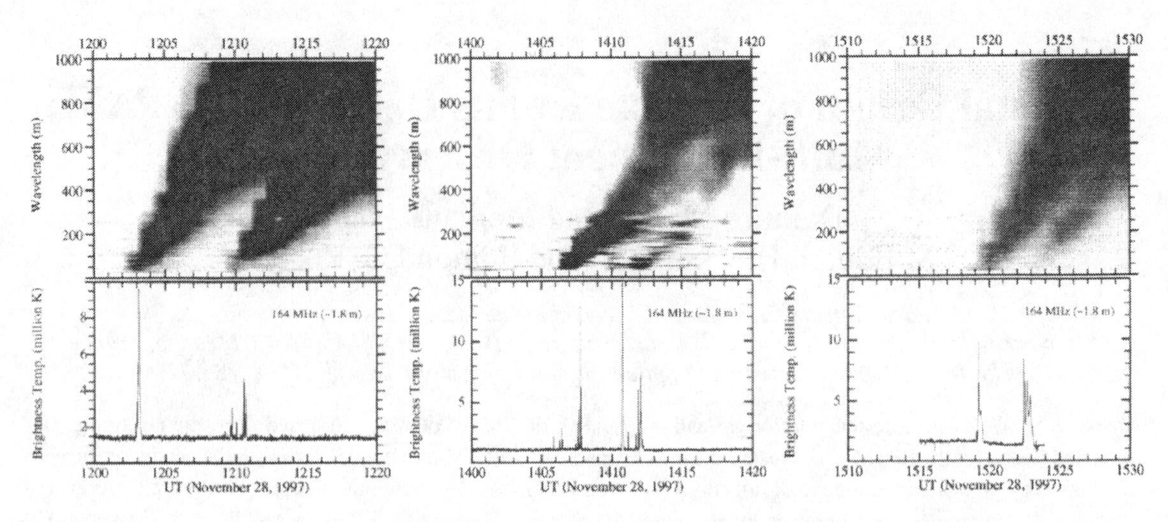

FIGURE 2. Radio flux plots at 164 MHz (measured by the Nançay radioheliograph) associated with the west limb type III like activity compared with simultaneous spectra from WIND/WAVES.

west direction. Figure 1 shows examples at 1203 (west limb) and 1230 UT (east limb) of the emission sites seen by the NRH at 164 MHz. The NRH on this day provided images at 5 frequencies (432, 410, 327, 236 and 164 MHz), with a cadence of an image every 0.5 seconds at each frequency. The eastern activity is of intermittent nature, with a large series of short weak bursts distributed along the day and two major occurrences around 1445 UT and 1510 UT; the 1510 UT burst is associated with a flare on active region 8113 (Solar Geophysical Data, 1997, 640, part I, p. 34). The western sources consist of three series of similar events. In each of these three radio events a first group of short bursts lasting from 1 to 2 minutes is followed by a second group a few minutes later; individual bursts within each group last only a few seconds. All these western bursts are also seen at 236 MHz and show a frequency drift typical of type III bursts. The WAVES (Bougeret et al. , 1995) experiment in the WIND spacecraft detected interplanetary type III bursts with a pattern of emission which is strikingly similar to the western sources seen by the NRH. Figure 2 shows clearly that the WAVES dynamical plots mimic the behavior of the western sources seen by the NRH, including the dual nature of the western activity. These western dual packets of type III bursts arise from 1203 to 1210 UT, from 1406 to 1412 UT, and from 1518 to 1523 UT. From the coronal and interplanetary sets of radio data one can unambiguously conclude that electrons beams originating in the western limb of the sun have readily access to open field lines and propagate in the interplanetary medium.

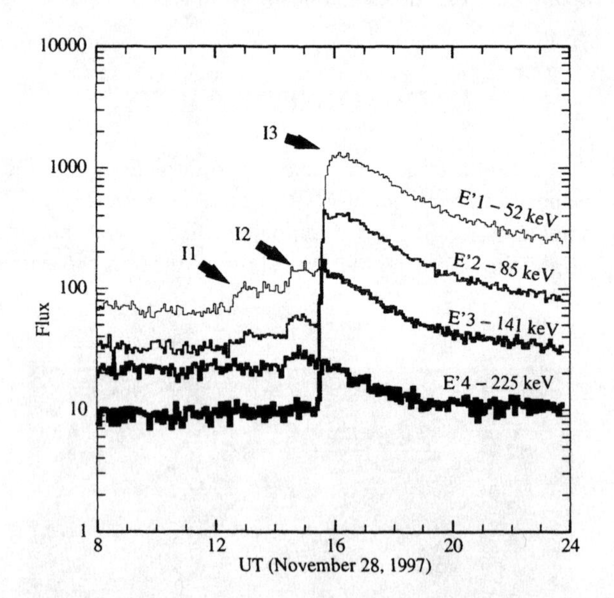

FIGURE 3. 260 second averages of the electron fluxes measured on November 28, 1997 by the LEFS60 particle telescope of the EPAM experiment on the ACE spacecraft. The different panels correspond to different energy channels. Three jumps in the electron intensities are identified by arrows and are labeled I1, I2 and I3.

IN SITU PARTICLE OBSERVATIONS

The Low Energy Foil Spectrometer oriented at 60° of the spin axis (LEFS60) of the EPAM experiment on the ACE spacecraft gives measures of electron fluxes in the energy range from 42 to 290 keV with better than 1 minute time resolution. The LEFS60 telescope is divided into 8 sectors, allowing measure of the electron fluxes over a wide range of directions. Figure 3 shows a plot of 260 second averages of the electron fluxes measured on November 28, 1997 for the 4 energy channels of the LEFS60 instrument. A series of three successive increases in the flux are identified in the LEFS60 E'1 channel (45–62 keV electrons), these are labeled I1, I2 and I3 in Figure 3. The I3 event is seen in the 4 energy channels; the relatively weaker event I2 is seen in E'1, E'2 (62–102 keV electrons) and barely in E'3 (102–175 keV electrons) but not in E'4 (175-290 keV electrons). The I1 increase is seen only in the E'1 channel. The time lag between the onsets of I1, I2 and I3 events corresponds very closely to the time separation of the packets of type III bursts seen in the low corona on the western limb.

The solar wind speed at the EPAM spacecraft at the time of our events remains close to 330 km s^{-1}, meaning that the archimedian spiral connecting magnetically the spacecraft to the sun has a length of 1.21 AU, and the spiral is anchored around 72° of heliolongitude. This spiral length defines the shape of the onset times versus energy curve for a beam of particles propagating scatter free from the solar surface with 0° pitch angle. LEFS60 energies are near-relativistic ($\beta \sim 0.4, 0.7$) and so we have to take that into account when computing the energy curves. In terms of onset time versus $1/\beta$ then onset times should follow a linear relation. The instant corresponding the interception of the time axis at $1/\beta = 0$ marks the release time at sun. The slope of the straight line in the time versus $1/\beta$ is simply the spiral length divided by the speed of light; for a spiral length of 1.21 AU this means a variation of roughly 10 minutes for every change of one unity in the $1/\beta$ values.

In order to get a meaningful slope we need pitch angle discrimination. Another necessary condiiton is for the error bars on the times to be much smaller than the dispersion in arrival times. The energy range of the LEFS60 channels corresponds roughly to one unity difference in $1/\beta$ values, thus an accurate estimate of the spiral length directly from ACE energetic electrons data will require errors to be only a small fraction of 10 minutes, one minute uncertainty means a 10% error in extrapolating the spiral length..

The I1 event shows a very weak signal to noise ratio and we can make a reliable estimate for the onset time of E'1 channel only. Hence we cannot estimate directly the spiral length from the time versus inverse β slope. What we can do for this channel is to infer a release time back at sun by assuming that the particles propagate scatter-free along the spiral length estimated from the solar wind speed. To compare with the metric radio data we add 500 s to the extrapolated time (to account for the electromagnetic radiation travel time). Using a spiral length of 1.21 AU we obtain for the I1 event an estimated release time around 1218 UT (± 5 min). This value is slightly higher the the times of the first package of radio bursts, which took place between 1203 to 1210 UT, the difference resides partly in the high error bars and partly in the bias arising from the lack of pitch angle discrimination. In fact, at the time of the I1 event all the LEFS60 sectors show a similar value for the value of the cosine of the pitch angle about -0.5, since each sector has an angular width of 53° this means that the lower limit for the sampled pitch angle is around 33°. Thus our bias can be as high as the cosine of this angle times the propagation time; for E'1 this corresponds to increase the travel time by about 5 minutes. This correction is enough to bring the inferred release time from the energetic particles detected by EPAM in good agreement with the type III bursts seen by the NRH.

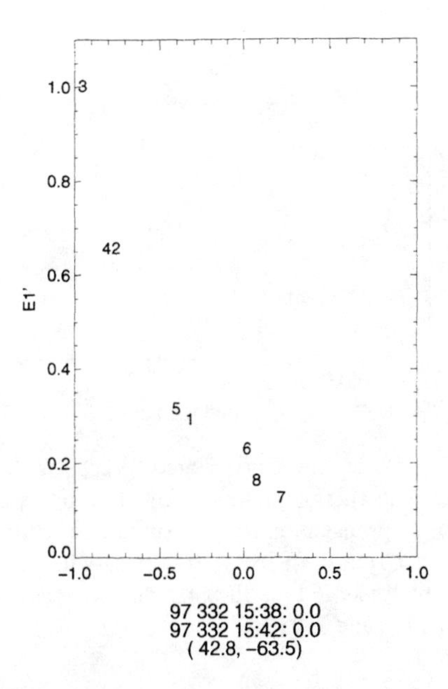

FIGURE 4. Normalized flux distribution as a function of the pitch angle distributions for the E'2 energetic electron channel during the onset of the I 3 event.

Altough slightly stronger than the I1 event, the signal to noise ratio in the I2 event still precludes an estimation of the spiral length, the uncertainties in the determination of the onset are around ± 2 minutes for E'1 and ± 2 for E'3. The inferred release time, assuming the particles have a trajectory length equal to 1.2 AU is 1414 UT ± 3, which agrees well with the second pack of radio bursts, which took place from 1406 to 1412 UT.

The strongest of our three events, the one we have labeled I3, has a very good signal to noise ratio at most energy channels, and the uncertainty in the determination of the onset times is very small, below 1 minute. From Figure 4, which shows the pitch angle distributions for the E'1 channel,averaged 4 minutes around the flux rise time of the I3 event, the anisotropy of the event is clearly evident. The flux peaks sharply around the magnetic field aligned sector. From Figure 4 the beam-like nature of

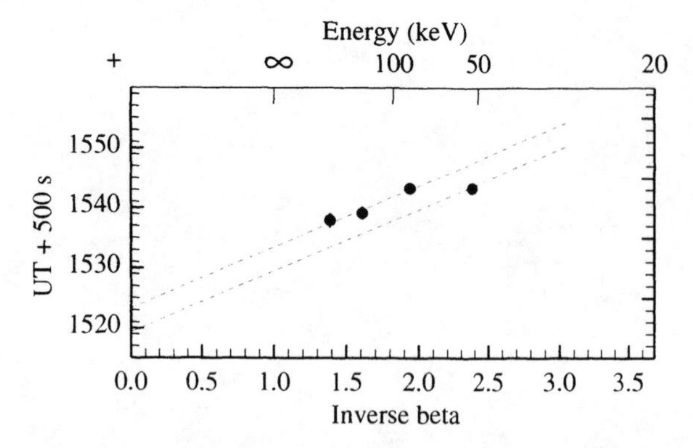

FIGURE 5. Onset times of energetic electrons at the ACE spacecraft versus energy for the I3 event. The curves shown correspond to the propagation curves expected for particles propagating scatter free from the solar surface with release times coinciding with metric type III bursts like sources on the west limb about 1518 and 1523 UT.

the I3 event is quite obvious. The lines that are plotted in Figure 5 are those one would expect for scatter-free transport from the two bursts. When one plots the measured onset times versus $1/\beta$, as shown in Figure 5, one sees that the slope of all four points is to small (much less than 10 minutes) for the electrons to be associated with the type III bursts. If, however, only the three highest points are used, the slope and the onset time values are approximately consistent with an origin in the second type III burst group from the third packet of metric type III bursts seen from 1518 to 1523 UT.

Given the very small uncertainties in the onset times, below 1 minute, the measured values are incompatible with a single sudden injection of the energetic electrons. The electron onset times obtained from the EPAM channel with energy below 50 keV agrees well with the curve giving an onset time at the sun around 1518 UT (once corrected by 500s). The onset times for electrons at higher energies are not compatible with a scatter free propagation from the solar surface with such a release time; requiring a release time at the sun a few minutes later, around 1523 UT.

DISCUSSION

We find an excellent association between 3 successive energetic electron events, 3 interplanetary type III bursts packets and 3 packets of metric type III bursts. The source at the sun has been unambiguously determined. The transit times correspond to a spiral length of 1.21 AU which is in agreement with the highly collimated nature of the pitch angle distributions.

The strongest in situ particle event shows a dispersion in the inferred release times as function of energy. The sense of the dispersion, with the low energies liberated before the high energies, agrees well with an event reported by Roelof et al. (this volume). In the case we report here the dispersion in the inferred release times is bounded by the occurrence of two groups of type III bursts observed in the low corona. Does this mean that we have two successive injections separated by a few minutes? The double injection seems in fact to be the most parsimonious explanation, yet not such feature is apparent in the case studied by Roelof et al in this volume. Nonetheless, the two groups of type III bursts closely separated in time seem to be an important feature of the series of recurrent events that we have presented here. Some questions arise, namely, how frequent are these recurrent double events? To what kind of structures are they related in the corona?

We finally emphasize the importance of radio images for establishing a correct association between metric and interplanetary type III bursts and the exact position of the source location. This is clearly demonstrated in the present study. The events we analyze happen with almost simultaneous activity coming from the opposite limb of the sun. Strong radiobursts are observed a few minutes preceding the third event, they are associated with the only flare reported in Solar Geophysical Data (1997, 640, part I, p. 34) which occurs in the eastern limb. Using flare locations from Hα or X-ray observations would have given in this case a faulty association.

REFERENCES

Bougeret, J. -L., et al. 1995, Space Science Reviews, 71, 231

Brueckner, G. E., et al. 1995, Sol. Phys., 162, 357

Gold, R. E., et al. 1998, Space Science Reviews, 86, 541

Kerdraon, A., Delouis, J. M. 1997, Coronal Physics from Radio and Space Observations, Lecture Notes Phys., vol. 483, p. 192.

Lin, R. P., Potter, D. W., Gurnett, D. A. & Scarf, F. L. 1981, ApJ, 251, 364

Lin, R. P. 1985, Sol. Phys., 100, 537

A Survey of 40-300 keV Electron Events with Beam-Like Anisotropies

S. E. Hawkins III, E. C. Roelof, R. E. Gold, D. K. Haggerty, G. C. Ho

The Johns Hopkins University Applied Physics Laboratory, Laurel, MD, 20723-6099

Abstract.

Beam-like solar energetic electron events provide new insight into acceleration and transport of ~40–300 keV particles from the solar corona to 1 AU. The EPAM instrument on the ACE spacecraft provides a unique set of measurements with its magnetically deflected "pure" electron channels, enabling us to unambiguously discriminate ions from electrons. With this identification, we can then also utilize our higher sensitivity "foil" electron spectrometers (which respond to ions with a low efficiency). This combination of "pure" and "foil" electron measurements then uniquely allows us to construct pitch-angle distributions over nearly a full 180-degree range. The electron observations used in this study have time resolution ~1 minute. We report on a survey of energetic electron events with large anisotropies in which the statistical significance of the intensity in the peak direction is much greater ($\gtrsim 10\sigma$) than the isotropic component. Beam-like pitch-angle distributions imply nearly scatter-free propagation. Because these electrons are almost relativistic ($0.4 \lesssim v/c \lesssim 0.8$) knowledge of the path length back to the Sun should in principle permit one to compute the release time in the corona to within a few minutes. Although impulsive injection produces the most obvious beam events, we often find periods with beams lasting many hours with nearly constant intensity, implying constant outflow from the corona with no apparent correlation to the 1–8 angstrom X-ray emission.

INTRODUCTION

The study of near-relativistic electrons in the interplanetry medium, and in particular the study of strongly anisotropic electron events, serves as an important diagnostic to the acceleration and transport of these particles from the Sun. Although there is a long history of the study of electrons emitted from the Sun (1, 2, 3), we now have observations of these energetic electrons at ~1 AU that can provide insight into the source, acceleration, and transport of these particles. Since the launch of the ACE spacecraft in August 1997, we are able to analyze the onset profile of ~40–300 keV electrons at ~1 AU with good angular resolution, high temporal resolution, and high sensitivity using the Electron Proton and Alpha Monitor (EPAM) instrument (4).

We characterize observations of strongly anisotropic electrons as beam-like when they are detected streaming mainly along the interplanetry magnetic field with pitch-angles near 0° or 180°. During beam-like events, the particles have pitch-cosine $\mu \sim 1$ implying $v_{\parallel} \approx v$. These beam-like electrons tend to propagate nearly scatter free. The transit time to travel a direct path along the nominal field length of ~1.2 AU from their origins in the corona to ~1 AU ranges from about 12 to 25 min for ~300-keV to ~40-keV electrons, respectively.

In this study, we provide an overview of the technique we have implemented to identify anisotropic electron events and the criterion we have chosen to select the beam-like events. Included in our general definition of beam-like events are the so-called prompt or impulsive events, e.g. (5). The design of EPAM's electron detectors, the favorable geometry of the telescope look-directions, and the temporal resolution of the instrument are ideally suited to the task of making an extensive survey of electron beam events.

INSTRUMENTATION

EPAM measures energetic particles ranging in energy from a few tens of keV to a few MeV. Its configuration consists of five solid-state detector systems on two stub telescopes. In this study we take advantage of two of the electron detector systems: the higher sensitivity "foil" electrons of the LEFS60 telescope, and the "pure" electrons magnetically deflected (DE) to separate them from any ions present.

The low-energy foil spectrometer or LEFS60 sensor is pointed 60° from the spin-axis of ACE, approximately oriented toward the Sun. In this position LEFS60 has an excellent viewing geometry sampling electrons stream-

CP528, *Acceleration and Transport of Energetic Particles Observed in the Heliosphere: ACE 2000 Symposium,* edited by Richard A. Mewaldt, et al.

© 2000 American Institute of Physics 1-56396-951-3/00/$17.00

ing along the nominal Parker field over a wide range (~180°) of pitch-angles. As ACE spins once every 12 s, the LEFS60 head sweeps out eight sectors and measures electron intensities in four energy channels spanning ~45–290 keV. An aluminized Parylene foil discriminates against detection of ions with energy below about 350 keV, while electrons above ~35 keV pass through to the solid-state detector.

A unique feature of EPAM is its ability to unambiguously identify electrons through its magnetically deflected sensor. Electrons entering an aperture 30° from the spacecraft spin axis are deflected by a rare-earth magnet into a solid-state detector, whereas the trajectory of a more massive ion is unaffected. The deflected electrons (DEs) are also detected in four energy channels from ~38–315 keV, in four sectors.

METHODOLOGY

We define an electron anisotropy indicator as

$$A_e = \frac{N_{pk} - N_{iso}}{\sigma_{rel}} \quad (1)$$

where N_{pk} is the number of electrons measured in the peak sector, N_{iso} represents the isotropic component (spin-averaged) number of electrons, and σ_{rel} is the statistical error of the numerator. This expression compares the statistical significance of the electron beam (the peak sector) to the pre-event electron population, which we estimate as the spin-average. The relative error in the denominator of (1) is given by the root-sum-square of the individual contributors:

$$\sigma_{rel} = \left[N_{pk} + \frac{1}{n(n-1)} \sum_i N_i \right]^{1/2} . \quad (2)$$

The second term in (2) is the variance of N_{iso}, corrected for bias introduced by the fact that N_{pk} is included in the spin-average. The index i runs over the n sectors: $n = 8$ for the LEFS60 telescope and $n = 4$ for DE. Although N_{pk} and N_{iso} are obviously not independent, A_e as defined in (1) adequately provides a relative measure of electron anisotropy, and tends to include all anisotropic events. The strength of the anisotropy and hence its beam-like characteristic must be inferred through additional tests such as examination of the electron pitch-angle distributions (PADs).

DATA ANALYSIS AND DISCUSSION

To demonstrate the application of (1) applied to the 45–62 keV electron channel of the EPAM, Figure 1 shows a time series spanning sixty days in 1997 from 07 October (day 280) to 06 December (day 340). Figure 1(a) shows spin-averaged electron intensities, hourly averaged from the LEFS60 telescope. The symbols plotted along the abscissa in this panel represent soft X-ray observations of class C and higher ($I > 10^{-6}$ Wm^{-2}) listed in Solar Geophysical Data. The points corresponds to class C, the triangles to class M, and the exes to class X. If an Hα flare was also observed with the X-ray emission, and it was on the western hemisphere, a 'W' is plotted below the X-ray flare symbol.

Figure 1(b) is plotted as in Figure 1(a) but using the pure electron DE channels from EPAM. By comparing the pure electron intensities plotted by the four channels DE1–4 in (b) with the intensities from the higher sensitivity detector measurements from LEFS60 in Figure 1(a), one can unambiguously distinguish pure electron events from events in which ions are present in LEFS60. The ordinate at the right corresponds to the trace at the bottom of Figure 1(b) showing the calculated anisotropy indicator A_e, using 5-min averaged data from the 45–62 keV electron channel from the LEFS60 telescope. The intensity maxima labeled 1–16 in Figure 1(a) correspond to individual electron beam events identified by $A_e \gtrsim 10$ and indicated by the arrows in Figure 1(b). These events are tabulated in Table 1.

We have selected a time range which includes the first large energetic particle events in Solar Cycle 23. The sixty day period shown in Figure 1 provides a representative sample of the frequency and duration of electron beam events we have continued to see since the turn-on of EPAM in August 1997. Anisotropic electron events occur frequently and the duration of the events lasts from minutes to many hours. We have not yet performed a detailed statistical study to correlate electron events at 1 AU with solar X-rays. However, our preliminary results suggest that smaller strong beam-like events are often seen without any obvious X-ray flares (which does not preclude the possibility that such flares may have occurred behind the limb).

We define the parameter S_b to give a measure of the strength of the electron beam anisotropy by comparing the peak intensity $J(\alpha_{pk})$ at at pitch-angle α_{pk}, to the intensity of the side of the distribution $J(\alpha_{sd})$, where $\alpha_{sd} \equiv$ the pitch-angle nearest 90°, and

$$S_b = \frac{J(\alpha_{pk}) - J(\alpha_{sd})}{J(\alpha_{sd})} . \quad (3)$$

Table 1 lists the events numbered in Figure 1(a) along with the approximate times of the electron beam onsets in which the anisotropy indicator $A_e \gtrsim 10$. The third column gives an indication of the beam strength S_b and the fourth column gives the two pitch-angles used in calculating S_b using (3).

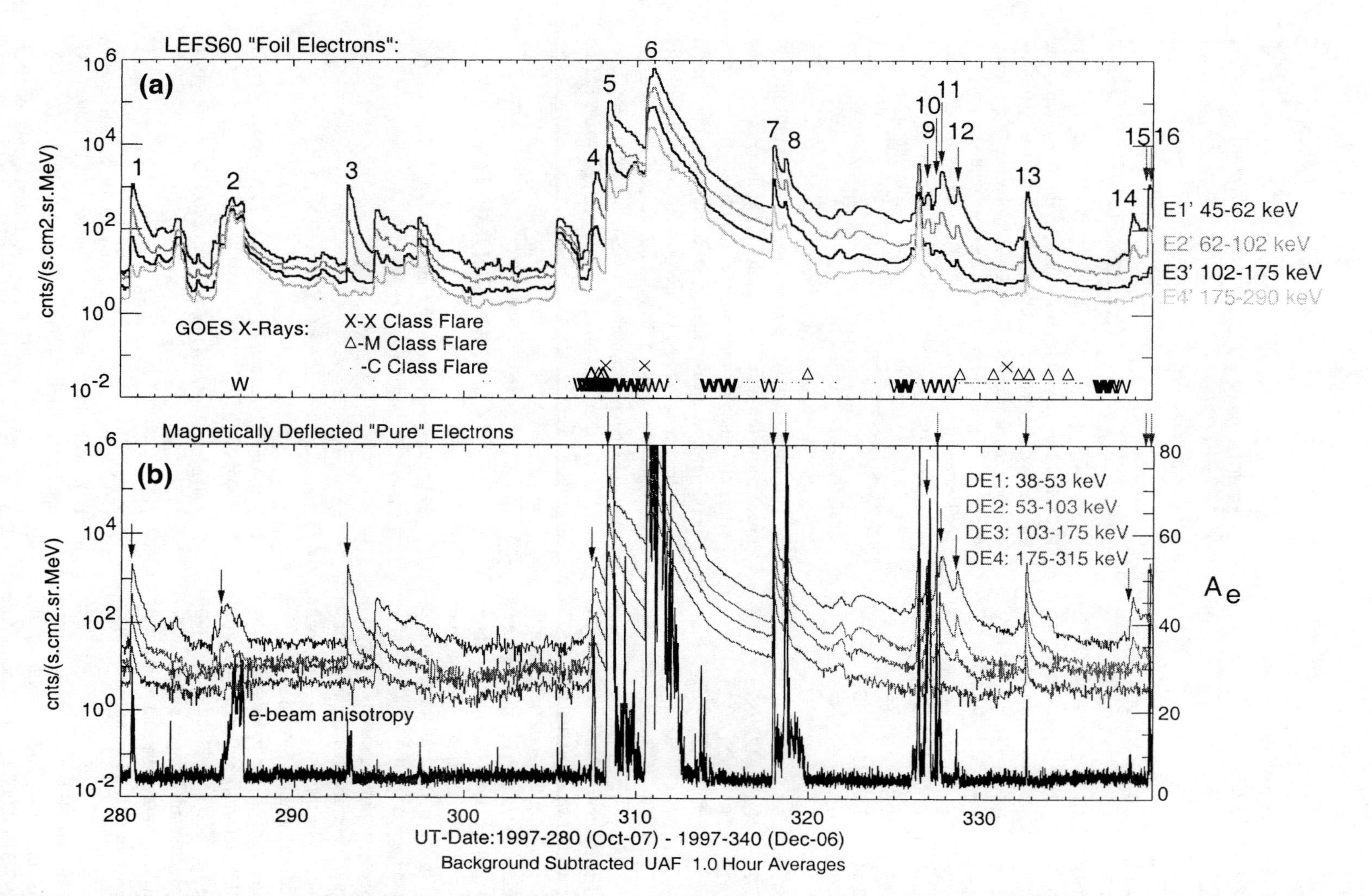

FIGURE 1. Example of method used to identify periods of strong beam-like electron anisotropy for a 60 day interval during 1997. (a) The four traces in this panel correspond to ~45–290 keV electrons plotted as intensity versus time (hourly averages). The numbered events (listed in Table 1) are electron beam events, determined from the heavy trace in (b) and identified as pure electron events by comparing with the magnetically deflected electron (DE) channels shown in (b). The plotted symbols near the bottom of the figure correspond to soft X-rays identified by the NOAA GOES satellite. If an Hα flare was identified and appeared on the western hemisphere it is denoted by a W. (b) The top four traces are plotted as in (a) but using the DE channels (~38–315 keV). The dark trace labeled e-beam anisotropy is A_e from (1), caclulated using the lowest energy channel in LEFS60. The arrows identify valid electron beaming events (see text).

Table 1. Approximate onset times for the beam-like electron events shown in Figure 1 during the 60-day period in 1997 from 07 Oct–06 Dec. The third column gives an estimate of the strength of the beam $S_b = J(\alpha_{pk})/J(\alpha_{sd}) - 1$, and α_{pk} and α_{sd} are the pitch-angles of the peak intensity and the nearest pitch-angle to 90° ($\mu = 0$).

Beam Event	≈Event Time (DOY-UT)	S_b	α_{pk}/α_{sd} (deg)
1	280-14:22	2.0	33.9/86.4
2	285-21:38	4.6	55.7/74.4
3	293-04:11	4.5	15.1/86.8
4	307-10:30	5.7	167.8/87.2
5	308-06:15	0.7	20.0/88.6
6	310-12:45	0.5	32.6/76.9
7	317-21:53	0.9	28.3/83.1
8	318-13:45	2.2	4.5/78.2
9	326-18:50	4.5	17.3/89.0
10	327-10:23	50.5	16.7/88.1
11	327-12:54	2.0	154.7/87.5
12	328-14:39	0.8	57.0/89.1
13	332-15:35	1.5	134.9/87.4
14	338-15:35	8.3	33.7/75.3
15	339-17:26	2.1	17.1/89.2
16	339-19:22	1.1	30.6/76.1

Event 7 from Figure 1 is shown in Figure 2 at higher time resolution, along with two representative PADs. The intensity measurements from the eight sectors of the LEFS60 ~45–62 keV electron channel are plotted versus time. The single trace at the bottom of Figure 2 shows $A_e > 10$. The anisotropy is evident in the intensity plot by the large separation in the individual sectors shortly after the onset of the event at ~317-21:53 UT.

The leftmost PAD labeled (a) in Figure 2 is taken from day 317-23:14 UT, about 1.5 hours after the initial onset of the electron beam event. For this PAD, $S_b = 4.4$ with $\alpha_{pk}/\alpha_{sd} = 32.0/88.1$. The symbols plotted in the PAD plot the normalized intensities of the individual sectors from the LEFS60 and DE detectors versus $\mu = cos(\alpha)$. The eight sectors from LEFS60 are plotted as the numbers 1–8 and the four sectors of DE are plotted as the letters A–D. In PAD (a) in Figure 2 the sectors are well-ordered and all directions lie in the forward hemisphere ($\mu > 0$).

The second PAD, labeled PAD (b), shows a very weakly anisotropic distribution. The strength of the anisotropy $S_b = 0$ for 97-318-02:03 UT, and one minute earlier measures $S_b = 0.05$ for comparision with the values in Table 1.

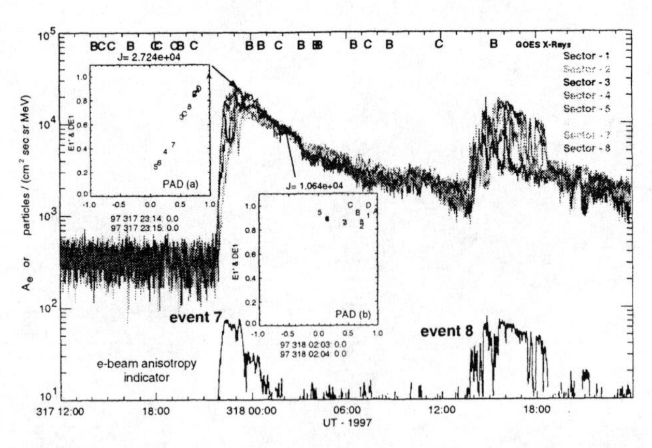

FIGURE 2. Higher time resolution of events 7 and 8 plotting intensity versus time for the eight sectors of the LEFS60 ~45–62 keV electron channel. The trace at the bottom of the figure shows A_e computed from 1 min averages plotted every 30 s. The leftmost inset figure labeled PAD (a) shows the very well-orderd pitch-angles near the peak of the event, and the rightmost PAD (b) shows the weakly anisotropic distribution just following the beam event.

The letters appearing near the top of the plot in Figure 2 show the soft X-ray activity for these days. The letters 'B' and 'C' indicate the identification of class B and C flares as measured by GOES. Note that the nearest X-ray flare observed by GOES occured from 317-20:09–20:18, more than an 1.5 hours before the particle event onset. There were a number of western hemisphere flares 8–10 days after this (cf. Figure 1, events 9–13), but no large flares were reported.

ACKNOWLEDGMENTS

This work was supported under NASA GSFC Contract NAS5-97271 task order 009.

REFERENCES

1. Van Allen, J. A., and Krimigis, S. M., *J. Geophys. Res.* **70**, (1965) 5737–5751.
2. Anderson, K. A., and Lin, R. P., *Phys. Rev. Letters* **16**, (1966) 1121.
3. Lin, R. P., *Space Sci. Rev.* **16**, (1974) 189–256.
4. Gold, R. E., Krimigis, S. M., Hawkins III, S. E., Haggerty, D. K., Lohr, D. A., Fiore, E., Armstrong, T. P., Holland, G., and Lanzerotti, L. J., *Space Sci. Rev.* **86**, (1998) 541–562.
5. Krucker, S., Larson, D. E., Lin, R. P., and Thompson, B. J., *Astrophys. J.* **519**, (1999) 864–875.

Heavy Ions in ^{3}He Enhanced Solar Energetic Particle Events

G.C. Ho[1], E.C. Roelof[1], G.M. Mason[2,3], R.E. Gold[1], S.M. Krimigis[1], J.R. Dwyer[2]

[1]*Applied Physics Laboratory, Johns Hopkins University, Laurel, MD 20723*
[2]*Department of Physics, University of Maryland, College Park, MD 20742*
[3]*Institute for Physical Science and Technology, University of Maryland, MD 20742*

Abstract. Prior to the launch of ACE, ^{3}He/^{4}He ratios in ^{3}He-rich solar energetic particle (SEP) events were typically reported at 10% or higher only from above a few MeV per nucleon because of limited instrument capabilities. However, with the new ULEIS instrument on ACE, we are able to explore the ^{3}He/^{4}He ratio below the 10% level and also at energy less than 1 MeV per nucleon. We have investigated a total of 45 ^{3}He-rich (^{3}He/^{4}He > 0.01 at 0.23 - 0.45 MeV per nucleon) SEP events from November 1997 to May 1999. We found the event average ^{3}He/^{4}He and Fe/C ratios are clearly correlated. This correlation suggests that the ^{3}He and Fe enrichments could be related and places further constraints on any ^{3}He-enhancement models.

INTRODUCTION

^{3}He-rich solar energetic particle (SEP) events represent a much more dramatic class of isotopic enhancements than those that are found in the solar wind. Average solar wind ^{3}He/^{4}He ratios were found to be 5×10^{-4} [1], but on rare occasions the ratio could be as high as ^{3}He/^{4}He = 7×10^{-3} within coronal mass ejections [2]. However, as was found first by Hsieh and Simpson [3], in certain impulsive SEP events the ^{3}He/^{4}He ratio can be 3-4 orders of magnitude higher than the solar wind value [4,5].

Following the discovery of these ^{3}He-rich SEP events, theories were developed to explore the underlying mechanism for the selective enhancement of the isotope [6,7,8]. Most theories involved some form of wave-particle resonant interaction with ^{3}He. Often, during ^{3}He-rich SEP events, the Fe/C ratio was also enhanced relative to its coronal value [4]. However, at least at energies greater than 1 MeV/nucleon, the Fe/C enhancements were found to be uncorrelated with the ^{3}He/^{4}He enhancements. Hence a separate wave mode was proposed that would preferentially accelerate the Fe/C independent of enhancing the ^{3}He/^{4}He [8]. In this paper, we investigate the Fe/C enhancement within ^{3}He-rich SEP events using new data from the Ultra-Low Energy Isotope Spectrometer (ULEIS) instrument on board the Advanced Composition Explorer (ACE) spacecraft.

INSTRUMENT DESCRIPTION

The ULEIS instrument is a high-resolution mass spectrometer that measures both elemental and isotopic ion compositions from 50 keV/nucleon to a few MeV/nucleon. It uses a relatively long time-of-flight path (50 cm) and position sensing micro-channel plates to achieve high mass resolution ($\sigma_m \sim 0.2$ amu). The overall instrument geometric factor is about 1.3 cm^2-sr. Mason *et al.* [9] describe the instrument in full detail.

OBSERVATIONS

The ^{3}He-rich events were selected using the ULIES pulse-height-analysis (PHA) data from November 1997 to May 1999. The high-mass resolution and relatively large geometric factor of ULEIS allow us to separate the helium isotopes easily. Figure 1 shows the mass histograms of 0.2 - 2.0 MeV/nucleon helium isotopes for two selected events during this study. A

CP528, *Acceleration and Transport of Energetic Particles Observed in the Heliosphere: ACE 2000 Symposium*, edited by Richard A. Mewaldt, et al.
© 2000 American Institute of Physics 1-56396-951-3/00/$17.00

total of 45 ^{3}He-rich events were detected during this 19-month period.

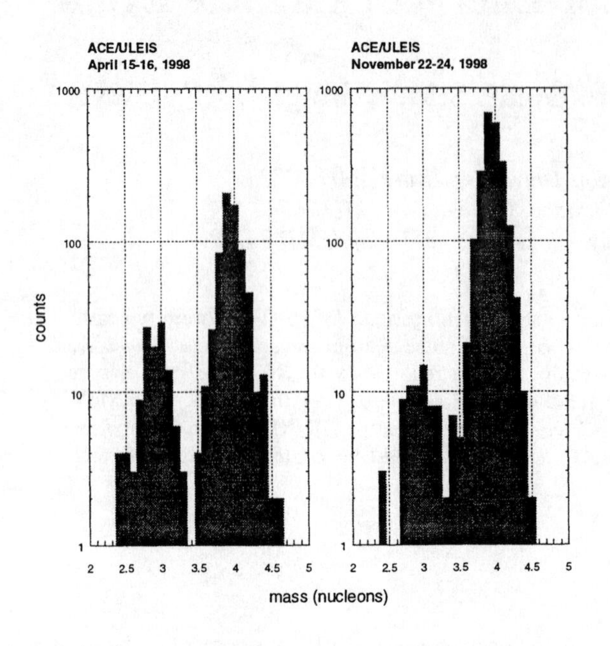

FIGURE 1. Mass histograms of 0.23 - 0.45 MeV/nucleon ACE/ULEIS data for the helium isotopes in two selected ^{3}He/^{4}He-rich SEP events.

Iron Enhancement

Figure 2 is a scatter plot of the event average ^{3}He/^{4}He (0.23 - 0.45 MeV/nucleon) versus Fe/C ratios (0.27 - 0.45 MeV/nucleon) from those ^{3}He-rich SEP events that have uncertainties below 50%. This plot extends the analysis of Reames *et al.* [5] to lower energy range and also to lower ^{3}He/^{4}He ratios (^{3}He/^{4}He < 0.1). We fitted a power law function (not shown) to all events in Figure 2 and obtained a power law index of 0.48 with correlation coefficient of 0.59. A better fit (power index of 0.56 and correlation coefficient of 0.65) was obtained if we remove the two events that have high ^{3}He/^{4}He and low Fe/C ratios (shown in filled circles). However, if we limited the fit to only those events that have ^{3}He/^{4}He > 0.1 to compare with the Reames *et al.* [5] analysis, the power law index became 0.35 and correlation coefficient of only 0.31. We conclude that for our data set a correlation is found between the ^{3}He/^{4}He and Fe/C enhancement.

We also examined the relationship between the C/O and ^{3}He/^{4}He enhancements within the same set of events. In Figure 3, we plot the event average C/O versus ^{3}He/^{4}He from those ^{3}He-rich SEP that have

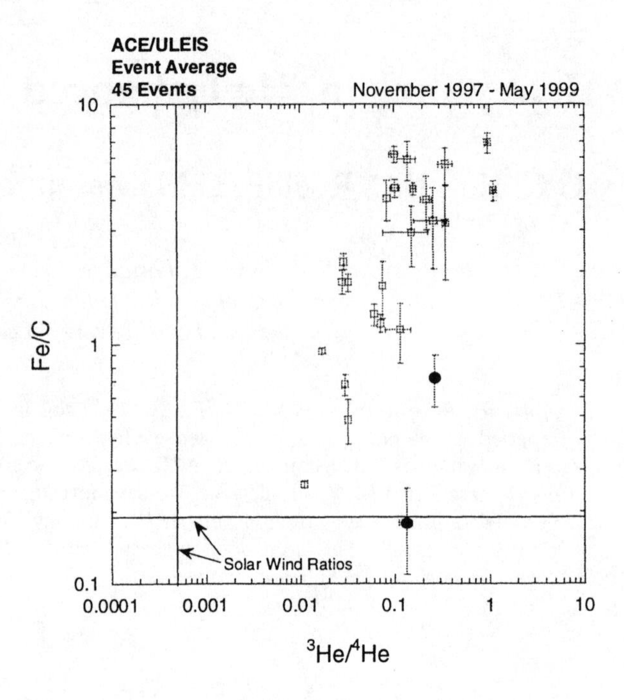

FIGURE 2. Event average Fe/C ratios as a function of the ^{3}He/^{4}He ratios. Only events that have ratio uncertainties below 50% are shown here. A positive correlation between the Fe and ^{3}He can be seen for events that have ^{3}He/^{4}He < 0.1

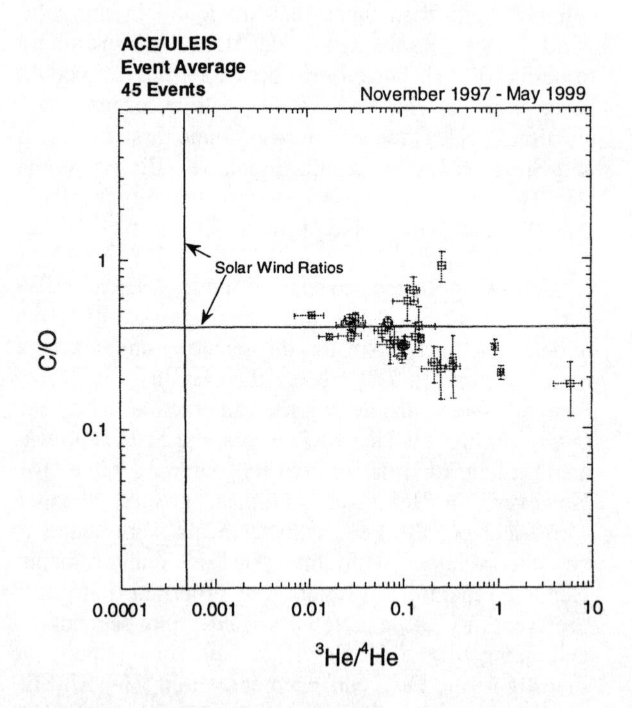

FIGURE 3. Event average C/O ratios as a function of the ^{3}He/^{4}He ratios. The C/O ratio is taking at 0.27 - 0.45 MeV/nucleon while ^{3}He/^{4}He is at 0.23 - 0.45 MeV/nucleon. No correlation is apparent in this plot.

both C/O and $^3He/^4He$ uncertainties below 50%. Although this study extends the C/O ratios to lower energies and lower $^3He/^4He$ then previous studies [4], no apparent correlation between the C/O and $^3He/^4He$ can be found in these ^3He-rich events.

Temporal Variation

The temporal variation of the $^3He/^4He$ within one single SEP event gives us clue on their injection, propagation and acceleration mechanism. However, due to limited instrument capabilities, the $^3He/^4He$ temporal variation within a single event was reported only in few cases before the launch of ACE [10]. We studied the temporal evolution of the $^3He/^4He$ ratio in our events and found that mostly it is constant throughout, indicating a common injection and propagation of the two helium isotopes. However, in some of the events, the ratios vary due to possible different injection and/or propagation history. In Figure 4, we plot the hourly average intensities profile of ^{12}C, ^3He (2-hour average) and ^4He from 0.27 to 0.45 MeV/nucleon for day of year (DOY) 197 to 202 of 1998. Using only the ^{12}C and ^4He intensities, one single SEP event is observed between DOY 197 to 202.

However, on closer examination, a small ^3He-rich injection is observed embedded inside this large SEP event. In Figure 5, each data point represents 2-hour averages of the ^3He and ^4He intensity during the embedded ^3He-rich SEP event from DOY 199 to 201. If the ratio of the $^3He/^4He$ were constant through out this smaller event, the points would more or less form a line with constant slope unity in the log-log plot. During the early part of the event, the $^3He/^4He$ ratios were constant for 14 hours during onset (the $^3He/^4He$ had a constant ratio = 0.3). However, during the latter portion of the event, the ratios changed abruptly and decreased steadily to lower values (0.3 - 0.05). The change of ratio probably indicates there were new injection of ^4He while the ^3He took on the decay profile from the smaller ^3He-rich SEP. This event illustrates the complexity of identifying separate ^3He-rich SEP event during active time periods.

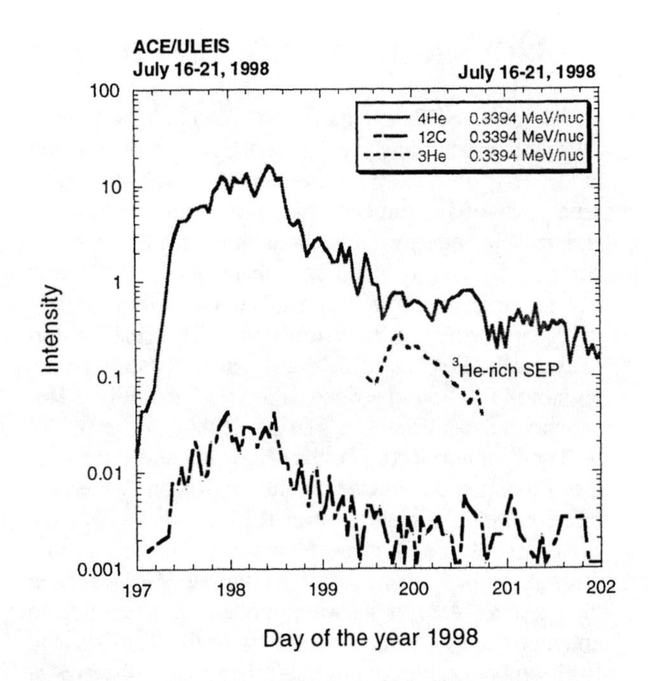

FIGURE 4. The time-intensity profiles of ^3He, ^4He and ^{12}C (0.27 - 0.45 MeV/nuc) from day of year 197 to 202 of 1998. A small ^3He-rich SEP is embedded within a larger SEP events.

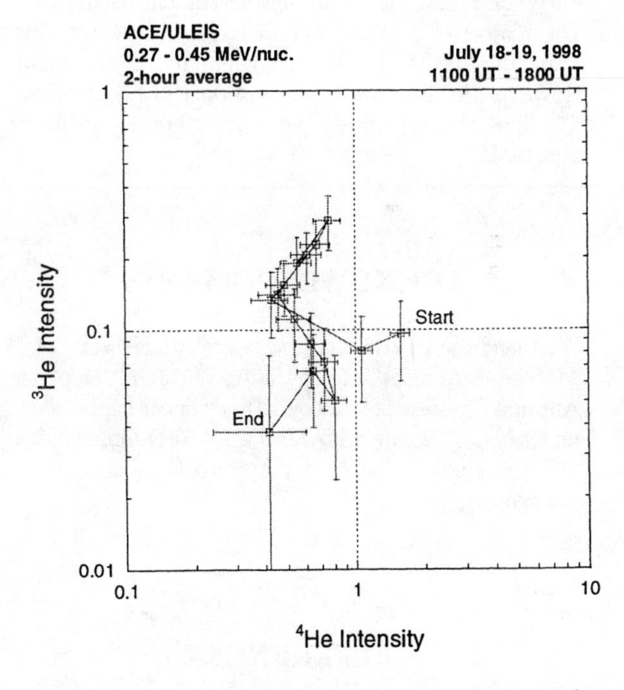

FIGURE 5. Temporal variations of the ^3He and ^4He intensities in one ^3He-rich SEP event. Each data point is a 2-hour average. A constant slope of unity in the data represents a constant $^3He/^4He$ ratio.

DISCUSSION AND CONCLUSIONS

^{3}He-rich SEP events have been studied with different instruments ever since they were discovered in the early 1970s. With the launch of ACE, we have a new set of instruments that have higher sensitivity and broader energy range than those available to us before. The corresponding enhancement of Fe and ^{3}He is surprisingly well correlated when the ^{3}He/^{4}He ratios are extended to below 0.1. The relationship between the Fe/C and ^{3}He/^{4}He in our study showed the possible relation between the Fe and the ^{3}He-enhancement mechanism. In particular, in the model by Temerin and Roth [7], the ^{3}He is enhanced through the fundamental mode of the hydrogen generated electromagnetic ion cyclotron (EMIC) waves. In the same model, Fe is also enhanced at the same time through higher harmonics of the same EMIC waves in the corona. A separate wave mode is not required to enhance the Fe. Hence, according to the Temerin and Roth model, it is not surprising to observe a correlation between the Fe/C and ^{3}He/^{4}He enhancement in our data, at least at lower energies.

The temporal evolution of the ^{3}He/^{4}He ratio in ^{3}He-rich events has not been studied extensively before at lower energies due to limited instrument capabilities. The majority of the events we examined so far have constant ^{3}He/^{4}He ratios throughout the entire event. However, in few cases (one is shown in Figure 4 and 5), the ^{3}He/^{4}He ratios can vary due to different injections.

ACKNOWLEDGMENTS

We thank the ULEIS team members at University of Maryland, Aerospace Corporation, and Johns Hopkins Applied Physics Laboratory. This work is supported under NASA contract NAS5-97271, Task Order 009 to Johns Hopkins University Applied Physics Laboratory.

REFERENCES

1. Geiss, J., Bler, F., Cerutti, H., Eberhardt, P., and Filleux, C., "Solar Wind Composition Experiment", Apollo 16 Prelim. Sci. Rep., *NASA SP-315*, Nat. Aeronaut. And Space Admin. (1972).

2. Ho, G.C., Hamilton, D.C., Gloeckler, G., and Bochsler, P., *Geophys. Res. Lett.,* **27**, 309-312 (2000).

3. Hsieh, K.C., and Simpson, J.A., *Astrophys., J. (Letters),* **162**, L191-L196 (1970).

4. Mason, G.M., Reames, D.V., Kleckler, B., Hovestadt, D., and von Rosenvinge, T.T., *Astrophys., J.,* **303**, 849-860 (1986).

5. Reames, D.V., Meyer, J.P., and von Rosenvinge, T.T., *Astrophys., J.,* **412**, 386-400 (1993).

6. Fisk, L.A., *Astrophys., J.,* **224**, 1048-1055 (1978).

7. Temerin, M., and Roth, I., *Astrophys., J.,* **391**, L105-L108 (1992).

8. Miller, J.A., and Viña, A.F, *Astrophys., J.,* **412**, 386-400 (1993).

9. Mason, G.M., Gold, R.E., Krimigis, S.M., Mazur, J.E., Andrews, G.B., Daley, K.A., Dwyer, J.R., Heuerman, K.F., James, T.L., Kennedy, M.J., Lefevere, T., Malcolm, H., Tossman, B., and Walpole P.H., *Space Sci. Rev.,* **86**, 409-448 (1998).

10. Möbius, E., Hovestadt, and Kleckler, B., *Astrophy., J.,* **238**, 768-779 (1980).

Measurements of Heavy Elements in ^{3}He-rich SEP Events

P.L. Slocum[1], M.E. Wiedenbeck[1], C.M.S. Cohen[3], E.R. Christian[2], A.C. Cummings[3], R.A. Leske[3], R.A. Mewaldt[3], E.C. Stone[3], and T.T. von Rosenvinge[2]

[1]*Jet Propulsion Laboratory, Pasadena, CA 91109*
[2]*NASA/Goddard Space Flight Center, Greenbelt, MD 20771*
[3]*California Institute of Technology, Pasadena, CA 91125*

Abstract. Using the Solar Isotope Spectrometer (SIS) on the Advanced Composition Explor
the properties of a selection of small ^{3}He-rich solar energetic particle (SEP) events with heavy
energy range ~11-22 MeV/nucleon. These events contain significantly increased ^{3}He/^{4}He r
value of 0.0004 in the energy range ~4.5-5.5 MeV/nucleon. In order to characterize the even
have been investigated. First, the heavy element content has been measured and compared to
of impulsive SEP events. Next, the simultaneous 38-53 keV electron flux, measured with t
Alpha Monitor (EPAM) on ACE, has been examined for possible activity near the ^{3}He-rich eve
list of measured solar X-ray flares, with corresponding H-alpha flares where possible, has been
correlations with these events. The results show an apparent correlation between event onset an
and a possible association with X-ray flares.

INTRODUCTION

The heavy element content of ^{3}He-rich SEP events may provide insight into the nature and origin of this type of occurrence. ^{3}He-rich SEP events were first observed in 1970 by Hsieh and Simpson (1), and their related heavy element enhancements were first noted by Hurford et al. (2). Since 1970, enhancements of ^{3}He have come to be associated with "impulsive" SEP events, one of the two main classifications of SEP events (impulsive and gradual). Impulsive events, which usually have a duration on the order of several hours, typically contain ^{3}He/^{4}He ratios which are 2-3 orders of magnitude greater than that of the solar wind. In addition, the Fe content of impulsive events is normally ~10 times that of the solar wind (3, 4, 5), while the Ne, Mg, and Si abundances are enhanced by a factor of ~3 (4, 5). Impulsive events have been associated with ~keV electron emission from the Sun (6), as well as with X-ray flares.

Gradual events, on the other hand, have been thought to contain material which reflects closely the composition of the solar wind and corona. However, recent measurements of the composition of gradual events by Mason et al. and Cohen et al. (3, 7) have revealed deviations from coronal abundances toward the impulsive event characteristics described above. Because of these measurements, one can no longer assume that gradual events do not have ^{3}He enhancements. This complicates the distinction between impulsive and gradual events. The material in gradual events is understood to have been accelerated out of the interplanetary medium by shocks which are driven by coronal mass ejections. Gradual events, unlike impulsive events, typically have a duration of days.

There are several observational differences between impulsive and gradual SEP events with respect to their composition and accompanying electromagnetic emissions. By examining the heavy element content and coincident solar activity of small ^{3}He-rich SEP events, it is possible to gain information relevant to the classification of these events and to the origin of the ejected material.

DATA ANALYSIS AND DISCUSSION

Using the Solar Isotope Spectrometer (SIS) (8), a set of eight small ^{3}He-rich SEP events with heavy element enhancements which occurred between September 6, 1998 and July 12, 1999 have been selected. The events have been defined by the following criteria: (1) The ^{3}He/^{4}He ratio is $\geq$4% in the energy range ~4.5-5.5 MeV/nucleon. (2) The heavy element ($Z\geq10$) flux in the energy interval ~11-22 MeV/nucleon is greater than that of the solar quiet time background. (3) Event onset times are chosen to coincide with a rise in the ^{3}He flux. (4) Event stop times are defined by the return of the heavy element fluxes to background levels, determined by inspection. All eight of the events are shown in Figure 1, including the heavy element counts, ^{3}He flux, coinci-

CP528, *Acceleration and Transport of Energetic Particles Observed in the Heliosphere: ACE 2000 Symposium*,
edited by Richard A. Mewaldt, et al.
© 2000 American Institute of Physics 1-56396-951-3/00/$17.00

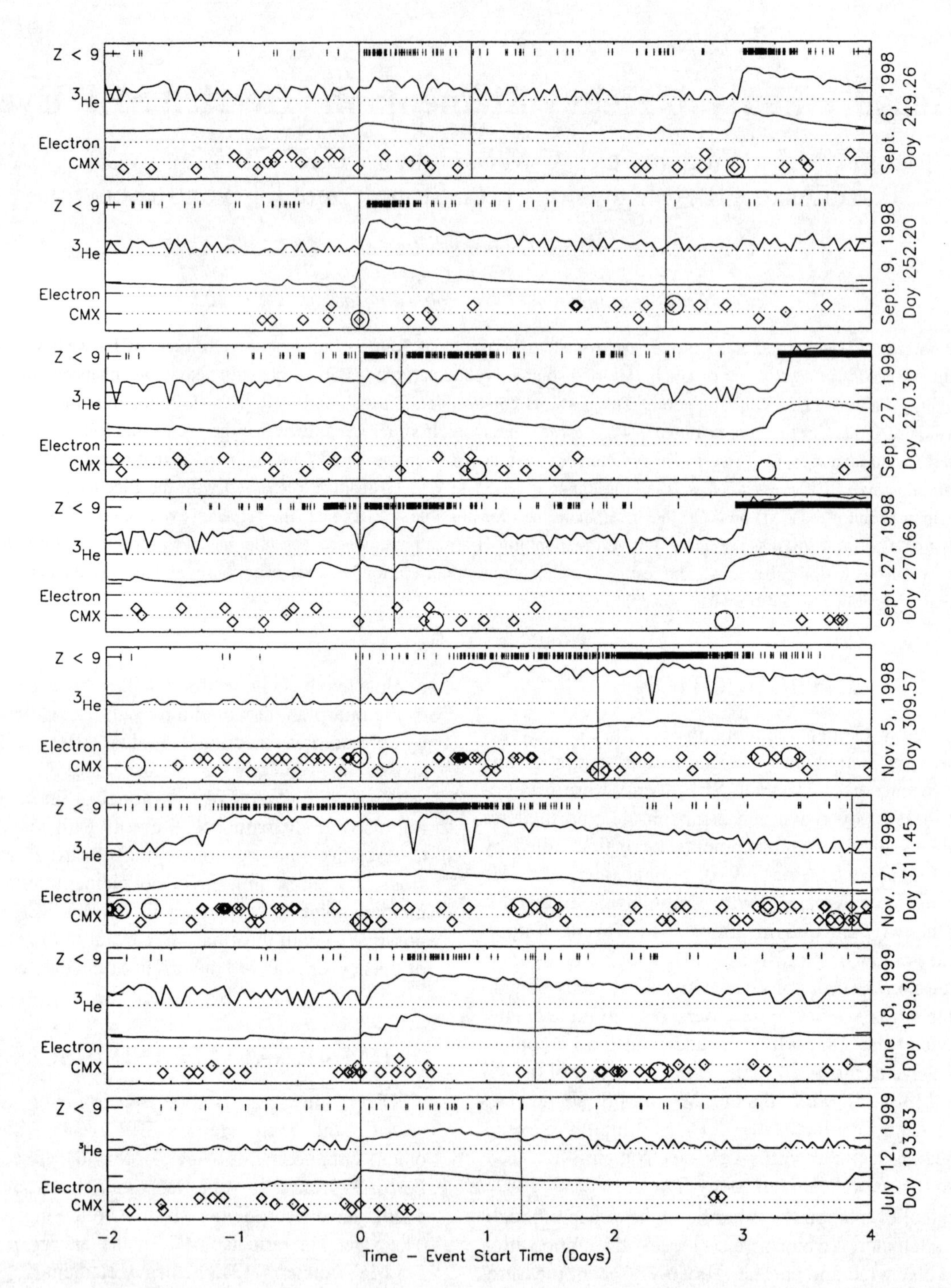

FIGURE 1. The eight events selected for this study. Each panel in the figure represents one event. The event start and stop times are denoted by the two vertical bars. The date of the event onset is noted at the far right side of the panel. The top row of ticks represents the arrival times of heavy nuclei with energy ~11-22 MeV/nucleon. The ^{3}He trace plot represents the ^{3}He flux measured using SIS for energy ~4.5-5.5 MeV/nucleon, plotted on a log scale where the first bin on the vertical axis equals one decade. The 38-53 keV electron flux, also shown on a log scale with a one decade indication, was measured using EPAM on ACE. The "CMX" plot represents the incidence of solar X-ray flares which are of C-Class or greater, as measured with the GOES-8 satellite. The diamond symbols denote C-Class flares, and circles M-Class. No X-Class flares were detected during these time periods. X-ray flares which could be associated with an H-alpha flare from the western (eastern) hemisphere of the Sun are shown above (on) the dotted line, while those which could not be associated with an H-alpha flare are plotted below the dotted line.

dent X-ray flares (C, M, and X-Class) and the 38-53 keV electron flux which has been provided by the Electron, Proton, and Alpha Monitor (EPAM) on ACE. The list of X-ray flares has been extracted from a GOES satellite database (*http://www.sel.noaa.gov/index.html*) at the NOAA Space Environment Center. Where possible, the X-ray flares have been associated with an H-alpha flare originating from a known location. The X-ray events in Figure 1 are therefore separated according to east, west, and unknown longitudinal origin on the Sun.

All eight of the events in Figure 1 have C or M-Class X-ray flares $\leq$ 2 hours before the onset time. If one assumes that the X-rays and energetic nuclei associated with a SEP event depart from the Sun simultaneously, then the minimum difference in transit times for X-rays and 5.5 MeV/nucleon ^{3}He over 1 AU is $\sim$1.2 hours. However, such an assumption is not clearly justified and could be a significant source of uncertainty in calculating expected X-ray flare onset times. Furthermore, because the eight events in this study are relatively small it is frequently difficult to determine their onset times to within $\lesssim$1 hour. Due to these two sources of uncertainty, and because of the abundance of X-ray flares shown in Figure 1, it is clearly difficult to deduce a unique association between the X-ray emissions and the ^{3}He-rich events. Therefore it is possible, but not definite, that these eight events are associated with X-ray flares.

Also shown in Figure 1 are the coincident electron fluxes, measured using EPAM. All of the events show increases in the electron flux within $\lesssim$ 2 hours of the event onset time. The average difference in transit times for 38-53 keV electrons and 5.5 MeV/nucleon ^{3}He over a distance of 1 AU is $\sim$1.1 hours. However, for the reasons stated above, the expected onset time of the electron flux relative to the ^{3}He flux is not well determined, and should not be used as a rigid criterion for correlation. Instead the apparent association can be seen by inspecting the electron fluxes at or very near the event onset times in Figure 1.

To classify these eight events further, the heavy element content has been investigated, as shown in Figure 2. In the figure, 17 average abundance ratio enhancements are plotted for the eight SEP events studied here, and are compared with those found by Reames et al. (4) in a study of 228 impulsive SEP events with the ISEE-3 spacecraft. The abundance ratio enhancements are defined relative to coronal values (9). From the figure it is clear that similar abundance enhancements are present in the two data sets. In this comparison it should be noted that the energy interval measured using ISEE-3 is 1.9-2.8 MeV/nucleon, somewhat lower than the $\sim$11-22 MeV/nuc in this study. In addition, while there have been many ^{3}He-rich events measured by SIS, most do not have large enhancements of heavy elements (Z$\geq$6) in the energy interval $\sim$11-

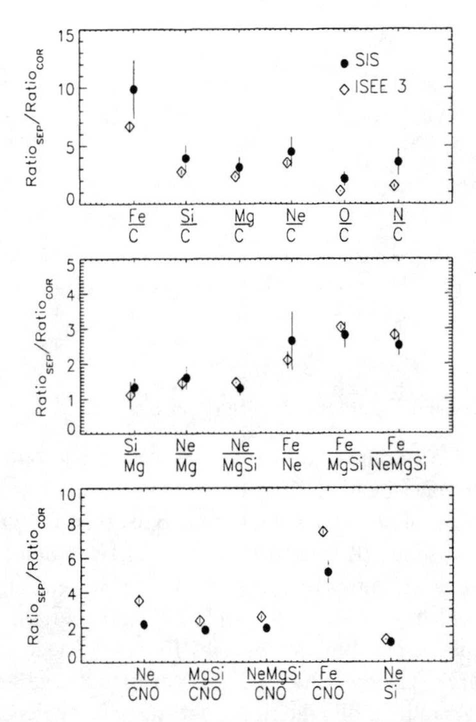

FIGURE 2. Average heavy element abundance ratios, normalized to coronal values, for the eight SEP events occurring between September 6, 1998 and July 12, 1999. Uncertainties in the ratios are statistical in nature, and are shown with one standard deviation.

22 MeV/nucleon. Consequently the ISEE-3 impulsive event sample is much larger than that of SIS.

The background contribution to the heavy element fluxes has been estimated using selected days of solar quiet time from three consecutive time periods. Between May 25, 1998 and November 13, 1998, 59 days of quiet time have been selected. From November 14, 1998 through July 22, 1999 there were 133 days of quiet time, and from July 23, 1999 through October 31, 1999 there were 60 days. Quiet times are defined by imposing a threshold on the He and O fluxes in the lowest three energy ranges of SIS. The thresholds are chosen such that the average O flux is less than 2×10^{-5} particles/(cm^{2}Sr sec MeV/nucleon) for the energy interval 7.1-15.6 MeV/nucleon, and the average He flux is less than 1×10^{-4} particles/(cm^{2}Sr sec MeV/nucleon) for 3.4-7.3 MeV/nucleon. This definition of solar quiet time is driven primarily by the He flux.

Also of interest in observing ^{3}He-rich SEP events is the possibility of a correlation between ^{3}He/^{4}He ratios and heavy element abundances. Such a correlation could place important restrictions on acceleration models for this type of event. Recent results by Ho et al. (10) report a correlation between the Fe/C and ^{3}He/^{4}He ratios, using data from the ULEIS instrument on ACE. Past investigations (11, 4) have not found a correlation between heavy

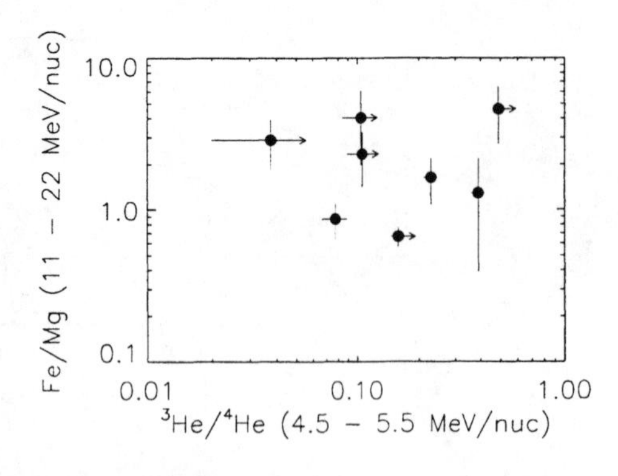

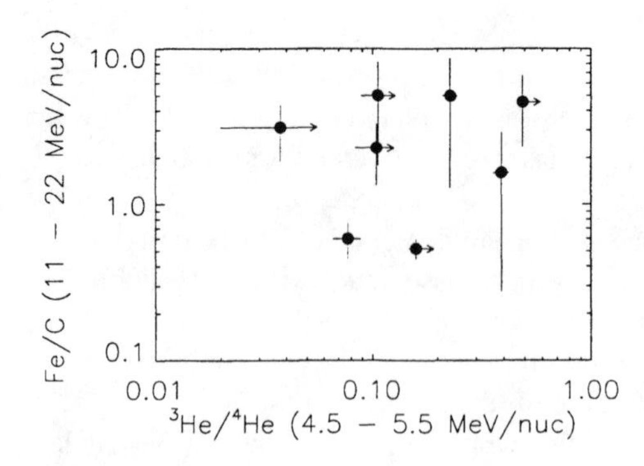

FIGURE 3. Plot of Fe/Mg (left panel) and Fe/C (right panel) against the ^{3}He/^{4}He ratio, for the eight events in this study.

element abundance ratios and ^{3}He/^{4}He ratio, but these studies have been limited to ^{3}He/^{4}He ratios $\gtrsim 0.1$ and have been at somewhat higher energies than the ULEIS measurements. The results by Ho et al. included events with ^{3}He/^{4}He ratios as low as 0.01. In a search for similar correlations using SIS, two heavy element ratios have been plotted as a function of ^{3}He/^{4}He ratio. The left panel of Figure 3 shows the Fe/Mg ratio plotted against the ^{3}He/^{4}He ratio, while the right panel depicts the Fe/C ratio as a function of ^{3}He/^{4}He ratio. In the Fe/Mg plot, there does not appear to be a correlation with the ^{3}He/^{4}He ratio. Similarly, the Fe/C plot does not show a correlation with the ^{3}He/^{4}He ratio, although this is difficult to determine without extending the ^{3}He/^{4}He measurements lower than the 4% limit possible with the SIS instrument in this energy range.

Measurement of the ^{3}He/^{4}He ratio is limited by several contributing factors. As mentioned above, the SIS instrument is not sensitive to ratios lower than $4\pm1\%$ in the energy range 4.5-5.5 MeV/nucleon, due to spillover contamination to the ^{3}He peak from the ^{4}He peak. A correction for this contribution to the ^{3}He/^{4}He ratio has been applied to the data shown in Figure 3. It should also be noted that no uncertainty has been added to account for contributions to the ^{4}He from neighboring SEP events. Therefore, the ^{3}He/^{4}He ratios in Figure 3 should be interpreted as lower limits where indicated.

SUMMARY

The heavy element content of these eight ^{3}He-rich events has been examined. The enhancements of Fe, Mg, Ne, and Si, which are standard indications as to the nature of the material which is present, are consistent within statistical and experimental limitations with the previous study of 228 impulsive SEP events by Reames et al. (4)

There does appear to be an association between electron emission and the onset of these events, which suggests that the events may be impulsive. This is also supported by the presence of X-ray flares at the event onsets, although the frequent occurrence of small X-ray events hinders any direct correlation. To combat this difficulty, future work may include studies of heavy ion velocity dispersion as a tool to extract the expected X-ray flare onset times. The combination of heavy element content, electron flux association and X-ray emission suggests that these eight events may be impulsive SEP events as defined by previous studies.

ACKNOWLEDGMENTS

This research was supported by NASA at the California Institute of Technology (under grant NAG5-6912), the Jet Propulsion Laboratory, and the Goddard Space Flight Center. We thank the EPAM science team for providing the electron data used in the first figure.

REFERENCES

1. Hsieh, K.C., Simpson, J.A., *Astrophys J. Letters*, **162**, L191-L196 (1970).
2. Hurford, G.J. et al., *Astrophys J. Letters*, **201**, L95-L97 (1975).
3. Mason, G.M. et al., *Astrophys J. Letters* **525**, L133-L136 (1999).
4. Reames, D.V. et al., *Astrophys J. Suppl.* **90**, 649-667 (1994).
5. Slocum, P.L. et al., *Proc. 26th Intern. Cosmic Ray Conf.* (Salt Lake City) **6**, 107-110 (1999).
6. Reames, D.V. et al., *Astrophys J.* **292**, 716-724 (1985).
7. Cohen, C.M.S. et al. "The Isotopic Composition of Solar Energetic Particles" in *ACE 2000 Symposium*, edited by R.A. Mewaldt et al., AIP Conference Proceedings this volume, New York: American Institute of Physics, 2000.
8. Stone, E.C. et al., *Space Sci. Rev.* **86**, 357-408 (1998).
9. Reames, D.V., *Space Sci. Rev.* **85**, 327-340 (1998).
10. Ho, G.C. et al. "Heavy Ions in ^{3}He Enhanced Solar Energetic Particle Events" in *ACE 2000 Symposium*, edited by R.A. Mewaldt et al., AIP Conference Proceedings this volume, New York: American Institute of Physics, 2000.
11. Mason, G.M. et al., *Astrophys J.* **425**, 843-848 (1994).

Enhanced Abundances of ^{3}He in Large Solar Energetic Particle Events

M. E. Wiedenbeck[1], E. R. Christian[2], C. M. S. Cohen[3], A. C. Cummings[3], R. A. Leske[3], R. A. Mewaldt[3], P. L. Slocum[1], E. C. Stone[3], and T. T. von Rosenvinge[2]

[1] *Jet Propulsion Laboratory, Pasadena, CA 91109, USA*
[2] *NASA/Goddard Space Flight Center, Greenbelt, MD 20771, USA*
[3] *California Institute of Technology, Pasadena, CA 91125, USA*

Abstract. Observations of a number of relatively large solar energetic particle (SEP) events that have occurred since the launch of ACE in August 1997 have shown that the ratio of ^{3}He/^{4}He can be enhanced over the solar wind value ($\sim 4\times10^{-4}$) by more than an order of magnitude in such events. Since particle acceleration in these "gradual" SEP events is thought to be caused by CME-driven shocks traveling through the solar corona and interplanetary medium, a source of ^{3}He in addition to the solar wind appears required to provide the seed material. Using data from the Solar Isotope Spectrometer on ACE, we have carried out a more detailed investigation of the characteristics of the ^{3}He enhancements at energies $>$ 5 MeV/nucleon in three large SEP events (4 Nov 1997, 6 May 1998, and 14 Nov 1998). We find that the ^{3}He/^{4}He ratios are essentially time-independent during the events, that the ^{3}He energy spectra are markedly harder than those commonly observed in impulsive events, and that the spectra of ^{3}He may be harder than those of ^{4}He.

INTRODUCTION

Particle acceleration in solar energetic particle (SEP) events is thought to be caused by two distinct mechanisms (1). First, the energy released by magnetic reconnection can cause acceleration at the site of a flare on the Sun, although the details of the mechanism remain uncertain. Second, the passage of coronal mass ejections (CMEs) thorough the corona and interplanetary space can drive a shock which will accelerate particles from the medium it is traversing.

Extensive efforts have been made to establish sets of observable characteristics of SEP events that can be used to distinguish which acceleration mechanism was operative in a particular event (2). Events classified as "impulsive" or "gradual" (based on the duration of the x-ray emission) are thought to be associated with acceleration by flares and CME-driven shocks, respectively. It is generally found that the SEP events with the highest particle intensities near Earth are of the gradual type.

Compositional signatures have also been extensively studied. In gradual events, heavy element abundances are generally consistent with coronal (or solar wind) values, with relatively modest fractionation which can be correlated with the mass-to-charge ratio (M/Q) of the ions (3). In impulsive events, more extreme deviations from coronal abundances are observed. Most notably the ^{3}He/^{4}He isotope ratio can exceed the solar wind value of $\sim 4\times10^{-4}$ (4) by as much as four orders of magnitude, and heavy elements exhibit a pattern of generally increasing enhancements with increasing atomic number, Z, with Fe/O reaching values $\sim 10\times$ the coronal value of 0.13 inferred from gradual SEP events (1). The extreme enhancements of ^{3}He are thought to result from selective heating by some resonant process acting on the pre-flare material. The degree of correlation between ^{3}He and heavy ion enhancements and the conditions under which such correlations occur are an active subject of debate and require further clarification.

In the prevailing view of the origin of SEP events one would expect very small values of the ^{3}He/^{4}He ratio in large, gradual events. Prior to the launch of the Advanced Composition Explorer (ACE) there were reports of a few large events with appreciable enhancements of this ratio and it had been suggested that such events could be of a "mixed" character containing both shock-accelerated material and flare-associated particles (see (2) and references therein). However, earlier instruments lacked the combination of sensitivity and resolution needed to detect ^{3}He in gradual events at levels much below $\sim$ 10% of ^{4}He under normal conditions.

Isotope spectrometers on ACE are able to resolve ^{3}He

CP528, *Acceleration and Transport of Energetic Particles Observed in the Heliosphere: ACE 2000 Symposium*, edited by Richard A. Mewaldt, et al.
© 2000 American Institute of Physics 1-56396-951-3/00/$17.00

at least down to levels of a few tenths of a percent of ^{4}He. At energies < 1 MeV/nuc, data from the Ultra-Low Energy Isotope Spectrometer (ULEIS) have shown the presence of significant ^{3}He enhancements in several gradual events (5, 6). At higher energies, $\gtrsim$9 MeV/nuc, observations with the Solar Isotope Spectrometer (SIS) indicate ^{3}He/^{4}He ratios exceeding $\sim 0.4\%$ in at least half of the 11 large SEP events that occurred between November 1997 and June 1999 (7, 8).

SIS OBSERVATIONS OF SEP ^{3}He

The SIS instrument (9) identifies the charge and mass of energetic nuclei using measurements of dE/dx, total energy, and trajectory in stacks of silicon solid-state detectors. For the present study where it is necessary to identify small fluxes of ^{3}He in the presence of significantly larger ^{4}He intensities, we consider particles which stop in the fourth, fifth, or sixth detectors in the stack (called ranges 2, 3, and 4). For each of these nuclei there are between 3 and 5 mass measurements which are required to be consistent to reduce backgrounds. Figure 1 shows He mass histograms for these three ranges from three of the large SEP events that have been studied (4 Nov 1997, 6 May 1998, and 14 Nov 1998). In each of these histograms there is a clearly resolved ^{3}He peak. In some of the other large events the SIS data do not show a distinct ^{3}He peak, but sensitivity is sometimes limited either by low statistics or residual spill-over from ^{4}He.

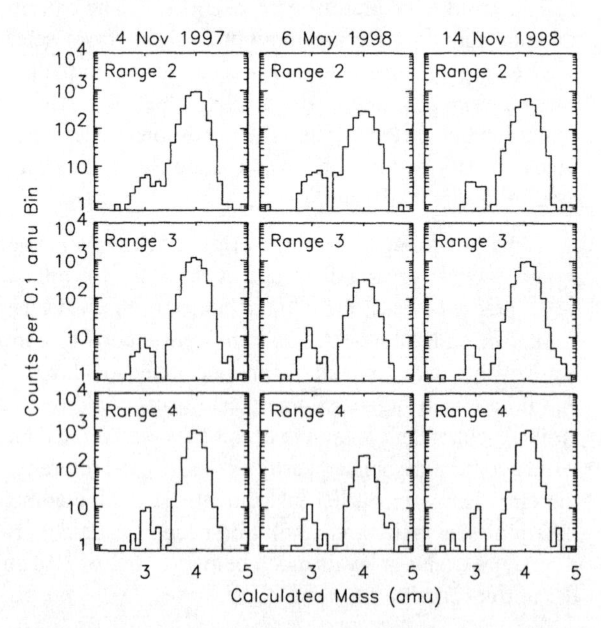

FIGURE 1. He mass histograms for SIS ranges 2 through 4 from three large SEP events. Energy intervals (MeV/nuc) for ^{4}He and ^{3}He in SIS are: Range 2, 6.5–7.6 and 7.7–9.0; Range 3, 7.7–10.1 and 9.1–11.9; Range 4, 10.2–13.8 and 12.0–16.3.

Figure 2 shows plots of ^{3}He and ^{4}He particle intensities vs. time for energies of 8.3 and 10.5 MeV/nucleon. For ^{3}He, these energies correspond to SIS ranges 2 and 3 (Fig. 1). The ^{4}He intensities were interpolated to the energies of the ^{3}He measurements.

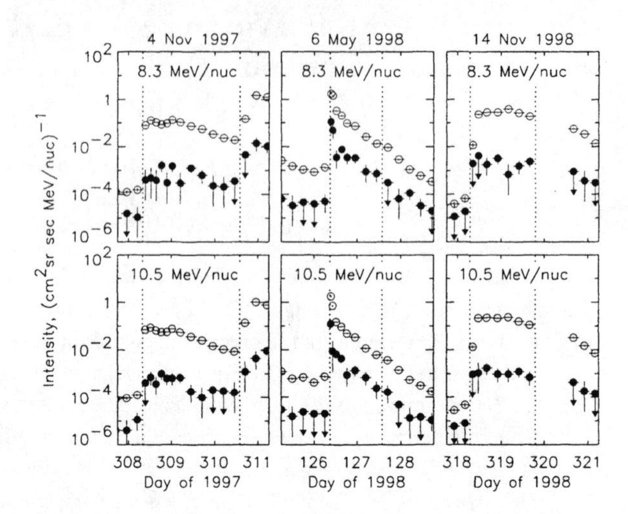

FIGURE 2. Time–intensity plots for He isotopes at 8.3 and 10.5 MeV/nucleon in three large SEP events. Filled points—^{3}He; open points—^{4}He. Dotted lines indicate start and stop times used for accumulating histograms in Fig. 1. A data gap occurred during the decline phase of the 14 Nov 1998 event.

The rise time of the particle intensities are short ($\lesssim$1–2 hr), as is commonly found when the Earth is magnetically well-connected to the acceleration site. These are consistent with the western-hemisphere locations which have been reported for the flares associated with these events (10). The intensity increases have relatively long durations (1–3 days) characteristic of gradual events, where particle acceleration continues as the shock propagates through the interplanetary medium. Shock associations have been reported for the events of 4 Nov 1997 (5) and 6 May 1998 (11). (The flare locations should also be the approximate locations from which CMEs driving the shocks were launched.)

In the 6 May 1998 event there is a brief "spike" in the intensities of ^{3}He and ^{4}He in the first hour or two of the event. A similar spike is also present for the heavy element fluxes in this event (10). The presence of this initial spike lasting less than 0.1 day is suggestive of a possible impulsive contribution to the event from flare-accelerated material. Temporal variations of abundances in this event are also discussed by Reames et al. (11) and von Rosenvinge et al. (12).

Since large enhancements of the ^{3}He/^{4}He ratio over the solar wind value of 4×10^{-4} are thought to be indicative of impulsive events, and impulsive event durations tend to be short (~hours), it is of interest to investigate whether this ratio varies over the course of the events we are investigating. In particular, does the intensity spike at the start of the 6 May 1998 event exhibit a distinctly greater enhancement of ^{3}He than the later, presumably gradual, phases of this event?

Figure 3 shows the time dependences of the ^{3}He/^{4}He values obtained from the ratios of the intensities shown in Figure 2. During the course of these events (delimited by the dotted vertical lines) no time variation of the ratio is evident, although the limited statistical accuracy of the ^{3}He measurements do not allow us to set very strong limits on this variation. We find that the ^{3}He enhancement in the 6 May 1998 event's initial spike is not appreciably greater than in the decay portion of the event.

The observation of time-independent ^{3}He/^{4}He ratios throughout these large events are similar to an example of time-independent helium isotopic composition reported by Mason et al. (6) at 0.7 MeV/nucleon in the event of 4 Jun 1999.

The average values of ^{3}He/^{4}He for the events studied here range from slightly less than 1% in the 14 Nov 1998 event to $\sim 5\%$ for the 6 May 1998 event. These values are all smaller than the 10% limit that traditionally has been used for identifying impulsive events on the basis of their ^{3}He/^{4}He enhancements. Nevertheless, they are 1 to 2 orders of magnitude greater than the solar wind ratio.

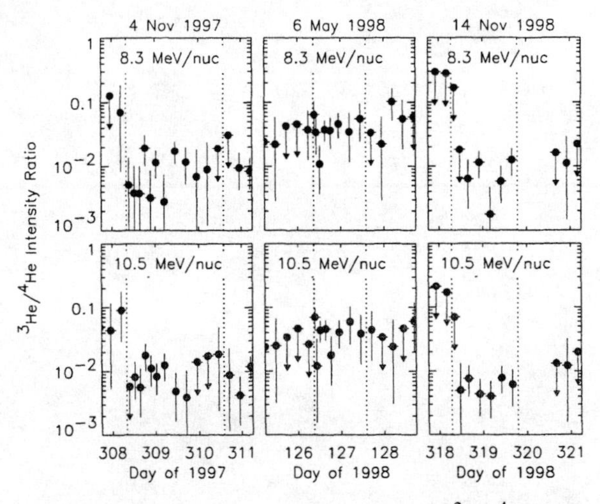

FIGURE 3. Time dependence of the measured ^{3}He/^{4}He intensity ratio at energies of 8.3 and 10.5 MeV/nucleon. Dotted lines indicate start and stop times used for the events. A data gap occurred during the decline phase of the 14 Nov 1998 event.

Figure 4 shows energy spectra for the helium isotopes obtained for the three large SEP events being investigated. The fluences shown are integrals of the intensities plotted in Figure 2 over the indicated event interval. (Note that the ^{3}He fluences have been multiplied by a factor of 50 to reduce the number of decades required for the plots.) The straight lines shown in Figure 4 result from power-law fits to the observed spectral points. The spectral indexes are summarized in Table 1. There is an apparent tendency for the spectra of ^{3}He to be somewhat harder than the corresponding spectra of ^{4}He, although the difference is greater than two standard deviations only in the 6 May 1998 event. Several authors have previously reported events with this characteristic (13, 14).

More significant is the comparison of the spectral indexes derived for ^{3}He in these large, presumably gradual, events with those found in events which have a clearly impulsive, ^{3}He-rich character. We have examined spectra of ^{3}He in ~ 40 such events observed between May 1998 and September 1999. Spectral indexes generally fell in the range -4.5 ± 1.5, with few instances of spectra as hard as those considered here (see also (13)).

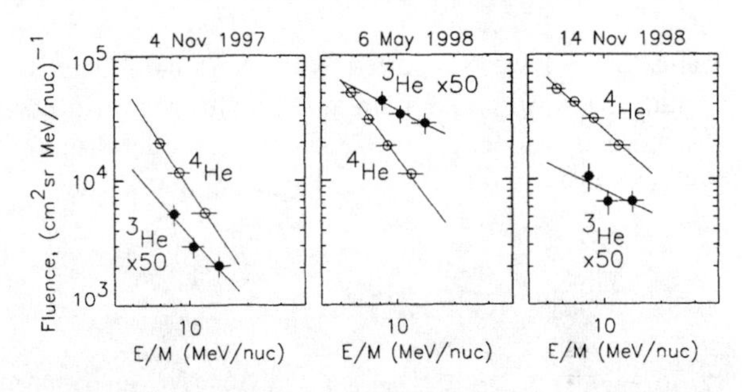

FIGURE 4. Energy spectra of ^{3}He and ^{4}He during the three large SEP events. Fluences are calculated as integrals over the time periods delimited by dotted lines in Fig. 2. Straight lines are least-squares fits of power laws to the observed spectra. Spectral slopes are summarized in Table 1.

Table 1. Slopes of Power-law Energy Spectra.

Event Start	Spectral Index [a] ^{3}He	^{4}He
4 Nov 1997	-1.77 ± 0.51	-2.41 ± 0.04
6 May 1998	-0.78 ± 0.50	-2.07 ± 0.05
14 Nov 1998	-0.72 ± 0.58	-1.40 ± 0.03

[a] Quoted uncertainties are 1σ statistical errors.

SUMMARY

The time and energy dependences of the abundances of ^{3}He and ^{4}He at energies > 5 MeV/nucleon have been examined in three large SEP events which were previously found to have significant enhancements of the ^{3}He/^{4}He ratio relative to its value in the solar wind. The isotope ratios show little, if any, variation throughout the course of the events, indicating that the ^{3}He enhancements are not the result of the occurrence of an impulsive, ^{3}He-rich event at the start of a larger gradual event with normal (i.e. solar-wind-like) helium isotopic composition. These observations do not rule out the possibility that the flare could be injecting material enriched in ^{3}He which is subsequently accelerated to MeV energies by the CME-driven shock (for further discussion see (12)).

The ^{3}He energy spectra are much harder than those found in typical, small, ^{3}He-rich events. In addition, there are indications that the spectra of ^{3}He may also be somewhat harder than the spectra of ^{4}He in the large events, although this result requires confirmation.

These observations further constrain models for the origin of ^{3}He abundance enhancements in SEP events and for the acceleration of particles in gradual events.

ACKNOWLEDGMENT

This research was supported by the National Aeronautics and Space Administration at the California Institute of Technology (under grant NAG5-6912), the Jet Propulsion Laboratory, and the Goddard Space Flight Center.

REFERENCES

1. Reames, D. V., *Sp. Sci. Rev.* **90**, 413 (1999).
2. Cliver, E. W., in *High Energy Solar Physics*, R. Ramaty et al. (eds.), AIP (Woodbury, NY), 1995, p. 45.
3. Breneman, H. and Stone, E. C., *Ap. J. Lett.* **299**, L57 (1985).
4. Gloeckler, G. and Geiss, J., *Sp. Sci. Rev.* **84**, 275 (1998).
5. Mason, G. M. et al., *Geophys. Res. Letters*, **26**, 141 (1999).
6. Mason, G. M. et al., *Ap. J. Lett.* **525**, 133 (1999).
7. Cohen, C. M. S et al., *Geophys. Res. Letters*, **26**, 2697 (1999).
8. Leske, R. A. et al., "Measurements of the Heavy-Ion Elemental and Isotopic Composition in Large Solar Particle Events from ACE", in *High Energy Solar Physics: Anticipating HESSI*, edited by R. Ramaty and N. Mandzhavidze, ASP Conf. Series, 2000, in press.
9. Stone, E. C., et al., *Sp. Sci. Rev.* **86**, 357 (1998).
10. von Rosenvinge, T. T., et al., *Proc. 26th Int. Cosmic Ray Conf.* (Salt Lake City) **6**, 131 (1999).
11. Reames, D. V., Ng, C. K, and Tylka, A. J., *Ap. J. Lett.* **531**, L83 (2000).
12. von Rosenvinge, T. T., et al., "The Solar Energetic Particle Event of 6 May 1998", in *Acceleration and Transport of Energetic Particles Observed in the Heliosphere*, edited by R. Mewaldt et al., New York, AIP Conf. Proceedings, 2000 (this volume).
13. Chen, J., Guzik, T. G., and Wefel, J. P., *Ap. J.* **442**, 875 (1995).
14. Mason, G. M., Dwyer, J. R., Mazur, J. E., Gold, R. E., and Krimigis, S. M., *Proc. 26th Int. Cosmic Ray Conf.* (Salt Lake City) **6**, 103 (1999).

The Solar Energetic Particle Event of 6 May 1998

T.T. von Rosenvinge[1], C.M.S. Cohen[2], E.R. Christian[1], A.C. Cummings[2], R.A. Leske[2], R.A. Mewaldt[2], P.L. Slocum[3], E.C. Stone[2], and M.E. Wiedenbeck[3]

[1]*NASA/Goddard Space Flight Center, Greenbelt, MD 20771*
[2]*California Institute of Technology, Pasadena, CA 91125*
[3]*Jet Propulsion Laboratory, Pasadena, CA 91109*

Abstract. The abundances of elements from helium to iron have been measured in more than a dozen moderate to large solar energetic particle (SEP) events using the Solar Isotope Spectrometer (SIS) on-board the Advanced Composition Explorer (ACE). Time variations within some of these events and from event to event have been reported previously. This paper presents an analysis of the event of 6 May 1998, for which relatively time-independent abundance ratios are found. This event has been considered to be an example of an impulsive event, a gradual event, and as a hybrid of the two. Difficulties with classifying this event are discussed.

INTRODUCTION

Well prior to the launch of the Advanced Composition Explorer (ACE) there was an established model of solar energetic particle (SEP) events which divided events into two classes, impulsive events and gradual events (1, 2). In addition, it had been proposed that some events are hybrid events which combine an initial impulsive event with a gradual event (3, 4). Since the launch of ACE there has been considerable debate as to whether certain events are gradual, impulsive, or hybrid events. In particular, the nature of the 6 May 1998 event has been extensively debated and will be the topic of this paper.

The characteristics of impulsive and gradual events have been summarized by Reames (1). In brief, impulsive events are events which are associated with acceleration by solar flares, whereas gradual events are associated with acceleration by shocks driven by Coronal Mass Ejections (CMEs). Starting in the corona, the strongest such shocks can continue to accelerate particles all the way out to 1 AU and beyond. Initially the terms impulsive and gradual referred to the durations of corresponding soft x-ray events, with 1 hour roughly marking the dividing line between the two. Subsequently the terms have evolved to refer to the durations of the particle events themselves, with impulsive events, at a few MeV/nuc, typically lasting for only hours, and gradual events lasting for several days (1). This of course leaves a wide window for events in between. In addition, no clear-cut criteria for measuring an event's duration have been established.

Impulsive events are typically ^{3}He-rich and have composition enhanced in heavy elements relative to coronal abundances. They are also associated with Type III radio bursts and low energy electrons. By contrast, gradual events are not ^{3}He-rich and are associated with Type II and Type IV radio emission. Their average abundances reflect coronal abundances (5). Mean Fe charge states were initially measured at ~1 MeV/nuc to be 20.5±1.2 in impulsive events and 14±1 in gradual events (6). Recent measurements have shown that 14 is perhaps an intermediate value (7), with some gradual events having a mean Fe charge state as low as 11 (7, 8). The relationship between these intermediate events and hybrid events, if any, has not been established. The charge states in gradual events are also, at least on occasion, energy dependent (9, 10, 11).

Due to the large scale of CMEs and their associated shocks, gradual event particles appear on magnetic field lines which connect back to the sun over a wide range of solar longitudes. By comparison, an observer must be well-connected magnetically to the flare site on the sun to observe an impulsive event. The observed range of good connection longitudes is from ~30 - 80 ° West (2).

Historically, events have been considered to be ^{3}He-rich when ^{3}He/^{4}He exceeded 10%, but this was largely due to the fact that early instruments could not reliably observe ^{3}He/^{4}He below about 10%. With improved measurement capabilities on ACE, it is apparent that many events have ^{3}He/^{4}He ratios which

CP528, *Acceleration and Transport of Energetic Particles Observed in the Heliosphere: ACE 2000 Symposium,* edited by Richard A. Mewaldt, et al.
© 2000 American Institute of Physics 1-56396-951-3/00/$17.00

are below 10% and yet are much enhanced over the average solar wind value of ~ 0.04%.

The Solar Isotope Spectrometer (SIS) on-board the Advanced Composition Explorer (ACE) has a large geometry factor (~ 40 cm^2-sr), enabling abundances in solar energetic particle (SEP) events to be observed on a time scale of hours or less. Substantial temporal variations of abundances from event to event and within events have been reported earlier using SIS data (12). For example, Fe/O varied from as low as ~ 0.1 times the average Fe/O value for gradual events to as high as ~ 8 times the average value in the events of 20 April 1998 and 6 November 1997, respectively. He/O varied from ~ 2 times average to ~ 0.1 times average within the 20 April 1998 event. (The average values for gradual events obtained from (5) are used as convenient reference levels for identifying enhancements and depletions). Ca/O enhancements are frequently similar to Fe/O enhancements. Si/O is rarely enhanced, although the event of 14 November 1998 had Si/O enhanced by ~ 2 times average. Time histories of the intensities and abundance ratios in these events are shown in (12).

Besides events with large temporal variations, some events have abundances which are essentially time-independent for all elements. The first four such events observed by SIS (6 November 1997, 2 May 1998, 6 May 1998, and 14 November 1998) all had abundances similar to the average abundances of impulsive events. These abundances are unlike the average abundances of gradual events (13, 12). However, other analyses (Tylka and Reames, private communication; also (14)) have led to the conclusion that all four of these events were gradual events. It has also been suggested that one of these events, the event of 6 May 1998, is a hybrid of an impulsive and a gradual event (15).

Ultimately it is essential to understand the temporal variations of solar energetic particle abundances in order to understand particle acceleration mechanisms and to reliably estimate the composition of the sun from direct observations of SEP abundances. To do so, each event needs to be separately understandable. The event studied here is sufficiently complex to challenge current thinking.

OBSERVATIONS

Solar observations for this period are given in NOAA's Solar-Geophysical Data Reports (16). There were two solar x-ray events, M class or greater, on 6 May 1998, both in a region nominally located at S11W65. The first of these, a class M2.9 event, started at 07:10, peaked at 07:25, and ended at 07:41. The second event, of class X2.7, started at 07:58, peaked at 08:09, and ended at 08:20. Type III emission started at 07:35 and continued intermittently to 08:13. Type II and IV radio emission started at 08:00 and at 08:03, respectively, presumably due to the second event. A CME was first observed in the LASCO C2 field of view (1.5-6 solar radii) at 08:04. The speed of the CME was measured to be 1053 km/sec (17). This CME is clearly associated with the second x-ray event and not the first. Type III emission is clearly associated with the first x-ray event and possibly with the second one as well.

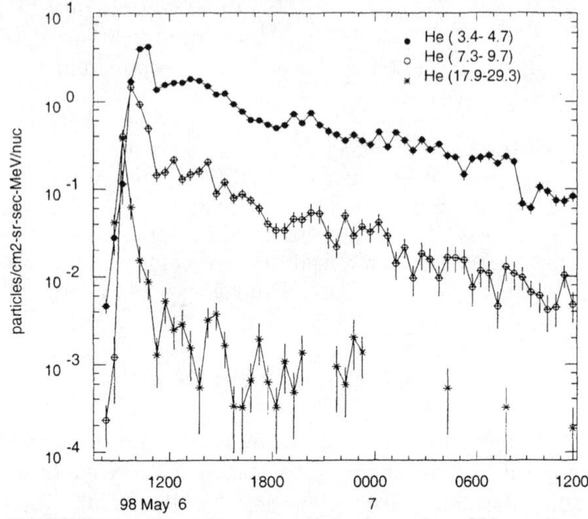

FIGURE 1. Time histories of the intensity of He at three different energies for the event of 6 May 1998.

Figure 1 shows the time history of He for this event at three different energies. The intensity of He in the energy interval 17.9 to 29.3 MeV/nuc fell by one and a half orders of magnitude in just over 2 hours and then declined further at a much slower rate. He in the energy interval 3.4 to 4.7 MeV/nuc shows a relatively gradual decline, with some evidence of the initial spike evident at higher energies. The latter is an unusual feature not present in the other events observed by SIS. Reames et al. (14) saw no sign of an initial spike in this event at low energies. Surprisingly, they also saw no evidence for a spike for H at 20 MeV/n.

The top panel of Figure 2 shows intensity-time histories for ^{4}He, ^{3}He, and O all at 10.5 MeV/nuc. The middle panel shows the corresponding ratio of ^{3}He/^{4}He, while the bottom panel shows the ratio of ^{4}He/O normalized to the gradual event average value of ^{4}He/O. SIS abundances have been analyzed previously at 14 MeV/nuc because the energy ranges of the SIS instrument are different for different elements; 14 MeV/nuc is the lowest energy that intensities are available for every element from He to Ni. Figure 2 corresponds to longer time averages than used in Figure 1 and somewhat lower energy than 14 MeV/nuc in order to improve the statistics for ^{3}He.

From the middle panel of Figure 2, the value of ^{3}He/^{4}He at the impulsive peak is the same as the value

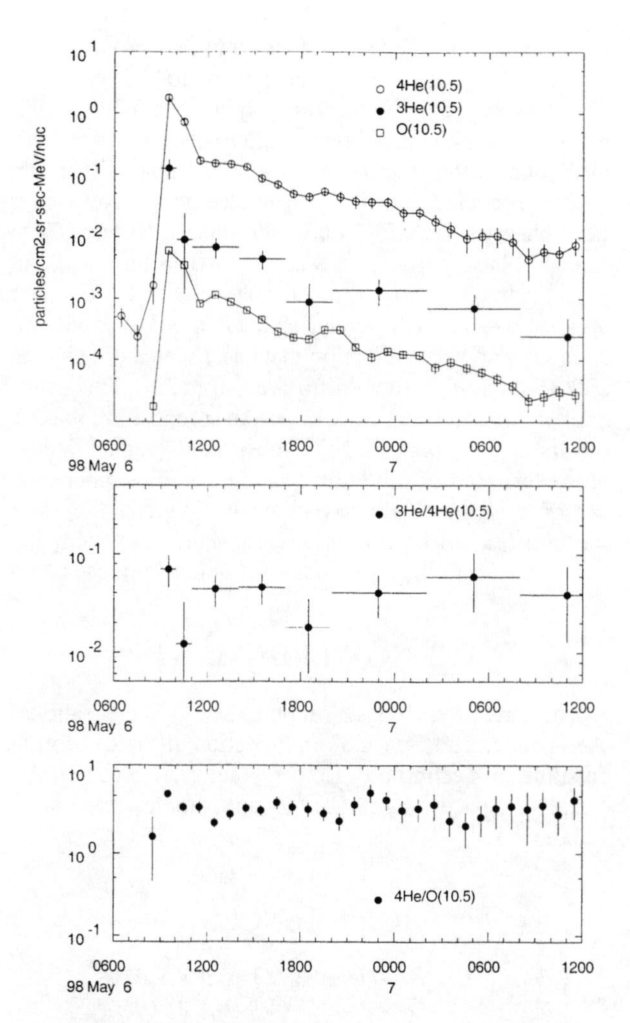

FIGURE 2. Top: 1-hour resolution time history profiles for ^{4}He, ^{3}He, and O at 10.5 MeV/nuc. Middle: ^{3}He/^{4}He versus time for 10.5 MeV/nuc. Bottom: ^{4}He/O versus time for 10.5 MeV/nuc. ^{4}He/O has been normalized to the average gradual event value from (5).

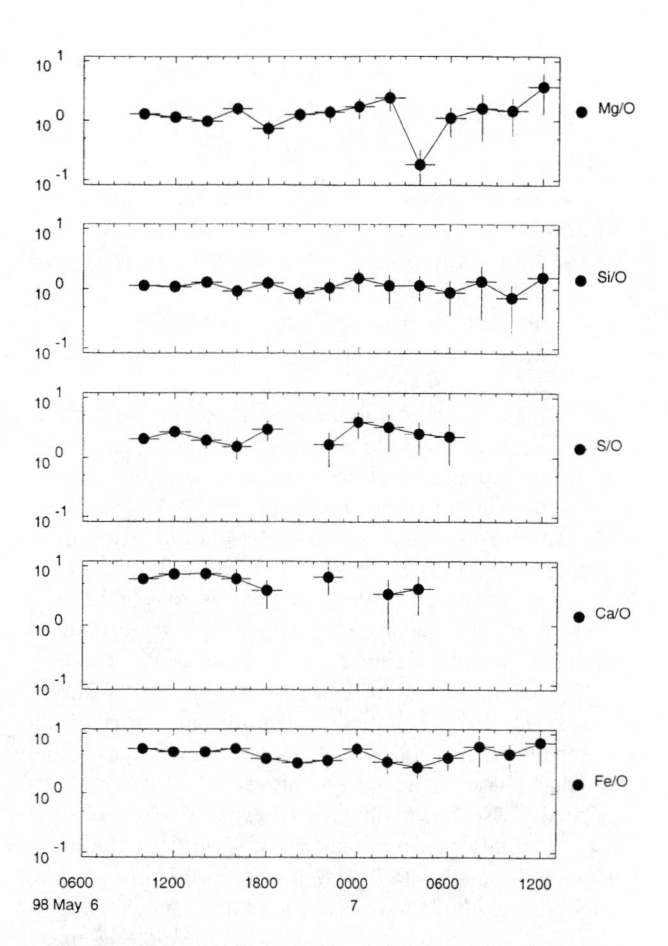

FIGURE 3. 2-hour averages of Mg/O, Si/O, S/O, Ca/O, and Fe/O at 14 MeV/nuc as a function of time for the event of 6 May 1998. These ratios are normalized to the average gradual event values from (5).

of subsequent points during the slow decay phase of the event. The value of ^{3}He/^{4}He averaged over the event is about 5%. The bottom panel of Figure 2 shows that the value of ^{4}He/O, while enhanced, changes relatively little during the event. Figure 3 shows the time histories of the abundance ratios Mg/O, Si/O, S/O, Ca/O, and Fe/O given as two hour averages at 14 MeV/nuc. The first data point of each panel coincides with the 'impulsive' peak at the leading edge of the event and subsequent data points correspond to the gradual decay phase. It is apparent that the ratios S/O, Ca/O, and Fe/O are all enhanced. It is also apparent that there is essentially no change in these ratios from the initial values for at least 12 hours.

DISCUSSION

The 6 May 1998 event has been described variously as an impulsive event (13), as a gradual event (14), and as a hybrid of an impulsive event plus a gradual event (15). These interpretations are complicated by the fact that the X2.7 x-ray event was preceded by an M2.9 x-ray event with Type III emission. This preceding x-ray event has been previously ignored, but it would clearly allow for yet other interpretations. For example, the first event could be impulsive and the second purely gradual. For the moment, it will be assumed that the first event can be ignored.

The 6 May 1998 event, then, appears to be impulsive because:

- the initial intensity spike has a duration of only about 2 hours
- its abundances are very similar to the average elemental abundances of impulsive events and not like those of gradual events (13).

- the event-averaged ratio of $^3He/^4He$, ~5%, is enhanced well above the coronal value even if it doesn't quite reach the historical standard of 10%.
- it is well connected to the associated flare region.

The 6 May 1998 event appears to be gradual because:

- the event profile at a few MeV/nuc looks like a normal gradual event.
- the event is associated with a fast CME.
- the event is associated with Type II and Type IV radio emission.
- there are time variations in abundances at a few MeV/nuc (not evident at SIS energies) which can be explained well in terms of waves generated in the vicinity of the shock (14).

The suggestion that the 6 May 1998 event is a hybrid event (15) is appealing because it allows an apparent resolution to the argument as to whether the event is either impulsive or gradual. That is, it removes the need to make a choice. In addition, Popecki, *et al.* (15) have shown that the charge states of Fe vary during the event, being initially high (an indicator of an impulsive event) followed by lower values during the gradual phase (more consistent with the charge states normally associated with gradual events). However, the measured differences are not so large: initially the mean Fe charge state is ~14.7 ±0.6, then dropping to ~13 (15). Moebius, *et al.* (7) have classified this event as being in the previously mentioned class of intermediate Fe mean charge states.

Both the $^3He/^4He$ and the elemental abundance ratios are consistent with having the same values during both the impulsive spike and the gradual decay phase. This seems highly unlikely to occur for a hybrid event in which the gradual phase is independent of the impulsive phase. First of all, one has to ask where the enhanced ^3He comes from during the gradual phase. If one argues that this comes from ^3He left over from other, previous impulsive events (18), such as possibly the M2.9 x-ray event mentioned before, then there is no reason to expect that the value of $^3He/^4He$ from these earlier events would match that from the impulsive portion of the hybrid event. This follows since it is well known that, even with a floor of 10%, the $^3He/^4He$ ratio varies by factors of 20 and more from event to event (e.g. 5). In addition, the acceleration of ^3He is due to a highly non-linear resonance process which is unlikely to produce the same $^3He/^4He$ ratio, even if repeating in the same region. Second of all, since the average elemental abundances are very different for impulsive and gradual events, it is difficult to see why the abundances of heavy elements shown in Figure 3 would also be so constant if the two phases were independent. One possibility then is that the event is a hybrid event but that the composition in the gradual phase is not independent of the impulsive phase. For example, perhaps the impulsive event injected particles into the coronal shock and the shock then accelerated them to SIS energies. This was suggested by (19) to explain the 3 June 1982 event. Cliver (4) proposed an alternative model, which also couples the impulsive and gradual phases. In this model, particles are first accelerated impulsively in a flare phase, populating a post-flare loop. Subsequently gradual phase particles are accelerated in the magnetic reconnection region behind the associated CME. Yet another possibility for coupling the two phases would be a single, impulsive injection that has a scatter-free initial spike with a subsequent diffusive wake (20). This could explain the common, impulsive-like composition in both the impulsive spike and the subsequent gradual decay. However, both of the latter two explanations ignore the presence of the CME driven shock. We conclude that the event could be a hybrid event, but, if so, with the two phases coupled by some as yet unknown means.

ACKNOWLEDGMENTS

This research was supported by the National Aeronautics and Space Administration at the California Institute of Technology (under grant NAG5-6912), the Goddard Space Flight Center, and the Jet Propulsion Laboratory.

REFERENCES

1. Reames, D.V., *Rev. Geophys.*, **33**, p. 585 (1995).
2. Reames, D.V., *Sp. Sci. Rev.*, **90**, p. 413 (1999).
3. Cane, H.V., et al., *Ap. J.*, **373**, p. 675 (1991).
4. Cliver, E.W., in *High Energy Solar Physics*, R. Ramaty, et al. (eds.), AIP (Woodbury, NY), 1995, p. 45.
5. Reames, D.V., *Adv. Space Res.*, **15**, p. 41 (1995).
6. Luhn, et al., *Ap. J.*, **317**, p. 951 (1987).
7. Moebius, et al., *Proc. 26th Int. Cosmic Ray Conf.* (Salt Lake City), **6**, 1999, p. 87.
8. Mason, G.M., et al., *Ap. J.*, **452**, p. 901 (1995).
9. Oetliker, et al., *Ap. J.*, **477**, p. 495 (1997).
10. Mazur, J.E., et al., *Geophys. Res. Lett.*, **26**, p. 173 (1999).
11. Möbius, E., et al., *Geophys. Res. Lett.*, **26**, p. 145 (1999).
12. von Rosenvinge, et al., *Proc. 26th Int. Cosmic Ray Conf.* (Salt Lake City), **6**, 1999, p. 131.
13. Cohen, C.M.S., et al., *Geophys. Res. Lett.*, **26**, p. 2697 (1999).
14. Reames, D.V., et al., *Ap. J. Lett.*, **531**, p. L83 (2000).
15. Popecki, et al., *Proc. 26th Int. Cosmic Ray Conf.* (Salt Lake City), **6**, 1999, p. 187.
16. NOAA Solar-Geophysical Data Reports, #646, Part I, p.6 and p. 30 (June, 1998); #647, Part I, pp. 121-122 (July, 1998); #651, Part II, p. 43 (November, 1998).
17. LASCO CME list, http://lasco-www.nrl.navy.mil/cmelist.html.
18. Mason, G.M., et al., Ap. J. Lett., **525**, p. L133 (1999).
19. Van Hollebeke, et al., *Ap. J. Suppl.*, **73**, p. 285 (1990).
20. Earl, J., *Ap. J.*, **188**, p. 379 (1974).

Variation in Solar Energetic Particle Elemental Composition Observed by *ACE* and *Wind*

Paul R. Boberg[2,1] and Allan J. Tylka[1]

[1]*E. O. Hulburt Center for Space Research, Naval Research Laboratory,* [2]*Consultant*

Abstract. We have used *ACE* and *Wind* data to study C/O and Fe/O ratios at ~2.6 – 15 MeV/nuc in nine large solar energetic particle (SEP) events between November 1997 and November 1998. Six events have event-integrated Fe/O ratios that are larger than the nominal coronal value by a factor of two or more. However, the energetic storm particle event of 25 August 1998 single-handedly restores the fluence-weighted Fe/O average to nearly the coronal value. We suggest that this balancing indicates a common acceleration mechanism in all of these events. We also compare SEP results to a recent ~300-day survey of slow solar wind by *Ulysses*. The average solar-wind C/O ratio is significantly larger (by ~40%) than in SEPs and in the photosphere. The origin of this difference in C/O is not understood. Finally, we also use 1-hour and 2-hour averaged data to examine intra-event correlations of SEP Fe/O vs. C/O.

INTRODUCTION

Historically, solar energetic particle (SEP) elemental abundances have been used to distinguish gradual and ^{3}He-rich events [1], to assess various charge-to-mass (Q/M) and first ionization potential (FIP) fractionation effects [2,3,4], and to derive coronal abundances that are difficult to determine through direct spectroscopic means [5]. Recent >10 MeV/nuc, event-averaged measurements from *ACE* [6-8] indicate that abundance-based distinctions between gradual and ^{3}He-rich events [5] may be less evident than observed at lower energies. Indeed, previous event-integrated measurements [3,4,9,10] showed that as energy increases, elemental abundances exhibit larger event-to-event variation.

In this paper, we use large-geometry-factor instruments *ACE*/SIS [11] and *Wind*/EPACT/LEMT [12] (~40 cm^2-sr and ~51 cm^2-sr, respectively) to examine energy-dependent C/O and Fe/O ratios in nine large SEP events that occurred between November 1997 and November 1998. We present ensemble-averaged ratios, event-integrated ratios, and intra-event correlations. We also compare these results to a recent survey of interstream solar-wind (SW) composition from *Ulysses*, taken in a ~300 day period nearly coincident with these SEP observations.

EVENT-ENSEMBLE AVERAGES

Table 1 shows event-ensemble-averaged C/O and Fe/O at several energies. Error bars are statistical only. In calculating these X/O ratios, we combined the X fluence from all nine events and then divided by the combined O fluence. This technique was suggested by [5] as the least-biased way in which to determine the source abundances; it can yield results quite different from averaging individual events' X/O ratios. (See Table 2 below.) These results are compared with an earlier survey [5], based on the ensemble average of 43 gradual events in Cycles 21 and 22. Also shown are the ^{3}He-rich-event average [5] and photospheric values [14].

TABLE 1. C/O and Fe/O from combined fluence of nine SEP events from LEMT (L) and SIS (S)

Energy, MeV/nuc	<C/O>	<Fe/O>
2.6 – 3.2 (L)	0.513 ± 0.002	0.166 ± 0.001
3.2 – 5.0 (L)	0.456 ± 0.001	0.211 ± 0.001
5 – 10 (L)	0.408 ± 0.001	0.157 ± 0.001
7 – 10 (S)	0.418 ± 0.001	--
10 – 15 (S)	0.475 ± 0.001	0.088 ± 0.001
5 – 12 [5]	0.465 ± 0.009	0.134 ± 0.004
^{3}He-rich [5]	0.434 ± 0.030	1.08 ± 0.05
Photosphere [14]	0.489	0.0468
Solar Wind [9]	0.670 ± 0.005	0.106 ± 0.003

CP528, *Acceleration and Transport of Energetic Particles Observed in the Heliosphere: ACE 2000 Symposium,* edited by Richard A. Mewaldt, et al.
© 2000 American Institute of Physics 1-56396-951-3/00/$17.00

Table 1 also shows *Ulysses*/SWICS solar wind (SW) measurements [13] for ~300 days from July 1997 to April 1998, "an exceptionally long and quiet period of slow solar wind with virtually no interference from CMEs or fast streams" [13] during which Ulysses was at ~5 AU and within 10° of the ecliptic. SW uncertainties in Table 1 are $1/\sqrt{300}$ of the one-sigma daily SW variations [13].

Small events, which we have neglected, would not significantly alter Table 1. But Table 1 may change as more *large* events are accumulated. We examined sensitivity of the present results to the limited number of events by re-evaluating ratios while excluding the large 20 April 1998 event. The Fe/O results were extremely sensitive to this change, ranging from a 35% decrease at 2.6-3.2 MeV/nuc to a factor of 2.6 increase at 10-15 MeV/nuc. Thus, another large event could significantly alter <Fe/O>, and interpreting Fe/O energy dependence or difference from solar wind is premature. However, note that photospheric Fe/O and a mean enhancement factor of ~4 at low FIP (as deduced by comparing Mg and Ne [5]) would suggest <Fe/O> ~ 0.18. Below 10 MeV/nuc, the present SEP averages in Table 1 are within ~25% of this value.

On the other hand, SEP <C/O> averages were relatively insensitive to omission of the 20 April 1998 event, increasing by only ~5-10% at the various energies. C/O shows less inherent variability than Fe/O (see Table 2 and Figure 1 below), so that even the small event sample to-date permits inferences about C/O. Thus, comparing SEP and photospheric C/O indicates that carbon (FIP=11.26 eV) should be considered a high-FIP, rather than an intermediate-FIP, element. Also, <C/O> in gradual-events, ^{3}He-rich events, and the photosphere agree to within $\pm$12%. But the average SW C/O is *higher* by ~40%.

EVENT-INTEGRATED ABUNDANCES

Table 2 shows event-integrated C/O and Fe/O for the nine events used in this study. Error bars are statistical only. For comparison purposes, values in this table have been normalized to the nominal SEP values given by [5], which are quoted in line 6 of Table 1. Unlike in Table 1, *equal weighting* of the events determined the multi-event SEP averages given in the penultimate row of Table 2.

Event-to-event variability (as quantified by the standard deviations in the penultimate row of Table 2) is larger at the higher energy. Also, two of the events (20 April 1998 and 30 September 1998) are cases in which the appellations "Fe-rich" and "Fe-poor" are clearly energy dependent.

A particularly striking point arises from comparing Fe/O ratios in Tables 1 and 2. In spite of the large energy dependence, the SEP ensemble-averages in Table 1 are within ~50% of the nominal SEP Fe/O value of 0.134 [5]. But at both energies in Table 2, six of the nine events are enhanced in Fe/O relative to this standard by more than a factor of two. A closer examination shows that the fluence from one event – 25 August 1998 -- almost single-handedly pulls the ensemble-averaged Fe/O values in Table 1 down to nearly the nominal coronal value.

Table 2 also shows solar-wind averages (normalized to [5]) with error bars corresponding to one-sigma daily-variations [13]. Except for the 25 August 1998 and lower-energy 30 September 1998 results, the SW <C/O> exceeds the event-integrated C/O values by more than 2 daily-variation sigmas. For the majority of events and energies, the SEP and SW Fe/O results also differ significantly.

TABLE 2. Event-integrated C/O and Fe/O, normalized to SEP-based coronal values [5], for nine large SEP events

SEP Event	[C/O] /<C/O>$_{SEP}$ (2.6-3.2 MeV/nuc)	[C/O] /<C/O>$_{SEP}$ (10-15 MeV/nuc)	[Fe/O] /<Fe/O>$_{SEP}$ (2.6-3.2 MeV/nuc)	[Fe/O] /<Fe/O>$_{SEP}$ (10-15 MeV/nuc)
04 Nov 1997	0.956 ± 0.017	0.931 ± 0.047	3.481 ± 0.056	3.40 ± 0.26
06 Nov 1997	0.827 ± 0.005	0.760 ± 0.030	2.313 ± 0.016	4.99 ± 0.15
20 Apr 1998	0.897 ± 0.003	1.013 ± 0.031	2.632 ± 0.009	0.237 ± 0.012
02 May 1998	0.815 ± 0.026	0.691 ± 0.051	3.095 ± 0.091	4.87 ± 0.46
06 May 1998	0.760 ± 0.015	0.721 ± 0.107	3.405 ± 0.056	3.97 ± 0.15
09 May 1998	1.020 ± 0.032	0.786 ± 0.105	1.440 ± 0.060	2.57 ± 0.36
25 Aug 1998	1.250 ± 0.010	1.552 ± 0.014	0.275 ± 0.006	0.133 ± 0.017
30 Sep 1998	1.191 ± 0.006	0.844 ± 0.015	0.899 ± 0.007	1.610 ± 0.034
14 Nov 1998	0.761 ± 0.009	0.786 ± 0.047	3.223 ± 0.034	3.72 ± 0.21
SEP Mean &. σ	0.94 ± 0.18	0.90 ± 0.27	2.31 ± 1.17	2.83 ± 1.83
SW & daily 1σ [13]	1.44 ± 0.18		0.79 ± 0.33	

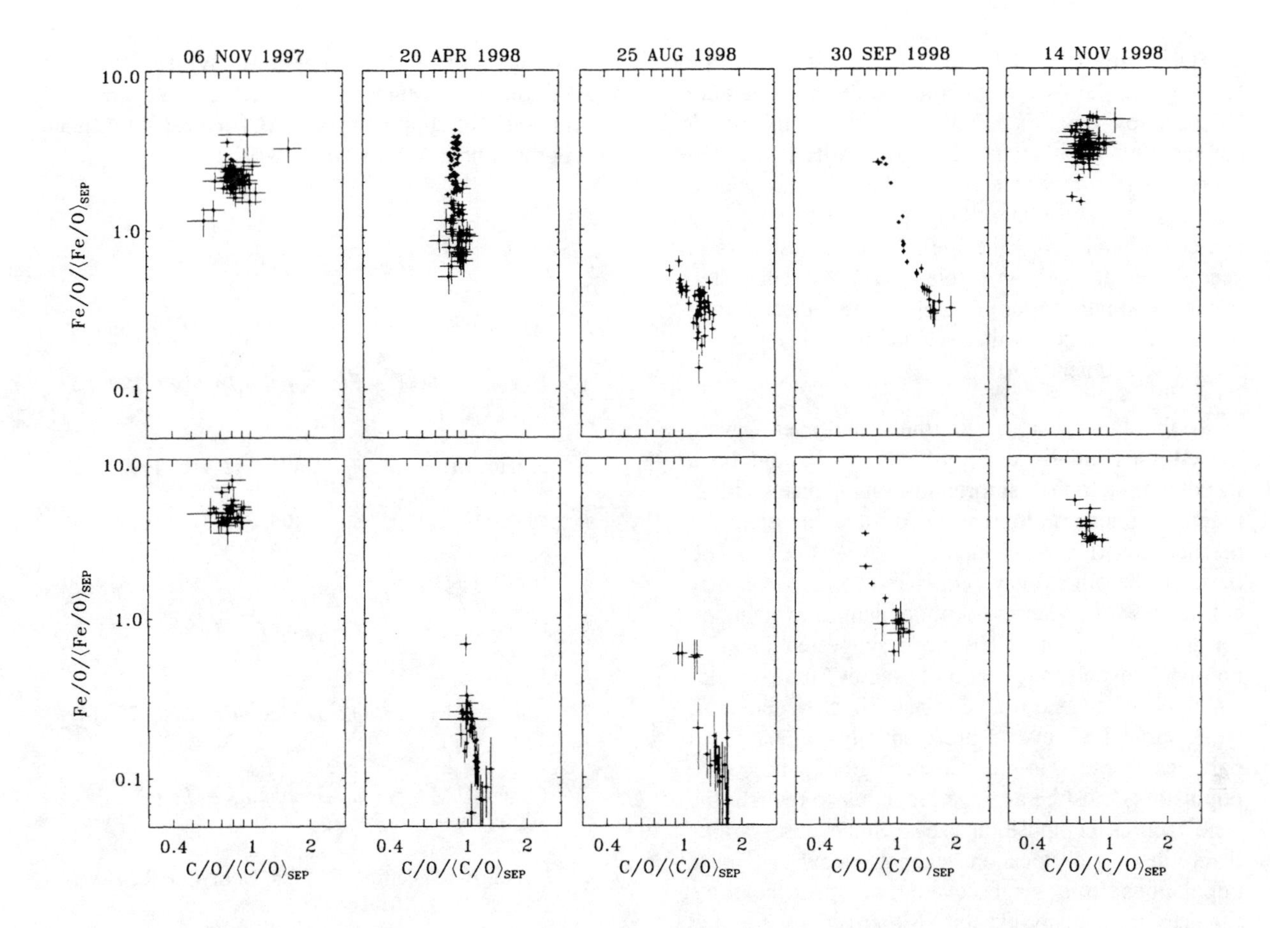

FIGURE 1. Fe/O vs. C/O, both normalized to SEP/coronal averages [5], for five events. Top row shows one-hour averaged *Wind*/EPACT/LEMT data at 2.6-3.2 MeV/nuc; the bottom row shows two-hour-averaged *ACE*/SIS data at 10-15 MeV/nuc. Error bars are statistical only. Points in which the statistical significance of either ordinate or abscissa is less than 1.5 standard deviations have been suppressed.

INTRA-EVENT VARIATION

Figure 1 shows correlations of Fe/O vs. C/O in two energy bins for the five events in which reasonable statistical errors could be obtained for one-hour or two-hour averages. Fe/O has more intra-event variation than C/O, as well as stronger energy dependence. There is a general trend of anti-correlation between C/O and Fe/O, with Fe/O larger than average when C/O is smaller than average, and vice versa. This anti-correlation is evident at both energies and is consistent with a Q/M-dependent fractionation process. In the 30 September 1998 event, the anti-correlation manifests itself over an especially wide dynamic range, particularly at the lower energy, encompassing much of the variation seen in the other events. In contrast to the other events, the lower energy C/O in 30 September 1998 extends to larger values than the higher energy C/O.

DISCUSSION

Previous discussions of events in this study have often revolved about the fact that most of these events are "Fe-rich" relative to the nominal composition given in [5]. Since enhanced Fe/O has been identified as a signature of flare acceleration, this Fe richness has been used to argue for a substantial flare component[1] in many of these events.

[1] Four of these Fe-rich events also have ^{3}He/^{4}He at levels of ~0.5 - 4% [8]. These enhancements are below the 10% threshold generally used in identifying ^{3}He-rich SEPs, but are nevertheless significantly above the average solar wind value of 0.05%. It has been suggested that these ^{3}He/^{4}He enhancements are another indication of a flare component. But another possible explanation has recently been advanced [20], involving a remnant population of flare-accelerated suprathermals. A small admixture of flare suprathermals in the solar wind might account for the additional ^{3}He without having a large impact on elemental abundance ratios.

But comparison of Tables 1 and 2 shows that the 'excess' Fe in seven of these events is largely balanced by the Fe 'deficit' that occurs in the 25 August 1998 event alone. This event was a so-called energetic storm particle ("ESP") event, associated with a near-central-meridian flare (N35E09) and a powerful shock, with the highest compression ratio seen so far at 1 AU in Cycle 23 [D. Berdichevsky, private communication]. Unlike the other events, most of the fluence in this event was collected while the shock was near 1 AU.

If the Fe-enrichment in the other events were indeed substantially due to flares, we would be a faced with a rather surprising coincidence: adding together fluences from two distinct acceleration mechanisms (i.e., a shock in the August event and flares in the others) have conspired to deliver nearly coronal/SW abundances. A more natural explanation for this 'coincidence' has been suggested in [15]: non-ESP events preferentially show us particles which have escaped from a distant shock accelerator, while large ESP events preferentially deliver to us particles trapped in the shock region. These two populations must be averaged together to recover the 'real' source composition [16]. Since observations from just one location cannot show us both populations in a single event, we must combine fluences from many events. Moreover, escape and trapping are inherently Q/M-dependent processes. Recent quantitative modeling efforts [17,18] confirm that these processes can account for a significant part of the variability in SEP elemental composition.

Additional observations and modeling are needed to determine the extent to which Fe/O suppression is a general characteristic of large ESP events.

Given the limited event sample to date, discussion of differences between SEP-average and SW-average Fe/O is not yet warranted. But the difference between SEP and SW C/O ratios in Table 1 is both highly significant and unlikely to be changed by additional data. It is not clear why this difference occurs. However, we note that C/O varies systematically with solar-wind speed in corotating interaction region (CIR) events [19] at a magnitude similar to the SEP C/O variations in Table 2 and Fig. 1. The origin of this effect is also unknown, perhaps suggesting a common missing ingredient in our understanding of CIR and SEP shock acceleration.

ACKNOWLEDGMENTS

We thank Don Reames, Tycho von Rosenvinge, Dick Mewaldt, Christina Cohen, and Rick Leske for their data and helpful discussions. We also thank the ACE Science Center for assistance. We gratefully acknowledge support by the ACE Guest Investigator Program under NASA DPR-W19501.

REFERENCES

1. Reames, D.V., *Adv. Sp. Res.* **13**, (9)331-(9)339 (1993).
2. Breneman, H.H. & Stone, E.C., *Ap.J.Lett.* **299**, L57-L61 (1985).
3. Mazur, J.E., *et al.*, *Ap. J.* **401**, 398-410 (1992).
4. Mazur, J.E., *et al.*, *Ap. J.* **404**, 810-817 (1993).
5. Reames, D.V., *Adv. Sp. Res.* **15**, (7)41-(7)51 (1995).
6. Leske, R.A., *et al.*, *Geophys. Res. Lett.* **26**, 153-156 (1999).
7. Cohen, C.M.S., *et al.*, *Geophys. Res. Lett.* **26**, 149-152 (1999).
8. Cohen, C.M.S., *et al.*, *Geophys. Res. Lett.* **26**, 2697-3000 (1999).
9. Tylka, A.J., Dietrich, W.F., & Boberg, P.R., *Proc. 25th ICRC* **1**, 101-104 (1997).
10. Tylka, A.J. and Dietrich, W.F., *Radiat. Meas.* **30**, 345-349 (1999).
11. Stone, E.C., *et al.*, *Sp. Sci. Rev.* **86**, 357-408 (1998).
12. von Rosenvinge, T.T., *et al.*, *Sp. Sci. Rev.* **71**, 155-206 (1995).
13. von Steiger, R., *et al.*, submitted to *J. Geophys. Res.*
14. Grevesse, N. & Sauval, A.J., *Sp. Sci. Rev.* **85**, 161-174 (1998).
15. Tylka, A.J., Reames, D.V., & Ng, C.K., *Geophys. Res. Lett.* **26**, 2141-2144 (1999).
16. Reames, D.V., *Proc. 26th ICRC*, Highlight & Rapporteur Papers, in press (2000).
17. Ng, C.K., Reames, D.V., & Tylka, A.J., *Geophys. Res. Lett.* **26**, 2145-2148 (1999).
18. Ng, C.K., Reames, D.V., & Tylka, A.J., *Proc. 26th ICRC* **6**, 151-154 (1999).
19. Mason, G.M., *et al.*, *Ap.J.Lett.* **486**, L149-L152 (1997).
20. Mason, G.M., Mazur, J.E., & Dwyer, J.R., *Ap.J. Lett.* **525**, L133-L136 (1999).

Gamma-Ray Evidence for Time-Dependent Heavy Ion Enhancement in a Solar Flare

R. Ramaty[a], G. Lenters[b] N. Mandzhavidze[c] and J. A. Miller[d]

[a]NASA/GSFC, Greenbelt, MD 20771, [b]NRC at NASA/GSFC, Greenbelt, MD 20771
[c]USRA at NASA/GSFC, Greenbelt, MD 20771, [d]UAH, Huntsville, AL 35899

Abstract. We briefly review the gamma-ray data of the 1991 June 1 flare and the analysis which provide evidence that the accelerated ion composition is that observed from impulsive flares and that the heavy ion enrichment, characteristic of impulsive flares, increased during the course of the flare as the gamma-ray fluxes decreased. We propose that this anti correlation is due to acceleration by cascading Alfvén turbulence. We also show that a large behind-the-limb flare, such as the June 1 event, could produce detectable gamma-ray line emission as the radioactive patch which it creates rotates onto the visible solar disk.

INTRODUCTION

A series of six X-class flares occurred in June 1991 (e.g. (1)). Of these, the June 1 flare, located 6 to 9° behind the East limb of the Sun, was probably the most powerful. It is this flare that we consider in the present paper. Even though behind the limb, the flare site was in full view of the hard X-ray detectors on Ulysses. These X-ray observations showed that as much as 10^{34} ergs in non relativistic electrons may have been generated by this very large flare (2). Gamma rays were observed with the Phebus instrument on Granat (3). These gamma rays must have been produced in the corona, because the flare site was occulted for an Earth orbiting instrument. Neutrons were also observed with OSSE on CGRO (4).

The gamma-ray data from the June 1 flare were analyzed in detail (5). It was shown that the appropriate interaction model is a thin target, meaning that the interacting particles in the coronal portions of the flare loops produce gamma rays in this interaction region without losing much of their energy before they escape to the foot point regions of the loops where they produce gamma rays by thick target interactions. Much more gamma-ray emission should have been produced in the occulted thick target source than in the visible thin target coronal region. For other flares on the visible disk of the Sun, the thick target component from the denser regions of the flare dominates the observed gamma-ray emission.

For a thin target source the ratio of the gamma-ray line emission from the heavy ions Ne, Mg, Si and Fe to that from C and O is enhanced relative to what is expected from a thick target source (5). This is because in a thin target all ions produce gamma rays as they traverse the same escape path, while in a thick target the heavier the ion the shorter the path is over which gamma rays are produced. Evidence for this effect was seen for the June 1 flare. Moreover, there was also evidence that the heavy ion enhancement increased with time during the course of the flare as the gamma-ray fluxes themselves decreased. As we suggest in the present paper, this effect could be due to the cascading of turbulence from long to short wave lengths.

We first review the gamma-ray data and its implications, and then discuss the acceleration. We also consider the delayed gamma-ray emission resulting from long lived radio isotopes produced in a large, behind the limb flare like the June 1 event. This delayed emission becomes observable as the radioactive patch created by the flare rotates onto the visible disk of the Sun.

THE GAMMA-RAY EMISSION

The Phebus detectors did not have enough energy resolution to resolve individual gamma-ray lines. But they provided evidence for two spectral features, between 1.1 and 1.8 MeV, and between 4.1 and 7.6 MeV. Theory tells us (e.g. (5)) that the lower energy feature is produced by deexcitations predominantly in Ne-Fe, while the feature at higher energy results mostly from C and O. Figure 1 (from (6)) shows the observed time profiles of the fluxes in the two channels (the two lower sets of data), and the ratio between these fluxes (the data set at the top of the figure). We see that as the fluxes decrease (after about 15:06 UT) the ratio increases.

CP528, *Acceleration and Transport of Energetic Particles Observed in the Heliosphere: ACE 2000 Symposium,* edited by Richard A. Mewaldt, et al.
© 2000 American Institute of Physics 1-56396-951-3/00/$17.00

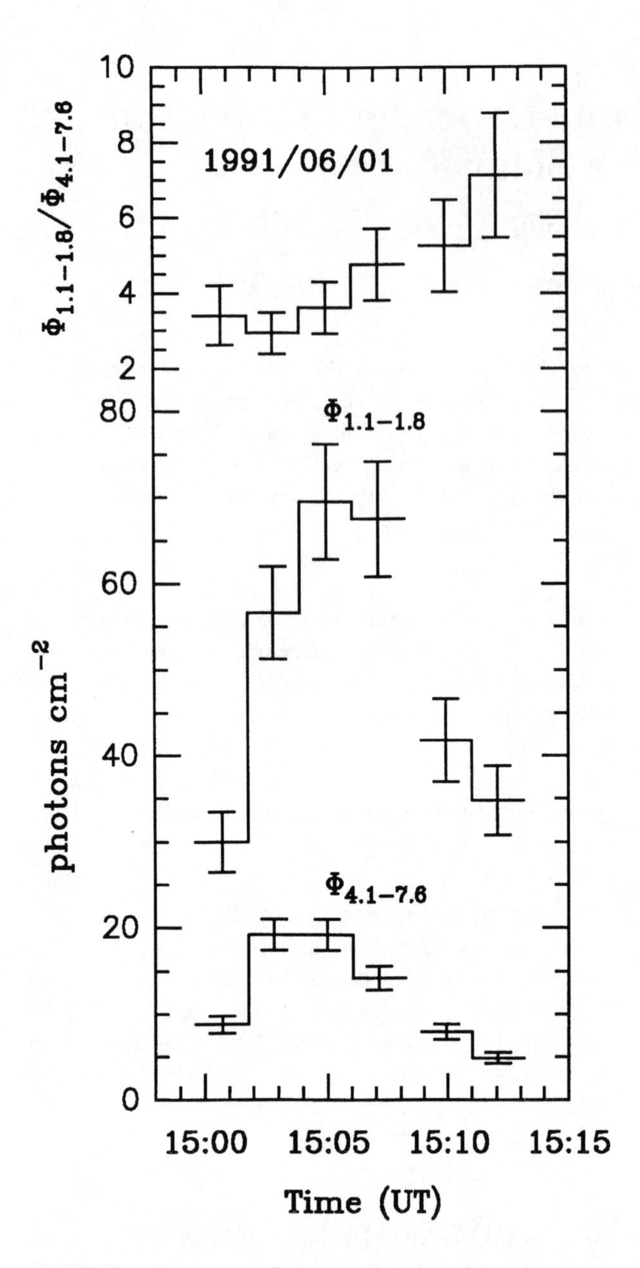

FIGURE 1. Gamma-ray fluxes as a function of time in two energy channels, and the ratio of the two as a function of time. The 1.1-1.8 MeV and 4.1-7.6 MeV channels are sensitive to the accelerated Ne-Fe and C-O, respectively.

The vertical bars in Figure 2 (from (5)) are the same observed ratios, also plotted as a function of time. The horizontal lines are the corresponding calculated ratios for various spectral indexes of the accelerated particle spectra, which are taken as power laws in kinetic energy per nucleon. Both thin and thick target calculations were performed (5), but here we only show the thin target results. The three panels in Figure 2 are for accelerated particles with gradual (G) and impulsive (I_1 and I_2) flare composition, and for ambient media of coronal composition (C_1 and C_2). I_1 and C_1, respectively, represent average impulsive flare and coronal compositions (7). I_2 and C_2 represent modifications to these averages, specifically enhanced Ne-Fe abundances relative to C and O. For example, Fe/O is an order of magnitude higher for I_2 than for I_1, but see (5) for all the details. For the G composition, α/O=90 maximizes the contribution of the narrow lines (lines from ambient heavy ions which have the C_2 composition). Nevertheless, as can be seen, the gradual composition fails to account for the data. Considering the impulsive compositions, we see that the average I_1 can account for the data, except at late times when extremely steep accelerated particle spectra would be required. Since such spectra are highly unlikely, a transition from the I_1 to the I_2 is suggested, perhaps at time interval 4 or 5, i.e. at the time when the gamma-ray fluxes themselves are declining. This means that there is evidence that as the gamma-ray fluxes decrease, the accelerated Ne, Mg and Si abundances relative to C or O increase by a factor of about 3, and Fe/O increases by an order of magnitude (the assumed I_2 abundances, see (5)).

We proceed now to discuss the fundamentals of stochastic gyroresonant acceleration, and to indicate how these abundance variations might be caused by acceleration due to cascading turbulence.

STOCHASTIC GYRORESONANT ACCELERATION

Magnetohydrodynamic (MHD) modes will be generated in the large scale restructuring of the magnetic field which occurs during the primary energy release in solar flares (8, 9). Three MHD modes can propagate in a flare plasma, namely, the fast mode wave, the intermediate mode or Alfvén wave, and the slow mode, which suffers from severe Landau damping and can be ignored. The mode of interest for ion acceleration is the parallel propagating Alfvén wave.

Parallel propagating Alfvén waves interact with the ions via a process known as gyroresonance (e.g. (10)). Gyroresonance of the particle with the wave electric field occurs when (i) the Doppler shifted frequency of the wave in the particle's guiding center rest frame, ω_{gc}, is equal to the gyrofrequency of the particle in that frame, Ω_{gc}, and (ii) the sense of rotation of the electric field and the particle about $\vec{B}_0$ are the same (e.g. (11)). Thus, the wave frequency and particle gyrofrequency must be transformed to the guiding center frame which moves along B_0 with the particle velocity given by $v_{\|}\hat{z} = c\beta_{\|}\hat{z}$. The Lorentz transformation gives $\omega_{gc} = \gamma_{\|}(\omega - v_{\|}k_{\|})$, where $\gamma_{\|} = (1-\beta_{\|}^2)^{-1/2}$ and ω and $k_{\|}$ are the wave fre-

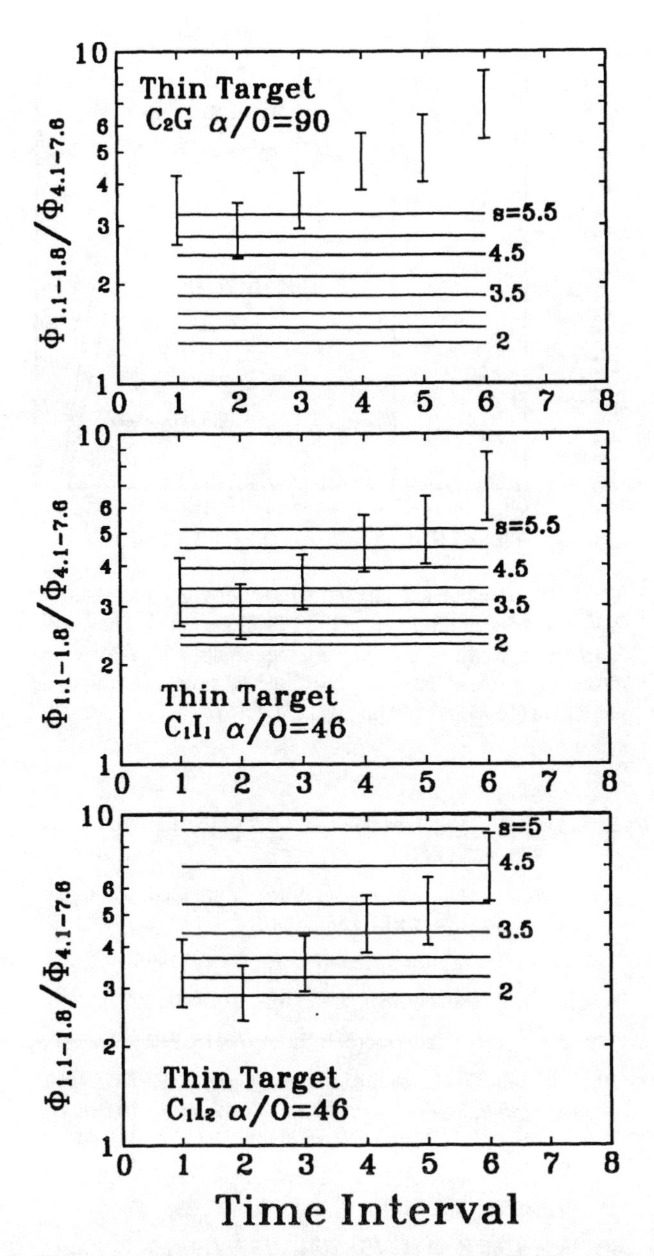

FIGURE 2. Vertical bars: observed ratios of the two fluxes as in Figure 1; horizontal lines: calculated ratios for thin target interactions for various power law spectral indexes s.

quency and parallel wavenumber of the wave, respectively. The particle guiding center gyrofrequency becomes $\Omega_{gc} = \gamma_{||}(\Omega/\gamma)$, where $\Omega = |q|B_0/mc$ is the particle gyrofrequency and γ is the particle Lorentz factor.

The first condition for gyroresonance then requires that $|\omega - k_{||}v_{||}| = |\Omega/\gamma|$. Assuming that the wave frequency is always positive, the second condition requires that $\omega - k_{||}v_{||} = \frac{\ell\Omega}{\gamma}$, where ℓ is an integer and is positive (negative) if the electric field vector and particle rotate in the same (opposite) sense. If $\ell = 0$ and the wave has a component of the electric field parallel to B_0, then the resonance condition yields the usual Landau resonance condition, $\omega - k_{||}v_{||} = 0$, which means if the wave's parallel phase speed is close to the particle's parallel speed, the wave can be Landau damped. The $\ell = 0$ and $\ell = \pm 1$ resonances are the most efficient for acceleration, and the higher harmonics can be neglected. This is because for the $\ell = 0, \pm 1$ resonances, the particle sees an electric field for the longest time (relative to the higher harmonics) during its orbit.

Once an ion gains energy from resonance with a wave, it will move out of resonance with that wave, so that a spectrum of waves is necessary for the ions to gain appreciable energy. This condition will be satisfied in the flare environment since Alfvén waves will be generated at long wavelengths (and low amplitude) comparable to the flare loop length and nonlinearly cascade to shorter wavelengths. This cascade can be viewed as an injection of energy into the acceleration region, which is transferred to the particles via gyroresonance. The cascade continues to shorter wavelengths until the energy is completely dissipated in the particles.

The wave frequency of the Alfvén waves is determined by the dispersion relation $\omega = k_{||}v_A$, where $k_{||} = 2\pi/\lambda_{||}$ is the component of the wave vector parallel to $\vec{B}_0$ and v_A is the Alfvén speed. The energy density per unit wave number of the waves $W(k)$ becomes a power law of $k_{||}^{-5/3}$, for continuous injection of wave energy and Kolmogorov cascading (e.g. (12)). Ions which have high charge-to-mass ratios (i.e. higher gyrofrequencies) will not only be the last ions accelerated in the flare, but also receive a smaller portion of the wave energy. Hence Ne-Fe, which are expected to be partially stripped and hence have lower charge-to-mass ratios than C and O, will be accelerated first and therefore enhanced relative to the latter, and as the energy injection rate decreases Ne-Fe will be even more strongly enhanced relative to the lighter C-O. This can explain the anti correlation indicated by the data and analysis described in the previous section.

THE OCCULTED THICK TARGET SOURCE AND ITS CONTRIBUTION TO DELAYED GAMMA-RAY EMISSION

As already mentioned, the thick target gamma-ray source of the June 1 flare, expected to be located near the foot points of the flaring loops at chromospheric or even photospheric heights, was not visible with Phebus/Granat because of the behind the limb location of the flare. However, a powerful flare like that on June 1 is expected to produce a large amount of radio nuclei whose decay leads to detectable gamma-ray emission as the radioactive patch rotates onto the visible disk. The most promising such line is at 0.847 MeV resulting from ^{56}Co decay into ^{56}Fe (half life 78.8 days). The ^{56}Co is produced in (p,n) reactions of both accelerated protons with ambient Fe and accelerated Fe with ambient H. Figure 3 shows the time dependence of the emission that would have been expected after the large flares of June 1991 . The two curves correspond to two different values (5) for the total prompt thick target line emission of the 1991 June 1 flare, which is obviously uncertain because of the occultation. The 0.847 MeV line is accompanied by additional lines, also resulting from ^{56}Co decay, at 1.238 and 2.599 MeV with fluxes of 68% and 17% relative to the 0.847 MeV line flux plotted in the figure. Such long duration lines from flare-produced radioactivity have not yet been observed. Their future detection with an imaging instrument such as HESSI (13) could measure the size of the radioactive patch, which would provide unique information on mixing and transport processes in the solar atmosphere.

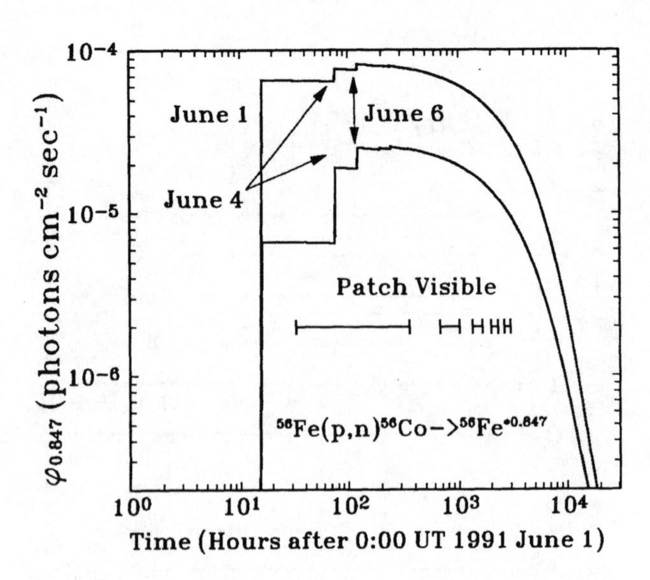

FIGURE 3. Calculated long term 0.847 MeV line flux resulting from ^{56}Co decay produced in the 1991 June flares. The curves correspond to two assumptions for the prompt 4.1-7.6 MeV line fluence which was only partially observed from this behind the limb flare (5600 and 560 photons cm^{-2} (5)).

CONCLUSIONS

We briefly reviewed the gamma-ray data of the 1991 June 1 flare and the analysis which provide evidence that the accelerated ion composition is that observed from impulsive flares, and that the heavy ion enrichment, characteristic of impulsive flares, increased during the course of the flare as the gamma-ray fluxes decreased. The abundance enhancements are thought (e.g. (14)) to be due to acceleration by cascading Alfvén turbulence from long to short wavelengths. We propose that when the level of turbulence decreases, either because of the weakening of its source or because turbulence is absorbed by the accelerated particles, fewer ions are accelerated, but also with less turbulence available, mostly the heavy ions are accelerated, causing the anti correlated gamma-ray fluxes and abundance enhancements.

We also suggest that the hidden portion of a large behind-the-limb flare, e.g. the June 1 event, could be studied by observing the radioactive patch created by the flare.

REFERENCES

1. Ramaty, R., Schwartz, R. A., Enome, S., and Nakajima, N., *ApJ*, **436**, 941 (1994).
2. Kane, S. R. et al., *ApJ.*, bf 446, L47)1995).
3. Barat, C. et al., *ApJ*, **425**, L109 (1994).
4. Murphy, R. J., Share, G. H., DelSignore, K. W., and Hua, X.-M., *ApJ*, **510**, 1011 (1999).
5. Ramaty, R., Mandzhavidze, N., Barat, C., and Trottet., G., *ApJ*, **479**, 458 (1997).
6. Trottet, G. et al., *High Energy Solar Physics*, eds. R. Ramaty et al., (AIP: New York), 153 (1996).
7. Reames, D. V., *Adv. Space Res.*, **15**, (7) 41 (1995).
8. Chiueh, T., and Zweibel, E. G., *ApJ*, **317** 900 (1987).
9. LaRosa, T. N., and Moore, R. L., *ApJ*, **418**, 912 (1993).
10. Karimabadi, H., Omidi, N., and Gary, S. P., *Geophysics Monograph, Solar System Plasmas in Space and Time*, eds. J. Burch & J. H. Waite, (Washington, D. C.: AGU), **84**, 221 (1994).
11. Steinacker, J. & Miller, J. A. 1992, ApJ, 393, 764
12. Verma, M. K. et al., *JGR*, **101**, 21619 (1996).
13. Lin, R. P. et al. *Proc. SPIE, Missions to the Sun II*, ed. C. M. Korendyke, **3442**, 2 (1998)
14. Miller, J. A. 1998, *Space Sci. Revs.*, **86**, 79 (1998).

Variable Fractionation of Solar Energetic Particles According to First Ionization Potential

R. A. Mewaldt[1], C. M. S. Cohen[1], R. A. Leske[1], E. R. Christian[2], A. C. Cummings[1], P. L. Slocum[3], E. C. Stone[1], T. T. von Rosenvinge[2], and M. E. Wiedenbeck[3]

[1]*California Institute of Technology, Pasadena, CA 91125*
[2]*NASA/Goddard Space Flight Center, Greenbelt, MD 20771*
[3]*Jet Propulsion Laboratory, Pasadena, CA 91109*

Abstract. The average composition of solar energetic particles (SEPs), like the solar corona, is known to be depleted in elements with first ionization potential (FIP) more than ~10 eV by a factor of approximately four. We examine evidence for event to event variations in the FIP-related fractionation of SEPs, following up a 1994 study by Garrard and Stone. In a survey of 46 SEP events from 1974 to 1999 the deduced FIP-fractionation varies by a factor of ~2 from event to event, with no apparent relation to charge-to-mass dependent fractionation patterns in these same events. These results are compared to similar variations observed in the solar wind.

INTRODUCTION

Variations in the elemental composition of solar energetic particles (SEPs) from event to event are usually ascribed to acceleration and transport processes that depend on the charge-to-mass (Q/M) ratio of the particles (1). In addition, it is well-known that the abundances of elements with first ionization potential (FIP) > 10 eV are depleted in SEPs by a factor of ~4 [see, e.g., (1)]. Indeed, it was this observation that first led to the realization that the solar corona and solar wind are depleted in high-FIP elements when compared to the photosphere (2, 1, 3). The depletion of high-FIP elements in the corona is viewed as evidence for ion-neutral separation processes that allow ionized species to be transported more efficiently from the photosphere to the corona [see, e.g., (4) and references therein]. In some models the relevant parameter is the "first ionization time" (FIT) rather than FIP [e.g., (5)].

In a 1994 paper, Garrard and Stone (6) presented evidence that the degree of FIP-fractionation in SEPs (the FIP depletion factor) varies by a factor of 2 or more from event to event, based on observations by Voyager, IMP-7&8 and Galileo from 1974 to 1989. Williams (7) also found the FIP-fractionation to vary in comparing two 1992 events. This paper re-examines evidence for event-to-event variations in the FIP-fractionation of SEPs by combining 1974-1989 data with more recent SAMPEX and ACE observations.

SOLAR PARTICLE DATA

The SEP events in this study span the years from 1974 to 1999 and include elements from C to Ni. Twenty-one events are from Breneman's study (8) of 5 to ~45 MeV/nucleon Voyager observations during 1977 to 1982. McGuire, von Rosenvinge and McDonald (9) reported 6.7 to 15 MeV/nucleon measurements for fifteen events observed by IMP-7&8 from 1974 to 1981. We include the nine events not measured by Breneman, in one case substituting higher-precision ISEE-3 data (10). Adding three 1989 events from Galileo (11), we obtain the 33 events studied by Garrard and Stone (6). To these we add two 1992 events from SAMPEX (7), and eleven events from the SIS instrument on ACE, where the energy interval is ~11 to ~40 MeV/nucleon [see, e.g., (12)].

Ionic charge-state measurements are available for only a few of the SEP events in this study, mostly at lower energies than the composition data. Following Garrard and Stone, we use ~1 MeV/nucleon mean

CP528, *Acceleration and Transport of Energetic Particles Observed in the Heliosphere: ACE 2000 Symposium*, edited by Richard A. Mewaldt, et al.
© 2000 American Institute of Physics 1-56396-951-3/00/$17.00

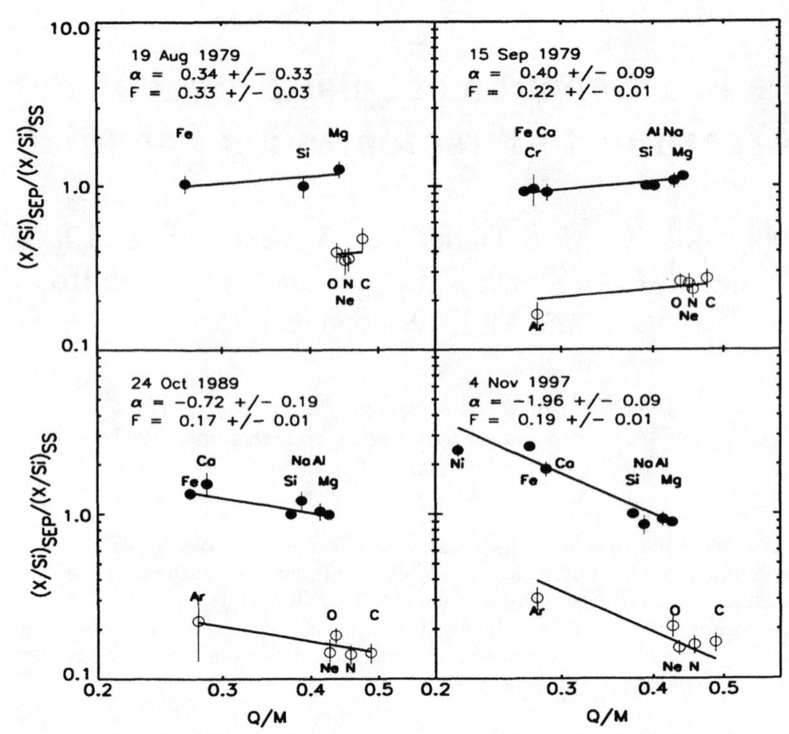

FIGURE 1. Examples of least squares fits to the normalized abundances of low-FIP (solid points) and high FIP elements (open points) in four SEP events. Best-fit vales for the Q/M fractionation index (α) and the FIP step factor (S) are indicated. Plotted uncertainties include uncertainties in the SEP abundances and in the tabulated solar system abundances (15).

charge states $\langle Q_Z \rangle$ from Luhn et al. (13), measured during 12 SEP events in 1978-1979, for Voyager, IMP-7&8, Galileo and ISEE-3 observations. For the SAMPEX and ACE events we use $\langle Q_Z \rangle$ values measured from ~15 to 50 MeV/nucleon by SAMPEX during two 1992 events (14).

To obtain normalized SEP abundances we define $\langle J_Z \rangle = (J_Z/J_{Si})/(P_Z/P_{Si})$ where J_Z is the fluence of element Z in a given SEP event and photospheric abundances (P_Z) are taken to be Anders and Grevesse "solar system" abundances (15). We assume (1) the fractionation of element Z relative to Si can be represented as a power-law in the ionic charge-to-mass ratio (Q_Z/M_Z) multiplied by a step-function FIP fractionation factor (F). Here F = 1 for elements with FIP ≤ 10 eV and F = S for FIP > 10 eV. Then

$$J_Z/J_{Si} = (P_Z/P_{Si})\, F(Z,S)[(Q_Z/M_Z)/(Q_{Si}/M_{Si})]^{\alpha}. \quad (1)$$

The values of S and α in a given SEP event are determined by a simultaneous least-squares fit to the available measured abundances (ranging from 7 to 12 species), assuming the same α and S for all species. Sulfur (FIP = 10.4 eV) is not included because its SEP abundance often corresponds to intermediate values of S (8). In the examples shown in Figure 1 note that both Fe-rich (α<1) and Fe-poor (α>1) events are included and that the fitted S values range from ~0.17 to ~0.33.

FIP-FRACTIONATION RESULTS

The scatter-plot of α and S values in Figure 2 indicates very little (if any) correlation between α and S, with a broad range of α values for both Fe-rich and Fe-poor events. It is perhaps not surprising that the fitted values of these parameters depend greatly on a few key elements with the smallest statistical uncertainties. For example, the best-fit values of S are highly correlated with the measured Mg/O ratio (see Figure 3) because the measured charge-to-mass ratios of these two abundant elements happen to be almost identical. We find $Q_{Mg}/M_{Mg} = 0.440 \pm .003$ and $Q_O/M_O = 0.437 \pm .001$, with ~ 3% average variation for the 12 SEP events measured by Luhn et al. (13).

Similarly, as shown in Figure 4, there is a very good correlation of α with the Fe/Si ratio in a given event because these two abundant low-FIP species have a large difference in Q/M [$Q_{Fe}/M_{Fe} = 0.267$ and $Q_{Si}/M_{Si} = 0.391$ using $\langle Q_Z \rangle$ from (13)]. The measured Mg/O and Fe/Si ratios therefore provide good proxies for S and α. A scatter-plot of S and Fe/Si is shown in Figure 5. There is no apparent correlation except possibly for very Fe-rich events. Note that event-to-event variations in S cannot be due to uncertainties in the photospheric abundances.

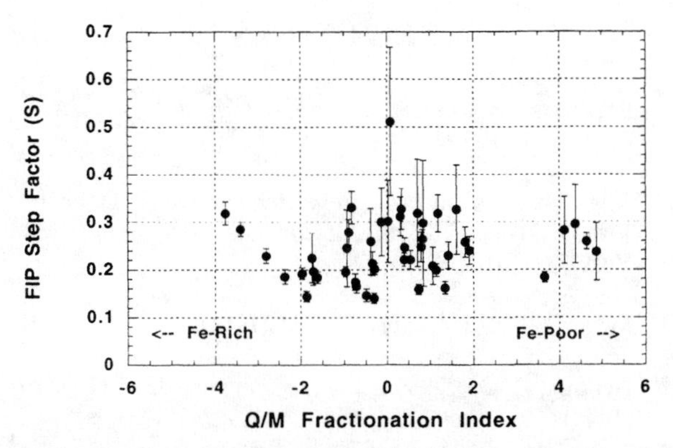

FIGURE 2. Scatter-plot of the Q/M-fractionation index (α) and the FIP fractionation step (S) for 46 SEP events.

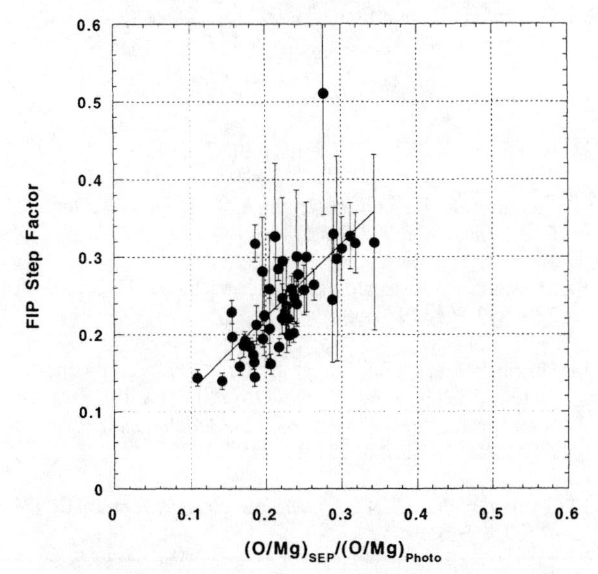

FIGURE 3. Comparison of the FIP step factor (S) with the Mg/O ratio measured in the same event.

A potential weakness in this approach is that in 44 of the 46 events (all but the SAMPEX events) it is necessary to assume that the $<Q_Z>$ values do not vary significantly from event to event or with energy. However, there is now evidence that $<Q_Z>$ values in some SEP events increase with increasing energy (16, 17, 18). Variations in $<Q_Z>$ will not affect the determination of S in events with little Q/M fractionation (small values of α). However, in very Fe-rich or very Fe-poor events the best-fit value of S can be affected if the Q/M values are considerably different than assumed. To assess this sensitivity we refit all events with two extreme assumptions: $<Q_Z>$ characteristic of (a) 1 MK and (b) 4 MK, based on $<Q_Z>$ calculations (19, 20). In these tests the best-fit values of α varied considerably, as expected. However, for events with $0.5 \leq <Fe> \leq 2$, only one of the values of S shifted by more than ±15%, and there was no significant difference in the distribution of S.

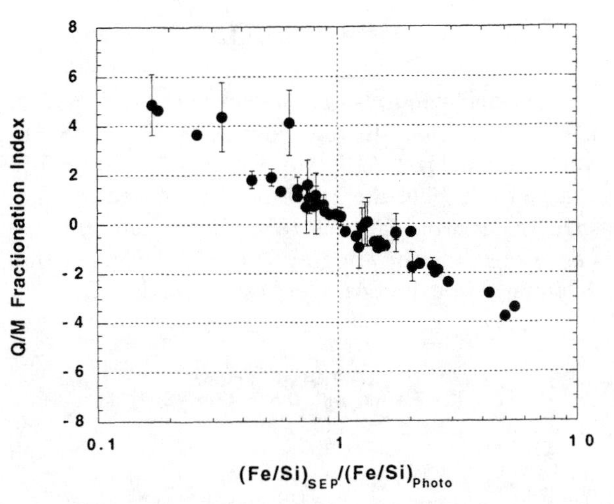

FIGURE 4. Comparison of the best-fit Q/M fraction indices (α) with the Fe/Si ratios in the same SEP events.

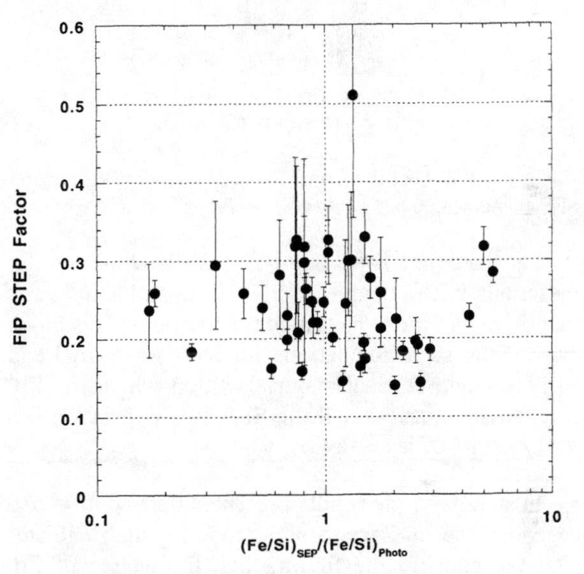

FIGURE 5. Comparison of the best-fit FIP step factors and the measured Fe/Si ratios for all 46 SEP events.

In order to minimize the possibility that uncertainties in $<Q_Z>$ are affecting the determination of S we confine the subsequent discussion to events with $<Fe>/<Si>$ ratio ranges from 0.5 to 2., a maximum variation of a factor of 2 from the solar system value. Figure 6 shows the distribution of S values for these 31 SEP events. The mean value is 0.254 with an rms deviation of 0.056, in reasonable agreement with the results of Garrard and Stone (6), where the mean was S = 0.23 with an rms deviation of 0.11. Differences from their study appear to be due to their use of a different set of photospheric abundances, with a greater abundance of Fe, and to our omission of very Fe-rich and Fe-poor events, which, in our opinion, provide less reliable FIP-fractionation estimates.

DISCUSSION

The solar wind also exhibits a variable degree of FIP fractionation. In the slow solar wind the FIP fractionation factor S is similar to the average SEP value of 0.25, but the FIP effect is reduced in high-speed solar wind streams [e.g., (5)]. There are also variations in the degree of FIP fractionation determined from coronal spectroscopy studies (21).

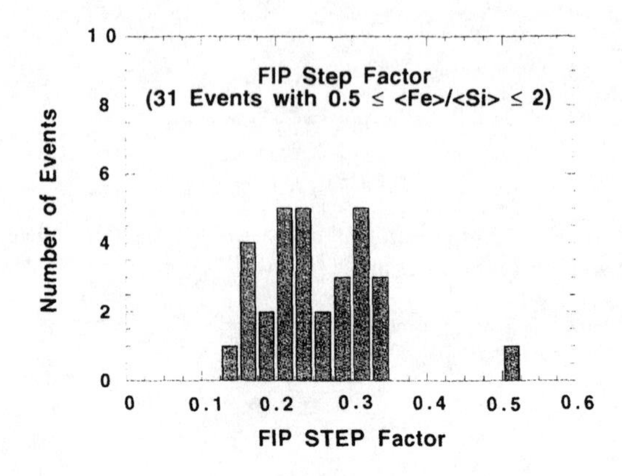

FIGURE 6. Distribution of FIP step factors (S) for the 31 SEP events with 0.5 ≤ <Fe>/</Si> ≤ 2.

The fact that solar wind and SEPs show similar variations in the degree of FIP fractionation suggests that these variations have a common origin. This might arise if the seed populations for both solar wind and SEPs include coronal material within which the FIP fractionation varies, or if the SEP seed population is a variable mix of fast and slow solar wind.

In summary, the results presented here confirm and extend those of Garrard and Stone (7) and Williams (8). We confirm the finding that the degree of FIP fractionation (the FIP depletion factor) varies by about a factor of two from event to event. The distribution of α and S values (Figure 2) shows no evidence for a correlation between the degree of FIP-fractionation and the degree of Q/M-dependent fractionation. It is interesting that variations in the degree of FIP fractionation of SEPs are similar to variations observed in the solar wind, suggesting that they may have a common origin.

ACKNOWLEDGMENTS

This work was supported by NASA at the California Institute of Technology (under grant NAS5-6912), the Jet Propulsion Laboratory, and Goddard Space Flight Center.

REFERENCES

1. Breneman, H., and Stone, E. C., *Astrophys. J. Letters* **299**, L57-L61 (1985).

2. Mewaldt, R. A., in *Proc. Conf. Ancient Sun*, R. O. Pepin, J. A. Eddy, and R. B. Merrill, eds, (Pergamon, New York), 81-101 (1980).

3. Meyer, J. P., *Ap. J. Suppl*, **57**, 151 (1985).

4. Henoux, J.-C., *Space Science Reviews* **85**, 215-226 (1998).

5. Geiss, J., *Space Science Reviews* **85**, 241-252 (1998).

6. Garrard, T. L., and Stone, E. C., *Adv. Space Res.* **14**, (10) 589-598 (1994).

7. Williams, D. L. PhD Thesis, Caltech (1996).

8. Breneman, H. H, PhD Thesis, Caltech (1985).

9. McGuire, R. E., von Rosenvinge, T. T., and McDonald, F. B., *Ap. J.* **301**, 938-961 (1986).

10. Mewaldt, R. A., Spalding, J. D., and Stone, E. C., *Ap. J.* **280**, 892-901 (1984).

11. Garrard, T. L. and Stone, E. C., "Heavy Ions in the October 1989 Solar Flares Observed on the Galileo Spacecraft", *Proceedings of the 22nd Internat. Cosmic Ray Conf.* (Dublin) **3**, 331-334 (1991).

12. Cohen, C. M. S., et al., *Geophys. Res. Letters* **26**, 2697-2700 (1999).

13. Luhn, A., et al. *Adv. Space Research* **4**, 161 (1984).

14. Leske, R. A., et al., *Ap. J.* **452**, L149-L152 (1995).

15. Anders, E., and Grevesse, N., *Geochim. Cosmochim. Acta* **53**, 197-214 (1989).

16. Oetliker, M., et al., *Ap. J.* **477**, 495-501 (1997).

17. Mazur, J. E., et al., *Geophys. Res. Letters* **26**, 173-176 (1999).

18. Moebius. E., et al., *Geophys. Res. Letters* **26**, 173-176 (1999).

19. Arnaud, M., and Rothenflug, R., *Astron. Astrophys. Suppl. Ser.* **60**, 425-457 (1985).

20. Arnaud, M., and Raymond, J., *Astrophys. J.* **398**, 394-406 (1992).

21. Feldman, U., *Space Science Reviews*, **85**, 227-240 (1998).

Abundance Variations and Fractionation Effects in a Gradual SEP Event

A. B. Galvin[1], E. Möbius[1], M. A. Popecki[1], L. M. Kistler[1], D. Morris[1], D. Heirtzler[1], D. Hovestadt[2], B. Klecker[2], A. T. Bogdanov[2], B. Thompson[3]

[1]*Institute for the Study of Earth, Oceans and Space and the Department of Physics, UNH, Durham, NH 03824 USA*
[2]*Max-Planck-Institut für extraterrestrische Physik, Postfach 1603, D-85740 Garching, Germany*
[3]*NASA Goddard Spaceflight Center, Greenbelt MD USA*

Abstract. Using data from the ACE SEPICA experiment, we examine elemental abundance variations for C, O, Ne, Mg, Si and Fe for the 1998 April 21-23 "pure" gradual SEP event in the energy ranges 0.6-0.8 and 0.8-1.0 MeV/nucleon. The high FIP element Ne has SEP abundances (Ne/O) consistent with nominal photospheric and coronal values. Low FIP elements Mg, Fe, and Si show enhancements (relative to O) over both photospheric and coronal abundances. Power law fits to the SEP abundances (normalized to coronal values) as a function of measured <Q>/A are performed and evaluated. It is found that for this event the single-parameter power law assumption does not yield a consistently good fit. The derived power law index γ varies with both time and energy.

INTRODUCTION

In-situ minor ion ($Z > 2$) measurements of solar energetic particles (SEP) are often used to infer coronal and even photospheric abundances when reliable spectroscopic measurements are not available (e.g., Neon and Argon: (1)). Indeed, gradual SEP composition measurements gave the first indication (2) that some coronal abundances varied systematically from photospheric values, organized according to the first ionization potential (FIP) of the element. The inclusion of SEP data greatly extends the number of resolved elements and isotopes over those reported in either the in-situ solar wind or spectroscopic coronal measurements.

Gradual event SEPs are correlated with the occurrence of coronal mass ejections (CMEs). The particles themselves are believed to be accelerated by the CME-driven coronal shock out of the plasma of the upper corona and perhaps continuing into the solar wind (3). The SEP heavy ion abundances are considered reflective of coronal abundances; however, the acceleration and transport processes may introduce a compositional bias dependent on the ion's rigidity, or equivalently (for the same energy per nucleon values) its ionic charge to mass ratio (Q/A). Breneman and Stone (4) found SEP relative abundances were well-ordered in Q/A, as represented by a single-parameter power law, $(Q/A)^{\gamma}$. The value of γ was determined from the power law fit and varied from one event to another. The assumed Q/A-fractionation is also energy dependent (5). The gradual SEP composition approaches the original coronal (and solar wind) composition at low SEP energies (energy range > 0.3 MeV/n).

Determining the Q/A dependence has at least three complications. First, while the mass (or Z) for an SEP species is well measured, normally the ionic charge is not. Hence, many abundance studies have assumed "nominal" charge states, using either averaged SEP event values (6, 7) or ionization equilibrium tables, or both. The implicit assumption in using averaged values is that charge states do not vary from event to event. Second, it has recently been shown that some (but not all) SEP events demonstrate an energy dependence on the ionic charge of a given species (8, 9). Hence, even if charge states are known within a given event at one energy range, it may or may not be appropriate to apply the same values to a vastly different energy

CP528, *Acceleration and Transport of Energetic Particles Observed in the Heliosphere: ACE 2000 Symposium*, edited by Richard A. Mewaldt, et al.
© 2000 American Institute of Physics 1-56396-951-3/00/$17.00

range. Third, there may be temporal variations within an event.

Measurements of Q/A and elemental abundances as a function of time and energy are necessary to remove ambiguities. Obtaining Q/A measurements for ions at SEP energies is non-trivial. Direct Q measurements were first made in the energy range 0.5-2.5 MeV/n by Hovestadt et al. (10). Because of low geometrical factor and limited resolution, these early results were averaged over an entire event or sometimes over several events. With the advent of the ACE mission, direct charge state measurements are again available through the SEPICA experiment for energies below ~1 MeV/n. For higher energies, indirect Q measurements are occasionally used (see review by Popecki (11)).

OBSERVATIONS

The Advanced Composition Explorer (ACE) was launched on 25 August 1997 and is currently in halo orbit about the Lagrangian L1 point (12). The Solar Energetic Particle Ionic Charge Analyzer (SEPICA) on ACE measures the energy (E) and nuclear charge (Z) of solar and heliospheric energetic particles in the overall energy range 0.1-15 MeV/nuc (specific energy range is species dependent). For energies below ~1 MeV/nuc (species dependent), the ionic charge (Q) is measured.

SEPICA combines a high precision mechanical collimator and electrostatic deflection with thin-window, multi-wire proportional counters to determine simultaneously the energy loss (ΔE) and the amount of deflection of incident ions. Ion-implanted solid state detectors measure the residual energy (E_{res}). From ΔE vs. E_{res} the nuclear charge (Z) and energy (E) are calculated, and from the deflection position the E/Q, and therefore Q, is determined. With its precision measurements and high geometrical factor, SEPICA provides both high charge resolution and high time resolution. A complete description can be found in Möbius et al. (13).

Event Periods

The SEP event in this study is a subset of the "pure" gradual SEP periods identified by Klecker et al. (14). Klecker et al. used the SEPICA measured Fe/O composition and the Fe mean ionic charge to differentiate among gradual, intermediate, and impulsive SEP populations. Periods were defined as "pure" gradual if the Fe/O ratio was low (Fe/O ~0.1 to 0.3 within the energy range 0.5-1.0 MeV/n) and the $<Q>_{Fe}$ was ~11 to 12 throughout the time interval.

The SEP event we have selected for detailed study is associated with the CME observed by SOHO on April 20th. An eruptive prominence was seen on the west limb (S43 W90) at 09:31 UT. An M1.4 flare observed by GOES peaked at 10:21 UT. The associated gradual SEP event was observed by ACE at L1 beginning a few hours later (doy 110, April 20th ~15 UT) and continuing for ~5 days (Fig 1). An interplanetary shock was observed on doy 113, April 23rd ~17:28 UT, however there was no associated increase in the low energy particle intensities measured by SEPICA. Fig. 1 shows the time profile (3-hr averages) for the Oxygen differential intensity (0.6-0.8 MeV/n) and abundance ratios relative to O (13.7 eV) for low FIP elements Mg (7.6 eV), Fe (7.9 eV), Si (8.2 eV), the intermediate FIP element C (11.3 eV), and the high FIP element Ne (21.6).

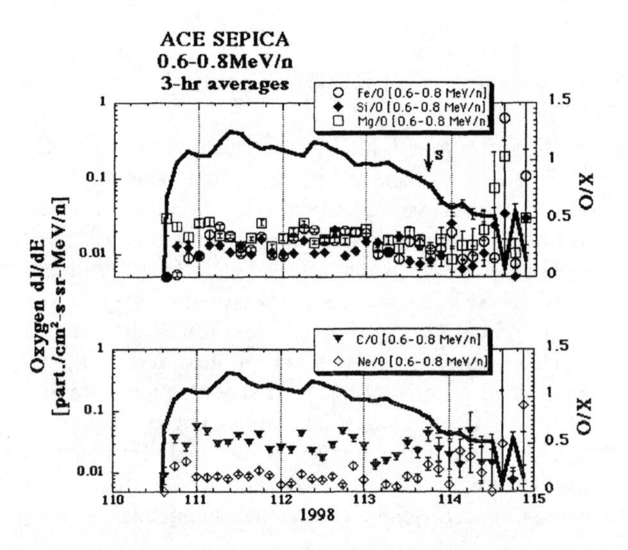

Figure 1. The gradual SEP event observed by ACE SEPICA from 20-24 April 1998 (day of year 110-114). Shown are the Oxygen differential intensities at 0.6-0.8 MeV/n and the abundance ratios relative to Oxygen for low FIP elements Fe, Si, Mg (top panel), and intermediate and high FIP elements C and Ne (bottom panel). The time of the interplanetary shock that was observed on April 23rd is shown in the top panel by an arrow.

Within this event, Klecker et al. identified three "pure" gradual periods (doy 111 0-24 UT, doy 112 0-24 UT, and doy 113 0-24 UT). We have derived SEP abundance ratios (Fe/O, Si/O, Mg/O, Ne/O and C/O) for these periods over two adjacent energy intervals: 0.6-0.8 MeV/n and 0.8-1.0 MeV/n (Table 1). In Fig. 2, the abundance ratios are shown normalized to photospheric values (15) and to nominal coronal values derived from SEP events (3). As expected from the FIP-effect, it is observed that the low FIP elements (Si, Fe, Mg) are much more overabundant relative to photospheric values than the high FIP element Ne, or the transitional element C. However, the level of overabundance (i.e., the "step factor") is quite variable (factor of 4 to 11).

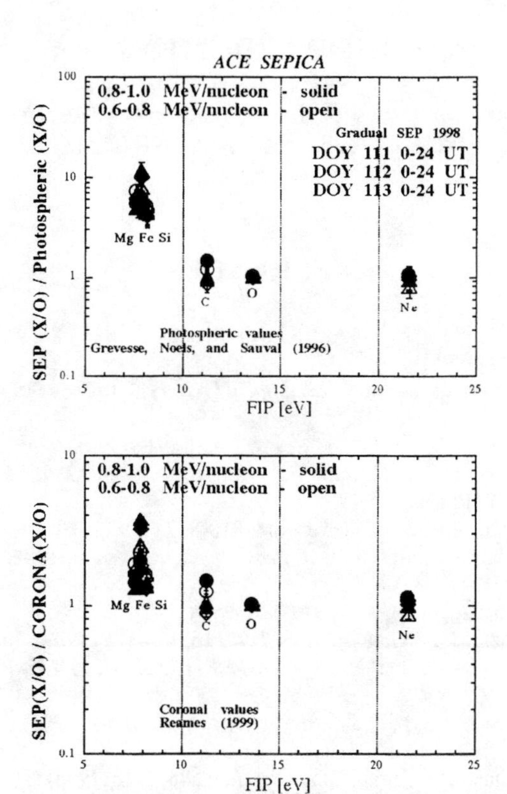

Figure 2. Top panel, the gradual SEP daily abundances normalized to nominal photospheric values for doys 111 (circles), 112 (triangles), and 113 (diamonds). The energy intervals are 0.6-0.8 MeV/n (open symbols) and 0.8-1.0 MeV/n (solid). Bottom panel, measured SEP abundances normalized to coronal values.

In the bottom panel, the SEP abundances are shown relative to nominal coronal values derived by (3). If only FIP-effect fractionation is present, one would expect the coronal normalization to approach unity. One could explain a given deviation from unity as due to a difference in the source coronal population from the assumed nominal coronal values. Variations in the size of the "step" of the FIP-effect may vary from event to event (16), but the step factor is typically ~3 to 5. Under an assumption of exclusively FIP-effect fractionation, additional daily variations require sampling different coronal populations during the SEP temporal evolution. Even with these possible scenarios, it is difficult to explain variations observed between energy intervals on the same day without assuming the presence of additional fractionation mechanism(s).

As previously discussed, additional fractionation may occur during acceleration or transport and is expected to be a function of Q/A (4). Using the measured ionic charges from SEPICA, we can investigate whether this additional fractionation in the current data set can be well-represented by a single-parameter power law, $(Q/A)^{\gamma}$.

For the three daily periods within the SEP event, <Q> has been determined individually for C, O, Ne, Mg, Si, and Fe (14). The selected measurement energy range for <Q> was optimized for each element (based on the SEPICA response function) and is therefore species dependent, inclusively covering 0.2 to 0.8 MeV/n. We have examined the measured charge states and observe no statistically significant variation with energy (Fig 3) within the stated sigma limits of (14). For comparison, the energy-dependent event reported by Möbius et al. (9) exhibited charge state increases for all measured elements, with especially noticeable increases in the Mg, Si, and Fe charge states ($\langle Q\rangle_{Fe}$ increase by ~3).

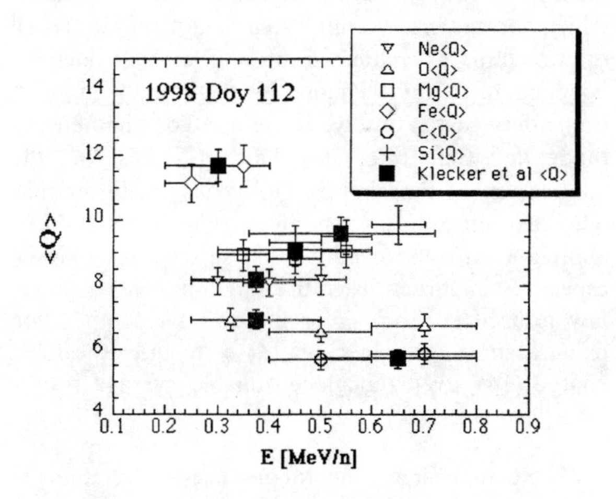

Figure 3. Measured Q as a function of energy for doy 112.

Using measured <Q>, we can directly determine the index γ from the relationship

$$\frac{SEP}{Coronal} = const. * (Q/A)^{\gamma} \quad (1)$$

TABLE 1. Event Summary

	Doy 111		Doy 112		Doy 113		Doys 111-113	
<E> MeV/n	0.6-0.8	0.8-1.0	0.6-0.8	0.8-1.0	0.6-0.8	0.8-1.0	0.6-0.8	0.8-1.0
Fe/O	0.29±.02	0.29±.03	0.34±.03	0.50±.04	0.25±.03	0.44±.06	0.30±.02	0.40±.02
Si/O	0.25±.02	0.21±.02	0.25±.02	0.21±.02	0.23±.03	0.20±.04	0.24±.01	0.21±.02
Mg/O	0.37±.03	0.28±.03	0.35±.03	0.25±.03	0.31±.04	0.31±.05	0.35±.02	0.27±.02
Ne/O	0.17±.02	0.17±.02	0.13±.01	0.15±.02	0.15±.02	0.16±.03	0.15±.01	0.16±.01
O/O	≡ 1	≡ 1	≡ 1	≡ 1	≡ 1	≡ 1	≡ 1	≡ 1
C/O	0.55±.04	0.66±.05	0.49±.03	0.45±.04	0.41±.04	0.67±.08	0.50±.02	0.57±.03
γ (χ^2_r)	-0.90±.11 (5.3)	-0.81±.15 (2.8)	-1.2±.1 (5.5)	-1.7±.1 (0.8)	-0.97±.18 (1.4)	-1.4±.2 (2.1)	-1.0±.1 (9.2)	-1.3±.1 (4.4)

This relationship is modified from that used by (4, 17) under the assumption of nominal coronal abundances (3). The results for all periods of this event (separate and combined) are given in Table 1. Gamma is determined from the best fit to the data, and the weighted reduced chi-squared is an indication of the goodness of fit. ($\chi^2_r \sim 1$ indicates a good fit.)

DISCUSSION

We have examined the composition ratios at <1 MeV/n for the "pure" gradual SEP event on 1998 April 21-23. We find the expected FIP-effect fractionation between coronal and photospheric values. There is evidence of additional fractionation which appears more complicated than a simple $(Q/A)^\gamma$ relationship, as indicated by the overall lack of goodness of fit. Furthermore, for this event a variation in the derived γ is observed over a rather small energy range and time interval. This indicates that the application of the simple $(Q/A)^\gamma$ has its limitations when extrapolating over large energies and should be applied with some caution. These cautions may be especially important when the single-parameter power law is used either to infer the coronal composition using assumed charge states (4) or to infer the ionic charge (16), even though reasonable average results have been obtained.

The complicated temporal and energy variations in the compositional behavior for this event have also been reported at higher energies (>2 MeV/n) by Tylka et al. (18). Their results are consistent with a rigidity-dependent escape of SEP particles from the acceleration (shock) region into interplanetary space. They conclude that time-dependence in the composition is a result of the evolution of the Alfvénic wave field near the shock.

ACKNOWLEDGMENTS

This work was performed under NASA NAS5-3226. SEPICA is a collaboration of UNH and MPE.

REFERENCES

1. Grevesse, N., and A.J. Sauval, Space Sci. Rev., **85**, 161-174 (1998).
2. Hovestadt, D., in *Solar Wind III*, ed. C.T. Russell, 1974, 2-25.
3. Reames, D., *Space Sci. Rev.*, **90**, 413-491 (1999).
4. Breneman, H. H. , and E. C. Stone, *ApJ. Lett.*, **299**, L57-L61 (1985).
5. Mazur, J.E., et al., *Ap. J.*, **404**, 810-817 (1993).
6. Luhn, A., et al., in *Conf. Proc 19th ICRC*, **Vol. 4**, 1985, p. 241-244.
7. Luhn, A., et al., *ApJ.*, **317**, 951-955 (1987).
8. Mazur, J. E., *Geophys. Res. Lett.*, **26**, 173-176 (1999).
9. Möbius, E., et al., *Geophys. Res. Lett.*, **26**, 145-148 (1999).
10. Hovestadt, D., et al., *Adv. Space Res.*, **1**, 61-64 (1981).
11. Popecki, M., this issue.
12. Stone, E. C., et al., *Space Sci. Rev.*, **86**, 1-22 (1998).
13. Möbius, E., et al., *Space Sci. Rev.*, **86**, 449-495 (1998).
14. Klecker, B., et al., in *Conf. Proc. 26th ICRC*, **6**, eds. D. Kieda, M. Salamon, and B. Dingus, 1999, 83-86.
15. Grevesse, N., Noels, A., and Sauval, A. J., in *Cosmic Abundances*, Astron. Soc. Pac., ed. S. S. Holt, G. Sonneborn, 1996, 117-126.
16. Garrard, T. L., and E. C. Stone, *Adv. Space Res.*, **14**, (10)589-(10)598 (1994).
17. Cohen, C. M. S., et al., *Geophys. Res. Lett.*, **26**, 149-152 (1999).
18. Tylka, A. J., D. Reames, and C. Ng, *Geophys. Res. Lett.*, **26**, 2141-2144 (1999).

Survey of Ionic Charge States of Solar Energetic Particle Events During the First Year of ACE

E. Möbius[1], B. Klecker[2], M. A. Popecki[1], D. Morris[1], G.M. Mason[3], E. C. Stone[4], A. T. Bogdanov[2], J.R. Dwyer[3], A. B. Galvin[1], D. Heirtzler[1], D. Hovestadt[2], L. M. Kistler[1], C. Siren[1]

[1]*Space Science Center and Department of Physics, University of New Hampshire, Durham, NH, USA*
[2]*Max-Planck-Institut für extraterrestrische Physik, Garching, Germany*
[3]*Department of Physics and IPST, University of Maryland, College Park, MD, USA*
[4]*Jet Propulsion Laboratory, Pasadena, CA, USA*

Abstract. The ionic charge state distributions of solar energetic particle events are determined with ACE SEPICA on an event by event basis, over the time period from launch through the end of 1998. Because of the large geometric factor of SEPICA the observations can be extended to events with very low fluxes. The study is confined to the most abundant species O, Ne, Mg, and Fe. Mean charge states for Fe are observed to vary between ≈ 11 for CME related events and ≈ 20 for small events that carry signatures of impulsive events. For these events all elements up to Mg, appear almost fully ionized. The charge states of all species follow the same trend as that of Fe in their variation from event to event. A comparison of observed mean charge states with a model assuming thermal equilibrium shows a general agreement with temperatures ranging from $1.2 - 10 \cdot 10^6$ K. However, noticeable deviations from charge states at a unique temperature for all species are seen for O at high and for Mg at both high and low charge states, which may suggest the presence of other processes. A distinct correlation is observed between the charge states and the overabundance of heavy ions in comparison with O. It remains puzzling that events with substantial deviations from coronal abundance accelerate almost fully stripped ions, which do not lend themselves easily to fractionation processes based on mass and charge.

INTRODUCTION

A widely accepted view holds that solar energetic particle events (SEPs) should be subdivided into two classes. Gradual solar events are characterized by high fluxes of energetic particles, generally with a low electron to ion ratio and a composition that reflects on average normal solar corona conditions (e.g. (1)). They are accompanied by long duration radio emissions (typically Type II and IV) and are generally associated with shocks in the corona and by a coronal mass ejection (CME). In contrast impulsive events generally show low fluxes of energetic particles in interplanetary space with a high electron to ion ratio and short time scale (several minutes) electromagnetic (X-ray and radio, typically Type III) emission. In these events substantial enhancements in the abundance of ions heavier than O are observed and frequently also dramatic enhancements of ^{3}He over ^{4}He (e.g. (2, 3)). In contrast to gradual events, whose charge states appear compatible with coronal temperatures in the neighborhood of $1 - 3 \cdot 10^6$ K (4), substantially higher mean charge states of Si (≈ 14) and Fe (≈ 20) have been reported for impulsive events (5, 6).

With the much larger sensitivity of the SEPICA sensor on ACE it is possible for the first time to determine ionic charge states of individual impulsive events and to study their variation. In a preliminary study based on 12 different SEP events, which included both gradual and impulsive events, we have demonstrated that the mean charge states are quite variable and cover a wide range (from ≈ 11 to ≈ 20 for Fe), including intermediate values (7). The limited sample also showed a distinct correlation between the observed Fe charge-state and the Ne/O and Fe/O ratios. In this paper we will expand the sample, test the charge states against the assumption of a thermal

CP528, *Acceleration and Transport of Energetic Particles Observed in the Heliosphere: ACE 2000 Symposium*, edited by Richard A. Mewaldt, et al.
© 2000 American Institute of Physics 1-56396-951-3/00/$17.00

equilibrium at the source, improve the composition results by including ACE ULEIS data, and discuss implications of the observed correlation.

INSTRUMENT AND OBSERVATIONS

The ACE spacecraft was launched on August 25, 1997, and injected into a halo orbit around the Lagrangian point L1 on December 17, 1997 (8). Within the complement of high-resolution spectrometers to measure the composition of solar and local interstellar matter, as well as galactic cosmic rays, SEPICA provides the ionic charge state distribution of energetic particles, while ULEIS provides the elemental and isotopic composition in the same energy regime with high resolution and sensitivity.

Instrumentation

The analysis of each ion starts with the determination of its energy/charge (E/Q) through electrostatic deflection in a collimator-analyzer assembly by measuring the impact position in a multi-wire proportional counter. The same counter is the energy loss (ΔE) element of a ΔE-E_{res} telescope, where the residual energy (E_{res}) is determined in an ion-implanted solid state detector. Z is determined from the specific energy loss of the particle. Combining all energy losses, including dead layers with E_{res}, provides the original energy E, which together with E/Q leads to the ionic charge state Q (detailed description in (9)). ULEIS determines the mass and energy of incoming ions by combining a time-of-flight measurement with the determination of the residual energy in a solid state detector (10).

TABLE 1: Selected SEP events (*ULEIS data)

#	Time Period	**Fe/O** (0.58-2.3 MeV/n)	**Q(Fe)** (0.18-0.44 MeV/n)
1	110 06:00 - 114 24:00	0.32	11.26
2	120 00:00 - 120 24:00	0.09	11.43
3	121 00:00 - 121 24:00	0.04	11.45
4	292 12:00 - 293 24:00	0.63	12.71
5	185 00:00 - 187 24:00	0.04	12.88
6	147 16:00 - 149 12:00	0.14	12.94
7	126 00:00 - 127 24:00	0.50	14.17
8	122 06:00 - 123 24:00	0.62	14.46
9	124 00:00 - 125 24:00	0.36	14.59
10	229 05:00 - 229 10:00	0.46*	15.86
11	269 12:00 - 270 09:00	0.86	17.19
12	270 09:00 - 271 12:00	0.95	18.06
13	149 12:00 - 150 24:00	0.74	18.25
14	252 09:00 - 252 19:00	1.60*	18.39
15	230 02:30 - 230 10:00	2.36*	18.55
16	136 08:00 - 136 12:00	1.67	18.75
17	251 12:00 - 252 09:00	2.65	18.90
18	249 12:00 - 251 12:00	3.98	19.66

Observations

For the present study we have selected 18 SEPs in 1998 that show a wide range of characteristics. The time intervals are shown in Table 1 with their average Fe/O ratios and the mean charge state of Fe. We have included both typical gradual events, with their generally low, but variable, Fe abundance, and impulsive events with much lower ion fluxes, but a substantial Fe enhancement. We have organized Table 1 by the charge state of Fe. It becomes apparent that the events seem to cover the entire range from low Fe abundance and $Q_{mean} \approx 11$ to high Fe/O-ratios ≥ 1 and $Q \approx 20$.

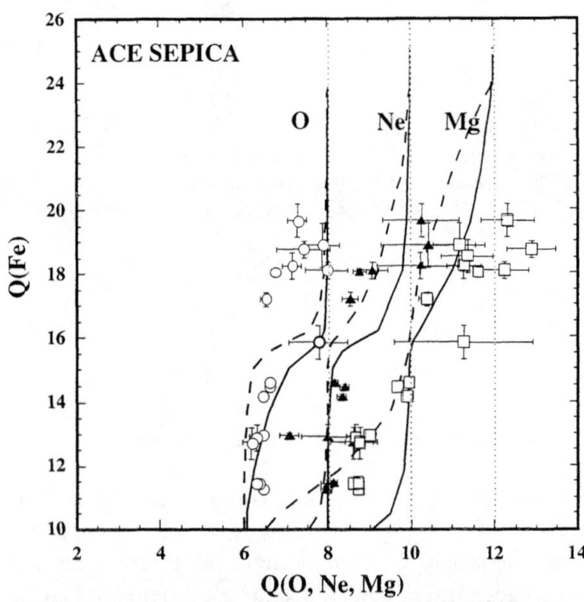

FIGURE 1: Observed mean charge state of Fe versus those of O, Ne and Mg together with modeled values. Solid lines: Mean charge states based on thermal equilibrium (11), (12). Dashed lines: T increased by a factor of 1.6 for Fe.

In Fig. 1 the mean charge states of O, Ne and Mg are shown as a function of the Fe charge state together with the standard error of the mean for each event. The charge states of these species follow a trend with Fe from low to high charge states. For most events with $Q_{Fe} \approx 18$ - 20, the ionic charge states of O, Ne and Mg are consistent with fully stripped ions. The same trend is also observed (not shown here) for Si, but only weakly for C, because it is close to fully stripped in all events. Also shown are two model curves (11, 12) for each element. Solid lines represent mean charge states according to thermal equilibrium with a single temperature. In the model with the dashed lines Fe is allowed to reach charge states typical for temperatures higher by a factor of 1.6. The motivation for this ad hoc assumption is that Fe with its many charge states reacts much more readily to a change in conditions and thus would be affected first by any deviations from a true thermal equilibrium. While the curves appear to be generally consistent with the trend, there are significant deviations, which will be discussed below.

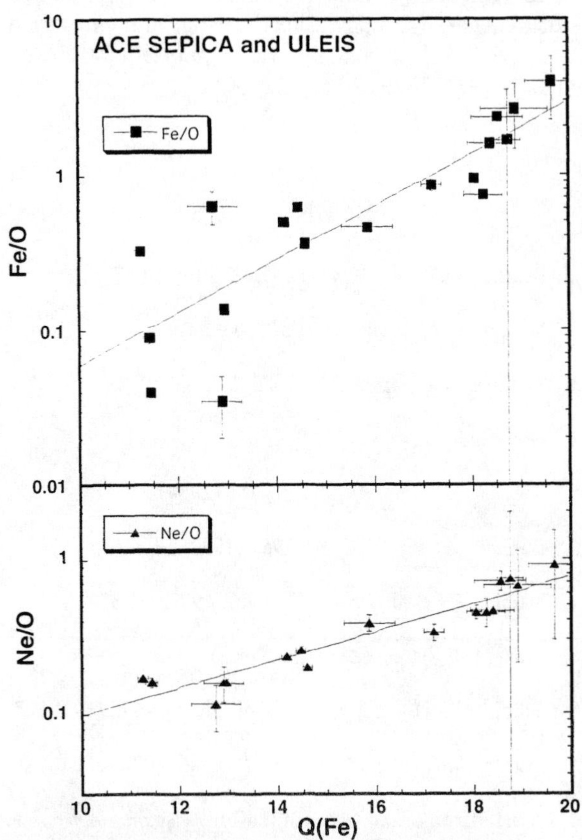

FIGURE 2: Fe/O and Ne/O ratio as a function of Q_{Fe}.

Breneman & Stone (13) have organized the abundance variations in SEPs in terms of M/Q, because various plausible fractionation processes can be invoked as potentially responsible for the observed variations. Reames et al. (14) pointed out a similar organization of elemental abundance in terms of M/Q for impulsive events. Because a change in Q directly affects the M/Q ratio, we have compiled in Fig. 2 Ne/O and Fe/O, two pairs of species with a large difference in M/Q, as a function of Q_{Fe}. Because of the substantially lower fluxes, the error bars for abundance ratios states and mean charge are generally larger for impulsive events. However, meaningful results can still be derived for individual events, while the earlier ISEE-3 results (5, 6) were averaged over all impulsive events. Both the Ne and Fe abundance increase as a function of the ionic charge state over the entire range. Yet Fe seems to exhibit a substantially larger abundance variation than Ne for events with low Q_{Fe}. Also shown in Fig. 2 are fits to the data that represent an exponential variation of the abundance ratios with Q_{Fe}. In both cases the correlation coefficient is somewhat higher for exponential, 0.95 and 0.91, than for linear variation, 0.89 and 0.79, respectively.

DISCUSSION

We have shown that ionic charge states vary widely from $Q \approx 11$ to ≈ 20 for Fe, and spread over the entire range. The charge states of the other species (O, Ne, Mg and Si) vary in lockstep with those of Fe. Events with the lowest Q values are clearly identified as gradual and are mostly compatible with equilibrium charge states for temperatures of $\approx 1.5 \cdot 10^6$ K, as discussed in detail by Klecker et al. (15). The ones with the highest charge states, which are consistent with impulsive events based on their charge and composition, require a temperature of $6 - 10 \cdot 10^6$ K, when compared with a thermal equilibrium model. While such a model follows the general trend of the data, there appear to be substantial deviations. The observations are represented best for Ne. For Mg they deviate substantially at low temperatures (see also (15)) and for O and Mg at high temperatures. Because small variations in T lead to relatively large variations in Q for Fe, we have varied the ionization state of Fe separately from the other species by allowing different values for T in order to explore this trend. Assuming T higher by a factor of 1.6 for Fe seems to produce better agreement at the low temperature end for Mg, but makes the deviations stronger at the high end. Therefore, it seems plausible to conclude that the observed charge states probably do not represent a thermal equilibrium. For example, the expanding corona and solar wind are thought to be the sources for CME related events, in which charge states of different species freeze in at different altitudes (e.g. (16)). Acceleration in impulsive events may occur under quickly varying conditions so that a dynamic model would be needed.

In Fig. 2 we have demonstrated that the abundance of heavy ions is correlated with the observed ionic charge states, where higher charge states of Fe appear to be synonymous with increased abundance of Ne and Fe. The scatter of Ne/O is much less than that of Fe/O. This is consistent with earlier observations that Fe/O itself varies substantially among gradual events, while the Ne/O ratio for these events is relatively constant (e.g. (14)). On the other hand both Ne/O and Fe/O show variations of about one order of magnitude over the entire range of charge states. Deviations from coronal abundance are most substantial in events for which we have confirmed that heavy ions through Mg are essentially fully stripped. While Fe could still be affected by M/Q dependent fractionation, when compared with O, Ne clearly cannot, and the difference in M/Q shrinks for events with the largest enhancements (see also (17)). Reames et al. (14) have suggested a two-stage model, in which the ions are only stripped after their acceleration out of a low T ($\approx 3 \cdot 10^6$K) environment. This would preserve the M/Q dependence for fractionation, but does not explain the observed correlation between charge states and abundance, because both are established in two independent processes.

It is possible that intermediate Q values and abundance can be related to variable contributions from two distinctly different source populations, as was also suggested by Mason et al. (18) based on ^{3}He observations. The lowest charge states would represent gradual events, composed of shock accelerated material with coronal composition. The highest Q values and substantial overabundance of heavy ions would be consistent with impulsive events. Recently observed temporal variations of charge states during individual events, related to differing contributions of such populations, as discussed by Popecki et al. (17), may be suggestive of such an explanation for the event to event variations. This also seems to be a natural explanation for the observed correlation, because mixtures would produce intermediate values for both Q and composition. However, such a model would predict a better correlation for a linear variation of the abundance ratios as opposed to the observed slight preference of an exponential dependence. Furthermore, Leske et al. (19) have shown for events 1), 7) and 8) of our sample that the abundance ratios of heavy elements and isotopes within a single event follow a $(M/Q)^\gamma$ dependence according to Breneman and Stone (13). This would be hard to achieve for events that contain a mixture of two different populations.

1. We have shown that the observed charge states in SEPs are probably not established in thermal equilibrium conditions. 2. The observed correlation between charge states and abundance ratios currently has no satisfactory explanation. Because several key species are fully stripped in high Q events, selective acceleration has been suggested to occur before and thus separate from the final charge formation by stripping. The otherwise attractive potential explanation with a mixture of two different populations leaves several questions open. It should be pointed out though that the current study is based on a small number of events and still contains relatively large statistical errors. An extended study with a substantially larger group of events will be needed to confirm the suggested correlation and its implications.

ACKNOWLEDGMENTS

The authors are grateful to many unnamed individuals at UNH and the MPE for their enthusiastic contributions to the completion of the ACE SEPICA instrument. They thank F. Gliem, K.-U. Reiche, K. Stöckner and W. Wiewesieck for the implementation of the S3DPU. The work was supported by NASA under NAS5-32626 and NAG 5-6912.

REFERENCES

1. Reames, D.V., *AIP Conf. Proc. 264*, 1992, p. 213.
2. Mason, G.M., et al., *Ap. J., 303*, 849 (1986).
3. Reames, D.V., *Ap. J. Suppl., 73*, 235 (1990).
4. Hovestadt, D., et al., *Adv Space Res., 1*, 61 (1981).
5. Klecker, B., et al., *Ap. J., 281*, 458 (1984).
6. Luhn, A., et al., *Ap. J., 317*, 951 (1987).
7. Möbius, E., et al., *Proc. 26th ICRC*, 1999, pp. 87-90.
8. Stone, E.C., et al., *Space Sci. Rev., 86*, 1-22 (1998).
9. Möbius, E., et al., *Space Sci. Rev.*, 86, 449-495 (1998).
10. Mason, G.M., et al., *Space Sci. Rev.*, 86, 409 (1998).
11. Arnaud, M., and R. Rothenflug, *Astron. Astrophys. Suppl. Ser., 60*, 425 (1985).
12. Arnaud, M., and J. Raymond, *Ap. J., 398*, 394 (1992).
13. Breneman, H.H., & E.C Stone, *Ap. J., 471*, L65 (1985).
14. Reames, D.V., J.P. Meyer, & T.T. Rosenvinge, *Ap.J. Suppl., 90*, 649 (1994).
15. Klecker, B., et al., *Proc. 26th ICRC*, 1999, p. 83-87.
16. Geiss, J., et al., Science, 268, 1033, (1995).
17. Popecki, M. et al., (2000) *this volume*.
18. Mason, G.M., J.E. Mazur & J.R.Dwyer, *Ap. J., 525*, L133, (1999).
19. Leske, R.A., et al., *Proc. 26th ICRC*, 1999, pp. 139-142.

Comparison of Ionic Charge States of Energetic Particles With Solar Wind Charge States in CME Related Events

B. Klecker[1], A. T. Bogdanov[1], M. Popecki[2], R. F. Wimmer-Schweingruber[3], E. Möbius[2], R. Schaerer[3], L. M. Kistler[2], A. B. Galvin[2], D. Heirtzler[2], D. Morris[2], D. Hovestadt[1], G. Gloeckler[4]

[1]*Max-Planck-Institut für extraterrestrische Physik, Postfach 1603, D-85740 Garching, Germany*
[2]*Space Science Center and Dept. of Physics, University of New Hampshire, Durham, NH, USA*
[3]*Physikalisches Institut, University of Bern, CH 3012 Bern, Switzerland*
[4]*University of Maryland, College Park, Md, 20742, USA*

Abstract. With SWICS and SEPICA onboard ACE, ionic charge measurements are now available for a wide energy range, covering the solar wind and suprathermal energetic particles. The high sensitivity of SEPICA has been utilized to examine the heavy ion composition and mean Fe ionic charge state as a function of time for several large, gradual, CME related solar energetic particle events. We selected those events for further analysis that clearly show small Fe/O ratios and low mean Fe ionic charge states (Q_{Fe} ~10–12) throughout the event, excluding by this method a possible admixture from impulsive events. We determine for these events with SEPICA the mean ionic charge states of C, O, and Fe in the energy range ~0.2-0.6 MeV/nuc and with SWICS the ionic charge distribution of C, O, and Fe ions of the solar wind. We find for this sample of 'pure' gradual events for suprathermal ions consistently low mean ionic charge states of ~5.5 (C), ~6.7 (O), and ~11–12 (Fe). We compare the suprathermal charge states with solar wind measurements. For events where local acceleration at an interplanetary shock is observed we find that C mean charge states are compatible with solar wind charge states and that the mean charge of O and Fe are ~0.6 and ~1–2 charge units larger for suprathermal ions than observed in the solar wind. We suggest that small systematic differences of ~0.6 (O) and ~1–2 (Fe) charge units between the solar wind source and suprathermal particles could be accounted for by Q/M dependent acceleration and propagation effects.

INTRODUCTION

Until recently, ionic charge measurements of suprathermal ions in gradual, CME or interplanetary (IP) shock related solar energetic particle events (SEP) were limited by counting statistics to event averages (Luhn et al, (1)). With the new high sensitivity instrumentation onboard the Advanced Composition Explorer (ACE), we are now able to investigate the ionic charge composition during individual large SEP events. This enables us to select 'pure' gradual events by their signature of elemental abundances and ionic charge distributions (Klecker et al., (2)), thus eliminating ambiguities due to the possible admixture of particles of different origin, e.g. from impulsive events with their much higher heavy ion charge states and elemental abundances (Möbius et al., (3)). Measurements of 'pure' gradual, CME or interplanetary shock related events consistently show low mean ionic charge states for C (~5.5–5.7), O (~6.6–6.9), Ne (~8), Si (~9.5), and Fe (~11–12) (Klecker et al., (2)). In this paper we present results of the analysis of ionic charge states of the solar wind and of suprathermal particles in the energy range ~0.2–0.6 MeV/nuc, obtained with the SWICS and SEPICA instruments onboard ACE, respectively. We make use of the high sensitivity of SEPICA to select short time periods of ≤ 1 day with large intensity increases near interplanetary shocks in CME related SEP events. We compare solar wind charge state distributions and elemental abundances with the ionic charge- and elemental abundances of suprathermal particles near the interplanetary shock, in order to investigate mass-per-charge dependent acceleration and propagation effects.

CP528, *Acceleration and Transport of Energetic Particles Observed in the Heliosphere: ACE 2000 Symposium,*
edited by Richard A. Mewaldt, et al.
© 2000 American Institute of Physics 1-56396-951-3/00/$17.00

INSTRUMENTS AND OBSERVATIONS

ACE is designed to study particle composition from solar wind energies to galactic cosmic rays. Within the ACE instrumentation the Solar Wind Ion Composition Spectrometer (SWICS) and the Solar Energetic Particle and Ionic Charge Analyzer (SEPICA) are the prime instruments to study the ionic charge distribution of the solar wind and of energetic particles in the energy range of ~0.2–1.0 MeV/n.

SWICS combines electrostatic deflection, post-acceleration, time-of-flight, and a total energy measurement to uniquely determine the mass, M, the charge, Q, and the energy, E, of solar wind ions, of suprathermal tails of the solar wind, and of pickup ions (Gloeckler et al., (4)).

SEPICA combines electrostatic deflection in a collimator-analyzer assembly for the determination of energy /charge (E/Q) with the measurement of residual energy (E_{res}) and energy loss in a solid state detector and multi-wire proportional counter, respectively, for the determination of the nuclear charge (Z) of the particles. From E_{res} and Z the incident energy can be derived, which, in combination with the E/Q measurement, provides the ionic charge Q of energetic ions. In order to combine high charge resolution with a large geometric factor, the experiment consists of three independent sensor units, one with high charge resolution ($\Delta Q/Q \sim 0.1$ for $E < 1$ MeV/Q) and small geometrical factor and the remaining two with their resolution reduced by a factor of three, but large geometric factor. A complete description of SEPICA may be found elsewhere (Möbius et al., (5)).

We have selected three time periods for our analysis when large increases in the intensity of low-energy ions ($E \leq 1$ MeV/nuc) have been observed in close connection with the passage of IP shocks at ACE. Figure 1 shows as an example the oxygen flux, the Fe/O ratio, and the mean ionic charge of O and Fe during the time period April 30 to May 3 (day 120–123), 1998. The dashed vertical line indicates the passage of an IP shock at ACE at 2123 UT on May 1 (Skoug et al., (6)). Note that the low mean ionic charge of Fe and the small Fe/O ratio on April 30 and May 1 are typical for large, gradual, CME or interplanetary shock related SEP events. The consistently low mean ionic charge of Fe during the two-day period preceding the shock indicates that there is no admixture from impulsive events during this time period. The large variations of the mean ionic charge of Fe and of the elemental abundances on May 2 1998 indicate contributions from different particle populations and are possibly related to the passage of a halo CME with

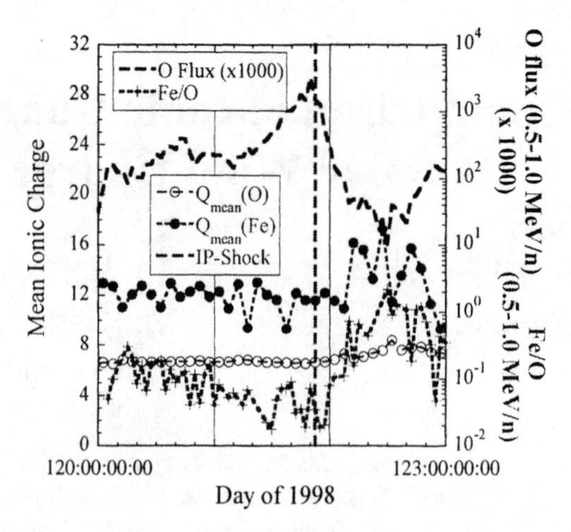

Figure 1. Oxygen flux, Fe/O-ratio, and mean ionic charge of O and Fe during the April 30 – May 1, 1998 interplanetary shock related SEP event.

unusual solar wind ionic charge composition (Gloeckler et al., (7)) and will be discussed in another paper in this volume (Popecki et al., this volume). We will concentrate our further analysis of this event on the 21.5 hour time period on May 1 preceding the shock.

In Figure 2 we compare the ionic charge distribution of solar wind Fe ions as measured on May 1 upstream of the IP shock with the corresponding distribution of Fe ions in the energy range 0.2-0.4 MeV/nuc. Figure 2 shows that both, the mean ionic charge and the relative abundance of higher charge states are larger for energetic Fe ions than for Fe ions in the solar wind.

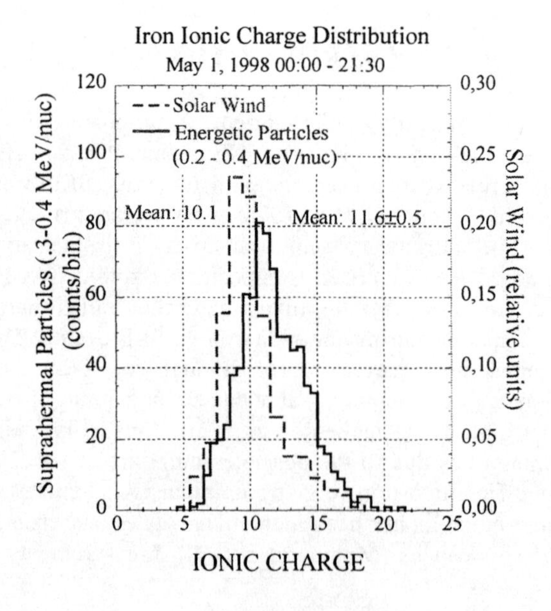

FIGURE 2. Charge distribution of Fe in the solar wind and of suprathermal particles (0.2-0.4 MeV/nuc) for May 1, 1998

TABLE 1. Mean Ionic Charge of Solar Wind and Suprathermal Particles

	Solar Wind			Suprathermal Particles		
Time Period	C	O	Fe	C	O	Fe
Day 121, 1998 00:00 – 21:30	5.50	6.20	10.1	5.53±0.3	6.8±0.3	11.6±0.5
Day 267, 1998 18:00 – 24:00	5.33	6.10	9.6	5.54±0.3	6.7±0.3	10.8±0.5
Day 311, 1998 00:00 – 24:00	5.48	6.17	9.8	5.64±0.3	6.8±0.3	11.7±0.5

In Table 1 we compare the mean ionic charge of C, O, and Fe in the solar wind with the corresponding mean ionic charge of suprathermal C (0.5-0.8 MeV/nuc), O (0.25-0.5 MeV/nuc), and Fe (0.2-0.4 MeV/nuc) for the May 1 1998 time period and two other interplanetary shock related SEP events. Table 1 shows that, within the 1 σ uncertainties, the mean ionic charge of C in the solar wind and in suprathermal particles is compatible, whereas the mean ionic charge of O and Fe of suprathermal particles are ~0.6 and ~1–2 charge units larger than the corresponding solar wind values.

DISCUSSION

The systematic differences between the mean ionic charge of iron in the solar wind and of suprathermal particles, as observed for IP shock related SEP events, provide constraints for the acceleration process and/or for the plasma parameters at the acceleration environment.

Energy Dependence of the Ionic Charge Distribution

Recent observations with SAMPEX and ACE showed in some large, gradual SEP events a significant energy dependence of the mean ionic charge of Fe (Oetliker et al., (8)) and of O, Ne, Si, and Fe (Möbius et al., (9)), Mazur et al., (10)) in the energy range of ~0.2–50 MeV/nuc. These measurements showed for Fe an increase of the mean ionic charge from ~11 at 0.5 MeV/nuc to ~18-20 at ~50 MeV/nuc. These large increases have recently been discussed in terms of additional stripping of electrons low in the corona. For equilibrium conditions it has been shown by Reames, et al. (11) that an increase of the mean charge by 3–4 charge units in the energy range ~0.1–1 MeV/ nuc can be expected (see also discussion in Bogdanov et al., this volume). At somewhat lower densities, i.e. if the acceleration occurs at higher altitudes, charge exchange equilibrium cannot be reached and the corresponding increase of the mean ionic charge will consequently be reduced (e.g. Barghouty and Mewaldt, (12)). However, in the May 1 1998 event, acceleration most likely occurs in interplanetary space where stripping effects are negligible. We will demonstrate this based on acceleration and transport time scales. The ~0.5 MeV/nuc oxygen flux upstream of the interplanetary shock shows an exponential increase with a very short time constant of $t^* = 6.2 \pm 0.5$ h. Using a simple planar diffusion-convection model at the shock, we can estimate the acceleration time scale t_c from $t_c \sim \kappa/V^2$, $t^* = x^*/V_s$, $x^* \sim \kappa/V$ (e.g. Toptyghin, (13)). x^*, κ, V, and V_s are the upstream e-folding distance of the intensity increase, the upstream diffusion coefficient, and the shock speed relative to upstream solar wind speed and spacecraft, respectively. With ~750 km/s for the local shock speed, estimated from the time difference between the observation of the shock at ACE and a magnetic storm sudden commencement at Earth (Solar Geophysical Data), and an average solar wind speed of ~360 km/s upstream of the shock, we obtain t_c ~12 h. This acceleration time scale is short compared to the transit time of the shock from the sun to Earth. Thus, local acceleration in the interplanetary medium, where stripping effects are insignificant, seems to be more likely in this event.

M/Q Dependent Acceleration Effects

Further processes that could result in small variations of the mean ionic charge are mass-per-charge dependent acceleration and propagation at coronal and interplanetary shocks. In a study of large, gradual SEP events Brenneman and Stone (14) showed that heavy ion abundances in individual SEP events, relative to average SEP abundances, are well ordered in terms of their charge-to-mass ratio, Q/M, and can be approximated by a power law in Q/M. This empirical relationship has recently been used very successfully to relate elemental and isotopic abundance variations (Leske et al., (15)) and to derive mean ionic charge states at ~12-60 MeV/nuc from elemental and isotopic abundances (Cohen et al., (16)). If Q/M–dependent effects are important for acceleration and propagation at interplanetary shocks, this will systematically change the mean ionic charge, in particular if the source distribution consists of many charge states as in the case of Fe. In order to quantify the influence of Q/M dependent effects, we compute in Figure 3 the expected mean charge of suprathermal particles sa a function of γ,

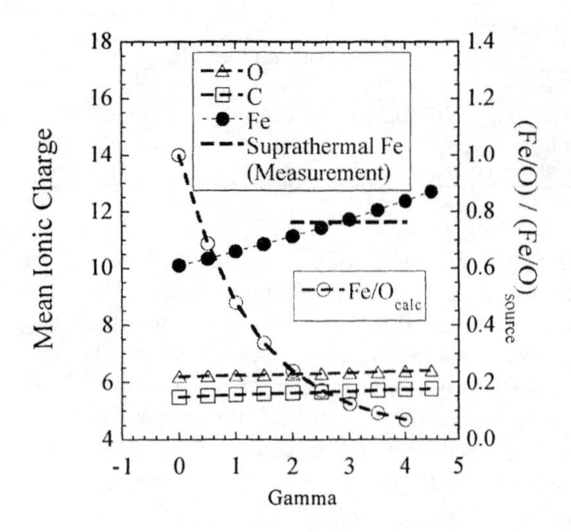

Figure 3. Variation of the mean ionic charge of C, O, and Fe and of the Fe/O abundance for Q/M dependent acceleration effects.

assuming a power law dependence of the acceleration and propagation bias for individual charge states Q_i:

$$N_{Qi} / N_{Qj} = S_{Qi} / S_{Qj} * (Q_i/Q_j)^{\gamma} \quad (1)$$

We start from the observed ionic charge distributions (S_Q) of C, O, and Fe in the solar wind (for Fe from Fig. 2). $\gamma \sim 2$–4 results in a shift of the mean ionic charge of Fe by 1–2 charge units, consistent with observations (dashed line). The effect on the mean ionic charge of C and O is much smaller ($\Delta Q \sim 0.1$-0.2), because for both, C and O, there is only one dominant charge state (+6) in the solar wind during this time period.

If the same mechanism is applied to elemental abundances, a difference between the elemental abundances of the source and suprathermal particles can also be expected for elements with different M/Q ratio, for example Fe/O. The corresponding Fe/O ratio as a function of γ, normalized to the source abundance, is also shown in Figure 3. For positive γ, the Fe/O abundance is expected to be reduced over the source. This is qualitatively in agreement with observation on May 1 1998. Averaged over a 10 h time period preceding the shock, the Fe/O ratio is ~0.048, i.e. a factor of ~2.5 smaller than the average Fe/O abundance of 0.12 ± 0.03 in the slow solar wind (v. Steiger et al., (17)).

In summary, the small differences of ~ 0.6 and 1–2 ionic charge units for O and Fe, respectively, between the solar wind source and suprathermal particles can, qualitatively, be attributed to Q/M dependent acceleration effects at interplanetary shocks. For a more quantitative treatment our observational results need to be compared with shock acceleration models. Q/M-dependent effects are, for example, introduced via a rigidity dependent scattering mean free path in models where heavy ions are scattered by Alfvén waves, amplified by the streaming protons (e.g. Ng et al (18) or Lee (this volume)), and accelerated at the shock.

ACKNOWLEDGMENTS

The authors are grateful to the many individuals contributing to the successful completion of the ACE SWICS and SEPICA instruments. The work on SWICS and SEPICA was supported by NASA contracts NAG 5-6912 and NAS 5-32626, respectively.

REFERENCES

1. Luhn, A., et al., *Adv. Space Res.* **4**, 161-164 (1984).
2. Klecker, B., Möbius, E., Popecki, M.A., et al., *Proc. 26th ICRC* (Salt Lake City) **6**, 1999, 83-86.
3. Möbius, E., et al., *Proc. 26th ICRC* (Salt Lake City) **6**, 1999, 87-90.
4. Gloeckler, et al., *Space Sci. Rev.* **86**, 497-539 (1998).
5. Möbius, E., et al., *Space Sc. Rev.* **86**, 449-495 (1998).
6. Skoug, R.M., et al., *Geophys. Res. Lett.* **26**, 161-164, (1999).
7. Gloeckler, G., et al., *Geophys. Res. Lett.* **26**, 157-160 (1999).
8. Oetliker, M., et al., *Ap.J. Lett.* **474,** L69-L72 (1997).
9. Möbius, E., et al., *Geophys. Res. Lett.* **26**, 145-148 (1999).
10. Mazur, J. E., et al., *Geophys. Res. Lett.* **26**, 173-176 (1999).
11. Reames, D.V., et al., *Geophys. Res. Lett.* **26**, 3585-3588 (1999).
12. Barghouty, A.F., and Mewaldt, R.A., *Astrophys. J. Lett.* **520**, L127-L130 (1999).
13. Toptyghin, I. N., *Space Sci. Rev.* **26**, 157-213 (1980).
14. Brenneman, H.H., and Stone, E.C., *Astrophys. J. Lett.* **471**, L65-L69 (1985).
15. Leske, R.A., Cohen, C.M., Cummings, A.C., et al., *Geophys. Res. Lett.* **26**, 153-156 (1999).
16. Cohen, C.M.S., et al., *Geophys. Res. Lett.* **26**, 149-152 (1999).
17. v. Steiger, R., Geiss, J., and Gloeckler, G., Composition of the solar wind, *in Cosmic Winds and the Heliosphere* eds J.R. Jokipii, C.P. Sonnett, and M.S. Giampapa, University of Arizona press, 581-616, (1997).
18. Ng, C.K., et al.., *Geophys Res. Lett.* **26**, 2145-2148, (1999).

Simultaneous High Fe Charge State Measurements by Solar Energetic Particle and Solar Wind Instruments

M. A. Popecki[1], T. H. Zurbuchen[2], R. M. Skoug[3], C. W. Smith[4], A. B. Galvin[1], M. Lee[1], E. Möbius[1], A. T. Bogdanov[5], G. Gloeckler[2], S. Hefti[2], L. M. Kistler[1], B. Klecker[5], N. A. Schwadron[2]

(1) Space Science Center and Department of Physics, University of New Hampshire, Durham, NH, USA; (2) The University of Michigan Space Research Laboratory, Ann Arbor, MI, USA; (3) Los Alamos National Laboratory, Los Alamos, NM, USA; (4) University of Delaware, Bartol Research Institute, Newark, DE, USA; (5) Max-Planck-Institut für extraterrestrische Physik, Garching, Germany

Abstract. During the May 2-3, 1998 CME event, iron charge state distributions were observed on the ACE spacecraft simultaneously in both solar energetic particles and the solar wind. Surprisingly, common signatures were found, even though the energy of these two particle populations differ by about two orders of magnitude. At the beginning of the event a substantial shift towards higher charge states ($Q \geq 14$) was detected for Fe nearly simultaneously in solar energetic particles by SEPICA, and in the solar wind by SWICS. The onset in the solar wind was somewhat more gradual, beginning approximately two hours before the rather sudden onset in the solar energetic particles. The high iron charge states coincided approximately with the magnetic cloud associated with the CME. Later, during the cloud passage, periods of high and unusually low, iron ionization states were observed in the solar wind. Just after the passage of the cloud, iron with high charge states again appeared in both energetic particles and the solar wind. Transport and acceleration scenarios will be discussed as possible explanations of this unusual event.

INTRODUCTION

With the launch of the Advanced Composition Explorer (ACE), it is possible for the first time to measure ionic charge states for the solar wind and solar energetic particles (SEPs) simultaneously. Charge states provide information about the source temperature, the SEP acceleration mechanisms and the propagation of SEPs through matter. Consequently, they can help identify the source population from which the SEPs are accelerated and can clarify details of the acceleration process.

In SEP events associated with a CME-driven interplanetary shock, known as gradual events (Reames (14)), the charge states of the SEPs are roughly similar to those in the solar wind (Klecker et al. (6), Gloeckler et al. (3)). The shock operates on ions in the corona or in the solar wind, where collisional effects are minimal during and after acceleration. In contrast, impulsive SEP events have been associated with flares (Reames (14)). Typically these have ion charge states much higher than the solar wind (Möbius et al. (11,12)), suggesting that they originate from a hotter source region than the gradual SEPs.

The May 2-3, 1998 magnetic cloud contained solar wind and SEP ions that had first high, then low, iron charge states. The presence of these ions in the cloud along with the plasma and magnetic field data provide insights into the structure of a cloud and the acceleration processes responsible for the SEPs, and additional ionization may play an important role.

OBSERVATIONS

The early May 1998 period was very disturbed. Four large CMEs occurred between April 27 and May 6. The May 2-3 1998 solar wind observations in this paper were associated with a CME that lifted off the

CP528, *Acceleration and Transport of Energetic Particles Observed in the Heliosphere: ACE 2000 Symposium*, edited by Richard A. Mewaldt, et al.
© 2000 American Institute of Physics 1-56396-951-3/00/$17.00

Sun on April 29, 1998. The details of the solar wind plasma parameters were presented by Skoug et al. (16). Using the SWEPAM and MAG instruments on ACE, they described a CME containing a cloud on May 2-3, 1998. They found an extended period of He+ enhancement during and after the cloud. X-ray flares with optical counterparts were observed on the solar disk. None appeared to be well-connected to the Earth.

Figure 1 shows, from top to bottom, the proton speed from SWEPAM, the solar wind iron charge state, the SEP iron charge state and the SEP Fe/O ratio at 0.47 MeV/nuc. This energy is approximately the same at which the SEP charge states are measured. Further details of the solar wind mass and charge state composition for this event can be found in Gloeckler et al. (3). Also noted in Figure 1 are the boundaries of the CME and cloud, as well as observations of bidirectional electrons and two shocks. One shock occurred at 2123 on May 1, before the CME arrived, and the other at 17 UT on May 3, just after the cloud passed. The CME itself is considered to have begun at 3-5 UT on May 2, and the cloud extended from 13 UT on May 2 to 12UT on May 3. The measurements discussed here were made inside the cloud, and then just outside.

Before the cloud arrived, a gradual event was observed at ACE. The iron charge states were compatible with solar wind iron (Gloeckler et al. (3)), with Q_{Fe}=9+, 10+ and 11+, as well as a tail toward higher charge states (Klecker et al. (7)). The first shock reached ACE during this event. The Fe/O ratio in the SEPs was low at this time, which is an indicator of a gradual event (Reames (14)). Very little SEP Fe was observed between this event and the cloud.

When the cloud arrived at ACE, a high charge component appeared in the solar wind and the SEPs at nearly the same time. The solar wind Fe, measured by SWICS, was mostly at Q_{Fe}=16+, with some even higher, but not shown because of a data processing artifact. SEP Fe, as observed by SEPICA, had a broad distribution centered around Q_{Fe}=15+, and extending from 10+ to 20+. These values were significantly higher than during the preceding gradual event on May 1. Although the charge states in both the solar wind and SEPs increased simultaneously, it is important to note that the SEP charge distribution was not an exact replication of the solar wind. The Fe/O ratio was also elevated, near one or two, in the SEP Fe at the beginning of the cloud. This was particularly so during the most intense SEP period of 13-14 UT.

At 21UT on day 122 (May 2, 1998), the charge state of iron decreased in the SEPs and more substantially in the solar wind. High and low iron charge states were present in the solar wind simultaneously, from 18 to 24 UT. A fresh injection occurred in the SEPs at 22 UT, as discerned from intensity vs. time curves (not shown). At the same time, an even lower charge iron population was measured in the solar wind (Q_{Fe} = 4 to 5+). The SEP charge state distribution in this low charge period extended down to Q=6+, and a Q=10-11+ component persisted until 16 UT on May 3. This drop in charge state occurred at approximately the same time as a break in the bidirectional electrons. The Fe/O ratio was variable in this period of mixed moderately high and very low SEP charge states, from as high as 1.3 to as low as 0.06.

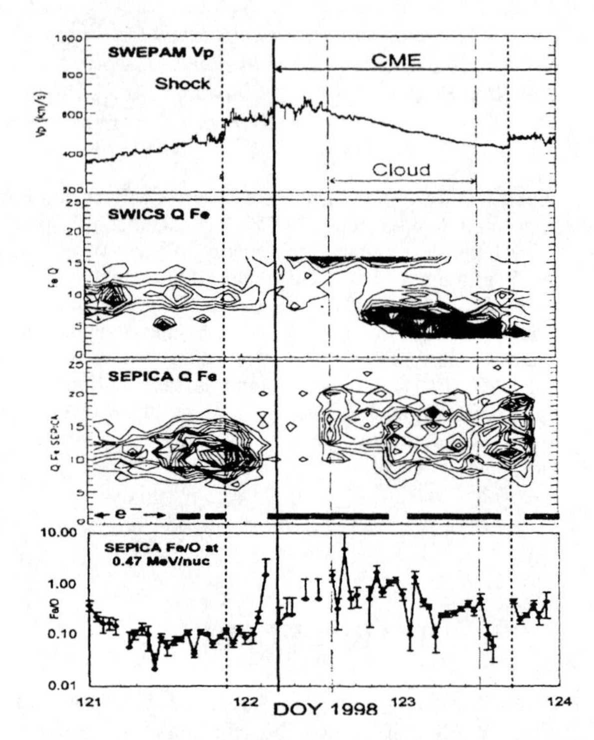

FIGURE 1. Solar wind proton speed, solar wind iron charge state, SEP iron charge state and SEP Fe/O are shown for the May 2-3, 1998 cloud CME period. Bidirectional electrons are indicated by horizontal bars in the SEPICA Q_{Fe} panel.

Again, at 16 UT on day 123, just after the cloud, a high charge component appeared in both the solar wind and SEPs, at the time of a shock passage. Iron 16+ is prominent in the solar wind, and a strong enhancement at Q_{Fe}=18-19+ occurred in the SEPs. This enhancement occurred during a break in the bidirectional electrons.

DISCUSSION

The charge state information suggests the presence of three separate iron populations, in three separate flux tubes, within and then just outside the cloud. The first population had a sudden onset of high charge state iron in both SEPs and the solar wind at the beginning of the cloud. These were accompanied by bidirectional electrons and a high Fe/O ratio in the SEPs. This simultaneous appearance of the SEPs and solar wind suggests that both were confined to a single flux tube. The high charge states, bidirectional electrons and the high Fe/O ratio are consistent with an active, heated source region at the Sun. Bidirectional electrons can indicate the presence of field lines that have both footpoints at the Sun, with at least one near an active region (Kahler & Reames (5), Gosling et al. (4)). Multiple flux tubes within a cloud have been discussed by Osherovich et al. (13), using the June 10-13 1993 Ulysses cloud observations. They inferred the cloud boundaries from solar wind proton density and the presence of ion acoustic waves.

The travel time for the solar wind is three to four days from the Sun to ACE, while it is only 6-7 hours for the SEP Fe. Therefore, their simultaneous appearance is not easy to understand in terms of propagation from a single event at the Sun. However, if a footpoint of the cloud remained magnetically connected to an active region on the Sun for several days, the following scenario is possible. The high charge solar wind could have left the Sun on this flux tube 3-4 days ago, while the high charge SEP Fe could have left from the same active region 6-7 hours ago, from a more recent eruption.

Previous observations have indicated that the footpoints of magnetic clouds may remain attached to the Sun long enough for a cloud to reach 1 AU. Mazur et al. (9) found SEP populations within magnetic clouds that had composition consistent with impulsive events. Kahler and Reames (5) observed proton SEPs that commenced during periods of bidirectional proton flows. Richardson and Cane (15) observed energetic particles in ejecta and concluded that their propagation was consistent with propagation along nonspiral, including looped, field lines. Larson, et al. (8) calculated field line lengths, using electron and Type III radio measurements for impulsive events that occurred while WIND was inside a cloud. They derived distances of 1.2 to 3 AU for propagation through the center and along the outside boundary of the clouds respectively. The May 2, 1998, SEP iron would take 6-16 hours to travel such distances, assuming a pitch angle of zero degrees. Six X-ray flares of at least class C occurred during this time, including two at E34-35 and two at W04-05 longitude. Three of these flares were in active region 8210, which was also active 3-4 days earlier.

Farrugia et al. (2) analyzed a magnetic cloud during which a rapid enhancement of 0.5-4 MeV $1 \leq Z \leq 2$ ions occurred. The presence of the cloud was established in part by bidirectional ion flows. Locally shock accelerated ions that passed the spacecraft prior to the cloud became less intense near the cloud, suggesting that they could not efficiently enter. Farrugia et al. (2) concluded that the 0.5-4 MeV ions were accelerated by a flare at the Sun a few hours earlier. Their access to the cloud suggested that the field lines remained connected to the Sun at least until the cloud passed the Earth.

We now turn to the second population of iron observed by SEPICA and SWICS in the May 2-3, 1998 period. The second population commenced nearly simultaneously in the solar wind and the SEPs. The charge states of iron were low in both the solar wind and SEPs. The solar wind iron contained lower charge states ($Q_{Fe} \geq 4+$) than the SEPs. SEPICA could measure an SEP iron charge state as low as 2.2 in this event, but the lowest observed was $Q_{Fe}=6+$. The low limit of Q=2.2+ is derived from the fact that SEPICA measures ionic charge via electrostatic deflection. A minimum deflection, and hence charge/energy ratio, is needed for complete measurement of the charge state distribution (Möbius et al. (10)).

It is conceivable that the low charge state SEP component was accelerated from the prominence material in the cloud by the shock that passed the spacecraft on May 3 at 17 UT. The He+ in this material was described by Skoug et al. (16). The observed difference between the minimum charge states of the solar wind and SEPs may be interpreted as a consequence of the acceleration process. It should be noted, however, that the SEPs and the solar wind observed simultaneously in a flux tube are not truly contemporaries because of their different speeds. It is possible that the solar wind source that underwent acceleration had a charge state distribution similar to that of the SEPs.

Shock acceleration of heated solar wind in the cloud may be an alternative explanation for the high charge state SEP component at the beginning of the cloud. In that case, a recent eruption from the same active region would not be necessary.

The abrupt onset of the third population of high charge iron observed by SEPICA and SWICS, commencing with the break in the bidirectional electron flows, is again suggestive of a separate magnetic confinement of both solar wind and SEPs. This flux tube may be detached from the Sun at one of the two footpoints, however. Larson, et al. (9) noted breaks and dropouts in bidirectional electrons, without disruption to the local cloud magnetic field rotation. They concluded that those field lines were disconnected at one or both ends from the Sun. Armstrong et al. (1) observed a unidirectional ion beam in a flux rope structure. They interpreted this as hot coronal source, traveling along a field line whose antisunward end was connected to the outer heliosphere.

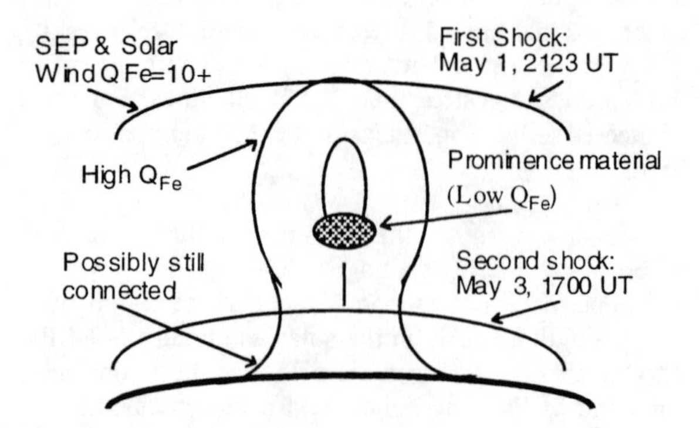

FIGURE 2. The May 2-3 1998 SEP and solar wind observations might be explained by a sequence in which a magnetic cloud passed the spacecraft, containing high charge state SEPs and solar wind on one flux tube. The SEPs either recently arrived from the same active region as the heated solar wind, or were shock accelerated from it. A second flux tube within the cloud, with lower charge state prominence material and SEPs derived from it, came next.

The third population of the solar wind and SEP ions occurred after the cloud was considered to have passed the spacecraft, possibly on a flux tube with footpoints separate from the cloud itself. This post-cloud flux tube had a mixture of high and low charge state iron in it, so it was not strictly derived from an active region. The presence of flare-related SEPs through the end of a cloud was also noted by Mazur et al. (9). Wimmer-Schweingruber et al. (17) observed high charge states for solar wind iron in the material trailing the magnetic cloud in the November 7-8, 1997, active period.

These three distinct iron populations, as identified by their charge state and their Fe/O ratio, and organized by breaks in the bidirectional electron flows, indicate that the spacecraft sampled at least three flux tubes within and then just outside a magnetic cloud. The high iron charge states in both solar wind and SEPs, just inside the cloud, are suggestive that the footpoints of the cloud were near an active region on the Sun for 3-4 days.

Finally, the possibility of shock acceleration on cloud field lines in the low charge state portion of the event may provide interesting insights on SEP acceleration from the solar wind. The acceleration of the SEPs, whether by an impulsive event or shock acceleration, could be better understood by including measurements of ^{3}He abundance and solar wind mass composition. The ^{3}He is an indicator of flare-associated (impulsive) acceleration. If it were observed in the initial portion of the cloud, where the iron charge states were high, then both the solar wind and SEP iron could be coming from active solar regions. However, if the mass composition of the solar wind was similar to that of the SEPs, then shock acceleration would be a plausible explanation.

ACKNOWLEDGMENTS

The authors are grateful to the many unnamed individuals at the University of New Hampshire, the Max-Planck-Institut fur extraterrestrische Physik and the Technical University of Braunschweig for their enthusiastic contributions to the completion of the ACE SEPICA instrument and the S3DPU. The work on the SEPICA instrument was supported by NASA under Contracts NAS5-32626 and NAG5-6912.

REFERENCES

1. Armstrong, T.P., et al., *GRL*, **21**, 1747-1750, (1994).
2. Farrugia, C.J, et al., *JGR*, **98**, 7621-7632, (1993).
3. Gloeckler, G., et al., *GRL*, **26**, 157-160, (1999).
4. Gosling, J.T. J. Birn and M. Hesse, *GRL*, **22**, 869-872, (1995).
5. Kahler, S.W. and D.V. Reames, *JGR*, **96**, 9419-9424, (1991).
6. Klecker,B., et al., *Proc. of the 26th Int'l Cosmic Ray Conf.*, eds. D. Kieda,et al., 1999, **6**, pp.83-86.
7. Klecker, B., et al., *ACE2000 Symp. Proc.s*, 2000.
8. Larson, D.E., et al., *GRL*, **24**, 1911-1914, (1997).
9. Mazur, J.E., et al., *GRL*, **25**, 2521-2524, (1998).
10. Mobius, E., et a., *Space Sci. Rev.*, **86**, 449-495 (1998).
11. Moebius, E., et al., *Proc. of the 26th Int'l Cosmic Ray Conf,*, eds. D. Kieda,et al., 1999, 6, pp. 87-90.
12. Möbius, E., et al., *ACE2000 Symp. Proc.*, 2000.
13. Osherovich V.A., et al., *GRL*, **26**, 401-404, (1999).
14. Reames, D.V., *Sp Sci Rev*, **90 (3/4),** 413-491 (1999).
15. Richardson, I.G. and H.V. Cane, *JGR*, **101**, 27521-27532, (1996).
16. Skoug, R.M., et al., *GRL*, **26**, 161-164, (1999).
17. Wimmer-Schweingruber, R.F., O. Kern and D.C. Hamilton, *GRL*, **26**, 3541-3544, (1999).

Energy Dependence of Ion Charge States in CME Related Solar Energetic Particle Events Observed with ACE/SEPICA and SOHO/STOF

A.T. Bogdanov[1], B. Klecker[1], E. Möbius[2], M. Hilchenbach[3], L.M. Kistler[2], M.A.Popecki[2], D. Hovestadt[1], J. Weygand[4]

[1]*Max-Planck-Institut für extraterrestrische Physik, Garching, D-85740, Germany*
[2]*Space Science Center, University of New Hampshire, Durham, NH, 03824, USA*
[3]*Max-Planck-Institut für Aeronomie, Katlenburg-Lindau, D-37189, Germany*
[4]*Physikalisches Institut der Universität Bern, CH-3012, Switzerland*

Abstract. We investigate the energy dependence of the charge states of heavy ions during a number of gradual solar energetic particle events related to coronal mass ejections. The data of two instruments with complementary energy ranges, the Solar Energetic Particle Ionic Charge Analyzer SEPICA on ACE and the time-of-flight mass and charge spectrometer STOF on SOHO are used to cover the energy range from ~ 0.02 to 0.7 MeV/amu for O and Fe. Measurements with SAMPEX and SEPICA/ACE contain evidence that some gradual CME related SEP events show energy dependent charge distributions. In the present paper we are closing the gap in the lower energy interval (STOF data) thus making future comparison to solar wind values - representative for coronal conditions - more reliable. We find that energy dependent charge states of Fe are almost ubiquitous in gradual SEP events. Very large differences from event to event are observed however. In cases of local acceleration at the CME driven shock the variation in the energy interval up to 0.5 MeV/amu is ~ 1-2 charge units. In alternative cases the mean ionic charge variation for Fe reaches ~4 charge units over the same energy interval. The respective charge spectra could be interpreted as the result of ionization near the sun, in the dense coronal plasma, with or without reaching charge state equilibrium.

INTRODUCTION

A set of particle instruments on the ACE and SOHO missions offer possibilities for new insights into the physics of gradual SEP events by gaining new and more complete data on their spectra and time evolution. Furthermore, they present an opportunity for better understanding the selection and acceleration mechanisms in CME related solar energetic particle events.

An interesting recent development concerning the physics of gradual SEP events was the observation that contrary, to expectations based on prevailing theories, at least some gradual CME related SEP events show energy dependent charge distributions. First observed for Fe ions by instruments on SAMPEX (Oetliker et al., [5]) it has later substantially evolved in energy and element range to include ions of C, O, Ne, Si, by observations made with SEPICA/ACE (Möbius et al., [4]) and again with SAMPEX by Mazur et al. [2]. We extend now the investigation of this aspect by adding data from the time-of-flight mass and charge spectrometer STOF on SOHO (Hovestadt et al., [1]). ACE and STOF, combined, cover the energy range from 0.02 to 0.70 MeV/amu for O and Fe. The investigation of newly available events and the extension of the energy range to lower values by including data points from STOF seems to confirm that the phenomenon of energy dependent charge states may be of more general nature than an exotic singular case. Thus we can go low enough in energy to enable future comparisons with values typical for the solar wind (see paper by Klecker et al. in this Volume). Moreover, comparison between experiment and existing models on particle acceleration in gradual CME events could give a clue to the correct interpretation of the formation of the observed charge spectra and help answer the questions on the seed populations, accelerations sites, stripping mechanisms and whether eventual contributions by impulsive flare sources could play a role.

CP528, *Acceleration and Transport of Energetic Particles Observed in the Heliosphere: ACE 2000 Symposium*, edited by Richard A. Mewaldt, et al.
© 2000 American Institute of Physics 1-56396-951-3/00/$17.00

TABLE 1. List of events studied in this paper			
Event number	Time interval	Event type	Fe/O ratio
A, fig. 1	1998: 121:00 – 121:24	Gradual + Shock	0.07
B, fig. 2	1999: 177:00 – 177:20	Gradual + Shock	0.16

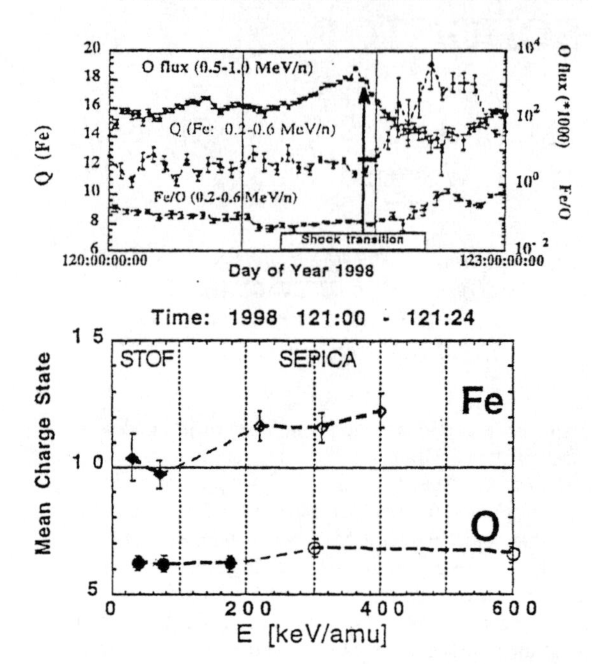

FIGURE 1. Fe and O charge spectra for an event with local shock acceleration. The peak intensity of the O-flux in the 0.5-1.0 MeV/amu range is almost coincident with the moment of the shock traversal.

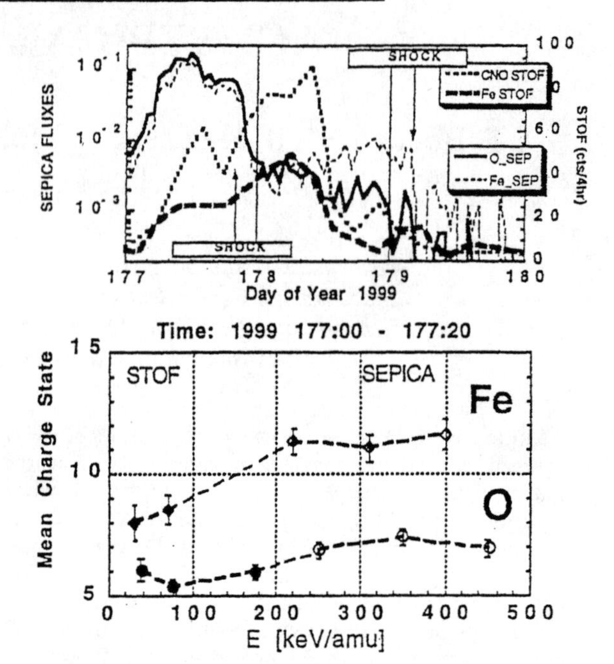

FIGURE 2. The case of strong energy dependence of the charge states. Acceleration and charge formation in the lower corona have to be considered according to the model discussed here.

OBSERVATIONS

Data from two instruments of complementary energy ranges have been used for the present analysis. The energy range between 0.20 and 0.70 MeV/amu is covered by SEPICA onboard ACE whereas the lowest energy interval (0.02 - 0.20 MeV/amu) is measured by STOF. Only the most substantial parameters of the instruments are mentioned below. The Solar Energetic Particle Ionic Charge Analyzer (SEPICA/ACE) is designed to measure the ionic charge state, the kinetic energy, and the nuclear charge of energetic heavy ions from ~0.2 to ~1 MeV/amu. It is a dE/dx - E telescope with a proportional counter - solid state detector combined with a collimator - electrostatic analyzer assembly. The combined entrance aperture geometry factor of the two instrument subsystems used in these measurements is 0.20 cm^2sr, Möbius et al. [3]).

The Suprathermal TOF sensor (STOF/SOHO) is an ion telescope with a geometry factor of 0.05 cm^2sr and an energy range from 0.03 to 0.73 MeV/e achieved by employing an electrostatic deflection system with cylindrical deflection plates, time-of-flight measurement and an SSD for the determination of the residual energy of the incident particles - Hovestadt et al. [1]).

We have studied 5 active periods in 1998 and 1999 and investigated the energy dependence of the mean ionic charge of heavy ions. We have taken only cases where inputs by impulsive SEP's are excluded. For this reason, we did not include here, for example, the event on day 178 of 1999 - see Fig.2. We present the results for the two most representative cases out of the full set, (Table 1). The events have been identified as belonging to the gradual class by various characteristics such as their energetic heavy ion average charge state signatures, Fe/O ratio and the accompanying shock transition in the solar wind. To reflect the principally possible cases of charge spectrum formation one should ideally study events: a) in which the energetic particles appear in close association with the shock (peak fluxes coincide in time with the shock transition) and b) in which they come in a prolonged interval well ahead of the CME-driven shock thus carrying a signature of an acceleration site near the Sun. Period (A) - Fig.1 and Table 1, reflects the case of local acceleration at the CME driven shock in the first of a sequence of events in May 1998 (Skoug et al., [7]) with the peak intensity of the suprathermal and energetic particles seen by both instruments coming immediately ahead of the shock transition. While in the near suprathermal range Q (Fe) remains essentially on the level of the solar wind sample, in the range of .2-.4 MeV/amu it increases by ~1.5-2 charge states. For O the variation over the energy span of 0.03 to 0.6 MeV/amu is statistically insignificant. During period (B) - Fig.2 - the energetic particles are accelerated also at a CME driven shock, this time however not locally but nearer to the sun. The case of non-local acceleration can be recognized here in the time shift of the intensity peak of energetic particles observed with both instruments. The maximum heavy ion flux values for SEPICA are reached some 8 hours before the shock traversal (at about 20 hr UT on day 177 of 1999). The local maximum of the fluxes measured with STOF at lower energy per nucleon is about 4 hours retarded with respect to the SEP's measured in SEPICA and another 4 hours ahead of the shock itself. The clear separation of the peak intensities of the SEP fluxes from the shock itself points to a much earlier time of acceleration. Here the mean charge of Fe ions increases by almost 4 charge units with energy, from extremely low level of about 8 at the lower suprathermal end up to ~12 at 400 keV/amu. The oxygen charge states are also very low in this event. Their variation is larger over the entire energy range studied but still of no statistical significance. See also Table 1 for details.

DISCUSSION

In view of the present observations and accounting for earlier results by Oetliker et al., [5]), Möbius et al., [4]) and Mazur et al., [2]), we can assert that energy dependent charge spectra are obviously no exception and seem to have a more general origin than assumed hitherto.

In principle, two extreme cases of charge spectrum formation are possible: I) Charge states determined by coronal plasma temperature. The acceleration to MeV/amu-energies takes place at sites where no further stripping is possible, i.e. in the outer corona and in interplanetary space (local interplanetary acceleration). In this case <Q> would be expected to be similar to solar wind values and independent from the energy of the accelerated particles. This is our case "A" of local shock acceleration where, for Fe, the <Q> variation with energy is only a minor effect and could be explained with M/Q-dependent acceleration (see the contributions by Klecker et al. and by Lee in this volume). II) Equilibrium <Q> spectrum is formed in an acceleration process by stripping in moderately dense coronal plasma. The charge states at ~1 MeV/amu reach their equilibrium values if the condition $\tau * N \geq 10^{10}$ s*cm^{-3}, is met, which is realistic only below some 2 solar radii (Reames, Ng and Tylka, [6]). τ (above) is the duration of the acceleration process (residence time) and N - the average ambient plasma density. In this case with acceleration close to the Sun a larger dispersion in the arrival times of particles of different energies is to be expected.

All intermediate cases between complete absence of energy dependence and the case of equilibrium Q(E) are also possible if the above condition is not fulfilled but the acceleration takes place where stripping is still possible.

If charge state equilibrium is reached during the acceleration the following semi-empirical formula can be used as shown by Reames, Ng and Tylka, [6], as a generalization for the case of non-zero electron temperature.

$$Q(Z,\beta_{ion}) = Z*[1-\exp(-125(\beta_{ion}+\beta_{e,th})/Z^{2/3})] \quad (1)$$

Here $\beta_{ion}=v/c$, $\beta_{e,th}=2kT_e/m_ec$ and Z is the atomic number. The authors have shown that this form based on the electron velocity in the Thomas-Fermi model ($\beta_{TF}= Z^{2/3}/137$) provides a reasonable fit to the data of the earlier measurements (Möbius et al., [4] and Mazur et al., [2]), where energy dependence of the mean ionic charge was observed, if one assumes an electron temperature of ~ 1.5 MK. The energy range in which the above relation is valid has a lower limit of ~100 keV/amu, hence it delivers unrealistically low Q-values for the solar wind energy range at ~ 1 keV/amu. In Fig. 3 we compare curves obtained by using relation (1) for two electron temperatures with the data from event "B". We can see that in this case of the relatively strong dependence of Q on energy, the experimental points are compatible with the curve for electron temperature of the ambient plasma of 1.0 MK. A plausible interpretation for the discrepancy between experiment and theory in the high energy range is that we are confronted here with the intermediate case, where no charge equilibrium has been reached before the particles have left the region where stripping is still possible.

A more accurate description of the energy dependent charge spectra origin can be achieved via more involved modelling [8, 9]. An illustration to this point is given in a paper by Stovpyuk & Ostryakov [8]), in which SEPICA data (Möbius et al., [4]) is explicitly compared with results from simulations assuming that the acceleration of the particles takes place at a parallel shock under realistic conditions of stripping and recombination, described with a system of diffusion equations for the full set of charge states of a given element. Within the constraints of their model a very good agreement with the experiment can be reached by taking into account stripping caused by both electrons and protons in a proton plasma of given density, temperature and wave turbulence level.

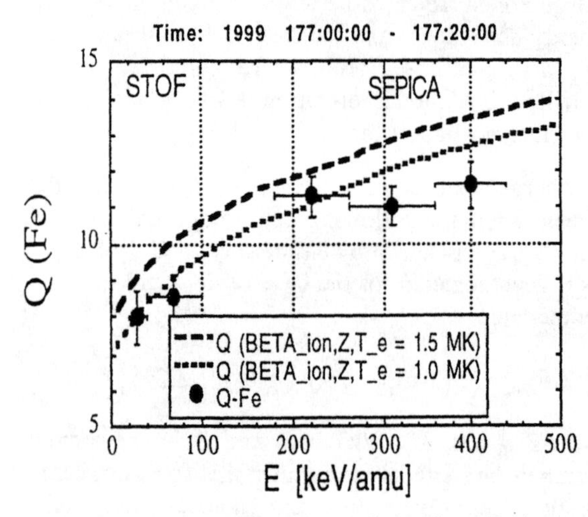

FIGURE 3. Comparison of experimental results (our case "B") with the predicted mean ionic charge values for Fe. See the discussion and equ. (1).

CONCLUSIONS

The mean ionic charge of Fe shows large event-to-event variations of the energy dependence. For events dominated by local interplanetary acceleration, the Fe mean ionic charge increases from solar wind values by 1-2 charge units within the energy range 30 - 450 keV/nuc. These small increases could be due to Q/M dependent acceleration effects. For the case of acceleration in the dense coronal plasma, near to the CME origins, the mean Fe charge state varies over a significantly larger interval of 4 charge units and this spreading can be explained in terms of stripping effects during acceleration in dense coronal plasma. With oxygen the variations observed in the events studied so far are not large enough in order to make a definitive statement and further work is needed.

ACKNOWLEDGEMENTS

The authors would like to thank the many unnamed contributors to the successful completion of the SEPICA and STOF/CELIAS instruments onboard ACE and SOHO, respectively. The work on SEPICA was supported under NASA Contract NAS5-32626 and the work on STOF was supported by DLR / Germany under contracts 50 OC 89056 and 50 OC 96059.

REFERENCES

1. Hovestadt, D. et al., *The SOHO Mission* (editors: Fleck, B. et al.), *Solar Physics* **162**, 441-481 (1995).
2. Mazur, J. E., G. M. Mason, M.D. Looper, et al., *Geophys. Res. Letters* **26**, 173-176 (1999).
3. Möbius, E., et al., *Space Science Reviews* **86**, 449-495 (1998)
4. Möbius, E., M. Popecki, B. Klecker, et al., *Geophys. Res. Lett.* **26**, 145-148 (1999).
5. Oetliker, M., B. Klecker, D. Hovestadt, et al., *Astrophys. J.* **477**, 495-501 (1997).
6. Reames, D.V., C.K Ng., and A.J Tylka, *Geophys. Res. Lett.* **26**, 3585-3588 (1999).
7. Skoug, R.M., S.J. Bame, W.C. Feldman, et al., *Geophys. Res. Lett.* **26**, 161-164 (1999).
8. Stovpyuk, M.F., Ostryakov, V.M., *Proceedings of the 26th ICRC*, vol. **6**, 66-69 (1999).
9. Barghouty, A.F., R.A.Mewaldt, *Proceedings of the 26th ICRC*, vol. **6**, 138-141 (1999).

Temporal Evolution in the Spectra of Gradual Solar Energetic Particle Events

Allan J. Tylka[1], Paul R. Boberg[2,1], Robert E. McGuire[3], Chee K. Ng[4,5], and Donald V. Reames[4]

[1]E.O. Hulburt Center for Space Research, Naval Research Laboratory, [2]Consultant, [3]National Space Science Data Center, NASA Goddard Space Flight Center, [4]Laboratory for High Energy Astrophysics, NASA Goddard Space Flight Center, [5]Dept. of Astronomy, University of Maryland

Abstract. We examine solar energetic particle (SEP) spectra in two very large "gradual" events (20 April 1998 and 25 August 1998), in which acceleration is caused by fast CME-driven shocks. By combining data from ACE/SIS, Wind/EPACT/LEMT, and IMP8/GME, we examine all major species from H to Fe, from ~2 MeV/nuc to the highest energies measured. These events last for several days, so we have divided the events into 8-hour intervals in order to study the evolution of the spectra. The spectra reveal significant departures from simple power laws. Of particular note is the behavior at high energies, where the spectra exhibit exponential rollovers. We demonstrate that the fitted e-folding energies reflect both ionic charge states and a complex but orderly temporal evolution. We speculate that this behavior may be related to evolving rigidity dependence in the near-shock diffusion coefficient, which is of potentially great importance for models of SEP acceleration and transport.

INTRODUCTION

In this paper, we focus on the two largest "gradual" SEP events seen so far in Solar Cycle 23: 20 April 1998, which had the largest fluence of >10 MeV/nuc particles; and 25 August 1998, which had the highest intensities of ~MeV/nuc particles. Both of these events were caused by fast CME-driven shocks. CMEs and shocks certainly slow down [1] and otherwise evolve as they move through the changing interplanetary plasma conditions of the inner heliosphere. This evolution should be reflected in the SEP spectra. Thus, unlike nearly all previous SEP studies [e.g., 2, 3], we do not simply examine event-integrated spectra. Instead, we take advantage of the large geometry factors of Wind/EPACT/LEMT [4] and ACE/SIS [5] to investigate how the spectral characteristics evolve during the event. We combine these data with extensive H and He measurements from IMP-8/GME [6]. As we shall demonstrate, the H and He spectra provide a critical 'calibration' for understanding the spectra of heavier ions.

Ellison and Ramaty [7] have discussed shock-theory expectations for SEP spectra. The differential energy spectrum of ion species X should follow a power-law, modulated by an exponential,

$$F_X(E) = C_X \cdot (E^2 + 2ME)^{-\gamma} \cdot \exp(-E/E_{0X}) \quad (1)$$

where C_X is a normalization factor and M = 938.3 MeV/nuc. The spectral index γ is determined by the shock compression ratio and is the same for all species. The exponential rollover at high energies is caused by finite shock-lifetime and/or finite shock-size effects. Moreover, if the diffusion coefficient at the shock has the form $\kappa \sim \beta R^{\alpha}$ (β = particle speed, R = rigidity) *with* $\alpha = 1$, then the e-folding energy E_{0X} varies from species to species, with a value which is directly proportional to the ion's charge-to-mass ratio, Q_x/A_x. Strictly speaking, equation (1) applies only at the shock. But we will nevertheless compare it to observed spectra, implicitly assuming that effects related to escape from the shock region and subsequent interplanetary transport are relatively small at sufficiently large rigidities.

CP528, *Acceleration and Transport of Energetic Particles Observed in the Heliosphere: ACE 2000 Symposium*, edited by Richard A. Mewaldt, et al.
© 2000 American Institute of Physics 1-56396-951-3/00/$17.00

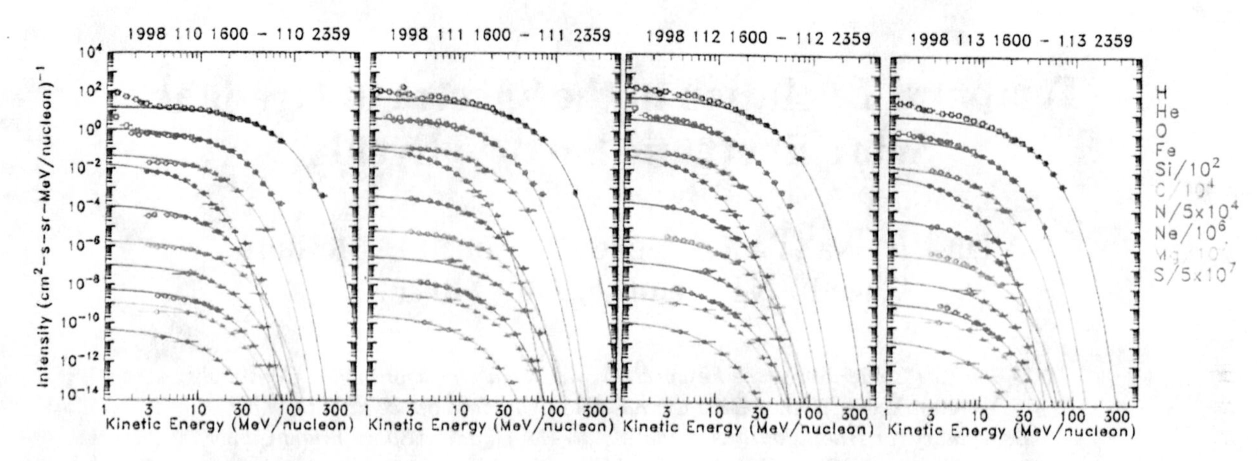

Figure 1. Examples of 8-hour-averaged spectra in the 20 April 1998 event. Each species is color-coded and shown in the same order from top to bottom as in the legend at the right. Note the scale factors used in plotting some elements. Data come from IMP-8/GME (■), Wind/EPACT/LEMT (●) and ACE/SIS (▲). Curves are exponential fits to the high-rigidity points. Open symbols denote low-rigidity points which were not used in the fits. Datapoints consistent with residual Galactic and/or anomalous cosmic-ray background are not shown.

20 APRIL 1998 EVENT

This SEP event was caused by a fast (~1600 km/s) western-limb CME which was first detected by SoHO/LASCO at 10:07 UT on 20 April 1998. The associated flare (~W87) was well removed from the footpoint of the Sun-Earth field line, and flare and CME activity were low in the preceding ~2 weeks. Thus, this event has provided an unusually "clean" baseline for studying gradual events. The event shows strong systematic variation in elemental composition [8] due to rigidity-dependent transport through proton-amplified Alfvén waves [9, 10].

Figure 1 shows a sample of the thirteen 8-hour averaged particle spectra starting at 16 UT on 20 April 1998 (DOY 110), ~6 hours after onset so as to avoid initial dispersion effects. Non-SEP backgrounds (as estimated from solar-quiet periods [11]) have been subtracted. In addition, slight adjustments have been made for normalization discrepancies (typically ~10-20%) among the instruments. These adjustments have been applied "globally" to all measurements in each 8-hour interval, so as to preserve the spectral shapes reported by each instrument.

Eight hours is long compared to Sun-Earth transit times at these energies, so residual velocity dispersion effects should be small. At low energies, these spectra are very flat, with spectral index $\gamma \sim 0$. This value is inconsistent with any physical shock compression ratio. In fact, this flattening is another reflection of rigidity-dependent escape from the shock region [9]. When these spectra are plotted vs. rigidity [8], assuming typical charge states, the plateaus extend to ~230 MV. Thus, in order to minimize the impact of such transport effects in this analysis, the exponential fits in Figure 1 have used only datapoints corresponding to rigidity >230 MV.

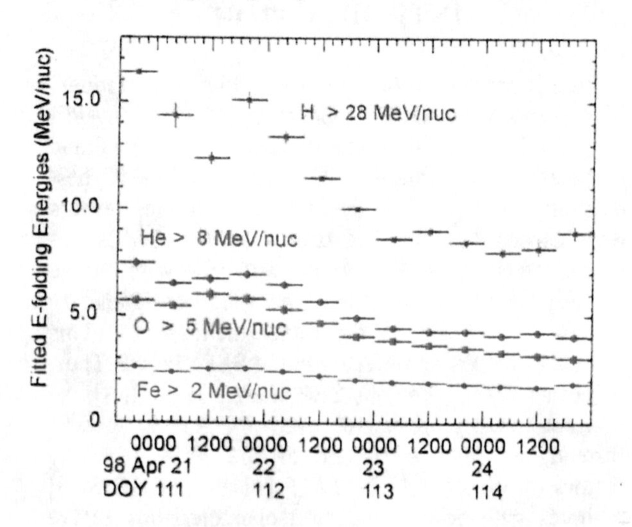

FIGURE 2. E-folding energies E_0 vs. time, from exponential fits of high-energy H, He, O, and Fe in Figure 1.

The steepness of the Fe rollover relative to that of other heavy ions is particularly striking. As seen in Figure 1, exponentials generally provide acceptable fits to the high-rigidity parts of the spectra throughout this event. Figure 2 shows temporal evolution of the fitted e-folding energies (E_0) for some species (H, He, O, and Fe). E_0 values decline more or less smoothly during the first ~3 days of the event[1]. However, starting shortly before shock arrival at ~17 UT on DOY 113, E_0 values become roughly constant. This behavior is consistent with onset of the invariant spectrum region [12, 13].

The fitted E_0 values in Figure 2 are roughly proportional to Q/A, as suggested by Ellison & Ramaty [7] for $\kappa \sim \beta R$. But careful examination shows that E_{0He}/E_{0H} differs by a small but significant amount from 0.5 early in the event. As a slightly more complicated alternative, we therefore consider e-folding energies proportional to a *power of Q/A*, i.e.,

$$E_{0x} = E_{0H} (Q_x/A_x)^{\delta} \quad (2)$$

Figure 3 shows the values of δ, as determined from the He E_0 values, assuming that the He ions are fully stripped. These δ values are generally, but not always, close to unity.

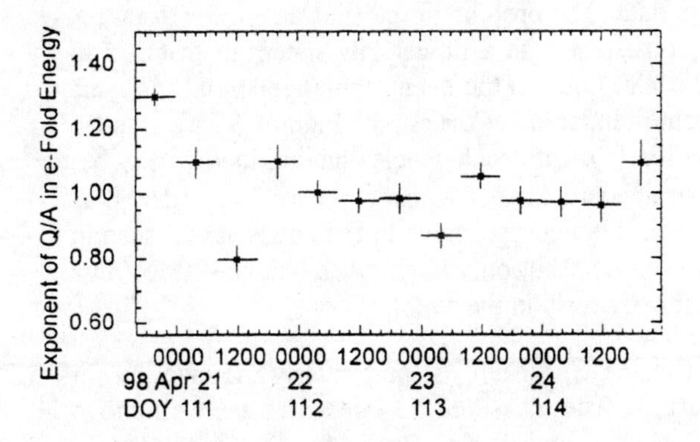

FIGURE 3. Value of the exponent δ in equation (2), as determined from H and He.

We have thus used the proton and alpha spectra to "calibrate" the Q/A dependence of E_0 in this event. We now employ equation (2), along with the E_{0H} values in Figure 2 and the δ values in Figure 3, to infer charge states for all the other species in each time interval. Figure 4 shows these results. There is little time dependence in these charge states. ACE/SEPICA also reports virtually no time dependence in their directly measured charge states at ~0.2-1.0 MeV/nuc in this event [14].

Table 1 compares the mean charge states (from averaging over all time intervals) with those reported by ACE/SEPICA for this event [14]. The values[2] agree remarkably well, except for Mg. In fact, all of the charge states in Table 1 are typical of the solar wind and consistent with a single source plasma temperature of ~1.5 MK, except for the ACE/SEPICA Mg result, which is low by ~1 charge unit [14].

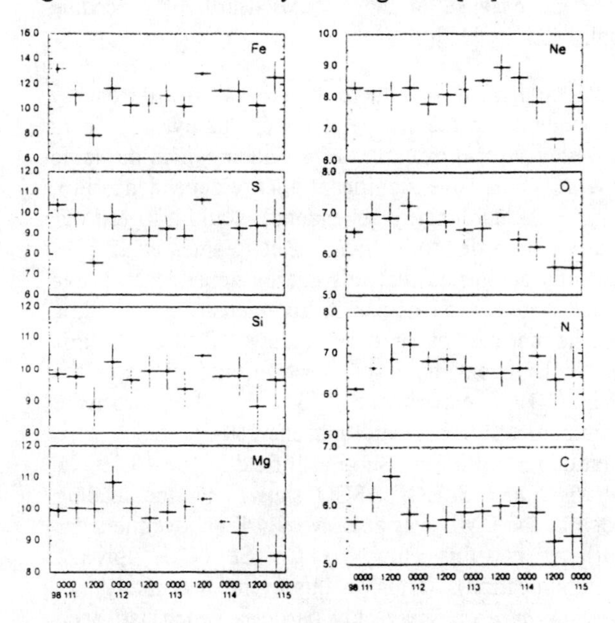

Figure 4. Inferred charge states vs. time in the 20 April 1998 event.

TABLE 1. Mean SEP Charge States: 20-25 April 1998

Element	ACE/SEPICA ~0.2 – 1.0 MeV/nuc	This Analysis ~2 – 60 MeV/nuc
C	5.68 ± 0.17	5.85 ± 0.17
N	--	6.6 ± 0.1
O	6.92 ± 0.17	6.59 ± 0.08
Ne	8.27 ± 0.23	8.20 ± 0.11
Mg	8.96 ± 0.26	9.84 ± 0.13
Si	9.67 ± 0.29	9.78 ± 0.14
S	--	9.5 ± 0.2
Fe	11.68 ± 0.29	11.4 ± 0.2

The internal consistency of these charge state determinations suggests that fitted E_0 values are indeed meaningful quantities. But it is important to understand *why* these spectral fits reflect the charge states so well in this particular event. For example, we

[1] The exception is a slight increase in E_0 values between Periods 3 and 4. This increase appears to be associated with abrupt changes in particle intensities and large fluctuations in magnetic field directions, perhaps indicating a change in our connection to the CME-driven shockfront.

[2] Table 1 shows only formal errors from the fitting procedures and error propagation. In order to assess potential *systematic* errors in this study, we repeated the charge-state analysis using the full equation (1) form (not just an exponential) and all datapoints above 2.4 MeV/nuc (not just those at high rigidity). This yielded slightly different E_0 values, but none of the average inferred charge states changed by more than 6%.

would not expect good fits to the Fe spectra if the Fe ions actually arose from a broad distribution of charge states. But according to ACE/SEPICA, this event has one of the narrowest SEP Fe charge distributions observed so far [M. Popecki, private communication], perhaps because of the relative dearth of preceding solar activity [15].

Similarly, why is there apparently so little energy dependence in the charge states of this event? There are now several examples of events in which the ionic charge states have significant energy dependence [16, 17, 18]. Such energy dependence would also foil our charge state determinations. But Reames et al. [19] recently suggested that this energy dependence arises when acceleration begins in high-density regions low in the corona, at altitudes below 0.2 Rs. In this particular event, however, electron onset times in the Wind/3DP experiment (S. Krucker, private communication) indicate that energetic particles were first deposited on the Sun-Earth field line at 10:19 UT *at the Sun.* SoHO/LASCO shows that the leading edge of the CME was already ~4.4 Rs above the solar surface at this time (O.C. St. Cyr, private communication). Thus, in this event, it is likely that we saw *here at Earth* only particles which had been accelerated in low-density regions. Hence, no further stripping – and no strongly energy-dependent charge states – should be expected in this event.

25 AUGUST 1998 EVENT

The 25 August (DOY 237) 1998 SEP event offers an instructive comparison to the 20 April 1998 event [8]. This was a central meridian event, associated with an X1.0 flare at N35E09 at 21:50 UT on DOY 236. A strong IP shock arrived at Earth ~34 hours later, corresponding to a mean transit speed of ~1200 km/s. SoHO/LASCO observations are unavailable, but this event was presumably associated with a fast halo CME. As in the 20 April 1998 event, the footpoint of the Sun-Earth field line was far removed from the flare site; it is thus unlikely that we saw flare-accelerated particles. But, whereas the 20 April 1998 event only showed us particles accelerated in the relatively weak shock at the far-eastern flank of the CME, in this case we observe particles coming from strong parts of the shock for most of the event. Also, unlike the relatively quiet conditions that preceded the April event, two smaller SEP events occurred in the week preceding this event.

Figure 5 shows a sample of 8-hour averaged particle spectra in this event. Exponential rollovers are difficult to detect early in the event, presumably because they occur at energies above the range of these data. The protons in the first panel, for example, are consistent with a power law spectrum from ~5 to 500 MeV. Later in the event, the exponential rollovers become apparent. Curves in Figure 5 are fits to equation (1), with each species having its own E_{0X} but constrained to keep the same power law index γ as protons. Flattening relative to the power law is seen in this event too, but only at energies below ~3 MeV/nuc and mostly early in the event.

The Fe spectrum is particularly noteworthy in Figure 5. The measured Fe intensities are well above GCR background levels. But in the first panel, the Fe spectrum is complicated and cannot be fit by the

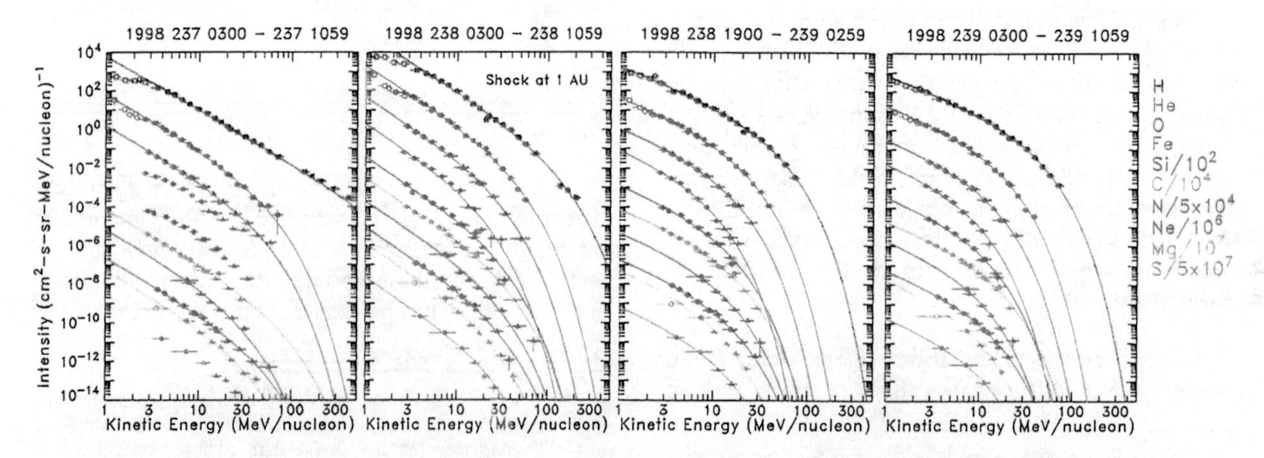

Figure 5. Examples of 8-hour-averaged spectra in the 25 August 1998 event. Symbols are as in Figure 1. Curves are fits to equation (1), using all datapoints above 2.4 MeV/nuc. Open symbols denote low-energy points that were not used in the fits. No satisfactory fits were found for Fe, S, and Si in the first panel. The shock arrived at 1 AU at the middle of the time period covered by the second panel.

simple functional form of equation (1). Similar problems appear for S and Si. Only later in the event does this form work reasonably well for these species. When integrated over the entire event duration, this event is "Fe poor" at ACE/SIS energies. But in the first time interval, the Fe/O ratio increases with energy and approaches unity (or perhaps even to exceeds it) at the highest energies. Other events have shown this behavior in event-integrated spectra [3].

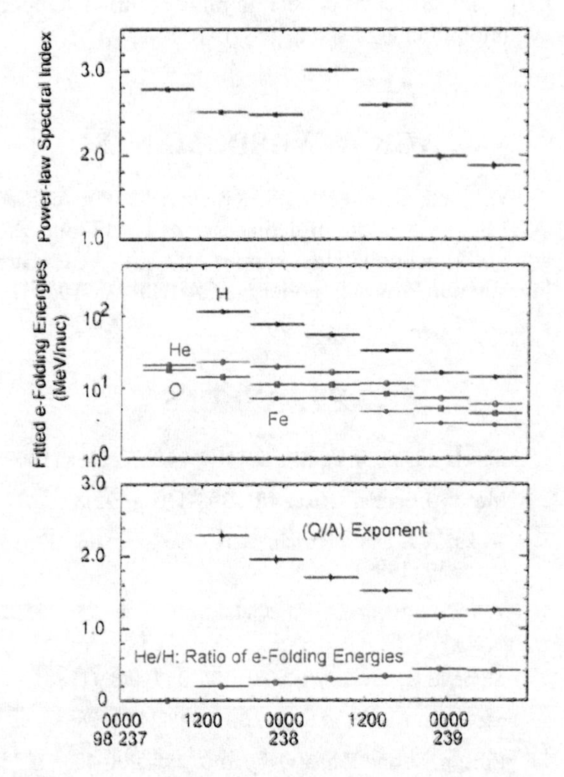

FIGURE 6. Evolution of fit parameters in the 25 August 1998 event. (a) top: power-law spectral index γ; (b) middle: E_0 for H, He, O, and Fe; (c) bottom: ratio E_{0He} to E_{0H}, and exponent δ for equation (2), as determined from H and He.

Figure 6a shows the power law spectral indices, as determined from the protons. Shock passage occurs in the middle of the fourth interval. The spectral indices are significantly smaller both before and after shock passage. This suggests that the apparent temporal evolution in the spectral index may be more related to transport effects, rather than changes in the compression ratio. Such transport effects are likely to be rigidity dependent. Our procedure of constraining all species to have the same γ as protons is therefore probably not strictly correct, even though it appears to work reasonably well.

Figure 6b shows the fitted E_0 energies for H, He, O, and Fe, except for H in the first time interval (where E_{0H} cannot be determined since no rollover is observed below 500 MeV) and Fe in the first two intervals (where fits to equation (1) fail). *The E_0 energies in this event are much larger and evolve more rapidly than in the April event.* Only late in the event, in the post-shock invariant spectrum region, do these E_0 values become similar to those of the April event.

Figure 6c plots the ratio of the fitted E_0 energies for H and He. This ratio does not attain ~0.5 until late in the event. The direct proportionality $E_0 \sim Q/A$ thus does not apply throughout most of this event. Figure 6c also shows the values of exponent δ for equation (2), derived from the fitted H and He E_0 values. Initially, $\delta \sim 2.4$ (and perhaps even larger, since we cannot determine E_{0H} in the first time interval) and evolves smoothly towards unity. Again, *this behavior is quite different from the relatively weak Q/A dependence shown in Figure 3 for the April event.*

Finally, Figure 7 shows the ionic charge states we deduce by applying equation (2) and the parameters from Figure 6 to other species. No charge states are shown for the first time interval, where E_{0H} could not be determined; nor for Fe in the second interval, where the Fe spectrum was inconsistent with equation (1). In this event, the inferred charge states are not consistent with a single plasma temperature. No ACE/SEPICA measurements are available for this event. However, these inferred charge states are similar to previous measurements of gradual events. In particular, the mean Fe charge state here agrees well with the result reported by SAMPEX/MAST in the Oct-Nov. 1992 event at ~15-70 MeV/nuc [20]. The higher Fe charge state here (compared to the April event) may also be indicative of a source population comprising a mixture of solar-wind and coronal [21] and/or remnant flare-accelerated suprathermals [15].

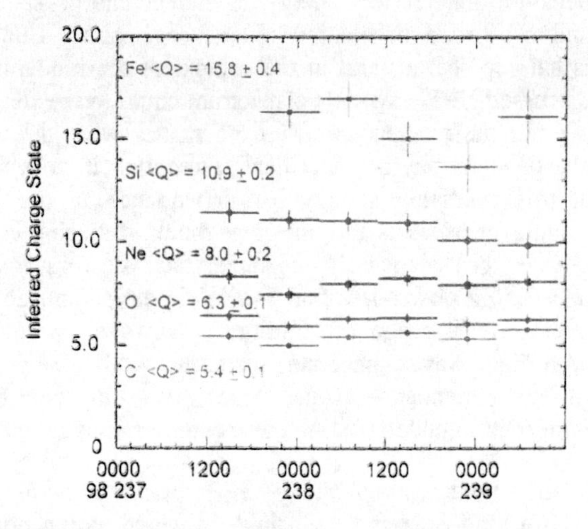

FIGURE 7. Inferred charge states vs. time in the 25 August 1998 event.

DISCUSSION AND CONCLUSIONS

The two events considered here show a complex but orderly pattern in the temporal evolution of SEP spectral shapes that was inaccessible to earlier instruments. The striking cogency of the charge state determinations suggests that the fitted e-folding energies do indeed contain fundamental information about the acceleration process. According to Ellison & Ramaty [7], we should expect $E_{0X} \sim Q_x/A_x$ when the near-shock diffusion coefficient is proportional to the first power of rigidity. It is therefore tempting to think that the stronger Q/A dependence in e-folding energies of the August event reflects even stronger rigidity dependence in the near-shock scattering. But what drives the temporal evolution of the scattering during the event? And why are scattering conditions so different in these two events? Indeed, there are other events in the historical record in which the spectral rollovers occur at much higher energies. For example, in the 29 September 1989 event, Lovell et al. [22] showed that the proton spectrum had $E_0 \sim 1$ GeV!

We now know that in very large SEP events, escape from the shock region and subsequent interplanetary transport is governed by Alfvén waves amplified by the streaming energetic protons themselves [8, 9, 10]. These same waves presumably also have significant impact on acceleration efficiency, through the increase in cross-shock scattering that they cause. Moreover, the spectrum of these proton-amplified waves is highly dynamic. Thus, it is also tempting to think that these waves play a key role in the observed spectral evolution.

The behavior of the Fe spectrum early in the August event is particularly puzzling. In part, this behavior may reflect energy-dependent charge states and/or a broad distribution of Fe charge states. But it is hard to see, at least in the context of the modeling discussed here, how the Fe spectrum could ever extend beyond the oxygen spectrum, no matter what the Fe charge state may be. A critical assumption in arriving at this relatively simple Q/A-dependence in the e-folding energies is that the near-shock scattering can always be adequately characterized as a *single, decreasing* power-law function in rigidity, applicable over a wide range of rigidities. However, proton-amplified waves generate scattering with complex rigidity dependence. Thus, these waves may also be important in understanding these complex spectra.

In summary, high-energy SEP spectra provide us with a kind of 'remote sensing', in which exponential rollovers probe scattering conditions in the shock region, even when the shock is still far from Earth. However, it should be noted that in many events, these rollovers cannot be adequately defined with the limited energy range of current instruments. This is especially true for protons and alphas. Future SEP experiments should extend spectral measurements to ~GeV/nuc energies. In addition, as we have shown here, high-energy charge states critically test our understanding of these spectra. Thus, new experiments should also provide a follow-on to SAMPEX [17,20] and LDEF [23], with sufficient collecting power to track temporal evolution in charge states at ~10-100 MeV/nuc.

ACKNOWLEDGMENTS

We thank B. Klecker, S. Krucker, M. Popecki, and O.C. St. Cyr for helpful discussions. AJT and PRB gratefully acknowledge support by the ACE Guest Investigator Program, under NASA DPR#W-19501.

REFERENCES

1. Sheeley, N.R. Jr. et al., *JGR* **104**, 24739-24768 (1999).
2. Mazur, J.E. et al., *Ap.J.* **401**, 398-410 (1992).
3. Tylka, A.J. and Dietrich, W.F., *Radiat. Meas.* **30**, 345-359 (1999).
4. von Rosenvinge, T.T. et al., *Sp.Sci.Rev.* **71**, 155-206 (1995).
5. Stone, E.C. et al., *Sp.Sci.Rev.* **86**, 357-408 (1998).
6. McGuire, R.E. et al., *Ap.J.* **301**, 938-961 (1986).
7. Ellison, D. and Ramaty, R., *Ap.J.* **298**, 400-408 (1985).
8. Tylka, A.J. et al., *GRL* **26**, 2141-2144 (1999).
9. Ng, C.K. et al., *GRL* **26**, 2145-2148 (1999).
10. Lee, M.A., *JGR* **88**, 6109-6119 (1983).
11. Leske, R.A. et al., *Proc. 26th ICRC* **7**, 539-542 (1999).
12. Reames, D.V. et al., *Ap.J.* **491**, 414-420 (1997).
13. Kahler, S.W. et al., *Proc. 26th ICRC* **6**, 26-29 (1999).
14. Klecker, B. et al., *Proc. 26th ICRC* **6**, 83-86 (1999).
15. Mason, G.M., et al., *Ap.J. Lett.* **525**, L133-L136 (1999).
16. Oetliker, M. et al., *Ap.J.* **477**, 495-501 (1997).
17. Mazur, J.E. et al., *GRL* **26**, 173-176 (1999).
18. Moebius, E. et al., *GRL* **26**, 145-148 (1999).
19. Reames, D.V. et al., *GRL* **26**, 3585-3588 (1999).
20. Leske, R.A. et al., *Ap.J. Lett.* **452**, L149-L152 (1995).
21. Boberg, P.R. et al., *Ap.J. Lett.* **471**, L65-L68 (1996).
22. Lovell, J.L. et al., *JGR* **103**, 23733-23742 (1998).
23. Tylka, A.J. et al., *Ap.J. Lett.* **444**, L109-L113 (1995).

Particle Acceleration at Coronal Mass Ejection Driven Shocks

W.K.M. Rice and G.P. Zank

Bartol Research Institute, University of Delaware, Newark, DE

Abstract. There is increasing evidence to suggest that energetic particles observed in "gradual" solar energetic particle (SEP) events are accelerated at shock waves driven out of the corona by coronal mass ejections (CMEs). Energetic particle abundances suggest too that SEPs are accelerated from in situ solar wind plasma rather than from high temperature material. We present here a a dynamical time dependent model of particle acceleration at a propagating interplanetary shock. The model includes the determination of the particles injection energy, the maximum energy of particles accelerated at the shock, energetic particle spectra at all spatial and temporal locations, and the dynamical distribution of particles that are trapped in the post shock flow and those that escape upstream and downstream from the evolving shock complex.

INTRODUCTION

Circumstantial evidence suggesting that energetic particles observed in large "gradual" solar energetic particle (SEP) events are accelerated at coronal mass ejection (CME) driven shocks has been accumulating for the last decade [see Reames, 1995, 1999; Cane, 1995; Kahler, 1992; Dryer, 1994; Gosling, 1993].

Energetic particle abundances [Mason et al., 1984; Reames and Stone, 1986] suggest too that low energy SEPs are accelerated from in situ solar wind plasma rather than from high temperature flare material. The most commonly accepted mechanism for particle acceleration at a shock wave is first-order Fermi acceleration, or diffusive shock acceleration [Axford et al., 1977; Bell, 1978]. We present here a dynamical, time dependent model in which particles are accelerated at a CME driven shock wave. Once accelerated the particles can either be trapped in the post shock flow (which is assumed to be turbulent) or can escape upstream or downstream into the less disturbed interplanetary medium. The model is able to produce realistic spectra at any spatial or temporal location and gives the temporal evolution of the particle intensity observed at a given spatial location.

BASIC METHOD

A numerical hydrodynamic code [Pauls et al., 1995] is used to produce a steady state solar wind solution between 0.1 and 3 AU. The magnetic field is added analytically using the standard Parker spiral [Parker, 1958]. A CME is injected at the inner boundary and produces a spherically symmetric, propagating shock wave. The model assumes that in a small time interval (Δt) the shock can be treated as steady and planar. As the shock propagates out into the solar wind, the region swept up by the shock is approximated by a series of nested shells, with shell i produced during time interval Δt_i. This is very similar to the onion shell models used to study particle acceleration at supernova remnant (SNR) shocks [Moraal & Axford, 1983; Bogdan & Völk, 1983; Völk et al., 1988] and is schematically represented in Figure 1.

During each small time interval the shock speed and compression ratio are calculated numerically. The injection energy (the energy required for particles to be injected into the diffusive shock process) is taken to be some fraction of the downstream thermal energy per particle. The maximum particle energy is calculated by equating the dynamical shock lifetime with the acceleration timescale [Drury, 1983]. The accelerated particles are then assumed to have a power law spectrum with the spectral index given by the compression ratio.

The number of accelerated particles is taken to be some small fraction of the number of solar wind particles swept up during the time interval considered. The particle distribution is given by the solution to the steady state convection diffusion equation [Drury,

CP528, *Acceleration and Transport of Energetic Particles Observed in the Heliosphere: ACE 2000 Symposium,* edited by Richard A. Mewaldt, et al.
© 2000 American Institute of Physics 1-56396-951-3/00/$17.00

1983]. For a given momentum, the intensity is constant in the shell created during the current time interval and decays exponentially upstream of the shock, with a length scale determined by the assumed momentum dependent diffusion coefficient. The diffusion coefficient is determined from the wave energy density [Bell, 1978]. In this case, the shock is strong enough to use the Bohm limit in which the particle's mean free path equals its gyroradius. The diffusion approximation is assumed to apply only within a certain distance upstream of the shock. Any particle that diffuses beyond this distance escapes and propagates in a scatter free fashion to the observer.

Unlike Ng et al. [1999], who carefully model the transport of particles that escape the shock but do not model particle acceleration dynamically, we treat the escaped particles very simply (using a telegrapher-like transport model) and concentrate on acceleration at the shock and the subsequent evolution of those particles trapped downstream of the shock complex.

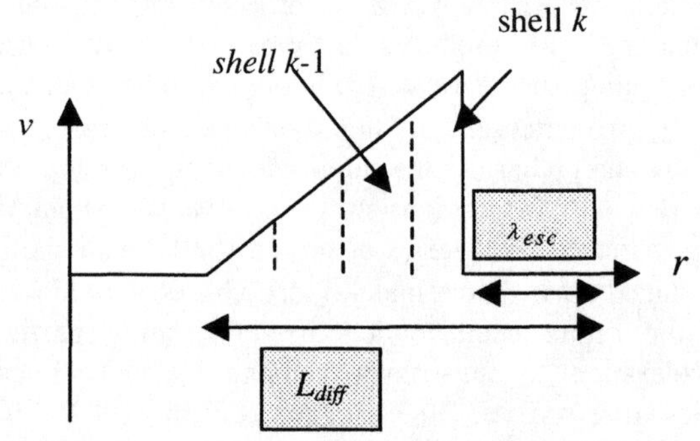

FIGURE 1. Schematic showing the shock profile, the nested shells, the escape length (λ_{esc}), and the region over which the diffusion approximation is taken to be valid (L_{diff}).

Particles that do not escape are trapped within the most recently created shell. Older shells contain particles accelerated at earlier times. These particles convect with the shells, cool as the shells expand, diffuse between shells, and may eventually escape both upstream and downstream of the shock complex. At all spatial locations, and at all times, we are therefore able to calculate the resulting particle distributions for all momenta. Full details of the model can be found in Zank et al. [2000].

RESULTS

For the simulation presented here, we consider a strong shock with a compression ratio of 4 at all times. Figure 2 shows the numerically computed shock speed and the resulting injection and maximum particle energies. The shock speed is almost constant at just above 1400 km s^{-1} out to about 0.5 AU. It then decreases to about 850 km s^{-1} by 2 AU. This profile is similar in shape, but not magnitude, to that of a flare associated shock shown by Cane [1995] (her Figure 3).

Since the shock has a large Mach number, the injection energy does not change significantly, dropping from 5 keV at 0.1 AU to ~ 2.3 keV at 2 AU. At small radial distances, where the magnetic field is large and the shock speed is high, the maximum energy is ~ 2 GeV. It has been suggested [Kahler, 1994; Reames et al., 1999] that shock acceleration starts at radial distances considerably less than 0.1 AU. Calculations of the maximum particle energy for an initial radial distance of 0.05 AU (with a shock speed of 1500 km s^{-1}) show that the maximum particle energy would be higher by a factor of ~ 3–4. As the shock propagates out into the weakening magnetic field, the maximum energy drops rapidly, tending to ~ 100 MeV at 2 AU.

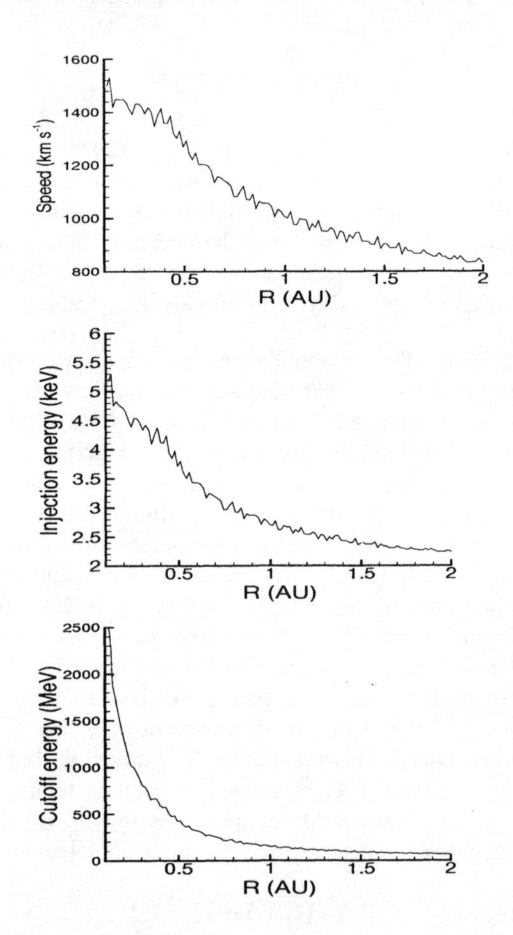

FIGURE 2. Numerically computed shock speed, and the resulting injection and maximum energies, plotted against radial distance.

Figure 3 shows the energetic particle spectra as the shock complex convects past an observer at 0.5 AU. The spectrum observed as the shock reaches the observer (shell 24) has a convex shape. Such a "broken" power law distribution was described briefly in Reames [1999] and an energetic particle spectrum upstream of an interplanetary shock was shown. This shape is a consequence of the higher energy particles (which have the largest diffusion coefficients) finding it easier than the lower energy particles to escape upstream of the shock.

As the post shock flow convects past the observer, the spectra become flatter and eventually concave. The lowest energy particles (with the lowest diffusion coefficients) are essentially trapped within their shells and simply cool as they convect out with the expanding solar wind. That the oldest shells (shell 3 is the oldest) have the lowest minimum energy is evidence of this cooling. The highest energy particles have diffusion coefficients large enough for them to mix easily across all or most of the shells. Thus the intensity at the highest energies is almost the same in all the shells. This reduces the intensity, at these high energies, in the newest shells (producing a convex shape) and increases it in the oldest shells (producing a concave shape). The concavity in the oldest shells is enhanced by the intermediate energy particles, in shells near the downstream boundary, being able to escape the shock complex. At these energies the diffusion coefficient is not large enough for these particles to mix across all the shells and hence their intensities are reduced in the oldest shells but not in the newest.

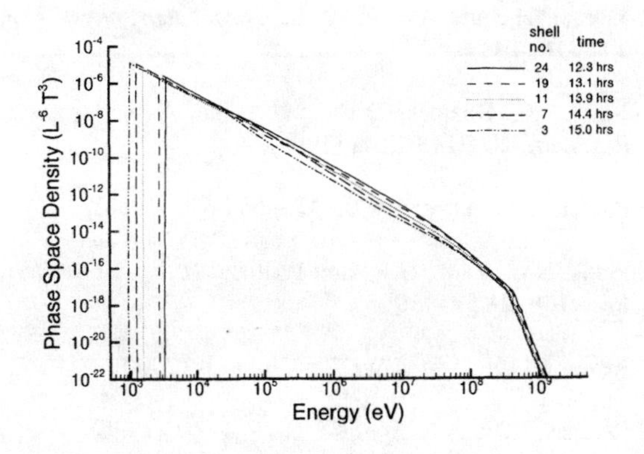

Figure 3. Particle distribution functions observed at 0.5 AU as the shock complex convects past the observer. An E^{-2} spectrum is included to guide the eye.

Since we keep track of the number of particles of all energies of interest that escape or that are trapped in the post shock flow, we can determine the intensity profiles of particles of various energies observed at a given location. Figure 4 shows the intensity profiles observed at 1 AU at 5 different energies. Particles of the highest energies (177 MeV and 46 MeV) arrive almost immediately after the shock is initiated at 0.1 AU. Their intensities are almost constant until they begin an exponential increase prior to the shock arrival. This increase begins once the shock is an escape length from the observer. The discontinuous jump in intensity at this point is a consequence of the assumption that particles beyond this escape length propagate in a scatter free fashion. The 177 MeV intensity starts decreasing before the arrival of the shock. From Figure 2 it can be seen that the shock is no longer accelerating particles of this energy and hence the observer is seeing particles that have been trapped in the post shock flow and subsequently leaked out gradually.

The lower energy particles (322 keV, 43 keV and 16 keV) arrive a few hours to a day after the shock is initiated at 0.1 AU, with the lowest energy particles arriving last. Their intensities increase when the shock arrives at the observer. They do not show any exponential intensity increase immediately prior to the shock arrival since their escape lengths (which are momentum dependent) are much shorter than the resolution of Figure 4. Once the shock has passed the observer, the intensities slowly decrease due to the effects of cooling. This is also evident for the 46 MeV particles but, at this energy, the effect of cooling is offset by their large diffusion coefficient allowing the particles to mix across a large portion of the post shock structure. Once the entire shock complex has passed the observer the intensities decrease and the observer measures particles that have escaped downstream of the post shock flow.

CONCLUSION

Using a numerical, hydrodynamic code to model a CME driven shock wave, together with analytic solutions to the cosmic ray transport equation, we are able to model particle acceleration at CME driven shocks in a dynamical, time dependent manner. We find that the energy required for particles to be injected into the diffusive shock acceleration mechanism is between 5 keV and 3 keV, and that the maximum energies can be as high as a few GeV at early times, decreasing to below 100 MeV by 2 AU. In the shock precursor, the energetic ions drive unstable Alfvén waves which scatter the particles, allowing them to return to the shock and undergo diffusive shock acceleration. In this particular case, we have chosen a shock strong enough that the Bohm limit applies. By solving for the wave energy density [Bell, 1978] this model can be applied to shocks of arbitrary strength. Using the Bohm limit we find that the radial mean free paths are ~ 2 orders of magnitude smaller than interplanetary values.

The accelerated particle spectrum experiences significant evolution as it is convected through the flow downstream of the shock. At low energies it maintains its original spectral index. At high energies the large diffusion coefficient mixes particles across all the shells producing a convexity in the newer shells and a concavity in the older shells. The concavity in the older shells is enhanced be the loss of intermediate energy ions through the inner boundary of the post shock flow.

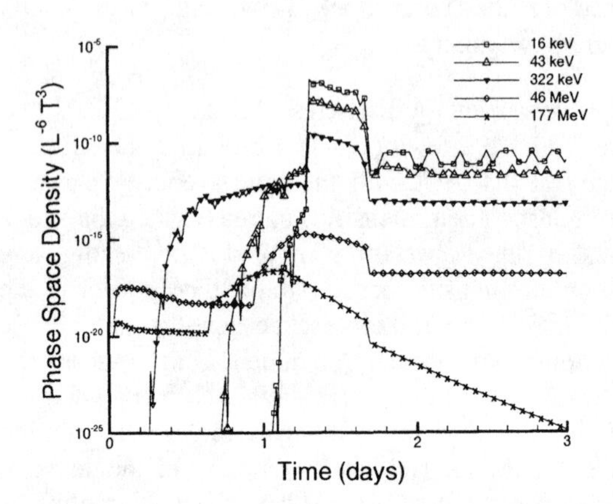

Figure 4. The temporal evolution of the intensity profiles observed at a given location (1 AU) at five different energies.

The intensity profiles as a function of time show a number of features. A sudden rise (almost instantaneous at the highest energies) followed by a plateau. At the lowest energies, the shock arrives before the plateau can form. Prior to the shock arrival there is an exponential growth in intensity. This is not visible at the lowest energies as their diffusion lengths are too short. At the highest energy, the intensity decreases before the arrival of the shock due to the shock no longer accelerating particles of this energy. The remaining profiles peak at the shock and decay slowly afterwards. Once the entire post shock flow has passed the observer, the intensity drops and particles that have escaped downstream of the shock material are observed.

The spectra and energetic particle intensity profiles presented here are, in certain cases, very similar to those observed (e.g. Reames [1999]). More detailed models (2-dimensional, different parameters, etc.) should allow us to consider further the effects of shock strength and observer location (i.e. magnetic connectivity to the shock front) on the observed spectra and particle intensities.

ACKNOWLEDGMENTS

The support of NSF grants ATM-9713432 and ATM-9713223 and NASA awards NAG5-7796 and NAG5-5054 is gratefully acknowledged.

REFERENCES

1. Axford, W.I., Leer, E., and Skadron, G., *Proc. 15th Int. Cosmic Ray Conf. (Plovdiv)*, **11**, 132 (1977).
2. Bell, A.R., *Mon. Not. Roy. Astron. Soc.*, 182, 147-156 (1978).
3. Bogdan, T.J., and Völk, H.J., *Astron. Astrophys.*, 122, 129-136 (1983).
4. Cane, H.V., *Nucl. Phys. B.*, **39A**, 35-44 (1995).
5. Drury, L.O'C., *Rep. Prog. Phys.*, **46**, 973-1027 (1983).
6. Dryer, M., *Space Sci. Rev.*, **67**, 363-419 91994).
7. Gosling, J.T., *J. Geophys. Res.*, **98**, 18,937-18,949 (1993).
8. Kahler, S.W., *Ann. Rev. Astron. Astrophys.*, **30**, 113-141 (1992).
9. Kahler, S.W., *Astrophys. J.*, **428**, 837-842 (1994).
10. Mason, G.M., Gloeckler, G., and Hovestadt, D., *Astrophys. J.*, **280**, 902-916 (1984).
11. Moraal, H., and Axford, W.I., *Astron. Astrophys.*, **125**, 204-216 (1983).
12. Ng, C. K., Reames, D.V., and Tylka, A. J., *Geophys. Res. Lett.*, **26**, 2145-2148 (1999).
13. Parker, E.N., *Astrophys. J.*, **123**, 664-676 (1958).
14. Pauls, H.L., Zank, G.P., and Williams, L.L., *J. Geophys. Res.*, **100**, 21,595 (1995).
15. Reames, D.V., *Space Sci. Rev.*, in press (2000).
16. Reames, D.V., *Adv. Space Sci.*, **15**, 41-51 (1995).
17. Reames, D.V., and Stone, R.G., *Astrophys. J.*, **308**, 902-911 (1986).
18. Reames, D.V., Ng, C.K., and Tylka, A.J., *Geophys. Res. Letts.*, **26**, 3585-3588 (1999).
19. Völk, H.J., Zank, L.A., and Zank, G.P., *Astron. Astrophys.*, **198**, 274-282 (1988).
20. Zank, G.P., Rice, W.K.M, and Wu, C.C., *J. Geophys. Res.*, submitted (2000).

Solar Energetic Particles vs. Global Cosmic-Ray Diffusion

J. Giacalone[1], J. R. Jokipii[1], and J. E. Mazur[2]

[1] *Lunar & Planetary Laboratory, University of Arizona, Tucson, AZ, 85721*
[2] *Aerospace Corporation, El Segundo, CA, 90245*

Abstract. We have addressed the problem of the discrepancy between observations of solar cosmic rays, often showing sharp-gradients, and observations of galactic and anomalous cosmic rays which are consistent with the diffusive-transport description. We have carried out numerical simulations of the propagation of energetic charged particles in a turbulent magnetic field similar to that observed in the solar wind. If the particles are released impulsively near the sun, in a region small compared with the field coherence scale (a solar flare, for example), they exhibit characteristic fluctuations in intensity at 1 AU (dropouts), associated with very-steep localized gradients. The field coherence scale near the sun is the size of a supergranulation cell, and is about 0.01 AU at the orbit of Earth. These numerical simulations are quantitatively very similar to recent observations by the ACE spacecraft. These fluctuations occur naturally as part of the particle transport in the same field which results in large-scale cross-field diffusion and which has previously been used to study the propagation of CIR-associated particles to high heliographic latitudes.

INTRODUCTION

Models of galactic and anomalous cosmic-ray propagation in the heliosphere are quantitatively in agreement with the diffusive-transport equation first derived by Parker(1). However, observations of solar cosmic rays (or solar-energetic particles) do not often fit this model. Particularly, sharp, localized, cross-field gradients are often observed in impulsive-flare-associated energetic-particle events. In order to have such steep gradients, the perpendicular diffusion coefficient must be much smaller than is needed to explain the global propagation of cosmic rays in the heliosphere.

Here we use numerical simulations of particle motions in a turbulent magnetic field to examine the transport of solar energetic particles. If particles are released impulsively in a small region near the sun, we find that they can simultaneously have significant large-scale diffusion normal to the average magnetic field and very-steep localized gradients in the same magnetic field configuration. We suggest that this result may help explain recent observations of impulsive flare events (2). We conclude that the apparently contradictory attributes of transient localized gradients normal to the local magnetic field and significant large-scale diffusion normal to the average magnetic field are actually compatible and can occur in the same turbulent magnetic field.

OBSERVATIONS OF SOLAR COSMIC RAYS

Recent observations by the ACE spacecraft show fine-scale variations in the intensity profiles of energetic particles associated with impulsive solar flares (2). These are consistent with the passage of alternatively filled and empty tubes of energetic-particle flux by the spacecraft. The magnetic field lines, for which these particles adhere closely, are mixed and braided on spatial scales that are much larger than the gyroradii of the particles. Similar observations have also been discussed (3).

On the other hand, Dwyer et al.(4) reported on observations of energetic particles associated with corotating interaction regions (CIRs) that show very large cross-field streaming. The inferred cross-field diffusion coefficient from these data is almost as large as the parallel diffusion coefficient during the peak of the event.

GLOBAL TRANSPORT OF GALACTIC AND ANOMALOUS COSMIC RAYS

The transport of charged particles across a turbulent magnetic field is not very well understood. There is currently no widely-accepted quantitative theory for the perpendicular diffusion coefficient. Therefore, it has been treated as a phenomenological parameter in modeling cosmic-ray transport in the heliosphere. Values of the ratio of perpendicular to parallel diffusion coefficients

CP528, *Acceleration and Transport of Energetic Particles Observed in the Heliosphere: ACE 2000 Symposium,*
edited by Richard A. Mewaldt, et al.
© 2000 American Institute of Physics 1-56396-951-3/00/$17.00

as large as .01 - 0.1, considerably larger than the value classical scattering would predict, have been used for the last 2 decades. Values this large seem to be needed to account for galactic and anomalous cosmic-ray observations (5, 6).

Observations from the Ulysses spacecraft at high heliographic latitudes have revealed that galactic and anomalous cosmic rays are affected by the lower-latitude CIRs. Moreover, lower-energy ions and electrons accelerated at CIRs revealed co-rotating effects up to high latitudes where there were no apparent effects of the CIRs on the magnetic field or plasma (7). One explanation for this involves significant cross-field diffusion, while another involves direct latitudinal magnetic connection via a non-Parker magnetic field (8). The quantitative numerical model of Kóta & Jokipii (9) was used to simulate the effects of CIRs and they found that values of the ratio $\kappa_{\perp}/\kappa_{\parallel}$ of the order 0.03-0.1 (the same order as that used in the modulation of galactic cosmic rays discussed above) were need to account for the Ulysses observations.

Although perpendicular diffusion is not well understood, CIR-related observations as well as the modulation of galactic and anomalous cosmic rays can be naturally interpreted in terms of sizable perpendicular diffusion of particles across the nominal interplanetary magnetic field. Simulations reported by Giacalone and Jokipii (10) suggest that even for very low-rigidity particles ($\sim 40-200$ MV), the ratio $\kappa_{\perp}/\kappa_{\parallel}$ can be as large as ~ 0.05 in fluctuating fields similar to those observed in the solar wind. On the other hand, there are observations of energetic particles showing large spatial gradients across the local magnetic field (3). These seem to imply very small perpendicular diffusion. It has not been clear how to reconcile these observations with large $\kappa_{\perp}/\kappa_{\parallel}$. We address this in the next section.

MODEL

We study the motion of solar-energetic particles under the influence of the forces associated with thge large-scale fluctuations in the interplanetary magnetic field by using a numerical simulation. The model is the same as the one described by Giacalone (11) which was used to study the propagation of CIR-associated particles to high heliographic latitudes. In this model, the interplanetary magnetic field is composed of a background field, which is the nominal Parker spiral, and a fluctuating component which is derived by assuming that the magnetic footpoints are anchored into the solar supergranulation network. The footpoints execture a random walk across the photosphere since the plasma motions associated with supergranulation are turbulent in nature (see (12) and (13)). The random motions of the magnetic footpoints embedded in the solar supergranulation network, lead to large-scale fluctuations in the interplanetary magnetic field. The temporal and spatial scales of the IMF fluctuations are related to those associated with supergranulation. The radial variation of the variance in the IMF fluctuations based on this idea has shown to be consistent with Ulysses observations (14). Note that other causes of field-line mixing such as, for example, reconnection near the sun would produce a similar magnetic field structure, and the particle transport would be similar to that reported here. Hence, our picture could apply to both the fast and slow solar wind.

We integrate the trajectories of charged particles interacting with a single realization of the magnetic field. The characteristic scale size for the supergranulation was taken to be 40,000 km, the characteristic time 1 day, and rms speed 0.6 km/s, consistent with observations of solar supergranulation (15). We follow 150,000 particles in order to obtain reasonable statistics. The form of the magnetic field, $\mathbf{B}(r,\theta,\phi,t)$ is given in (11). The electric field is determined from $\mathbf{E} = -\mathbf{V}_w \times \mathbf{B}/c$, where $\mathbf{V}_w = V_w\hat{r}$ is the solar wind velocity. The particles are followed in a frame that is fixed with respect to the rotating sun. Other details of the numerical model, such as how the particles are released and how pitch-angle scattering is incorporated can be found in (16). A typical mean-free path that is used in the simulations presented here is 1 AU for 1-MeV/nuc. particles (the actual functional form mean-free path is described in (16)).

Shown in left panels of Figure 1 are the positions of all particles projected onto the solar-equatorial plane for two different source sizes (as indicated). The two sets of simulations are identical in every respect except the spatial size of the injection region. The orbit of Earth is indicated as the solid curve. The event rotates with the angular velocity of the sun and is observed for a brief period as the filled flux tubes sweep past a spacecraft. Some structures which correspond to filled and empty flux tubes are clearly evident for the case of a smaller source. Shown in the right panels of Figure 1 are plots of the arrival time versus energy for the particles observed by a spacecraft placed at 1 AU. For the case of a small source size, there are distinct dropouts in particle intensity. On the other hand, the dropouts are nearly absent for the larger source, and the event lasts longer. This explains why larger, or non-impulsive, events do not show the dropouts (2).

The existence of the large, cross-field gradients in the intensity of the solar energetic particles would seem to be evidence for a small rate of perpendicular diffusion. However, it is interesting to note that the parameters of the model magnetic field used in the simulations discussed above are the same as those used previously to demonstrate significant latitudinal transport (11). That is,

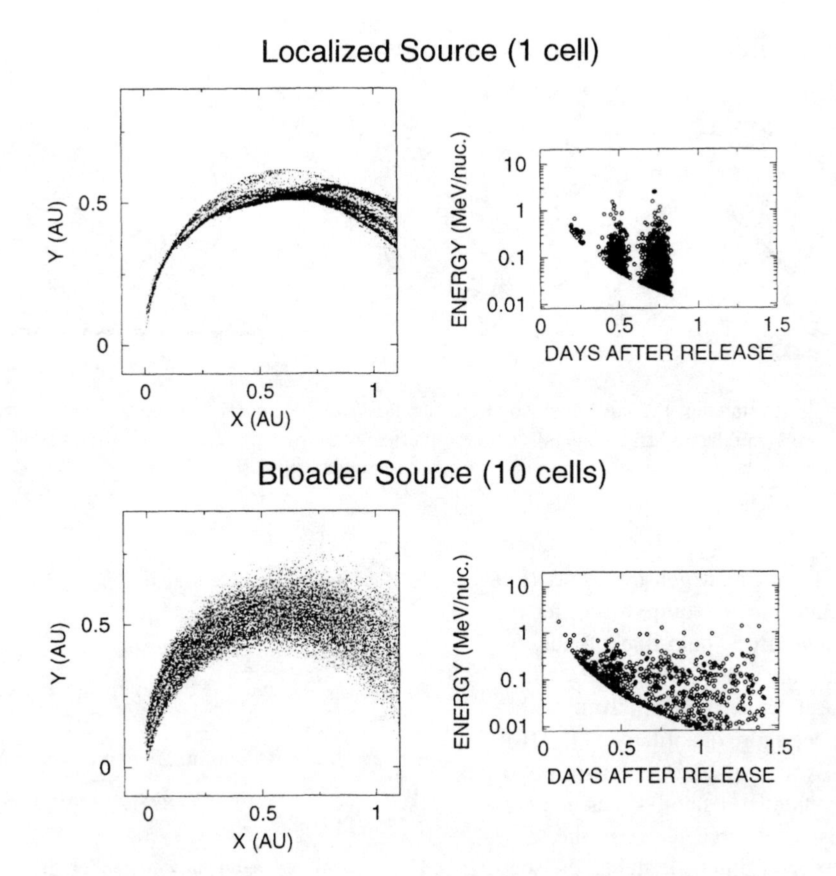

FIGURE 1. (Left panels) Positions of energetic ions associated with a simulated impulsive flare projected onto the ecliptic plane. Shown are the positions of the particles 1 day after their release. (Right panels) Energy vs. time scatter plots of the particles detected by an arbitrarily placed spacecraft, corresponding to the simulations shown on the left. The upper and lower panels are for different source sizes as indicated.

the models suggest that the observed large cross-field gradients (manifested in the observations as the dropouts in Figure 2a), are actually quite consistent with the simultaneous large-scale diffusion.

DISCUSSION

Our simulations reveal that energetic-particle intensities can have simultaneously large localized gradients normal to the local magnetic field and relatively large cross-field diffusion coefficients on large scales (much larger than the small-scale gradients would suggest). It is well known that the transport of particles normal to the magnetic field involves a significant contribution from the spatial meandering of magnetic field lines. Giacalone & Jokipii (10) specifically discussed this contribution to the perpendicular transport. For the same magnetic-field configuration, as the energy of the particles is reduced, one expects that particles which are initially placed closer together than the field coherence scale will remain close for some time, depending on the rate of separation of "neighboring" lines of force. This time should be at least of the order of the scattering time. For MeV protons this is on the same order as the time it takes the particles to reach 1 AU from the sun. Hence, if the particle acceleration/injection occurs in a small-enough region, the perpendicular motion will be initially non-diffusive, and injected particles will remain close together for some time. On the other hand, for phenomena having much longer characteristic time scales, such as CIRs, significant diffusion can take place. The same simulation model applied to the propagation of CIR-associated energetic ions confirms this (11).

Figure 2 illustrates our interpretation of the simulations and observations. For a localized source of energetic particles, it takes many scattering times for the particles to effectively diffuse through a large region (as shown in the middle panel). If the time is on the order of the scattering time of the particles, they are essentially

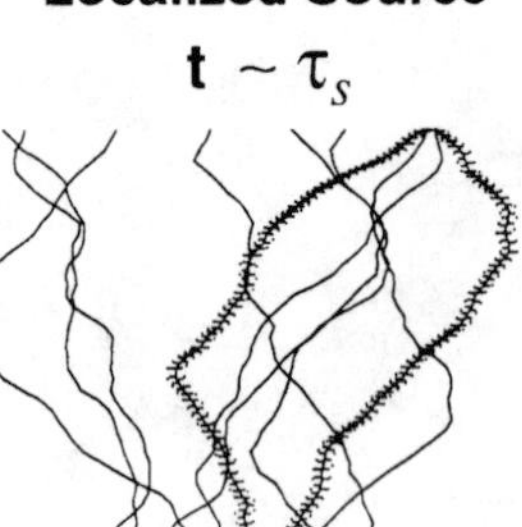

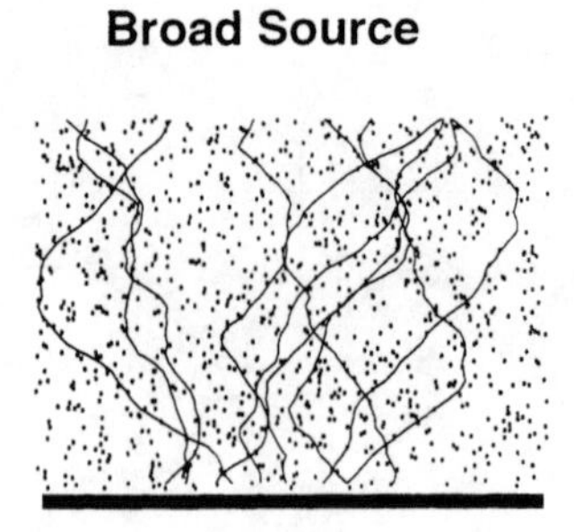

FIGURE 2. Cartoon sketch illustrating how the same magnetic field can have simultaneously large localized gradients (far left panel), and significant particle diffusion (middle and right panels). τ_s is the particle scattering time (the mean-free path measured along the magnetic lines of force divided by the particle speed) and **t** is the time of the observation.

tied to the magnetic field lines in which they start (left panel). We note that scattering is required in order for particles to be able to move off of field lines. Thus, for times larger than the scattering time, the gradients will be much smaller (as we expect, for example, in the outer heliosphere). This is discussed more quantitatively in (16).

We conclude that the explanation for the possibility of both steep cross-field gradients on small scales and significant cross-field diffusion at large scales in the same system lies in the coherence of magnetic fields, the scattering time of the particles of interest, and the spatial size of the regions of acceleration.

ACKNOWLEDGEMENTS.

This work was supported, in part by NASA under grants NAG5-7793 and NAG5-6620, and by the National Science Foundation under Grant ATM 9616547.

REFERENCES

1. Parker, E. N., *Planet. Space Sci.*, 13, (1965) 9-49.
2. Mazur, J. E., G. M. Mason, J. R. Dwyer, J. Giacalone, J. R. Jokipii, and E. C. Stone, 1999, *Astrophys. J.*, 532, (2000) L79-L82.
3. Roelof, E. C., R. B. Decker, & S. M. Krimigis, *J. Geophys. Res.*, 88, (1983) 9889-9909.
4. Dwyer, J. R., Mason, G. M., Mazur, J. E., Jokipii, J. R., von Rosenvinge, T. T., and Lepping, R. P., *Astrophys. J. Lett.*, 490, (1997) L115-L118.
5. Kóta, J. & J. R.. Jokipii, *Astrophys. J.*, 265, (1983) 573-581.
6. Haasbroek, L. J. & M. S. Potgeiter, *Space Sci. Rev*, 72, (1995) 385-390.
7. Simnett, G. M., et al., *Space Sci. Rev.*, 83, (1998) 215-258.
8. Fisk, L. A., and J. R. Jokipii, *Space Sci. Rev.*, 89, (1999) 115-124.
9. Kóta, J. & J. R.. Jokipii, *Space Sci. Rev.*, 83, (1998) 137-145.
10. Giacalone, J. & J. R. Jokipii, *Astrophys. J.*, 520, (1999) 204-214.
11. Giacalone, J., *Adv. Space Res.*, 23, (1999) 581-590.
12. Jokipii, J.R. & E.N. Parker, *Phys. Rev. Lett.*, 21, (1968) 44.
13. Jokipii, J.R., & J. Kóta, *Geophys. Res. Lett.*, 16, (1989) 1-4.
14. Jokipii, J.R., J. Kóta, J. Giacalone, T.S. Horbury, and E.J. Smith, *Geophys. Res. Lett.*, 22, (1995) 3385-3388.
15. Wang, H., 1988, *Solar Phys.*, 17, 343-358.
16. Giacalone, J., J. R. Jokipii, and J. E. Mazur, Small-scale gradients and large-scale diffusion of charged-particles in the heliospheric magnetic field, *Astrophys. J. Lett.*, 532, (2000) L75-L78.

Solar Energetic Particle Propagation in 1997-99: Observations from ACE, Ulysses, and Voyagers 1 and 2

R. B. Decker, E. C. Roelof, and S. M. Krimigis

Applied Physics Laboratory, The Johns Hopkins University, Laurel, MD 20723

Abstract. Solar energetic particles injected during intense activity in April-May 1998 and observed at ACE and Ulysses are identified in 0.5-1.5 MeV proton data from the LECP instrument on Voyager 2 (56 AU, 20°S) and Voyager 1 (72 AU, 33°N). A shell of ≈1 MeV protons ~60 days wide reached peak intensity at Voyager 2 on 1998.7 and at Voyager 1 on 1998.8, some 6 months after the April-May 1998 solar activity. We interpret these results in terms of infrequent scattering during propagation from 1 to 72 AU in helioradius and over ~50° in latitude.

INTRODUCTION

The intercomparison of directional intensities of a given particle species measured at comparable energies and at widely separated positions within the heliosphere has proven invaluable for investigating energetic particle (SEP) propagation in the heliosphere. The Low Energy Charged Particle (LECP) instruments on the Voyager 1 and 2 spacecraft in the distant heliosphere each possess a subset of energetic particle channels well-matched with selected channels from instruments on key spacecraft in the inner heliosphere, including the EPAM instrument on ACE and the HI-SCALE instrument on Ulysses.

SOLAR ACTIVITY IN 1997-1999

Figure 1 shows intensities of ≈1 MeV protons, during 1997.7-1999.3, from the energetic particle instruments EPAM on ACE, HI-SCALE on Ulysses, LECP on Voyager 2 (V2), and LECP on Voyager 1 (V1). Inverted triangles along the top axis indicate approximate occurrence times of major flares and/or CME-associated SEP enhancements at 1 AU during (A) 1997 DOY 308 and 310, (B) 1998 DOY 110, 122, and 126, and (C) 1998 DOY 239. We focus on these three periods because of the likely correspondence between these bursts of SEP activity and variations in energetic proton intensities observed at V1 and V2 in 1998 and 1999. A preliminary version of this paper appears in the proceedings of the 26th ICRC (1).

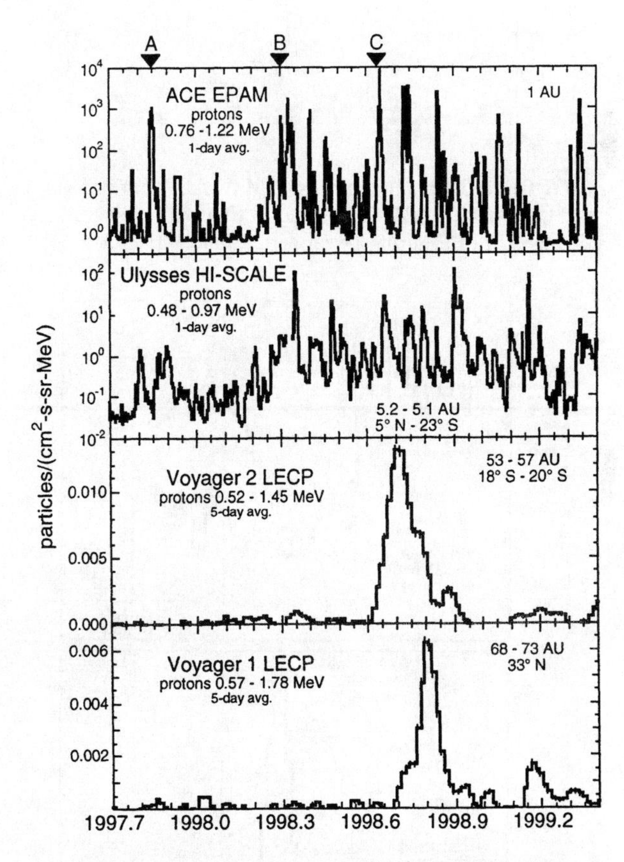

FIGURE 1. Intensities of ≈1 MeV protons at (top to bottom) ACE (1 AU), Ulysses (5 AU), Voyager 2 (56 AU), and Voyager 1 (72 AU) associated with enhanced solar activity during 1997-99. Note that the intensity axis is logarithmic in the top two panels, and linear in the bottom two panels.

CP528, *Acceleration and Transport of Energetic Particles Observed in the Heliosphere: ACE 2000 Symposium,* edited by Richard A. Mewaldt, et al.
© 2000 American Institute of Physics 1-56396-951-3/00/$17.00

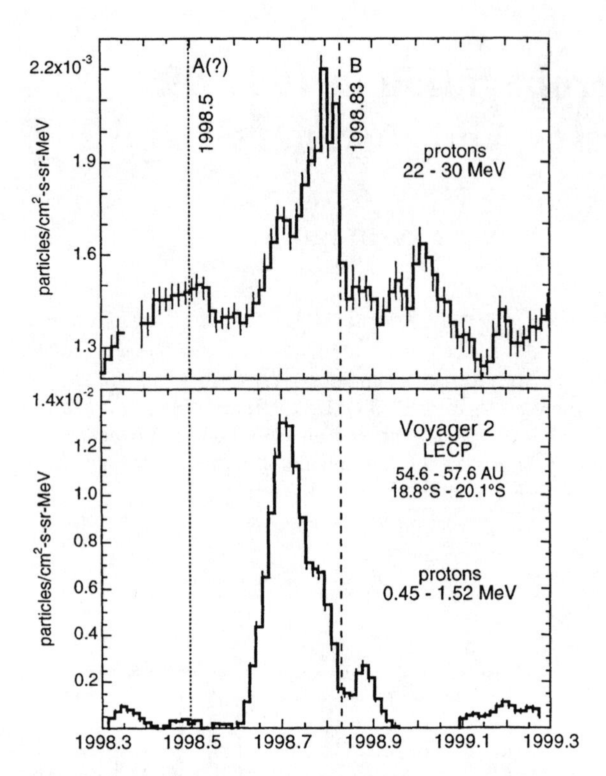

FIGURE 2. Intensities of 22-30 MeV ACR protons (top) and 0.45-1.52 MeV SEP protons (bottom) observed by Voyager 2 LECP during 1-year period 1998.3-1999.3.

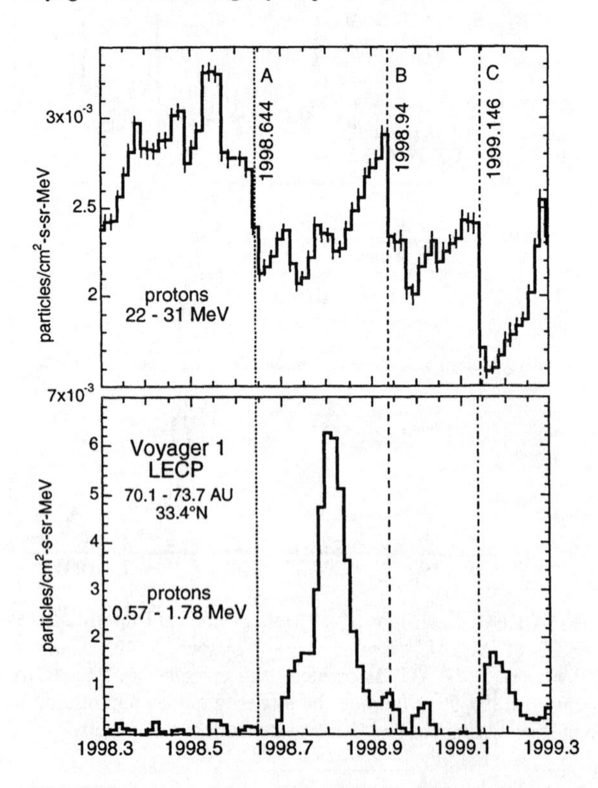

FIGURE 3. Intensities of 22-31 MeV ACR protons (top) and 0.57-1.78 MeV SEP protons (bottom) observed by Voyager 1 LECP during 1-year period 1998.3-1999.3.

Sustained intensity increases of protons ≈1 MeV from the April-May 1998 activity (B) are identifiable at 1 and 5 AU. The V2 and V1 data show that features of individual solar events are washed out and that intensities are greatly reduced at 56 and 72 AU. The SEPs at V2 and V1, which we here associate with protons injected during period B, comprise a shell with peak intensity at V2 $\sim 10^{-2}$/cm^2-s-sr-MeV, reduced by a factor $\sim 10^4$ from that at 1 AU, and a FWHM ~0.15 yr., or ~14 AU for a solar wind speed of 435 km/s. At V1, separated from V2 by 16 AU in radius, 50° in latitude, and 40° in longitude, the intensity is half that at V2.

VOYAGER OBSERVATIONS

Figures 2 and 3 show energetic proton intensities at V2 and V1 during 1998.3-1999.3. The top panels of each figure show protons ≈20-30 MeV, predominantly anomalous cosmic ray (ACR) hydrogen nuclei. The bottom panels show protons ≈0.5-1.5 MeV, which originate mainly at or near the Sun or are accelerated from a lower energy source by heliospheric shocks, plasma turbulence, or both. Intensity-time profiles of ACR protons 3-17 MeV and galactic cosmic ray (GCR) protons >70 MeV are similar to those of the ≈20-30 MeV ACR protons. Vertical lines labeled A, B, and C, indicate where in the >3 MeV proton data we see evidence for the passage of large-scale heliospheric disturbances, such as merged interaction regions (MIRs), that have a plausible association with solar active periods A, B, and C in Fig. 1 (1).

Our focus here is event B, which we designate 'Disturbance B.' Intensities of >3 MeV protons increase steadily as Disturbance B approaches, peak right before its arrival, and then drop abruptly after its passage. This is consistent with magnetic sweeping of the pre-existing ACR and GCR populations by the enhanced magnetic field of an MIR, as suggested by Burlaga et al. (2), based on their analysis V2 magnetic field data during this period. However, the ≈1 MeV protons represent a distinct population, peaking ≈45 days prior to the passage Disturbance B by V2 and V1. We contend that the ≈1 MeV protons are SEPs, that through relatively scatter-free propagation including scatter-free energy loss, arrive at 56 and 72 AU well ahead of Disturbance B. We do not share the view that the ≈1 MeV protons may be the low-energy portion of the swept-up ambient particle population (2,3).

ANALYSIS

We believe that the remarkably similar shapes of the ≈1 MeV proton intensity increases that peak at V2 and V1 ≈45 days before the ACR decrease (line B in

Fig. 4) are the consequence relatively infrequent scattering and continuous energy loss as SEP protons propagate from the inner to the outer heliosphere.

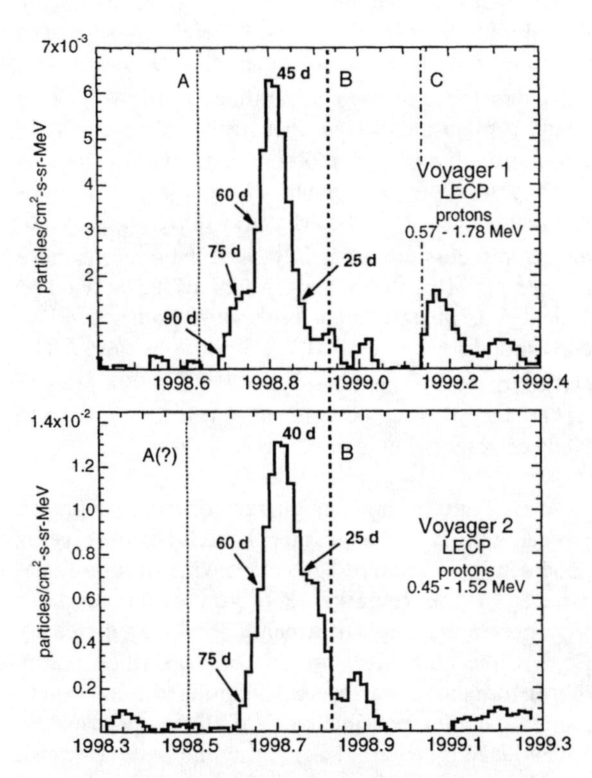

FIGURE 4. Intensity plot of ≈0.5-1.5 MeV SEP protons at V1 (top) shifted in time relative to that at V2 (bottom) so that Disturbance B is lined up at the two spacecraft.

For weak scattering, the gyro-averaged motion of the protons will be the vector sum of the guiding center velocity $v_{\parallel}$ and $V_{\perp}$ (the $E \times B$ drift velocity in the solar wind). Then the radial component of the total velocity is $dr/dt = (v_{\parallel})_r + (V_{\perp})_r$. In an archimedean spiral field beyond 3 AU, we can write to good approximation

$$\frac{dr}{dt} = \frac{\mu v V}{\Omega r \sin\Theta} + V, \tag{1}$$

where μ is the pitch cosine, V the solar wind velocity, Θ the co-latitude, and Ω the sidereal rotation frequency of the Sun. Since our observations are at latitudes <30°, we set $\sin\Theta = 1$. For the energy loss, we will use the relation for guiding center motion in the equatorial heliospheric magnetic field. Even in the absence of pitch-angle scattering, there is a continual momentum loss (independent of mass or charge) that has been shown (4) to be

$$\frac{dv}{dt} \cong -\left(\frac{vV}{2r}\right)(1+\mu^2). \tag{2}$$

The familiar expression for "adiabatic cooling" is recovered by replacing $(1+\mu^2)$ by its isotropic average of 4/3.

We could completely pose the problem by adding an additional relation giving μ as a function of r, such as the conservation of the magnetic moment $(1-\mu^2)v^2/B = const.$ However, we know that there is some pitch-angle scattering, although we believe it to be occasional. Even so, magnetic focussing will tend to keep an outward particle going outward. Thus, we believe that we can obtain a good indication of the essential nature of particle propagation beyond 5 AU by holding $\mu \cong const.$ in Eqs. (1) and (2) and considering it an "equivalent" pitch-cosine for the particle's transit from $r = r_0$ to $r > r_0$. Since the RHS of Eq. (2) varies only by a factor of 2 for $0 \le \mu^2 \le 1$, we will also consider $\mu \cong const.$ therein.

A very useful integral for our discussion is immediately obtained by substituting v in terms of dv/dt from Eq. (2) into Eq. (1) and then integrating from time t_0 (when the particles leave the disturbance at r_0 with velocity $v = v_0$) until they arrive at radius r at time t (which is before the disturbance arrives):

$$r - r_0 = -\frac{2\mu(v - v_0)}{\Omega(1+\mu^2)} + V(t - t_0). \tag{3}$$

Since Eq. (2) shows that the particle is always losing energy (regardless of the sign of μ), we always have $v < v_0$ in the first term on the RHS of Equation (3). Thus Eq. (3) implies that the radial distance δr that the particle can "run ahead" of the solar wind is given by

$$\delta r = \frac{2\mu(v_0 - v)}{\Omega(1+\mu^2)} < \frac{2\mu v_0}{\Omega(1+\mu^2)}. \tag{4}$$

This is because the distance along each winding of the spiral eventually becomes so long that the particle moves out radially faster by field line convection ($V_{\perp}$) than by running along the field ($v_{\parallel}$). Interestingly, because of the geometry of the archimedean spiral, δr is independent of the solar wind velocity V.

Eq. (4) gives us the time separation at some distance r between the arrival of the particles with "equivalent" pitch-cosine μ as $\delta t = \delta r / V$. We do not have space here to present the complete solution of Eqs. (2) and (3) for v_0 as a function of μ, v, r, and r_0, but we summarize the results in Fig. 5. Protons that arrive at V2 and V1 with E=0.5 MeV were launched

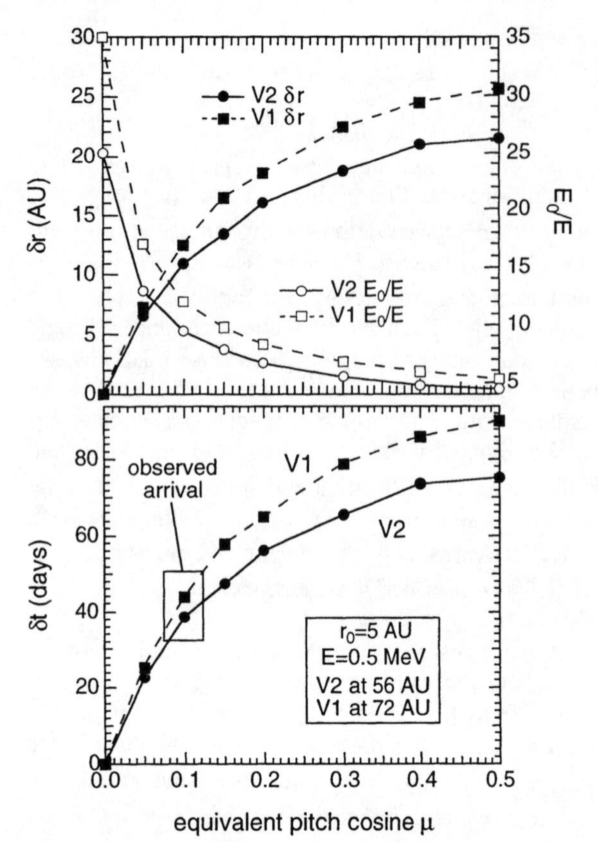

FIGURE 5. Top: Distance $\delta r(AU)$ ahead of Disturbance B (left axis), E_0/E (right axis). Bottom: Arrival time $\delta t(days)$ prior to the arrival of Disturbance B.

with energy E_0>0.5 MeV from Disturbance B when it had begun to coalesce at r_0=5 AU. For each part of the distribution that propagates with an "equivalent" value of the pitch cosine μ, Fig. 5 shows, for both V2 and V1, the distance $\delta r(AU)$ ahead of Disturbance B, E_0/E, and the arrival time $\delta t(days)$ prior to the arrival of Disturbance B.

DISCUSSION AND CONCLUSIONS

The results in Fig. 5 predict that if the proton population is confined to the vicinity of the shock at r_0 in the inner heliosphere at time t_0, the protons will be broadly distributed over distances δr ahead of the disturbance when they arrive at radius r in the outer heliosphere at a time interval δt before the disturbance arrives. Particles whose propagation can be characterized by small "equivalent" pitch cosines ($\mu \cong 0$) will arrive with the shock, while particles with $\mu \cong 0.5$ (60° "equivalent" pitch-angle) will arrive months before the disturbance. We have run Monte Carlo simulations with occasional weak scattering and the results can indeed be characterized by the "equivalent"- μ formalism.

Compare the values of δt in Fig. 5 with the intensity histories in Fig. 4. The general dependence of δr (and δt) upon μ is rather similar at the two Voyagers (despite their separation of 16 AU). This would explain the similar appearance of the events in Fig. 4. Relatively few protons arrive within the few weeks just before Disturbance B. This lack of protons with equivalent μ=0 is consistent with some degree of magnetic focussing, especially since the effect is more marked at V1. The intensity peaks at 40d at V1 and 45d at V2, in agreement with δt=38.5d and 43.8d corresponding to μ=0.10 at both Voyagers. The earliest detected intensities were 75d and 90d prior to Disturbance B, in good agreement with δt=75.3d and 90.0d corresponding to μ=0.5 at both Voyagers.

We conclude that our characterization of proton propagation by an "equivalent" pitch-cosine gives a good semi-quantitative agreement with all significant aspects of the remarkable proton events at both Voyagers preceding Disturbance B. As predicted by Eq. (4), the >0.5 MeV proton population had outrun the disturbance from 5 AU, attaining the maximum radial separation possible by field-aligned propagation (constrained by energy loss) during the ~50 AU transit to the Voyagers. This event confirms both the concepts of relatively scatter-free propagation (with magnetic focussing) and scatter-free energy loss (4).

ACKNOWLEDGMENTS

This work was supported by Voyager Interstellar Mission under NASA Grant NAG5-4365 and by Ulysses HI-SCALE under NASA Grant NAG5-6113.

REFERENCES

1. Decker, R.B., Roelof, E.C., and Krimigis, S.M., "Solar Energetic Particles from the April 1998 Activity: Observations from 1 to 72 AU," in *Proc. 26th Internat. Cosmic Ray Conf., Vol. 6*, 1999, pp. 328-331.
2. Burlaga, L.F., McDonald, F.B., and Ness, N.F, "Intense Magnetic Fields Observed by Voyager 2 during 1998," in *Proc. 26th Internat. Cosmic Ray Conf., Vol. 7*, 1999, pp. 107-110.
3. McDonald , F.B., Burlaga, L.F., McGuire, R.E., and Ness, N.F, "The Onset of Cosmic Ray Modulation (Cycle 23) at 1 AU Coupled with a Transient Low Energy Cosmic Ray Increase in the Outer Heliosphere," in *Proc. ACE 2000 Symposium*, 1999, submitted.
4. Roelof, E.C., *Space Sci. Rev. 89*, 238-240 (1999).

Solar Energetic ^{3}He Mean Free Paths: Comparison Between Wave-Particle and Particle Anisotropy Results

B. T. Tsurutani[a], L. D. Zhang[a], G. Mason[b], G. S. Lakhina[c], T. Hada[d], J. K. Arballo[a], and R. D. Zwickl[e]

[a]*Jet Propulsion Laboratory, California Institute of Technology, Pasadena, California*
[b]*Department of Physics, University of Maryland, College Park, Maryland*
[c]*Indian Institute of Geomagnetism, Colaba, Mumbai/Bombay, India*
[d]*ESST, Kyushu University, Kasuga, Japan*
[e]*Space Environment Laboratory, NOAA, Boulder, Colorado*

Abstract. Energetic ^{3}He particle mean free paths are calculated using in-situ wave amplitudes. The wave polarization (outward propagating, arc-polarized, spherical) and wave **k** directions (outward hemispherical) are included in a first-order cyclotron resonant calculation. Values for λ_{W-P} are ~ 0.2 AU. This is roughly ~ 5 times smaller than the particle mean free path as determined from modeling applied to measured front-to-back ^{3}He particle anisotropies. It is suggested that this difference is due to much slower pitch angle diffusion through 90°.

INTRODUCTION

In the past, particle transport from the solar corona to 1 AU has been studied by inferring the amount of pitch angle scattering that has taken place from an analysis of the particle distributions themselves, or by taking a characteristic interplanetary wave spectrum and theoretically calculating the amount of scattering that should have taken place assuming that the spectrum is representative (1, 2, 3). For a detailed discussion of the two methods, see (4) and (5). Calculation of the energetic particle scattering mean free paths using the magnetic field data and a quasi-linear theory of the field fluctuations has led to a long-standing discrepancy wherein this calculated mean free path is generally much smaller than the mean free paths derived from particle measurements. Some recent theoretical studies (6, 7) have obtained improved results (i.e., larger calculated particle scattering mean free paths) by using more complex models for the waves. Wanner et al. (8) presented evidence showing that the "slab" turbulence approximation was fundamentally flawed, and this was followed by Bieber et al. (9), who showed that two-dimensional (2D) turbulence was playing a major role. Bieber et al. (9) applied a 2D model to ~ 10 MeV proton observations from *Helios*, and found good agreement between the mean free paths calculated from the turbulence and from the energetic particle observations.

It is known that the amount of wave power present in the interplanetary medium can vary by orders of magnitude (10). It is the purpose of this paper to examine the simultaneous 1 AU LF wave properties (at frequencies near the particle cyclotron resonance) during ^{3}He rich events taken from (11). These solar energetic particles have energies near 1 MeV/nucleon, much lower than the $\sim 10-20$ MeV energies considered in recent studies (5, 9). These results will be compared to the mean free paths for the same events as determined from modeling applied to the measured 1 AU particle anisotropies.

RESULTS

^{3}He-Rich Events

Figure 1 shows the May 17, 1979 energetic ion event. Three He energy channels are given in the top panel. Velocity dispersion is clearly present, with the highest energy particles arriving first, as expected for propagation from a remote source. The magnetic field is given in the next four panels. The field is relatively quiet during the particle onset. The fluctuations in the three components are small, and the field magnitude is $\sim 3-5$ nT. An examination of the solar wind velocity (bottom panel) indicates that this particle event occurred in the far trailing part of a high velocity stream. This general region is noted for a lack of large amplitude Alfvén waves (12).

To quantify the characteristics of the interplanetary fluctuations present during this particle event, we have made power spectra of the magnetic field components and

CP528, *Acceleration and Transport of Energetic Particles Observed in the Heliosphere: ACE 2000 Symposium*, edited by Richard A. Mewaldt, et al.
© 2000 American Institute of Physics 1-56396-951-3/00/$17.00

the magnitude. We have used a field-aligned coordinate system to determine the power due to transverse fluctuations and the power due to compressional variations. The transverse wave power is responsible for resonant pitch angle scattering and is the important quantity for the calculations presented in this paper.

The transverse power spectra for the nine (11) particle events have been calculated and compared with power spectra for "quiet", "intermediate" and "active" periods (from (13)). It is found that the interplanetary medium is typically "quiet" during the ^{3}He-rich events.

A proper description of interplanetary Alfvén waves is that they are phase-steepened, arc-polarized spherical waves (10). For purposes of the calculations here, we can assume that they can be approximated as linearly polarized waves with equal power present in right- and left-hand rotations.

The **k** direction of rotational discontinuities, the phase-steepened edges of interplanetary Alfvén waves has been shown to be isotropic (14, Fig. 15). Since Alfvén waves are outwardly propagating (14, Fig. 6), the wave **k** distribution is an outward hemisphere.

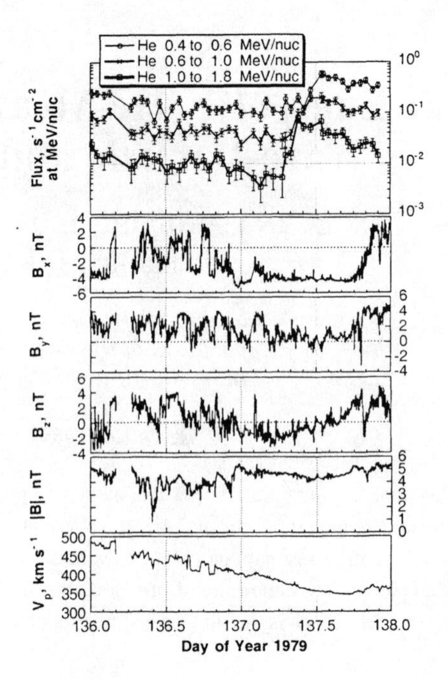

FIGURE 1. Particle flux, magnetic field and solar wind velocity for 16-17 May, 1979.

Scattering Mean Free Paths Determined by Particle Measurements

The scattering mean free paths for the nine ^{3}He events were obtained by comparing the event time/intensity profiles and anisotropies with the predictions of a Boltzmann equation model of interplanetary scattering which includes the effects of particle pitch-angle scattering and adiabatic defocusing as the particles move through magnetic fields of varying strength (15, 16, 17). Mason et al. (3) published numerical solutions of this equation based on the technique of (18) for observations from the *ISEE-3* ULEWAT instrument for nominal values of solar wind speed. We use these solutions here to estimate the scattering mean free paths. The results are given in Table 1.

Resonant Wave-Particle Interaction Calculation of Mean Free Paths Using IMF Power Spectra

The particle pitch angle diffusion coefficient (i.e., pitch angle scattering rate) has been derived in (19) and (20). The condition of first order cyclotron resonance between the waves and the anti-sunward directed particles can be written as:

$$\omega - k_{||}V_{||} = -\Omega \qquad (1)$$

In the above, ω and **k** are the wave frequency and wave vector, Ω is the particle cyclotron frequency in the ambient magnetic field. The particle velocity **V** has a parallel component $V_{||} = \mu V_0$, where μ is the cosine of the particle pitch angle.

For Alfvén waves propagating in the solar wind plasma frame, the phase velocity is V_A. Taking the angle between **k** and B_0 to be θ and the angle between **k** and $\mathbf{V}_{SW}$ to be ψ, Equation 1 now becomes:

$$2\pi f\left(1 - \frac{\mu V_0}{V_{SW}\cos\psi}\cos\theta\right) = -\Omega \qquad (2)$$

If the particles of interest are He^{++} and of 0.4 MeV/nucleon energy, $V_0 = 8.8 \times 10^8$ cm/s is much larger than the solar wind speed, V_{SW}. The ions are resonant with right-hand polarized waves, therefore in the final estimate of mean free paths, the effective wave transverse power should be $(P_1 + P_2)/2$, where 1 and 2 indicate two transverse directions to the ambient field.

Following equation (3.9) of (19), the pitch angle scattering rate for a given resonant velocity due to interactions with waves in a wave-number band of width Δk about resonance is:

$$D = \frac{(\Omega^{++})^2}{2\pi}\left(\frac{V_{SW}\cos\psi}{\mu V_0\cos\theta}\right)\frac{P_{res}}{B_0}, \quad P_{res} = \left.\frac{(B')^2}{\Delta f}\right|_{res},$$
$$\Delta f = \frac{1}{2\pi}\Delta k V_{SW}\cos\psi \qquad (3)$$

Table 1. Mean Free Paths for 1 MeV/nucleon ^{3}He-rich "Scatter-free" Events

Event (Date)	B_0 (nT)	$\Omega_{^3He_{++}}$	$P_{transverse}$ (nT^2/Hz)	Scattering Rate (s^{-1})	λ_{W-P} (AU)	λ_{He} (AU)
23 Oct 1978	6.30	0.402	$3.26\times10^{-3}f^{-1.7}$	1.03×10^{-3}	0.09	1.0
26 Dec 1978	8.11	0.517	$2.81\times10^{-3}f^{-1.7}$	5.77×10^{-4}	0.16	1.0
17 May 1979	4.63	0.293	$6.58\times10^{-4}f^{-1.8}$	6.05×10^{-4}	0.15	0.5
14 Dec 1979	9.93	0.634	$7.43\times10^{-3}f^{-1.7}$	1.08×10^{-3}	0.08	2.0
13 Jan 1980	6.25	0.399	$1.89\times10^{-3}f^{-1.7}$	6.05×10^{-4}	0.15	0.5
9 Nov 1980	11.47	0.732	$4.82\times10^{-3}f^{-1.7}$	5.49×10^{-4}	0.17	0.3
14 Nov 1980	7.41	0.473	$4.91\times10^{-3}f^{-1.8}$	1.94×10^{-3}	0.05	0.5
31 Jul 1981	9.66	0.616	$3.60\times10^{-3}f^{-1.6}$	3.45×10^{-4}	0.27	0.5
12 Feb 1982	15.56	0.993	$1.08\times10^{-2}f^{-1.7}$	7.33×10^{-4}	0.13	0.5

where B' is the wave amplitude in resonance with the particle. Assuming the wave power spectra to have a power spectral index of α, that is, $P_{res} = Af_{res}^{-\alpha}$ and $\mu V_0 \cos\theta \gg V_{SW}$, the effect of averaging over the ψ (at 1 AU) and θ angles is:

$$\langle D\rangle_\theta \cong \frac{1.1}{\alpha} D|_{\theta=0} = \frac{(\Omega^{++})^2}{2\pi} \frac{1.1}{B_0^2} \frac{1}{\alpha} A(f^{++})^{-\alpha} \left(\frac{\mu V_0}{V_{SW}}\right)^{\alpha-1} \quad (4)$$

Averaging over the cosine of the particle pitch angle gives:

$$\langle D\rangle_{\theta,\mu} = \frac{1.1}{\alpha^2} \frac{(\Omega^{++})^2}{2\pi} \frac{V_{SW}}{V_0} \frac{1}{B_0^2} A \left(\frac{V_{SW} f^{++}}{V_0}\right)^{-\alpha} \quad (5)$$

The time for scattering one radian in pitch angle T is $\sim 1/D$, and the particle mean free path is: $\lambda_{W-P} = TV_{He}$, where λ_{W-P} stands for wave-particle interaction estimate of the mean free path.

In this paper, we have considered only the first order cyclotron resonance term ($n = -1$). Use of higher order cyclotron resonance terms is more theoretically complete, but should only change the results slightly.

The results of the calculations are shown in Table 1. λ_{W-P} ranges between 0.05 and 0.30 AU while λ_{He} ranges between 0.3 and 2.0 AU.

DISCUSSION

Although the wave polarization, wave normal distributions, and in-situ transverse power spectra were included in this study, there are still substantial differences between the calculated λ_{W-P} and λ_{He} values. Previous works (21) have noted even greater discrepancies between the two values.

We note that $\langle D\rangle_{\theta,\mu}$, the average diffusion rate, is the typical quantity calculated. The value $1/\langle D\rangle_{\theta,\mu}$ and the particle velocity are used to derive λ_{W-P}. However, λ_{He} is the mean free path derived from particle pitch angle scattering across 90° pitch (where $D = 1/\tau = 0$). The time τ to diffuse across 90° pitch from quasilinear theory is infinite. Could this be the primary difference between the λ_{W-P} and λ_{He} values?

To examine this further, we perform a test particle simulation, in which ion orbits are integrated in time under the influence of static magnetic field turbulence, which is given as a superposition of parallel, circularly polarized Alfvén waves with equal propagation velocities (slab model). In this model, the ion energy in the wave rest frame is constant, thus there is no energy diffusion of ions. Both right- and left-hand polarized waves are included. Although each mode is non-compressional, superposition of the waves yields a ponderomotive compressional field, which may act to mirror-reflect the ions. In the simulation, we assume a power-law distribution of wave power with a spectral index γ when $k_{min} < k < k_{max}$, and zero otherwise, where k, k_{min}, k_{max} are respectively the wave number, the minimum, and the maximum wave numbers included in the simulation. The wave phases are assumed to be random.

Figure 2 shows the time evolution of the distribution of ion pitch angle cosine, μ. For each panel, the horizon-

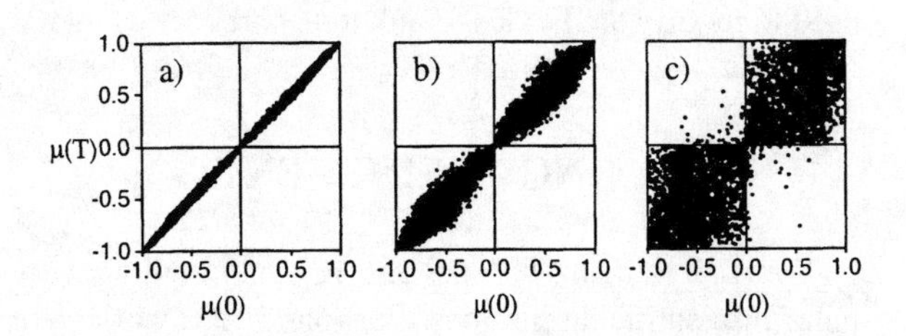

FIGURE 2. Time evolution of the particle pitch angle distribution.

tal axis represents the initial distribution, $\mu(0)$, and the vertical axis denotes the distribution at some later times, $\mu(T)$, with (a) $T = 40$, (b) $T = 640$ and (c) $T = 10240$. Each dot represents a single test particle. Parameters used are: the ion velocity, $v = 10, \gamma = 1.5, k_{min} = 6.13 \times 10^{-3}$, $k_{max} = 3.14$, and the variance of the normalized perpendicular magnetic field fluctuations, $\langle B_{perp}^2 \rangle = 4 \times 10^{-4}$.

At $T = 40$, the distribution of μ has not much evolved, and so the dots are almost aligned along the diagonal line in panel (a). Later, at $T = 640$, pitch angle diffusion is more evident, represented by thickening of the diagonal line (panel (b)). It is also clear that the diffusion is absent in essentially two regimes, $\mu \approx 0$ and $|\mu| \approx 1$. The former is due to the lack of waves which resonate with near 90° pitch angle ions. And the latter is due to geometry (i.e., the Jacobian), which appears as the pitch angle is transformed to its cosine, vanishes at $|\mu| = 1$, representing that a small deviation of the pitch angle from an exactly parallel direction does not give rise to a deviation of μ of the same order.

Clearly, the majority of the ions stay within the hemisphere in which they began. However, we should also note that a few ions did escape into the opposite hemisphere (see also (22)). More detailed analysis on test particle simulations will be done in a future study.

FINAL COMMENTS

What is the physical process of scattering particles across 90° pitch angle? The presence of large amplitude waves with $\delta B/B_0 \approx 1$ could lead to large, single-encounter pitch angle scattering across 90° (see (23)). This is a nonresonant interaction which involves large amplitude waves and is not included in the present quasi-linear theories. A second process is particle mirroring via interaction with $|\mathbf{B}|$ variations (24, 25, 26).

Random superposition of small amplitude waves may produce the $|\mathbf{B}|$ power spectra shown in Figure 2, and lead to mirroring across 90°. Computer simulations using particle-in-cell (PIC) codes should be useful to determine the relative effectiveness of the above two processes. Analytical expressions could then be derived which could be used to modify the Fokker-Plank transport coefficients.

ACKNOWLEDGMENTS

We wish to thank F. Jones and F. V. Coroniti for very helpful scientific discussions. Portions of this work were performed at the Jet Propulsion Laboratory, California Institute of Technology under contract with NASA, and at the University of Maryland supported by NASA.

REFERENCES

1. Jokipii, J. R., and Coleman, P. J. , Jr., *J. Geophys. Res.*, **73**, 5495, (1968).
2. Zwickl, R. D., and Webber, W. R., *Solar Phys.*, **54**, 457, (1977).
3. Mason, G. M., Ng, C. K., Klecker, B., and Green, G., *Astrophys. J.*, **339**, 529, (1989).
4. Palmer, I. D., *Rev. Geophys. Space Phys.*, **20**, 335, (1982).
5. Wanner, W., and Wibberenz, G., *J. Geophys. Res.*, **98**, 3513, (1993).
6. Schlickeiser, R., *Astrophys. J.*, **336**, 243, (1989).
7. Schlickeiser, R., and Miller, J. A., *Astrophys. J.*, **492**, 352, (1998).
8. Wanner, W., Jaekel, U., Kallenrode, M.-B., et al., *Astron. Astrophys.*, **290**, L5, (1994).
9. Bieber, J. W., Wanner, W., and Matthaeus, W. H., *J. Geophys. Res.*, **101**, 2511, (1996).
10. Tsurutani, B. T., Ho, C. M., Smith, E. J., et al., *Geophys. Res. Lett.*, **21**, 2267, (1994).
11. Kahler, S., Reames, D. V., Sheeley, N. R., Jr., et al., *Astrophys. J.*, **290**, 742, (1985).
12. Tsurutani, B. T., Gonzalez, W. D., Gonzalez, A. L. C., et al., *J. Geophys. Res.*, **100**, 21717, (1995).
13. Siscoe, G. L., Davis, L., Jr., Coleman, P. J., Jr., et al., *J. Geophys. Res.*, **73**, 61, (1968).
14. Tsurutani, B. T., Ho, C. M., Arballo, J. K., et al., *J. Geophys. Res.*, **101**, 11027, (1996).
15. Roelof, E. C., in *Lectures in High Energy Astrophysics, NASA SP-199*, edited by H. Ogelmann and J. R. Wayland, NASA, 1969, 111.
16. Earl, J. A., *Astrophys. J.*, **188**, 379, (1974).
17. Earl, J. A., *Astrophys. J.*, **252**, 739, (1981).
18. Ng, C. K., and Wong, K.-Y., in *Proc. 16th Internat. Cosmic Ray Conf.*, Kyoto, 1979, 252.
19. Kennel, C. F., and Petschek, H. E., *J. Geophys. Res.*, **71**, 1, (1966).
20. Tsurutani, B. T., and Lakhina, G. S., *Rev. Geophys.*, **35**, 491, (1997).
21. Tan, L. C., and Mason, G. M., *Astrophys. J. Lett.*, **L29**, 409, (1993).
22. Terasawa, T., in *Plasma Waves and Instabilities at Comets and in Magnetospheres*, edited by B. T. Tsurutani and H. Oya, Washington D.C., AGU, 1989, pp. 41-49.
23. Yoon, P. H, Ziebell, L. F., and Wu, C. S., *J. Geophys. Res.*, **96**, 5469, (1991).
24. Ng, C. K., and Reames, D. V., *Astrophys. J.*, **453**, 890, (1995).
25. Ragot, B. T., *Astrophys. J.*, **518**, 974, (1999).
26. Kuramitsu, Y., and Hada, T., *Geophys. Res. Lett.*, **27**, 629, (2000).

Charged particle composition in the inner heliosphere during the rise to maximum of Solar Cycle 23

C. G. Maclennan[1], L. J. Lanzerotti[1], L. A. Fisk[2], and R. E. Gold[3]

[1]*Bell Laboratories, Lucent Technologies, Murray Hill, NJ 07974*
[2]*Space Physics Laboratory, University of Michigan, Ann Arbor, MI 48109*
[3]*Johns Hopkins University, Applied Physics Laboratory, Laurel, MD 20723*

Abstract. Flux distributions and abundances relative to oxygen of interplanetary ions ($Z > 1$) are statistically studied and compared for measurements made at 1 and at $\sim$ 5 AU on the ACE and the Ulysses spacecraft near the ecliptic plane. Over the nearly two year interval studied, the distributions of the relative abundances and the fluxes of particles at the two locations are found to be approximately log normal. The statistical distributions of the relative abundances are found to be similar at the two helioradii. On a statistical basis, the fluxes at Ulysses times the distance of the measurements appear to be proportional to the fluxes at ACE. This radial dependence of the fluxes is consistent with the interpretation that, statistically, the ion parallel diffusion coefficient is large.

INTRODUCTION

Between the launch of the ACE spacecraft in late 1997 and mid-1999, the Ulysses (ULS) spacecraft (launched in October 1990 into a polar orbit of the Sun) was within about 30° of the ecliptic plane and at $\sim 4.5 - 5$ AU distance from the Sun. Ulysses was just ending its Solar Minimum Mission and beginning its Solar Maximum Mission. The essentially identical low energy charged particle instrumentation (EPAM on ACE and HI-SCALE on Ulysses: (1, 2)) that is flying on the two space probes provides an ideal opportunity to study the near-ecliptic distribution of interplanetary particles in the inner heliosphere during the beginning of the rise of solar cycle 23.

The charged particle detectors on the two space probes each consist of five solid state detector telescopes that are oriented to cover almost 4π steradians of the sky on the spin-stabilized spacecraft (spin rate of 5 rpm on both ACE and ULS). Of central relevance to the study reported herein is the composition aperture (CA) telescope of each instrument system. Each CA is a three element solid state telescope that consists of a $5\mu m$ first detector followed by two $100\mu m$ second and third detectors. Particle atomic species (hydrogen to iron) and energies ($\sim$ 0.5–8 MeV/nucl) are identified by energy loss and total energy measurements. A priority scheme is included to enhance the counting statistics of less abundant atomic species.

The Earth (and thus ACE, located sunward along the Earth-Sun line near the Lagrangian point) traveled nearly two complete orbits of the Sun during the time interval examined herein. That is, ULS only infrequently occupied a region of space that included a flux tube that might connect it to the ACE spacecraft near Earth. Therefore, this study of atomic species at 1 and 5 AU is made on a statistical, rather than an event-by-event, basis.

A number of spacecraft (most significantly Pioneers 10 and 11 and Voyagers 1 and 2) have traveled between the Earth and the orbit of Jupiter ($\sim$ 5 AU). Several investigators have used instrumentation on these space probes to measure the helioradius distribution of particle fluxes. Generally, past investigations have studied the propagation of protons from distinct solar events, e.g. (3), or anomalous cosmic rays, e.g. (4, 5). A few papers have determined a radial dependence for ions, with values ranging from $r^{-.5}$ to r^{-2}, depending on factors such as location and solar activity. We concentrate here on the statistical abundances at these two helioradial distances for atomic species with Z between 1 and 26.

MEASUREMENTS

Plotted in Figure 1 are oxygen (O) fluxes measured at $\sim$1 and $\sim$5 AU for the interval day 244, 1997, to day 190, 1999. The upper two panels show the O fluxes in the range 0.57–1.0 MeV/nucl for ACE and 0.5–1.0 MeV/nucl for ULS. The lower two panels contain the fluxes for 2.8–6.0 MeV/nucl oxygen. All sectors sampled by the spin of the spacecraft are averaged over one day intervals, which are themselves plotted as sliding 5-day averages every one day. The heliolatitude of ULS throughout the interval is given at the top of the Ulysses data panels; this latitude

CP528, *Acceleration and Transport of Energetic Particles Observed in the Heliosphere: ACE 2000 Symposium,* edited by Richard A. Mewaldt, et al.
© 2000 American Institute of Physics 1-56396-951-3/00/$17.00

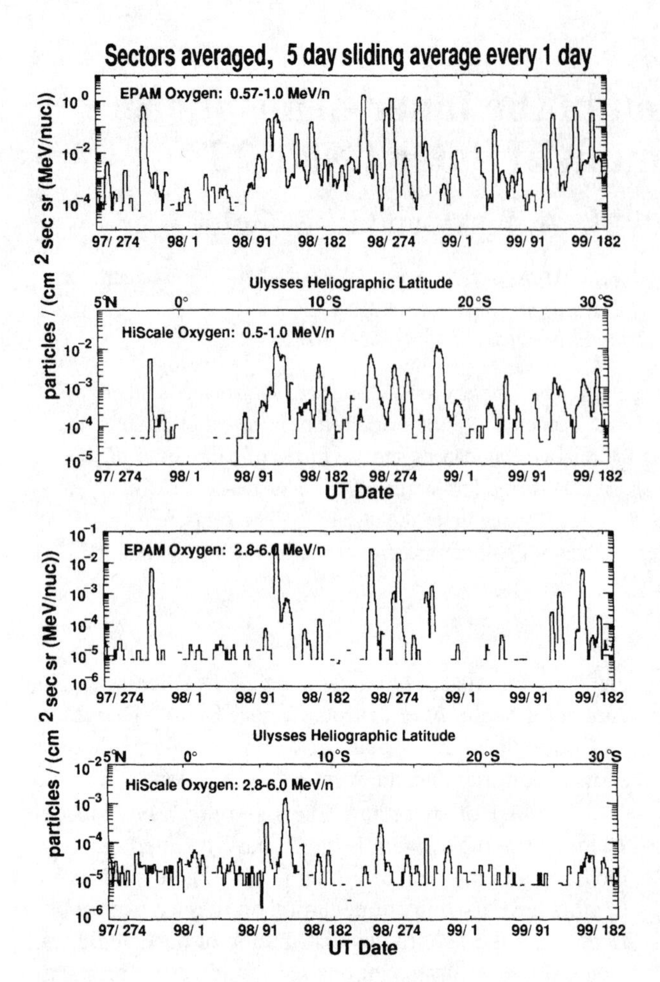

FIGURE 1. ACE and Ulysses oxygen fluxes in two energy ranges

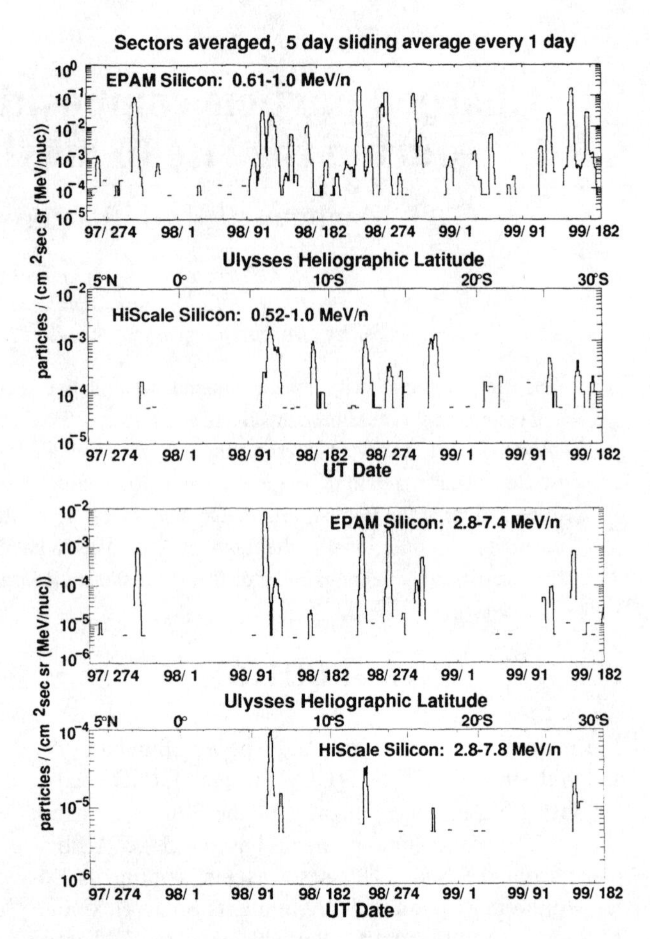

FIGURE 2. ACE and Ulysses silicon fluxes in two energy ranges

changes from ~ 5° N to ~ 30° S over the time shown as ULS passed through its aphelion distance of ~ 5.4 AU from the Sun.

The fluxes of the higher energy (2.8–6.0 MeV/nucl) O measured by Ulysses from the beginning of the time plot to about day 91, 1998, are anomalous oxygen at 5 AU. Beginning about day 91, 1998, to about day 1, 1999, the oxygen fluxes measured at both spacecraft became significantly more variable, especially in the lower energy range plotted. These continuing changes in the fluxes with time are produced by the increasing solar activity.

The abundance of the higher energy O particles in the time interval around day 91, 1999 (at a time of lower solar activity), is significantly lower than in the interval leading up to day 91, 1998. A discussion of these data in the context of anomalous cosmic ray oxygen and the implications for their removal from the inner heliosphere is contained in (6).

Shown in Figure 2 are the fluxes of silicon (Si) as measured at both spacecraft in approximately the same energy ranges as in Figure 1. This Si should be largely of solar origin, although some contributions from the sputtering of interplanetary grains from comets and other sources cannot be ruled out entirely. The fluxes of Si are larger at 1 AU than at 5 AU. The increases in the fluxes due to the increase in solar activity are clearly evident, especially at the lower energies and beginning around day 91, 1998.

COMPOSITION COMPARISONS

Plotted in Figure 3 are statistical comparisons of the fluxes of hydrogen and helium (0.5–1.0 MeV/nucl) that were measured at the two spacecraft locations for the time interval day 100, 1998 to day 262, 1999 (ACE fluxes as heavy solid lines; ULS fluxes as light lines). This time interval excludes the interval from day 244, 1997, to day 99, 1998, when most, if not all, of the O measured at 5 AU was anomalous oxygen (see discussion of Figure 1). The fluxes in each bin are normalized to the total number of daily averages for each spacecraft. The top two pan-

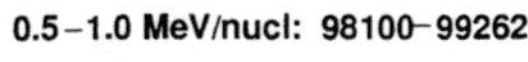

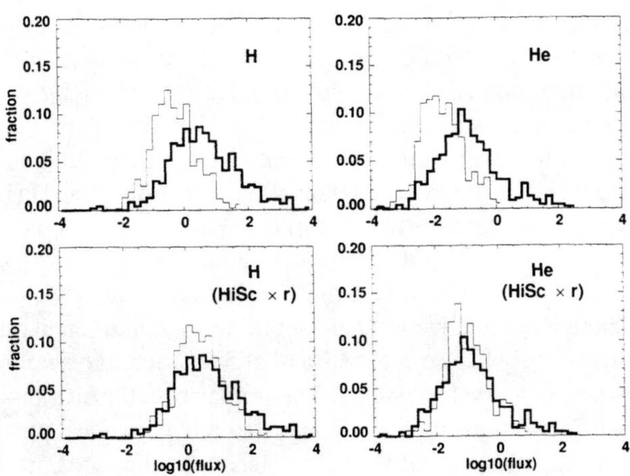

FIGURE 3. Distributions of 0.5–1.0 MeV/nucl H and He for ACE and Ulysses (heavy and light lines, respectively)

els contain the distributions of measured flux values; the lower two panels contain the distributions after multiplying each ULS daily flux value by the radial distance of ULS on that day. Thus, the ACE fluxes are the same in the upper and the lower panels.

The hydrogen and helium flux distributions in Figure 3 are seen to be approximately log-normal. (Lanzerotti et al. (7) remark on the log normal distribution of Voyager-measured hydrogen particles in the heliosphere beyond the orbit of Jupiter.) It is also evident that, on a statistical basis, the fluxes at each location are comparable when the ULS-measured flux values are scaled by the radial distance at the time of the measurement.

Figure 4 shows a comparison of the fluxes of atomic abundances (0.7–1.0 MeV/nucl) for carbon, nitrogen, oxygen, neon, magnesium, silicon, sulfur, and iron. The ACE distributions are plotted as heavy dark lines; the ULS distributions are the light lines and have been multiplied by the radial distance of the measurements before compiling the statistics. The distributions are seen to be similar at the higher flux levels. The $Z > 2$ fluxes are not log normally distributed, having low flux intensity cutoffs of the distributions that correspond to the measurement (or not) of a single count during an averaging interval. For this reason, the distributions are not similar at the lowest fluxes.

Plotted in Figure 5 are comparisons of the abundances of atomic species relative to oxygen (0.7–1.0 MeV/nucl) as measured at ACE (heavy dark lines) and ULS (light lines) for the same time interval as in Figure 3. These abundance distributions are approximately log normal and are similar at the two helioradii. That is, the statistical abundances do not appear to be affected by any

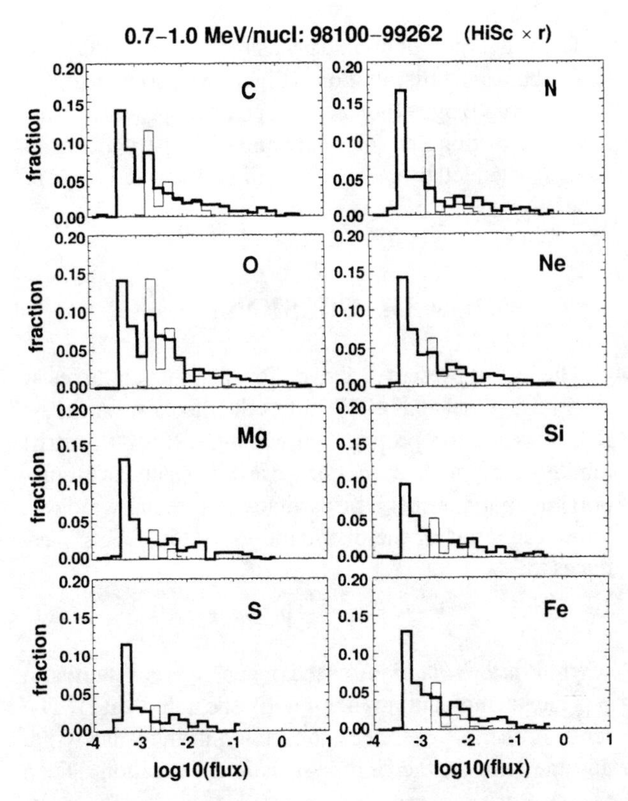

FIGURE 4. Distributions of heavy elements (0.7–1.0 MeV/nucl) for ACE and Ulysses (heavy and light lines, respectively). HiScale measurements are multiplied by the radial distance of the spacecraft

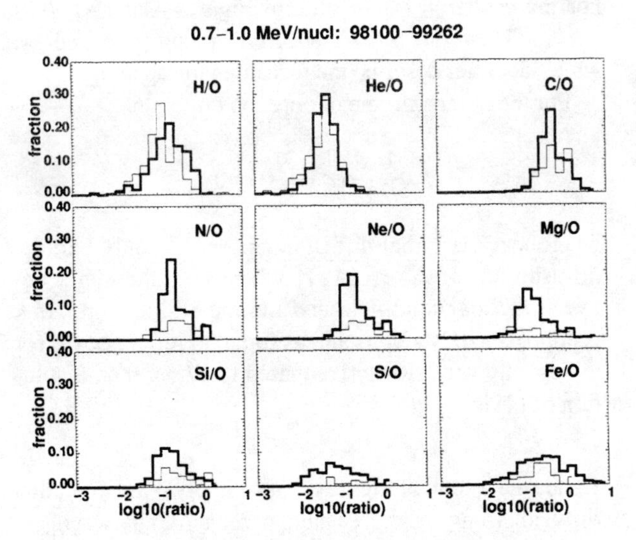

FIGURE 5. Abundance ratios to oxygen (0.7–1.0 MeV/nucl) for ACE and Ulysses (heavy and light lines, respectively)

interplanetary dynamics that might occur between 1 and 5 AU.

The suggestion of double peaks in the N/O, Ne/O, and Fe/O abundance distributions (Figure 5) also appears in other energy ranges (not shown), and may provide information to distinguish solar flare and CME particles from those accelerated by CIRs. This will be discussed in more detail elsewhere.

DISCUSSION

The results presented above show that the ion fluxes at r = 5 AU, j_5, scale as $\sim 1/r$ times the fluxes at Earth, j_1. If we assume ion propagation without diffusion or drift (and no additional sources or losses), the standard transport equation for energetic particles in the solar wind (8), written in terms of the distribution function $f = p^2 j$ reduces to:

$$V\frac{\partial f}{\partial r} = \frac{1}{3r^2}\frac{\partial}{\partial r}[r^2 V]p\frac{\partial f}{\partial p} \quad (1)$$

where V is the solar wind speed (taken to be constant) and p is the ion momentum (non-relativistic in this case). The term on the left describes convection in the solar wind, and the term on the right, adiabatic deceleration. Then (1) becomes

$$\frac{\partial f}{\partial r} = \frac{2}{3r}p\frac{\partial f}{\partial p} \quad (2)$$

which has a solution $f \propto r^a p^b$ where a and b are related as

$$a = \frac{2}{3}b \quad (3)$$

For the case $a = -1$ (results of Figures 3 and 4), $b = -3/2$. Then $j \propto p^{1/2}$, a spectrum rising with energy, which does not describe the measured ion spectra.

For the other extreme of pure ion diffusion,

$$\frac{1}{r^2}\frac{\partial}{\partial r}(r^2 \kappa \frac{\partial f}{\partial r}) = 0 \quad (4)$$

where κ is the radial diffusion tensor. For only parallel diffusion $\kappa_{\|}$, $\kappa = \kappa_{\|}(\cos\psi)^2$ where ψ is the angle between the magnetic field and the radial direction. If κ is approximately a constant, as found in reference (3) (or varies only very slowly compared to $1/r^2$), then a solution of (4) is

$$f \propto 1/r \quad (5)$$

Hence the results of Figures 3 and 4 imply rapid ion radial diffusion (long mean free path, order 1 AU) between $r \sim$ 1 AU and 5 AU. This is not an unreasonable description of particle propagation in the inner heliosphere, although the statistical results in (3) reported mean free paths < 0.1 AU between 1 and 5 AU.

The foregoing conclusion of rapid ion diffusion cannot eliminate, however, the possible effect of interplanetary statistical acceleration off-setting the adiabatic deceleration in a pure convection/diffusion regime (which incorporates adiabatic deceleration). A preliminary examination of this possibility shows that statistical acceleration is not significant in the context of the diffusion term. This will be explored further in a future work, as will the implications for possible CIR acceleration.

As shown in Figure 5, on a statistical basis the ion abundances relative to O are found to be similar and approximately log normal at 1 and at 5 AU over the nearly two year interval examined. The similarity of the distributions at both locations indicates that any dynamical processes that are operative in the solar system beyond Earth to the orbit of Jupiter, such as acceleration or deceleration by traveling shock waves and statistical acceleration, operate statistically equally on all ions. On a statistical basis, therefore, this implies that the charge state of the ions examined at the two locations is the same, probably a charge of one.

ACKNOWLEDGMENTS

We thank Dr. B. Klecker for helpful comments and our EPAM and HI-SCALE colleagues for their continuing contributions to the success of these investigations.

REFERENCES

1. Gold, R.E., S.M. Krimigis, S.E. Hawkins III, D.K. Haggerty, D.A. Lohr, E. Fiore, T.P. Armstrong, G. Holland, and L.J. Lanzerotti, *Space Science Reviews* **86**, 541-562 (1998).
2. Lanzerotti, L. J., et al., *Astron. Astrophys.* **92**, 349-363 (1992).
3. Zwickl, R. D., and W. R. Webber, *Solar Physics* **54**, 457-504 (1975).
4. McDonald, F. B., B. J. Teegarden, J. H. Trainor, and W. R. Webber, *Ap. J.* **187**, L105-L108 (1974).
5. Webber, W. R., F. B. McDonald, T. T. von Rosenvinge, and R. A. Mewaldt, *17th Inter. Cosmic Ray Conf.* **10**, 92 (1981).
6. Lanzerotti, L. J., and C. G. Maclennan, *Ap.J.Lett.* in press (2000).
7. Lanzerotti, L. J., R. E. Gold, D. J. Thomson, R. E. Decker, C. G. Maclennan, and S. M. Krimigis, *Ap. J.* **380**, L93-L96 (1991).
8. Parker, E. N., *Planet. Space Sci.* **13**, 9-49, (1965).

Propagation of Inclined Solar Neutrons: Scattering, Energy Decrease, Attenuation, and Refraction Effect

Lev I. Dorman[1], Irina V. Dorman[2], J.F. Valdes-Galicia[3]

[1]*IZMIRAN, Technion and Israel Cosmic Ray Center, affiliated to Tel Aviv University; Current address: P.O. Box 2217, Qazrin 12900, Israel; e-mail: lid@physics.technion.ac.il*
[2]*Institute of History of Science and Technology, Russian Ac. of Science, Staropansky 1/5, Moscow 103012, Russia*
[3]*Institute of Geofisica of UNAM, Mexico D.F., 04510 Mexico*

Abstract. We develop our simulation and analytical calculation results of solar neutron propagation in the atmosphere for different initial zenith angles, taking into account not only scattering and attenuation, but also neutron energy decrease (what leads to increase both of cross-section of interaction and scattering angles, what is especially important for small energy solar neutrons). We test the usually used suggestion that solar neutron propagation through the atmosphere of depth h at some initial zenith angle θ_o is the same as for vertical direction, but for depth $h/\cos\theta_o$. Our calculations of multi-scattering of neutrons on small angles with attenuation and energy change for different initial zenith angles, show that this suggestion is not correct. Taking into account the neutron energy change shows that with decreasing of solar neutron energy the asymmetry in solar neutron propagation and refraction effect became stronger. We show that during the propagation through the atmosphere the effective zenith angle of solar neutron flux sufficiently decreases. These decreases are especially great for the big initial zenith angles what gives expected solar neutron fluxes many times bigger than in the frame of previous theory of solar neutron propagation. We show also that the optimum direction of solar neutron telescope must be not the direction on the Sun (as for gamma-ray telescope), but between the Sun and vertical in dependence of Sun's zenith angle and effective energy of neutrons.

INTRODUCTION

In [1-3] we extended the investigation of solar neutron propagation which was done in [4] only for vertical initial incidence. We considered solar neutrons arriving at different initial zenith angles 0^0, 15^0, 30^0, 45^0, 60^0, 75^0 and did calculations of the angular distribution of arriving neutrons at different atmospheric levels by taking into account neutron scattering and attenuation. Here we will take into account the energy decrease of neutrons in scattering processes. We will show that the energy decrease of solar neutrons leads to increase the refraction effect. We calculate here expected solar neutron angle distribution, effective zenith angle of arriving neutrons and expected integral multiplicities for different initial zenith angles on different depth in the atmosphere for different energies of solar neutrons.

DECREASE OF SOLAR NEUTRON ENERGY DURING PROPAGATION

In each elastic scattering with O and N atoms in the atmosphere the energy of neutrons E_n decreases, as an average, proportionally to the coefficient 0.8793 (according to data in [4]). If solar neutron arrives to the boundary of the atmosphere with initial energy E_{no} and initial zenith angle θ_o, the energy of neutron at a level h and at zenith angle θ will be

$$\ln(E_n) = \ln(E_{no}) + (L_e(\theta_o,\theta,h)/\lambda)\times\ln(0.8793), \quad (1)$$

where $L_e(\theta_o,\theta,h)$ (in g/cm^2) is the effective average path of neutrons propagating from the boundary of atmosphere (h=0) to the level h at zenith angle θ (determined in [2,3]), and λ is the average path for neutron scattering and attenuation. According to [4] λ is about 110 g/cm^2 and is practically independent from the energy of neutrons.

CP528, *Acceleration and Transport of Energetic Particles Observed in the Heliosphere: ACE 2000 Symposium,* edited by Richard A. Mewaldt, et al.
© 2000 American Institute of Physics 1-56396-951-3/00/$17.00

DEPENDENCE OF SINGLE EFFECTIVE SCATTERING ANGLE FROM NEUTRON ENERGY

On the basis of data reviewed in [4], the dependence of single effective scattering angle $\delta(E_n)$ (in radians) for $-4 \le \ln(E_n) \le 2.3$ (where E_n is in *GeV*), may be approximated (with correlation coefficient 0.9975) as:

$$\delta(E_n) = 0.01034(\ln E_n)^2 - 0.0470 \ln E_n + 0.0686 \quad (2)$$

SOLAR NEUTRON ANGULAR DISTRIBUTION

Taking into account (1) and (2) we determine, according to [2], the expected solar neutron angular distribution in the atmosphere at depth h from 50 up to 1050 g/cm^2 in steps of 50 g/cm^2 for initial zenith angles θ_o from 0^0 to 88^0 and initial energies E_{no} from 0.1 *GeV* to 15 *GeV*.

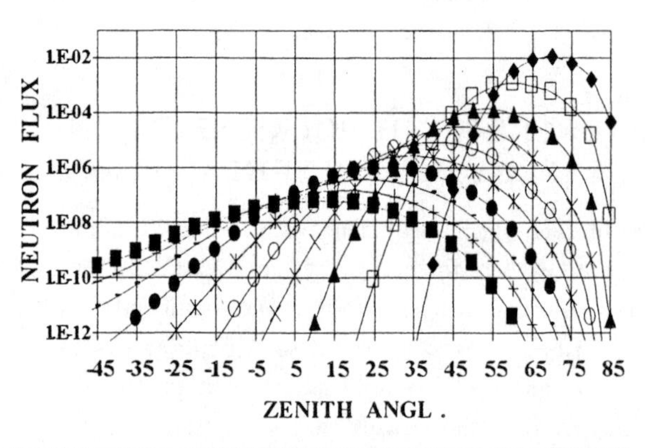

Figure 1. Expected angular distributions for $E_{no} = 1\,GeV$ and $\theta_o = 75^o$ at different levels h from 100 to 1000 g/cm^2; the values on the ordinate axis show the expected flux per one initial neutron per 5 degrees of zenith angle.

As an example, we show in Figure 1 the zenith angle distributions in the plan contains vertical and direction to the Sun at different h for $E_{no} = 1\,GeV$ and $\theta_0=75^0$ (negative zenith angles mean that arriving neutrons have azimuth angles opposite than azimuth angle of the Sun). Figure 1 shows that with increasing h the median zenith angle moved from about 70^0 at h=100 g/cm^2 to about 10^0 at 1000 g/cm^2. Taking into account energy change increases this effect because with decreasing of neutron energy during propagation $\delta(E_n)$ increases according to (2).

EFFECTIVE ZENITH ANGLE OF ARRIVING NEUTRONS

On the basis of determined angular distributions $F(\theta, h, \theta_o, E_{no})$ (see an example in Figure 1) we calculate expected effective zenith angle as:

$$\langle\theta\rangle = \int \theta F(\theta, h, \theta_o, E_{no}) d\theta / \int F(\theta, h, \theta_o, E_{no}) d\theta \quad (3)$$

In Figure 2 are shown dependencies of $\langle\theta\rangle$ from level h and initial zenith angle θ_o for $E_{no} = 3\,GeV$.

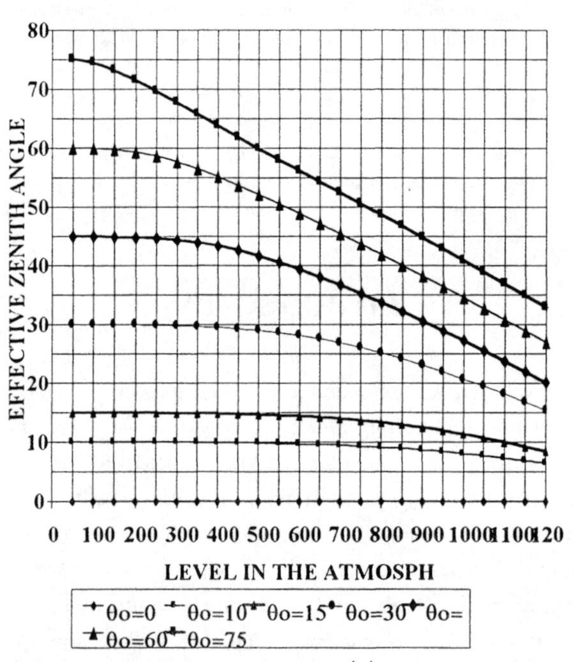

Figure 2. Effective zenith angle $\langle\theta\rangle$ versus level h for different initial zenith angle θ_o and $E_{no} = 3\,GeV$.

INTEGRAL MULTIPLICITIES FOR SOLAR NEUTRONS

We determine expected total neutron flux per one solar neutron, i.e. integral multiplicity $m(E_{no}, h, \theta_o)$ for E_{no} from 0.1 *GeV* to 15 *GeV* at different depth h in dependence of initial zenith angle θ_o with taking into account the change of neutron energy according to (1) and effective scattering angle according to (2). As an example, results for the initial neutron energy $E_{no} = 3\ GeV$ are shown .in Figure 3.

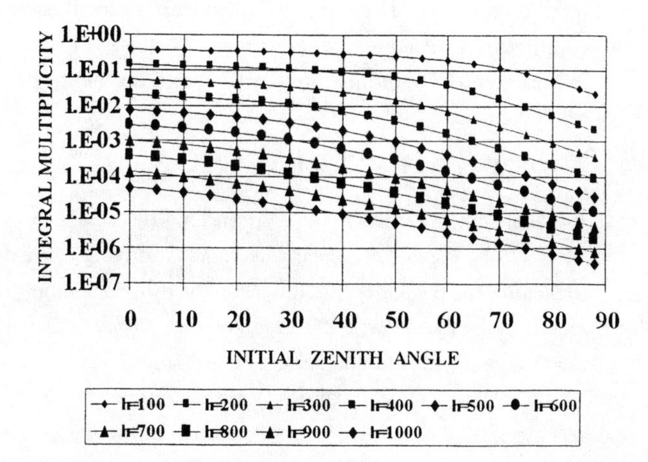

Figure 3. Expected integral multiplicity $m(E_{no}, h, \theta_o)$ for solar neutrons in dependence of initial zenith angle at different h from 100 to 1000 g/cm^2 for $E_{no} = 3\ GeV$.

REFRACTION EFFECT

Let us determine the value of refraction effect $R(E_{no}, h, \theta_o)$ as ratio of expected solar neutron integral multiplicity with taking into account attenuation, scattering and neutron energy change (see an example in Figure 3) to the usually used as integral multiplicity for vertical initial direction but on the depth $h/\cos\theta_o$ [4]. As an example, results for $E_{no} = 3\ GeV$ are shown in Figures 4 and 5.

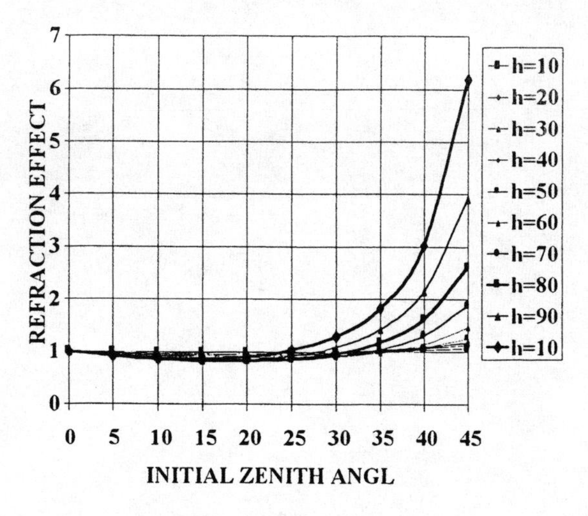

Figure 4. $R(E_{no}, h, \theta_o)$ vs. θ_o from 0^0 to 45^0.

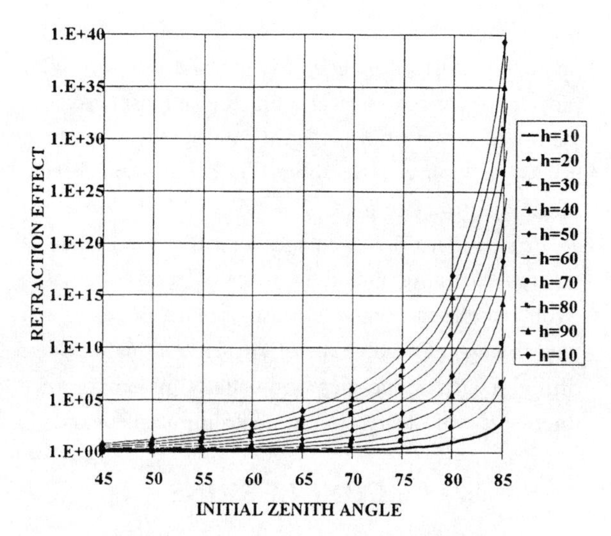

Figure 5. $R(E_{no}, h, \theta_o)$ vs. θ_o from 45^0 to 85^0.

COMPARISON OF RESULTS WITH AND WITHOUT INCLUDING OF THE NEUTRON ENERGY CHANGE

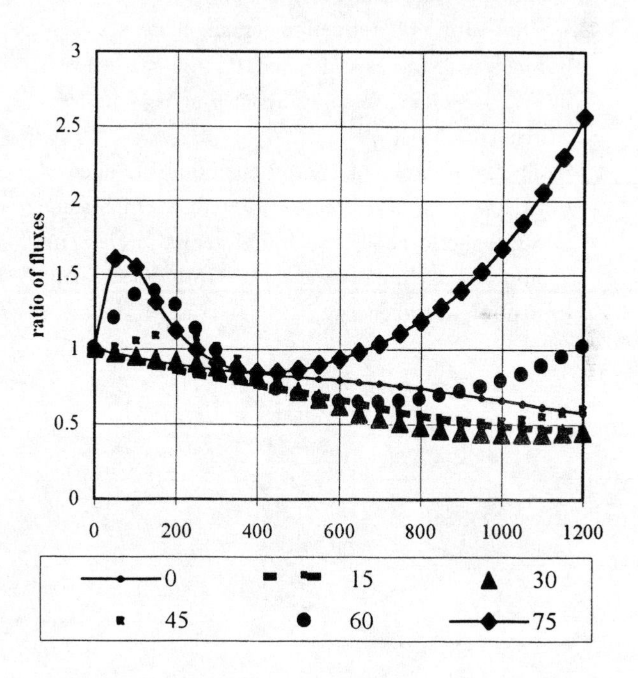

Figure 6. Ratios between fluxes of solar neutrons computed by taking into account neutron energy change and not taking into account the change of neutron energy. These ratios are given for initial energy $E_{no} = 3\ GeV$ and for different initial zenith angles from 0^0 to 75^0.

To show the importance of the including of the neutron energy change in calculations of solar neutron propagation through the atmosphere, we compare results for integral multiplicity for both cases. In Figure 6 we show this comparison for the initial neutron energy $E_{no} = 3\ GeV$.

From Figure 6 can be seen that the importance of energy changing including in calculations increased with increasing both of atmospheric depth and initial zenith angle. Calculations of the same ratios for other initial neutron energies show that this importance increases with decreasing of neutron initial energy.

CONCLUSIONS

1. The effective zenith angle of arriving neutrons decreases sufficiently with increasing h; this effect is especially important for great initial zenith angles θ_o.
2. Integral multiplicity of solar neutrons decreases sufficiently with increasing h and θ_o.
3. The value of refraction effect increases very much with increasing h and θ_o: for sea level for $E_{no} = 3\ GeV$ it reaches about 6 at $\theta_o = 45^0$ and about 10^{10} at $\theta_o = 75^0$.
4. The importance of energy changing included in calculations increases with increasing atmospheric depth and initial zenith angle. This importance increases also with decreasing neutron initial energy.

Acknowledgements. LID thanks the organizers of ACE 2000 for the kind invitation to take part in the Conference, and Marty Lee and Giorgio Villoresi for interesting discussions. LID thanks also the referee for useful comments (according to these comments it was added a special Section on the importance of including in computations the neutron energy changes).

REFERENCES

1. Dorman, L.I., Dorman, I.V., and Valdes- Galicia, J.F., "Simulation of solar neutron scattering and attenuation in the Earth's atmosphere for different initial zenith angles". *Proc. 25- th Intern. Cosmic Ray Conference*, Durban (South Africa), **1**, 1997, pp 25-28.
2. Dorman, L.I., and Valdes-Galicia, J.F., "Analytical approximation to solar neutron scattering and attenuation in the Earth's atmosphere". *Proc. 25- th Intern. Cosmic Ray Conference*, Durban (South Africa), **1**, 1997, pp 29-32.
3. Dorman, L.I., Valdes-Galicia, J.F., and Dorman, I.V., "Numerical simulation and analytical description of solar neutron transport in the Earth's atmosphere". *J. Geophys. Res.* **104**, No.A10, 22417-22426 (1999).
4. Shibata, S., *J. Geophys. Res.*, **99**, No. A4, 6651 (1994).

On the Heliospheric ^{3}He Observations and Their Astrophysical Implications

Ilan Roth*

Space Sciences Laboratory, University of California at Berkeley, CA ,94720

Abstract. Analysis of heliospheric observations indicating enhanced abundances of rare isotopes, modified abundance ratios of elements and isotopes, or elements with unusual charge states present challenges with astrophysical connotations. Besides processes involving nuclear reactions one must envision physical mechanisms which involve atomic processes at the outer solar layers as well as plasma processes in the presence of solar or interplanetary magnetic field. The most spectacular heliospheric enhancement with an anusual abundances involves the ^{3}He isotope and heavy elements (mainly Mg, Ne, Si and Fe) in impulsive solar flares. Primordial nucleosynthesis and galactic evolution confines the coronal ratio of the He isotopes, ^{3}He/^{4}He, to several times 10^{-4}; this ratio is enhanced during impulsive flares by several orders of magnitude due to resonant wave-particle interaction with electromagnetic ion cyclotron (emic) waves. Recent observations on the ACE satellite may help in explaining two unrelated astrophysical phenomena: (1) Observations of enhanced abundances of ^{3}He in planetary nebulae, (2) Formation of radioactive ^{26}Al isotope, which decays into the observed fossil element ^{26}Mg, due to the energetic flare-related ^{3}He in the early solar system.

INTRODUCTION

One of the cornerstones of modern astrophysics includes the study of the evolution of different elements since their origin to the present time. Stellar nuclear reactions modify the relative abundances of the light elements from the primordial values created by the Big Bang, while important local changes in abundances may occur in stellar magnetized plasmas due to atomic and electromagnetic processes. The evolution of astrophysical plasmas is determined to a large extent by their interaction with electromagnetic waves. These waves are excited either by unstable, nonisotropic (in the momentum space) distributions of particles, or result from large-scale inhomogeneous configurations. The wave-particle interactions may accelerate and heat various populations and significantly modify the abundances of nonthermal particles. Abundances of ions measured in situ by satellites (heliosphere) or deduced remotely by spectroscopic methods (interstellar medium), reveal various nuclear, atomic, and plasma processes which affect their evolution. The main processes which form the light elements occur in the primordial nucleosynthesis which creates ^{4}He, ^{2}H, ^{3}He and ^{7}Li [15] and in the spallation by the galactic cosmic rays which synthesizes ^{6,7}Li, ^{9}Be, and ^{10,11}B [20]. The cosmic (solar) abundances change during transport from the photosphere to the corona due to the first-ionization potential effect [12]. In-situ observations of impulsive-flare-accelerated ions in interplanetary space indicate abundance enhancement up to four orders of magnitude in the ^{3}He/^{4}He ratio and up to one order of magnitude in heavy ions (above CNO, up to Fe) vs coronal abundances [19]. The present paper surveys the astrophysical context of these abundances emphasizing the importance of ^{3}He measurements to the analysis of the galactic evolution, describes the excitation of low-frequency waves due to electron fluxes on coronal field lines and the resonant interaction of emic waves with selective ions, and using the recent ACE observations, suggests a possible scenario to explain the large abundance of ^{3}He in planetary nebulae and of the fossil radioactive element ^{26}Mg in extinct meteorites.

EVOLUTION OF THE HE ISOTOPES

^{2}H, ^{3}He, ^{4}He and ^{7}Li are the light nuclides produced in the primordial homogeneous Big-Bang nucleosynthesis (BBN). When the temperature of the Universe falls below 10^{10} K, the disintegration of deuterons by the photons slows down and the strong interaction starts creating the light nuclides, with a possibility of extremely low traces of heavier elements. Since the cross sections at the BBN epoch with temperatures 0.1 MeV $< T <$ 1 MeV

* In parts supported by Grants NAG5-3182, NAG5-3596 and NAG5-6985

CP528, *Acceleration and Transport of Energetic Particles Observed in the Heliosphere: ACE 2000 Symposium,* edited by Richard A. Mewaldt, et al.
© 2000 American Institute of Physics 1-56396-951-3/00/$17.00

are well known, the predicted abundances of the light elements which are produced in the BBN depend on: (i) baryon-to-photon ratio $\eta = n_b/n_\gamma$, (ii) universal expansion rate, which relies on the number of neutrino flavours (determined to be 3) and (iii) neutron lifetime, which is known accurately (887 sec). The resulting abundances are compared to different observations in the Galaxy, resulting in powerful conclusions relating the evolution of light elements and the status of the Universe (e.g., [21]).

The calculated abundance ratios of the BBN nuclei vs η (e.g. [25]) show the weak dependence of the ^{4}He abundance on the baryon-to-photon ratio with ^{4}He mass fraction of ~ 0.23, and the monotonous and relatively strong dependence of ^{2}H and ^{3}He abundance on η, making them very good baryonometers. The main change in the light nuclides abundances during the galactic evolution is the augmentation of ^{3}He in the pre-main sequence phase of stellar nucleosynthesis by a complete burning of the fragile ^{2}H to ^{3}He, such that ^{2}H and (^{3}He + ^{2}H) can be used, respectively, to assess the primeval upper and lower bounds of η [29]. The consistency of the observed abundances of the light nuclides in low metallicity astrophysical regions (weak galactic processing) over nine orders of magnitude, together with the Hubble expansion and the cosmic background microwave radiation, are the basis of the Big Bang cosmology. The concordance values, based on observational overlap of measured light nuclides abundances, and the known photon density in the microwave background radiation, place constraints on the baryon density, $2 \times 10^{-10} < \eta < 6 \times 10^{-10}$, indicating missing mass in the Universe. From the BBN and models of galactic evolution the deduced value for the primordial ^{3}He/^{4}He ratio emerges as $\sim 1.5 \times 10^{-4}$, while the solar ^{3}He/^{4}He ratio, which increases mainly due to ^{2}H burning, becomes $\sim 4.0 \times 10^{-4}$. These values are consistent (i) with the nucleosynthesis result for the primordial abundances [29] and the observations of the hot, ionized gas clouds [27] and (ii) with the solar photospheric and the solar wind values.

RESONANT INTERACTION BETWEN CORONAL IONS AND EMIC WAVES

There exists an interesting analogy between physical processes on active auroral and flaring coronal field lines, as was suggested by [26] and investigated by [13] and [23]. Both environments consist of very low β plasmas, are dominated by two majority ion species (H and O at Earth, H and He at Sun), and are subjected to large electron fluxes due to magnetic field reconfigurations. In the corona these electron fluxes are deduced from X-ray emissions, while in the aurora they can be measured in situ by spacecraft. Auroral observations on the S3-3, Freja, Viking, Polar and Fast satellites indicate that electromagnetic ion cyclotron waves are associated with accelerated electrons which are responsible for the discrete aurora. Similarly, one may postulate that the fluxes of streaming electrons along the flaring coronal field lines that generate the solar flare X-ray emissions, are also correlated with several low-frequency plasma modes. The emic waves are of particular interest due to their effective interaction with selective ions and charge states of heavier elements.

In the presence of two or more ion species, the oblique emic wave propagates in a frequency range below the hydrogen gyrofrequency and above the two-ion-hybrid frequency. The mechanism of ion acceleration is based on the gyroresonant interaction of ^{3}He and heavy ions. When the wave which propagates along the inhomogeneous magnetic field passes through Doppler-shifted gyrofrequency of ^{3}He or higher gyroharmonic of the heavier ions, these ions are resonantly accelerated. The residence time in the resonance region is determined by the local gradient of the magnetic field which affects the mirror force, and by the propagation direction of the waves with respect to the magnetic field. Waves propagating into stronger magnetic fields produce a parallel ion acceleration that balances the effects of the mirror force and increases the residence time, thereby enhancing the ion heating.

The waves are damped near the H and ^{4}He gyrofrequencies; at the high coronal temperatures above 4 MK ^{4}He and the main isotopes of CNO are fully ionized with a charge-to-mass ratio of 0.5 (in units of the H), hence they are not significantly affected by the waves. The lack of O^6 enhancement in impulsive flares, in contrast to its observation in the solar wind, indicates that the source of the accelerated ions is located at high temperatures. Therefore CNO nuclei, which are among the main products of stellar nucleosynthesis due to nuclear processes, are underabundant in the impulsive flares due to atomic processes (which are determined by the coronal temperature) and by plasma processes (which result from interaction with the emic waves). Similarly, heavy elements and isotopes which are not fully ionized at the coronal temperatures may be enhanced in impulsive flares.

The acceleration rate of a thermal ($\sim$100eV), resonant heavy ions is smaller by factor of $k_\perp v_\perp/\Omega_i \ll 1$ as compared to ^{3}He [23], since the resonant waves resonate with higher harmonics of the heavy ion gyrofrequency. An alternative scenario for acceleration of heavy ions involves cascading Alfven waves which resonate more effectively with the heavier ions with lower gyrofrequencies [14]. The data are inconclusive regarding the relative enhancements of ^{3}He and heavy ions; some recent measurements may indicate a positive correlation between the two enhancements [9]. In this scenario, although the initial acceleration of the Fe ions is smaller than of ^{3}He, fraction of both acceleration is due to the same mechanism.

ASTROPHYSICAL ^{3}HE PROBLEM

The abundance of ^{3}He, which can serve as a probe of cosmology, the evolution of low-mass stars, and the chemical evolution of the Galaxy, can be determined beyond the local solar neighbourhood only via measurements of the 3.46 cm (8.665 GHz) hyperfine transition of $^3He^+$. H II regions and planetary nebulae (PN) are the main observable sources of this transition (e.g. [4]). ^{3}He measurements in the protosolar material [7], the local interstellar medium [8] and galactic H II regions [2] indicate that the ^{3}He/H abundance ratio is similar to the non-processsed galactic value of $\sim 2 \times 10^{-5}$. However, in a series of very long observations [1;3] several PN sources were detected with $^3\mathrm{He/H} = 10^{-4} - 10^{-3}$, i.e. more than an order of magnitude larger than those found in any H II region, local interstellar medium or protosolar system. The high ^{3}He/H poses an important question with respect to the galactic evolution of the light elements. The disagreement between the observed abundances of ^{3}He throughout the Galaxy and the chemical evolution models, together with observations of ^{3}He enhancement around PN was term the "^{3}He problem" [6], which has stimulated studies of many nonstandard models for both stellar and galactic chemical evolution.

Recent ACE measurements show enhancements in ^{3}He/^{4}He (or equivalently ^{3}He/H) ratio in the range from MeV/nucleon all the way down to tens of keV/nucleon [28], confirming the previous observations [24;18]. More interestingly, the ^{3}He/^{4}He enhancements have been observed in gradual events lasting several days, when the isotopic ratio has increased by a factor of five or more above the solar ratio [11] and even during quiet times ^{3}He spectrum shows a low-energy turn-up below 10 MeV [28]. Some gradual events show also an increased Fe/O ratio, indicating that the impulsive events which are embedded into the gradual events and which are able to enhance ^{3}He/^{4}He by several orders of magnitude, increase the baseline (average) heliospheric ^{3}He/^{4}He. Although the impulsive events last over relatively short period of time, their frequent appearance during the active solar periods increases the abundance of ^{3}He in the interplanetary medium and this source population is accelerated again (presumanbly by the fast CME shock), resulting in an increased average isotopic ratio. Therefore, in addition to the specific requirements from the stellar evolution [22;16] to produce ^{3}He in low-mass stars in PN and deplete it in H II regions, as required for consistency with observations, one may consider a planetary nebula with its very hot central star and very intense magnetic activity as a source of steady enhancement of ^{3}He over a broad range of energies. The resulting mixing of the He isotopes results in an average higher value of ^{3}He/^{4}He and ^{3}He/H in the surrounding environment. The large H II regions which consist of many stars with on the average much less intense magnetic activity will not reach such high baseline values.

ENHANCED ABUNDANCE OF ^{26}MG IN METEORITES

The observations of fossil radioactive elements in meteorites is connected intimately to the formation of the solar system. The discovery of several radionuclides with lifetimes smaller by several orders of magnitude than the age of the solar system indicates that these elements were present 4.5×10^9 years ago in rocky planetesimals and were formed either due to a supernova events or by solar cosmic ray bombardments from the young Sun. Of particlar interest are the observations of the Ca-Al inclusions, millimiter-sized assemblies of texturally related Al-rich minerals, with an enhanced abundance of ^{26}Mg, formed by the decay of radioactive ^{26}Al. The galactic ratio of ^{26}Al/^{27}Al (radioactive/stable) is 3×10^{-6}, while the deduced ratio in the Ca-Al inclusions is 5×10^{-5}. ^{26}Al decays with the emission of 1.809 MeV photons and a lifetime of 1.1×10^6 years. This is one of the strongest emissions from the galactic center, indicating that presently ^{26}Al is created in supernovae with a formation rate of 3 solar masses/My [17]. It is generally believed that the birth of our solar system occurred when a molecular cloud was seeded by a nucleosynthesis event; however, supernovae cannot make enough ^{26}Al [5]. The bombardment of presolar grains by solar cosmic rays (H, ^{4}He) which were accelerated in gradual events may give the correct yield of the radioactive ^{41}Ca and ^{53}Mn, but too low for ^{26}Al. A possible solution may involve reactions with energetic ^{3}He ions. Since young solar-like stars emit copious amount of X-rays it is clear that they are magnetically very active. Therefore the young Sun was a source of a very large number of impulsive flares and the heliosphere was filled with a huge amount of energetic ^{3}He. These MeV ions interact through several channels like ^{24}Mg(^{3}He,p)^{26}Al, ^{25}Mg(^{3}He,pn)^{26}Al and ^{27}Al(^{3}He,α)^{26}Al [10] to form the radioactive element ^{26}Al. Although these reactions are not yet well calibrated and the preliminary requirements for ^{3}He/H seem high [10], they have a sufficiently large cross sections to increase significantly the ^{26}Al before solidification. Therefore the selective acceleration of a rare isotope in impulsive flares may help to understand the selective formation of radioactive elements in the early Sun which are observed as radioactive fossil elements in meteorites. ACE measurements indicate a significant, steady increase in the heliospheric energetic ^{3}He during active periods, allowing us to conclude that at the early Sun there was a significant likelihood for the ^{3}He-induced reactions resulting in the formation of ^{26}Al and the subsequent decay into ^{26}Mg.

CONCLUSIONS

The BBN isotopic ratio ^{3}He/^{4}He $\sim 1.5 \times 10^{-4}$, similar to the observed values at several primitive, low-metallicity astrophysical objects, changes during galactic chemical evolution mainly due to (a) the almost complete burning of the fragile deuterium into ^{3}He, which increases the He isotopic ratio towards the solar value of 4.1×10^{-4} and (b) the solar reactions. Electromagnetic interaction in the solar corona increases this ratio spectacularly in impulsive solar flares.

Impulsive solar flares generate fluxes of electrons with power-law halo and positive gradients in their distribution function, which may drive a variety of wave modes. The resonant interaction between ^{3}He ions and low-frequency ion cyclotron waves on coronal field lines accelerates efficiently these ions to energies of tens of keV to several MeV/nuc. The ^{3}He ion is unique, having a gyrofrequency at the range of the emic waves, hence easily accelerated. The ACE satellite observes enhanced ^{3}He abundances due to impulsive flares as well as increased average isotopic ratio ^{3}He/^{4}He over long periods of days during gradual events. Similar effect may exist in planetary nebulae where the spectroscopically observed He isotopic ratio is increased by an order of magnitude. The intense magnetic activity at the early Sun with the numerous impulsive flares and enhancement of MeV ^{3}He fluxes may explain the formation of the radioactive nuclei ^{26}Al and the decay product of a fossil ^{26}Mg.

REFERENCES

1. Balser, D.S., Bania, T. M., Rood, R.T., and Wilson, T.L., *Astrophys. J.*, **483**, 320, 1997.

2. Balser, D.S., Bania, T.M., Rood, R.T., and Wilson, T.L., *Astrophys. J.*, **510**, 759, 1999a.

3. Balser D.S., Rood R.T., and Bania T.M., *Astrophys. J.*, **522**, L73, 1999b.

4. Bania, T.M., Balser, D.S., Rood, R.T., and Wilson, T.L., *Astrophys. J. Supp.*, **113**, 353, 1997.

5. Cameron, A.G., Hoflich P., Myers P.C., and Clayton, D.D., *Astrophys. J.*, **447** , L53, 1995.

6. Galli, D., Stanghellini L., Tossi M., and Palla, F., *Astrophys. J.*, **477**, 218, 1997.

7. Geiss, J., in *Origin and Evolution of the Elements*, ed. N. Prantzos, E. Vangioni-Flam and M. Casse, (Cambridge: Cambridge University Press), 1993.

8. Gloeckler, G. and Geiss, J., *Nature*, **381**, 210, 1996.

9. Ho, G.C., ACE Workshop, Indian Wells, CA, 2000.

10. Lee T., Shu F.H., Shang, H., Glassgold, A.E. and Rehm, K.E., *Astrophys. J.*, **506**, 898, 1998.

11. Mason, G.M., ACE Workshop, Indian Wells, CA, 2000.

12. Meyer, J.P., *Astrophys. J. Sup.*, **57**, 173, 1985.

13. Miller, J.A. and Vinas A.F., *Astrophys. J.*, **412**, 386, 1993.

14. Miller, J.A., *Space Science Reviews*, **86**, 79, 1998.

15. Olive, K.A., Schramm D.N., Steigman, G., and Walker, T.P., *Phys. Lett.*, **B236**, 454, 1990.

16. Olive, K.A., Rood, R.T., Schramm D.N., and Vangioni-Flam E., *Astrophys. J.* , **444**, 680, 1995.

17. Prantzos, N. and Diehl, R., *Phys. Reports* **267**, 1, 1996.

18. Reames, D.V., Barbier, L.M., von Rosenvinge, T.T., Mason, G.M., Mazur, J.E., and Dwyer, J.R., *Astrophys. J.*, **483**, 515, 1997.

19. Reames, D. V., Meyer, J. P., and von Rosevinge, T. T., *Astrophys. J. Sup.*, **90**, 649, 1994.

20. Reeves, H., Fowler W.A., and Hoyle F., *Nature*, (London) **26**, 727, 1970.

21. Reeves, H., The saga of light elements, in *Origin and Evolution of the Elements*, ed. N. Prantzos, E. Vangioni-Flam and M. Casse, Cambridge: Cambridge University Press), 1993.

22. Rood, R.T., Steigman D.G., and Tinsley, B.M., *Astrophys. J.*, **207**, L57, 1976.

23. Roth, I. and Temerin, M., *Astrophys. J.*, **477**, 940, 1997.

24. Serlemitsos, A. T. and Balasubrahmanyan, V. K., *Astrophys. J.*, **198**, 195, 1975.

25. Schramm, D.N., in *Origin and Evolution of the Elements*, ed. N. Prantzos, E. Vangioni-Flam and M. Casse, Cambridge: Cambridge University Press), 1993.

26. Temerin, M., and Roth, I., *Astrophys. J.*, **391**, L105, 1992.

27. Vangioni-Flam, E., Olive K.A., Prantzos N., *Astrophys. J.*, **427**, 618, 1994.

28. Wiedenbeck, M.E., ACE Workshop, Indian Wells, CA, 2000.

29. Yang J., Turner M.S., Steigman, G., Schramm, D.N., and Olive K.A., *Astrophys. J.*, **281**, 493, 1984.

Measurement of Accelerated Particles at the Sun

Gerald H. Share and Ronald J. Murphy

E.O. Hulburt Center for Space Research, Naval Research Laboratory, Washington, DC 20375

Abstract. Solar γ-ray lines and continua provide information on flare-accelerated particles that interact at the Sun. We primarily discuss observations of flare spectra made by the OSSE experiment on *CGRO* and the gamma-ray spectrometer on *SMM*. Continuum γ-ray spectra reflect the MeV electron population. These spectra show various shapes above 0.1 MeV: single power laws, broken power-laws (both hardening and softening > 0.1 to 0.3 MeV), and spectra that harden ≥ 1 MeV. The spectra and directionality of accelerated protons and α–particles are revealed in the narrow lines from excited ambient nuclei and α-^{4}He fusion. These measurements imply power-law spectral indices between ~ -3 and -5 for energies >5 MeV/nucleon, evidence for both broad angular distributions and directionality in the interactions, and high accelerated α/p ratios (~0.5) in many flares. The presence of accelerated ^{3}He is revealed in weak line features suggesting ^{3}He/^{4}He ratios of ~0.1 or greater. Strongly Doppler-broadened lines reveal the composition and directionality of heavy accelerated ions. Fe appears to be enhanced by about the same ratio found in impulsive SEPs. We present some preliminary results on flares that individually were not detected at energies ≥ 1 MeV; their summed spectra exhibit hardening >1 MeV that may be due to nuclear radiation. There appears to be equipartition of energy between accelerated electrons and ions in flares with strong line emission.

LINK BETWEEN SOLAR GAMMA RAYS AND PARTICLES AT THE SUN

Figure 1 shows the γ-ray spectrum of the 1991 June 4 X12+ solar flare (N30E70) observed by the *Compton Gamma Ray Observatory (CGRO)* OSSE experiment.[1] The captions describe how γ-ray line and continuum studies reveal the physics of flares[2-5]. We use this figure to illustrate what has been learned about ion acceleration, transport, and interaction in flares.

Electron bremsstrahlung

We represent the electron bremsstrahlung by a power law(s) in this fit to the data. The shape of the continuum and its strength is variable from flare-to-flare and within flares. We find single power laws, broken power-laws (both hardening and softening >0.1 to 0.3 MeV), and spectra that harden > 1 MeV[6]. The acceleration process in flares is capable of producing particles that are energetically dominated by ions[7] or by electrons, as in electron-dominated episodes of some flares[8]. The nuclear contribution in these latter events is at least an order of magnitude below that typically found for events in which nuclear lines were observed. A tenfold decrease in the >1 MeV electron bremsstrahlung/nuclear line ratio was observed over a one hour period during the 1991 June 4 flare[1,9].

Narrow Nuclear Lines

About 30% of all flares with emission > 0.3 MeV exhibit characteristic features of ion interactions. Narrow γ-ray line observations are key to understanding the characteristics of accelerated protons and α particles at the Sun. They also provide information on ambient composition, temperatures, and densities. To date 17 distinct and relatively narrow de-excitation lines have been identified in solar flares[6]. These originate from proton and α-particle interactions on ambient material and include lines at 0.847 MeV (^{56}Fe), 1.238 MeV (^{56}Fe), 1.317 MeV (^{55}Fe), 1.369 MeV (^{24}Mg), 1.634 MeV (^{20}Ne), 1.778 MeV (^{28}Si), 4.439 MeV (^{12}C), and 6.129 MeV (^{16}O). Share & Murphy[10] found flare-to-flare variations in relative line fluxes suggesting that the abundance of elements in the flare plasma is grouped with respect to first ionization potential (FIP). They also showed that the Ne/O line ratio (see Figure 1) suggests that power-laws fit the accelerated particle spectra better than the

CP528, *Acceleration and Transport of Energetic Particles Observed in the Heliosphere: ACE 2000 Symposium,* edited by Richard A. Mewaldt, et al.
2000 American Institute of Physics 1-56396-951-3

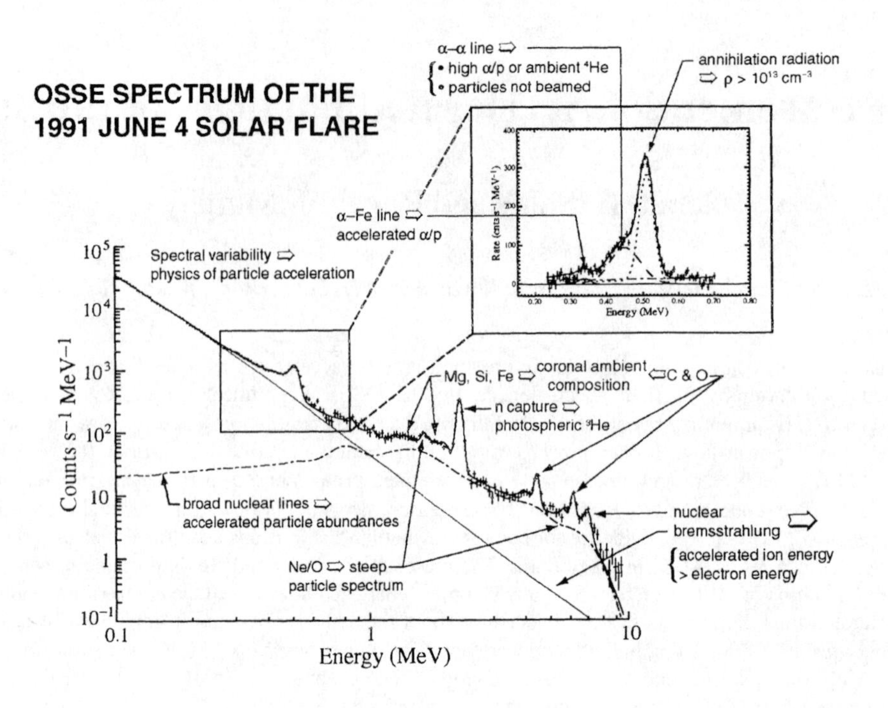

FIGURE 1. OSSE spectrum of the 1991 June 4 solar flare summarizing the physics revealed by γ-ray spectroscopy.

harder (at low energy) Bessel functions used previously. Using the published line fluences, measured cross sections, and kinematical calculations, Ramaty *et al.* showed that the composition of the flare plasma is, on average, close to coronal[11]. However, the flare on 1988 December 16 was depleted in low FIP emission lines which suggests a composition similar to that of the photosphere[10]. This is consistent with a local density of $>10^{14}$ cm^{-2} found in our study of the annihilation line and continuum[12] (see insert Figure 1). This suggests that flare particles may interact in regions with compositions ranging from those found in the upper photosphere to those in the corona. Recently reported spectroscopic measurements of flares with OSSE and *Yohkoh* suggest that the ambient composition may also change within flares[1,13]. This implies that ions accelerated in different flares and at different times in flares may interact at significantly different depths in the solar atmosphere. This could happen, for example, if the height of the magnetic field mirroring point varies.

Recently we have performed detailed spectroscopic studies on the narrow lines that reveal red shifts in the energies for flares near the center of the solar disk[6]. This suggests a dominance for interactions of particles moving in the downward direction.

The delayed 2.223 MeV neutron-capture line is the most intense line observed in spectra of flares that are not too close to the limb. Its narrow width and strength makes it an excellent indicator of the presence of ions in flares. Measurement of its intensity and temporal variation relative to prompt de-excitation lines has provided information on the spectra of ions above ~10 MeV[2] and on the concentration of ^{3}He in the photosphere[14]. The latter measurements are possible because ^{3}He nuclei capture neutrons in competition with photospheric hydrogen and this affects the decay time of the 2.223 MeV capture line. Observations[1,13] with *SMM*/GRS, *GRANAT*, *CGRO*/OSSE, and *Yohkoh*/GRS suggest ^{3}He/H ratios of $\sim 2 - 4 \times 10^{-5}$. The ratios are dependent on assumptions concerning the depth of interaction and solar atmospheric model, however.

Helium Composition

The inset of Figure 1 shows a detail of the region containing α-He fusion lines at 0.429 MeV (^{7}Be) and 0.478 MeV (^{7}Li); this complex is particularly sensitive to the angular distribution of flare-accelerated α particles. Its shape was found to be consistent with either isotropic or fan-beam distributions of accelerated α particles[15]. In contrast, a downward beam of accelerated particles was ruled out at high confidence [(99.99% and 99.8%)] for the two most intense disk-centered flares.

We have found high fluxes in the α-He fusion lines relative to the de-excitation lines in the 1991 June 4 flare[1] and in the *SMM*/GRS flares[15]. This led us to conclude that the accelerated α/proton ratio typically had to be large, ~0.5, for an assumed ambient ^{4}He/H abundance ratio of 0.1. Mandzhavidze *et al.* suggested that the ambient ratio might be higher in some flares and described a way in which γ-ray spectroscopy could distinguish between the two explanations[16]. This required the measurement of other lines that result from interactions of α-particles on ^{56}Fe. There is evidence for a weak line at 0.339 MeV from such interactions (see inset of Figure 1) in the *SMM* and *CGRO*/OSSE spectra. Based on this we concluded that, on average, the ambient ^{4}He abundance is consistent with accepted photospheric values and a high accelerated α/p ratio is needed[17]. Mandzhavidze *et al.* performed studies of individual flares and concluded[18] that there is evidence for both a higher accelerated α/p ratio and enhanced ambient ^{4}He.

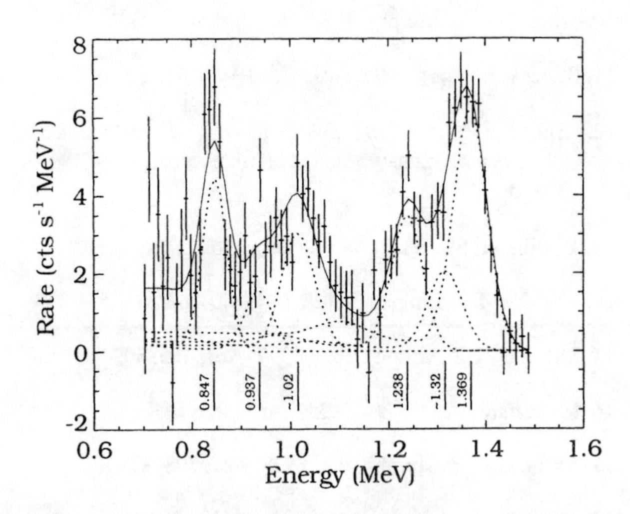

FIGURE 2. Gamma-ray spectrum revealing lines between 0.7 and 1.5 MeV observed in the sum of 19 *SMM*/GRS flares. Line energies are identified.

These same spectral studies have provided information on the accelerated ^{3}He/^{4}He ratio in flares. Shown in Figure 2 is the summed spectrum of 19 *SMM* flares[17] revealing the 0.847 and 1.238 MeV lines from ^{56}Fe, the weak newly-observed line from ^{55}Fe at 1.317 MeV, and the strong ^{24}Mg line at 1.369 MeV. For clarity the best-fit bremsstrahlung, highly broadened lines, and instrumentally degraded radiation have been subtracted before plotting. The key line features for understanding the ^{3}He abundance appear near 0.937 MeV and ~1.02 MeV. The relative strength of the ~1.02 MeV feature suggests a high accelerated α/p ratio from interactions on ^{56}Fe or high ^{3}He/^{4}He ratio from interactions on ^{16}O, or both. There is evidence for ^{3}He in the shift of the data points to higher energies in comparison with a model for ^{3}He/^{4}He = 0. On the other hand the weakness of the 0.937 MeV line generally precludes a flare-averaged ^{3}He/^{4}He-ratio close to 1. Studies of individual *SMM* flares[18] suggest that a ^{3}He/^{4}He ratio of 0.1 was consistent with all the flares and that a ratio as high as 1 could occur in some flares. Therefore the accelerated ^{3}He/^{4}He ratio is often $10^3 \times$ that found in the photosphere.

Accelerated Heavy Ions

We have recently demonstrated the ability to spectroscopically reveal the broad γ-ray lines from interactions of accelerated ions with ambient H and ^{4}He in data obtained by *CGRO*/OSSE and *SMM*/GRS[19]. Broad lines near 0.847 MeV (^{56}Fe) and 4.439 MeV (^{12}C) are individually resolved as can be seen in Figure 3. Broad lines from ^{24}Mg, ^{20}Ne, and ^{28}Si are not resolved from each other. The ^{16}O lines are also blended.

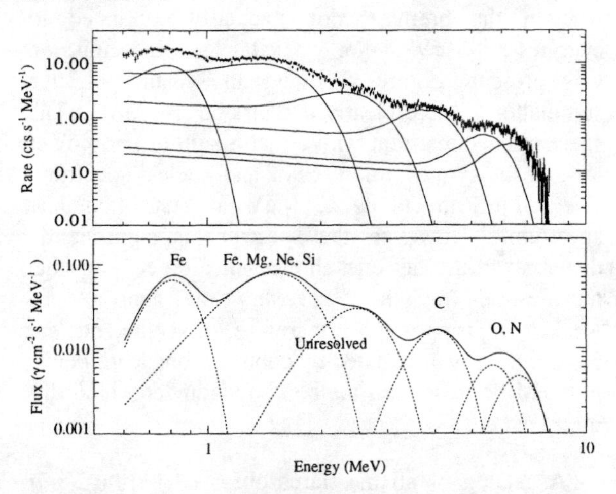

FIGURE 3. γ-ray spectrum revealing broad lines from accelerated heavy ions. Count spectrum after subtracting narrow lines and bremsstrahlung continuum (top panel). Inferred photon spectrum with lines identified (bottom panel).

Measurements of the widths and energies of these broad lines imply that the particles interact over a broad range of incident angles and suggest that they preferentially interact in the sunward direction. Comparisons of broad-line fluxes from individual accelerated nuclei with the respective fluxes in narrow lines from the ambient material measure the relative enhancements in the accelerated particles. We find that the accelerated ^{56}Fe abundance is enhanced over its ambient concentration by a factor consistent with that measured in solar energetic particles (SEP) in space from impulsive flares.

FIGURE 4. Summed spectrum from 40 flares with emission ≤ 1 MeV.

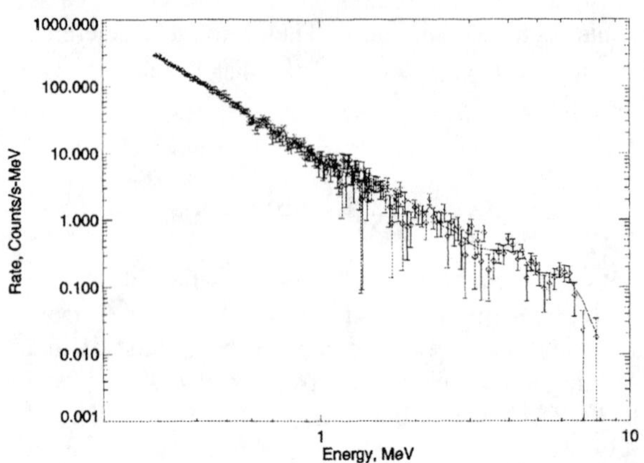

In a related study, we summed spectra from 40 flares observed by the *SMM*/GRS with no significant emission ≥ 1 MeV. We were interested in determining whether the bremsstrahlung actually continued to energies >1 MeV and/or weak nuclear line emission was present. Figure 4 shows the results of this summation. There clearly is emission >1 MeV. The spectrum is consistent with either a sum of two power laws or the sum of a power law and nuclear spectrum. The fall off in counts >7 MeV suggests a nuclear spectrum. However, the spectrum is significantly different from the one shown in Figure 1 in that narrow lines (e.g. the 2.223 MeV line) appear to be weak. It is possible, therefore, that the nuclear contribution is dominated by contributions from heavy ions. Miller discusses the conditions under which this might occur[20].

A catalog of all the flares observed by the *Solar Maximum Mission* (*SMM*)/GRS spectrometer has been published[21] and a compilation of high-energy flares observed by the *CGRO*/OSSE instrument is online at http://gamma.nrl.navy.mil/solarflare/flarelib.htm.

ACKNOWLEGMENTS

This work was supported by NASA grants DPR W-18,995 and W-19,498.

REFERENCES

1. Murphy, R.J., *et al.* 1997, ApJ, 490, 883
2. Ramaty, R., *et al.*1979, ApJS, 40, 487
3. Ramaty, R. & Murphy, R.J. 1987, Sp. Sci. Rev., 45, 213
4. Chupp, E.L. 1990, Science, 250, 229
5. Hudson, H. & Ryan, J. 1995, ARA&A, 33, 239
6. Share, G.H. & Murphy, G.H. 2000, in *Anticipating HESSI*, Ramaty and Mandzhavidze eds., ASP Conference Proceedings, Vol. 206
7. Ramaty, R. & Mandzhavidze, N. 1999, in *Highly Energetic Physical Plasmas,* Martens & Tsuruta, eds., IAU Symp 195, ASP Conference Proceedings
8. Rieger, E., *et al.* 1998, Solar Physics, 183, 123
9. Ramaty, R., *et al.* 1994, ApJ, 436, 941
10. Share, G.H. & Murphy, R.J. 1995, ApJ, 452, 933
11. Ramaty, R., *et al.* 1995, ApJ, 455, L193
12. Share, G.H., *et al.* 1996, in *High-Energy Solar Phenomena*, Ramaty, Hua & Ramaty eds., AIP Proc. 374, p.172.
13. Yoshimori, M., *et al.* 1999a/b, Proc. 26th ICRC, 6,5/30
14. Hua. X.-M. & Lingenfelter, R.E. 1987, ApJ, 319, 555
15. Share, G.H. & Murphy, R. J. 1997, ApJ, 485, 409
16. Mandzhavidze, N., *et al.* 1997, ApJ, 489, L99
17. Share, G.H. & Murphy, R.J. 1998, ApJ, 508, 876
18. Mandzhavidze, N., *et al.* 1999, ApJ, 518, 918
19. Share, G.H. & Murphy, R.J. 1999, Proceedings 26th ICRC, 6, 13
20. Miller, J. 1998, Space Sci. Rev., 86, 79
21. Vestrand, W.T., *et al.* 1999, ApJS, 120, 409

CGRO observations of gamma-ray flares associated with ACE particle events

D.J. Morris[1], L.M. Kistler[1], B. Klecker[2], E. Möbius[1], M.A. Popecki[1] and J. Ryan[1]

[1]*Space Science Center, University of New Hampshire, Durham, NH 03824-3525 USA*
[2]*Max-Planck-Institut für extraterrestrische Physik, D-85740 Garching, Germany*

Abstract. During the period in which the ACE spacecraft has been operating, the Compton Gamma-Ray Observatory (CGRO) has continued monitoring hard X-ray and γ-ray emission from the Sun, among other sources. The high-energy photons provide information about both the electrons, through bremsstrahlung continuum emission, and nuclei, through nuclear line emission, while they are still near the site of their acceleration. The question of the common origin of solar photons and particles is addressed through comparison of the photon observations with associated ACE events. Among many hard-X ray flares associated with particle events seen by ACE, three had particularly interesting spectra from the CGRO Burst and Transient Source Experiment (BATSE). Two, on 4 Nov. 1997 and 22 Nov. 1998, had very hard continuum spectra up to MeV energies. The third event, on 28 Nov. 1998, showed substantial 511-keV line emission, above the ever-present instrumental background; this could be due to the production either of short-lived positron emitters, such as ^{11}C, or pions.

INTRODUCTION

ACE is providing the most detailed observations of solar energetic particles ever made. However, a full understanding of particle acceleration and transport requires a coordinated analysis of the particle data together with information obtained through other channels. Hard X-ray and γ-ray photons, and solar neutrons all convey information from near the acceleration site, in close coincidence with the acceleration process. The four instruments on the Compton Gamma-Ray Observatory (CGRO), shown in Figure 1, detect photons at energies from 10 keV to 30 GeV, and neutrons from 10 to 150 MeV.

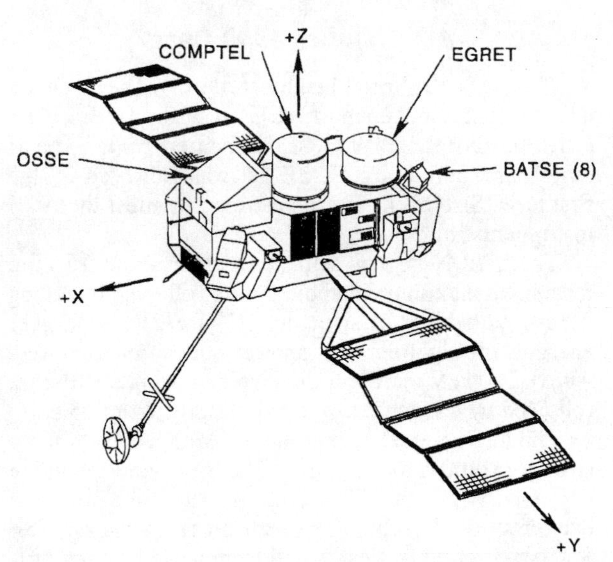

FIGURE 1: The CGRO spacecraft, showing the four instruments aboard: BATSE, OSSE, COMPTEL and EGRET

Following its launch in April 1991, CGRO made important observations of a number of large flares, near the end of the last solar maximum. All four instruments continue to operate, and can make an even bigger contribution to the understanding of energetic solar events during the current solar maximum. The capabilities of each instrument are discussed below. Following that are some CGRO observations of three interesting X-class flares associated with particle events observed by ACE.

CGRO AS A SOLAR OBSERVATORY

Each of the CGRO instruments provides unique information on the high-energy photons and neutrons produced in solar flares.

BATSE. The primary mission of the Burst and Transient Source Experiment (BATSE) is to study γ-ray bursts, but it is also well suited to study other high-energy transient phenomena, such as solar flares [1]. It is composed of eight modules, oriented like the faces of an octahedron to provide a full-sky field of view. BATSE provides about 50% solar coverage, whatever the orientation of the spacecraft.

CP528, *Acceleration and Transport of Energetic Particles Observed in the Heliosphere: ACE 2000 Symposium,* edited by Richard A. Mewaldt, et al.
© 2000 American Institute of Physics 1-56396-951-3/00/$17.00

Each BATSE module has two components: a Large Area Detector (LAD) and a Spectroscopy Detector (SD). Each LAD is a 20" × 0.5" NaI(Tl) disk, shielded by a thin (0.25") plastic charged particle detector (CPD). The LADs provide high time-resolution light curves in 16 energy channels from 20 keV to 8 MeV. In very intense flares, which can saturate the LADs and other CGRO instruments, the CPD count rates have provided light curves for the γ-ray emission above 1 MeV, though with no energy resolution [2]. Each SD is a 5" × 3" NaI(Tl) cylinder, providing better energy resolution than the LADs over a range that can extend from 10 keV to 100 MeV, depending on the gain settings of the SD modules. Normally, two SD modules are operated with a low gain (threshold near 200 keV), to provide good spectra of the nuclear line range.

COMPTEL. The Compton Imaging Telescope, COMPTEL, is a double-scatter γ-ray telescope, which uses Compton scattering in an upper layer of detector modules, followed by absorption in a lower layer of modules, to determine a photon's energy and constrain its incident direction to a circle on the sky [3]. The upper detector consists of seven 27.6 cm × 8.5 cm cylinders containing NE 213A liquid scintillator. The lower detector consists of 14 cylindrical NaI(Tl) modules with dimensions 28 cm × 7.5 cm. COMPTEL sees photons in the energy range 0.8-30 MeV over a field extending about 30° from the telescope axis, but intense transient sources may be seen out to 60°.

Using the pulse shape in the liquid scintillator to distinguish the scattering of photons and neutrons, COMPTEL can also serve as a neutron telescope: following elastic scattering of a neutron off a H nucleus in the NE 213A, the energy of the scattered neutron is determined from its time-of-flight to the lower detector. Neutrons can be detected over a range of 10-150 MeV and out to 60° from the telescope axis.

In addition, two of the NaI modules are operated as burst detectors to provide spectra covering two overlapping ranges of 0.1-1.5 MeV and 1-10 MeV. Their unobstructed field of view is about 2.5 sr.

OSSE. The Oriented Scintillation Spectrometer Experiment (OSSE) is composed of four identical 330-mm diameter NaI(Tl)-CsI(Na) phoswich detectors, each with a collimated field of view of 3.8° × 11.4° FWHM for the energy range 0.1–10 MeV; OSSE's full energy range for photons extends to 250 MeV [4]. The detectors can be rotated in a 192° range about the spacecraft y-axis; they are normally operated in pairs which oscillate between a source and background direction. Because of its limited independent pointing capability, OSSE can often be reoriented to look at the Sun when it receives a solar flare trigger signal from BATSE.

Pulse shape discrimination in the NaI portion of the phoswich is used to distinguish energy deposits by neutrons in the 10-250 MeV range.

Another paper in this volume [5] presents a detailed analysis of some solar flare observations by OSSE, illustrating it capabilities.

EGRET. The upper part of the Energetic Gamma-Ray Experiment Telescope (EGRET) is a stack of spark chamber modules interleaved with pair-conversion plates [6]. Photons with energies of 30 MeV - 30 GeV are converted to electron-positron pairs which are then tracked through the spark chamber. The remaining energy of particles which reach the bottom of the spark chamber is measured in the Total Absorption Shower Counter (TASC), a 76 cm × 76 cm × 21 cm NaI crystal. The telescope's field of view extends about 20° from its axis.

Due to depletion of the spark chamber gas, the EGRET telescope has been turned off at most times since the launch of ACE. It is turned on for a brief period after it receives a solar flare trigger from BATSE if the sun is in the field of view. The TASC system provides pulse height spectra, over the range 1-200 MeV, even when the spark chamber is turned off.

CGRO OBSERVATIONS OF SOLAR FLARES ASSOCIATED WITH ACE EVENTS

Using the ISTP/IACG Workshop list [7] and a list of ULEIS events (J. Dwyer, priv. comm.), the BATSE data has been searched for gamma-ray flares during 1997 and 1998 that may be associated with SEP events seen by ACE. Most BATSE flares show only hard X-ray emission, up to about 100 keV. Three exceptions were flares on 4 Nov. 1997, 22 Nov. 1998 and 28 Nov. 1998.

4 November 1997 flare

This was the third brightest flare to be seen by CGRO since the launch of ACE. It was a GOES X2.1 flare located at S14W33, accompanied by a Type II radio burst and a large CME. It was followed by the first large SEP event which was seen by most the ACE instruments [8].

The hard X-ray emission lasted for about 15 min, reaching a maximum in about 2 min, followed by a long decay, with hints of structure (Figure 2). At higher energies the light curve is shorter but more structured. Above 500 keV there is a sharp spike lasting about 25 s, followed by a second, lower peak lasting about 65 s.

The instrumental background at MeV energies has a strong orbital modulation. The high-energy spike occurred near a minimum in the instrumental background. To obtain a spectrum from the BATSE SD, background was estimated from periods about one-half orbit (45 min) preceding and following the spike, when the instrumental background also went through a minimum. Above 500 keV the spectrum (Figure 3) is very hard, with a power-law index of −1.3 for the period

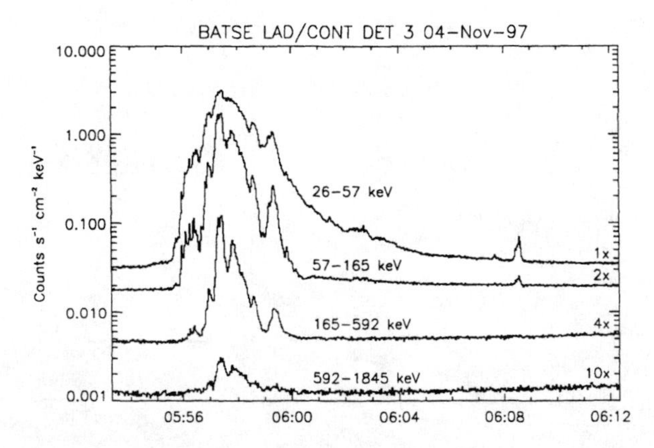

FIGURE 2. BATSE LAD light curves for the 4 Nov. 1997 flare in four energy ranges.

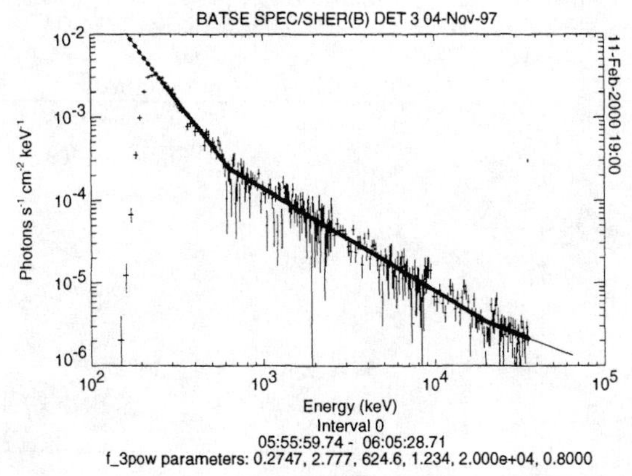

FIGURE 3. BATSE SD spectrum for the 4 Nov. 1997 flare, integrated over both peaks and most of the tail, with a broken power-law fit.

05:56:00 to 06:05:29. The emission extends into the nuclear line range, but no individual lines can be distinguished in the BATSE SD spectrum.

During this flare, the Sun was almost directly below the CGRO spacecraft (zenith angle 179°). Still it was seen by the COMPTEL burst modules, through an estimated 13 g/cm^2 of material. The COMPTEL light curve above 500 keV is similar to that seen by BATSE. The count spectrum for the first spike extends to 7 MeV; the spectrum for the second peak is somewhat softer, but shows a clear deuterium production line at 2.2 MeV. Thus it appears that the very hard spectrum seen by BATSE is due in part to nuclear line emission.

22 November 1998 flare

This event was a GOES X3.1 flare associated with AR8384 which was located at S27W82. It was accompanied by Type II and IV radio bursts.

The hard X-ray emission for this event lasted over 15 min; the BATSE light curve (Figure 4) ends at spacecraft sunset. The γ-ray light curve is again shorter and more structured, with two nearly equal peaks, each lasting about 1 min. The BATSE SD spectrum for the

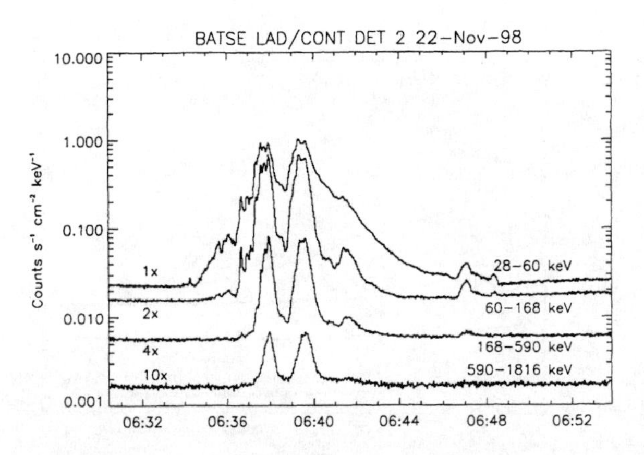

FIGURE 4. BATSE LAD light curves for the 22 Nov. 1998 flare in four energy ranges.

period including the γ-ray peaks is hard, though not as hard as the 4 Nov. 1997 flare, with a power law index of about –2.0, and no indication of line emission.

This flare was also seen by YOHKOH (Yoshimori, private comm.). The YOHKOH spectrum is also hard, extending above 10 MeV, and featureless. The YOHKOH hard X-ray spectrum had an index of –2.25.

Though OSSE was not pointed at the Sun, it still saw the γ-ray emission, extending to energies above 10 MeV, which penetrated its anti-coincidence shields [9]. The OSSE lightcurve is similar to those seen by BATSE and YOHKOH. The flare was far outside the COMPTEL field of view, and blocked by over 30 g/cm^2 as seen from the burst modules.

This flare occurred about 6 hours after the start of a two-day event period from the ULEIS list. There was no SEPICA event associated with this flare.

28 November 1998 flare

This event was a GOES X3.3 flare associated with AR8395 which was located at S14W33. It produced Type II and IV radio bursts as well as a CME. It was followed by a small, gradual event in SEPICA and other ACE instruments.

The BATSE hard X-ray light curve (Figure 5) begins with a sharp peak, with a rise time of about 5 min, followed by a slow decay lasting over 30 min. The γ-ray emission is nearly all in a 2-min period around the initial sharp peak. The BATSE SD data for the low-gain module viewing the Sun was not transmitted. However, the spectrum from the high-gain module pointed nearest the Sun shows a possible 511-keV line. Since there is an ever-present 511-keV line in the instrumental background, such a line could well be due to inadequate background correction, and so this result should be treated with caution. However, evidence for the line is also seen in a simple comparison of uncorrected SD light curves. Figure 6 shows SD light curves for a 68-keV wide energy window around the 511-keV line together light curves for three neighboring energy windows of similar width. The

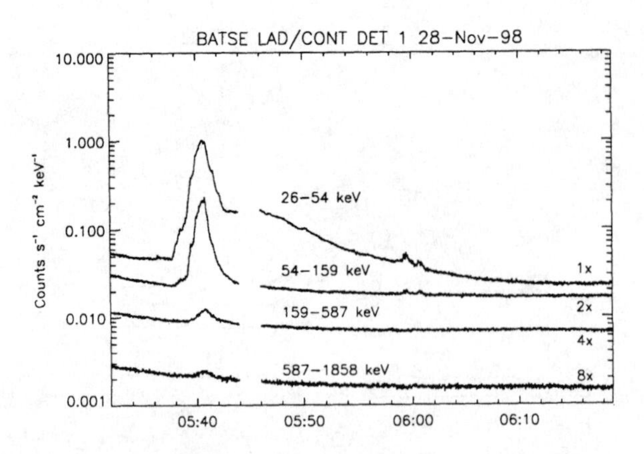

FIGURE 5. BATSE LAD light curves for the 28 Nov. 1998 flare in four energy bands. There is a gap in the data at approximately 05:44-05:46.

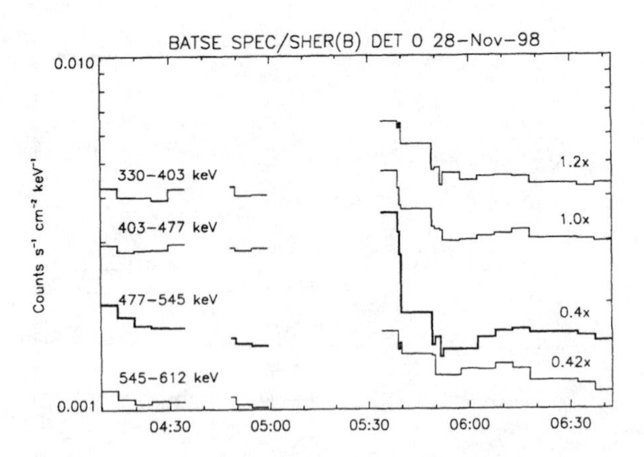

FIGURE 6. BATSE SD light curves for the 28 Nov. 1998 flare in an energy window around the 511-keV line and three neighboring energy windows; the 511-keV window is third from the top. There are large data gaps at 04:34-04:48 and 04:59-05:34.

large data gap preceding the flare was produced when that data was overwritten to allow data with better time resolution to be transmitted for the period following the flare trigger. The light curve for the 511-keV window shows an emission enhancement in the first few minutes of the flare which is much larger than that seen in the other energy windows.

This flare was located at a CGRO zenith angle of 58°. Nothing is visible in the COMPTEL telescope data, but spectra were obtained from the two burst modules; the 511-keV line is seen in the high-gain spectrum (Figure 7). While no individual lines can be identified in the low-gain spectrum, there is an excess in the nuclear-line range of 3-7 MeV. Both the 511-keV line and the 3-7 MeV emission are coincident with the γ-ray peak seen in the BATSE LADs.

This event was also seen by YOHKOH. The YOHKOH hard X-ray spectrum was relatively soft (index –4.70), and YOHKOH saw no indication of γ-ray emission above 1 MeV.

This flare occurred when CGRO was near the SAA where the temporal variations in the background can be particularly complex. Thus, it is possible that both BATSE and COMPTEL may simultaneously see an enhancement in the 511-keV background line. However, no such enhancement is seen in COMPTEL spectra near the same point in the orbit on the preceding or following day. This feature is under further study.

CONCLUSION

In the coming year, CGRO is scheduled to be pointed at the Sun nearly half the time. During an additional six weeks the Sun is a secondary target of the OSSE instrument, observed when the primary target is occulted by the Earth. If the mission is allowed to continue, despite the recent loss of one gyroscope, there is the potential to obtain a wealth of data on solar γ-ray and neutron emissions.

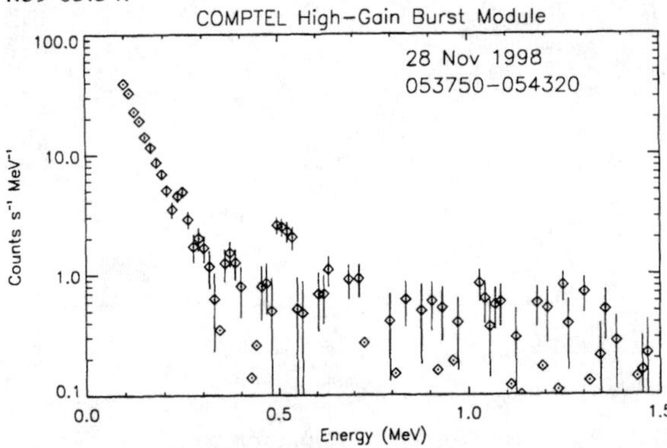

FIGURE 7. COMPTEL count spectrum for the 28 Nov. 1998 flare during the period 05:37:50-05:43:20.

ACKNOWLEDGMENTS

We thank Richard Schwartz for valuable advice on the analysis of the BATSE data.

REFERENCES

1. Bond, D.L. et al., *Exp. Astr.* **2**, 307 (1992).
2. Ramaty, R. et al., *Ap. J.* **436**, 941 (1994).
3. Schönfelder, V. et al., *Ap. J. Suppl.* **86**, 657 (1993).
4. Johnson, W.N. et al., *Ap. J. Suppl.* **86**, 693 (1993).
5. Share, G.M. and R.J. Murphy, these proceedings.
6. Thompson, D.J. et al., *Ap. J. Suppl.* **86**, 629 (1993).
7. ISTP/IACG Workshop Event List (1999) http://orpheus.nascom.nasa.gov/istp/typeII_eva.html
8. Mason, G.M. et al., *Geophys. Res. Lett.* **26**, 141 (1999).
9. Murphy, R.J. and G.M. Share, OSSE Solar Flare Observations website (1999). http://osse-www.nrl.navy.mil/solarflare

Particle Acceleration in the 6 November 1997 Event as viewed from Gamma Rays and Solar Energetic Particles

M.Yoshimori, K.Suga, A.Shiozawa, S.Nakayama and H.Takeda

Rikkyo University, Toshima-ku, Tokyo 171-8501, Japan

Abstract. An intense solar event was observed on 6 November 1997 with Yohkoh and solar energetic particle (SEP) satellites. The Yohkoh gamma-ray spectrum and soft / hard X-ray images exhibit a characteristic of an impulsive event, while the SEP energy spectra, elemental abundances and charge states indicate features of a mixed-gradual event. These observations suggest that two different types of particle populations were produced during this event. We discuss possibilities of particle acceleration processes for these two populations. For the gamma-ray producing-particles and impulsive SEPs stochastic aceleration through cascading Alfven waves and Fermi stochastic acceleration at fast shocks have been proposed. These plasma waves are generated at a magnetic reconnection site above a top of the flaring loop. On the other hand, for the gradual SEPs a large amount of coronal particles and solar wind are accelerated by a coronal and CME-driven shocks in the higher corona and interplanetary space.

INTRODUCTION

There are two types of high energy solar particle events, impulsive and gradual (Reames, 1995). Impulsive events are associated with solar flares and gradual events are with CMEs. Gamma-rays and solar energetic particles (SEPs) provide clue on the particle acceleration process related to high energy solar phenomena. A variety of data of X-rays, gamma-rays, solar energetic particles (SEPs), CME and CME-driven shock were obtained from a large solar event on 6 November 1997. Yohkoh observed an impulsive solar flare which exhibited a small flaring magnetic loop with double hard X-ray sources and strong gamma-ray emission extending above a few tens of MeV. The IMP-8, SAMPEX and ACE measurements provide the energy spectra, elemental composition and charge states of SEPs. Further, a strong CME and its related shock were recorded with SOHO and ACE, respectively (Mason et al., 1999) . This event is the most suitable for study of high energy particle production because the detailed data of high energy photons and particles are available. In this paper we discuss the possible particle acceleration processes based on the observed results of gamma-rays, X-ray images and SEPs.

OBSERVATIONS

Yohkoh observed a gamma-ray spectrum and soft / hard X-ray images from the 6 November flare (X9/2B, S18W64) (Yoshimori et al., 2000). The gamma-ray data show impulsive time profiles with a duration of about 4 min for both electron bremsstrahlung and gamma-ray line emissions. The count rate time profile at 4-7 MeV (dominated by C and O lines) is shown in Figure 1. The neutron capture line at 2.22 MeV and nuclear deexcitation lines of C, O and heavy nuclei are superposed on the electron bremsstrahlung continuum. The gamma-ray spectrum extends to a few tens MeV. The X-ray images indicate a compact magnetic loop structure (loop length is ~15,000 km) and two hard X-ray sources which are located at both footpoints of the loop. The temporal variation in hard X-ray images at 53-93 keV is shown in Figure 2. A distance between two sources is nearly constant (about 10,000 km) during the rising and maximum phase of the flare but gradually increases during the decay phase. This increasing separation of hard X-ray sources suggests that a magnetic reconnection (energy release / particle acceleration) site could have gradually moved up during the decay phase (Sakao et al.,

CP528, *Acceleration and Transport of Energetic Particles Observed in the Heliosphere: ACE 2000 Symposium,* edited by Richard A. Mewaldt, et al.
© 2000 American Institute of Physics 1-56396-951-3/00/$17.00

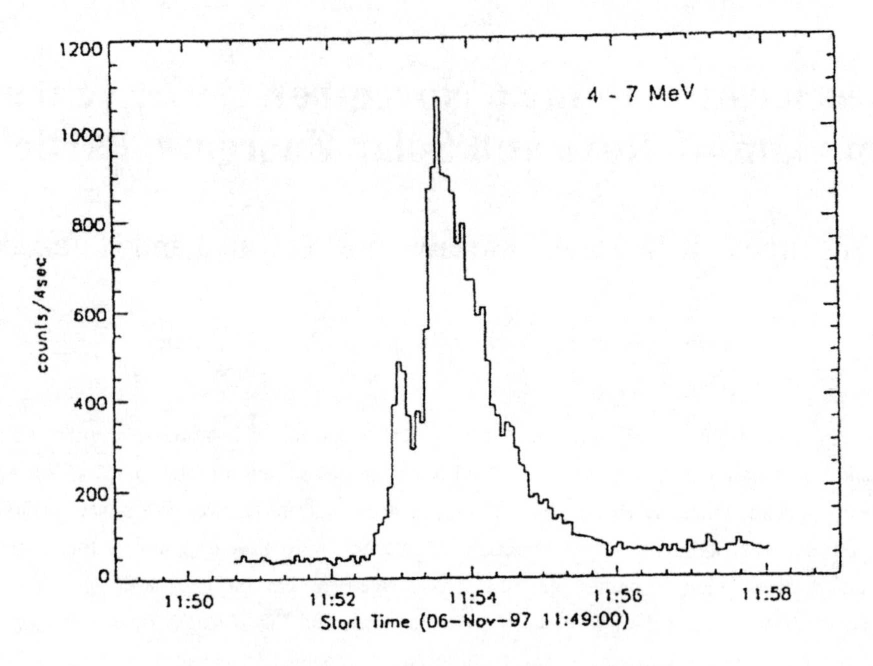

FIGURE 1. Count rate time profile at 4-7 MeV .

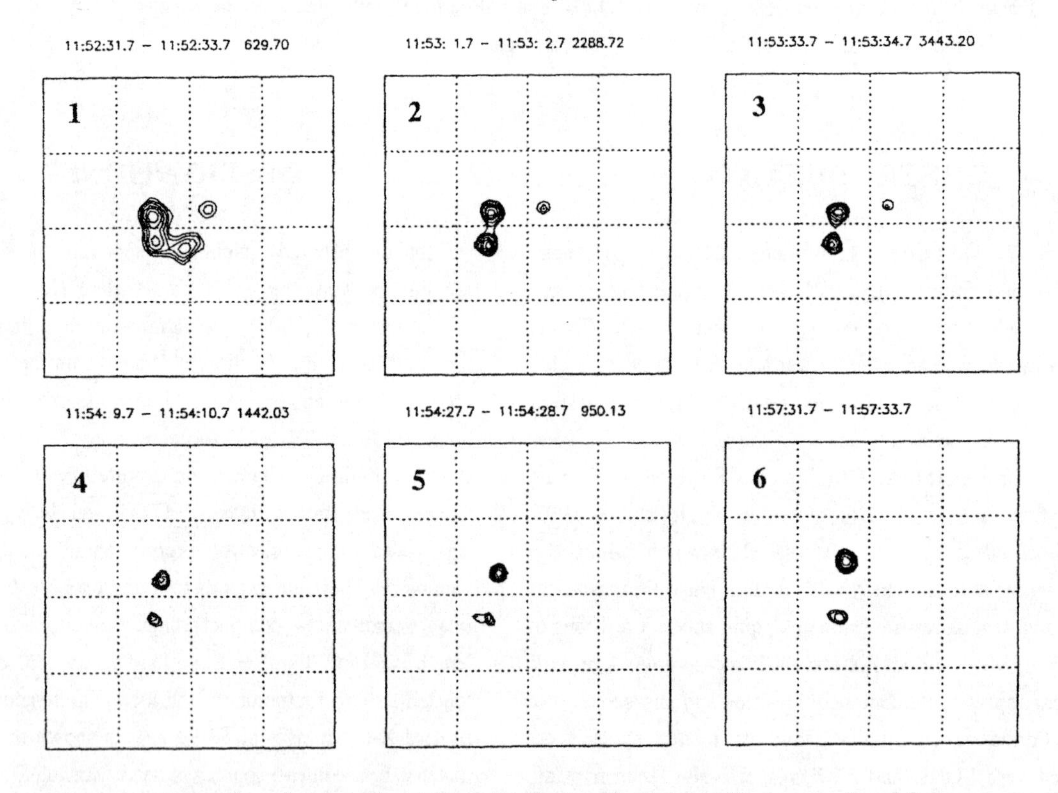

Figure 2. Temporal variation in hared X-ray images at 53-93 keV. The images 1 and 2 were measured at the rising phase, 3 at the maximum phase and 4-6 at the decay phase.

1998). The Yohkoh observations imply that electrons and ions were simultaneously accelerated to high energies on a short time scale and streamed down to the chromosphere to produce X- and gamma-rays. This flare exhibits a characteristic of impulsive events.

The energy spectrum of accelerated protons (10-100 MeV) can be derived from a time-diferentiation of the flux of neutron capture line at 2.22 MeV (Gan, 1998). In order to perform it, we need the yields of neutrons and C and O lines, conversion factors from neutron to 2.22 MeV line and the decay constant

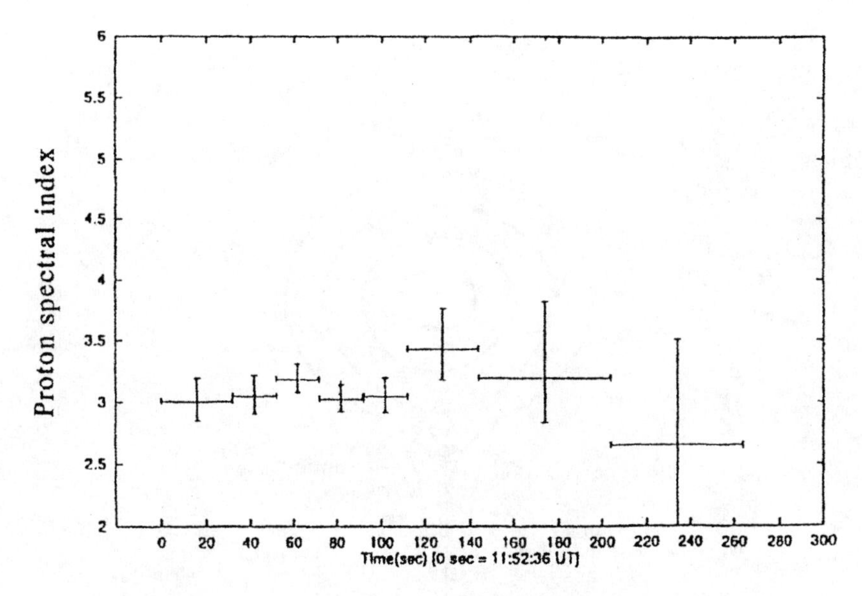

FIGURE 3. Temporal variation in the power law index of of proton spectrum

of the 2.22 MeV line emission. The first two parameters are available from Hua and Lingenfelter (1987) and the third one from Yoshimori et al. (1999). The temporal variation in the power law indices of the proton spectrum is shown in Figure 3. The average spectral index is 3.2+/-0.3, though the errors of the indices are large in the decay phase.

The SEP observations demonstrate that the proton and heavy ion energy spectra are exceptionally hard (spectral index is 1.5-2.5) at 1-100 MeV/nuc (Mason et al., 1999; Dietrich and Lopate, 1999) and the mean charge of Fe ions are 11-14 at 0.1-1 MeV/nuc, increasing with ion energies (Moebius et al, 1999; Mazur et al., 1999). In particular, the IMP-8 data shows that the spectral index of the proton spectrum is 1.3+/-0.2, which is much different from that derived from the gamma-ray data. These SEP results indicate the characteristic of a gradual event. However, the abundance ratio of Fe/O is 0.5-1.0 (Mason et al., 1999), which is much larger than the coronal value of 0.134 (Reames, 1999). Moreover, the charge distribution of Fe ions has a tail extending to about 20 (Moebius et al., 1999). These are the characteristics of impulsive SEP events. The SEP data suggest that the 6 November event is a mixed-gradual one.

DISCUSSION

The Yohkoh and SEP observations suggest that two types of high energy particle populations, impulsive and gradual, were produced during the 6 November 1997 event. The impulsive population is responsible for gamma-ray production and impulsive SEPs. This population is accelerated on a short time scale and most of the particles stream down to the chromosphere, but a part escapes into interplanetary space and forms the impulsive event. In general, the impulsive SEP event is much smaller than the gradual SEP event because most of impulsive SEPs are trapped by small and strong magnetic field and can not escape into the higher corona. The gradual population is responsible for gradual SEP event. A number of gradual SEP events are related to CMEs and the particles are thought to be accelerated by CME-driven shocks. A strong CME and its driven shock associated with this event were recorded by SOHO and ACE. A possible explanation for these observations is that two different types of particle acceleration mechanisms effectively operate at two sites and produce the impulsive and gradual particles. As the possible acceleration processes for the impulsive population, stochastic acceleration through cascading Alfven waves (Miller, 1998) and Fermi stochastic acceleration at fast shocks (Tsuneta and Naito, 1998) have been proposed. The Alfven waves and fast shocks are generated at a magnetic reconnection site above the top of the flaring loop and accelerate plasma particles on a short time scale. The downward particles produce gamma-rays in the chromosphere, while the upward ones become the impulsive SEPs. On the other hand, as the acceleration mechanism for the gradual population, a large number of coronal and solar wind particles are thought to be energize by a coronal and CME-driven shocks. These particles are observed as the CME-related gradual SEP event which has very hard energy spectra and coronal element abundances. We show a cartoon in

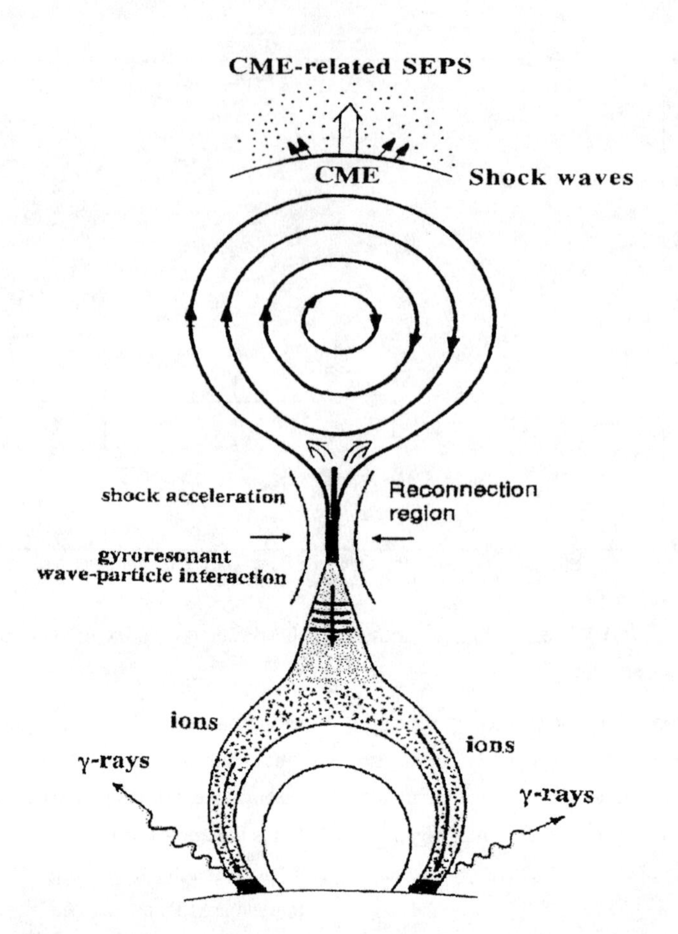

Figure 4 to outline the acceleration processes during the 6 November event. This explanation seems to explain the Yohkoh and SEP satellite results. The elemental and charge state composition of solar wind from the 6 November event was reported by Wimmer-Schweingruber et al. (1999). The unusually broad and high charge state distributions of Fe ions and enhanced Fe/O ratio were observed, indicating the observed compositional similarities in the solar wind and SEPs. There is the possibility that both kinds of particles were accelerated in a common reservoir. Further, very high energy particles accelerated to more than 10 GeV were measured with the ground-based water Cherenkov detector (Ryan et al., 1999). Its duration lasted for a few hours, implying that an extraordinary powerful particle accelerator existed.

ACKNOWLEDGMENTS

This work was supported by Grant-in-Aid for Scientific Research of Minustry of Education, Science and Calture C-11640290. We wish to thank Dr. H. Suzuki for his drawing up the manuscript.

REFERENCES

Dietrich, W,. and Lopate, C., 26 ICRC. **6**, 71 (1999).

Gan, W.Q., ApJ. **496**, 992 (1998).

Hua, X-,M. and Lingenfelter, R.E., Sol. Phys. **107**, 351 (1987).

Mason, G.M. et al., Geophys. Rev. Lett. **26**, 141(1999).

Mazur, J.E. et al., Geophys. Rev. Lett. **26**, 173 (1999).

Moebius, E. et al., Geophys. Rev. Lett. **26**, 26 (1999).

Miller, J.A., Space Sci. Rev. **86**, 79 (1995).

Reames, D.V., Rev. Geophys. Suppl. **33**, 585 (1995).

Reames,D.V., Space Sci. Rev. **90** (1999) 413.

Ryan, J..M. et al., 26 ICRC. **6**, 378 (1999).

Sakao, T. et al. , *Observational Plasma Astrophysics*, Kluwer Academic Publishers, 1998, p.273.

Tsuneta, S., and Naito, T., ApJ. **495**, L67 (1998).

Wimmer-Schweingruber et al., Geophys. Res. Lett. **26**, .3541 (1999).

Yoshimori, M. et al., 26ICRC. **6**, 5 (1999).

Yoshimori, M. et al., Adv. Space Res. **25** (9), 1801 (2000).

Detection of the 6 November 1997 Ground Level Event by Milagrito

Abe D. Falcone, for Milagro Collaboration

University of New Hampshire, Space Science Center, Morse Hall, Durham, NH 03824 USA

Abstract. Solar Energetic Particles (SEPs) with energies exceeding 10 GeV associated with the 6 November 1997 solar flare/CME (coronal mass ejection) have been detected with Milagrito, a prototype of the Milagro Gamma Ray Observatory. While SEP acceleration beyond 1 GeV is well established, few data exist for protons or ions beyond 10 GeV. The Milagro observatory, a ground based water Cherenkov detector designed for observing very high energy gamma ray sources, can also be used to study the Sun. Milagrito, which operated for approximately one year in 1997/98, was sensitive to solar proton and neutron fluxes above ~4 GeV. Milagrito operated in a scaler mode which was primarily sensitive to muons, low energy photons, and electrons, and the detector operated in a mode sensitive to showers and high zenith angle muons. In its scaler mode, Milagrito registered a rate increase coincident with the 6 November 1997 ground level event observed by Climax and other neutron monitors. An analysis, based on preliminary effective area estimates, indicates the presence of >10 GeV particles.

INTRODUCTION

Particle acceleration beyond 1 GeV due to solar processes is well established (1), but its intensity and energy still amazes researchers. However, few data exist demonstrating acceleration of protons or ions beyond 10 GeV (2, 3). The energy upper limit of solar particle acceleration is unknown but is an important parameter, since it relates not only to the nature of the acceleration process, itself not ascertained, but also to the environment at or near the Sun where the acceleration takes place. The Milagro instrument, a water Cherenkov detector near Los Alamos, NM, is at 2650 m elevation with a geomagnetic vertical cutoff rigidity of ~4 GV. It is sensitive to hadronic cosmic rays from approximately 5 GeV to beyond 1 TeV. These primary particles are detected via Cherenkov light, produced by secondary shower particles, as they traverse a large (80 × 60 × 8 m) water-filled pond containing 723 photomultiplier tubes (228 PMTs for the prototype, Milagrito). This energy range overlaps that of neutron monitors (in the region < 10 GeV) such that Milagro complements the worldwide network of these instruments. These ground-based instruments, in turn, complement spacecraft cosmic ray measurements at lower energies. This suite of instruments may then be capable of measuring the full energy range of solar hadronic cosmic rays, with the goal of establishing a fundamental upper limit to the efficiency of the particle acceleration by the Sun. Milagro's baseline mode (air shower telescope mode) of operation measures extensive air showers above 300 GeV from either hadrons or gamma rays. A description of Milagro's capabilities as a VHE gamma ray observatory is available elsewhere (4). Milagro measures not only the rate of these events but also the incident direction of each event, thereby localizing sources. While performing these measurements, the instrument records the rate of photomultiplier hits (the scaler mode), with an intrinsic energy threshold of about 4 GeV for the progenitor cosmic ray to produce at least one hit. The scaler mode provides data that are similar to those of a neutron monitor, while the telescope mode can significantly reduce background by pointing. With a proposed fast data acquisition system (DAQ) and modified algorithms for determining incident directions of muons, the energy threshold of Milagro's telescope mode will be reduced to ~4 GeV by detecting the (~300 kHz) single muons and mini muon showers. For now, this low energy threshold can only be achieved by using Milagro in the scaler mode, which is not capable of localizing

CP528, *Acceleration and Transport of Energetic Particles Observed in the Heliosphere: ACE 2000 Symposium*, edited by Richard A. Mewaldt, et al.
© 2000 American Institute of Physics 1-56396-951-3/00/$17.00

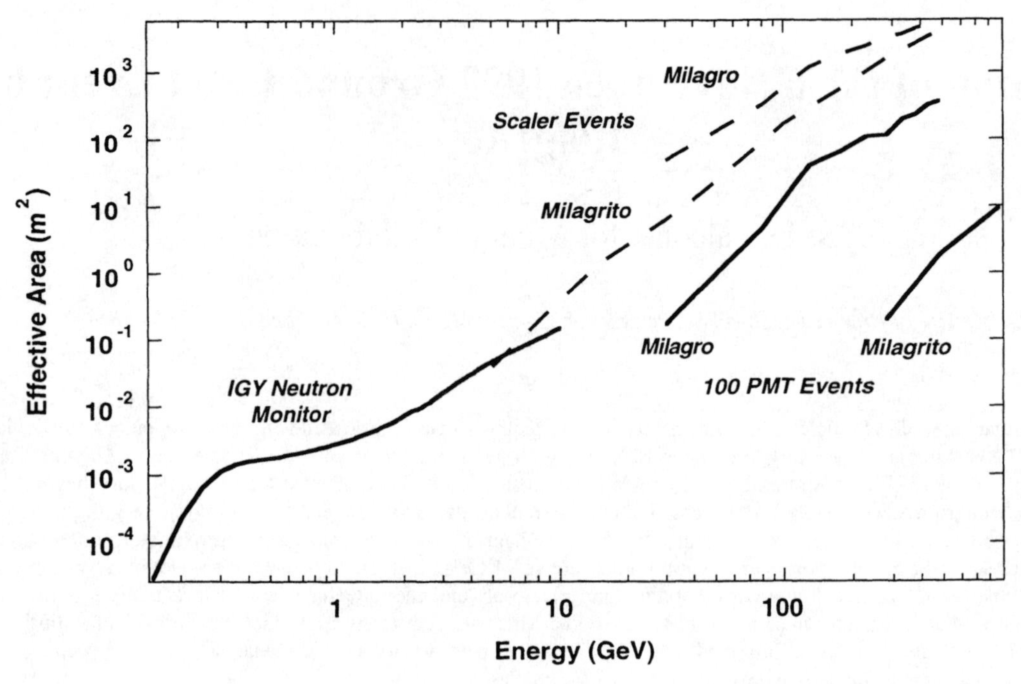

Figure 1. Preliminary, lower-limit effective area curves for Milagro and Milagrito, with a sea-level IGY neutron monitor for comparison. (The Milagro shower trigger is no longer set at 100 PMTs, but Milagrito shower data was recorded with a 100 PMT threshold.)

sources. A description of the Milagro solar telescope mode can be found in another publication (5).

SOLAR MILAGRO/MILAGRITO SCALER MODE

In the scaler mode, a substantial portion of the rate recorded by Milagro (and Milagrito) is due to muons, and an integral measurement above threshold is performed. These data provide an excellent high energy complement to the network of neutron monitors, which has been, and continues to be, a major contributor to our understanding of solar energetic particle acceleration and cosmic rays. With Monte Carlo calculations, we estimated the effective areas of Milagrito to protons incident on the atmosphere isotropically, at zenith angles ranging from 0°-60° (Figure 1). The effective area curves for Milagro, which have been plotted for the sake of comparison, are for beamed protons. At 10 GeV, Milagro's scaler mode is at least an order of magnitude greater than the effective area of a sea level neutron monitor, with the effective area rising rapidly with energy, while Milagrito had approximately 4 times the effective area of a neutron monitor at 10 GeV. Based on current work, these effective area curves appear to be lower limits to the instruments' actual effective area. The plotted areas were calculated using Monte Carlo events whose shower cores were thrown at, or near, the Milagro pond. We are currently calculating the complete response of the instrument by taking into account hits on the pond which are caused by hadronic showers with cores very far (> 5 km) from the detector. This effect will increase the calculated effective area, but the general shape of the curve is expected to remain unchanged.

In order to apply this area to an analysis of an observed rate, pressure, temperature, and other diurnal corrections must be applied to the ground level scaler rate (6). We have begun to determine these correction factors for Milagro/Milagrito, and we find them to be reasonably consistent with past work with muon telescopes (7). However, these corrections are less important for transient (i.e. solar) events that rise above background quickly and have short durations.

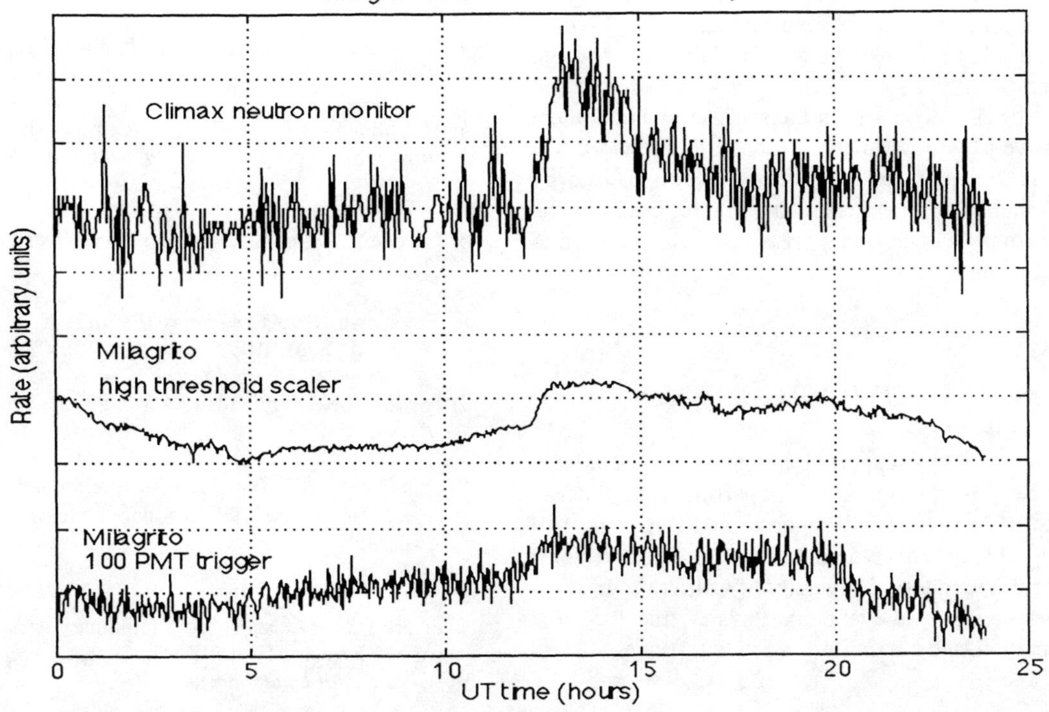

Figure 2. Milagrito registered a rate increase coincident with that of Climax during the GLE of Nov. 6, 1997. The y-axis units have been scaled and shifted for each plot to make comparison easier. (Climax data courtesy of C. Lopate, Univ. of Chicago)

6 NOVEMBER 1997 GROUND LEVEL EVENT

On 6 November 1997 at approximately 12:00 UT, an X-class flare with an associated coronal mass ejection occurred on the Sun. This produced a nearly isotropic (8) ground level event registered by many neutron monitors. A preliminary analysis of neutron monitor data for this proton event yields a spectral index of ≈5.5 at event maximum in the 1-5 GV rigidity range, assuming a power law rigidity spectrum for protons (9). Climax, located in nearby central Colorado, is the closest of these neutron monitors to Milagro/Milagrito. Milagrito, a prototype version of Milagro with less effective area, registered a scaler rate increase coincident, within error, with that measured by Climax (Figure 2). If one accounts for meteorological fluctuations, the event duration and time of maximum intensity, as seen by Milagrito, are also consistent with that of Climax. The magnitude of this rate increase was ≈22 times the RMS fluctuations of the instrument's background (RMS of fluctuations is used since the scaler background fluctuations are nearly twice that expected from Poisson statistics). It is likely that this signal is overwhelmingly due to protons, but there could be some contribution from other sources, such as iron (10) and other high Z ions.

The high threshold scaler rate increase can be used to derive characteristics of the primary proton spectrum. This is done by folding a trial power law spectrum of protons through the effective area of the detector. The parameters of the trial spectra are then varied until a good fit to the measured rate increase is achieved. When compared to the neutron monitor network's spectrum for protons < 4 GeV, the preliminary results of this analysis indicate the presence of protons in excess of 10 GeV. Further work, particularly on the effective area of the detector, is necessary to better determine the spectrum at these energies.

The 100 PMT shower trigger rate also experienced an increase, although the significance is not as great as that in scaler mode. It is not yet clear which of several possible mechanisms initiated the signal in the 100 PMT shower trigger, so the detector's sensitivity to various mechanisms is being investigated. This increase could have been caused by high energy primaries (> 100 GeV, see Figure 1) and/or secondary muons arriving from a nearly horizontal direction. If horizontal secondary muons contributed to this signal, they would have been the result of high energy proton primaries, but the effective area of the detector would be significantly different from that assumed here and cannot be used to constrain the spectrum without more extensive Monte Carlo calculations. Future work will address this issue by considering events caused by horizontally incident secondary muons, revising the preliminary effective area calculations, and recalculating the spectrum.

ACKNOWLEDGEMENTS

This work is supported in part by the National Science Foundation, U.S. Department of Energy Office of High Energy Physics, U.S. Department of Energy Office of Nuclear Physics, Los Alamos National Laboratory, University of California, Institute of Geophysics and Planetary Physics, the Research Corporation, and the California Space Institute.

REFERENCES

1. Parker, E.N., *Physical Review* **107**, 830 (1957).

2. Chiba, N., et al., *Astroparticle Physics* **1**(1), 27-32 (1992).

3. Lovell, J.L., Duldig, M.L., Humble, J.E., *Journal of Geophysical Research* **103**(A10), 23733 (1998).

4. McCullough, J.F., et al., "Status of the Milagro Gamma Ray Observatory," *Proc.26th Int. Cosmic Ray Conf, 1999.*

5. Falcone, A.D., et al., *Astroparticle Physics* **11**(1-2), 283-285 (1999).

6. Hayakawa, S., *Cosmic Ray Physics*, New York: John Wiley and Sons, 1969.

7. Fowler, G.N., Wolfendale, A.W., S.Flügge, eds., *Cosmic Rays I*, 1961.

8. Duldig, M.L. & Humble, J.E., "Preliminary Analysis of the 6 November 1997 Ground Level Enhancement," *Proc.26th Int. Cosmic Ray Conf, 1999, Vol. 6, pp. 403-406.*

9. Smart, D.F & Shea, M.A., "Preliminary Analysis of GLE of 6 November 1997A," *Proc. Spring American Geophysical Union Meeting, 1998.*

10. Mason, G.M., et al., Geophysical Research Letters, **26**(2), 141 (1999).

Milagro as a Solar Energetic Particle Observatory

J.M. Ryan for the Milagro Collaboration

University of New Hampshire
Durham, NH 03824

Abstract. We describe the existing Milagro Gamma-Ray Observatory and discuss how proposed enhancements will make it a unique facility for studying solar phenomena at GeV energies.

INTRODUCTION

In the last two solar cycles we have obtained a new appreciation of the ability of the Sun to accelerate cosmic rays. Proposed enhancements to the Milagro Gamma-Ray Observatory will make it a unique tool for extending solar neutron and proton observations to the multi-GeV regime with only a modest investment to an existing experimental program. The proposed enhancements will, among other things:

- Extend the energy regime for studying Ground Level Events and solar flares to the 5-100 GeV energy regime
- Facilitate the study of both impulsive and extended emission at high-energies with fine time resolution
- Provide both spatial and temporal data for the study of particle isotropy- and spectral-evolution
- Enable unique multi-wavelength, and multi-particle, studies of solar phenomena in collaboration with other instruments
- Enhance our ability to study the nature of particle acceleration at the highest energies

SCIENTIFIC MOTIVATION

Particle acceleration at, or near, the Sun produces energetic protons (and neutrons) both directly and indirectly. Understanding the particle acceleration process is of both fundamental scientific interest and of practical importance. Since 1956 it has been clear that cosmic rays exceeding 1 GeV could be accelerated in the solar environment [1]. With the neutron and high-energy gamma-ray measurements performed with the Solar Maximum Mission and the Compton Gamma Ray Observatory, we have gained a new insight into the solar-flare phenomenon. We now know that GeV ions (including both protons as well as heavier ions) can be accelerated on rapid timescales (< 1 minute) in flares [2], while in some flares, the acceleration process can last for several hours [3, 4]. Our understanding of high-energy solar flare particles was recently reviewed by Hudson and Ryan [5].

FIGURE 1. Aerial view of the Milagro Gamma-Ray Observatory site. Visible are the large covered pond and the counting house, pond utility building, and access roads

The transient nature of solar proton phenomena, combined with the limited physical size of the acceleration region, suggests that there must exist an effective upper limit to

CP528, *Acceleration and Transport of Energetic Particles Observed in the Heliosphere: ACE 2000 Symposium*, edited by Richard A. Mewaldt, et al.
© 2000 American Institute of Physics 1-56396-951-3/00/$17.00

the energy spectrum of particles produced by the Sun, whether they are accelerated in a flare or by an interplanetary shock. Although, solar proton phenomena have regularly amazed researchers by their intensity and energy, we would expect that the "upper limit" often occurs in the broad energy range of a few GeV to above 100 GeV. This corresponds to an energy range over which the modified Milagro would be sensitive. If the spectrum energy limit occurs above 300 GeV, it may be detected with Milagro in its baseline air-shower configuration. This would provide a third integral measure of the spectrum. A far more likely prospect, though, is that the energy upper limit frequently occurs at lower energies. By measuring the spectrum of solar proton emissions at the highest energies, we would be collecting data to aid in restricting the location of the acceleration site (via time of flight) and the mechanism involved (via energy cutoff and spatial distributions). Acceleration models are most severely tested at the extremes and it is the extreme part of the spectrum that we wish to investigate with Milagro.

The Milagro Gamma-Ray Observatory

Milagro is the first water-Cherenkov detector built specifically to study extensive air showers. Its primary science goals are to survey the sky for TeV gamma-ray emission from astrophysical sources. Located 2650m above sea level in the Jemez Mountains of New Mexico, Milagro is a covered, light tight 60 m × 80 m × 8 m pond filled with ~5 million gallons of purified water and instrumented with photomultiplier tubes (PMTs) (Fig. 1). Milagro has a number of advantages over previous ground based extensive air shower (EAS) arrays. Milagro's location above sea-level, combined with its ability to detect a large fraction of the shower particles that reach ground level, leads directly to a relatively low energy threshold for air-shower detection (median triggered energy ~2-3 TeV). In addition, its high duty-cycle for observations (~100%) makes it well suited for the study of transients and long-term continuous monitoring of astrophysical sources.

Relativistic, charged air-shower particles produce Cherenkov radiation as they traverse a detection medium — in this case water. In addition to charged secondaries, shower photons can also be detected via any pair-produced and/or Compton scattered electrons that are produced in the detector. The water Cherenkov technique allows Milagro to instrument a large area with a sparse spacing of photo-sensors keeping costs low. Techniques such as these have been used for decades in high-energy physics. A total of 723 20-cm PMTs arrayed on a 3m grid are used to detect the Cherenkov radiation. Existing electronics provides for the acquisition of both PMT pulse-heights and relative PMT hit times with nanosecond accuracy.

Milagro has two planes of PMTs. The first layer of 450 upward-facing tubes, the *air shower layer*, views the top 2-m of water. This layer is used to measure the charged particle air-shower wavefront (via Cherenkov radiation), providing the information needed to reconstruct the primary particle direction. A second layer of 273 upward-facing PMTs, the *muon layer*, is located at a depth of 6.5 m. The layered design of Milagro provides a modest calorimetric measure that may aid the identification of cosmic-ray induced showers, thereby improving the TeV gamma-ray sensitivity of the instrument. The full Milagro instrument was constructed during 1998-1999 and is now operational.

Milagro As A Solar Observatory

Milagro will study the high-energy solar emission by indirectly detecting solar neutron and proton interactions in the Earth's atmosphere. At high-energies these particles produce cascades that propagate to ground level. However, at energies of interest to solar physics investigations (<100GeV) these secondary cascade particles range out leaving very few particles at ground level. Quite often the only remnant of these low-energy showers is a penetrating muon. The muon events at ground level represent an important probe of solar high-energy phenomena. The background to the solar signature is the large flux of muons associated with cosmic ray induced showers. This large flux requires high-throughput acquisition electronics and event processing to reconstruct individual muon arrival directions, characterize the spatial distributions of solar transients and the cosmic-ray background, and study the temporal evolution of both the spectra and spatial distributions. The Milagro experiment, as it currently exists, is not able to address this physics, since it was designed to measure the global properties of air-showers not individual particles. We have therefore proposed a new high-throughput data acquisition system to facilitate the study of solar phenomena at GeV energies.

GeV particle detection is not a unique capability of Milagro. Neutron Monitors (NMs) routinely monitor the solar-terrestrial neighborhood at similar energies. However these instruments simply measure particle count rates — there is no arrival direction information available on an event-by-event basis. The degree of particle isotropy (pitch angle distributions relative to the magnetic field lines) is determined using event

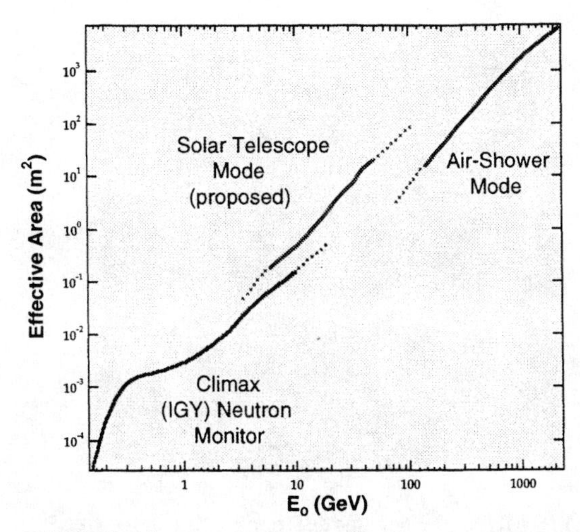

FIGURE 2 Predicted Milagro effective area to neutrons and protons (green). Also shown are the effective areas for an IGY neutron monitor and the existing Milagro air-shower trigger mode

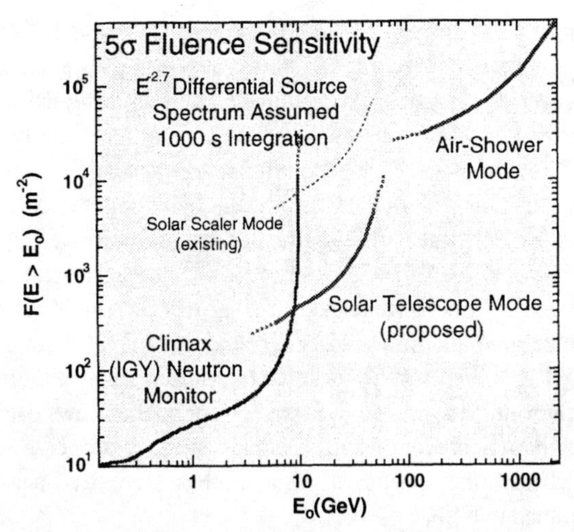

FIGURE 3 Predicted sensitivity of the proposed Milagro Solar trigger mode. Also shown are the sensitivities for an IGY neutron monitor and the existing Milagro air-shower trigger mode

rates measured at different geomagnetic locations around the world.

Milagro is unique since it allows high-energy solar phenomena to be imaged. Imaging is achieved by reconstructing the arrival direction of individual triggered events. Once fit, the solar origin of flare protons and/or neutrons could be established by identifying localized event-rate enhancements and correlating with solar flare measurements at other wavelengths, e.g., optical, radio, or X-ray. The solar origin for neutrons would be directly inferred because they come from the solar direction and would be consistent with the instrumental point spread function. For a beamed component of interplanetary protons the situation is different. Here the reconstructed direction need not point back to the Sun since these particles travel along magnetic field lines. This raises the sensitivity level because of the increased number of degrees of freedom used in the analysis. In addition, a beamed component of protons is unlikely to be consistent with the instrumental point spread function due to intrinsic anisotropies in the pitch angle distributions relative to the field lines. The degree of anisotropy and its evolution with time is an important diagnostic tool in the study of particle acceleration and transport. Event reconstruction and source imaging are important and unique features of this instrument that lead directly to a sensitivity improvement by many orders of magnitude, depending on the instrumental angular resolution achieved.

The wide field-of-view and imaging capability of Milagro can significantly aid solar monitoring during outburst periods. The most intense high-energy solar events exhibit 1 GeV gamma-ray and neutron emission for long periods of time, sometimes in excess of 2 hours [6, 7]. Interplanetary particle events are frequently of long duration. Measuring such long solar emissions is difficult in low Earth-orbiting spacecraft because of the satellite occultations every ~90 minutes. Milagro will permit uninterrupted tracking of the Sun across the sky for many hours.

Typically, the anisotropy of a solar particle event decays over the course of the event. This means that the leading edge of the event will often be detected with the best signal-to-noise ratio and, thus, sensitivity. The leading edge is also important for measuring the onset time from which one can compute the production time or production altitude of the particles. The evolution of the high-energy particle anisotropy provides the only data on the corresponding evolution of the mean free path of these protons above ~10 GV.

In addition to independent analyses, Milagro solar physics data can be combined with other instruments for more a complete analysis of a given event. Observations across multiple wavelengths and particle types provide context for data interpretation and model building. When used in conjunction with the Climax neutron monitor (500 km to the north) and with the appropriate temperature corrections, Milagro could serve as a surrogate for a neutron monitor at a different geomagnetic latitude. The scaler rate increases due to solar events can augment the data from the worldwide network of neutron monitors in the same manner as underground muon detectors but with a threshold that is more representative of solar flare particles.

Milagro, in its present configuration, is capable of triggering and recording these low-energy events. However, the existing data acquisition system designed to operate at ~1 kHz event rates, is incapable of handling the high muon rates. The experimental challenge is to distinguish single muons, produced by energetic solar particles, from the large number of muons produced by the isotropic background cosmic rays in the atmosphere. Cosmic-ray muons, unrelated to solar emission, will lead to an azimuthally uniform background rate of ~30 kHz in the Milagro detector. To perform the various science tasks briefly outlined above, we have designed a new data acquisition and event reconstruction system. A proposal to implement the proposed system has recently been submitted to NSF.

Expected Performance

Monte Carlo simulations have been used to study reconstruction algorithms, compute effective areas, and estimate flux sensitivities. The complete simulation of the detector response is done in two steps: (1) initial interaction of the primary particle (proton or neutron) with the atmosphere and the subsequent generation of secondary particles, and (2) detector response to the secondary particles reaching the detector level. The effective area and sensitivity estimates of Milagro in are illustrated in Figures 2 & 3; the effective area and sensitivity of an IGY neutron monitor is also shown for comparison. At the multi-GeV energies of interest here the spectrum of protons or neutrons may be falling rapidly at these energies, however, the increased effective area of Milagro compensates significantly, as does the important imaging capability.

Simulations have also been used to begin development of event reconstruction algorithms. The principal task of reconstruction is determining directionality. To do this we use a χ^2-minimization applied to PMT pulse-heights and relative timing distributions. Figure 4 shows preliminary results from this effort. The peak in the angular resolution corresponds to ~5°, and should represent the width (1σ) of a Gaussian point spread function. The distribution, however, is not the purely the projection of a 2-D Gaussian, but it also has a long tail. Experience with air-shower reconstruction suggests that further improvements (i.e., significant reduction of distribution tail) are expected when PMT pulse-height weighting, iterative fitting algorithms, and other optimizations are incorporated. The purpose of this preliminary algorithm development was to illustrate that muon events can be reconstructed at multi-kHz rates using data from the muon-layer PMTs only.

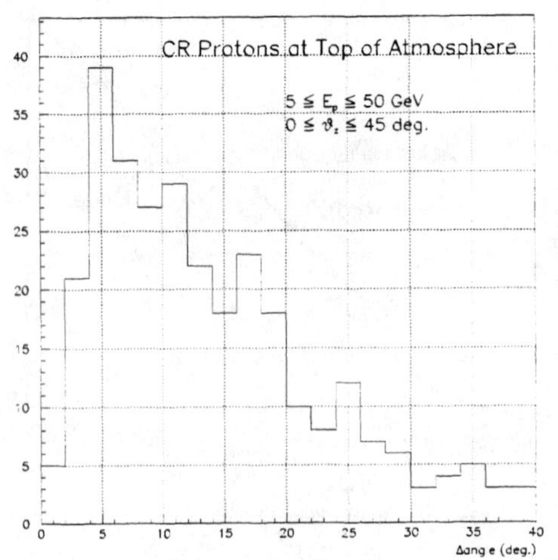

FIGURE 4 The angular difference between the fitted muon direction and initiating primary particle.

Summary

Proposed enhancements to the Milagro Gamma-Ray Observatory will make it a unique instrument for studying high energy particle emission from solar flares and Coronal Mass Ejections. These enhancements include new high throughput (>30kHz event rate) electronics and event reconstruction infrastructure. With these enhancements Milagro will be capable of reconstructing and studying the spatial and temporal distributions of secondary muons produced by solar energetic neutron and proton events

REFERENCES

1. E. N. Parker, *Physical Review* **107**, 830-836 (1957).
2. E. L. Chupp, *Annual Review of Astronomy and Astrophysics* **22**, 359-387 (1984).
3. G. Kanbach, et al., *Astronomy and Astrophysics (Suppl. Series)* **97**, 349-353 (1993).
4. G. Rank, PhD, Technical University of Munich (1996).
5. H. Hudson, J. Ryan, *Annual Review Astronomy and Astrophysics* **33**, 239-82 (1995).
6. G. Rank, *CESRA Workshop on Coronal Explosive Events*, 61 (1998).
7. E. J. Schneid, et al., *Astronomy and Astrophysics (Supplement Series)* **120**, 299-302 (1996).

The High Energy Solar Spectroscopic Imager (HESSI) Mission

R. P. Lin and the HESSI team[1]

Physics Department and Space Sciences Laboratory, University of California, Berkeley

Abstract. The primary scientific objective of the High Energy Solar Spectroscopic Imager (HESSI) Small Explorer mission selected by NASA is to investigate the physics of particle acceleration and energy release in solar flares. The HESSI instrument utilizes Fourier-transform imaging with 9 bi-grid rotating modulation collimators and cooled germanium detectors to make observations of X-rays and γ-rays from ~3 keV to ~17 MeV. It will provide the first imaging spectroscopy in hard X-rays, with ~2 arcsec angular resolution, time resolution of 2s for full image (tens of ms for crude image), and ~1 keV energy resolution; the first solar γ-ray line spectroscopy with ~1-5 keV energy resolution; and the first solar γ-ray line and continuum imaging, with ~36-arcsec angular resolution. The instrument is mounted on a Sun-pointed spin-stabilized spacecraft, and is planned to be launched in July 2000 into a 600 km-altitude, 38° inclination orbit. HESSI will provide detailed information on the energetic particle populations at the Sun, for comparison to measurements of solar energetic particles by ACE, WIND, and other interplanetary spacecraft.

INTRODUCTION

The Sun is the most powerful particle accelerator in the solar system. Both solar flares and fast Coronal Mass Ejections (CMEs) appear to accelerate ions up to tens of GeV and electrons to hundreds of MeV. Solar flares release up to 10^{32}-10^{33} ergs in 10^{2}-10^{3} s with accelerated 10-100 keV electrons (and possibly >1 MeV ions) containing a significant fraction, perhaps the bulk, of this energy. How the Sun releases this energy, presumably stored in the magnetic fields of the corona, and how it rapidly accelerates electrons and ions with such high efficiency, and to such high energies, is presently unknown. CMEs involve the ejection of up to ~10^{15}-10^{16} gms of solar material, also with total energy release of ~10^{32}-10^{33} ergs. It is believed that Large Solar Energetic Particle (LSEP) events seen at 1 A.U. are due to acceleration by the shock waves driven by fast CMEs as they travel through the corona.

High-energy emissions are the most direct signature of energetic particles near the Sun. Hard X-ray/γ-ray continuum is produced as bremsstrahlung by energetic electrons. Nuclear collisions of energetic protons and heavier ions with the ambient solar atmosphere result in a complex spectrum of narrow and broad γ-ray lines that contain unique information on not only the accelerated ions but also the ambient solar atmosphere.

It is not known how the electrons and ions that produce the X-ray and gamma-ray emissions in solar flares are related to solar energetic particles observed in the interplanetary medium.[1] Gamma-ray flare events have similarly large e/p ratios as impulsive (so-called because the soft X-ray imaging burst is < 10 minutes duration) SEP events, and may have similar enrichments of ^{3}He and heavy ions.[2] On the other hand, there is evidence indicating that LSEP events may be predicted by flare hard X-ray bursts which exhibit systematic hardening of the spectrum through the event.[3] HESSI hard X-ray and γ-ray spectroscopy and imaging, together with observations of coronal shocks, SEPs, and CMEs by other spacecraft such as ACE, WIND, etc., are needed to understand the relationship of the various acceleration processes.

[1] See Acknowledgements

CP528, *Acceleration and Transport of Energetic Particles Observed in the Heliosphere: ACE 2000 Symposium,* edited by Richard A. Mewaldt, et al.
© 2000 American Institute of Physics 1-56396-951-3/00/$17.00

Electron Acceleration and Energy Release

Bursts of hard (>20 keV) X-rays are the most common signature of the impulsive phase of a solar flare (Fig. 1). These X-rays are bremsstrahlung produced by accelerated electrons colliding with the ambient solar atmosphere. The Yohkoh Hard X-ray Telescope (HXT) often observes double footpoint structures, with the two footpoints brightening simultaneously to within a fraction of a second.[4] These coincide, spatially and temporally, with H_α and white-light brightenings, implying the X-ray emitting electrons have energy $E>kT$ of the ambient gas. Then the energy lost by the electrons to bremsstrahlung is only ~10^{-5} of their energy lost to Coulomb collisions with ambient thermal electrons. This inefficiency means that, for many flares, the energy in accelerated >20 keV electrons must be comparable to the total flare radiative and mechanical output.[5] Thus, the acceleration of electrons to tens of keV may be the most direct consequence of the basic flare-energy release process.

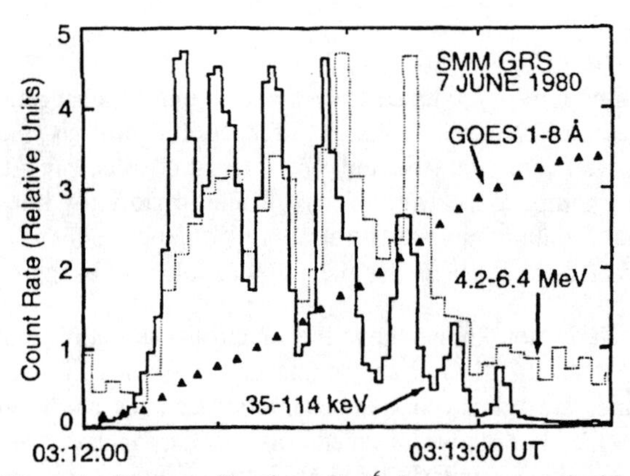

Figure 1. Time profiles for a flare[6] showing near-coincident impulsive peaks in 35-114 keV hard X-rays (from energetic electrons) and 4.2-6.4 MeV γ-rays (from energetic ions).

For some flares occurring near the solar limb, HXT has detected a co-temporal, weaker, hard X-ray source in the corona[7] above the soft X-ray loop linking the hard X-ray footpoints. This source has been interpreted as evidence for energy release by magnetic reconnection in a region above the soft X-ray loop.

HESSI's hard X-ray imaging spectroscopy allows the photon spectrum in each spatial and temporal element to be directly inverted to obtain N(E, **r**, t), the X-ray producing electron number density, as a function of energy (E), and position (**r**), and time (t). Context observation by the fleet of spacecraft (SOHO, Wind, ACE, Ulysses, TRACE, GOES, Yohkoh, SAMPEX, CGRO, etc.) that will already be in place, and by ground based instruments, will provide information on the ambient density, magnetic field strength and topology, etc. Then the electron loss processes can be directly evaluated, and with transport calculations (using a spatially dependent continuity equation including loss processes), the spatially and temporally resolved *accelerated* electron source distribution, F(E, **r**, t), can be obtained. This can be used to test detailed quantitative models of the acceleration, energy release, and energy propagation.

Ion Acceleration

Nuclear collisions of accelerated nuclei with the ambient solar atmosphere result in a rich spectrum of gamma-ray lines.[8] γ-ray line emission has been observed from many solar flares.[9,10] Energetic protons and α-particles colliding with carbon and heavier nuclei produce narrow de-excitation lines (widths of ~few keV to ~100 keV), while energetic heavy nuclei colliding with ambient hydrogen and helium produce much broader lines (widths of a few hundreds keV to an MeV). Neutron capture on hydrogen and positron annihilation produce narrow lines (at 2.223 MeV and 0.511 MeV, respectively) which are delayed.

The bulk of the γ-ray line emission is produced by ions with energies of 10-100 MeV/nucleon that contain only a small fraction of the energy in the >20 keV electrons. However, systematic study of SMM γ-ray line flares[10] shows that the 1.634 MeV ^{20}Ne line is unexpectedly enhanced. Because the cross section for ^{20}Ne has an unusually low energy threshold (~2.5 MeV), this effect may be due to large fluxes of low-energy ions with a total energy content comparable to that in accelerated electrons.[11] HESSI's high spectral resolution allows the identification of accelerated ^{3}He induced lines[12] in the broad group of lines around 1 MeV. It also provides high sensitivity measurements of the narrow neutron capture lines and detailed line shapes (a measure of the energetic particle anisotropy) for the α-α lines around 400 keV.[13]

HESSI will provide the first localizations of solar γ-rays and the most precise imaging ever achieved in γ-ray astronomy. For large flares, HESSI should be able to image in narrow γ-ray lines, e.g. the 2.223 MeV neutron capture and 0.511 MeV positron annihilation lines, where line counts dominate over the background. Although γ-ray lines have never been directly imaged, the 2.223 MeV line was once detected in a behind-the-limb flare.[14] This line is formed when thermalized neutrons are captured by ambient protons in the photosphere, so the neutrons must have been produced by charged particles interacting on the visible hemisphere of the Sun. Thus, either the acceleration site was far removed from the optical flare site, or the charged particles were transported over large distances.

INSTRUMENTATION

The HESSI scientific objectives will be achieved with a single instrument consisting of an Imaging System and a Spectrometer.[15] The Imaging System is made up of nine Rotating Modulation Collimators (RMCs), each consisting of a pair of widely separated identical grids in front of an X-ray/γ-ray detector (Fig. 2), mounted on a rotating spacecraft. Each grid consists of a planar array of equally-spaced, X-ray-opaque slats separated by transparent slits. The transmission through the grid pair depends on the direction of the incident X-rays. For slits and slats of equal width, the transmission is modulated from zero to 50% and back to zero for a change in source angle to collimator axis (orthogonal to the slits) of p/L where p is the pitch and L is the separation between grids. The angular resolution is then defined as p/(2L).

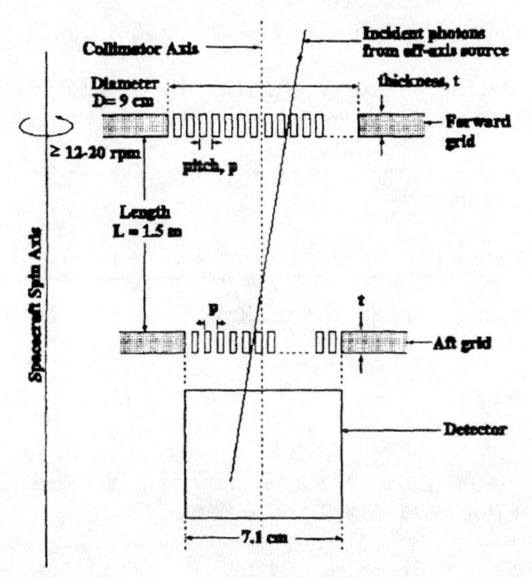

Figure 2. Schematic showing the instrument parameters that define the imaging capability.[15]

For a parallel incident beam, the modulated waveform generated by a smoothly rotating spacecraft has a distinctive quasi-triangular shape whose amplitude is proportional to the intensity of the beam and whose phase and frequency depend on the direction of incidence. For complex sources, and over small rotation angles, the amplitude and phase of the waveform provide a direct measurement of a single Fourier component of the angular distribution of the source.[16] Different Fourier components are measured at different rotation angles and with grids of different pitches.

For HESSI, the separation between grids in each RMC is L = 1.5 m and the grid pitches range from 34 μm to 2.75 mm in steps of $\sqrt{3}$. This gives angular resolutions that are spaced logarithmically from 2.3 arcsec to ~3 arcmin, allowing sources to be imaged over a wide range of angular scales. The chosen grid thicknesses provide imaging from 3 keV to ~100 keV with ~2 arcsec resolution, up to several hundred keV with 20 arcsec, and > 10 MeV with 36 arcsec.

The detector records the arrival time and energy of individual photons, allowing the modulated counting rate to be determined as a function of rotation angle. HESSI uses hyperpure (n-type) closed end coaxial germanium detectors (GeDs), cooled to provide high spectral resolution (~1-5 keV FWHM). The inner electrode is segmented into two contacts that collect charge from two electrically independent detector segments. The front segment thickness of ~1.5 cm is chosen to stop photons up to ~150 keV, where photoelectric absorption dominates. Photons with energies from ~150 keV to ~20 MeV, including all nuclear γ-ray lines, stop primarily in the thick rear segment The intense 3-150 keV X-ray fluxes that usually accompany large γ-ray line flares are absorbed by the front segment, so the rear segment will always count at moderate rates to provide optimal spectral resolution and high throughput for γ-ray line measurements. The GeDs are cooled on-orbit to ~75 K by a single Sunpower Inc. single stage, integral (counterbalanced) Stirling cycle mechanical cooler.

In a half rotation (2 s) the 9 RMCs measure amplitudes and phases of ~1100 Fourier components compared to 32 for the Yohkoh HXT. Detailed simulations show that HESSI can obtain accurate images with a dynamic range (defined as the ratio of the brightest to the dimmest feature reliably seen in an image) of up to 100:1 compared to ~10:1 for Yohkoh HXT. High-resolution X-ray spectra can be obtained for each location in the image, thus providing true high-resolution imaging spectroscopy.

The Solar Aspect System (SAS) provides six precise measurements of the solar limb every 10 ms, yielding Sun center position to ~1.5 arcsec. A star scanner Roll Angle System (RAS) samples the roll orientation at least once per rotation.

Spacecraft and Mission

The instrument field of view (~1°) is much larger than the Sun (0.5°) so spacecraft pointing is relaxed (within ~0.2° of Sun center). The energy and arrival time of every photon, together with SAS and RAS aspect data, are recorded in the spacecraft's on-board 4-Gbyte solid-state memory and telemetered when the spacecraft goes over the ground station. All of the photon data for the largest flare can be stored in the spacecraft memory and downlinked in <~48 hours, so flare data will rarely, if ever, be lost. Consequently, HESSI is planned for an automated store-and-dump operation, with normally no real-time access.

The HESSI spacecraft weighs a total of ~290 kg, uses ~280 watts, is passively spin-stabilized at 15 rpm, and points continuously at the Sun. It will be launched into a 600 km circular, 38° inclination orbit with a Pegasus XL rocket. A single 11-meter dish ground station at Berkeley provides the required data downlink rate and command uplink.

The complete data output of the HESSI mission will be made available promptly to the scientific community, without restriction, together with a fully documented analysis package, consisting of the same software used by the PI team.

HESSI is planned, and presently on schedule for, launch in mid-2000 near the predicted next solar maximum. A two-year nominal mission (a third year is highly desirable) will provide observations of tens of thousands of microflares, thousands of hard X-ray flares, and of order a hundred γ-ray line flares.

ACKNOWLEDGMENTS

The HESSI team includes Co-I's G. Hurford, and N. Madden at UCB; B. Dennis, C. Crannell, G. Holman, R. Ramaty and T. von Rosenvinge at GFSC; A. Zehnder at the Paul Scherrer Institute (PSI) Switzerland; F. van Beek in The Netherlands; P. Bornmann at NOAA; R. Canfield at Montana State Univ.; G. Emslie at U. Alabama Huntsville; H. Hudson at UCSD; A. Benz at Switzerland; J. Brown at U. of Glasgow, Scotland; S. Enome at NAO, Japan; T. Kosugi at ISAS, Japan; and N. Vilmer at Observatoire de Paris, Meudon, France. HESSI associated Scientists are D. Smith, J. McTiernan, I. Hawkins, S. Slassi-Sennou, A. Csillaghy, G. Fisher, C. Johns-Krull at UCB; R. Schwartz, L. Orwig, D. Zarro at GFSC; E. Schmahl at U. of Maryland; and M. Aschwanden at Lockheed-Martin. The engineering team at UCB is led by P. Harvey, D. Curtis, and D. Pankow; at GSFC, D. Clark and R. Boyle; at PSI, R. Henneck, A. Mchedlishvili, and K. Thomsen. The HESSI spacecraft was developed by Spectrum Astro Inc. The project manager at the GSFC Explorer Office is F. Snow. This research is supported by NASA contract NAS5-98033 to the UCB.

REFERENCES

1. Cliver, E., these proceedings.
2. Reames, D. V., Meyer, J. P., and von Rosenvinge, T. T., *Astrophys. J. Suppl.* **90**, 649-667 (1994).
3. Kiplinger, A. L., *Astrophys. J.* **453**, 973-986 (1995).
4. Sakao, T., Kosugi, T., Masuda, S., Yaji, K., Inda-Koide, M., and Makashima, K., *Proc. Kofu Symp.* **360**, 169- (1994).
5. Lin, R. P., and Hudson, H. S., *Solar Phys.* **50**, 153-178 (1976).
6. Chupp, E. L., *AIP Conf. Proc.* **77**, 363-381 (1982).
7. Masuda, S., Kosugi, T., Hara, H., Tsuneta, S., and Ogawara, Y., *Nature* **371**, 495-497 (1994).
8. Ramaty, R., and Murphy, R. J., *Space Sci. Rev.* **45**, 213-268 (1987).
9. Chupp, E. L., *Ann. Rev. Astron.* **22**, 359- (1984).
10. Share, G. H., and Murphy, R. J., *Astrophys. J.* **452**, 933-943 (1995).
11. Ramaty, R., and N. Mandzhavidze, *IAU Symp. V.* **195**, (1999).
12. Mandzhavidze, N., R. Ramaty, and B. Kozlovsky, *Astrophys. J.* **518**, 918-925 (1999).
13. Share, G. H., and R. J. Murphy, *Astrophys. J.* **485**, 409-418 (1997).
14. Vestrand, W. T., and Forrest, D. J., *Astrophys. J.* **409**, L69-L72 (1993).
15. Lin, R. P. and The Hessi Team, *SPIE Conf. on Missions to the Sun II* **3442**, 2-12 (1998).
16. Prince, T. A., Hurford, G. J., Hudson, H. S., and Crannell, C. J., *Solar Phys.* **118**, 269-290 (1988).

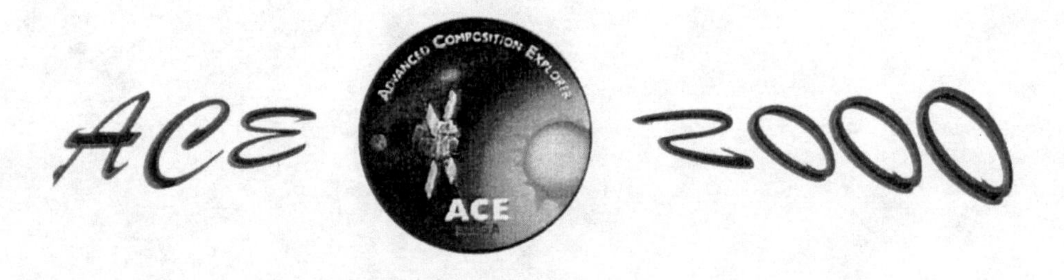

Interplanetary Acceleration

Electron Heating and Acceleration at Collisionless Shocks in the Solar Wind

J. T. Gosling

Los Alamos National Laboratory, MS D466, Los Alamos, New Mexico 87545

Abstract. This paper provides a brief synthesis of certain aspects of the electric and magnetic field structure of a collisionless shock in the solar wind and outlines how electron heating and acceleration at the shock is related to that structure. Emphasis is on measurements obtained in the vicinity of Earth's bow shock, although the results should have broad applicability to other shocks in the heliosphere.

INTRODUCTION

A shock in a gas is best characterized as a large and nearly discontinuous change in pressure that travels faster than the characteristic speed (the sound speed) with which small amplitude pressure waves propagate through the gas. In any shock rest frame the flow normal to the shock surface changes from supersonic upstream to subsonic downstream. In an ordinary gas, where particle collisions are essential for effecting the shock transition, this slowing of the flow crossing the shock is produced by a steep pressure gradient. The Rankine-Hugoniot equations, which express conservation of mass, momentum, and energy, together with an equation of state determine the magnitude of the changes in density, flow speed, temperature, and pressure that occur at the shock.

For a shock propagating through the ionized and nearly collisionless solar wind, the characteristic speed is either the fast mode speed or the slow mode speed and the slowing of the flow must be produced primarily by electric and magnetic forces. Moreover, in addition to satisfying the ordinary Rankine-Hugoniot equations, the changes that occur at the shock must be consistent with Maxwell's equations. In the absence of measurements within the shock layer, early theoretical studies (1) suggested that plasma microinstabilities would play the role that collisions play in an ordinary gas in heating the plasma and in producing irreversible dissipation. These instabilities were generally thought to be associated with currents flowing perpendicular to the magnetic field in the shock layer (such currents are necessary to produce the jump in magnetic field strength at the shock). The possible importance of the macroscopic electric and magnetic fields within the shock layer in producing gross distortions in electron and ion velocity distributions, f(v), was not generally appreciated. It is now understood, however, that such distortions are of fundamental importance in initiating the heating process for both electrons (2) and ions (3,4) at strong shocks in the solar wind. The role of microinstabilities is to smooth out these distortions and thereby make the shock transition irreversible. Our purpose here is to describe how electron heating and acceleration occur at a collisionless shock in the solar wind and to relate that heating and acceleration to the electric and magnetic structure of the shock layer, which is primarily determined by the ion dynamics. We concentrate on observations made at the Earth's bow shock, where shock motions relative to a spacecraft such as ISEE-2 are relatively small and the shock layer can be resolved by the observations. The bow shock observations have general applicability to relatively strong shocks observed elsewhere in the solar wind even though the bow shock is more highly curved than most interplanetary shocks and field line connection times are correspondingly shorter.

ELECTRON HEATING MECHANISM

Figure 1 illustrates the distortions in the solar wind electron f(v) produced by the electric field within the Earth's bow shock layer (5). The figure shows cuts parallel to the magnetic field, **B**, obtained as ISEE-2 progressed through the shock from up to downstream. The field was directed in the upstream direction so that negative speeds in the figure correspond to the downstream direction. An offset maximum in f(v) developed within the shock layer that moved to ever higher speeds as the spacecraft penetrated deeper within the shock layer. With increasing penetration the amplitude of the peak decreased and the overall distribution flattened and broadened. The offset peak

CP528, *Acceleration and Transport of Energetic Particles Observed in the Heliosphere: ACE 2000 Symposium*, edited by Richard A. Mewaldt, et al.
© 2000 American Institute of Physics 1-56396-951-3/00/$17.00

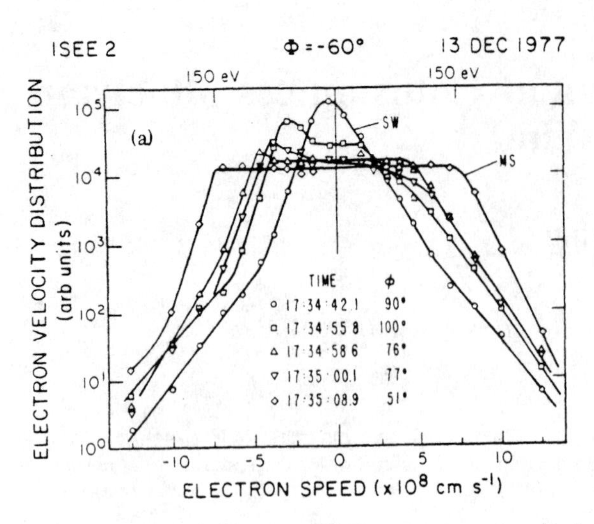

FIGURE 1. Cuts, parallel to **B**, through electron velocity distributions measured during a crossing of Earth's bow shock by ISEE-2. The significant part of each cut was measured in less than 100 ms. SW and MS refer to solar wind and magnetosheath, respectively. Adapted from (5).

was the result of the bulk acceleration of the electrons parallel to **B** through the cross-shock electric potential. In this case the potential drop experienced by the electrons was ~150 eV. The general decrease in amplitude of the beam and the redistribution of the particles to other pitch angles presumably were the result of microinstabilities generated by the field-aligned beam. These and similar measurements clearly indicate that electron heating at strong shocks in the solar wind is initiated by acceleration of the electrons parallel to **B** through the cross-shock potential. On the other hand, at weaker shocks the electron temperature increase is almost entirely a result of the adiabatic compression of the plasma (6).

MAGNETIC FIELD ROTATION OUT OF THE PLANE OF COPLANARITY

The magnetic structure of a collisionless shock, which is determined primarily by the more massive ions, causes the cross-shock potential to be frame-dependent, and thereby limits the heating experienced by the electrons at the shock. Conservation of mass and momentum and Maxwell's equations lead to the following relationships, which are a subset of the modified Rankine-Hugoniot equations:

$$[\rho v_n] = 0 \tag{1}$$

$$[B_n] = 0 \tag{2}$$

$$[v_t] = B_n[\mathbf{B}_t] / 4\pi\rho v_n \tag{3}$$

$$[v_t] = [v_n\mathbf{B}_t] / B_n. \tag{4}$$

Here ρ is the mass density, **v** is the flow velocity, **B** is the magnetic field, the subscripts n and t refer respectively to components normal and tangential to the shock surface, and the brackets indicate the change across the shock. By combining (3) and (4) one obtains the well known coplanarity result, which states that the vector changes in the magnetic field and flow velocity at the shock must lie in the plane determined by the upstream magnetic field and the shock normal.

The change in tangential velocity embodied in (3) and (4) arises from magnetic stresses and has no counterpart in ordinary gas dynamic shocks. The internal magnetic structure of a collisionless shock is, to a large degree, determined by the need to provide this change in tangential velocity for the ions (7-9). Although the average upstream and downstream magnetic fields and the shock normal must lie in a common plane, the field is free to rotate out of this plane within the shock layer. In fact it can be shown (7) that, except for exactly perpendicular shocks, the field must rotate out of the coplanarity plane within the shock layer in order to provide the transverse deflection of the ions at the shock. We define a coordinate system such that the +x-axis is anti-parallel to the shock normal, the +z-axis is parallel to the projection of the upstream field on the shock surface (the x-z plane is thus the coplanarity plane), and the +y axis completes a right-handed system. Figure 2 demonstrates the rotation of the magnetic field out of

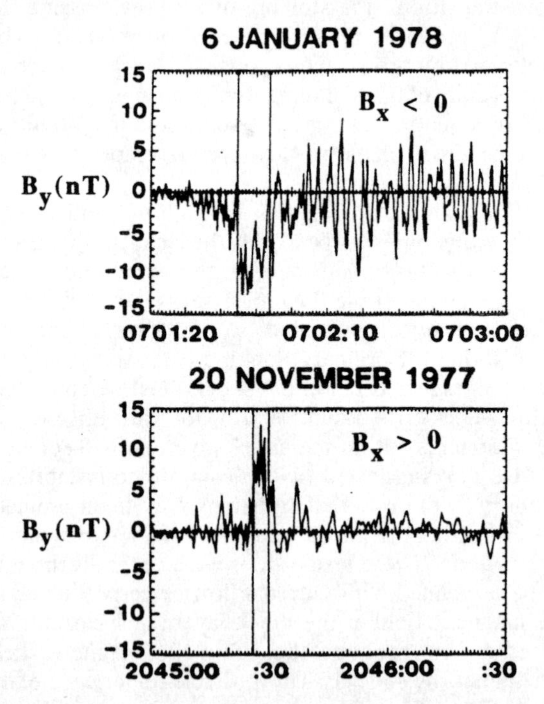

FIGURE 2. Non-coplanarity component (B_y) of the magnetic field measured by the ISEE-1 magnetometer during two crossings of Earth's bow shock. In each case the upstream region is to the left of the shock ramp, which is indicated by the vertical lines. Adapted from (10).

the coplanarity plane for two crossings of the Earth's bow shock (10). As is required to provide the proper transverse deflection of the ions, the non-coplanar component within the shock layer always has the same sign as the x-component of the magnetic field.

FRAME DEPENDENCE OF THE CROSS-SHOCK POTENTIAL

The rotation of the magnetic field out of the coplanarity plane within the shock layer causes the cross-shock potential to be frame dependent (11). The two most commonly used shock frames are the normal incidence (NI) frame, in which the upstream flow is parallel to the shock normal, and the deHoffman-Teller (HT) frame, in which $\mathbf{v}$ and $\mathbf{B}$ are parallel both upstream and downstream. The normal components of the electric field in the two frames are related by

$$\mathbf{n}\cdot\mathbf{E}^{HT} = \mathbf{n}\cdot\mathbf{E}^{NI} + \mathbf{n}\cdot[\mathbf{V}^{T} \times \mathbf{B}_{y}] / c \qquad (5)$$

where c is the speed of light, $\mathbf{n}$ is the shock normal, and $\mathbf{V}^{T}$ is the transformation velocity from the normal incidence (NI) to the deHoffman-Teller (HT) frame, given (12) by

$$\mathbf{V}^{T} = \mathbf{n} \times (\mathbf{v}_{nu} \times \mathbf{B}_{u}) / \mathbf{n}\cdot\mathbf{B}_{u} \qquad (6)$$

where u denotes the upstream region. The frame-dependent difference in cross-shock potential is therefore given by

$$\Delta\phi = v_{xu}B_{zu}\int_{0}^{L} B_{y}\, dx / cB_{x} \qquad (7)$$

where L is the shock thickness. This shows that the potential drop in the HT frame is always less than that in the NI frame for a fast mode shock because v_{xu} and B_{zu} are always positive by definition, and the integral in (7) and B_x always have the same sign within the shock layer (see Figure 2). Figure 3 shows the frame dependence of the cross-shock potential for a crossing of Earth's bow shock (13). Examination of a large number of bow shock crossings shows that the potential drop in the HT frame typically is a factor of 2-6 smaller than that in the NI frame (10). As the potential actually experienced by the electrons crossing the shock is that in the HT frame, the electrons crossing the shock receive a smaller acceleration (and thus also less heating) than might be expected on the basis of the observed deceleration of the normal component of the ion flow across the shock.

PRODUCTION OF SUPRATHERMAL ELECTRON TAILS AT NEARLY PERPENDICULAR SHOCKS

Electron heating experienced at a shock is strongly correlated with the solar wind speed and quantities derived from it. The change in electron temperature at the shock correlates best with the change in the ion bulk flow energy across the shock (14), as shown in Figure 4. This dependence is consistent with a heating process associated with acceleration through the cross-shock potential in the HT frame, as noted above. Nevertheless, the electron temperature increase at the shock is also a function of Θ_{Bn}, the angle between the

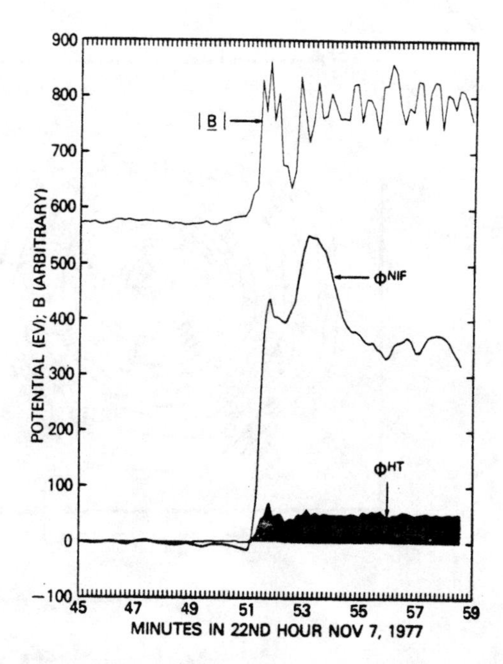

FIGURE 3. Normal incidence frame (NIF) and deHoffman-Teller frame (HT) cross-shock potentials determined by ISEE-1 during a crossing of Earth's bow shock. The trace of the magnetic field magnitude is shown at the top. Adapted from (13).

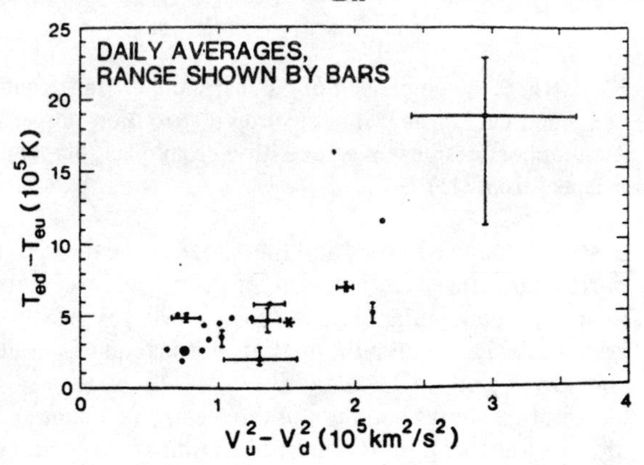

FIGURE 4. Observed increase in electron temperature across the shock as a function of the decrease in the square of the bulk flow speed, which is a measure of bulk flow energy dissipated by the shock. Bars indicate the range of values for multiple crossings of the shock. Adapted from (14).

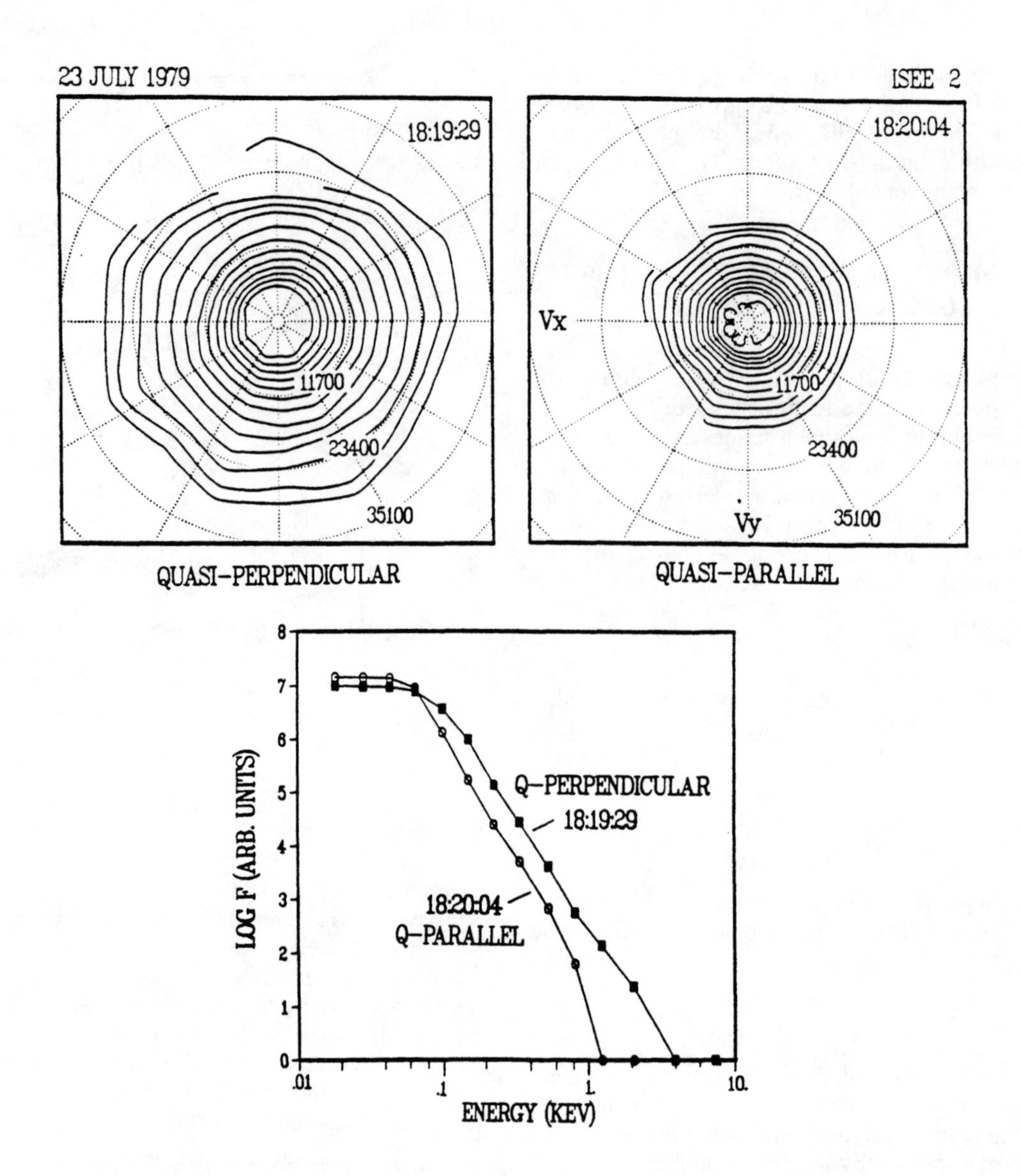

FIGURE 5. A comparison of 3s snapshots of (top) contours of two-dimensional electron distribution functions and (bottom) one-sided cuts through the electron distribution perpendicular to **B** obtained downstream from Earth's bow shock by ISEE-2 on opposite sides of a transition from quasi-perpendicular to quasi-parallel upstream field geometry on 23 July 1979. Adapted from (15).

upstream magnetic field and the shock normal (15). In particular, the electron temperature increase across nearly perpendicular shocks ($\Theta_{Bn} \sim 90°$) typically is considerably greater than that across quasi-parallel shocks ($\Theta_{Bn} < 45°$).

Figure 5 shows contours of the electron f(v) and one-sided cuts through f(v) perpendicular to **B** obtained immediately downstream from the Earth's bow shock. The snapshots and cuts are separated in time by less than a minute and bracketed a rotation in the upstream magnetic field orientation from the nearly perpendicular geometry to the quasi-parallel geometry while the other upstream parameters remained nearly constant. Below ~60 eV, where the cross-shock potential most affects the distributions, both distributions are roughly flat-topped as in Figure 1, with the phase space density at these energies being greater for the quasi-parallel geometry. However, above ~60 eV the spectra diverge considerably. At energies above ~300 eV, f(v) for the nearly perpendicular shock exceeds that for the quasi-parallel shock by a factor greater than 10 in this case.

Figure 6 further contrasts the changes in electron f(v) that occur at the quasi-perpendicular and quasi-parallel shocks. The top panel compares 1-sided cuts perpendicular to B immediately up and downstream of a quasi-perpendicular shock crossing, while the lower panel compares similar cuts for a quasi-parallel shock crossing. Whereas both downstream distributions were

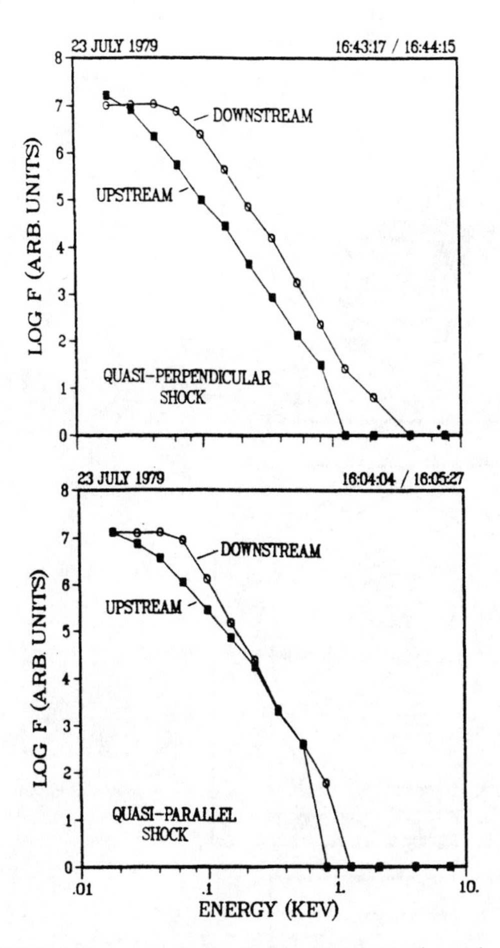

FIGURE 6. A comparison of one-sided cuts through the electron distribution function perpendicular to **B** upstream and downstream from crossings of Earths bow shock when the shock was (top) quasi-perpendicular and (bottom) quasi-parallel on 23 July 1979. Adapted from (15).

flat-topped at energies below ~60 eV, at energies above ~300 eV there was no downstream enhancement in f(v) for the quasi-parallel shock. In contrast, the downstream enhancement in f(v) at energies greater than 300 eV was approximately a factor of 20 for the quasi-perpendicular shock crossing. To summarize, extended high energy tails are commonly observed downstream of quasi-perpendicular shocks but are rarely, if ever, observed downstream of quasi-parallel shocks. The suprathermal tails can cause the downstream electron temperature to be about 50% greater at the nearly perpendicular shock than at the quasi-parallel shock for otherwise similar upstream conditions. Measurements show that the tails often extend upward to energies greater than 100 keV (16,17), although the measurements shown here extend up only to ~10 keV.

The flux of suprathermal electrons usually peaks immediately downstream from the nearly perpendicular shock at all energies. Figure 7 shows representative snapshots of the 2-dimensional electron f(v), together with the overall electron density and temperature profiles, for a crossing of the nearly perpendicular bow shock. The times of the f(v) snapshots are indicated on the temperature trace. Suprathermal electrons with speeds greater than 20,000 km s^{-1} (~1.14 keV) were first observed as a broad, field-aligned beam directed upstream along **B** within the shock ramp (snapshot 2) at a time when the bulk of the electrons had been only partially heated. This escaping beam of electrons was much more energetic than the nominal backstreaming heat flux observed over most portions of the bow shock (2). The flux of suprathermal electrons maximized within the shock overshoot, at which point the suprathermal electron distribution was roughly isotropic. Farther downstream the suprathermal flux decreased and the distributions developed a pronounced anisotropy perpendicular to **B** (not shown). Similar sequences, including the escaping beams, have been observed for a number of other crossings of the nearly perpendicular shock.

The above-described pattern clearly indicates that a portion of the solar wind electron distribution is accelerated to relatively high energy as the plasma crosses a nearly perpendicular shock. Such acceleration decreases as Θ_{Bn} decreases, and is insignificant at quasi-parallel portions of the shock. The most field-aligned particles quickly leak back upstream along **B**, thus producing the energetic electron beams commonly observed upstream from nearly perpendicular portions of the shock (18). As these energetic electron beams travel upstream they generate a variety of plasma emissions (19). (It seems likely that interplanetary type II radio emission is produced by similar electron beams escaping upstream from nearly perpendicular portions of CME-driven shocks (20). Indeed, the well known intermittency of the type II radio emission is probably related to the fact that energetic (> several keV) electron beams are only present where the shock normal is nearly perpendicular to the upstream **B**.) The upstream escape of suprathermal electrons with large parallel speeds eventually produces perpendicular anisotropies well downstream from quasi-perpendicular portions of the shock (15,21).

It is presently uncertain what produces the intense fluxes of energetic electrons observed downstream of nearly perpendicular shocks. It is relatively straight forward to show (15), however, that they are not simply a consequence of a phase space mapping from upstream to downstream in which individual electrons traversing the shock suffer negligible scattering, conserve both their magnetic moment and their total energy in the HT frame, and obey Liouville's theorem. The observed flux levels are far greater than predicted by such a mapping; moreover, the roughly isotropic suprathermal distributions observed throughout the shock layer (except for the escaping beam) and immediately downstream from the shock indicate that the electrons scatter significantly in pitch angle within

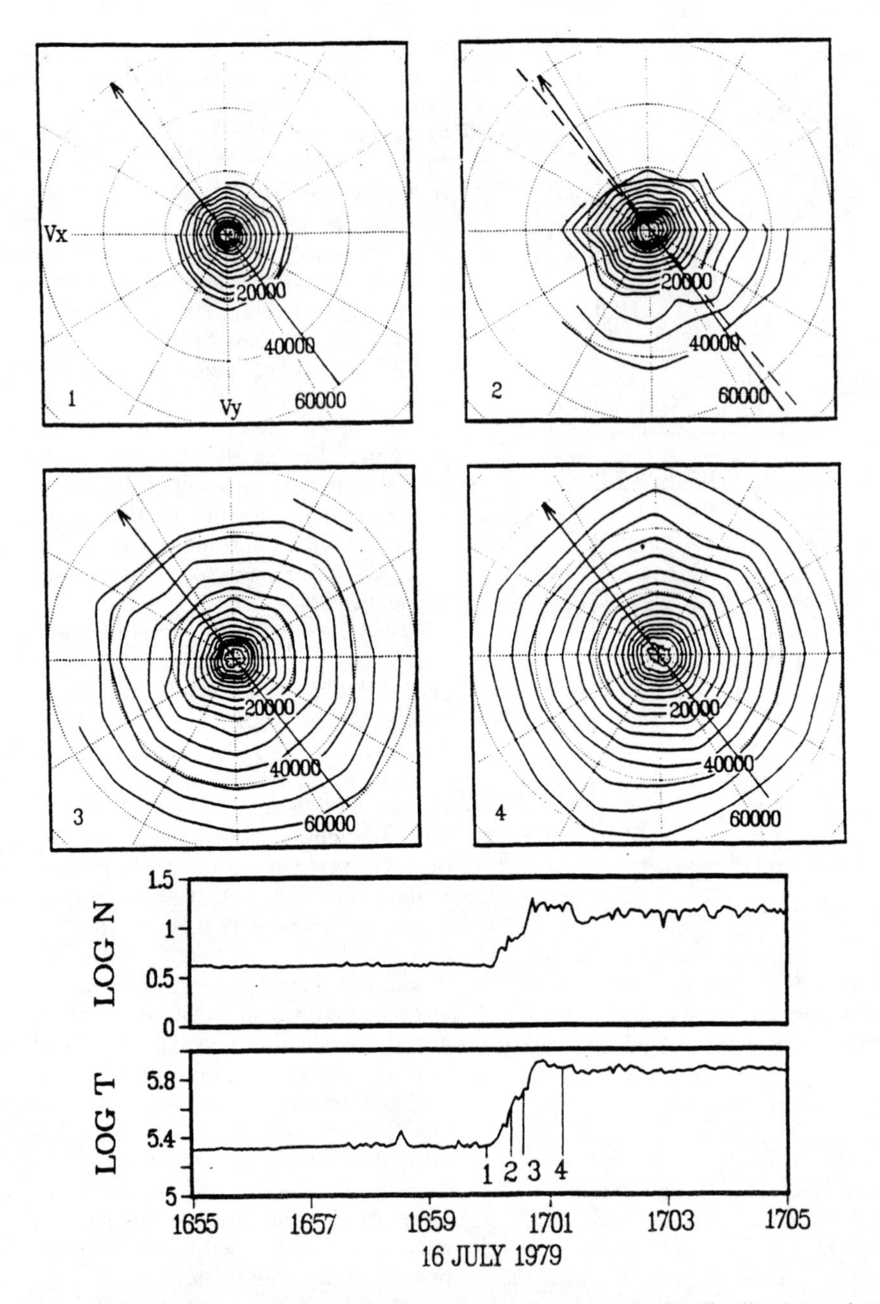

FIGURE 7. (Top) Representative 3s snapshots of 2-dimensional electron velocity distributions obtained by ISEE-2 during a crossing of Earth's bow shock when the upstream field was nearly perpendicular to the shock surface (orientation indicated by the dashed line in the second panel). The distributions are shown as contours of constant phase space density. Vectors drawn indicate the ecliptic projection of **B** at the time of the measurement. (Bottom) Electron density (cm^{-3}) and temperature (K) profiles for the shock crossing. Numbered vertical lines refer to the times of the spectral snapshots shown above. Adapted from (15).

the shock layer. Early theoretical work on electron acceleration at shocks (22,23) concentrated on the upstream region and assumed that the backstreaming electrons observed upstream from the nearly perpendicular shock were energized by reflection at the shock (conservation of energy in the HT frame); however, reflection can not explain the downstream fluxes of suprathermal electrons, which always are more intense than the upstream fluxes. Hybrid simulations (24) suggest that a combination of shock drift and scattering within the shock layer may be responsible for the particle acceleration even though electrons at least initially have gyroradii comparable to or smaller than the scale of the magnetic field gradients at the shock.

SUMMARY

Fluid considerations are essential for determining the relationship between upstream and downstream quantities at a shock via the Rankine-Hugoniot relations and Maxwell's equations. However, the actual shock transition in the nearly collisionless solar wind is more complex than would be inferred from fluid considerations alone, and fluid considerations do not specify the relative amount of heating experienced by the ions and the electrons at the shock. Collisionless shock complexity arises because electron and ion motions are effectively decoupled on the spatial scale of the shock layer and because the electrons and ions have considerably different masses. Because of their greater mass, it is the ion dynamics that primarily determine the macroscopic electric and magnetic structure of collisionless shocks. For example, the electric field within the shock layer is directed upstream in order to help slow the ions crossing the shock. (Note: the magnetic field also contributes to the slowing (4).) The electrons, on the other hand, are accelerated parallel to the magnetic field by the cross-shock potential. It is this acceleration that initiates the electron heating process at a strong shock in the solar wind. In addition, the magnetic field rotates out of the coplanarity plane within the shock layer in order to provide the transverse acceleration of the ions required by the modified Rankine-Hugoniot relations. That rotation causes the cross-shock potential to depend on reference frame; the cross-shock potential in the deHoffman-Teller frame, which determines the amount of acceleration and heating experienced by the electrons, is typically 2-6 times lower than that in the normal incidence frame.

To zeroth order, electron heating at the shock scales as the change in bulk flow energy experienced by the ions (where most of the solar wind energy resides) crossing the shock, a result of the role of the cross-shock potential in heating the electrons. However, for similar upstream conditions, the electrons are heated more at quasi-perpendicular portions of a shock than at quasi-parallel portions. This additional "heating" is actually the result of the formation of an extensive suprathermal tail at quasi-perpendicular portions of the shock that is absent at the quasi-parallel shock. The suprathermal tail often extends upward to energies greater than 100 keV downstream of the nearly perpendicular shock, where the flux of suprathermal electrons maximizes. Magnetic field-aligned portions of the suprathermal electrons that are energized at the shock escape back upstream, thus producing beams of energetic electrons on field lines intersecting the nearly perpendicular shock. Although the process for accelerating electrons to high energies at the nearly perpendicular shock is presently uncertain, shock drift in conjunction with scattering within the shock layer appears to be a likely candidate mechanism.

ACKNOWLEDGMENTS

This work was performed under the auspices of the U. S. Department of Energy with support from the National Aeronautics and Space Administration.

REFERENCES

1. Kennel, C. F., Edmiston, J. P., and Hada, T., *In Collisionless Shocks in the Heliosphere: A Tutorial Review*, R. G. Stone and B. T. Tsurutani, editors, Washington D.C.: AGU, 1985, pp 1-36.
2. Feldman, W. C., Anderson, R. C., Bame, S. J., Gary, S. P., Gosling, J. T., McComas, D. J., Thomsen, M. F., Paschmann, G., and Hoppe, M. M., *J. Geophys. Res.*, **88**, 96-110 (1983).
3. Gosling, J. T., and Robson, A. E., *In Collisionless Shocks in the Heliosphere: Reviews of Current Research*, B. T. Tsurutani, and R. G. Stone, editors, Washington D.C.: AGU, 1985, pp 141-152.
4. Goodrich, C. C., *In Collisionless Shocks in the Heliosphere: Reviews of Current Research*, B. T. Tsurutani, and R. G. Stone, editors, Washington D.C.: AGU, 1985, pp 153-168.
5. Feldman, W. C., Bame, S. J., Gary, S. P., Gosling, J. T., McComas, D., Thomsen, M. F., Paschmann, G., Sckopke, N., Hoppe, M. M., and Russell, C. T., *Phys. Rev. Lett.*, **49**, 199-201 (1982).
6. Feldman, W. C., Anderson, R. C., Bame, S. J., Gosling, J. T., Zwickl, R. D., and Smith, E. J., *J. Geophys. Res.*, **88**, 9949-9958 (1983).
7. Goodrich, C. C., and Scudder, J. D., *J. Geophys. Res.*, **91**, 7135-7136 (1986).
8. Jones, F. C., and Ellison, D. C., *J. Geophys. Res.*, **92**, 11,205-11,207 (1987).
9. Gosling, J. T., Winske, D., and Thomsen, M. F., *J. Geophys. Res.*, **93**, 2735-2740 (1988).
10. Thomsen, M. F., Gosling, J. T., Bame, S. J., Quest, K. B., Winske, D., Livesey, W. A., and Russell, C. T., *J. Geophys. Res.*, **92**, 2305-2314 (1987).
11. Goodrich, C. C., and Scudder, J. D., *J. Geophys. Res.*, **89**, 6654-6662 (1984).
12. Schwartz, S. J., Thomsen, M. F., and Gosling, J. T., *J. Geophys. Res.*, **88**, 2039-2047 (1983).
13. Scudder, J. D., Mangeney, A., Lacombe, C., Harvey, C. C., and Aggson, T. L., *J. Geophys. Res.*, **91**, 11,053-11,073 (1986).
14. Thomsen, M. F., Mellott, M. M., Stansberry, J. A., Bame, S. J., Gosling, J. T., and Russell, C. T., *J. Geophys. Res.*, **92**, 10,111-10,117 (1987).
15. Gosling, J. T., Thomsen, M. F., Bame, S. J., and Russell, C. T., *J. Geophys. Res.*, **94**, 10,011-10,025 (1989).
16. Fan, C. Y., Gloeckler, G., and Simpson, J. A., *Phys. Rev. Lett.*, **13**, 149-152 (1964).
17. Anderson, K. A., Harris, H. K., and Paoli, R. J., *J. Geophys. Res.*, **70**, 1039-1050 (1965).
18. Anderson, K. A., Lin, R. P., Martel, F., Lin, C. S., Parks, G. K., and Reme, H., *Geophys. Res. Lett.*, **6**, 401-404 (1979).
19. Kasaba, Y., Matsumoto, H., Omura, Y., Anderson, R. R., Mukai, T., Saito, Y., Yamamoto, T., and Kokubun, S., *J. Geophys. Res.*, **105**, 79-103 (2000).
20. Bale, S. D., Reiner, M. J., Bougeret, J.-L., Kaiser, M. L.,

Krucker, S., Larson, D. E., and Lin, R. P., *Geophys. Res. Lett.*, **26**, 1573-1576 (1999).
21. Potter, D. W., *J. Geophys. Res.*, **86**, 11,111-11,116 (1981).
22. Leroy, M. M., and Mangeney, A., *Ann. Geophys.*, **2**, 449-456 (1984).
23. Wu, C. S., *J. Geophys. Res.*, **89**, 8857-8862 (1984).
24. Krauss-Varban, D., *J. Geophys. Res.*, **99**, 2537-2551 (1994).

Observations of Non-Thermal Properties of Heavy Ions in the Solar Wind

Thomas H. Zurbuchen, Lennard A. Fisk,
Nathan A. Schwadron, and George Gloeckler

Space Research Laboratory, University of Michigan, Ann Arbor, MI USA

Abstract. Heavy ions in the solar wind are ideal for studying injection processes in the solar wind. We use composition data from Ulysses, ACE, and Wind to examine the properties of heavy ions from thermal energies to several 100 keVs. We show that these particles are observed to gain energy without any association with shocks. This paper provides a survey of recent observations of non-thermal properties of solar wind heavy ions which are consistent with the following picture: At thermal energies coherent wave-particle interactions preferentially heat and accelerate heavy ions with collisional processes limiting subsequent non-thermal properties. At higher energies heavy ion distribution functions are characterized by ubiquitous suprathermal tails. We argue that solar wind heavy ions are a good tracer for acceleration processes which are not directly associated with shocks. These stochastic processes are observed to be relevant for pre-disposing ions for shock acceleration.

INTRODUCTION

Shocks are often associated with energetic particle populations in the heliosphere. They are therefore an important source of particle acceleration, especially at higher energies. In fact, experimental data are in good agreement with theoretical predictions based on a combination of shock acceleration and field aligned diffusion (see, e.g., Fisk and Lee (1)). It is therefore not surprising that acceleration mechanisms are often considered to happen exclusively in association with shocks. However, it has been pointed out that shock acceleration processes have an injection threshold energy, E_{in}. For energies $E>E_{in}$, shock acceleration is effective but it is inhibited if this condition is not fulfilled. The exact value of E_{in} depends on the details of the acceleration mechanism. In the case of diffusive shock acceleration it is easily expressed as a function of solar wind speed, V, the shock speed, V_s, and the angle between the shock and the magnetic field, θ. A particle will only participate in shock drift acceleration if the particle moves upstream after its first interaction with the shock. The particle speed after the interaction with the shock, v_f, therefore has to exceed the solar wind speed relative to the shock:

$$v_f \cos\theta > V - V_s. \quad (1)$$

For a quasi-perpendicular shock, this injection speed is typically 1.3-2 V, measured in the solar wind rest frame. Solar wind heavy ions, if in thermal equilibrium with solar wind protons, should therefore have no chance of being directly injected into shock acceleration.

This injection process was examined by Chotoo (2) using Wind data associated with Co-rotating Interaction Regions (CIRs). He used a combination of three sensors, the Supra-Thermal Energetic Particle telescope (STEP), the Supra-Thermal Ion Composition Spectrometer (STICS), and the Ion Mass Spectrometer (MASS), which cover the energy range from solar wind thermal energies of up to MeVs. The result is summarized in Figure 1. The data in Figure 1 are plotted in the spacecraft frame. The data can, in principle, be transferred into the solar wind reference frame using a standard Compton-Getting correction (2). The reduced distribution functions, f, are three-dimensional distribution functions ϕ averaged over the full solid angle, e.g.

$$f(v) = \int d\Omega \cdot \phi(|\vec{v}| = v) \quad (2)$$

CP528, *Acceleration and Transport of Energetic Particles Observed in the Heliosphere: ACE 2000 Symposium*,
edited by Richard A. Mewaldt, et al.
© 2000 American Institute of Physics 1-56396-951-3/00/$17.00

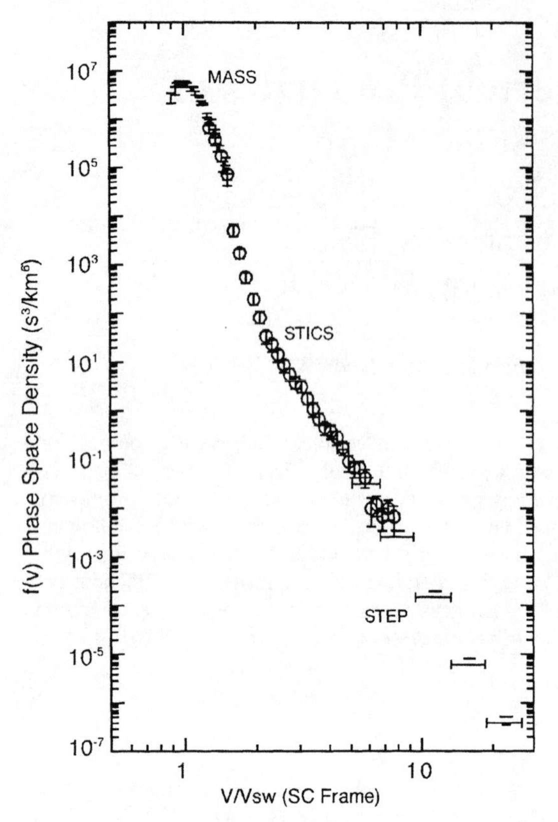

FIGURE 1. Reduced distribution function of He^{2+}combined from three sensors on Wind (figure copied from reference 2 with permission of author).

Apparently, solar wind α-particles are readily injected into shock acceleration. A single power-law spectrum can be fitted for speeds greater than 2-3V in the spacecraft rest frame. However, for smaller energies, $1.5V<v<2.5V$, another power-law–type spectrum can be seen. It would be difficult to directly inject the thermal core of the solar wind alphas into diffusive shock acceleration. The injected particles must first undergo a pre-acceleration process that prepares them for the shock injection by exceeding the injection threshold as determined by equation (1). If this process is of statistical nature, particles can be involved for which the particle speed is larger than the rms-speed of the scattering centers in the solar wind rest frame. For Alfvenic turbulence, this rms-speed is approximately equal to the Alfven speed.

$$v >> V_{rms} \approx V_A \quad (3)$$

Injection processes into statistical acceleration should therefore be more efficient for particles which are moving relative to the background plasma, or, which have large kinetic temperatures.

It is the goal of this paper to investigate the injection of heavy ions by following their acceleration out of the thermal background of the ambient solar wind into shock acceleration. We will first describe their average properties at thermal energies and then concentrate on the formation of supra-thermal tails and their relevance for injection into shock acceleration. We will use results from solar wind instruments on a number of spacecraft including ACE, Wind, Ulysses, and SOHO. For a detailed discussion on the relevance of these measurements for questions related to the source populations of ACRs refer to Gloeckler et al. in this issue. For a discussion of the observations reported here in the context of statistical acceleration, refer to Fisk et al. in this issue.

AVERAGE THERMAL PROPERTIES

General Comments

Heavy ions in the solar wind are generally not in thermal equilibrium with the solar wind protons. They exhibit significant differential speeds as compared to the protons and kinetic temperatures that exceed the proton temperature by large factors. However, the detailed interpretation of these data is complicated by a couple of instrumental constraints which are discussed here.

First, three-dimensional distribution functions are only available for α-particles (3, 4). All experimental data on rare ions in the solar wind published so far are based on reduced distribution functions as defined in equation (2). Differential speeds are therefore more easily observed when the magnetic field is predominantly radial. Also, the kinetic temperature along the magnetic field direction may not be the same as perpendicular to the magnetic field (4) introducing another dependence on the magnetic field direction. A detailed data-analysis of solar wind kinetic properties should therefore include, in detail, the orientation of the magnetic field relative to the instrument geometry.

Secondly, reduced distribution functions are measured during a time scale which is given by the instrument cycling time and the count-rate required for a statistical estimate of the moments. For an accurate determination of bulk speed and kinetic temperature of O ions at 1 AU, this takes on the order of 10-30 minutes. It is evident that there is some chance of broadening of these (averaged) distribution functions due to changes faster than the measurement cycle.

We will describe below the current status of the research on these non-thermal properties using recently published data.

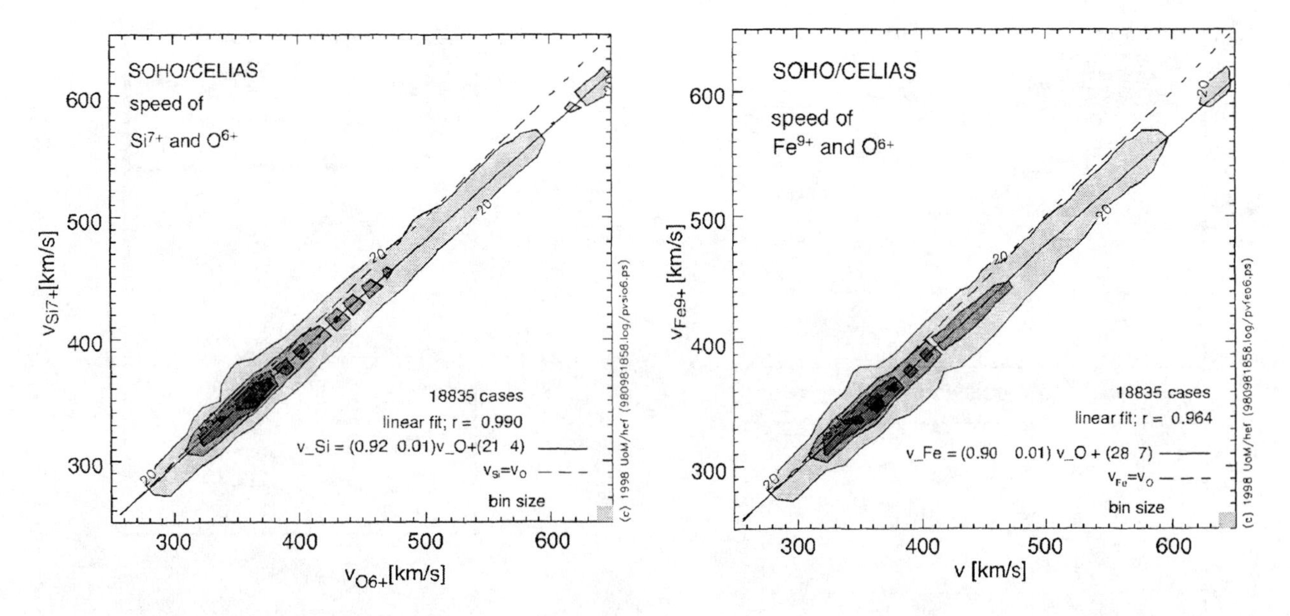

FIGURE 2. Ion bulk velocities for Si and O ions. Even though the speeds are rather identical for low-speed solar wind, there are significant differences for high-speed wind (after, 5).

Differential Speed

The differential speed of solar wind heavy ions has been extensively discussed based on observations of α-particles (3,4). In most cases α-particles are found to be faster than protons. The speed difference is typically $\Delta V \leq V_A$, the local Alfven speed. It has been pointed out that this can be understood in the framework of coherent wave-particle interactions in the heliosphere in which particles interact with outward propagating Alfven waves, being scattered in this wave-frame (4). However, Coulomb collisions are observed to play an important role in determining ΔV. When averaging over long periods of time, ΔV is well ordered by the ratio of the expansion time scale and the collision time scale. This has recently been tested (5) using a set of heavy ions with different M/Q at one particular instant. Small but significant differences are observed between different ion species, particularly at higher solar wind speeds. Figure 2 shows bulk speeds measured for a set of ions. At low solar wind speeds there are no significant differences, but in fast, coronal-hole–associated wind, there are significant differences.

The bulk speeds differ by a speed which is of the order of V_A relative to the proton speed. At any one instant, speed differences are observed to be ordered by their Q^2/M ratio. This confirms the importance of collisions. This is particularly important during time periods where the solar wind exhibits low-charge states, such as CME periods (6). Determinations of composition assuming equal speeds of all the species might therefore be misleading.

Kinetic Temperature

It has been pointed out that the kinetic temperature of heavy ions in the solar wind does not exhibit thermal equilibrium conditions (7). Enhanced temperatures are observed for heavy ions, often exhibiting constant thermal speeds independent of mass, resulting in a kinetic temperature $T \propto M$. That relationship is observed in fast, low-density solar wind where differential speeds are most significant. In low-speed solar wind, the distribution is much closer to thermal equilibrium conditions, consistent with the increasing influence of collisions on solar wind heavy ions.

This balance has recently been illustrated using Ulysses-SWICS data from a heliospheric distance of 2.5 AU (8) as shown in Figure 3. A detailed analysis of a large number of heavy ion species has been performed for both coronal hole and interstream solar wind. In the case of low-speed solar wind the distribution of kinetic temperatures are indeed close to thermal equilibrium conditions. However, for fast solar wind, a clear ordering is observed on Q^2/A and therefore the collision frequency. For a large Q^2/A the respective kinetic temperatures are closer to thermal equilibrium conditions.

The kinetic temperatures, as well as the flow speed differences, therefore seem to be controlled by a balance of coherent wave-particle interaction

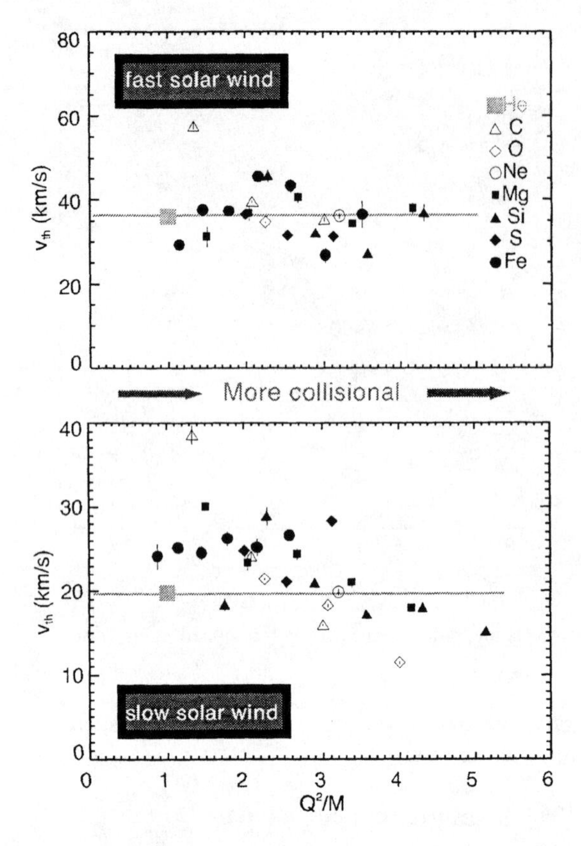

FIGURE 3. Thermal speeds of heavy ions in the solar wind. Upper panel: fast, coronal hole associated wind; lower panel: low-speed solar wind.

processes, as described before, and collisions. The enhanced relative speed and large kinetic temperatures increase the probability for injection into statistical acceleration according to equation (3).

SUPRATHERMAL PROPERTIES

Observations

The high-energy portion of distribution functions is often associated with long, extended suprathermal tails (2,8). These tails are ubiquitous and particularly obvious in low-speed solar wind.

A typical example of a reduced solar wind distribution function is shown in Figure 4. It is obvious that these data cannot be fit with a Maxwellian distribution function. Kappa-functions are more successful to fit the high-energy extensions as shown in Figure 4. In addition to using three fit parameters (density, speed, temperature), one more variable (κ) is fit which describes the structure of the high-energy tail. For $\kappa \rightarrow \infty$, κ-functions are identical to

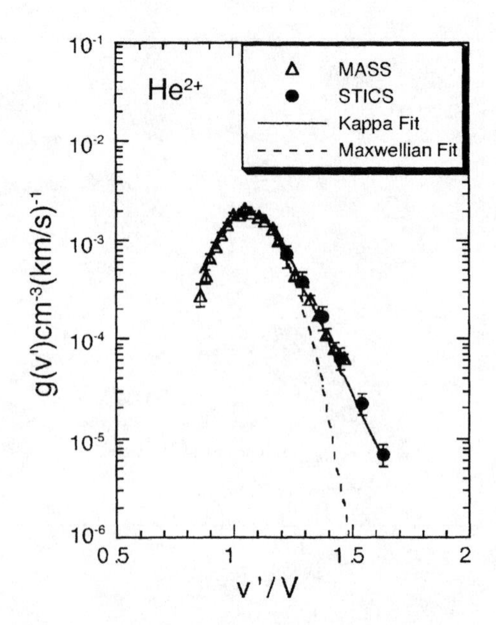

FIGURE 4. Suprathermal tails of solar wind heavy ions. The distribution function extends to larger speeds (from 2).

Maxwellians. Typical values for solar wind heavy ions are found to be $2.3<\kappa\leq 4.3$ (2,8). It has been pointed out (10) that similar supra-thermal tails are observed in distribution functions of pickup ions beyond $2V$. The exact structure of these tails, particularly their dependence on pitch-angle, is currently unknown.

Relevance of Collisions

We have indicated that Coulomb collisions are an important contribution in shaping the thermal properties of the distribution functions of heavy ions. It may be worth quantifying the collisional effects on the structure of the suprathermal extension of these distribution functions. Figure 5 shows the collisional mean-free path (mfp) for a particle of speed v normalized with the solar wind bulk speed V. The mfp is calculated for an O^{6+} ion in average slow solar wind at 1 AU (11).

This brings up the interesting possibility of a solar origin of these suprathermal tails. The situation with heavy ions would then be very similar to high-energy solar wind electrons (12). These electrons propagate scatter-free along magnetic field lines. Their signatures are therefore a direct measure for the thermal properties of their source region, the corona. The scale-height temperatures of our heavy ion distribution functions are similar to these electron distribution functions. However, due to instrumental constraints mentioned above, the angular extent of the suprathermal tail cannot be determined.

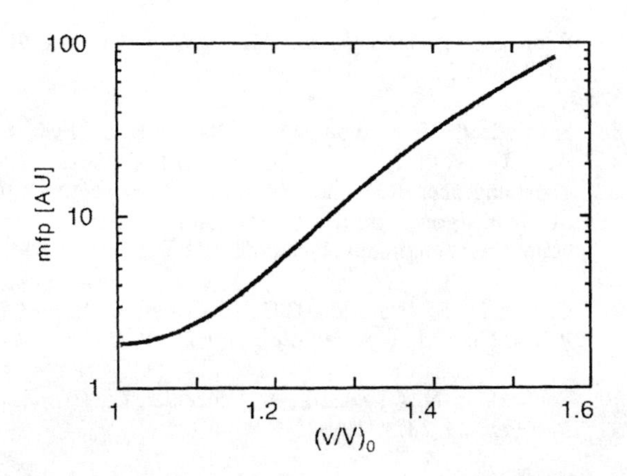

FIGURE 5. Collision mean-free path (mfp) for an O^{6+} ion in average slow solar wind around 1 AU. Even though the mfp is around 1 AU for ions in the thermal core of the distribution, it quickly increases to large distances for ions in the suprathermal tails.

Note that similar suprathermal tails are observed in distribution functions of pickup ions (10). If these tails are a signature of one and the same process, this would suggest interplanetary acceleration and not a solar source of these tails. However, the solar origin of suprathermal tails of heavy ions cannot be excluded based on the structure of the distribution function. A more detailed comparisons of the acceleration of pickup ions and solar wind heavy ions remains to be done.

Statistical acceleration is often invoked to explain the observed suprathermal tails (10, Fisk et al. this issue). The observations shown here do not specify the details of the relevant process. They have to be combined with an analysis of the observed wave properties to ascertain whether transient time damping of magnetosonic waves, second order Fermi acceleration, or another statistical process is operable.

Implications for Injection

The existence of these suprathermal tails has profound implications for the injection of these ions into any type of acceleration mechanism with a low-energy injection threshold. This can be quantified by comparing the phase space density available for acceleration at one particular injection energy for both, Maxwellian and κ-distributions, as shown in Figure 6. For this comparison it has been assumed that density, average speed, and average temperature of both functions are identical. The differences only come from a finite κ. The ranges of κ-values found in the literature are shown as well.

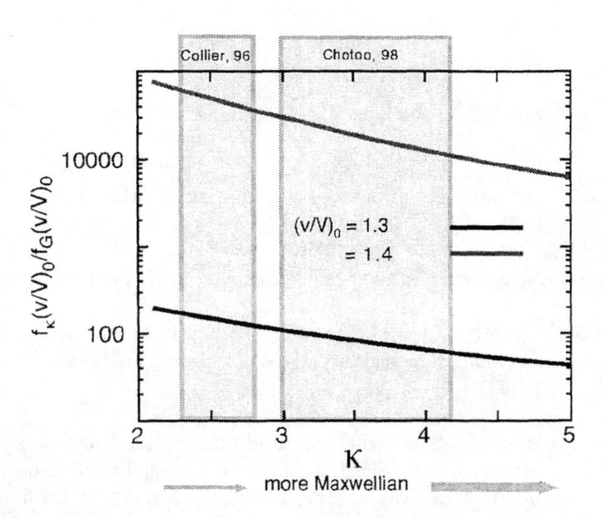

FIGURE 6. Comparison of phase space density at certain injection thresholds. The shaded areas indicate ranges of κ-values as found in the literature.

It is evident from Figure 6 that there is an enhancement of several orders of magnitude of phase space density. The ubiquitous suprathermal tails are therefore very important for the injection into shock acceleration in the heliosphere as shown in Figure 1.

SUMMARY

Heavy ions in the solar wind are observed to be a relevant source of shock-accelerated particle populations around 1 AU. However, they gain injection energy without any association with shocks. We present a survey of recent observations of non-thermal properties of heavy ions in the solar wind. These observations can be interpreted to support the following picture: At their lowest energies, particles undergo coherent wave-particle interactions, scattering in the frame of outward propagating waves. The resulting acceleration and heating provides them with enough energy to be efficiently injected into statistical acceleration. This causes ubiquitous extended tails in distribution functions with significant phase-space densities above the shock injection thresholds.

ACKNOWLEDGMENTS

This work has been supported, in part, by NASA grants NAG5-6912 and NAG5-7111, and by JPL grant #955460. We acknowledge helpful discussions with S. Hefti.

REFERENCES

1. Fisk, L. A., and Lee, M., *Astrophys. J.* **237**, 620 (1980).

2. Chotoo, K., Measurements of H^+, He^{2+}, and He^+ in corotating interaction regions at 1 AU, Thesis, University of Maryland, 1998.

3. Neugebauer, M., Fundam. *Cosmic. Phys.* **7**, 131 (1981).

4. Marsch, E., Muehlhaeuser, K.-H., Rosenbauer, H., Schwenn, R., Neubauer, F. M., *J. Geophys. Res.* **87**, 35 (1982).

5. Hefti, S., Solar wind freeze-in temperature and fluxes measured with SOHO CELIAS CTOF and calibration of the CELIAS sensors, Thesis, University of Bern, Switzerland, 1997.

6. Gloeckler, G., Fisk, L. A., Hefti, S., Schwadron, N. A., Zurbuchen, T. H., Ipavich, F. M., Geiss, J., Bochsler, P., and Wimmer-Schweingruber, R. F, *Geophys. Res. Lett.* **26**, 157 (1999).

7. Bochsler, P., Geiss, J., Joos, R., *J. Geophys. Res.* **90**, 10779 (1985).

8. Schwadron, N. A., von Steiger, R., Hefti, S., Fisk, L. A., Gloeckler, G., Ortland, D., Wimmer-Schweingruber, R. F., and Zurbuchen, T. H., University of Michigan, Internal document (http://solar-heliospheric.engin.umich.edu/publications.html), 1999.

9. Collier, M. R., Hamilton, D.C., Gloeckler, G., Bochsler, P., and Geiss, J., *Geophys. Res. Lett.* **20** (1995).

10. Schwadron, N. A., Fisk, L. A., Gloeckler, G., *Geophys. Res. Lett.* **23**, 2871 (1996).

11. Schwenn, R., in *Physics of the Inner Heliosphere I*, edited by R. Schwenn and E. Marsch, Berlin: Springer-Verlag, 1990, p. 99.

12. Marsch, E., in *Physics of the Inner Heliosphere II*, edited by R. Schwenn and E. Marsch, Berlin: Springer-Verlag, 1991, p. 45.

Sources, Injection and Acceleration of Heliospheric Ion Populations

George Gloeckler[1,2], Lennard A. Fisk[2], Thomas H. Zurbuchen[2] and Nathan A. Schwadron[2]

[1]Department of Physics and IPST, University of Maryland, College Park, Maryland 20742
[2]Department of Atmospheric, Oceanic and Space Sciences, University of Michigan, Ann Arbor, Michigan 48109-2143

Abstract. A variety of heliospheric ion populations -- from Anomalous Cosmic Rays (ACRs) to particles accelerated in Corotating Interaction Regions (CIRs) -- have been observed and studied for several decades. It had been commonly assumed that the solar wind was the source for all of these populations, except for the ACRs, and that shock acceleration produced the energetic particles observed, including the ACRs. For the ACRs the source that had been proposed a long time ago was the interstellar gas that penetrates deep into the heliosphere. Recent measurements of the composition and spectra of suprathermal ions, primarily from Ulysses, ACE and Wind, indicate that pickup ions are likely to be an important source not only of the ACRs but for other heliospheric ion populations as well. In particular, the newly discovered "Inner Source" pickup ions may be a significant source for particles accelerated in the inner heliosphere and may also be the seed material for ACR C, Mg, Si and Fe. Furthermore, the omnipresent suprathermal tails seem to tell us that shock acceleration may not be the primary mechanism energizing particles to ~0.1 MeV in the heliosphere. Explaining the origin of these persistent high velocity tails remains one of our challenges.

INTRODUCTION

Observational tools available to us today make it now possible to begin detailed exploration of acceleration mechanisms that produce a variety of heliospheric particle populations that were discovered during the last four decades. The key questions addressed, in addition to the acceleration processes, concern sources of accelerated particles and injection mechanisms. There is a clear advantage to observe these processes where they take place. What we learn in our heliosphere will have important implications to processes in other astrophysical settings for which detailed local observations are not possible.

Instruments on ACE and Ulysses provide us with the most comprehensive elemental and charge state composition measurements to date. The energy range covered by these instruments extends from typical solar wind energies (~1 keV/amu) to tens of MeV/amu. Thus velocity distributions of both the source material and the accelerated particles may be simultaneously obtained. In this report we confine ourselves to observations made with the SWICS instruments(1,2) on ACE and Ulysses, that measure the source populations from 1 to 5.4 AU, in the ecliptic and high latitudes, and cover the important velocity range where injection and the initial acceleration (pre-acceleration) occurs. SWICS measures elemental abundances, charge-state composition and speed distribution functions of ions from 0.6 to 60 keV/e on Ulysses and from 0.6 to ~100 keV/e on ACE.

OBSERVATIONS OF SOURCES

In order to determine which populations are the source of energetic particles produced in an acceleration process it is necessary first to establish the characteristics of these populations. Each source population has special characteristics which separate one from the other.

Solar Wind during Quiet Times

Shown in Figure 1 are the speed distribution functions (in the spacecraft frame) for hydrogen and helium averaged over a 65 day time period observed at 1 AU with SWICS on ACE. Time intervals around shocks and when magnetic clouds, counter-streaming electron events and magnetic holes were observed have been specifically excluded from these averaged distributions. It is remarkable that all three spectra, including He^+ pickup ions, have pronounced high-speed tails that approach a power law and persist even during time periods when no shocks and other disturbances are observed locally.

CP528, *Acceleration and Transport of Energetic Particles Observed in the Heliosphere: ACE 2000 Symposium*, edited by Richard A. Mewaldt, et al.
© 2000 American Institute of Physics 1-56396-951-3/00/$17.00

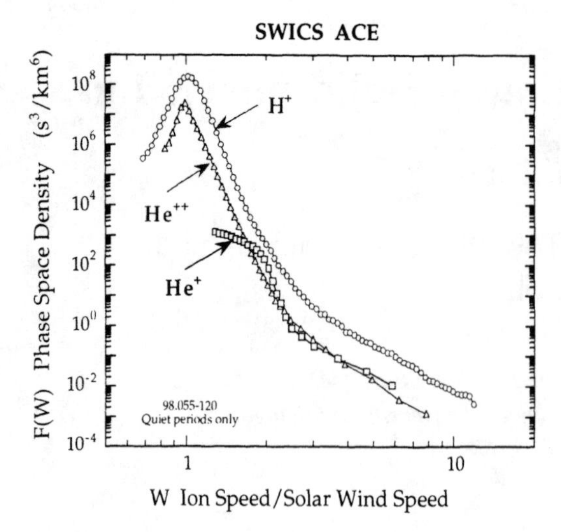

FIGURE 1. Speed distributions of H^+, He^{++} and He^+ observed in the quiet in-ecliptic solar wind at 1 AU. These spectra were measured with the SWICS instrument on ACE and averaged over the 65 day observation period (24 February to 30 April, 1998). Time periods around shocks or other disturbances were excluded from these averaged spectra (times of disturbances in the solar wind can be found at http://www.bartol.udel.edu/~chuck/ace/ACElists/obs_list.html).

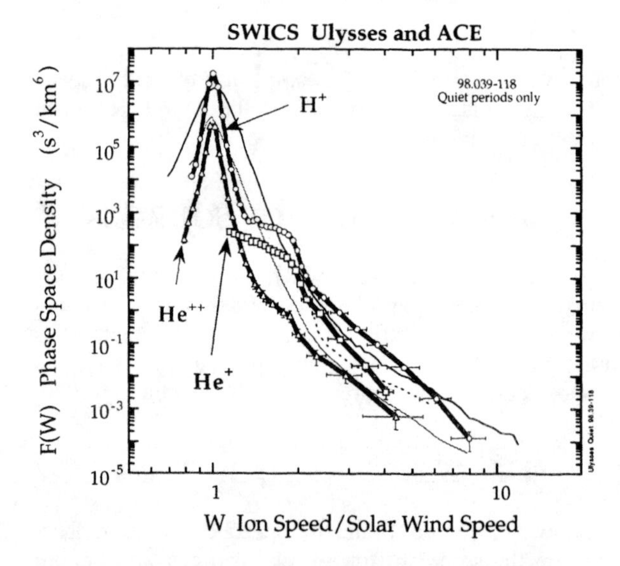

FIGURE 2. Speed distributions of H^+, He^{++} and He^+ (bold curves) observed in the quiet low-latitude solar wind at 5.4 AU with the SWICS instrument on Ulysses for approximately the same time period as in Figure 1. Ulysses and ACE were nearly radially aligned during this time. Disturbed time periods (e.g. around shocks, Forsyth and Gosling, private communication) are excluded from these averaged distribution functions. Shown for comparison from Figure 1, as the two thin curves, are the speed distribution functions of H^+ and He^{++}, reduced by 29.2 to account for the R^{-2} decrease in the solar wind density with heliocentric distance R, and of He^+, as the dashed curve, reduced by 8.5 in order to match the Ulysses spectrum below $W \approx 2$.

Figure 2 shows the velocity distributions of H^+, He^{++} and He^+ observed at 5.4 AU with SWICS on Ulysses during a time interval comparable to that in Figure 1. Ulysses, nearly radially aligned with ACE and the Sun during this time period, was sampling more or less the same solar wind as ACE. Again, time intervals around shocks and other disturbances were excluded from these averaged distributions. At 5.4 AU one observes, not only the solar wind ions and pickup He^+ seen at 1 AU, but, in addition, interstellar pickup H^+ and He^{++} at W (ion speed divided by the solar wind speed) above ~1.5. Again, each of the velocity distributions at 5.4 AU has a well-developed tail. Shown for comparison are also the scaled distribution functions from Figure 1 (see caption of Figure 2) of H^+ and He^{++}. Comparing the 1 AU spectra to the corresponding 5 AU distributions, we find that at ~5 AU the densities in the tail region ($W >$ ~2) relative to the bulk of the distribution (around $W \approx 1$) are about the same as at 1 AU, despite significant cooling clearly evident in the solar wind distributions. This implies that the tails are continuously regenerated in the out-flowing solar wind to overcome cooling of the distributions in the expanding solar wind. It should also be noted that at 5.4 AU, unlike at 1 AU, the density of He^+ for W above ~1.25 is much larger than that of He^{++}.

Interstellar Pickup Ions

Interstellar pickup ions are created by ionization of slow moving interstellar atoms inside the heliosphere(3,4). The pickup process produces suprathermal velocity distributions with a characteristic cut-off at twice the solar wind speed, $W \approx 2$. At 1 AU only inter-

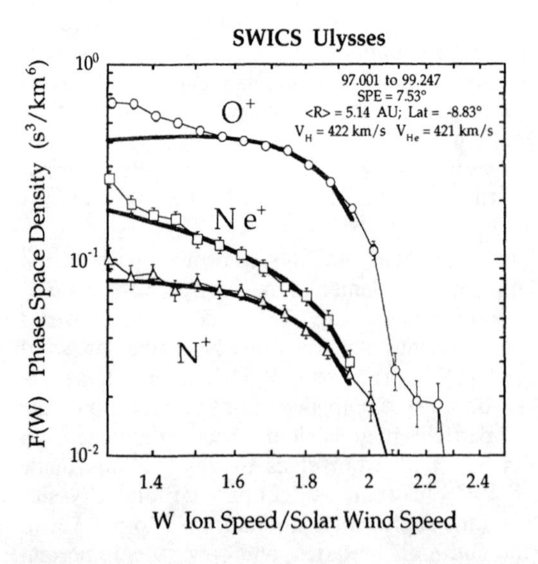

FIGURE 3. Time-averaged speed distributions of N^+, O^+ and Ne^+ observed at 5.4 AU with SWICS on Ulysses during the 2.68 year time period starting 1 Jan., 1997. The bold curves are model calculations of the predicted spectra.

stellar pickup He^+ (see Figure 1) is readily observed because atomic He, being difficult to ionize, penetrates closer to the Sun than ~0.5 AU(3,4). Beyond several AU, however, other interstellar pickup ion species begin to appear. At 5.4 AU interstellar pickup H^+ and He^{++} are visible above the suprathermal extension of the respective solar wind distributions (see also refs. 5 and 6). The velocity distributions of the heavy interstellar pickup ions, N^+, O^+ and Ne^+, are presented in Figure 3. The spectra of these ions all have the characteristic cut-off at $W \approx 2$. It is fairly straightforward to model the velocity distributions and predict the spatial profiles of the density of interstellar pickup ions(3,6,7). The shape of the velocity spectrum of a particular pickup ion species depends on the loss rate of the corresponding interstellar atoms, and the pickup ion density is the product of the production rate and the interstellar atomic density at the heliospheric termination shock(5,6). For distribution functions that are averaged over time periods of about one year or more the production and loss rates become approximately equal. In Figure 3 the fits to the data (bold curves) in the interval $\sim 1.5 < W < \sim 2$ were used to determine ionization loss rates. The densities of the corresponding interstellar atomic species at the termination shock are then found using the production rates that are assumed to be equal to the corresponding loss rates(6).

Pickup Ions from the Inner Source

Creation of interstellar pickup ions requires the presence of interstellar atoms deep in the heliosphere. Elements already ionized in the interstellar medium (for example C^+) flow around the heliopause and do not enter the heliosphere. Interstellar pickup ions of these elements are therefore not expected and have not been found. However, Geiss et al.(8) found amounts of pickup C^+ comparable to O^+ in the mass spectrum of singly-charged heavy ions with speeds between ~0.8 and 1.2 times the solar wind speed. This indicated another source for pickup ions which they called the "inner source" (8,9).

Clear evidence for the existence of the inner source is provided in Figure 4 which shows the distribution functions of O^+, C^+ and Ne^+ observed with SWICS on Ulysses at low and middle latitudes at an average heliocentric distance of ~1.5 AU(10). This close to the Sun, the density of interstellar pickup O^+ and Ne^+ is small because of removal of interstellar oxygen and neon atoms by ionization. Yet all three spectra show orders of magnitude higher densities at W around 1 then near the $W \approx 2$ cut-off. While C^+ has no detectable interstellar component, as is expected, its density at W between ~0.8 and ~1.2 is about the same as that of O^+.

The distribution functions of inner source pickup ions have certain characteristics: (1) they peak at or below the solar wind peak at $W = 1$, and (2) they are relatively cold, in that their effective thermal speed would be small compared to the solar wind speed, in clear contrast with the usual interstellar pickup ion distributions. The interstellar pickup ion distribution is quite broad and extends with nearly equal phase space density up to $W \approx 2$, beyond which the density drops(3,4). Interstellar pickup ions gyrate about the magnetic field in the solar wind immediately following their ionization(3). Although initially they will propagate inward in the frame of the solar wind, upon scattering they can reverse direction, in which case they can acquire speeds (in the spacecraft reference frame) up to the sum of twice the solar wind speed and the initial speed of the slow moving (compared to the solar wind) neutrals: that is, $W \approx 2$.

The same process should occur for the inner source pickup ions. However, they are most likely produced from neutrals emitted from heliospheric grains(10,11). Because neon is clearly present in the inner source pickup ions it is believed that the emitted neutrals

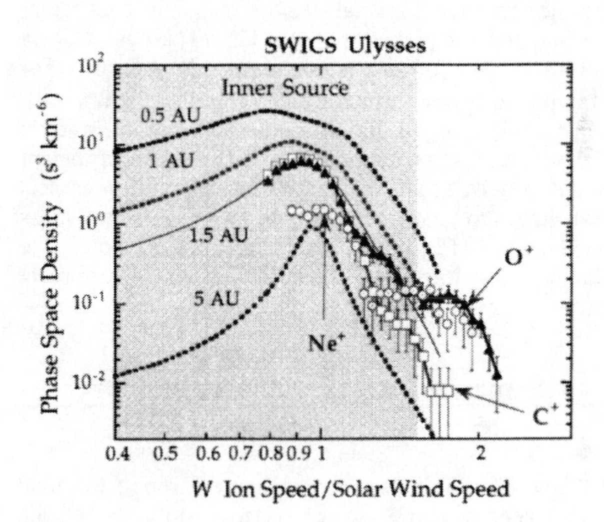

FIGURE 4. Time-averaged speed distributions of C^+, O^+ and Ne^+ observed from November 21, 1994 to May 30, 1995 at low and middle latitudes (<60°) and 1.5 AU with SWICS during the fast latitude scan of Ulysses. The distributions are plotted versus W, the speed of the ions relative to the solar wind speed, in a stationary reference frame relative to the spacecraft. The four curves are model calculations for O^+ and show the effects of strong cooling as evidenced by the narrowing of the peaks at $W \sim 0.9$ with increasing distance.

include 're-cycled' solar wind material. Solar wind ions are constantly absorbed by the grains and then re-emitted as slow moving atoms and molecules. Schwadron et al.(11) have shown that the absorption, re-emission, and ionization processes that produce the observed inner source pickup ion distributions occur near the Sun, and hence the inner source pickup ions suffer substantial cooling. The distribution functions in Figure 4 are clearly weighted toward speeds $W < 1$. The particles thus suffer relatively little scattering and retain their inward motion in the frame of the solar wind(7). The expansion of the solar wind, however, tends to reduce the particle speeds perpendicular to the magnetic field and substantially cool the distributions(11).

Modeling of the expected velocity distributions of the inner source pickup ions and matching them to the spectra observed at some fixed distance from the Sun allows us to estimate the radial profiles of the density of the neutral particles that produce a given inner source pickup ion species(11). Then we can predict the pickup ion distribution functions at other distances and the radial profile of pickup ion density. The model curves shown in Figure 4 represent the predicted velocity distributions of oxygen at the indicated radial distances.

Other pickup ion populations that could potentially become a source of accelerated energetic particles include planetary and cometary pickup ions. These are, however, very local sources, with their extended plasma tails confined to within several degrees of a line connecting the Sun and the planet or comet(12). The composition of these pickup ion populations is highly variable. For example, there is virtually no C^+ in the Earth's tail, while lowly charged sulfur ions are abundant in Jupiter's plasma tail. The composition of pickup ions in distant cometary tails seems to be the same for the few comets that have so far been studied with in situ mass spectrometers(12). For cometary pickup ions the abundance of nitrogen is low, and neon is extremely rare.

Comparison of Abundances of Source Populations

Except for the inner source, Figure 5 shows the total abundance of each source relative to its oxygen abundance. For the inner source, the abundance is relative to neon and divided by 10 for ease of comparison. Relative abundances of the three source populations(6,10) and of cometary tail pickup ions(12) are compared in Figure 5 to the sample-averaged composition of energetic (~2 to 20 MeV/amu) particles accelerated in Corotating Interaction Regions, CIRs(13,14). From comparison of the composition alone it is clear that a cometary source can be ruled out as a major contributor to the CIR accelerated energetic particles. This is not surprising because cometary sources are localized and hence do not add much to the source populations of CIR particles. On the other hand, some combination of roughly equal amounts from the three other sources (solar wind, interstellar and inner source pickup ions) could reproduce the CIR composition within a factor of two. We would therefore argue that using the similarity in compositions as the only criterion for deciding on the source material for accelerated particles is insufficient.

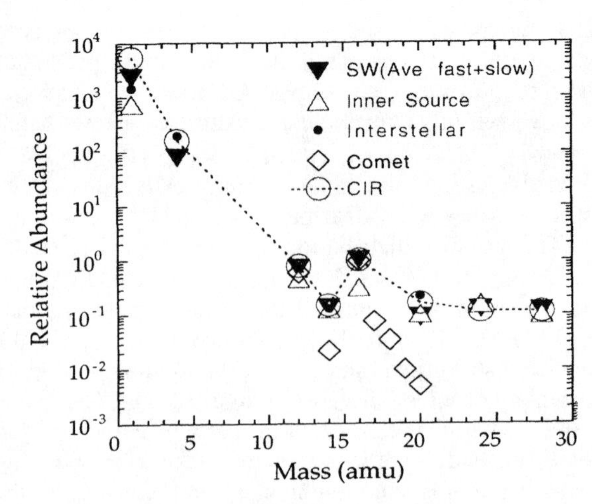

FIGURE 5. Relative abundances from four potential sources of energetic particles compared to the average composition of CIR accelerated particles.

Velocity Distributions in the Turbulent Solar Wind

Speed distributions of H^+, He^+ and He^{++} were measured(15) with SWICS/Ulysses downstream of the forward (FS) and reverse (RS) shocks in the CIR of late December 1992. Here we summarize several of the remarkable features apparent from the speed spectra shown in Figure 7 of reference 15. First, it was found that more He^+ than He^{++} is accelerated even though solar wind alpha particles are at least a factor of 10^3 more abundant than pickup He^+. Second, the spectral shapes in the high-speed ($W > 2.4$) tails behind the reverse shock were identical (within experimental uncertainties) to those behind the forward shock, even though the two shocks were different. The FS was weaker than the RS, and one was quasi-perpendicular while the other quasi-parallel(16). It was also evident from the speed distributions that the solar wind behind the reverse shock was heated more than the wind downstream of the weaker forward shock.

Downstream of the CIR reverse shock the shapes of the speed distributions above $W \sim 2.4$ were found to be identical (within experimental uncertainties) for all three species. This, combined with the fact that the FS and RS spectra also had the same shapes, implies that the injecting/pre-acceleration mechanism depends primarily on ion speed. In addition, the spectral shapes were found not to be simple power laws as would be predicted by standard shock acceleration models, providing further evidence that the high-velocity tail distributions that had been observed were not produced by a simple shock acceleration mechanism(17).

Figure 6 shows the speed distribution of H^+, He^+ and He^{++} measured with SWICS/ACE downstream of the large forward shock which occurred at ~0215 on May

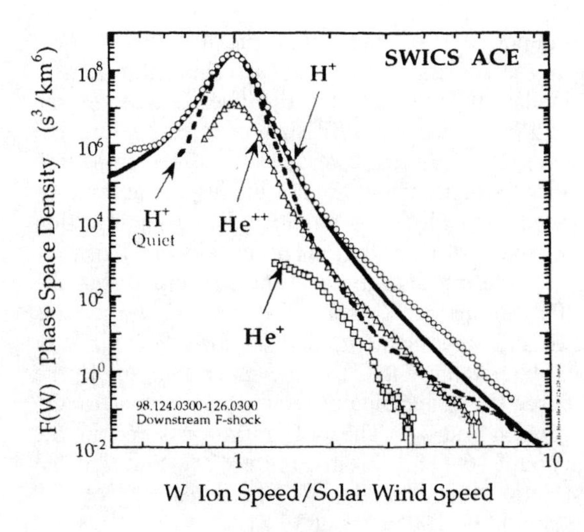

FIGURE 6. Speed distributions of H^+, He^{++} and He^+ observed during a two-day period (4 May 0300 to 6 May 0300, 1998) behind the large forward shock of 4 May, 1998. The quiet-time proton spectrum from Figure 1 is shown for comparison (dashed curve). The density of He^+ is below that of He^{++} for all W and its tail is steeper than that He^{++}.

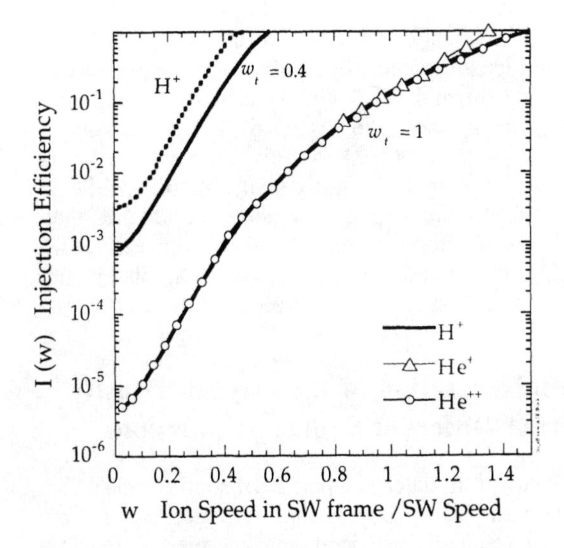

FIGURE 7. Injection efficiency vs. normalized ion speed in the solar wind frame for H^+, He^{++} and He^+ derived from speed distributions given in Figure 6 in the downstream region of the 4 May, 1998 shock. See text for description of the method used to compute the efficiencies.

4, 1998. The kappa function(6) shown as the solid curve, ($\kappa = 3.2$) is an excellent fit to the observed H^+ spectrum for $\sim 0.6 < W < \sim 1.5$. Outside this range, the proton density is above the kappa fit. The solar wind behind this strong shock is both hotter and has well developed tails compared to the solar wind during quiet times (see Figure 1 and dashed curve of Figure 6). The density of interstellar He^+ is well below that of the solar wind for all W. This is in contrast to the quiet-wind cases both at 1 AU, where the He^+ and He^{++} densities are comparable in the tail regions, and at 5.4 AU, where the He^+ density above $W > \sim 1.25$ is larger than the He^{++} density. The tail of He^+ downstream of the shock is steeper than the solar wind H^+ and He^{++} tails.

Injection Efficiency

Knowledge of the injection efficiencies and of the distribution functions (and not just the composition) of the source populations is required in order to determine the origin of accelerated particles. Injection efficiencies are not easily defined and difficult to measure or compute. Here we use a simple definition of the injection efficiency which will allow us to use measurements of the speed distribution of ions in the turbulent solar wind (e.g. downstream of a shock) to estimate injection efficiencies. Consider an ion of a certain species moving at some given speed, w_i, where w_i is the ion speed in the solar wind frame divided by the solar wind bulk speed. We define this ion to be injected for further acceleration (e.g. by a shock) if its speed is increased to a value above some given threshold speed, w_t. The injection efficiency, $I(w_i)$, is thus defined to be the integrated phase space density of all ions of that species above the threshold speed w_t, divided by the product of the phase space density of ions at w_i, times the number of w_i steps below w_t. With this definition we assume that the initial distributions have much lower density above w_t than the final distributions. This seems to be the case if the initial distributions are similar to the quiet-time spectra (e.g. compare in Figure 6 the quiet-time H^+ spectrum to the corresponding H^+ distribution in the downstream region).

Figure 7 shows the injection efficiencies of H^+, He^+ and He^{++} for $w_t = 1.0$, and also H^+ for $w_t = 0.4$, computed as defined above using the speed distribution functions shown in Figure 6. The turbulent solar wind behind the shock was chosen because ions from distributions in this region are most likely to be injected for further acceleration at the shock. For injection threshold speeds of $w_t = 0.4$ we were able to estimate $I(w_i)$ for protons moving inward in the solar wind frame ($W < 1$, toward the Sun, dotted curve) as well as for those moving outward ($W > 1$, solid curve). It appears that the inward-moving protons are injected more efficiently than those moving outward. For a 0.4 injection threshold speed the injection efficiency for thermal solar wind protons (those with $w < \sim 0.1$, a typical value of the thermal speed divided by the bulk speed) is a fraction of one percent.

For $w_t = 1$ and ions moving outward, $I(w_i)$ was computed for H^+ (solid curve), He^+ (solid triangles) and He^{++} (open circles). The three efficiency curves are nearly identical, implying a purely speed-dependent injection process with no dependence on the mass or mass/charge of the ion. With the injection threshold speed equal to the bulk solar wind speed (~630 km/s in this case) the injection efficiency of thermal ($w < \sim 0.1$) solar wind ions is about 10^{-5}. This very low efficiency for injecting thermal solar wind ions is consistent with our observation (shown in Figure 6) that the density of He^+ in the tail ($W > \sim 2$) is only about an order of magnitude lower than the tail density of He^{++} while at $W = \sim 1$ the density of He^{++} is about 10^4 higher than that of He^+. Because all efficiencies increase with increasing w, suprathermal ions, such as interstellar pickup ions, are expected to be preferentially injected.

Radial Variation of the Oxygen Density from Different Source Populations

The amount of material a particular source contributes to accelerated particle populations will depend on how far from the Sun the acceleration takes place. Because the density of interstellar pickup ions peaks at large heliocentric distances (~5 AU or more, except for He), interstellar pickup ions will be an important source of particles accelerated in the outer heliosphere. Conversely, because the density of the solar wind and of inner source pickup ions is highest close to the Sun these ions will contribute most of the material for particles accelerated in the inner heliosphere. The radial dependence of the density of oxygen from the three sources is shown in Figure 8. The density profile of inner source O^+ is derived from best fits of the model distribution function to the observed velocity distribution of O^+ (solid curve in Figure 4). The solar wind O^{6+} density (dashed line) has been reduced by a factor of 2000 to aid in comparison between the three distributions. The interstellar O^+ density was computed using ionization loss rate and atomic oxygen density obtained from best fits to the velocity distribution of O^+ shown in Figure 3.

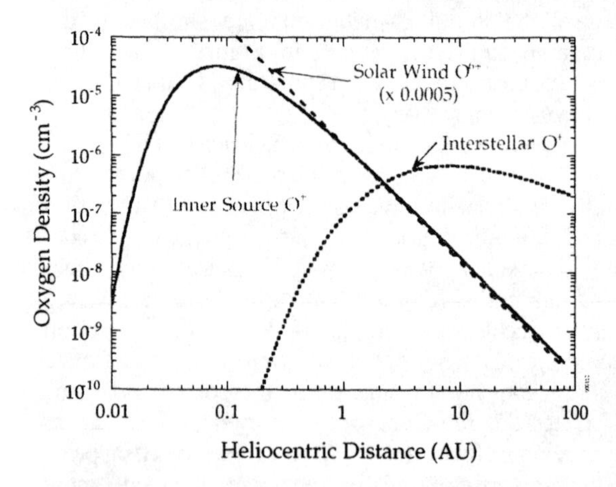

FIGURE 8. Predicted variation with heliocentric distance of the densities of inner source O^+, interstellar pickup O^+ and solar wind O^{6+}. For better comparison we reduced the solar wind oxygen density by a factor of 2000.

The importance of interstellar pickup ions beyond several AU is illustrated in Figure 8. There the density of interstellar O^+ exceeds that of the inner source and the solar wind, provided that efficiency of injecting thermal solar wind O^{6+} is less than $5 \cdot 10^{-4}$. Such injection efficiencies for thermal ions are not unreasonable, and, as can be seen from Figure 7, would indicate an injection threshold speed of $w_t \approx 0.5$. These calculated density profiles in Figure 8 also indicate the likelihood that the source for Anomalous Cosmic Ray (ACR) carbon (as well as Si and other elements that are already ionized in the Local Interstellar Cloud) is the inner source. In the inner source the abundance of C^+ is comparable to O^+ (see Figures 4 and 5). The O/C ratio measured in the ACR is about 160(18). From Figure 8 we find that the density ratio of interstellar O^+ to inner source C^+ (or O^+) should be 160 at about 30 AU.

Comparison of Velocity Distributions of Source Populations

In deciding on the source of a given energetic particle population it is not sufficient to compare the relative abundance of the accelerated ion population to that of one of the source populations. When two sources have a similar composition one must examine their velocity distributions at a given location in order to decide which is preferentially injected. In Figure 9 we compare the predicted distribution functions of inner source O^+ (from Figure 4) and interstellar O^+ at 1 AU with the velocity distribution of solar wind O^{6+} measured with ACE at 1 AU. The phase space density of solar wind oxygen exceeds that of the other two sources for $W > \sim 0.6$. For $W < \sim 0.6$ the inner source O^+ density is larger than that

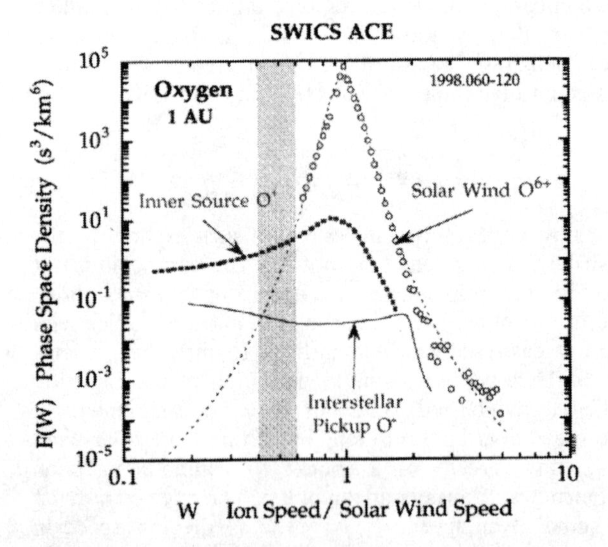

FIGURE 9. Comparison of the predicted velocity distributions, at 1 AU, of inner source O^+ and the interstellar O^+ with the solar wind O^{6+} spectrum measured with SWICS on ACE during the same time period as in Figure 1.

of the other two sources. For particles accelerated at 1 AU, interstellar O^+ is not a significant source since its density is below the other two sources for all speeds. The relative contributions of the other two sources to energetic particles accelerated at 1 AU will depend critically on the injection efficiency, which depends on ion speed W (see Figure 7). Thus, assuming for simplicity that injection efficiency is 0 for $\sim 0.6 < W < \sim 1.5$ and 1 for all other W, then inner source O^+ will be the principal source of accelerated particles. However, if the injection efficiency is zero only inside the smaller interval $\sim 0.75 < W < \sim 1.3$ then the solar wind will be the dominant source.

The situation is entirely different at other locations in the heliosphere. In Figure 10 we show what happens at 5 AU. Again we compare the predicted spectra of inner

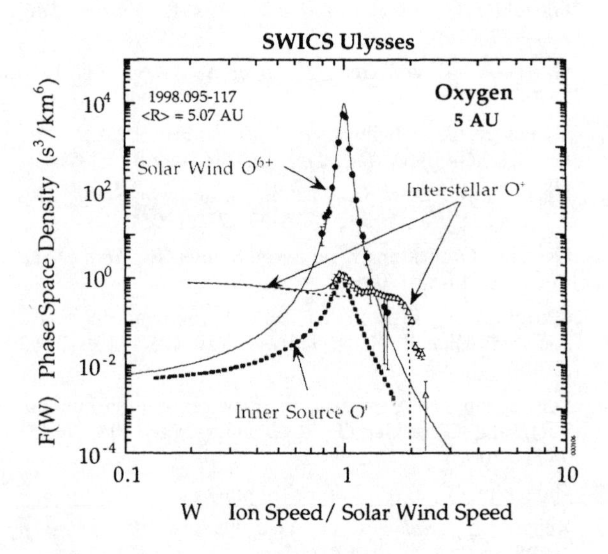

FIGURE 10. Comparison of the predicted speed distributions of inner source O^+ and interstellar O^+ at 5 AU with the solar wind O^{6+} spectrum measured with SWICS on Ulysses at 5.07 AU from April 5 to April 27, 1998.

source and interstellar O^+ with that of solar wind O^{6+}, measured by Ulysses at ~5.1 AU. Now the density of the interstellar pickup O^+ exceeds that of solar wind O^{6+} for $W > \sim 1.4$ and $W < \sim 0.6$. For ions, such as C^+, not present in the interstellar source, the inner source may still contribute even at 5 AU if the injection efficiency is small for $W > \sim 0.1$.

SUMMARY AND CONCLUSIONS

One of our most important findings is that the speed distributions of H^+, He^+ and He^{++} in the in-ecliptic solar wind free of shocks and other disturbances have well developed high-speed tails indicating the presence of a population of highly suprathermal ions at all times. These tails are not formed in the corona since they are seen in the distributions of interstellar pickup He^+. Furthermore, the density in the tails decreases no faster than R^{-2} with distance R from the Sun, as is most clearly seen in Figure 2 by comparing the He^{++} tails measured at 1 and 5 AU. This implies that the quiet-time high-speed tails are continuously regenerated to overcome continuous cooling with increasing distance. Such cooling is clearly visible in the bulk solar wind which at 5 AU has narrower peaks than at 1 AU (see Figure 2).

One of the open questions then is how these ubiquitous ions, with speeds nearly ten times that of the solar wind, are continuously created in the low-latitude solar wind, and apparently not the high-latitude fast wind(15), at times when there are no shocks, waves or other solar wind disturbances observed locally. Because of their slow mobility, it is unlikely that these ions were accelerated by distant shocks and then transported to the shock free regions of the low-latitude heliosphere, although this possibility cannot be excluded. Fisk et al.(19) address this problem.

Downstream of shocks, suprathermal tails increase in strength by one to two orders of magnitude(15). The amount of the increase most likely depends on the strength of the shock. Our observation that the acceleration process produces spectra of similar shapes despite the large difference in the shock parameters once more suggests that the process responsible for the formation of tails is not due to the shocks themselves but rather to the turbulence in the downstream regions of the shocks as was also reported previously(15,17,20). We suggest that the initial formation of the high-speed tails in the quiet solar wind and the subsequent strengthening of these tails in the turbulent solar wind downstream of shocks is the first step leading to acceleration of particles to energies above ~0.1 MeV/amu. In this sense, this initial step that energizes ions to many tens of keV/amu could be considered the injection process.

We now come to the question of what material is injected, and then further accelerated to MeV energies. The fact that more He^+ than He^{++} is observed above $W \approx 2$ both in the quiet in-ecliptic solar wind and in the turbulent regions of CIRs at 4 - 5 AU(15) excludes the bulk solar wind, that is, around $W \approx 1$, as a significant source of CIR accelerated energetic particles. As shown in Figure 7, interstellar pickup ions, in particular He^+, are efficiently injected and accelerated in CIRs. On the other hand, the injection efficiency of the thermal solar wind (ions within one thermal speed of the bulk speed) seems to be quite small ($< \sim 10^{-3}$). Just how much of each of the three source populations (interstellar pickup ions, inner source pickup ions and suprathermal solar wind ions) is injected for further acceleration will depend on the relative fraction of these populations in the suprathermal range, where the injection efficiency becomes large, around $w > 0.5$.

The mix of sources will also change with heliocentric distance (see Figures 8, 9 and 10) and with solar wind speed. Thus, the interstellar source should become dominant in the outer heliosphere at distances beyond 5 to 10 AU. The inner source may be as important as the suprathermal solar wind in the inner heliosphere, from ~0.5 to ~ 2 to 4 AU. Also, when the solar wind speed is high, interstellar pickup ion distributions extend to higher speeds, and at twice the solar wind speed they may again have densities comparable to or higher than those of suprathermal solar wind ions at $W \approx 2$. It seems to be the case that in CIRs at ~3 - 6 AU, the heated solar wind distributions for H, He and O and the corresponding interstellar pickup ion distributions have roughly comparable densities near $W \approx 2$ and therefore relatively small changes in either the solar wind thermal speed or its bulk speed will affect the proportion of solar wind versus pickup ions accelerated in CIRs. Observations of the larger relative abundance of energetic (MeV) H and He particles in CIRs (e.g. ref. 14) compared to that in the solar wind support our conclusion that pickup ions form a significant part of the seed population of energetic particles accelerated in these turbulent regions.

Finally, we come to the question of acceleration and of the source population of ACRs. It is well established that the source of the main components of ACRs are interstellar pickup ions(21). Inner source pickup ions, however, could well be the source for the rare ACRs(18) (C, Mg, Si, S and Fe). While much of the acceleration to hundreds of MeV is likely to take place at the heliospheric termination shock, some acceleration of pickup ions inside the heliosphere is required in order to compensate for cooling with increasing heliocentric distance. The same mechanisms that produce suprathermal tails in the quiet low-latitude solar wind in the absence of shocks could also accelerate interstellar and inner source pickup ions in the outer heliosphere right up to the termination shock. Another possibility, suggested by Fisk(22), is acceleration at distances beyond ~10 AU at the interface between the low-speed solar wind in the ecliptic and the high-speed wind from the polar coronal holes at higher latitudes. Presumably, pickup ions in the turbulence associated with this interface would be accelerated as efficiently as in CIRs. Both process should be solar-cycle dependent. High-velocity tails seem to be most pronounced in the slow solar wind, which dominates the heliosphere at solar maximum. The slow-fast wind interface, on the other hand, is most extensive at solar minimum, characterized by the permanent high-speed wind from the large coronal holes.

ACKNOWLEDGMENTS

We gratefully acknowledge the essential contributions of the many individuals (see Gloeckler et al.(1,2)) at the University of Maryland, the University of Bern, the Max-Planck-Institute für Aeronomie and the Technical University of Braunschweig which assured the success of the SWICS experiments on ACE and Ulysses. Of particular benefit have been the many illuminating discussions with Johannes Geiss. We thank Christine Gloeckler for her help with data reduction. This work was supported in part by NASA/Caltech grant NAG5-6912 and NASA/JPL contract 955460.

REFERENCES

1. Gloeckler, G., *et al.*, *Space Sci. Revs.* **86**, 495-537 (1998).
2. Gloeckler, G., *et al.*, *Astron. Astrophys. Suppl. Ser.* **92**, 267-289 (1992).
3. Vasyliunas, V. M., and Siscoe, G. L., *J. Geophys. Res.* **81**, 1,247-1,252 (1976).
4. Gloeckler, G., *Space Sci. Revs.* **78**, 335-346 (1996).
5. Gloeckler, G., Fisk, L. A., and Geiss, J., *Nature* **386**, 374-377 (1997).
6. Gloeckler, G., and Geiss, J., *Space Sci. Revs.* **86**, 127-159 (1998).
7. Gloeckler, G., Schwadron, N. A., Fisk, L. A., and Geiss, J., *Geophys. Res. Lett.* **22**, 2665-2668 (1995).
8. Geiss, J., Gloeckler, G., Fisk, L. A., and von Steiger, R., *J. Geophys. Res.* **100**, 23,373-23,377 (1995).
9. Geiss, J., Gloeckler, G., and von Steiger, R., *Space Sci. Revs.* **78**, 43-52 (1996).
10. Gloeckler, G., Fisk, L. A., Geiss, J., Schwadron, N. A., and Zurbuchen, T. H., *J. Geophys. Res.* **105**, 7459-7463 (2000).
11. Schwadron, N. A., Geiss, J., Fisk, L. A., Zurbuchen, T. H., and Gloeckler, G., *J. Geophys. Res.* **105**, 7465-7472 (2000).
12. Gloeckler, G., *et al.*, *Nature* (in press).
13. Keppler, E., *Surveys in Geophysics* **19**, 211-278 (1998).
14. Mason, G. M., and Sanderson, T. R., *Space Sci. Revs.* **89**, 77-90 (1999).
15. Gloeckler, G., *Space Sci. Revs.* **89**, 91-104 (1999).
16. Balogh, A., *et al.*, *Space Sci. Revs.* **72**, 171-180 (1995).
17. Gloeckler, G., *et al.*, *J. Geophys. Res.*, **99**, 17,637-17,643 (1994).
18. Cummings, A. C., Stone, E. C., and Steenberg, C. D., "Composition of Anomalous Cosmic Rays and Other Ions from Voyager Observations", in *Proceedings of the 26th International Cosmic Ray Conference* **7**, Salt Lake City, Utah, 1999, 531-534.
19. Fisk, L. A., Gloeckler, G., Zurbuchen, T. H., and Schwadron, N. A., "Statistical Acceleration of Suprathermal Particles in the Solar Wind", in these *Proceedings of the ACE 2000 Symposium*, 2000.
20. Schwadron, N. A., Fisk, L. A., and Gloeckler, G., *Geophys. Res. Lett.* **23**, 2871-2874 (1996).
21. Fisk, L. A., Kozlovsky, B., and Ramaty, R., *Astrophys. J. Lett.* **190**, L35-L38 (1974).
22. Fisk, L. A., *Space Sci. Revs.* **78**, 129-136 (1996).

Ubiquitous Statistical Acceleration in the Solar Wind

L. A. Fisk, G. Gloeckler, T. H. Zurbuchen, and N. A. Schwadron

Department of Atmospheric, Oceanic, and Space Science
University of Michigan, Ann Arbor, MI 48109

Abstract. One of the more interesting observations by ACE is the ubiquitous presence of higher energy tails on the distribution functions of solar wind and pickup ions. The tails occur continuously in the slow solar wind, but less so in fast wind. Their presence is not correlated with the passage of shock waves. It is pointed out that statistical acceleration by transit-time damping of propagating magnitude fluctuations in the magnetic field of the solar wind is a likely mechanism to yield the observed tails.

INTRODUCTION

In companion papers in this Proceedings, Gloeckler et al. (1) and Zurbuchen et al. (2) provide stunning new observations of the ubiquitous presence of pronounced tails on the distributions of ions in the solar wind, both of solar wind origin and pickup ions. Shown in Figure 1 is a summary of these observations, for hydrogen and singly and doubly charged helium, as seen by the SWICS instrument on ACE. Similar distribution functions are observed from Ulysses/SWICS during its passage through the solar equatorial plane. There are clear tails on all three distributions, extending from essentially solar wind thermal energies, up to the limits of these instruments; in the case of hydrogen up to ten times the solar wind flow speed. From ACE at 1 AU, the hydrogen tail is due to accelerated solar wind protons. From Ulysses at 4-5 AU, interstellar pickup hydrogen is present and accelerated. Singly charged helium is an interstellar pickup ion, and is accelerated at the locations of both ACE and Ulysses. The acceleration of the pickup ions indicates that the tails on the distribution functions are created in the solar wind, and are not the remnant of some coronal process.

The most significant point concerning these tails is that in the slow solar wind, near the solar equatorial plane, the tails on the thermal distributions of solar wind and pickup ions are always present. They do not appear to occur as commonly at the higher latitudes seen by Ulysses (3). However, they are ubiquitous in the slow solar wind. Most important for deciding on an explanation for their origin, they are unrelated to the presence or passage of any shock wave. These tails are continuous from thermal speeds of order the solar wind flow speed, up to ten times that value. At the lower speeds, the particles cannot propagate far in the solar wind, but rather they are transported simply by convection with the solar wind. Such particles do not arrive at earth from some distant shock such as the standing shocks that bound Co-rotating Interaction Regions (CIRs) since they lack the upstream speed to do so. We are forced to conclude, therefore, that these particles are not generated by shocks.

The only alternative acceleration mechanism is statistical acceleration, in which the particles interact with random electric fields present in the solar wind. We will conclude in this paper that only transit time damping of propagating magnitude fluctuations in the solar wind, as proposed by Fisk (4), is likely to cause the observed acceleration. There are continuous sources of magnitude fluctuations in the solar wind, some of which are described, and the transit time damping mechanism is quite efficient.

With regard to higher-energy particles, in the MeV range, shock acceleration is undoubtedly the preferred mechanism. There is a well known correlation of accelerated particles with shocks in CIRs (5). Although the particles observed in the tails of the thermal distributions do not originate in shocks, they could serve as the particles that are injected into a shock acceleration mechanism. Indeed, the particles

CP528, *Acceleration and Transport of Energetic Particles Observed in the Heliosphere: ACE 2000 Symposium*, edited by Richard A. Mewaldt, et al.
© 2000 American Institute of Physics 1-56396-951-3/00/$17.00

well up on the tails, with higher speeds, should be able to be readily injected into a standard diffusive shock acceleration mechanism.

We begin by reviewing the general properties of statistical acceleration, and the requirements for the efficiency of the acceleration that is needed to produce the observed tails. We then consider the statistical acceleration that accompanies pitch-angle scattering by Alfven waves, and from transit-time damping of magnetosonic waves, and conclude only the latter is viable. Finally, we discuss possible sources of magnetosonic waves in the solar wind.

GENERAL PROPERTIES OF STATISTICAL ACCELERATION

Statistical acceleration can always be expressed as a diffusion in momentum space, by, e.g., the following equation:

$$\mathbf{V}\cdot\nabla f = \nabla\cdot\kappa\cdot\nabla f + \frac{1}{p^2}\frac{\partial}{\partial p}\left(p^2 D_{pp}\frac{\partial f}{\partial p}\right) + \frac{p}{3}\nabla\cdot\mathbf{V}\frac{\partial f}{\partial p} \qquad (1)$$

The equation is written as a steady state. The term on the left side describes convection with the solar wind velocity, **V**. The first term on the right describes spatial diffusion; the second term on the right is diffusion in momentum, or the statistical acceleration; and finally the third term on the right is due to adiabatic deceleration in the solar wind. The spatial diffusion term needs to be treated with some caution in the case of the low velocity particles shown in Figure 1. Spatial diffusion is usually derived in the limit that the particle speed is large compared to the plasma flow speed. The statistical acceleration term is probably reasonably accurate so long as the particle speed is larger than the characteristic speeds of the waves or turbulence with which the particles interact.

Statistical acceleration is governed by a diffusion coefficient in momentum space D_{pp}. The statistical acceleration must be expressed only in terms of derivatives of the distribution function in momentum, because the Vlasov or Liouville's equation that was the origin of equation (1) contained only derivatives of the distribution with momentum.

Statistical acceleration can also be described as a diffusion in energy space, with diffusion coefficient, D_{TT}. The two diffusion coefficients are of course related to each other, as $D_{TT}=v^2D_{pp}$. In the case of diffusion in energy, there is also a mean change in energy. There are no negative energies, and so as particles diffuse in momentum, the mean energy increases at the rate $D_{TT}/2T$. When evaluating a statistical acceleration mechanism any of these various forms can be used since they are all interrelated.

To create the tails shown in Figure 1, the diffusion in momentum or energy must have a characteristic time of order the solar wind convection time, or equivalently,

$$\frac{T^2}{D_{TT}} \approx \frac{r}{V} \approx 3.75\times 10^5 \text{ sec} \qquad (2)$$

As can be seen in equation (1), the time change in the distribution function due to statistical acceleration will then be comparable to the time during which particles are convected through the heliosphere, and a significant tail can be produced at earth.

STATISTICAL ACCELERATION BY PITCH ANGLE SCATTERING

One possibility for a statistical acceleration mechanism is pitch-angle scattering from moving magnetic irregularities, as was discussed by, e.g. Jokipii (6) and Wibberentz and Beuermann (7). The concept here is quite simple. Particles are scattered in

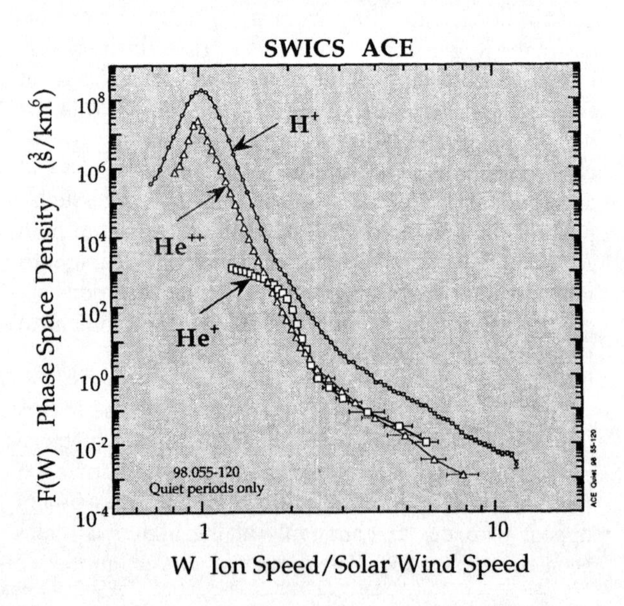

Figure 1. The distribution functions of H^+, He^+, and He^{++}, as seen by the SWICS instrument on ACE (1).

pitch angle presumably from Alfven waves, which move randomly. Like tennis balls bouncing off a moving wall, the reversal of the particle's velocity in the frame moving with the wave causes an energy change in the solar wind frame. As discussed by Jokipii (6) and by Fisk (4), the relevant diffusion coefficient in energy space is

$$D_{TT} = V_{Alfven}^2 T^2 / \kappa_{parallel} \qquad (3)$$

Here V_{Alfven} is the Alfven speed and $\kappa_{parallel} = v\lambda/3$ is the diffusion coefficient for pitch-angle scattering parallel to the mean magnetic field, with λ the mean free path.

The difficulty with this mechanism is the mean free path required for this diffusion coefficient to meet our criteria of being important on the time scale for convection. The value for the mean free path required to satisfy equation (2) is

$$\lambda = \frac{3V_{Alfven}^2 r}{vV} < 5\times10^{-2}\,\text{AU} \qquad (4)$$

Recall that pick up ions in particular are known to have mean free paths of order 1 AU (8), i.e., very little pitch angle scattering. Statistical acceleration by pitch angle scattering is not a viable acceleration mechanism in this instance.

STATISTICAL ACCELERATION BY TRANSIT-TIME DAMPING

An alternative statistical acceleration mechanism results from transit-time damping of fluctuations in the magnitude of the magnetic field in the solar wind, as was discussed by Fisk (4). This mechanism was applied by Schwadron et al. (9) to create tails on the distributions of pickup ions, as seen by Ulysses. The physics here is also straightforward. With magnitude fluctuations there is a moving gradient in the magnitude of the magnetic field, which exerts a force and alters the particle speed,

$$\frac{dv_{parallel}}{dt} \approx -\frac{v_\perp^2}{B_o}\frac{\partial B}{\partial z} \qquad (5)$$

or, $|\Delta v_{parallel}| \approx v_\perp^2 \eta \Delta t / L_{parallel}$ where $\eta = \delta B / B$, and $L_{parallel}$ is the characteristic scale length parallel to the mean field.

The gradient, which is produced by a wave, moves parallel to the mean field direction at the parallel phase speed of the wave. Unlike Alfven waves, which have a constant phase speed parallel to the mean field, the magnetosonic waves that generate the magnitude fluctuations have a parallel phase speed that depends on the angle the wave makes with the mean field, or $u_{phase,parallel} = u_{phase}(L_{parallel} / L_\perp)$, where $L_\perp$ is the characteristic scale length normal to the mean field. The resulting change in momentum is then the phase speed parallel to the mean field times the change in the particle speed, or

$$\frac{\Delta p}{p} \approx \frac{u_{phase,parallel}\Delta v_{parallel}}{v^2} \qquad (6)$$

The resulting diffusion coefficient in momentum is

$$\frac{D_{pp}}{p^2} \approx \left(\frac{v_\perp}{v}\right)^4 \frac{\eta^2 u_{phase}^2}{v_{parallel}} \frac{L_{parallel}}{L_\perp^2} \qquad (7)$$

It is also possible to do a formal quasi-linear calculation of the diffusion coefficient in momentum, as was done by Fisk (4). The result is qualitatively the same as equation (7). However, the quasi-linear calculation does reveal that the relevant length scale normal to the field, $L_\perp$, is not the perpendicular correlation length of the turbulence. Rather, turbulence in the solar wind has a relatively hard power spectrum, generally falling off at less than (wave-number)$^{-2}$. As a result, the level of power falls off less steeply than do the length scales. In a weighted average, then, it is the short length scales that matter most, and, in fact, $L_\perp$ is weighted closer to the particle gyro-radius than to the correlation length, which increases the acceleration substantially.

Schwadron et al. (9) performed a study of the tails seen on the distribution functions of interstellar pickup ions, as seen by Ulysses when it was at low latitudes, i.e., in regions where Co-rotating Interaction Regions (CIR) in the solar wind occur. Figure 2, which shows the observed distribution function of interstellar hydrogen, is repeated from Schwadron et al. (9). The quantity η^2 is the square of the ratio of the amplitude of the magnitude fluctuations in the magnetic field to the mean field. The observed distribution function of pickup hydrogen is averaged over 150 days in 1992. The magnitude fluctuations in the field are assumed to be inhomogeneous, with two levels possible, but with the average of the two levels equal to the average observed in the 150-day interval. The diffusion coefficient in momentum space was calculated by Schwadron et al. (9) using the formal quasi-linear

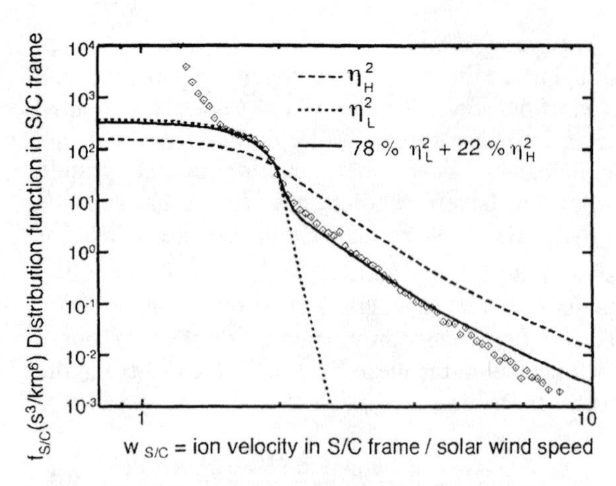

Figure 2. The average proton distribution functions observed by Ulysses/SWICS between day 50 and 200 in 1992, in the spacecraft reference frame. Model results are plotted in the solid curve (9). The mean value of η^2=0.03.

derivation of Fisk (4). As can be seen in Figure 2, the resulting acceleration yields an excellent fit to the observed tails. The presence of inhomogeneous fluctuations permits a detailed fit to the data, including the knee in the distribution function at twice the solar wind speed. Inhomogeneous turbulence is reasonable over the long time interval used here in particular since turbulence can be expected to be generated within the CIRs itself, but less so in the regions outside the CIR.

Schwadron et al. (9) also did a detailed examination of the correlation of the presence of strong tails on the distribution function with the occurrence of shock waves, e.g., the standing shocks surrounding CIRs. No correlation was found, and indeed the strongest correlation was with enhanced magnitude fluctuations, such as occur interior to the CIR.

Statistical acceleration by transit-time damping is thus a very viable mechanism for producing tails on the distribution function of both solar wind and pickup ions. It is a mechanism which can account in detail for the tails on interstellar pickup ions observed by Ulysses (9) and is a very likely candidate for the tails observed near Earth by ACE, e.g. as seen in Figure 1 (1).

MECHANISMS FOR GENERATING MAGNETIC FIELD MAGNITUDE FLUCTUATIONS

The strength of statistical acceleration by transit-time damping is also its weakness. It is a very efficient acceleration mechanism, and thus presumably will readily damp the magnitude fluctuations that are necessary for a reasonable acceleration rate, i.e. the diffusion coefficient in momentum varies as the square of the amplitude of the field magnitude fluctuations. Indeed, the observed amplitudes of the magnitude fluctuations are quite small, presumably because they are readily damped. As is seen in Figure 2, however, even with small amplitudes, the calculated acceleration rate can account for the observed tails.

To be a viable acceleration process it will be necessary for there to be a continuous source of field magnitude variations. In the case of acceleration in CIRs, as described by Schwadron et al. (9), the stream-stream interactions of the solar wind are presumably a continuous source of field magnitude fluctuations. For a continuous source of field magnitude fluctuations near Earth, there are also some possibilities.

There are several arguments that suggest that the footpoints of magnetic field lines random walk in position at the Sun. Jokipii [e.g., (10)] invokes such a random walk to account for cross-field diffusion of energetic particles in the solar wind. Fisk et al. (11) argue that such random walk must occur in the corona that yields slow solar wind. The large scale transport of magnetic flux in the corona in latitude, introduced by Fisk (12), requires a region near the solar equatorial plane where the flux can be transported in longitude, presumably by a reconnection process that in turn yields a random walk of the footpoints. Further, Fisk et al. (13) argue that the emergence of new magnetic flux, in the form of small bipolar loops, and their subsequent reconnection, can account for the acceleration of the solar wind. These loops and their reconnection will also result in random walk of the footpoints.

Inherent in any random walk of field lines is the introduction at the Sun of a component of the field normal to the heliocentric radial direction. As the field line is transported outward with the solar wind, the radial component of the field will decline as $1/r^2$. The normal component will decline as $1/r$. Since normal components of various strengths are possible, and thus various angles for field lines are possible, the magnetic strength on adjacent field lines will vary. Equivalently, magnitude variations in the magnetic field in the solar wind are an inherent consequence of the evolution of field lines that random walk back at the Sun.

CONCLUDING REMARKS

One of the more interesting recent observations from ACE has been the ubiquitous presence of pronounced tails on the distribution functions of both solar wind and pickup ions in the slow solar wind. The tails show no correlation with the passage of shock waves, and so appear to be the result of a continuous acceleration in the solar wind. Statistical acceleration is a likely candidate, particularly statistical acceleration by transit-time damping of magnitude fluctuations in the magnetic field. Transit-time damping is a very efficient mechanism; so long as propagating magnitude fluctuations are presence, it should provide any needed acceleration.

Issues remain, however. It will be important to identify the source of the magnitude fluctuations and to show that adequate energy is available to account for the observed tails. It will be important to determine that no other species, e.g. solar wind electrons, are accelerated, such that they would damp and thus remove the available energy in the magnitude fluctuations. It will be important to have a conclusive argument as to why tails occur in the slow solar wind but less so in the fast wind. And finally, as in the case of statistical acceleration in CIRs (9), it will be important to perform detailed fits to observed distribution functions with observed levels of magnetic field magnitude fluctuations.

ACKNOWLEDGEMENTS

This work was supported, in part, by NASA contract #NAG5-6912.

REFERENCES

1. Gloeckler, G., Fisk, L. A., Zurbuchen, T. H., and Schwadron, N. A., in *The Acceleration and Transport of Energetic Particles Observed in the Heliosphere*, AIP, 2000.

2. Zurbuchen, T. H., Fisk, L. A., Schwadron, N. A., and Gloeckler, G., in *The Acceleration and Transport of Energetic Particles Observed in the Heliosphere*, AIP, 2000.

3. Gloeckler, G., Geiss, J., and Fisk, L. A., in *Results of the Ulysses Mission*, edited by R. Marsden, A. Balogh, and E. Smith, in press, 2000.

4. Fisk, L. A., *J. Geophys. Res.* **81**, 4641 (1976).

5. Barnes, C. W., and Simpson, J. A., *Astrophys. J.* **210**, L91 (1976).

6. Jokipii, J. R., *Phys. Rev. Lett.* **26**, 666 (1971).

7. Wibberentz, G., and Beuermann, K. P., in *Cosmic Plasma Physics*, edited by K. Schwindler, New York: Plenum, 1972, pp. 339.

8. Gloeckler, G., Schwadron, N. A., Fisk, L. A., and Geiss, J., *Geophys. Res. Lett.* **22**, 2665 (1995).

9. Schwadron, N. A., Fisk, L. A., and Gloeckler, G., *Geophys. Res. Lett.* **23**, 2871 (1996).

10. Jokipii, J. R., and Parker, E. N., *Phys. Rev. Lett.* **21**, 44 (1969).

11. Fisk,. L. A., Zurbuchen. T. H., and Schwadron, N. A., *Astrophys. J.* **521**, 868 (1999).

12. Fisk, L. A., *J. Geophys. Res.* **101**, 15547 (1996).

13. Fisk, L. A., Schwadron, N. A., and Zurbuchen, T. H., *J. Geophys. Res.* **104**, 19765 (1999).

Composition and Energy Spectra of Ions Accelerated in Corotating Interaction Regions

G. M. Mason

Department of Physics and I.P.S.T., University of Maryland, College Park, MD 20742

Abstract. Corotating Interaction Regions (CIRs) arise from the interaction of fast- and slow-solar wind streams. Ions accelerated in association with CIRs are one of the primary heliospheric energetic particle populations, often reaching energies of >10 MeV/nucleon and intensities near 1 MeV/nucleon comparable to large solar energetic particle events. New instruments on the WIND, *Ulysses*, and ACE spacecraft have for the first time probed the CIR population in the suprathermal energy range. These new observations show that the 1 AU CIR abundances of suprathermal particles are locally accelerated, most often without any corresponding shock. In addition, enormous abundance enhancements of He^+ (rare in the solar wind) and details of heavy ion spectra and abundance ratios show that the source population for the energetic particles is not the solar wind itself but is most likely the suprathermal ion population in the speed range ~1.8-2.5 times that of the solar wind.

INTRODUCTION

Corotating interaction regions were discovered in the 1960s when observations showed increases of ~1 MeV energetic particle intensities at 1 AU that recurred each solar rotation (27 days) (1, 2). Observations on deep space probes showed that these intensities were higher outside Earth orbit, peaking at several AU (3, 4). The discovery of inward flowing intensities of the ions at 1 AU (5) was consistent with this overall picture, and simple models (6) reproduced the spectral features down to ~0.5 MeV/nuc. Fig. 1 shows a typical geometry and solar wind speed associated with CIRs. With a few puzzling exceptions such as He and C, the abundances in CIRs were similar to other heliospheric ion populations, and so it was generally assumed that the solar wind was the source population (7, 8). A complete review of CIRs has recently been published in the ISSI Space Science Review Series (9).

Figure 2 shows the place of CIRs in a flow diagram of inner heliospheric source material and energization processes based on observations through the 1980s (10). Starting at the top, the solar photosphere is the original source of material which moves by a fractionating process to the corona, which forms the source population for the solar wind, CME accelerated particles, and CIR accelerated particles. After acceleration, particle populations are transported to 1 AU where we observe them. In this view, the CIR population is one of several similar energetic particle populations in the heliosphere.

This basic picture explained many observed features but was derived without detailed knowledge of the solar wind composition, or of the suprathermal ion particle composition. Below we review recent observations of CIRs and suprathermals and examine their major impact on this basic picture.

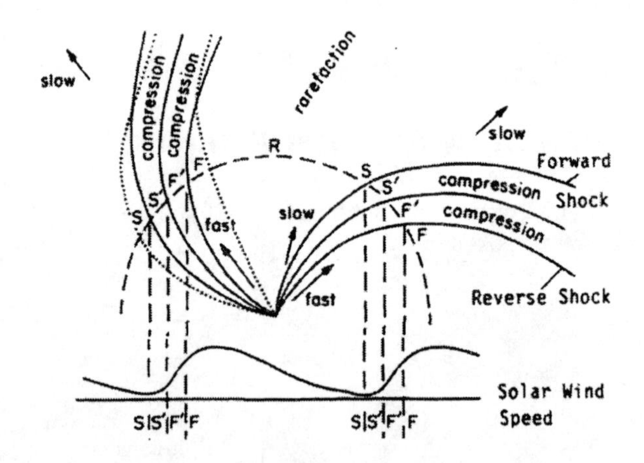

FIGURE 1 Schematic representation of CIR geometry adapted from Richardson *et al.*,(11). Dashed line represents typical spacecraft trajectory through the CIR. Solar wind regions are: S, slow solar wind; S', compressed, accelerated solar wind; F', compressed, decelerated fast solar wind; F, fast solar wind. Dotted lines show typical magnetic field lines.

CP528, *Acceleration and Transport of Energetic Particles Observed in the Heliosphere: ACE 2000 Symposium,* edited by Richard A. Mewaldt, et al.
© 2000 American Institute of Physics 1-56396-951-3/00/$17.00

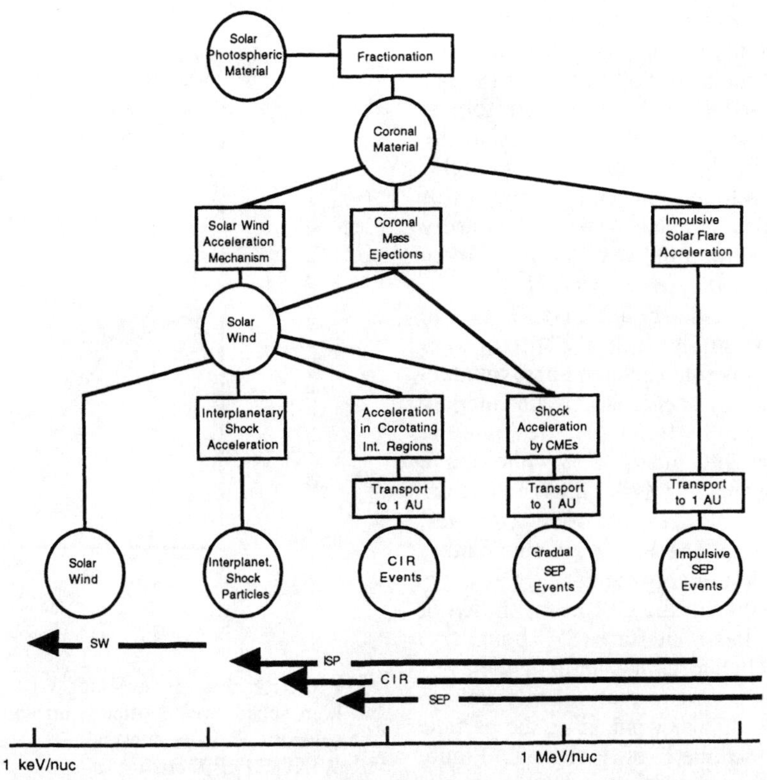

FIGURE 2 Inner heliospheric material sources (circles) and physical mechanisms (rectangles) that produce energetic particle populations at 1 AU. Approximate energy scale for the particles is at bottom. Adapted from Stone *et al.* (10).

OBSERVATIONS

Figure 3 shows solar wind and energetic particle intensities for a CIR event beginning Dec. 6, 1994 (12). Referring to the CIR regions shown in Fig. 1, the sharp rise in solar wind speed, simultaneous with the rise in suprathermal He intensities, occurs virtually simultaneously with the crossing of the stream interface (SI) between the low- and high-speed streams (S'-F' interface in Fig. 1). There were no shocks observed in this CIR, as is usually the case at 1 AU; presumably shocks have formed in this case at several AU (13) as is often observed in other CIRs. Notice that the lowest energy particles (60 keV/nuc) reach their peak intensity first, around 1500UT on day 340, while at higher energies the time of maximum is progressively later until at ~6 MeV/nuc, it occurs around 0600UT on day 341, 15 hours later than the lowest energies. Referring to figure 1, the magnetic field line connections in this case map out as follows: The lowest energy particle peak intensities occur *inside* the CIR, in the F' region, and are therefore not connected to possible shocks in the outer heliosphere, since the shocks form at the F'-F interface. At the time of peak intensity at ~6 MeV/nuc, the spacecraft has moved into the high speed solar wind (F region) and is connected to the F'-F interface outside 1 AU. However, since this intensity peak occurs only

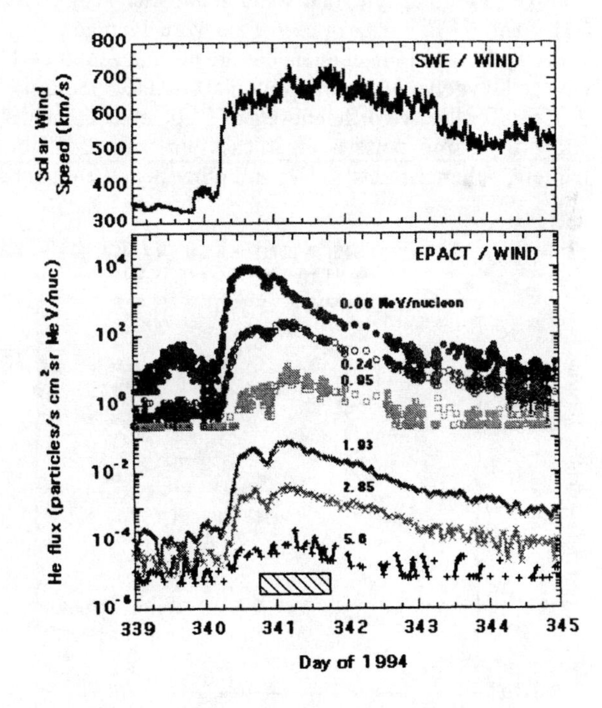

FIGURE 3 A CIR observed shortly after launch of the WIND spacecraft (12). *Top panel:* solar wind speed. *Bottom panel:* Helium intensities from suprathermal through energetic particle range

6-8 hours after the spacecraft passes into the high speed solar wind stream, a simple model shows that the field line the spacecraft is on near 0600UT on day 341 connects to the CIR within ~0.5 AU. Therefore if particles are following magnetic field lines as in the nominal picture shown in Figure 1, all the CIR particles shown in Figure 3 come from heliocentric distances of ~<1.5 AU. This is also consistent with outwardly flowing anisotropies seen for ~300 keV ions in a survey of 1978-79 CIRs by *Richardson and Zwickl* (14)

Figure 4 shows energy spectra of H, He, CNO, and Fe during the CIR shown in Figure 3; the spectra were computed over a 24 hour period centered approximately at the peak intensities at higher energies, as indicated by the hatched box in Figure 3. Below ~1 MeV/nuc the spectra are power laws with index ~2.5, while above this energy they steepen appreciably. Although there are slight differences in slope between the different species, their similarity is evident. The heavy ion abundances are roughly solar in the low energy regime.

The particle intensities in the CIR event shown in Figure 3 peak after the stream interface (S'-F' boundary) as is typical at 1 AU. Therefore, in examining the average abundances, the appropriate thermal reservoir to consider is the high speed solar wind, since the particle energization takes place in the F' or F regions. Figure 5 shows CIR abundances at ~150keV/nucleon from the WIND/STEP and ACE/ULEIS instruments (12, 15) relative to high speed solar wind abundance (16). The STEP and ULEIS data in each case were averaged over several CIRs, and the abundances are normalized in each case to Oxygen. While the proton abundances are quite different for the two different sets of CIRs, other elements show the same pattern for these two sets of CIRs, namely, enhancements in He, and other heavy elements

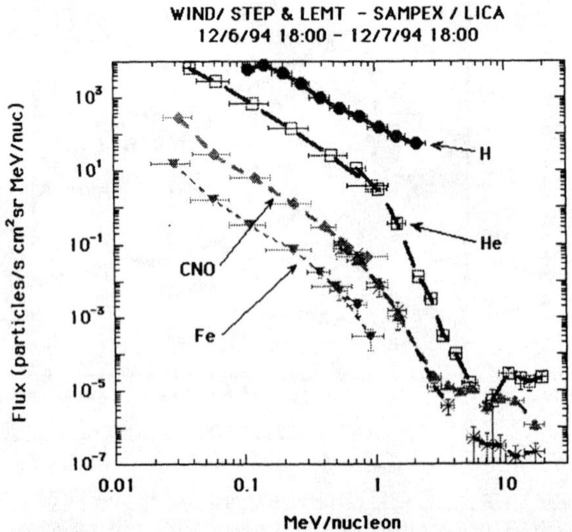

FIGURE 4 Energy spectra during peak of CIR shown in figure 3 (12). Note similarity of spectral forms for different elements, and roll over above ~1 MeV/nuc. Flat spectra above ~8 MeV/nuc are anomalous cosmic rays.

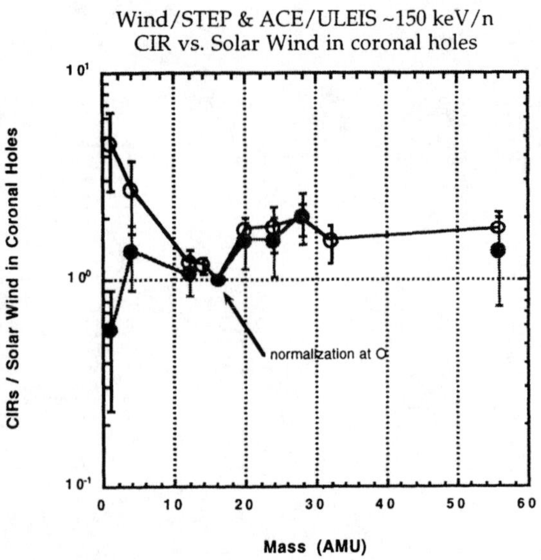

FIGURE 5 150 keV/nuc CIR abundances relative to high speed stream solar wind abundances averaged over several CIR event, normalized to Oxygen. *Filled circles:* WIND/STEP; *open circles:* ACE/ULEIS observations (12, 15).

by roughly a factor of 2. If we had chosen another material sample for comparison, for example the slow-speed solar wind (that is often observed in the ecliptic plane), the reference reservoir would have had higher abundances of elements with low first ionization potential. This would lower the CIR/SW ratio of Mg, Si, S, and Fe, changing the apparent "excess" shown in Figure 5 to a depletion of roughly a factor of 2. Since He and Ne are high FIP elements, and since the C/O ratio is very similar in both the high- and low-speed solar wind, the points for those elements would not change significantly.

The spectra shown in Figure 4 go down to ~30 keV/nucleon, well above solar wind energies. Figure 6 shows the He distribution function spectra all the way down to the solar wind, measured in the compressed solar wind (F') region in a CIR observed at 1 AU on May 30, 1995 (17). Three instruments on the WIND spacecraft (MASS, STICS, and STEP) were required to cover this energy range. Notice that the higher energy spectra continue to rise in a power law fashion down to speeds of only 2-3 times the solar wind speed Vsw.

The power law form of the spectra shown in Figures 4 and 6, going down to a few times the solar wind speed is not expected according to simple models of CIR acceleration and transport, such as that of Fisk and

Lee (6). In this model, particles are accelerated at CIR shocks at 3-4 AU, and propagate into the inner solar system along field lines such as that sketched in Figure 1. While this model is consistent with many features of the multi-MeV CIR particles, it breaks down in the suprathermal energy range discussed here.

Figure 7 shows predictions of the Fisk and Lee model for particle spectra observed at 1 AU in cases where the particles originated at shocks located at 1.0, 1.5, 3, and 6 AU. Notice that even for particles coming from shocks as close at 1.5 AU, there is a distinct flattening of the intensity below a few 10s of keV/nucleon; yet comparing to figure 6, such a flattening is not seen even down to energies corresponding to 2-3 times the solar wind speed, i.e. energies of 5-10 keV/nuc. Thus, even if shocks are producing suprathermal ions at several AU, they will not propagate into the inner heliosphere in significant numbers. The physical reasons for this effect are a combination of low particle speed, scattering, and adiabatic deceleration -- all of which are included in the Fisk and Lee model.

We now consider the time evolution of abundances in CIRs, as observed in the slow solar wind region (prior to the stream interface) and the high speed region. One of the key new results from *Ulysses* has been the discovery that low-and high-speed solar wind streams show markedly different enhancements for elements with low first ionization potential (FIP), wherein in the fast stream the enhancement is less pronounced by a factor of 3-4 (16). Figure 8 shows this effect for a *Ulysses* survey of a series of CIRs with the data points arranged according to the CIR regions in order: pre-forward shock, forward-shock to stream interface, stream-interface to reverse-shock, post reverse-shock (corresponding to S,

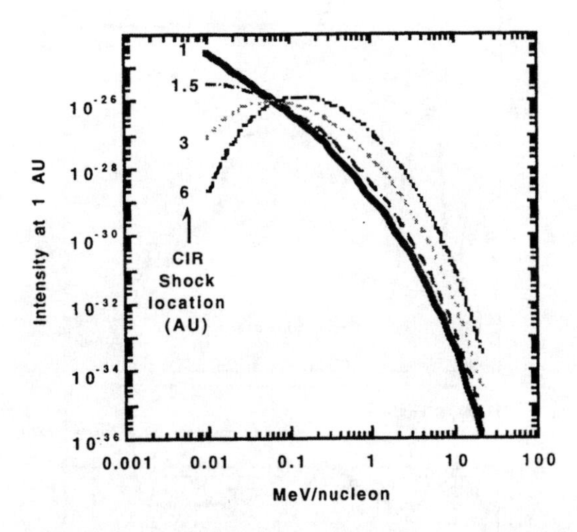

FIGURE 7 Calculation of spectral forms using the model of Fisk and Lee (6), predicting that below a few hundred keV/nucleon the low energy particle spectra will turn over (18).

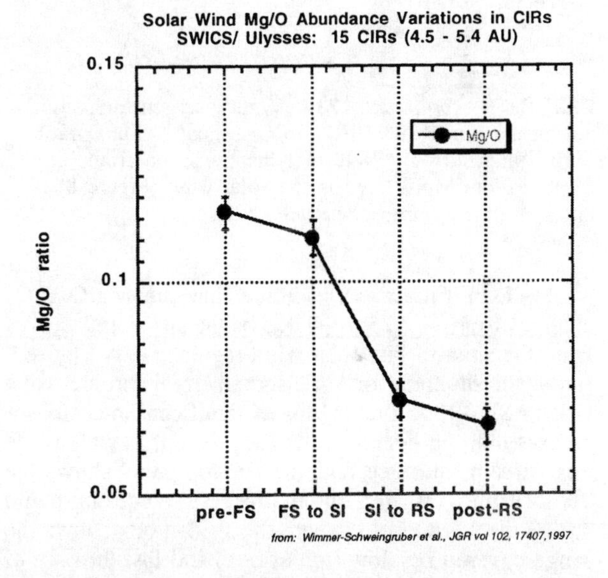

FIGURE 8 Epoch analysis of *Ulysses* solar wind data up- and down-stream from CIR stream interfaces, showing the transition from FIP-enhanced (slow) solar wind to fast solar wind Mg/O abundances (19).

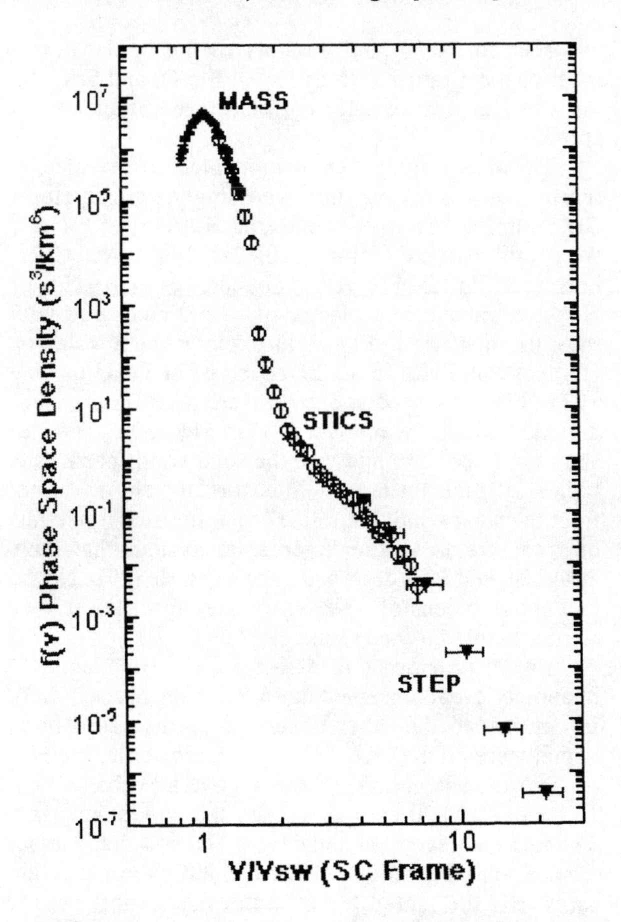

FIGURE 6 Helium distribution function from solar wind to suprathermal to energetic particle regime, observed during a CIR on May 30, 1995 (17). Notice the smooth joining of the suprathermal tail to the energetic particle regime.

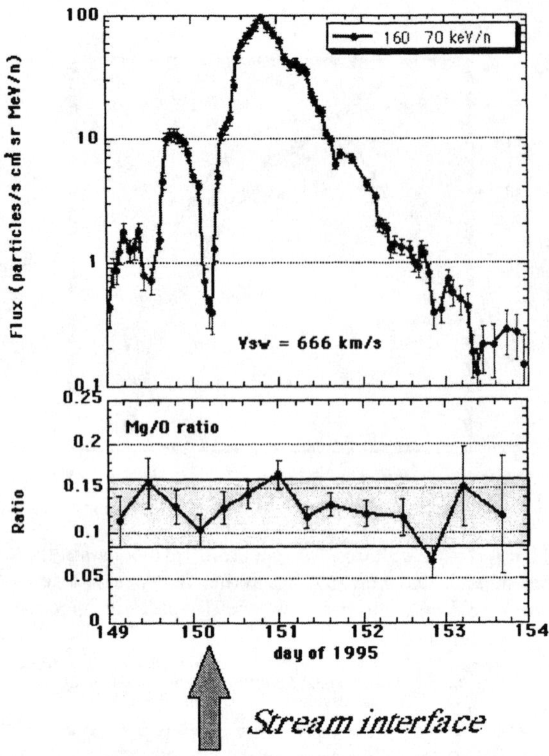

FIGURE 9 *Top panel:* 70 keV/nuc Oxygen intensities during the May 1995 CIR. *Bottom panel:* Mg/O ratio during this period. Note that the transition from high Mg/O to low Mg/O seen in the solar wind (Figure 8) is not seen in the energetic particles (12).

S', F', F in Figure 1). Notice that the Mg/O ratio changes from ~0.125 to ~0.065 as the spacecraft moves from the slow to fast solar wind regions (19). Figure 9 shows the situation for 1 AU suprathermal particles for a CIR in May 1995 that exhibited significant intensity increases in the pre-stream interface region, as well as the post-stream interface region. The top panel shows the 70 keV/nuc Oxygen intensities; the bottom panel shows the Mg/O ratio, where the shaded box shows the range covered by slow (top of box) and fast (bottom of box) solar wind. The suprathermal Mg/O shows no evidence of a transition at the stream interface, and in fact during the entire period it fluctuates around a value roughly equal to the average of the fast- and slow-solar wind Mg/O ratios.

Figure 10 shows an additional feature of suprathermal ion abundances for 1 AU CIRs: in a series of 12 CIRs observed during 1994-1995, the C/O ratio showed a significant variation with the speed of the solar wind at the time of peak intensity. The variation seen was more than a factor of 2, much greater than seen in the solar wind C/O ratio, which shows only small variations from high- to low-speed solar wind streams. In this series of CIRs a similar increase with solar wind speed

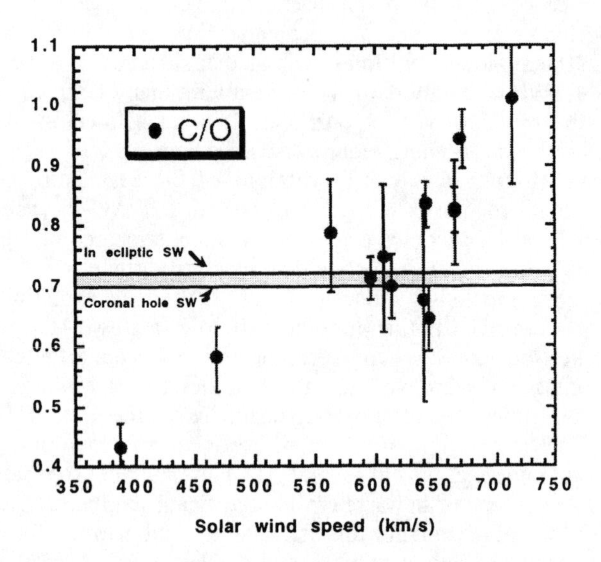

FIGURE 10 C/O ratio at 150 keV/nucleon for 12 CIRs observed on WIND. Note that the C/O ratio varies from below to far above the solar wind value (yellow box) depending on solar wind speed (12).

was seen for Ne/O, and possibly for Fe/O. However, other element ratios such as He/O, Mg/O, and Si/O did not show any systematic correlation with solar wind speed.

The most striking compositional feature revealed by recent instruments has involved singly ionized He in CIRs. In the bulk solar wind, the $He^+/He^{++} < 5\times10^{-5}$, with only rare exceptions (20, 21). In 1 AU CIRs, however, He^+ is observed at abundances of 10-15% of He^{++}, an enhancement factor of >1000 compared with the bulk solar wind. Figure 11 explores the low energy H^+, He^{++}, and He^+ in a CIR observed at 1 AU in May 1995 (17). The spectra shown were taken in the compressed fast solar wind region (F' in Figure 1). The He^+ shows a "knee" around twice the solar wind speed, as is typical of pick up ions -- thus the He^+ shown is not from the solar wind, but rather from interstellar neutrals or from grains in the inner solar system that have absorbed, and later desorbed solar wind He (20). Notice that above speeds of ~3Vsw, the He^+ spectrum is very similar to that for the H^+ and He^{++}.

Figure 12 shows CIR associated H^+, He^{++}, and He^+ in another event, observed at ~4.5 AU on *Ulysses* (22), in contrast to the other observations discussed here, which were all at 1 AU. In the *Ulysses* case, the He^+ *exceeds* the energetic He^{++}, for an overall enhancement factor of ~10,000 compared with the bulk solar wind. The relatively larger abundance of He^+ at 4.5 AU arises from a combination of effects, including the ionization rates of the interstellar neutrals, inner source contributions, and a $1/r^2$ fall off of solar wind He^{++}.

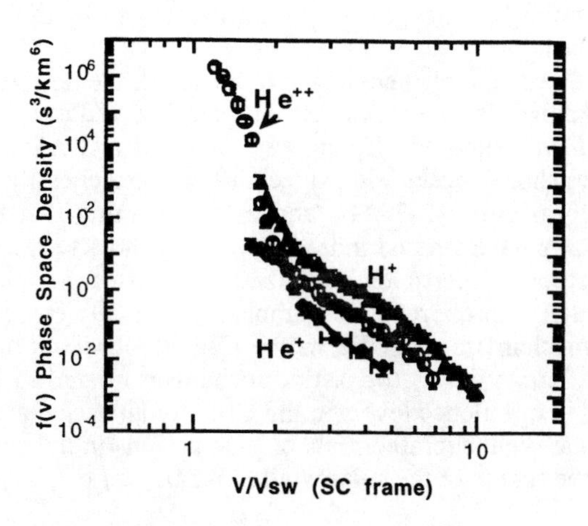

FIGURE 11 Solar wind and suprathermal H^+, He^{++}, and He^+ observed in the compressed fast solar wind stream in a CIR on May 30, 1995 (17). Notice that the He^+ to He^{++} ratio is close to unity for ion speeds close to twice the solar wind speed.

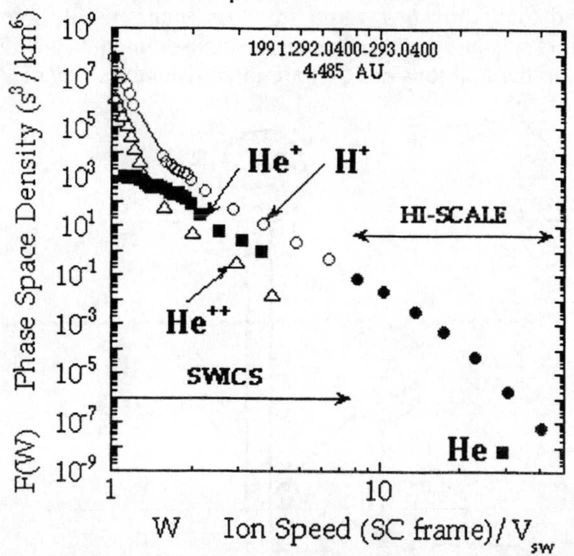

FIGURE 12 *Ulysses* observations of H^+, He^{++}, and He^+ in a CIR at 4.5 AU (22). Note that He^+ exceeds the He^{++} abundance above ~2x the solar wind velocity, an enhancement of >10000 compared to thermal abundances.

DISCUSSION

The suprathermal ion properties we have described may be summarized as follows:

- Particle intensities peak within the CIR itself, in the compressed, fast-solar wind stream where the nominal magnetic connection to the F'-F interface is *inside* 1 AU.
- The spectra are power laws down to ~10 keV/nucleon, and do not exhibit turnovers (change of sign of spectral slope) at any energy or time.
- The relative composition of ions is roughly similar to the solar wind, with distinct overabundance of He and Ne, and possibly for other elements as well depending on whether the slow- or fast-solar wind is chosen as the reference.
- He^+ is present in CIRs with the same spectral form as H^+ and He^{++} over the entire range of available measurements (~10 to 50 keV/nucleon).
- The suprathermal particle composition does not change in the CIR between the pre- and post-stream interface regions, even though a distinct change is seen in these regions for the solar wind.
- In a series of CIRs, the C/O and Ne/O ratios increased with solar wind speed in the fast stream, with the variation seen being well outside the range seen in the solar wind for slow- and fast-streams.

Taken in the context of prior research at higher energies including observations in the inner and outer heliosphere (e.g., 9), we interpret these new observed features of suprathermal ions as follows:

First, the suprathermal ions observed at 1 AU are "locally" accelerated: their source lies within a heliocentric distance of ~1.5 AU, and may even be *at or inside* 1 AU. The observational features supporting this are: (a) The timing of the peak intensities, which are within the CIR, not outside it in the high speed stream. This leads to a magnetic connection to the F'-F interface inside 1 AU. We note that several studies have concluded that certain observational features of CIRs require non-field aligned transport or cross-field diffusion at a much greater rate than generally assumed (23, 24). While such transport would change the magnetic connections such as those shown in Figure 1, in general they would tend to allow connection to shocks *closer to* the observer at least in the F region. (b) The lack of a turn-over or severe flattening of the spectra below a few 10s of keV/nucleon as predicted by CIR models. (c) The factor of 10 difference in He^+/He^{++} ratio observed at 1 AU (~0.20) versus 4.5 AU (~>2), showing that the particle population at 4.5 AU is not propagating into 1 AU.

Second, the bulk solar wind is *not* the source of CIR particles; rather the source material consists of ions in the range ~1.8-2.5 times the solar wind speed, which have several possible sources which may contribute varying fractions of the ion population at these energies. Known candidates for these sources are: (a) the suprathermal tail of the solar wind, as enhanced, e.g., by turbulence, shocks or other transients, (b) pick-up ions whose parent population is interstellar neutrals, or (c) "inner source" neutrals emitted from interplanetary grains. The primary feature leading to this conclusion is the enormous enhancement of He^+ in the CIRs -- factors of ~1000 at 1 AU, and closer to 10,000 at 4.5 AU. Other abundance features observed in the suprathermal population are consistent with this picture, namely the long-known enhancements of He and C in CIRs, and the Ne enhancement clearly detected in the suprathermal energy range. Additional features that are *inconsistent* with a bulk solar wind source include: the lack of change of ratios such as Mg/O from the slow- to fast-stream regions, and the factor of ~2 variation of the C/O ratio with solar wind speed. These latter features point to the importance of seed particles with properties notably different from those seen in the bulk solar wind: i.e., slow- versus fast-stream abundance changes, and a C/O ratio that changes only slightly from slow- to fast-streams.

Third, shock acceleration is *not* the energizing mechanism for the suprathermal ions observed at 1 AU. The *local* origin of CIR suprathermal ions at 1 AU precludes shock acceleration since CIR shocks generally do not form until ~2-3 AU, and even assuming that the CIRs we have studied indeed have such shocks present, suprathermal particles energized at them do not flow into 1 AU in appreciable numbers. Thus, the energizing mechanism must lie in the CIR as observed at 1 AU: it may be a stochastic mechanism related to increased turbulence levels in the CIRs, or an acceleration of solar wind suprathermals or pick up ions in the compressed region of the solar wind (25-27).

CONCLUSIONS

Energetic particle observations in CIRs at energies > 1 MeV fit reasonably well into the "classic" picture sketched in Figure 2: namely, the particles are accelerated from a solar wind source by shocks at several AU, and then they propagate into the inner heliosphere. Notwithstanding this success, measurements of CIR suprathermal ions at 1 AU are inconsistent with *all* of

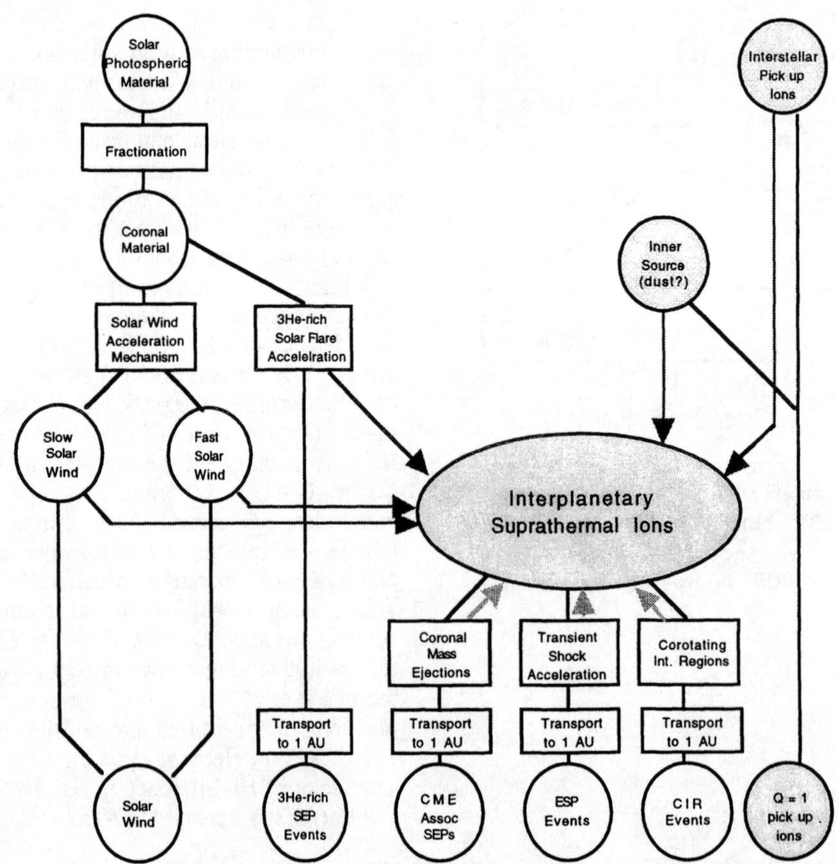

FIGURE 13 Modified version of figure 1, illustrating the role of interplanetary suprathermal ions as the seed population for many particle populations observed in the heliosphere. Shaded areas are new source populations (compare figure 2).

the basic assumptions in the classic picture: the source is the suprathermals including pickup ions, not the solar wind; the acceleration is not by shocks, but by some other mechanism; the acceleration is close (~<0.5 AU) to 1 AU, not at several AU; and the particles have not propagated far.

These results combined with other studies lead to a significant revision of the basic picture of particle sources and processes sketched in Figure 2, where the solar photosphere/corona was the source for all the energetic particle populations energized in the inner heliosphere. Rather, we have the following particle populations contributing to a reservoir of interplanetary suprathermal ions that apparently lie in the proper energy range to play a key role in subsequent energization processes:

- suprathermal tails of the fast and slow solar wind
- interstellar and "inner source" pick-up ions that fill the heliosphere at different intensities depending on location (20)
- impulsive solar energetic particle events, whose remnants can fill substantial portions of the inner heliosphere with a distinct class of suprathermal ions (28).

These particle sources are shown schematically in Figure 13, which is a modification of Figure 2 taking account of recent progress based on WIND, *Ulysses,* and ACE. Shaded areas represent particle populations whose existence was known, but whose role in contributing to the seed population of energetic particles is only now being revealed. The interplanetary suprathermal ion reservoir is not a fixed quantity, but has multiple, time-dependent inputs as shown. Interplanetary transients and CIRs interact with this seed population, producing the energetic particles we observe in particular events. These same transients, i.e. shocks and CIRs, themselves leave enhanced levels of suprathermals that may be energized further by a subsequent event if it occurs in a reasonably short period of time.

ACKNOWLEDGMENTS

We gratefully acknowledge kind permission from the authors and following organizations to publish copyrighted material: the American Astronomical Society/University of Chicago Press (Figs. 3,4,9,10); the American Geophysical Union (Figs. 1, 8, 12); Kluwer Academic Publishers (Figs. 2,7).

REFERENCES

1. Bryant, D.A., *et al.*, *Phys. Rev. Lett.* **14**, 481 (1965).
2. McDonald, F.B. and U.D. Desai, *JGR* **76**, 808 (1971).
3. McDonald, F.B., *et al.*, *ApJ* **203**, L149 (1976).
4. Van Hollebeke, M.A.I., *et al.*, *JGR* **83**, 4723 (1978).
5. Marshall, F.E. and E.C. Stone, *JGR* **83**, 3289 (1978).
6. Fisk, L.A. and M.A. Lee, *ApJ* **237**, 620 (1980).
7. Gloeckler, G., *et al.*, *ApJ* **230**, L191 (1979).
8. Reames, D.V., *et al.*, *ApJ* **382**, L43 (1991).
9. Balogh, A., *et al.*, eds. *Corotating Interaction Regions*. Space Science Series of ISSI. 1999, Kluwer Academic Publishers: Dordrecht.
10. Stone, E.C., *et al.*, *Space Sci. Rev.* **86**, 1 (1998).
11. Richardson, I.G., *et al.*, *JGR* **98**, 13 (1993).
12. Mason, G.M., *et al.*, *ApJ* **486**, L149 (1997).
13. Gosling, J.T., *et al.*, *JGR* **81**, 2111 (1976).
14. Richardson, I.G. and R.D. Zwickl, *Planet. Space Sci.* **32**, 1179 (1984).
15. Dwyer, J.R., *et al.* in *ACE-2000: The acceleration and transport of particles observed in the heliosphere*. 2000. Indian Wells, CA: Caltech, SRL.
16. von Steiger, R., *et al.*, in *Cosmic Winds and the Heliosphere*, J.R. Jokipii, *et al.*, Editors. 1997, Univ. of Arizona Press: Tucson. p. 581.
17. Chotoo, K., Ph.D. thesis, Department of Physics, University of Maryland, College Park, MD, (1998).
18. Mason, G.M., *et al.*, *Space Sci. Rev.* **89**, 327 (1999).
19. Wimmer Schweingruber, R.F., *et al.*, *JGR* **102**, 17407 (1997).
20. Gloeckler, G. and J. Geiss, *Space Sci. Rev.* **86**, 127 (1998).
21. Gloeckler, G., *et al.*, *GRL* **26**, 157 (1999).
22. Gloeckler, G., *et al.*, *JGR* **99**, 17637 (1994).
23. Dwyer, J.R., *et al.*, *ApJ* **490**, L115 (1997).
24. Intriligator, D.S. and G.L. Siscoe, *JGR* **100**, 21605 (1995).
25. Richardson, I.G., *Planet. Space Sci.* **33**, 557 (1985).
26. Giacalone, J. and J.R. Jokipii, *GRL* **24**, 1723 (1997).
27. Schwadron, N.A., *et al.*, *GRL* **23**, 2871 (1996).
28. Mason, G.M., *et al.*, *ApJ* **525**, L133 (1999).

High Latitude Observations of Corotating Interaction Regions: Remote Sensing Using Energetic Particles

Edmond C. Roelof

Johns Hopkins University, Applied Physics Laboratory, Laurel, MD 20723-6099

Abstract. Ulysses observations revealed a strong diminution of 26-day-recurrent energetic ion and electron increases as its latitude increased above 20°. There was a systematic several day lag of the 38-53 keV electron peak intensity behind that of the 480-966 MeV protons. The recurrences persisted to the highest latitudes ~80° in both the southern and northern polar passes, well above the most poleward *in situ* detection of corotating interaction regions (CIRs) in the solar wind. These effects have been explained in terms of remote magnetic connection from Ulysses to the reverse shock forming the poleward boundary of the CIR. We present theoretical arguments that the dominant mechanism in producing the velocity ordering of the particle intensities is energy loss as the particles propagate from the CIR inward towards the spacecraft. The velocity ordering (regardless of species) follows immediately from the striking result that the energy loss process is depends only upon particle velocity (and not on mass or charge) in the limit of very weak scattering.

INTRODUCTION

One of the striking observational results from the first southern pass of the Ulysses spacecraft was the continuation of the appearance of 26-day recurrent energetic particle events all the way up to the maximum latitude attained (80.22°S heliographic on September, 1994, day 257). The sensitivities of the instruments were such that electrons with energies ~50 keV could be tracked right through the highest latitudes (1) and protons with energies ~1 MeV almost that high (2). The recurrent electron event at 80°S was number 32 in a series that had begun in early July 1992 when Ulysses had only reached a latitude of 13°S (3). At low to moderate latitudes, the energetic electron and ion events were clearly associated (4,5) with corotating interaction regions (CIRs). These are stream-stream interactions in which the low-speed solar wind (from the streamer belt) interacts with high-speed solar wind (from the equatorial extension of polar coronal holes) to compress the interplanetary magnetic field (IMF).

The CIR phenomenon had been identified and interpreted (6) from Pioneer 10 and 11 observations beyond 1 AU in the ecliptic during the decline of Solar Cycle 20 in 1974-1975. They were observed a decade later during the decline of Solar Cycle 21 while Voyager 1 and 2 were still at relatively low latitudes after their encounters with Jupiter. Ulysses found them once again at low- to mid-heliolatitudes during the decline of Solar Cycle 22. However, the surprise was that the recurrent particle events continued after the last reverse shock was observed at 46°S (7), because the reverse shock had been identified with the poleward boundary of the CIR (8). The energetic electron and ion events had accompanied the forward and reverse shocks of the CIRs up to mid-latitudes, yet they continued to reappear every 26 days, long after the *in situ* shock signatures had ceased to appear. Since the CIRs were the only likely acceleration site in the outer heliosphere for the recurrent electrons and ions, and it was understood that these low energy particles must essentially follow field lines as they propagate along them. Simnett and Roelof (9) argued that Ulysses at mid-latitudes had to be connected magnetically to the CIRs at helio-radii well beyond Ulysses.

However, as the recurrent events continued to be observed up to high latitudes, it became clear that an ideal Parker magnetic field could not do this, because its field lines lie on cones of constant helio-latitude. Thus motivated to explain these high latitude recurrent energetic particle events, Fisk (10) produced a new model of the heliospheric magnetic field based on

CP528, *Acceleration and Transport of Energetic Particles Observed in the Heliosphere: ACE 2000 Symposium*, edited by Richard A. Mewaldt, et al.
© 2000 American Institute of Physics 1-56396-951-3/00/$17.00

photospheric differential rotation and non-radial motion of the near-sun solar wind. It predicts that some high-latitude field lines from the Sun will connect to CIRs at latitudes as low as 30° at distances ~15 AU. Later, Kota and Jokipii (11) argued that Parker field lines that wandered stochastically in latitude could accomplish the same connection and also explain the Ulysses energetic particle observations.

Recurrent energetic particle events did not immediately reappear as Ulysses climbed into the northern hemisphere after its perihelion in early 1995. It was not until after the northern polar pass, when Ulysses was descending through 71°N helio-latitude in late 1995, that first the 26-day recurrent electron events reappeared above instrument background (12). Energetic ~0.5 MeV protons reappeared three solar rotations later on 1 January 1996. Yet there was no in situ measurement of a CIR reverse shock until 6 August 1996 at 30°N. Clearly, the reappearance of the 26-day recurrent energetic particle events at 71°N implied their magnetic connection to CIRs at ~30°N at helioradii well beyond Ulysses at 5 AU.

As this paper is being written, Ulysses is again in latitudes ~40°S, but we are now in the rise-to-maximum of Solar Cycle 23. Although there are tendencies toward 26-day recurrences in the energetic particles and plasma, they have not exhibited the dramatic, strict 26-day period that we saw during the decline of Solar Cycle 22.

This paper will mainly be concerned with how the energetic particles propagated from the CIRs to Ulysses to form the 26-day recurrent events observed at high latitudes during the first southern polar pass. We will show that there was a general pattern in the way that the intensity histories changed as the latitude increased. We will argue that this pattern is ordered essentially by particle velocity, and we will offer a theoretical interpretation based on energy loss during nearly scatter-free propagation.

OBSERVATIONS

A truly comprehensive summary of observations and interpretations has been published both as a book *Corotating Interaction Regions* (13) and as a special combined issue of *Space Science Reviews* (13). These are the Proceedings of an International Space Science Institute Workshop that was held 6-13 June 1998 in Berne, Switzerland. Detailed data plots and discussions thereof relevant to the material we shall discuss here will be found therein, particularly in the Report of Working Group 4 "Corotating Interactions in the Outer Heliosphere." The energetic particle measurements essential to our discussion here are summarized in Figure 1. Each of the three panels presents a 26-day plot of the unidirectional differential intensities of three particle species measured by the HI-SCALE detectors on Ulysses (14). The species shown are: 47-68 keV ions (essentially protons); 480-966 keV protons; and 38-53 keV electrons (with intensities scaled by 10^{-5} for visibility relative to the protons). The main point of the figure is that the intensity histories are generally ordered by the velocities of the particles (regardless of their mass or charge). The velocities (specified by $\beta = v/c$) for each of these species are 0.011, 0.038, and 0.394, respectively. Other proton channels with $0.04 < \beta < 0.10$ could have been plotted (13), but they would simply fill in the trend that is already apparent in Fig. 1.

The three 26-day periods chosen are: CIR 9 when Ulysses was at 5.0 AU and at helio-latitude 24°S; CIR 18 (4.3 AU, 39°S); and CIR 25 (3.4 AU, 56°S). The periods are aligned to the nearest day by the vertical line indicating the times of the reverse shocks on CIRs 9 and 18, and a weak reverse wave on CIR 25. In the low-latitude CIR 9, the reverse shock at 02:57:44 UT on day 22 of 1993 is strong (actually super-critical), so it exhibits the characteristic signatures of CIR particle acceleration: a shock spike in the electrons and a plateau in the low-energy ions that drops precipitously at the reverse shock. The narrower particle signatures of the preceding forward shock at 04:59:49 UT on day 20 are shock spikes for all three species. The intensities decay much faster at the lower velocities after the reverse shock, with the fast electrons decaying most slowly of the three. In the mid-latitude CIR 18, the reverse shock at 19:16:28 UT on day 261 of 1993 is much weaker, and there are no intensity spikes in the particles. The low-velocity proton intensity maxima are now approximately centered on the reverse shock, while the electrons actually do not rise until after it. Again, the low-velocity ions decay rapidly, while the electrons do not decay at all for nearly a week. Finally, in the high-latitude CIR 25, the low-velocity ions peak several days after the weak reverse wave at ~1200 UT on day 69 of 1994 (with the β=0.04 ions peaking a few days later than the β=0.01 protons), while the β=0.40 electrons exhibit a broad maximum at least 10 days after the reverse shock.

INTERPRETATION

The attenuation of the ion and electron intensities with latitudes above 20° and the systematic lag of the 38-53 keV electron peak behind the 480-966 MeV

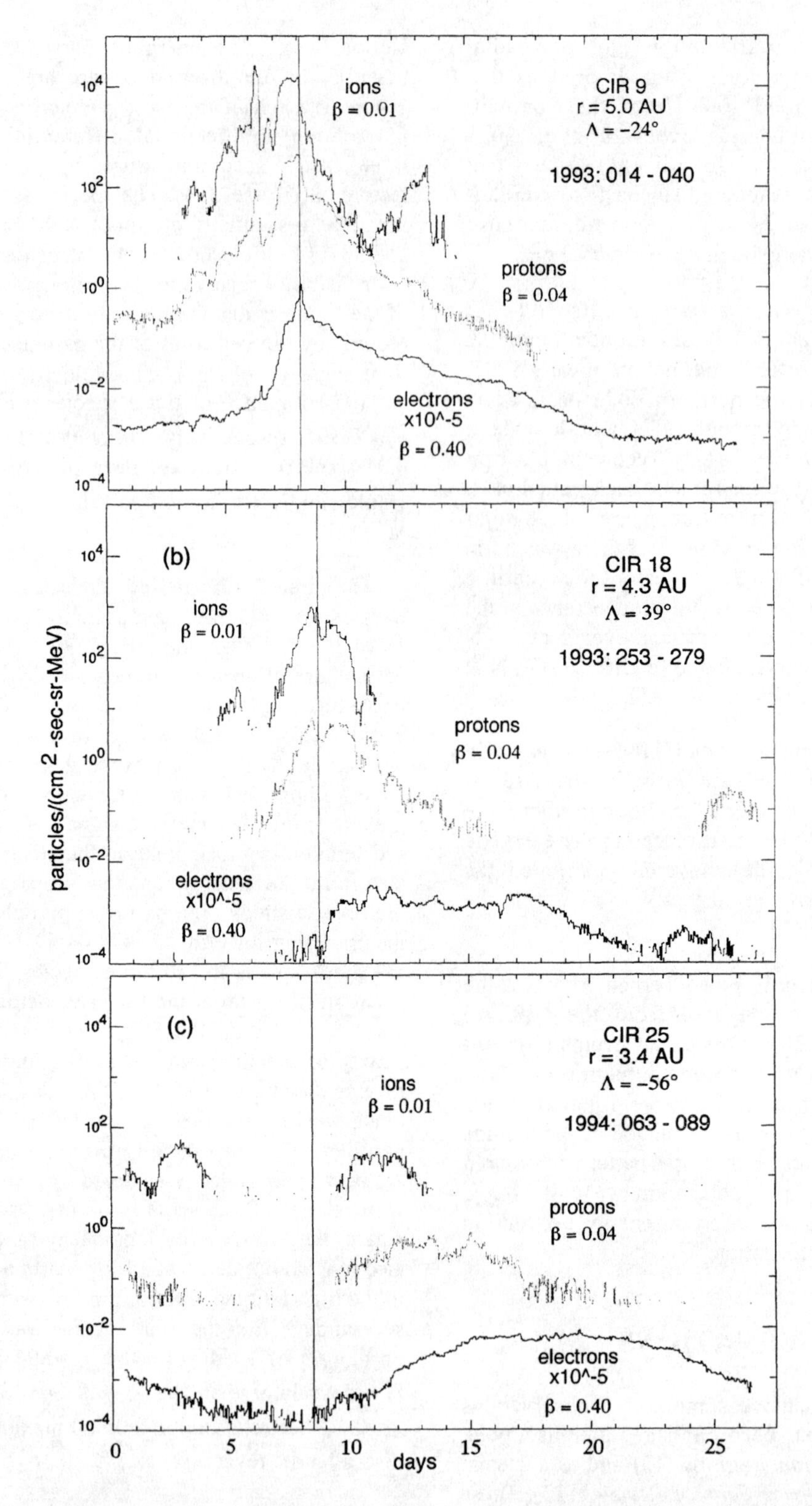

FIGURE 1. Intensity histories measured by Ulysses/HI-SCALE during three separate 26-day solar rotations (dates indicated in each panel) during the first south latitude pass: 47-68 keV ions (mostly protons), 480-966 keV protons, and 38-53 keV electrons (intensities scaled by 10^{-5} for clarity of presentation). Vertical lines give the times of observed reverse shocks for CIRs 9 and 18 and a reverse wave on CIR 25. The evolution of the intensity histories is systematically ordered by particle velocity ($\beta = v/c$), independent of charge or mass.

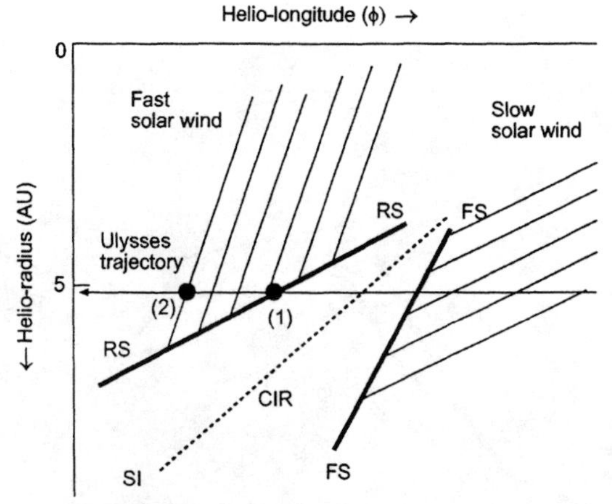

FIGURE 2. Corotating streamlines of solar wind on cone of constant latitude containing Ulysses' trajectory. Distance of magnetic connection from spacecraft to reverse shock (RS) increases rapidly after RS passage over spacecraft.

proton peak was noted and explained in terms of remote magnetic connection from Ulysses to the CIR (9). The concept is illustrated by the sketches in Figures 2 and 3. Fig. 2 represents solar wind stream-lines on a corotating cone of constant latitude that contains the Ulysses trajectory. The CIR is bounded by a forward (FS) and reverse (RS) shock and contains a stream interface (SI). Ulysses appears to move through the CIR, and after the RS is observed the distance from the spacecraft to the RS rapidly increases (9). For a solar wind velocity V=800 km s^{-1} and a velocity V_R=700 km s^{-1} for the RS, the radial distance $\Delta r = VV_R(t_2 - t_1)/(V - V_R) \cong 14$ AU for $t_2 - t_1$=4 days.

We shall now present theoretical arguments that the dominant mechanism in producing the velocity ordering of intensities apparent in Fig. 1 is energy loss as the particles propagate from the CIR inward towards the spacecraft. This process is sketched in Fig. 3, which emphasizes the necessity of magnetic connection from Ulysses at high latitudes outward to the CIR at lower latitudes. The velocity ordering of the intensity histories (regardless of species) follows immediately from the striking result that the energy loss process depends only upon particle velocity (and not on mass or charge) in the limit of very weak scattering.

THEORY

The basic equation for guiding-center transport along solar-wind convected magnetic field lines (unit vector $\mathbf{b} = \mathbf{B}/|\mathbf{B}|$) is

$$d\mathbf{r}/dt = \mu v \mathbf{b} + \mathbf{V}_\perp \quad (1)$$

where v is the particle velocity, μ is its pitch cosine, and $\mathbf{V}_\perp = \mathbf{V} - \mathbf{b}(\mathbf{b}\cdot\mathbf{V})$ is the $\mathbf{E}\times\mathbf{B}$ drift velocity in a perfectly conducting plasma moving with velocity $\mathbf{V}$. Defining $\cos\psi = \mathbf{b}\cdot\mathbf{r}/r$ where r is the helio-radius, and assuming that the solar wind velocity is radial, so that $\mathbf{V} = (V/r)\mathbf{r}$, the radial component of Eq. (1) is

$$dr/dt = \mu v \cos\psi + V_\perp \sin^2\psi \quad (2)$$

The equation for the energy loss in the guiding center approximation has been given by Northrop (14) and recast by Roelof (15):

$$\gamma/(\gamma+1) d\ln T/dt = (1/2)(1-\mu^2)(\partial \ln B/\partial t + \mathbf{V}_\perp \cdot \nabla \ln B) + \mu^2 \mathbf{V}_\perp \cdot (\mathbf{b}\cdot\nabla\mathbf{b}) \quad (3)$$

where $mc^2 = \gamma m_0 c^2$, $\gamma = (1-\beta^2)^{-1/2}$, and $\beta = v/c$. Actually, the LHS of Eq. (3) is an exact derivative.

$$\gamma/(\gamma+1) d\ln T/dt = d\ln(\beta\gamma)/dt \quad (4)$$

The RHS is therefore exactly the time rate of change of the logarithm of the (relativistic) momentum

Faraday's law in perfectly conducting plasma is $\partial\mathbf{B}/\partial t = \nabla\times(\mathbf{V}\times\mathbf{B})$ which may be expanded to give

$$\partial\mathbf{B}/\partial t = \mathbf{V}(\nabla\cdot\mathbf{B}) - (\mathbf{V}\cdot\nabla)\mathbf{B} - \mathbf{B}(\nabla\cdot\mathbf{B}) + (\mathbf{B}\cdot\nabla)\mathbf{V} \quad (5)$$

Then, because $B\partial B/\partial t = \mathbf{B}\cdot\partial\mathbf{B}/\partial t$, we have

$$\partial \ln B/\partial t = -\nabla\cdot\mathbf{V} + \mathbf{b}\cdot(\mathbf{b}\cdot\nabla)\mathbf{V} - \mathbf{V}\cdot\nabla\ln B - \mathbf{b}\cdot(\mathbf{V}\cdot\nabla)\mathbf{b} \quad (6)$$

where the term containing $\nabla\cdot\mathbf{B} = 0$ in Eq. (5) does not appear in Eq. (6) and the last term in Eq. (6) vanishes because it is the component of the derivative of a unit vector parallel to itself. Because only $\mathbf{V}_\perp$ appears in the cross product $\mathbf{V}\times\mathbf{B}$, we can replace $\mathbf{V}$ with $\mathbf{V}_\perp$ everywhere in Eq. (6). When Eqs. (4) and (6) are substituted into Eq. (3), we obtain

$$d\ln(\beta\gamma)/dt = (1/2)(1-\mu^2)[\mathbf{b}\cdot(\mathbf{b}\cdot\nabla)\mathbf{V}_\perp - \nabla\cdot\mathbf{V}_\perp] + \mu^2[\mathbf{V}_\perp\cdot(\mathbf{b}\cdot\nabla\mathbf{b})] \quad (7)$$

There are two remarkable properties of Eq. (7). To start with, even though no time derivative of the magnetic field appears on the RHS of Eq. (7), it is nonetheless correct for a time-dependent magnetic field (in the ideal plasma approximation for Faraday's law). Much more importantly, however, the expression for the fractional rate of momentum change given by

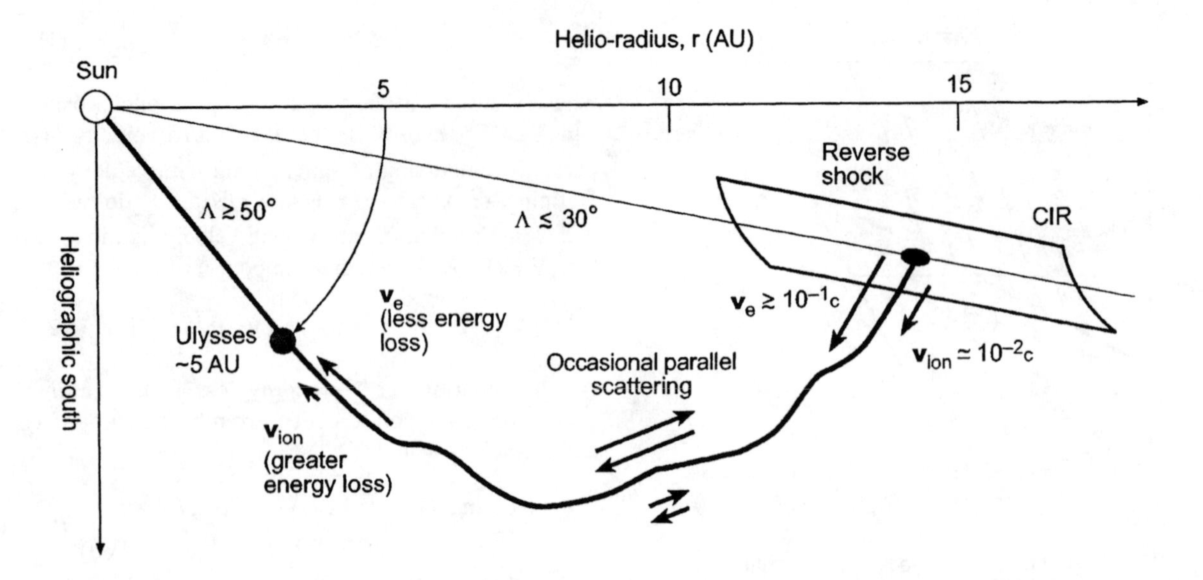

FIGURE 3. Sketch of three-dimensional magnetic connection outward from Ulysses at high latitudes to reverse shock bounding CIR. Fast electrons and slow ions lose energy as they traverse the distance (possibly with some weak parallel scattering), but both fractionally lose energy at the same rate. Consequently, the net fractional energy loss of the slow ions is considerably greater than that of the fast electrons.

the RHS is independent of particle mass, energy, *and* charge (positive or negative). The RHS depends only upon the *square* of the particle pitch-cosine and the local spatial dependence of the plasma velocity and the magnetic field. Consequently *any* two particles with the same pitch cosine (whether moving inward or outward along the field line) will have the *same* instantaneous fractional rate of change of momentum. For example, a 60 keV (non-relativistic) proton and a 1 MeV (relativistic) electron, each having a pitch-cosine equal to $\pm\mu$, will change their momenta at the same fractional rate. This means that the *net* energy loss in propagating from an initial position $\mathbf{r}_0$ to a final position $\mathbf{r}$ depends mainly on the time spent by the particle in getting from $\mathbf{r}_0$ to $\mathbf{r}$. Actually, the final positions will be slightly different, because each particle's magnetic field drift velocity depends on the particle mass, velocity, and charge. Ignoring this second-order effect for a direct guiding-center transit from $\mathbf{r}_0$ to $\mathbf{r}$, the time difference can be calculated by integrating Eq. (1) along with Eq. (7), as we shall now demonstrate for a simple configuration of the solar wind and magnetic field.

In the limit of a Parker spiral field in the distant equatorial heliosphere ($r >> V/\Omega$), where Ω is the sidereal rotation frequency, we find $\cos\psi \rightarrow V/\Omega r$ and $\sin\psi \rightarrow 1$ so that $\mathbf{B}$ is nearly azimuthal. Therefore $\mathbf{V}_\perp \rightarrow (V/r)\mathbf{r}$ so that $\nabla\cdot\mathbf{V}_\perp \rightarrow 2V/r$. Also $\mathbf{V}_\perp\cdot(\mathbf{b}\cdot\nabla\mathbf{b}) \rightarrow \mathbf{V}\cdot(-\mathbf{r}/r^2)$, so that in the outer equatorial heliosphere the limiting forms of Eqs. (3) and (7) are

$$dr/dt = \mu\beta cV/\Omega r + V \tag{8}$$

$$d\ln(\beta\gamma)/dt = (1+\mu^2)(-V/2r) \tag{9}$$

Let us pause here for an aside. This set of equations describing guiding center motion in the outer heliosphere contains only two parameters, the solar wind velocity V and the characteristic distance for a Parker field $R_P = c/\Omega$=715 AU for Ω=2π/(26days). This scaling distance R_P is the equatorial helio-radius at which the corotating Parker field would reach the speed of light (according to Newtonian physics). If we were to invoke the third guiding center relation (thus forming a complete set of equations for r, v, μ, t), we would add the first adiabatic invariant $p^2(1-\mu^2)/B$= constant. However, this would not introduce any additional parameters, because $p \propto \beta\gamma$ and $B \propto 1/r$. Thus R_P and V are the two basic parameters in determining energy loss rates in an outer (ideal) Parker heliosphere.

Although these equations describe a distant equatorial Parker field, they should also be roughly correct for estimating energy loss during scatter-free particle propagation in a mid-latitude Fisk field (10) or alternatively, a stochastic Parker field (11). The ideal

Parker field lines lie on cones of constant heliographic latitude, whereas the field lines must actually cross cones of constant latitude in order to connect Ulysses (at high latitudes) to the only likely site for interplanetary particle acceleration (the CIRs at low latitudes. The Fisk (or stochastic Parker) field lines connecting different latitudes are longer than the ideal Parker field lines that lie on cones of constant latitude. However, the equatorial ideal Parker spirals are also longer (at the same radius) than the mid-latitude Parker spirals. Since we only need approximate estimates for our discussion, we will therefore use the mathematically simpler distant equatorial Parker field expressions in Eqs. (8) and (9), even though we are applying these equations to mid-latitude fields that are more appropriately describable by either the Fisk or stochastic Parker model.

We now calculate the energy loss of a charged particle in moving outward from an initial radius r_0 to a final radius r. It will be sufficient for our purposes here to integrate Eqs. (8) and (9) in their non-relativistic form, so we set $\gamma = 1$. This is because the fastest particles we will be discussing are ~50 keV electrons which have $\gamma = 1.1$. For the situation of inward propagation of particles from the CIRs at low latitudes to Ulysses at high latitudes, we will be interested in the minimum possible energy loss. We are then rigorously justified in setting $\mu = -1$ in Eqs. (8) and (9) because we thus obtain the exact energy loss for field-aligned travel (which takes the least amount of time).

Here is another aside. We will not immediately set $\mu = -1$ and $\mu^2 = 1$ in Eqs. (8) and (9) because it will be interesting to see what is predicted by just holding μ =constant in the integrations. Our motivation for this approach is that we believe there is very weak scattering of the parallel velocity that violates the first adiabatic invariant. Because of the focussing due to the decreasing magnetic field, this scattering can produce a local "equilibrium" pitch-cosine distribution that can be characterized by a mean value that we identify with "equivalent" values of the functions of μ that appear in the equations. To be at all useful, these values should depend only weakly on position. The values have to be estimated on a case-by-case basis (depending on the propagation conditions), and the estimates should be validated from either observations or simulations. For example, if the pitch-cosine distribution is nearly isotropic (so that the equivalent value of μ is small compared to unity), then the equivalent value assigned to μ^2 would be 1/3. Substituting this value into Eq. (9), we recover the familiar "two-thirds" law for momentum loss due to adiabatic deceleration. We have found that this "equivalent-μ" approximation, although not in general mathematically rigorous, can offer convenient insight (owing to the ease of computation) into the propagation of energetic particles in the heliosphere. However, it must be applied judiciously and consistently and requires independent crosschecks on its validity. The paper by Decker et al. (16) in these Proceedings offers an independent example of how the "equivalent-μ" approach can be applied to the analysis of energetic particle propagation in the outer heliosphere.

Returning to our calculation, we first obtain an "orbit" equation (in which time does not appear explicitly), by taking the ratio of Eq. (9) to Eq. (8). We then consider r to be a function of v, so that $(dr/dt)/(dv/dt) = dr/dv$, obtaining

$$dr/dv = -k(\mu/\Omega + r/v) \qquad (10)$$

where $k = 2/(1+\mu^2)$ is a weak function of μ ($1 \le k \le 2$). Note that the solar wind velocity has cancelled out of this orbit equation. Next, we substitute the expression for v in terms of dv/dt from Eq. (9) into Eq. (8) and integrate with respect to time to obtain

$$r = r_0 + Vt - (1/\Omega)\int dt\, \mu k\, dv/dt \qquad (11)$$

We now can integrate both Eqs. (10) and (11) (holding μ, k, and μk constant at their equivalent values) to obtain

$$r = r_0(v_0/v)^k + (\mu k v/\Omega)/(k+1)[(v_0/v)^{k+1} - 1] \qquad (12)$$

$$r = r_0 + Vt + (\mu k/\Omega)(v_0 - v) \qquad (13)$$

If we specify the observed velocity v at r in the inner heliosphere, then the initial velocity v_0 that the particle had when it left the CIR at r_0 is determined by Eq. (12). As we anticipated above, this velocity ratio v_0/v does not depend on the solar wind velocity V. However, when the set of values (r, r_0, v, v_0) is then substituted into Eq. (13), the time can be extracted as the product Vt (the distance traveled outward by the solar wind while the particle is traveling inward from the CIR).

We now set $\mu = -1$ (the particles are moving inward toward r from the CIR boundary at r_0) so that $k = 1$ and $\mu k = -1$. Then Eq. (13) becomes quadratic in the velocity ratio v_0/v with the solution

$$v_0/v = \Omega r_0/v + [1 - 2\Omega r/v + (\Omega r_0/v)^2]^{1/2} \qquad (14)$$

We excluded the negative sign before the square root because it always yields a solution with $v_0/v<1$. Let us examine the behavior of the positive square root solution in the two limits of high and low velocities (relative to the corotation velocities Ωr and Ωr_0).

$$v_0/v \to 1+(\Omega/v)(r_0-r)+(1/2)(\Omega/v)^2(r_0^2-r^2)+\dots,\quad \Omega r/v \ll 1 \text{ and } (\Omega r_0/v)^2 \ll 1 \tag{15}$$

$$v_0/v \to 2\Omega r_0/v - r/r_0 + (v/\Omega r_0)(1-r^2/r_0^2)+\dots,\quad \Omega r_0/v \gg 1 \tag{16}$$

For high energies, Eq. (15) states that there is very little velocity change for the minimum-time trajectory, it being equal to the difference in the corotation velocities divided by the final velocity. We can look at this result another way. The distance Δs along a Parker spiral in the outer heliosphere from r to r_0 is $\Delta s \cong (\Omega/2V)(r_0^2-r^2) = (1/2)(r_0+r)(\Omega/V)(r_0-r)$, so this limit may be rewritten as $v_0/v \to 1+2V/(r_0+r)(\Delta s/v)+\Omega V(\Delta s/v^2)$. This result is therefore in agreement with Eq. (9), because $\Delta s/v=\Delta t$ is the transit time along the field, and $\langle r\rangle=(r_0+r)/2$ is the average radius, so $v_0/v \to 1+V\Delta t/\langle r\rangle$ or $\Delta v/v\Delta t=-V/\langle r\rangle$. The next correction term is then $(\Omega\Delta t)(V/v)$. On the other hand, in the low-velocity limit, Eq. (16) states that the particle has to start out from the CIR boundary with at least twice the corotation velocity at r_0 in order to reach the spacecraft at radius r with a residual velocity that is very much less than the original velocity.

We can get a better feeling for the behavior predicted from Eq. (14) by writing $\Omega r_0/v = r_0/(\beta R_P)$ and $\Omega r/v = r/(\beta R_P)$, where we have defined $R_P = c/\Omega$=715 AU for $\Omega/2\pi/(26\text{days})$. For 40 keV electrons, β=0.4, so for r_0=15AU and r=5AU, these dimensionless parameters are $\Omega r_0/v$=0.052 and $\Omega r/v$=0.01. These electrons are described by the limit in Eq. (15). On the other hand, 50 keV electrons with β=0.01 take the values $\Omega r_0/v$=2.1 and $\Omega r/v$=0.70. They therefore are not described by the just the first term (twice corotation) of the low velocity limit of Eq. (16), but require several more terms in the expansion.

Finally, the minimum transit time is immediately obtained by substituting Eq. (14) into Eq. (12)

$$Vt = r - r_0 + (v/\Omega)(v_0/v - 1) \tag{17}$$

The limiting values, corresponding to Eqs. (15) and (16), are

$$Vt \to (r_0^2-r^2)(\Omega/2v)+\dots,\quad \Omega r/v \ll 1 \text{ and } (\Omega r_0/v)^2 \ll 1 \tag{18}$$

$$Vt \to (r+r_0)(1-v/\Omega r_0)+(v/\Omega r_0)^2(r_0-r^2/r_0),\quad \Omega r_0/v \gg 1 \tag{19}$$

The minimum transit time in the high velocity limit agrees with $\Delta s/v$, as derived in the discussion following Eq. (15). This is because there is very little energy loss, so $v \cong v_0 \cong$constant. However, in the low velocity case, the minimum time for the particle to come in from r_0 to r is $t \to (r+r_0)/V$ which exceeds by $2r/V$ the transit time $(r_0-r)/V$ of the solar wind outwards from r to r_0.

DISCUSSION AND SUMMARY

The full relationships that give v_0/v and Vt (in AU) as functions of the final velocity (expressed as $\beta = v/c$) from Eqs. (14) and (17) are plotted in Figures 4 and 5 for the non-relativistic range β<0.5. We can now get the exact values of the minimum energy loss and transit time for 50 keV electrons and protons. The 50 keV electrons (β=0.4125) have v_0/v=1.035 and Vt =0.345 AU, while the 50 keV protons (β=0.0103) have v_0/v=3.98 and Vt=12.0 AU. Thus the 50 keV electrons lose ~7% of their energy during a minimum transit time of ~1.4 days (for V=420 km/s) that it takes to cover ~100 AU along the spiral. However, the 50 keV protons started out at 15 AU with at least 800 keV of energy, and their minimum transit time is 48 days!

These numbers immediately explain the drastic attenuation of 50 keV (and even the 500 keV) protons within a few days after the passage of the vestigial reverse wave on CIR 25. The ion intensity at the RS is much less at the initial energy of 800 keV than it is at the final energy of 50 keV at Ulysses, because of the steep ion energy spectrum observed in CIRs by Voyagers 1 and 2 near 14 AU (17). The gradual rise in the electron intensity at Ulysses after the RS passage (that produces the lag behind the low-energy proton), must mean that the 50 keV electron intensity at the RS near 15 AU must be greater than it is at 5 AU.

There is another remarkable aspect to these results. Figs. 4 and 5 show that 0.5 MeV protons observed at 5 AU begin leave 15 AU energies greater than 3 MeV, but they take at least 20 days in transit. This means that 1 MeV protons at Ulysses were probing the release of >2 MeV protons at the CIR more than one half a solar rotation period *earlier*. In other words, the configuration of the CIR when they left it was more like it had been on the solar rotation *previous* to the one when they were observed. The fact that the 26-day-recurrent energetic electron events at Ulysses were so "clock-like" (9) at high latitudes during 1993 and 1994 argued for a remarkably stable structure of the heliosphere at this time. The recurrence of the ions is actually an even more sensitive test of the stationarity of the field structure (because of their ~1 solar rotation transit times) than the much faster electrons. The lack of strict 26-day recurrences in the current Ulysses southern pass can be assigned to the higher solar activity that disrupts CIR structure.

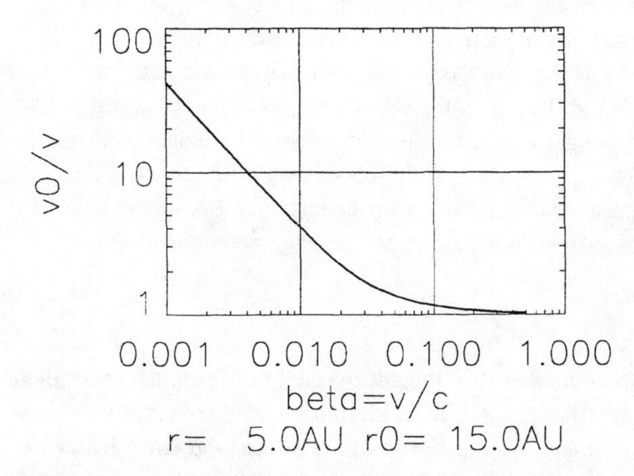

FIGURE 4. Ratio of initial velocity v_0 at CIR (15 AU) to final velocity v at Ulysses (5 AU) as a function of initial velocity expressed as $\beta = v/c$.

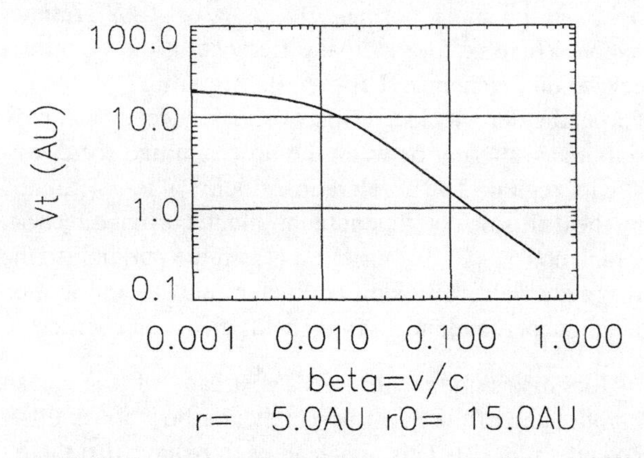

FIGURE 5. Particle transit time as a function initial velocity $\beta = v/c$ from the CIR (15 AU) to Ulysses (5 AU) expressed in terms of the distance Vt (AU) that would be traveled by the solar wind during the particle transit.

In summary, the process of energy loss explains the dominant trends observed in the systematic variation of energetic particle intensities observed by Ulysses at high latitudes, in both the southern and northern hemispheres of the heliosphere. A fully quantitative comparison with data still awaits detailed models for CIR structure and evolution, magnetic connection to the CIRs, and particle acceleration efficiency as a function of radius and latitude.

ACKNOWLEDGMENTS

I am indebted to Dr. and Mrs. Walker Fillius for their hospitality and technical support during the nascence of this paper and to my valued colleague Robert Decker for his insightful comments during its development. This work supported in part by the Ulysses HI-SCALE effort under NASA Grant NAG5-6113 and by the Voyager Interstellar Mission under NASA Grant NAG5-4365.

REFERENCES

1. Roelof, E.C., Simnett, G.M., and Tappin, S.J., *Astron. Astrophys.*, **316**, pp. 481-486 (1996).
2. Sanderson, T.A. *et al.*, in *Solar Wind 8*, edited by D. Winterhalter *et al.*, *AIP Conf. Proceedings,* 1996, pp. 411-414.
3. Bame, S.J. *et al., Geophys. Res. Lett.*, **20**, pp. 2323-2326 (1993).
4. Sanderson, T.A. *et al., Geophys. Res. Lett.*, **21**, pp. 1113-1116 (1994).
5. Simnett, G.M., Sayle, K., Roelof, E.C., and Tappin, S.J., *Geophys. Res. Lett.*, **21**, pp. 1561-1564 (1994).
6. Smith, E.J., and Wolfe, J.H., *Geophys. Res. Lett.*, **3**, pp. 137-140 (1976).
7. Gosling, J.T. *et al., Geophys. Res. Lett.*, **20**, pp. 2789-2792 (1993).
8. Pizzo, V.J., *J. Geophys. Res.*, **89**, 4173-4183 (1994).
9. Simnett, G.M., and Roelof, E.C., *Space Sci. Rev.*, **72**, pp. 303-308 (1995).
10. Fisk, L.A., *J. Geophys. Res.*, **101**, pp. 15547-15553 (1996).
11. Kota, J. and Jokipii, J.R., *Space Sci. Rev.*, **89**, pp. 240-242 (1999).
12. Roelof, E.C. *et al., J. Geophys. Res.*, **102**, pp. 11251-11262 (1997).
13. *Corotating Interaction Regions*, Kluwer Academic Publishers (Dordrecht), reprinted from *Space Sci. Rev.*, **89**, nos. 1-2, 1999.
14. Northrop, T.G., *The Adiabatic Motion of Charge Particles*, Interscience Publishers (John Wiley & Sons), New York, 1963.
15. Roelof, E.C., *Space Sci. Rev.*, **89**, pp. 238-240 (1999).
16. Decker, R.B. *et al., This Conference*, (2000).
17. Gold, R.E. *et al., J. Geophys. Res.*, **93**, pp. 991-996 (1987).

Injection and Acceleration of Ions at Collisionless Shocks: Kinetic Simulations

M. Scholer[1], H. Kucharek[1], V. V. Krasnosselskikh[2], and K.-H. Trattner[3]

[1]*Max-Planck-Institut für extraterr. Physik, 85740 Garching, Germany*
[2]*LPCE-CNRS, 45045 Orleans, France*
[3]*Lockheed Martin Missiles and Space, Palo Alto, CA 94304*

Abstract.
Kinetic simulations of collisionless shocks have provided a wealth of information on injection and acceleration of thermal ions into a diffusive acceleration process. At quasi-parallel shocks upstream diffuse ions and the induced upstream turbulence are an integral part of the collisionless shock structure. Before injected into a diffusive acceleration process thermal ions are trapped near the shock and are accelerated to higher energies. The injection and acceleration process for thermal ions at quasi-perpendicular shocks depends on the possibility of these ions to recross the shock many times. A viable mechanism for injection is cross-field diffusion of the specularly reflected ions after they have crossed the shock into the downstream region. Determination of the cross-field diffusion coefficient in strong turbulence suggests that specularly reflected ions can recross the quasi-perpendicular shock and can get further accelerated. At more oblique shocks the same injection process as at quasi-parallel shocks can work: particles gain high enough velocities during their first shock encounter so that they can escape the shock along the magnetic field in the upstream direction. Because of the form of the pickup ion distribution in velocity space there seems to be no problem for accelerating these ions at either quasi-parallel, quasi-perpendicular, or perpendicular shocks.

INTRODUCTION AND BACKGROUND

Shocks ara an important site for particle acceleration in astrophysical settings. In particular the first order Fermi acceleration mechanism as formulated by Krimsky (1), Axford et al. (2), Blandford and Ostriker (3), and Bell (4) has been widely employed in cosmic ray physics. The diffusive shock acceleration theory is not concerned with the question how a certain part of the ambient particles is injected into the acceleration process, but starts with suprathermal seed particles, which can either be already present in the upstream flow or are injected at the shock. One of the important questions in ion acceleration at collisionless shocks is how ions are extracted from the thermal population and are injected into a subsequent acceleration process, which usually is assumed to be diffusive shock acceleration. By thermal population we mean here and in the following the upstream plasma, i.e., in case of the heliosphere the solar wind. This so-called injection problem is of great interest, since once we understand injection we may be able to predict the flux in the suprathermal energy range at the shock. Standard diffusive shock acceleration theory, with possible complications as adiabatic deceleration in the solar wind etc, will then predict spectral shapes and absolute fluxes of the accelerated particles. Most detailed information on ion acceleration comes from in situ measurements at the Earth's bow shock and, to a lesser extent, at interplanetary traveling shocks and shocks bounding corotating interaction regions. Shocks are usually divided into quasi-perpendicular and quasi-parallel depending on whether the angle between the upstream magnetic field and the shock normal, Θ_{Bn}, is $>$ or $<$ 45°, respectively. There is no principal difference in the Fermi acceleration mechanism between quasi-parallel and quasi-perpendicular shocks: diffusive shock acceleration only requires scattering between the upstream and the downstream region. The acceleration efficiency is determined by the diffusion coefficient in the shock normal direction, which consists of a contribution from the parallel (to the magnetic field) diffusion coefficient and a contribution from the perpendicular (cross-field) diffusion coefficient.

Ion distributions observed upstream of the quasi-parallel and of the quasi-perpendicular bow shock differ considerably (5). This may lead us to believe that the injection mechanism also differs for the two regimes. There are two problems with using in situ particle observations obtained upstream of the Earth's bow shock in order to delineate the seed particles: firstly, bow shock observations are complicated by the fact that the ion distribu-

CP528, *Acceleration and Transport of Energetic Particles Observed in the Heliosphere: ACE 2000 Symposium*, edited by Richard A. Mewaldt, et al.
© 2000 American Institute of Physics 1-56396-951-3/00/$17.00

tions from the region upstream of the quasi-perpendicular shock, like field-aligned beams, are convected by the solar wind into the quasi-parallel regime and can thus interact with the quasi-parallel bow shock. Secondly, observations can not give us information about the initial distribution which is subsequently further accelerated; one observes a more or less final state and it is not possible to distinguish between ions just having been injected or those which have been scattered between the upstream and downstream medium many times and partake in a Fermi acceleration process. The latter also holds for interplanetary shocks.

However, the question arises whether a distinction between injection and acceleration is at all possible. According to the view held by Malkov and Völk (6) "the problem of injection consists not so much in the source of the particles to be injected, but in the way to describe their acceleration out of the thermal distribution". A natural assumption for acceleration out of the thermal distribution is an extension of the diffusive acceleration theory down to thermal energies, i.e., that also part of the shock heated solar wind plasma can freely scatter across the shock many times. In the case of quasi-parallel shocks the upstream leaking thermal particles are simply the hot downstream ions which have an upstream directed velocity which exceeds the shock velocity (both taken in the upstream rest frame). This model has first been developed by Ellison (7) using Monte Carlo particle simulations, wherein particle trajectories are determined from a prescribed scattering law (elastic scattering). These simulations can model thermal particle injection and acceleration and determine simultaneously the average shock structure, including shock mediation by the accelerated particles. Predictions of this model for spectra and abundances agree favorably well with the observations at the quasi-parallel bow shock, which is maybe not too surprising, considering the fact that the scattering law contains several free parameters, as rigidity dependence and absolute value, which have to be adjusted. Ellison et al. (8) have extended their Monte Carlo model for oblique shocks by assuming that particles are scattered parallel and perpendicular to the field according to a hard sphere scattering law, i.e., the perpendicular diffusion coefficient is given by $\kappa_\perp = \kappa_{||}/(1+\eta^2)$ where η is related to the parallel mean free path $\lambda_{||} = 3\kappa_{||}/v$ by $\lambda_{||} = \eta r_g$ with r_g and v as the particle's gyroradius and velocity, respectively. The parameter η determines the scattering strength, and injection and acceleration becomes efficient when scattering is strong ($\eta << 10$).

The same idea, i.e., that also part of the shock heated solar wind plasma can freely scatter across a shock many times, has been put on an analytical basis by Malkov and Völk (6). They have extended the theory of diffusive particle acceleration to low particle energies, where the difference between the upstream and the downstream fluid frame is essential and where the particle distribution at the shock may become highly anisotropic. In their model for parallel shocks wave excitation and pitch angle scattering are treated self-consistently by assuming that pitch angle scattering is due to self-excited MHD waves propagating along the ambient magnetic field. These waves are excited in cyclotron resonance due to the pitch angle anisotropy of the backstreaming ions, i.e., by an electromagnetic ion/ion beam instability. Quasi-linear theory results in two coupled equations for the evolution of the particle distribution and the wave spectrum. The solution for the particle spectrum contains the source of the particles to be injected. As the source of the injected particles Malkov and Völk (6) also assume downstream heated particles with a velocity exceeding the shock velocity, although other sources can, in principle, be incorporated. To summarize: the Monte Carlo model (7) assumes that the heated downstream solar wind scatters according to a hard sphere scattering law and thus circumvents the injection problem; the model predicts absolute fluxes and spectra. Theory (6) is an extension of the diffusive acceleration theory valid in the suprathermal energy range.

INJECTION AND ACCELERATION AT QUASI-PARALLEL SHOCKS

First results on the direct injection of ions out of the incident thermal plasma at a collsionless shock into the Fermi acceleration mechanism were based on hybrid simulations of an exacactly parallel shock (9). In hybrid simulations the ions are treated as macroparticles and the electrons are represented by an inertialess electron fluid. Thus frequencies of the order of the ion gyrofrequency and smaller, and length scales of the order of the ion inertial length are treated correctly. Following the work by Quest (9), Scholer (10), Kucharek and Scholer (11) and Giacalone et al. (12) have performed hybrid simulations of quasi-parallel shocks which resulted in upstream diffuse proton densities of 3-4% of the incident solar wind proton density. Giacalone et al. (12) introduced artificially a high level of upstream magnetic field fluctuations, so that the injected ions were efficiently scattered back and forth between the upstream and downstream region. This led to the build up of a power law distribution of the diffuse particles in the low energy region as predicted by steady state diffusive acceleration theory. Trattner and Scholer (13, 14) included self-consistently He^{2+} in their shock simulations and found that a few percent of the upstream He^{2+} ions are extracted at the shock and are subsequently further accelerated.

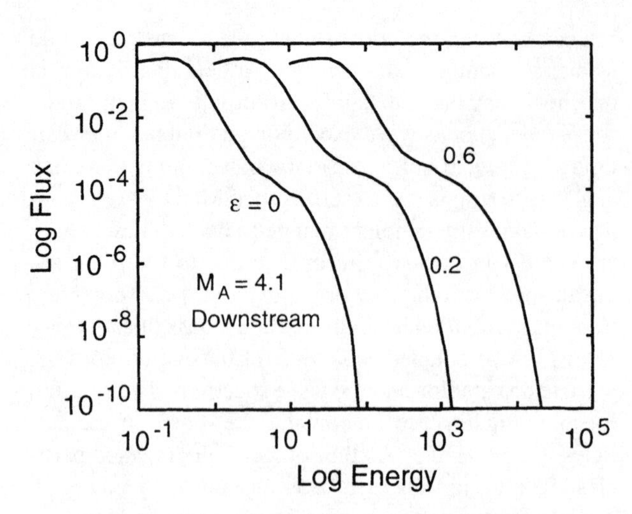

FIGURE 1. Differential flux (arbitrary units) of protons downstream of simulated shock ($\Theta_{Bn} = 5°$, Alfvén Mach number = 4.1)for 3 different values of upstream imposed magnetic field turbulence (from 15).

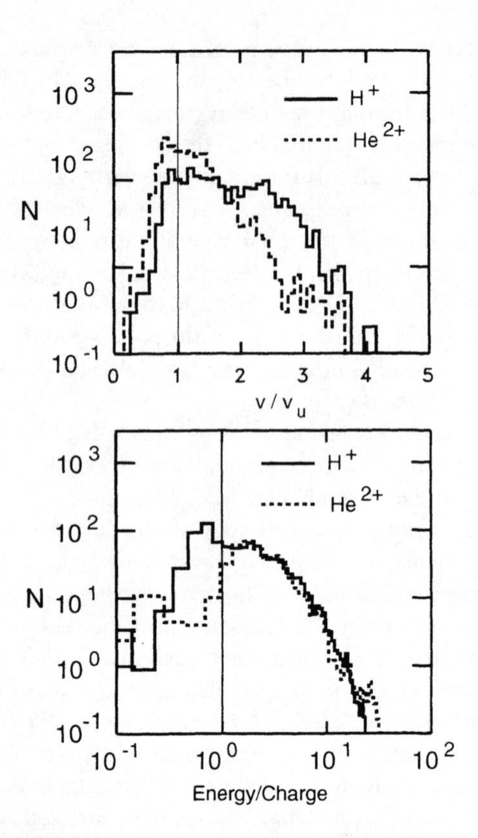

FIGURE 2. Histogram of number of H and He^{2+} ions versus velocity in units of shock velocity (upper panel) and energy per charge in units of shock ram energy (lower panel), when the particles cross for the first time a boundary 10 ion inertial length upstream of the shock (from 15).

Scholer et al. (15) performed quasi-parallel shock simulations for a wide range of parameters. In particular, they were concerned with the spectral shape of diffuse ions. Figure 1 shows downstream energy spectra (differential flux) in a log-log representation for 3 different values of upstream imposed turbulence (ε = total integrated upstream power in the magnetic field fluctuations). Energy is in units of the shock ram energy $E_p = m_p v_u^2/2$, where v_u is the upstream bulk speed relative to the shock and m_p the proton mass. One can see the heated downstream plasma at low energies and a second population with a high energy cut-off, which is simply due to the fact that the acceleration time is limited. The spectra of the accelerated particles can be expressed as exponentials in energy; the e-folding energy (normalized to the shock ram energy) increases with time (which can be transformed at the bow shock into magnetic field connection time) and with ε, the level of upstream turbulence. At the same solar wind ram velocity the e-folding energy increases with shock Mach number. Spectra of diffuse protons and alpha particles are known to exhibit e-folding energies which are about equal in energy per charge. In order to explain this in terms of the Fermi acceleration model it has been assumed in the past that there is no or very weak scattering beyond some distance upstream of the bow shock (free escape boundary). If the diffusion coefficients for protons and He^{2+} ions are identical at the same energy per charge steady state Fermi theory with a free escape boundary predicts for the spectra equal e-folding energies in energy per charge. However, Scholer et al. (15) have argued that the introduction of a free escape boundary is an artifact introduced in order to obtain exponential spectra by a steady state theory, since steady state shock acceleration theory in an infinite medium predicts power law distributions. In particular a Monte Carlo simulation is intrinsically steady state; in such a theory a spectral cut-off can only be obtained by a free escape boundary or some other loss process. Lee (16) has actually proposed cross-field diffusion to the flanks of the bow shock as a possible loss process leading to exponential spectra.

What can we learn from the hybrid simulations as far as injection is concerned? In order to obtain the "injection spectrum" we imagine a boundary close to the shock. We then determine the distribution of the solar wind ions which later in their life become diffuse upstream ions when they cross this boundary the first time in the upstream direction. Figure 2, taken from Scholer et al. (15), shows in the upper panel the number of backstreaming ions versus velocity in units of the shock ram velocity v_u (upstream bulk velocity in the shock frame) and versus energy per charge $(E/E_p)/Q$ (lower pane). In this run He^{2+} has been included self-consistently. The upper panel shows the following: a large contribution to the spectrum consists of ions, which have in the shock frame about shock ram velocity ($v/v_r \sim 1$), i.e., which are specularly reflected at the shock. However, the injection spectrum extends to several tens of the shock ram energy and is almost identical for both species when evaluated at equal energy per charge. Firstly, it is not necessarily the subsequent diffusive shock acceleration process, which determines the ordering in energy per charge; the injection spectrum exhibits already such an ordering. Secondly, injection is not from the hot downstream distribution. If it were so, the injection spectrum should not drop to zero at zero velocity. If heated downstream ions escape upstream the maximum intensity is expected to be at $v = 0$ in the shock frame which, as can be seen from Figure 2, is clearly not the case. Such a simplified kinematic leakage model does not take into account that the downstream particles are trapped in the large amplitude downstream waves originating upstream from the turbulence excited by the backstreaming ions. Malkov (17) has suggested a model where the large amplitude downstream waves efficiently trap the thermal ions and regulate their upstream leakage. We will come back to this model later.

One can gain more insight by investigating individual particle orbits from a typical quasi-parallel shock run. Figure 3 shows in the top panel the trajectory of a proton in $x-t$ space which ends up as a diffuse upstream ion. x is the direction normal to the shock. The heavy line is the shock position; upstream is to the left, downstream to the right. The unit of distance is the upstream proton inertial length λ_o (equal to the gyroradius for a beta =1 plasma) and the unit of time is the inverse proton gyrofrequency Ω^{-1}. Because of technical reasons the simulation frame is the downstream rest frame. In this frame the shock moves in the upstream direction. The thin line is the trajectory of a solar wind ion which moves toward the shock. It reaches the shock at $\Omega t \sim 80$ and stays for about 3 gyroperiods in the vicinity of the shock before escaping upstream. The lower panel shows a plot of perpendicular velocity $v_\perp$ versus parallel velocity $v_{||}$ during this time interval. The particle reaches the shock and is reflected. Subsequently it's perpendicular velocity increases by about a factor 3. It then bounces back and forth between the shock and the upstream region, while the perpendicular energy continues to increase.

Analysis of many trajectories like the one shown in Figure 3 suggest that the acceleration process in the thermal and suprathermal energy range is not well described by a diffusive process. It is still true that the problem of injection consists not so much in the source of the particles to be injected, but in the way to describe their acceleration out of the thermal distribution; however, the simulations seem to tell us that diffusive theory is not an adequate description. This is not to say that there is no diffusive shock acceleration. The backstreaming ions are

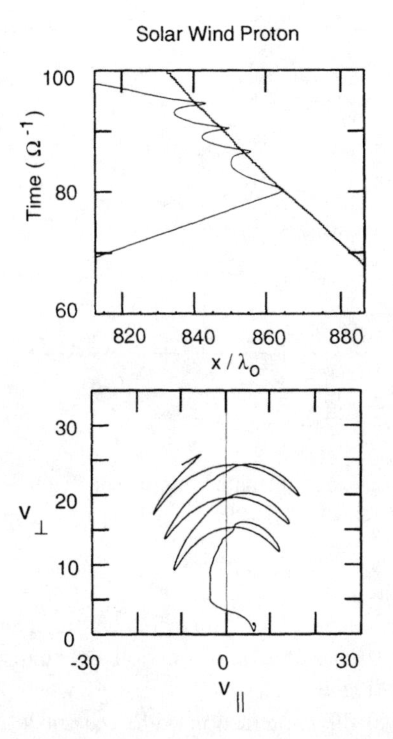

FIGURE 3. Trajectory of a solar wind proton which becomes energized at the shock in $x-t$ space (upper panel), and the perpendicular versus parallel velocity during this time interval (lower panel)

actually scattered in the upstream wave field, return to the shock, and are further accelerated. Figure 4 shows a trajectory of a proton over an extended time interval. After initial contact with the shock the particle moves upstream, interacts again with the shock for an extended time and moves again upstream. This stage can be described by standard diffusion theory, eventually taking into account the particles anisotropy. But what seems to be also important is the fact that the particle is not scattered from downstream, but is reflected at the shock ramp. In this energy regime the forces acting on the particles at the shock ramp, like shock potential and magnetic mirror forces, and thus the physics at the shock transition is of importance. A simple boundary condition, as the constancy of the particle streaming, does not necessarily describe correctly the ongoing physics. Likewise, a Monte Carlo model which assumes a step-like change in the velocities of the scattering centers, eventually mediated by the energetic particles, can not correctly describe the physics at the shock ramp.

The simulations discussed above suggest that the problem of injection and acceleration can be divided into three tasks: 1. Given the shock conditions, one has to determine the distribution of particles originating from the upstream flow which are reflected from the shock or es-

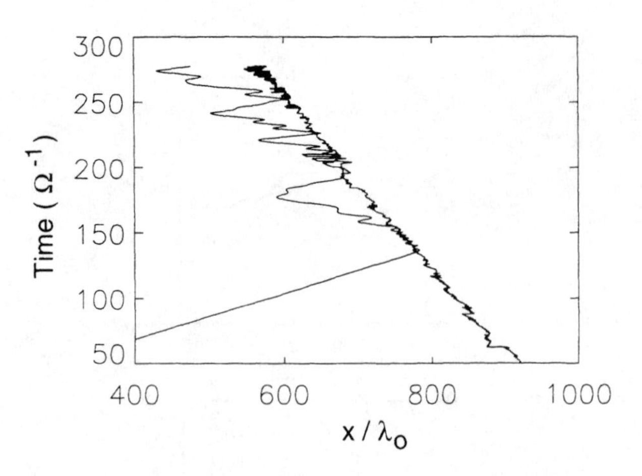

FIGURE 4. Trajectory of a diffuse ion showing multiple interactions with the shock and scattering in the upstream medium

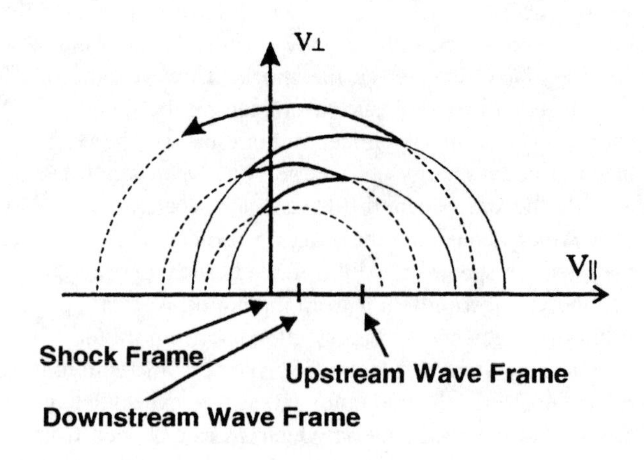

FIGURE 5. Schematic of the velocity space trajectory of a particle during repeated crossing of the shock front (from 18).

cape from downstream. 2. One follows trajectories of these particles when they have multiple encounters with the shock. 3. The particles have an energy which is high enough so that their interaction withe the shock can be described by standard diffusive shock acceleration theory, possibly modified due to anisotropies. A model for steps 2 and 3 has been developed within the framework of quasi-linear theory in (6). However, since phase-mixing is assumed in (6), this theory does probably not correctly describe step 2. A description of the possible physics concerned with step 2 based on hybrid simulation results has been given by Sugiyama and Fujimoto (18). The large-amplitude upstream and downstream waves lead to non-linear phase-trapping of the particles injected in step 1: the pitch angle of non-resonant particles changes over a wide range within one gyroperiod. This leads to rapid crossing from upstream to downstream and vice versa. Since the energy in each wave frame is constant and the phase velocity changes drastically from upstream to downstream a particle moves, depending on whether it is upstream or downstream, on circles with different centers in $v_\perp - v_{||}$ velocity space and moves to higher and higher velocities. This is schematically shown in Figure 5 (from 18). However, so far no theory has yet been developed for step 2 which results in a source function, which is then the input for a diffusive theory (step 3).

The theory developed in (17) for step 1 is a thermostat model, i.e., particles escape into the upstream region out of a hot downstream plasma. The large amplitude downstream waves filter the hot plasma in its leakage upstream by resonant interaction. The model is self-consistent: when the beam of escaping ions is weak, it excites only small amplitude waves and the leakage will be increased to produce stronger waves. Escaping ion intensity and turbulence amplitude rest at some definite and unique level. The theory developed by Malkov (17) is so far the only theory which quantitatively predicts the injection rate and the turbulence level as a function of Alfvén Mach number and shock compression ratio.

We propose as an alternative possibility to the model of leakage from a downstream thermostat a model for step 1, which is rather similar to the shock surfing mechanism developed for quasi-perpendicular shocks (19, 20). We assume that a solar wind ion is trapped at the shock ramp in its motion along the magnetic field and is accelerated perpendicular to the magnetic field by resonance in the upstream circularly polarized wave. The resonance condition is $\omega - k_{||}v_{||} = \pm\Omega$, where ω is the wave frequency, $k_{||}$ the wave vector parallel to the upstream magnetic field, and $v_{||}$ the parallel velocity of the particle. The acceleration can exist at the shock as long as the particles is trapped, i.e., $v_{||}$ is constant and zero in the shock frame. In order for this to happen the force of the electromagnetic wave in the parallel direction toward the shock has to be balanced by some other force at the shock. This force can be either the cross-shock potential or the electromagnetic force in the shock ramp due to the increase in the tangential magnetic field component in quasi-parallel shocks. In order to have $v_{||}$ constant a specific spatial dependence of the trapping force is required. Eventually the particle is detrapped and leaves the shock in the upstream or downstream direction. The maximum energy gain by this process is limited by the time the particle stays trapped. So far, this theory is not self-consistent, since it does not predict the injection rate and the relation to the upstream wave amplitudes. To summarize: the Malkov theory (17) for step 1 depends on the downstream region being a thermostat. It predicts which particles are detrapped in the downstream turbulence and escape upstream, leading to upstream turbulence and determining the downstream wave amplitudes and is a leakage model. In our model it is assumed that some solar wind ions perform a non-adiabatic motion when they reach the shock so that their parallel velocity is small. The particles stay trapped at the shock by the electric and magnetic forces in the shock ramp and by the electromagnetic force exerted by the incoming wave, and are accelerated in gyroresonance with the upstream waves convected into the shock.

We have only briefly mentioned composition. Theory (17) makes also predictions about the dependence of the injection rate on the mass to charge ratio of different species. However, since this theory relies on the downstream region being a thermostat, assumptions have to be made about the downstream thermal distribution function of the minor ions. Simulations have shown that depending on Mach number and beta (upstream thermal to magnetic field pressure) the minor ions are not thermalized on the same scale as the protons. Thus a thermostat model is not appropriate. In simulations of quasi-parallel collisionless shocks where the upstream plasma contains a few percent alpha particles as a minor component it is found that the alpha particle of the incident solar wind gyrate downstream as a coherent beam before being finally thermalized far downstream of the shock ramp (14). During the gyration this beam reaches occasionally the shock front and bunches of alpha particles escape upstream. They are subsequently scattered in the upstream waves and can interact again with the shock, thus gaining higher energies.

A final question is from where in velocity space the injected and accelerated particles originate. In the simulations in (14) the ions were sorted according to their initial velocity in the upstream rest frame (peculiar system). The thermal speed of the upstream plasma is v_{th}. Ions were subdivided into bins with equally spaced thermal speed and the ratio of diffuse ion density to the ion density in each subpopulation was determined. No diffuse ions originate from the core, i.e., from $0 < v < 0.5v_{th}$, whereas the ratio of diffuse ions to incident ions in the $v > 2v_{th}$ bin is, depending on the plasma beta, of the order of 40%. Thus diffuse ions originate from the outer region (in the peculiar system) of the distribution in velocity space.

INJECTION AT QUASI-PERPENDICULAR SHOCKS

An important question in shock acceleration studies is whether upstream thermal ions can be efficiently injected and accelerated at quasi-perpendicular shocks, i.e., shocks with $\Theta_{Bn} > 45°$. Backstreaming ions have been observed at the Earth's quasi-perpendicular bow shock; however, these ions seem to originate from locations such that $\Theta_{Bn} < 70°$ (21). When $\Theta_{Bn} > 70°$ specularly reflected ions at the Earth's bow shock are accelerated by the interplanetary electric field parallel to the shock surface and turned around under the influence of the Lorentz force. When they return to the shock they have gained sufficient energy to surmount the effective potential barrier that originally caused their reflection and they end up downstream. On the other hand energetic particles have actually been observed at quasi-perpendicular interplanetary traveling shocks and at the quasi-perpendicular foreward and reverse shocks bounding the corotating interaction regions. In order for charged particles to be efficiently accelerated by quasi-perpendicular shocks they must encounter the shock several times. One way for this to happen is by scattering rapidly enough that they diffuse against the downstream convection. Since the magnetic field is nearly perpendicular to the flow, the diffusion is across the magnetic field. Another way for particles to encounter the shock several times is by being trapped between the electrostatic potential and the upstream Lorentz force. The latter mechanism (shock surfing) has been proposed for acceleration of pickup ions at quasi-perpendicular shocks (19, 20). This mechanism can work for pickup ions, but not easily for thermal solar wind ions, since it requires that the ions have in the shock frame a velocity normal to the shock which is much smaller than the solar wind bulk velocity. This is the case for a fraction of the pickup ions at traveling interplanetary shocks as well as at the heliospheric termination shock.

Self-consistent particle simulations of collisionless shocks should results in the appropriate parallel and perpendicular scattering of thermal solar wind ions and possible injection and acceleration. However, these simulations have been performed almost exclusively in one or two spatial dimensions in such a way as to ignore the coordinate normal to the plane containing the asymptotic magnetic field. Jokipii et al. (22) and Jones et al. (23) have presented a general theorem according to which charged particles in fields with at least one ignorable spatial coordinate is effectively forever tied to the same magnetic line of force, except for motion along the ignorable coordinate. Thus, 1-D and 2-D kinetic simulations of quasi-perpendicular shocks can not give results on the injection and acceleration of thermal ions. Since at present long time 3-D simulations of shocksare not computationally feasible, Giacalone et al. (24) have introduced an ad hoc perpendicular diffusion in an 1-D hybrid simulation of perpendicular shocks. In these simulations pickup ions have been included self-consistently. Assuming a scattering time of about 20 times the inverse ion gyrofrequency upstream of the shock these authors found that only pickup ions are injected efficiently, whereas thermal solar wind ions are not.

The turbulence behind quasi-perpendicular shocks is produced by the combined distributions of transmitted solar wind ions and transmitted specularly reflected ions. The specularly reflected ions gyrate downstream as a gyrophase-bunched distribution. The combined distribution is susceptible to the Alfvén ion cyclotron instability and to the mirror mode instability. In order to determine the parallel and cross field diffusion coefficient in such a turbulence field Scholer et al. (25) have performed 3-D simulations of a system consisting initially of a core distribution and a gyrophase-bunched distribution. From the temporal development of the spatial variance over many particle trajectories a ratio of the perpendicular to parallel diffusion coefficient of $\kappa_{\perp}/\kappa_{||} \sim 0.1$ has been derived, which is by an order of magnitude larger than the value predicted by hard sphere scattering, i.e., parallel scattering is considerably more effective than perpendicular scattering. Thus, it is expected that specularly reflected ions are rapidly pitch-angle scattered onto a sphere in velocity space. The perpendicular diffusion coefficient results in a perpendicular scattering time constant of $\Omega t \sim 20$. In the 1-D quasi-perpendicular shock simulations by Giacalone et al. (24) pitch angle scattering was prohibited, since in the 1-D setup the instability with k vectors parallel to the magnetic field can not be excited. However, since specularly reflected ions should rapidly pitch-angle scatter in a 2- or 3-D setup, they should behave as far as cross-field diffusion is concerned similarly as pickup ions. Thus, the cross-field scattering in the turbulence generated by the specularly reflected ions may be sufficient to inject and accelerate these ions at quasi-perpendicular shocks.

In oblique shocks, that is in a Θ_{Bn} range between $\sim 50° - 60°$, particles can get injected by the same mechanism that works at quasi-parallel shocks: we have pointed out that particles are trapped near the shock and gain energies of up to 10 times the shock ram energy before they move upstream and are possibly further accelerated by a diffusive acceleration mechanism. At these high energies particles can leave the shock in the upstream direction for shock normal - magnetic field angles exceeding 45°, can subsequently produce upstream waves, and can get backscattered. 1-D simulations of oblique shocks ($\Theta_{Bn} = 60°$) have indeed resulted in high energy backstreaming ions. These simulations have to be repeated at least in 2-D: in 1-D simulations only waves with **k** vectors parallel to the shock nomal are allows whereas **k** of maximum growth is expected to be parallel to the magnetic field.

INJECTION OF PICKUP IONS

The injection/acceleration of pickup ions by the shock surfing mechanism (19 20) was already mentioned earlier. Such a process cannot be modeled by hybrid simulations since the mechanism relies on the spatial scale of the shock potential being of the order of the electron inertial length. Full particle simulations are necessary to verify this acceleration mechanism. Pickup ions have a velocity distribution which is close to a spherical shell with a radius of solar wind speed around the solar wind velocity. Pickup ions are easily reflected from quasi-parallel shocks. The potential in a quasi-parallel shock helps to decelerate the solar wind; a large part of the pickup ion distribution has a velocity between zero and solar wind speed and can get reflected at the shock, as has been seen in 1-D hybrid simulations of Scholer and Kucharek (26). Injection at quasi-perpendicular shocks by cross-field diffusion is another possibility. Since a large part of the pickup ion distribution has a very small velocity in the shock normal direction these ions are able to recross the shock due to cross-field diffusion many times. This has been demonstrated in the hybrid simulations of Giacalone et al. (24) and in the Monte Carlo simulations of Ellison et al. (27).

SUMMARY

Self-consistent quasi-parallel collisionless shock simulations have shown that $\sim 2-4\%$ of the upstream ion are extracted during their interaction with the shock and are subsequently accelerated to higher energies. This may not sound as an important result, but in view of the large number of sceptics of such a process it is actually a rather important result. The simulations have shown that the injected and accelerated particles are an integral part of the quasi-parallel shock structure: the upstream diffuse ions excite waves via the electromagnetic ion/ion beam instability. These waves are convected into the shock and lead to dissipation. Detailed analysis of particle trajectories during the simulations allows determination of the processes which lead to injection and acceleration. Once we know these processes, we may be able to construct theoretical models. The models by Malkov (17) for the injection from downstream and the model by Malkov and Völk (6) for the diffusive process in the suprathermal energy range are particularly noteworthy.

The injection and acceleration process for thermal ions at quasi-perpendicular shocks depends on the possibility of these ions to recross the shock many times. A viable mechanism for injection is cross-field diffusion of the specularly reflected ions after they have crossed the shock into the downstream region. Determination of the cross-field diffusion coefficient in strong turbulence suggests that specularly reflected ions can recross the quasi-perpendicular shock and can get further accelerated. One-dimensional simulations of shocks cannot give results on injection of thermal particles, even when an ad hoc perpendicular scattering is introduced. The specularly reflected particles have to scatter fast in pitch angle, so that they have small velocities in the shock normal direction. Pitch angle scattering is absent or only weak in 1-D simulations of perpendicular shocks since the excitation of waves with wave vectors parallel to the magnetic field is excluded. We have to await fully 3-D simulations of quasi-perpendicular shocks in order to verify such an injection and acceleration mechanism. At more oblique shocks the same injection process as at quasi-parallel shocks can work: particles gain high enough velocities during their first shock encounter so that they can escape the shock along the magnetic field in the upstream direction.

We may almost state that pickup ions like to be accelerated, not only by shocks, but also by turbulent fields. In magnetic field turbulence, as for instance behind shocks, transit time damping can accelerate pickup ions, since these ions have velocities exceeding the Alfvén speed, the minimum speed needed for transit time damping to work (28). Pickup ions are reflected from quasi-parallel shocks: hybrid simulations of quasi-parallel shocks have shown that the reflection rate of pickup ions is large. They can subsequently partake in a diffusive acceleration process. At quasi-perpendicular shocks the surfing mechanism is a possible mechanism. Cross-field diffusion at quasi-perpendicular shocks can inject and accelerate ions at these shocks. As far as pickup ions are concerned the problem seems to be the determination of the dominant acceleration mechanism in a particular situation. Simulations and theory have to come up with better predictions on the dependence of the injection efficiency on the charge to mass ratios. Comparison with the excellent ACE composition observations may eventually discriminate between the various injection and acceleration processes for pickup ions.

ACKNOWLEDGEMENTS

Part of this work has been performed during the Collisionless Shock Workshops at the Intern. Space Science Institute, Bern. The work at Lockheed was supported by NASA contracts NAS5-30302 and NAG5-8072.

REFERENCES

1. Krimsky, G. F., *Dokl. Akad. Nauk. SSRR* **234**, 1306 (1977).
2. Axford, W. I., Leer, E., and Skadron, G., in *Proc. Int. Conf. Cosmic Rays 15th* **11**, 132 (1977).
3. Blandford, R. R., and Ostriker, J. P.*Astrophys. J.* **221**, L29 (1978).
4. Bell, A. R., *Mon. Not. R. Astron. Soc.* **182**, 147 (1978).
5. Gosling, J. T., et al., *Geophys. Res. Lett.* **5**, 957 (1978).
6. Malkov,M. A., and Völk, H. J., *Astron. Astrophys.* **300**, 605 (1995).
7. Ellison, D. C., *Geophys. Res. Lett.* **8**, 991 (1981).
8. Ellison, D. C., Baring, M. G. and Jones, F. C.,*Astrophys. J.* **473**, 1029 (1996).
9. Quest, K. B., in *Proc. 6th Intern. Solar Wind Conf.*, NCAR Techn. Note **306**, 503 (1988).
10. Scholer, M., *Geophys. Res. Lett.* **17**, 1153 (1990).
11. Kucharek, H., and Scholer, M., *J. Geophys. Res.* **96**, 22195 (1991).
12. Giacalone, J. et al., *Geophys. Res. Lett.* **19**, 433 (1992).
13. Trattner, K.-J., and Scholer, M., *Geophys. Res. Lett.* **18**, 1817 (1991).
14. Trattner, K.-J., and Scholer, M., *J. Geophys. Res.* **99**, 6637 (1994).
15. Scholer, M., Kucharek,H., and Trattner, K.-H., *Ann. Geophys.* **17**, 583 (1999).
16. Lee, M. A., *J. Geophys. Res.* **87**, 5063 (1982).
17. Malkov, M. A., *Phys. Rev. E.* **58**, 4911 (1998).
18. Sugiyama, T., and Fujimoto, M., in *Proc. Int. Conf. Cosmic Rays 25th* **4**, 423 (1999).
19. Lee, M. A., Shapiro, V. D., and Sagdeev, R. Z., *J. Geophys. Res.* **101**, 4777 (1996).
20. Zank, G. P., et al., *J. Geophys. Res.* **101**, 457 (1996).
21. Paschmann, G., et al., *J. Geophys. Res.* **86**, 4355 (1981).
22. Jokipii, J. R., Kota, J., and Giacalone, J., *Geophys. Res. Lett.* **20**, 1759 (1993).
23. Jones, F. C., Jokipii, J. R., and Baring, M. G., *Astrophys. J.* **509**, 238 (1998).
24. Giacalone, J., Jokipii, J. R., and Kota, J., *J. Geophys. Res.* **99**, 2929 (1994).
25. Scholer, M., Kucharek, H., and Giacalone, J., in press *J. Geophys. Res.*, (2000).
26. Scholer, M., and Kucharek, H., *Geophys. Res. Lett.* **26**, 29 (1999).
27. Ellison, D. C., Jones, F. C., and Baring, M. G., *Astrophys. J.* **512**, 403 (1999).
28. Fisk, L. A., *J. Geophys. Res.* **81**, 4633 (1976).

Particle transport, composition, and acceleration at shocks in the heliosphere

Joe Giacalone

Lunar & Planetary Laboratory, University of Arizona, Tucson, AZ, 85721

Abstract. The physics of energetic charged-particle transport in the inner heliosphere is discussed in the context of spacecraft observations, and their connection to anomalous cosmic rays and the outer heliosphere. It will be argued that although there may be a variety of sources of energetic nuclei in the inner heliosphere, it is likely that the contribution from interstellar pickup ions dominates in the outer heliosphere. These ions may be further energized by the termination shock and contribute to anomalous cosmic rays. At corotating interaction regions, pickup ions (either of interstellar or inner-source origin) are naturally accelerated to high energies. It will be shown that their acceleration is strongly favored at the reverse shock over the forward shock. The physics of the injection problem (the difficulty in accelerating near thermal-energy particles) will also be discussed. It will be shown that long-wavelength magnetic fluctuations enhances the probability that particles are accelerated to high energies.

INTRODUCTION

There are a number of open issues related to the subject of energetic ions in the inner heliosphere and their importance in the global heliosphere. For instance, their appears to be a wide variety of sources of these nuclei. These include accelerated "pickup ions" which represent a non-thermal distribution of ionized neutrals originating from, for example, dust grains (known as the "inner source"), or interstellar atoms entering the solar system. Other sources of energetic ions are solar cosmic rays which consist of, for example, ions associated with solar flares which are assumed to be accelerated near the sun, or those associated with coronal mass ejections which are assumed to be accelerated by the large-scale shock waves associated with them. Another open issue is how low-energy pickup ions can be accelerated by these shocks. Shock-acceleration theories have discussed a problem of accelerating these low-rigidity particles and this has been commonly referred to as the "injection problem."

In this paper, we discuss issues related to the acceleration, transport, and composition of energetic ions in the inner heliosphere. These will be discussed in the context of their connection to anomalous cosmic rays, which are heliosphere-accelerated ions having primarily the same composition as that of interstellar pickup ions. One of the open issues related to anomalous cosmic rays is the injection problem. We argue in this paper that there is not an injection problem. We show that there are two possible resolutions to this. One is that interstellar pickup ions are accelerated in the inner heliosphere and are then transported to the outer heliosphere, already with high energies, where they are further accelerated at the termination shock to anomalous cosmic-ray energies. The other explanation that we discuss is the acceleration of previously unaccelerated pickup ions locally at the termination shock itself. We do not attempt to deduce which explanation is the more likely since spacecraft observations do not confine the model parameters to this extent.

SOURCES OF ENERGETIC NUCLEI IN THE INNER HELIOSPHERE

It is widely known that energetic-particle intensities are enhanced by interplanetary disturbances corotating with the sun (1, 2). These corotating interaction regions (CIRs) are bound by two shocks: the forward and reverse shock pairs. Shocks are known to accelerate particles. However, there are still some unanswered questions regarding the source of the particles which are to be accelerated.

Figure 1 shows particle fluxes at various heliocentric distances associated with corotating interaction regions. The observational data from Imp-7, Helio-2, and Pioneers 10 and 11 are from (3). Also shown are Voyager observations from (4). Also shown in Figure 1 is a dashed curve which represents the expected radial dependence assuming the adiabatic cooling and convection of a source that has a power-law dependence on energy and is normalized to the largest flux shown. Energy spectra of energetic nuclei observed to be associated with CIRs are usually

CP528, *Acceleration and Transport of Energetic Particles Observed in the Heliosphere: ACE 2000 Symposium*, edited by Richard A. Mewaldt, et al.
© 2000 American Institute of Physics 1-56396-951-3/00/$17.00

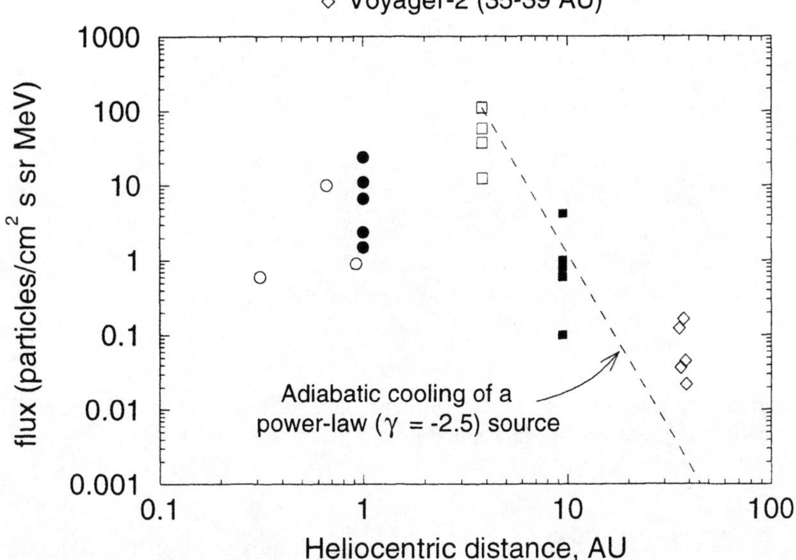

FIGURE 1. Intensity of CIR-associated energetic particles (~ 0.5-2 MeV) observed by several spacecraft at various heliocentric distances. The Voyager observations were from the period 1992-93 (from (4)), while all of the other were in the 1975-76 period (from (3)). The curve shows the expected radial dependence assuming the adiabatic cooling/convection of a source coinciding with the largest intensity shown, having a power-law energy dependence with a spectral index of -2.5.

exponential (5). We note that the convection/cooling of a source with an exponential energy dependence would result in a much more rapid decline with heliocentric distance than a source with a power-law energy dependence.

Figure 1 also shows that the intensity of CIR-associated energetic-particle streams increases from the inner heliosphere and is peaked at about 4 AU. After which, for larger heliocentric distances, the intensity declines but with a much smaller gradient than would expected from simple convection/cooling of a source in the inner heliosphere.

One explanation for the peaked intensity of CIR particles at about 4 AU is that this is where the shocks are the strongest. It is known that the shocks associated with CIRs do not form until beyond about 1-2 AU. However, it is intersting to note that the flux of the interstellar pickup ions H^+ and O^+ also peaks at about 4 AU (as shown in Figure 2). As will be further discussed in later sections, pickup ions are naturally accelerated to high energies by corotating shocks. We also note that the flux of interstellar pickup ions falls off as $1/r$ in the outer heliosphere. Thus, the large number of CIR particles seen by Voyager are likely accelerated interstellar pickup ions. Note also that pickup ions of the "inner source" origin and solar-energetic particles may also contribute in the outer heliosphere. However, we feel that because these ions will have a $1/r^2$ radial dependence, they will be much less abundant than interstellar pickup ions.

Figure 2 shows the radial dependence of interstellar pickup ions based on the model of Vasyliunas and Siscoe (6). Hydrogen and Oxygen peak at about 4-6 AU, which is coincidentally the peak in the energetic particle fluxes associated with CIRs shown in Figure 1. Helium and Neon, on the other hand, peak much closer to the sun.

As suggested by Figures 1 and 2, it is difficult to separate the abundances of pickup ions (either inner source or interstellar) and solar particles when the observations are made at 1 AU. This is because there are so few interstellar pickup H and O while thermal solar wind is difficult to accelerate. Helium, on the other hand, should be dominated by interstellar pickup ions. Ne may also be dominated by interstellar pickup ions, but this is not as clear. In the outer heliosphere we believe that the observed fluxes of CIR-related energetic particles are dominated by interstellar pickup ions. Solar energetic particles, which may dominate in many 1 AU observations, are likely to be much less abundant in the outer heliosphere due to adiabatic losses.

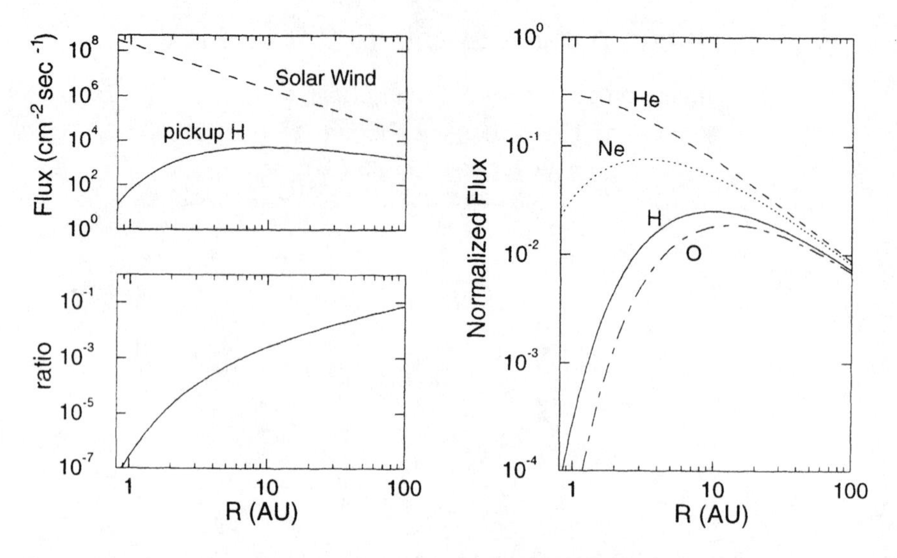

FIGURE 2. Upper left: Pickup hydrogen and solar wind flux versus heliocentric distance. Lower left: the ratio of the pickup-ion to solar wind flux. Right: The normalized outward-directed pickup-ion flux for H, He, Ne, and O.

CONNECTION TO THE OUTER HELIOSPHERE AND ANOMALOUS COSMIC RAYS

The current widely accepted theory of anomalous cosmic rays is that they are the consequence of acceleration of interstellar pickup ions at the nearly perpendicular termination shock of the solar wind (lying between $\sim 60 - 100$ AU). The theory naturally accounts for the observed energy spectra, spatial distributions, and charge state. However, indications are that some initial acceleration of low-energy pickup ions needs to occur since it is very difficult to accelerate them locally at the termination shock. Giacalone et al. (7) have shown that there are enough low-energy ions observed by Voyager 2 at 40 AU heliocentric distance to be further accelerated at the solar-wind termination shock and account for the observed levels of anomalous cosmic rays. They used a widely accepted model for the global transport and acceleration of cosmic rays (8). The key assumption was that the preaccelerated pickup ions (accelerated in the inner heliosphere) were of high enough energy to be adequately described by the diffusive transport equation (?). If in fact the Voyager 2 observations of CIR-related particles (4) are accelerated interstellar pickup ions, as suggested by Gloeckler et al.(10), then the acceleration of pickup ions in the inner heliosphere is an important stage in the formation of anomalous cosmic rays.

Figure 3 is adapted from (7). It shows the simulation results (curves) normalized using Voyager 2 data (stars) and compared to anomalous cosmic-ray data (squares). The simulations used diffusion coefficients which were based on consensus values. The assumed source was located at 10 AU although the placement of the source is not an important parameter since the simulations are normalized to the low-energy observations made at 40 AU. It should also be pointed out that the low-energy portion of the spectrum decreases with increasing radial distance, while the peak at anomalous cosmic-ray energies increases with radial distance. These are consistent with other Voyager observations in the outer heliosphere of other species (11).

ACCELERATION OF PICKUP IONS AT CIRS

Because the shocks associated with CIRs are moving only slightly faster (or slower in the case of the reverse shock) than the solar wind speed, pickup ions which were ionized in the flow into which the shock is propagating are very mobile. That is, they can readily encounter the shock several times and are naturally accelerated. The SWICS instrument on board the Ulysses spacecraft has shown that pickup ions are efficiently accelerated by interplanetary shocks (10).

However, there is observational evidence reported by Schwadron et al.(12) that interplanetary shocks may not efficiently accelerate unaccelerated pickup ions. These authors have proposed that the pickup ions are accelerated by a stochastic process, called transit-time damping of magnetic fluctuations. They suggest that this is the initial stage of a two-step process. The fluctuations nec-

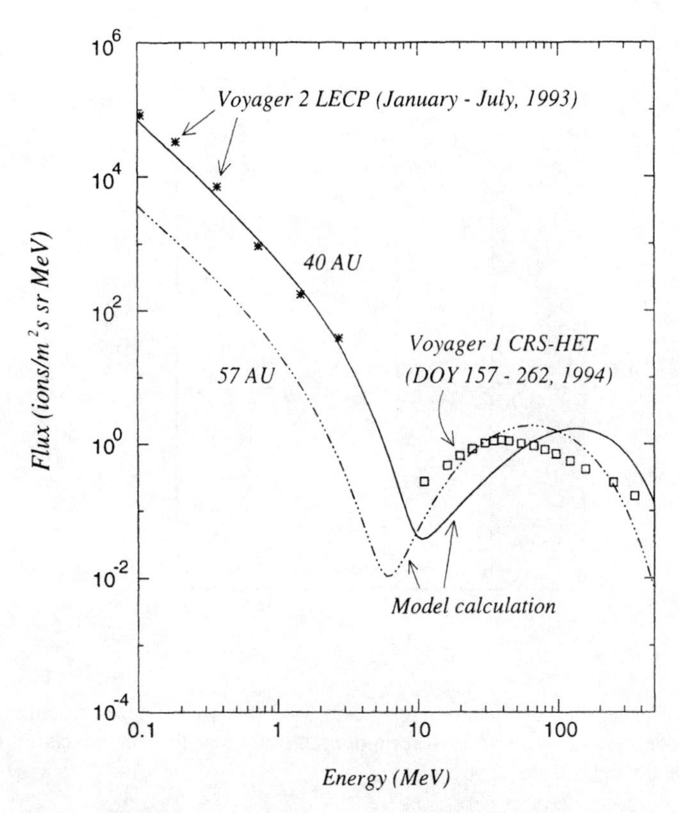

FIGURE 3. Simulated anomalous cosmic-ray proton spectra at 40 AU and 57 AU using a source located in the inner heliosphere (10 AU). The star symbols are Voyager 2 observations of low-energy ions when Voyager 2 was at 40 AU and the square symbols are Voyager 1 observations of anomalous cosmic-ray protons reported by Christian et al.(13) when Voyager 1 was at 57 AU. This figure is adapted from (7).

essary for this process to work are contained within the turbulent regions within the corotating stream. Once the particles have been pre-accelerated in this manner, they may be further accelerated by the shocks.

There exists an asymmetry in the interaction between pickup ions with the forward and reverse shocks bounding a CIR, depending on whether they were first ionized in either the fast or slow solar wind. Giacalone and Jokipii (14) shows that this naturally explains why there are typically more particles accelerated at the reverse shock than the forward shock (3, 15). Since the pickup ions are moving much faster in the fast wind, acceleration in the fast wind should be more efficient because these particles start with approximately four times as much energy as those in the slow wind. This idea is illustrated in Figure 4 for the case where the accelerated spectrum is an inverse power law (having a v^{-4} dependence which would occur for shock acceleration at a strong shock). Note that a steeper spectrum would produce a larger effect. Moreover, this general conclusion should be independent of the acceleration mechanism.

The form of the energy spectra of CIR-associated energetic particles was also discussed by Fisk and Lee (5). They included an adiabatic loss term in the diffusive transport equation and obtained approximate analytic solutions. They found that the spectra have an exponential-like energy dependence at 1 AU. The spectral index depends inversely on the flow speed, so that the spectra associated with a reverse shock are harder than those associated with the forward shock, also in agreement with observations (3, 15).

We have modeled the acceleration of pickup ions at a CIR by integrating the trajectories of test particles moving in the fields associated with a CIR. The CIR model is a crude approximation to the expected global structure associated with a CIR. In our model, the flow speed, density, and magnetic field are spherically symmetric. Initially a disturbance propagates radially from the sun. The disturbance is specified as a gradual change in flow speed from a fast solar wind to a slow solar wind over a distance which decreases as the disturbance moves outward (the profiles were obtained by using combinations of hyperbolic tangent functions). At a prechosen radial dis-

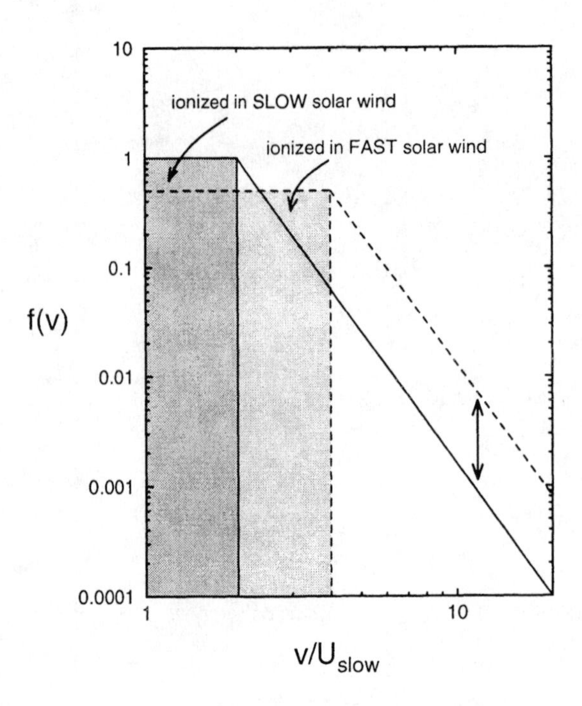

FIGURE 4. Cartoon illustration of pickup ion distributions for pickup ions that were ionized in either the fast or slow wind, as indicated. The high-energy tails arise from some unspecified form of acceleration mechanism which, in this case, is independent of the local wind speed so that the spectral index is the same.

tance, the disturbance is at a minimum thickness where two shocks are created. Subsequently, one of the shocks moves outward in a frame moving with the initial disturbance (the forward shock) and the other moves inward (the reverse shock). Note that both shocks are moving outward in the fixed frame. The position of the stream interface is exactly between the forward and reverse shocks for distances beyond the shock-formation location.

The magnetic field in our model is determined from the plasma density which is also a prespecified function like the flow speed. The plasma density has the same jump at both shocks (2.5), and also increases at the stream interface. The radial component of the magnetic field is proportional to $1/r^2$, while the azimuthal component is given by

$$B_\phi = B(r_0)\frac{\rho}{\rho(r_\odot)}\frac{r\Omega_\odot \sin\theta}{V_w(r_\odot)} \quad (1)$$

where r and θ are spherical coordiantes, ρ is the plasma density, $r_\odot$ is the radius of the sun, B is the field strength, $\Omega_\odot$ is the solar rotation frequency, and V_w is the flow speed. There is an infinitesimal current sheet located at a polar angle of 90 degrees. Finally, the electric field is determined from the ideal MHD approximation.

A source of interstellar pickup ions were followed in these fields. The spatial variation of the source is based on the model of Vasyliunas and Siscoe (6). The particles are treated as test particles. Pitch-angle scattering of these ions was modeled using ad-hoc scattering. A mean-free path of 0.05 AU was assumed. Other parameters that were used were typical of CIRs (fast wind speed of 800 km/s, slow wind speed of 400 km/s, intermediate speed of 600 km/s, density jump across each shock of 3).

Results of the model are shown in Figure 5. The left panel shows the case where the shocks form at 0.75 AU and the observation of the ions is made at 3.5 AU. The intensity of the particles is peaked near the reverse shock, in agreement with the theory discussed above. When the shock formation location is much closer to where the observation is made, the peak in intensity occurs between the two shocks (right panels). The reason for this is that the particles are capable of diffusing through the velocity gradient associated with the CIR formation, before the shocks form, and are efficiently accelerated. This gradient is fairly large since the flow speed changes from 400 km/s in the slow wind to 800 km/s in the fast wind. Consequently, acceleration of pickup ions may occur before the shocks form. This will lead to a peak in the intensity within the CIR which will diminish relative to the intensity at the shocks as the CIR evolves. This is partly due to adiabatic energy losses, but also because the intensity at the shocks increase due to shock acceleration. This idea may be a possible alternative explanation of the observations by Schwadron et al.(12).

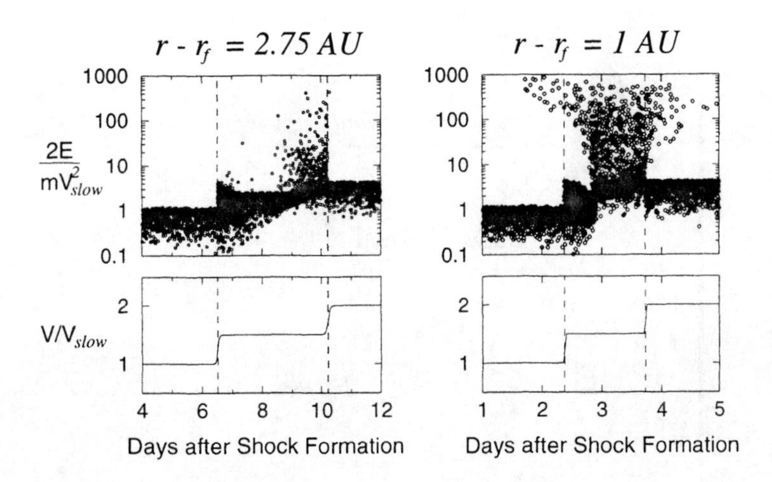

FIGURE 5. Upper panels: Scatter plots of the particle energy as a function of the time they were observed for an ensemble of test particles encountering a model CIR (see the text for a description of the model). Lower panels: the flow-velocity profile of the CIR. Left panels are for the case in which the CIR forms very far from the point of observation, whereas the right panels are for the case in which the point of observation is very near the shock formation location.

THE PHYSICS OF DIFFUSIVE SHOCK ACCELERATION AND THE ISSUE OF INJECTION

Diffusive shock acceleration (DSA) is a widely known theory describing the acceleration of charged particles by collisionless shock waves. Subsets of this general theory are often discussed separately. For instance, shock-drift acceleration in the limit of no scattering has received much attention (16), as has the limit of strong scattering (17). Both of these are contained within the general framework of diffusive shock acceleration provided the distribution function is quasi-isotropic.

It is well known that the energy spectrum of charged-particles, which collectively are well described by the diffusive transport equation, downstream of a collisionless shock is a power law (in the absence of either adiabatic losses or a free-escape boundary). However, diffusive shock acceleration only describes the acceleration of particles whose distribution function is quasi-isotropic. Low-energy particles are not isotropic in all frames of reference. This is especially true for particles whose energy is near the thermal peak. Consequently, DSA is not adequate to describe how the particles are injected. To address this issue, it is of interest to examine the conditions under which the diffusive transport equation is valid. It is reasonable to expect the theory to be valid only when the diffusive-streaming anisotropy in the plasma frame of reference is small. The general form of this condition was given in (?, 19). For the case of a perpendicular shock (the average magnetic field is perpendicular to the shock-normal direction), the condition is given by

$$\Upsilon = \frac{3U_1}{v}\sqrt{1+(\kappa_A/\kappa_\perp)^2} \ll 1 \tag{2}$$

where U_1 is the flow speed upstream of the shock, v is the plasma-frame particle velocity, κ_A is the antisymmetric component of the diffusion tensor, and $\kappa_\perp$ is the component perpendicular to the average magnetic field.

One may regard Υ as an indicator of the injection efficiency. This can be understood in terms of the speed at which the process becomes diffusive (i.e., the speed at which (1) is satisfied). If the distribution function for speeds larger than this injection speed, w_{inj}, is, for instance, a power law, then the intensity at high energies will depend critically on w_{inj}. Consequently, large values of Υ imply that the injection is likely to be very inefficient, whereas, smaller values indicate that the injection will be more efficient.

It is useful to consider special cases of Equation (1). In the case of classical scattering, $\kappa_\perp$ and κ_A can be written in terms of the ratio of the scattering mean-free path, $\lambda_\parallel$ to the particle gyroradius, r_g. For this case, Equation (1) becomes,

$$\Upsilon = \frac{3U_1}{v}\left[1+\left(\frac{\lambda_\parallel}{r_g}\right)^2\right]^{1/2} \tag{3}$$

Equation (2) confirms that the stronger the scattering (and hence, the smaller the mean-free path) the more efficient the particle injection. This is in agreement with the Monte-Carlo simulations by Ellison et al. (20).

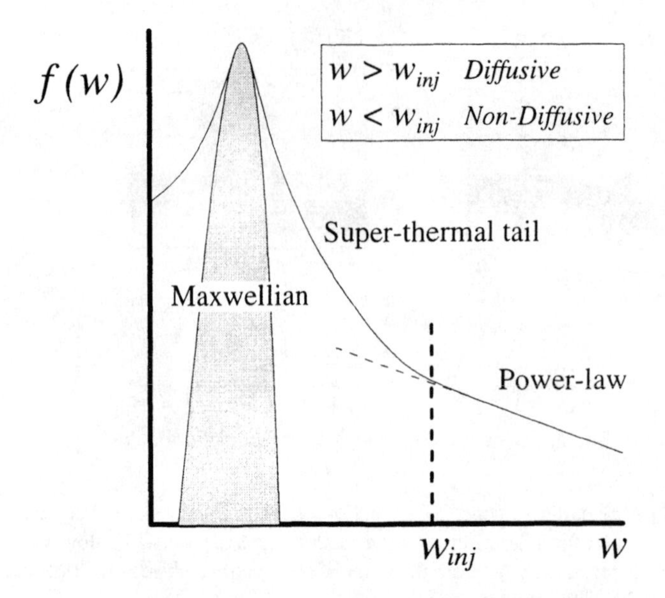

FIGURE 6. Cartoon sketch of a distribution function showing a thermal peak, and suprathermal tail which connects to a power-law tail at high energies. The velocity at which the process is diffusive is indicated with the speed w_{inj}.

Role of Large-Scale Field-Line Meandering

Classical scattering theory may not give a valid description of $\kappa_\perp$. The test-particle simulations of Giacalone and Jokipii (?) for charge-particles moving in time-stationary spatially isotropic magnetic turbulence revealed that $\kappa_\perp$ does not agree with the theory of classical-scattering. There is currently no consensus theory for $\kappa_\perp$, thus we can consider other cases which seem reasonable. For instance, assuming that the ratio of perpendicular to parallel diffusion, $\kappa_\perp/\kappa_\parallel$, is a constant, which may be reasonable in the case of large-scale magnetic fluctuations (those with a typical scale size which far exceeds the particle gyroradii), we find

$$\Upsilon = \frac{3U_1}{v}\left[1+\left(\frac{\kappa_\parallel}{\kappa_\perp}\right)^2\left(\frac{\lambda_\parallel/r_g}{1+(\lambda_\parallel/r_g)^2}\right)^2\right]^{1/2} \quad (4)$$

where still use $K_A = (1/3)vr_g$, which is a widely accepted representation (see also (21)).

Equation (3) shows that Υ decreases with increasing scattering mean-free path. This implies that the injection efficiency increases with $\lambda_\parallel$ which is contrary to the result discussed above for the case of classical scattering. Physically, the interpretation is that the particles with long mean-free paths will be able to sample the large-scale magnetic field and hence have a larger perpendicular diffusion coefficient than they would have by simply moving a single gyroradius across the field each time they scatter.

Is There Really an Injection Problem?

The injection problem is evident in the theories of shock acceleration. However, shocks are observed to accelerate particles. So, is there really an injection problem? Or is it just that we are unable to explain the acceleration analytically? Shown in Figure 6 is a cartoon sketch of a distribution function. The speed at which the process becomes diffusive is w_{inj}. The distribution above this point is power law (resulting from a steady-state planar shock and no losses). The thermal peak is also shown. Our interpretation of the injection problem is that the physics which describes the suprathermal tail, i.e. its velocity dependence and density is not known. There are a number of processes described in the literature which address this physics, but there is currently no consensus theory. Examples of some important factors contributing to the formation of suprathermal tails are the microphysics of the shock layer (22, 23, 24) and also statistical mechanisms such as transit-time damping (12).

DISCUSSION

Energetic ions in the inner heliosphere stem from a variety of sources and it is difficult to separate the various contributions arising from, for example, interstellar pickup ions, inner source pickup ions, and solar energetic particles. We have concentrated here on the contributions from interstellar pickup ions and their connection to anomalous cosmic rays and the global heliosphere. We find that although the composition of solar energetic particles associated with corotating interaction regions may not always resemble interstellar pickup ions, they are likely to be the most abundant contribution to inner-heliosphere related particles in the outer heliosphere. The reason for this is simply that (A) interstellar pickup ions have fluxes which fall off as $1/r$ as opposed to other sources which fall off more rapidly like $1/r^2$, and (B) pickup ions, of either interstellar or inner source origin, are naturally favored for acceleration at propagating shocks in the inner heliosphere over thermal solar wind.

Anomalous cosmic rays are the result of the acceleration of interstellar pickup ions, at least in part, by the termination shock of the solar wind. We have previously shown that this may occur via a two-step process by which interstellar pickup ions are first accelerated in the inner heliosphere, and are then transported to the termination shock where they are accelerated further to anomalous cosmic-ray energies. There is currently no reason to discount this possibility. However, we have also shown that the acceleration of low-energy, previously unaccelerated pickup ions can occur in the presence of long-wavelength magnetic fluctuations. Consequently, the two-step mechanism may not be necessary. Currently this issue is not resolved and it is likely that both effects contribute.

ACKNOWLEDGEMENTS.

This work was supported, in part by NASA under grants NAG5-7793 and NAG5-6620, and by the National Science Foundation under Grant ATM 9616547.

REFERENCES

1. Barnes, C. W., and J. A. Simpson, *Astrophys. J.*, 210, (1976) L91-L96.
2. McDonald, F. B., B. J. Teegarden, J. H. Trainor, and T. T. von Rosenvinge, *Astrophys. J.*, 203, (1975) L149-L154.
3. Van Hollebeke, M. A. I., F. B. McDonald, H. H. Trainor, and T. T. von Rosenvinge *J. Geophys.Res.*, 83, (1978) 4723-4731.
4. Decker, R. B., S. M. Krimigis, S. M., and M. Kane, *Proc. 24th Int. Cosmic Ray Conf. (Rome)*, 4, (1995) 421-424.
5. Fisk, L. A., and M. A. Lee, *Astrophys. J.*, 237, (1980), 620-626.
6. Vasyliunas, V. M., and G. L. Siscoe, *J. Geophys.Res.*, 81, (1976) 1247-1252.
7. Giacalone, J., J. R. Jokipii, R. B. Decker, S. M. Krimigis, M. Scholer, and H. Kucharek, *Astrophys. J.*, 486, (1997) 471-476.
8. Jokipii, J. R., *J. Geophys. Res.*, 91, (1986) 2929-2932.
9. Parker, E. N., *Planet. Space Sci.*, 13, (1965) 9-49.
10. Gloeckler, G., J. Geiss, E. C. Roelof, L. A. Fisk, F. M. Ipavich, K. W. Ogilvie, L. J. Lanzerotti, R. von Steiger, and B. Wilken, *J. Geophys. Res.*, 99, (1994) 17,637-17,643.
11. Hamilton, D. C., M. E. Hill, R. B. Decker, and S. M. Krimigis, *Proc. 25th Int. Cosmic Ray Conf. (Durban)*, 2, (1997) 261-264.
12. Schwadron, N. A., L. A. Fisk, and G. Gloeckler, *Geophys. Res. Lett.*, 21, (1996) 2871-2874.
13. Christian, E. R., Cummings, A. C., and E. C. Stone, *Astrophys. J. Lett.*, 446, (1995) L105-L108.
14. Giacalone, J., and J. R. Jokipii, *Geophys. Res. Lett.*, 24, (1997) 1723-1726.
15. Pesses, M. E., J. A. Van Allen, B. T. Tsuratani, and E. J. Smith, *J. Geophys. Res.*, 89, (1984) 37-46.
16. Armstrong, T. P., M. E. Pesses, and R. B. Decker, Shock drift acceleration, in *Collisionless Shocks in the Heliosphere: Reviews of Current Research, Geophys. Monogr. Ser.*, vol. 35, edited by B. T. Tsuratani and R. G. Stone. pp. 271-286, AGU, Washington, D. C., 1985.
17. Jones, F.C., and D.C. Ellison, *Space Sci. Rev.*, 58, (1991) 259-346.
18. Giacalone, J. & J. R. Jokipii, *Astrophys. J.*, 520, (1999) 204-214.
19. Giacalone, J., and D. C. E.Ellison, J. Geophys. Res., (2000), in press.
20. Ellison, D.C., M. G. Baring, and F. C. Jones, *Astrophys. J.*,473, (1996) 1029-1050.
21. Giacalone, J., J. R. Jokipii, and J. Kota, *Int. Conf. Cosmic Rays 26th*, 7, (1999) 37-40.
22. Giacalone, J., T. P. Armstrong, and R. B. Decker, *J. Geophys. Res.*, 96, (1991) 3621-3626.
23. Zank, G. P., H. L. Pauls, I. H. Cairns, and G. M. Webb, *J. Geophys. Res.*, 101, (1996) 457-477.
24. Lee, M. A., V. D. Shapiro, and R. Z. Sagdeev, *J. Geophys. Res.*, 101, (1996) 4777-4789.

Two Distinct Plasma and Energetic Ion Distributions within the June 1998 Magnetic Cloud

D. K. Haggerty[1], E. C. Roelof[1], C. W. Smith[2], N. F. Ness[2], R. M. Skoug[3], R. L. Tokar[3]

[1] *The Johns Hopkins University Applied Physics laboratory, Laurel MD 20723*
[2] *Bartol Research Institute, University of Delaware, Newark DE 19716*
[3] *Los Alamos National Laboratory, Los Alamos, NM 87545*

Abstract. On June 24-25, 1998 a magnetic cloud was observed by ACE near the L1 Lagrangian point. The cloud contained what at first appears to be a single interplanetary magnetic field (IMF) flux rope. However, within this flux rope we found two distinct distributions of plasma and energetic particles. The first region of the cloud was populated with anti-sunward streaming energetic ions and exhibited a low alpha to proton ratio, typical of the solar wind. Halfway into the magnetic cloud a second and distinct spatial region was encountered, demarcated by a significant drop in the proton temperature, increase in the proton density, and increase in the alpha to proton ratio. This second region was also populated with "pancake" energetic ion pitch angle distributions (PADs). Therefore not only were the two regions within the flux rope occupied by two different plasma regimes, implying different coronal origins, but the ion anisotropies were completely different, (unidirectional vs. bidirectional), implying markedly different global topologies for the magnetic field lines.

INTRODUCTION

The temporal and spatial evolution of a coronal mass ejection (CME) -- from the injection, through the convection or expansion, to the observation at 1 AU -- is of great scientific interest. Often these CME's are observed at Earth as magnetic clouds and an understanding of the processes involved in cloud formation is of fundamental significance to heliospheric physics. Both general models of magnetic clouds (2) and statistical properties of magnetic clouds (7) have been well reported in the literature. Changes in plasma properties during the passage of a magnetic cloud have been reported in the past. An unusual composition was detected by instruments on ACE for a magnetic cloud that was observed in May, 1998 (4,10). For the May cloud, changes occurred in composition during the middle of the event. Interesting changes during a magnetic cloud were also reported by Christon (3); however these changes were identified as temporal, not spatial changes. We report on a magnetic cloud observed by instruments on the ACE spacecraft near the L1 Lagrangian point on DOY 175-176, 1998 where simultaneous changes in both plasma and energetic particle distributions suggest a spatial change that challenges our understanding of these interplanetary structures.

OBSERVATIONS

On June 24, 1998 a magnetic cloud, identified by Smith (11), moved past the L1 point. Figure 1 shows the magnitude |B|, longitude (δ), and latitude (λ), of the interplanetary magnetic field (IMF) in a GSE coordinate system during the passing of the cloud as measured by the MAG experiment on ACE (12). An increase in the magnitude of the IMF began near 1648 UT on day 175. The rotations in both latitude and longitude associated with the increase in the IMF, traditional signatures of magnetic clouds (2), indicate the arrival of the flux rope observed at 1 AU. The rotations in the IMF and the enhancement in the IMF magnitude continued for over a day, until at 1550 UT

CP528, *Acceleration and Transport of Energetic Particles Observed in the Heliosphere: ACE 2000 Symposium*, edited by Richard A. Mewaldt, et al.
© 2000 American Institute of Physics 1-56396-951-3/00/$17.00

on day 176 a weak forward shock was observed at 1 AU, followed by yet another rope-like magnetic structure that convected past the spacecraft.

The last four panels of Figure 1 show the He++/H+ ratio, number density, kinetic temperature, and proton speed respectively, all measured by the Solar Wind Electron Proton Alpha Monitor (SWEPAM) (8). The velocity of the plasma increased prior to noon on day 175, followed by a long period of decreasing proton velocity. Although pressure balance (not shown) was maintained throughout the cloud, there were abrupt changes in the number density, kinetic temperature, and the He++/H+ ratio at a region we have identified as the plasma separatrix, 176:0314.

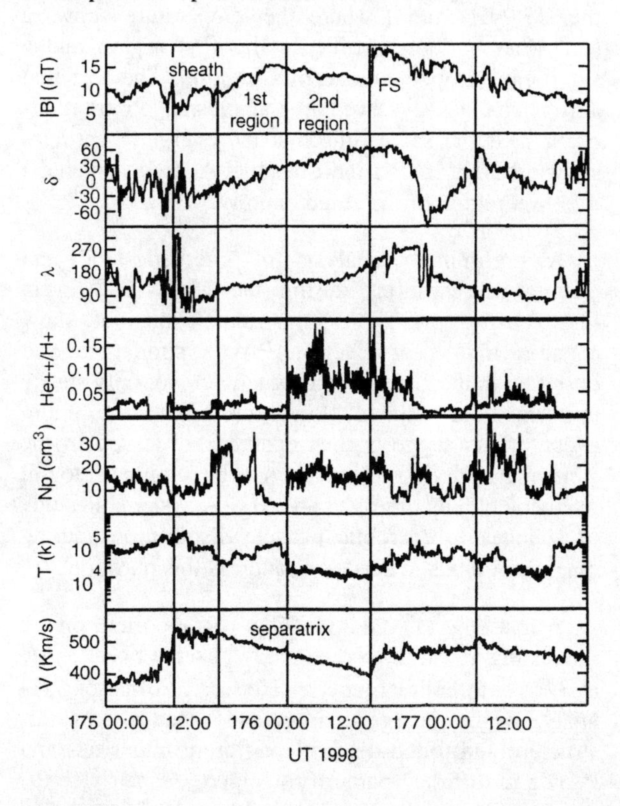

FIGURE 1. IMF magnitude, longitude, and latitude. Measured in a GSE coordinate system, along with the He++/H+ ratio, number density, kinetic temperature, and proton velocity. Five vertical lines indicate the solar wind increase and beginning of the sheath, the end of the sheath and beginning of the magnetic cloud, the plasma separatrix, the forward shock, and the end of the energetic particle event respectively.

Figure 2 shows spin averaged ion intensities and ion pitch angle distributions measured by the ACE/EPAM instrument (5). Ion intensities from channels centered at 55 keV, 149 keV, 427 keV, 1412 keV are plotted in panels from top to bottom. These energy channels come from the LEMS30 detector system that has four 90° sectors and whose angle from the ACE sun-oriented spin axis is 30°. The pitch angle distributions (PAD's) shown at the bottom of Figure 2 contain observations from both the LEMS30 system (described above with the four sectors labeled A-D) as well as the LEMS120 system that contains eight 45° sectors (labeled 1-8) and is oriented at 120° from the ACE spin axis. The energy channels from the two systems are essentially the same. For each sector displayed in the pitch angle distributions below, the cosine of the particle pitch angle is plotted on the abscissa and the intensity (normalized to the maximum sector) is plotted on the ordinate. The upper row of PAD's are 0.046-0.067 MeV ions while the lower row of PAD's are 1.060-1.880 MeV ions. The peak flux (J) measured in particles/(cm^2 sr sec MeV) is indicated above each PAD.

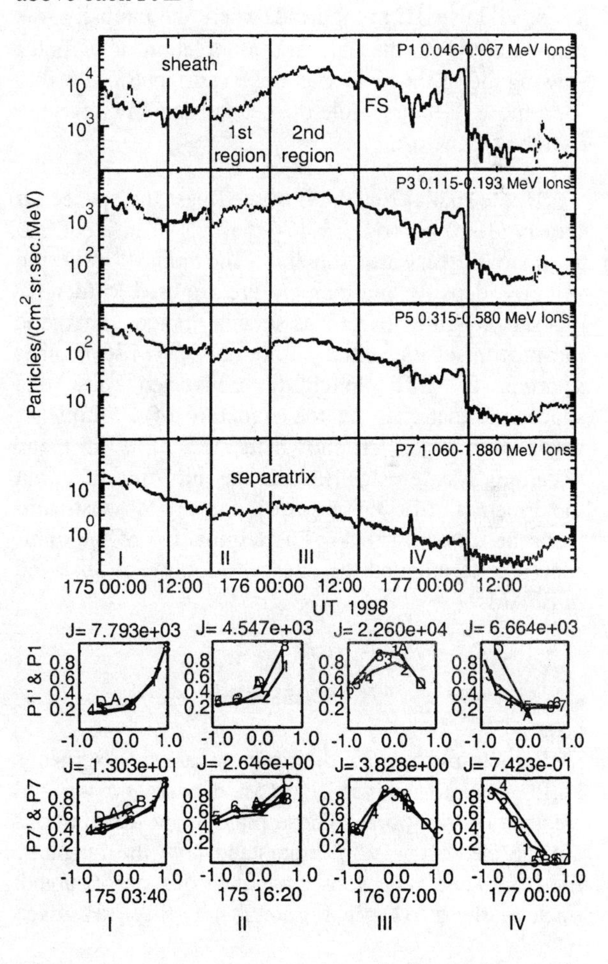

FIGURE 2. Ion intensities are shown in the top four panels. The boxes below represent 20 minute averaged ion PAD's. The top row of boxes are channels P1 and P1' (0.046-0.067 MeV). The four 90° P1 sectors are indicated with letters A-D while the eight 45° P1' sectors are indicated with numbers 1-8. The lower row of boxes are from P7 and P7' (1.060-1.880 MeV), labeled accordingly to the same scheme.

Prior to the arrival of the magnetic cloud the energetic ions exhibited strong field aligned flow. This distribution persisted for well over a day before the arrival of the magnetic cloud at 1 AU. The region identified as a possible sheath was observed at 1012 on day 175, just after the jump in proton speed from 400 km/sec to well over 500 km/sec. The profile of the solar wind speed became less erratic at the end of the sheath region when at 175:1648; the structure we identify as the first region appeared at ACE. Inside this first region, that lasted through the second half of day 175, the energetic ions exhibited strong outward (anti-sunward) flows along the magnetic field as seen in the ion PAD's (II) in Figure 2. This distribution persisted at all energies until at 176:0220 a transition to a trapped (pancake) distribution was observed. These pancake distributions were well established by 176:0314 where a plasma separatrix was observed. These PAD's (III) in Figure 2, where the intensity was peaked at 90° to the field and a depletion of particles flowing along the field was observed, appeared stable for almost 12 hours while the second region convected past the spacecraft.

A new spatial region convected past the spacecraft shortly after noon on day 176 that has been identified as a weak forward shock. The jump in proton velocity, density, and temperature are used to identify this shock. Little effect was seen in the local energetic ion profiles at that time. Late on DOY 176 another rotation in IMF longitude convected past the spacecraft that may be the signature of a "mini" or partial flux rope. As shown in the PAD's (IV), ion streaming was re-established in the mini-flux-rope, but the direction of flow was sunward, away (downstream) from the forward shock. The termination of the mini-rope also terminated the energetic ion event observed in EPAM.

Discussion

Particle pitch angle distributions are the "surgeons knife" used to analyze changes in energetic ion and electron distributions. These pitch angle distributions are the clearest way to understand how the magnetic field and energetic ions and electrons are coupled. Each of the PAD's in Figure 2 have been averaged over twenty minutes, during this time the particles have traveled significant distances following the interplanetary magnetic field lines. For the specific case of the June magnetic cloud the distributions were remarkably constant over the period of hours, far longer then the averaging interval. Since the 1.5 MeV protons have a velocity of 0.40 AU/h, this implies that the magnetic structure observed at ACE extends a significant fraction of an AU in either direction beyond the spacecraft along the IMF.

The energetic particles are not the only indication that there are different structures within the cloud. The plasma observations also show a significant change in the number density, kinetic temperature, and the He++/H+ ratio, where the temperature shows a somewhat variable profile in the first region and a significant drop in the second region. The fact that abrupt changes observed in the plasma observations occur near the same time that the energetic particles show different PAD's indicates that ACE has entered a different region of the cloud entirely.

A preliminary analysis of energetic ion and electron observations during the May 1998 event (4,10) using the EPAM instrument do not show changes in ion or electron PAD's similar to that observed in the June 1998 magnetic cloud, only steady uniform field aligned flows throughout the entire event. This suggests the seperatrix observed in the June magnetic cloud might not be common to all magnetic clouds observed at 1AU. A systematic study of changes in energetic particle distributions during magnetic clouds in needed to address this question.

Armstrong (1) and Gosling (6) reported on an interesting and well-studied cloud observed near 5 AU at 32° south heliolatitude by Ulysses. Armstrong (1) analyzed the energetic ions and electrons in this structure and found distinctly different intensities and PAD's in different parts of the cloud. Osherovich (9) interpreted the Ulysses observed cloud as a complex structure of multiple helices embedded in a cylindrically symmetric flux rope. Although other explanations that could explain the different distributions observed in the energetic particles during magnetic clouds need to be investigated, a multi-tube model may provide part of the explanation for the observed distributions.

ACKNOWLEDGMENTS

Thanks to Ed Santiago for providing the ACE/SWEPAM datasets. The work performed at the Johns Hopkins University Applied Physics Laboratory has been supported by NASA under Task 009 of Contract NAS5-97271. Support of NFN and CWS was provided by CIT subcontract PC251439 to the Bartol Research Institute under NASA grant NAG5-6912. Work at Los Alamos was performed under the auspices of the U.S. Department of Energy with financial support from the NASA ACE program.

REFERENCES

1. Armstrong T. P., D. Haggerty, L. J. Lanzerotti, C. G. Maclennan, E. C. Roelof, M. Pick, G. M. Simnett, R. E. Gold, S. M. Krimigis, K. A. Anderson, R. P. Lin, E. T. Sarris, R. Forsyth, and A. Balogh, Observation by Ulysses of hot (~270 keV) coronal particles at 32(south heliolatitude and 4.6 AU, Geophys. Res. Lett., 17, 1747, 1994.

2. Burlaga, L. F., R, Lepping, and J. Jones, Global configuration of a magnetic cloud, in Physics of Magnetic Flux ropes, Geophys. Monogr. Ser., vol, 58, edited by C. T. Russell, E. R. Priest, and L. C. Lee, p. 373, AGU, Washington, D. C., 1990

3. Christon S. P., C. S. Cohen, G. Gloeckler, T. E. Eastman, A. B. Galvin, F. M. Ipavich, Y. -K Ko, A. T. Y. Lui, R. A. Lundgren, R. W. McEntire, E. C. Roelof, and D. J. Williams, Concurrent observations of solar wind oxygen by Geotail in the magnetosphere and Wind in interplanetary space, Geophys. Res. Lett., 15, 2987, 1998.

4. Gloeckler G., L. A. Fisk, S. Hefti, N. A. Schwadron, T. H. Zurbucken, F. M. Ipavich, J. Geiss, P. Bochsler, and R. F. Wimmer-Schweingruber, Unusual Composition of the solar wind in the 2-3 May 1998 CME observed with SWICS on ACE, Geophys. Res. Lett., 2, 157, 1999.

5. Gold. R. E., S. M. Krimigis, S. E. Hawkins, III, D. K. Haggerty, D. A. Lohr, E. Fiore, T. P. Armstrong, G. Holland, L. J. Lanzerotti, Electron, Proton, and Alpha Monitor on the Advanced Composition Explorer Spacecraft, Space Sci. Rev., 86, 541, 1998.

6. Gosling J. T., S. J. Bame, D. J. McComas, J. L. Phillips, E. E. Scime, V. J. Pizzo, B. E. Goldstein, and A. Balogh, A forward-reverse shock pair in the solar wind driven by over-expansion of a coronal mass ejection: Ulysses observations, Geophys. Res. Lett., 3, 237, 1994.

7. Lepping, R. P., J. A. Jones, L. F. Burlaga, Magnetic Field Structure of Interplanetary magnetic Clouds at 1 AU, J. Geophys. Res., 95, 11,957, 1990.

8. McComas, D. J., S. J. Bame, P. Barker, W. C. Feldman, J. L. Phillips, R. Riley, J. W. Griffee, Solar Wind Electron Proton Alpha Monitor (SWEPAM) for the Advanced composition Explorer, Space Sci. Rev., 86, 563, 1998.

9. Osherovich V. A., Fainberg J., Stone R. G., Multi-Tube Model of Interplanetary Magnetic Clouds, Geophys. Res. Lett., 3, 401, 1999.

10. Skoug R. M., S. J. Bame, W. C. Feldman, J. T. Gosling, D. J. McComas, J. T. Steinberg, R. L. Tokar, R. Riley, L. F. Burlaga, N. F. Ness, and C. W. Smith, A prolonged He+ enhancement within a coronal mass ejection in the solar wind, Geophys. Res. Lett., 2, 161, 1999.

11. Smith C. W., R. J. Leamon, N. F. Ness, L. F. Burlaga, R. M. Skoug, R. L. Tokar, Magnetic Turbulence Properties Associated with Interplanetary Magnetic Clouds, Trans. Am. Geophys. U., 80, S267, 1999.

12. Smith C. W., J. L'Heureux, N. F. Ness, M. H. Acuna, L. F. Burlaga, J, Scheifele, The ACE Magnetic Fields Experiment, Space Sci. Rev., 86, 613, 1998.

Is there a record of interstellar pick-up ions in lunar soils?

R. F. Wimmer-Schweingruber and P. Bochsler

Physikalisches Institut, Universität Bern, Switzerland

Abstract. Solar wind noble gases and nitrogen implanted in the surface layers of lunar grains have frequently been studied to infer the history of the solar wind. In sub-surface layers, and thus presumably from particles with higher energies than solar wind, a mysterious population, dubbed "SEP", accounts for most of the implanted gas. This "SEP" population is mysterious for at least four reasons: i) In the case of neon it accounts for several tens of percent of the total amount of implanted gas, completely disproportionate from what is expected from solar wind particles; ii) its isotopic composition is distinct from solar; iii) while the heavy neon isotopes are enriched relative to ^{20}Ne, ^{15}N is depleted relative to ^{14}N, signatures which are unexpected from known fractionation processes in particle acceleration; iv) the elemental abundance of N with respect to the noble gases (e.g. Ar) is inconsistent with solar abundances. Many attempts to explain the origin and nature of this mysterious component seem unsatisfactory. In this work, we propose that pick-up ions from interstellar neutrals, accelerated in the heliosphere and subsequently implanted into grains of the lunar regolith might account for the large amount of non-solar "SEPs". The solar system must have encountered various dense interstellar clouds throughout its history. If this scenario is correct, lunar soils serve as a "travel diary" for the voyage of the solar system through the galaxy, preserving records of the isotopic and elemental composition of dense interstellar clouds.

INTRODUCTION

The investigation of volatiles in lunar soils has flourished since the first samples were returned to Earth and made available for scientific investigation. A multitude of applications has emerged from these studies involving interest from cosmochemistry, planetology, astrophysics, and solar physics. The paucity of natural volatiles in the lunar crust and the relative simplicity of handling noble gases in the laboratory have made such work particularly rewarding. Using stepwise etching and stepwise heating techniques (e.g. (1, 2)) it has been possible to isolate solar wind-implanted and solar energetic particle (SEP) volatiles in lunar soils. At solar wind energies, particles penetrate typically a fraction of a micron into a mineral grain such as ilmenite ($FeTiO_3$). In the range from 10 to 1000 keV/amu, penetration depths are some microns and in stepwise heating and etching experiments the release behavior of such particles differs substantially (3).

ISOTOPIC SIGNATURES

Figure 1 is a three-isotope plot of neon. The pattern is typical for stepwise heating and etching experiments with surface implanted volatiles: At low temperatures, a component is released which is isotopically very similar to that found in the solar wind as determined by in situ observations with the foil technique (4), or with spacecraft born mass spectrometers (5, 6). With increasing extraction temperatures gradually neon enriched in ^{22}Ne is released and the turning point in Figure 1 (label "SEP") is approached. Finally, approaching the melting temperature of the sample, the galactic cosmic ray produced neon ("GCR") with roughly equal amounts of ^{20}Ne, ^{21}Ne, and ^{22}Ne is released. Although the isotopic composition of the lunar "SEP" component still seems compatible with some long-term averages of solar energetic particles and although this component has apparently been implanted with energies consistent with presently observed suprathermal and energetic solar particles, their abundance relative to the solar wind concentrations retained in lunar grains exceeds what one would expect from contemporaneously observed solar particle populations by three to five orders of magnitude. This immediately leads to the question whether "SEP" might be a misnomer, and whether the origin of these particles may be non-solar. Another malaise with the identification of solar particles in lunar grains arises from the fact that surface correlated nitrogen in lunar soils if related to the noble gas abundances, exceeds the amount expected from a solar wind origin by a factor of three or even an order of magnitude (7). Furthermore, large variations, up to 50%, have been found in the nitrogen isotopic composition (e.g. (8)). Mathew et al. (9) have found a correlation between the relative deviation of ^{15}N with respect to its terrestrial atmospheric nitrogen content, δ^{15}N, and

CP528, *Acceleration and Transport of Energetic Particles Observed in the Heliosphere: ACE 2000 Symposium*, edited by Richard A. Mewaldt, et al.
© 2000 American Institute of Physics 1-56396-951-3/00/$17.00

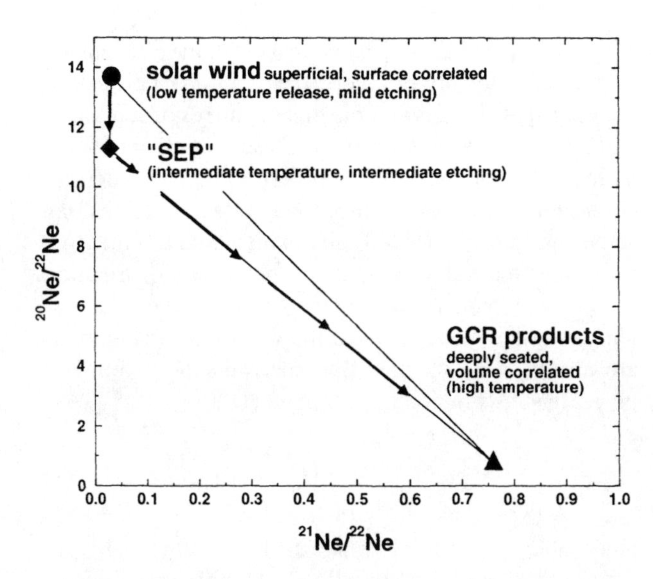

FIGURE 1. Schematics of neon isotope release from lunar grains (see e.g. (3) for more details).

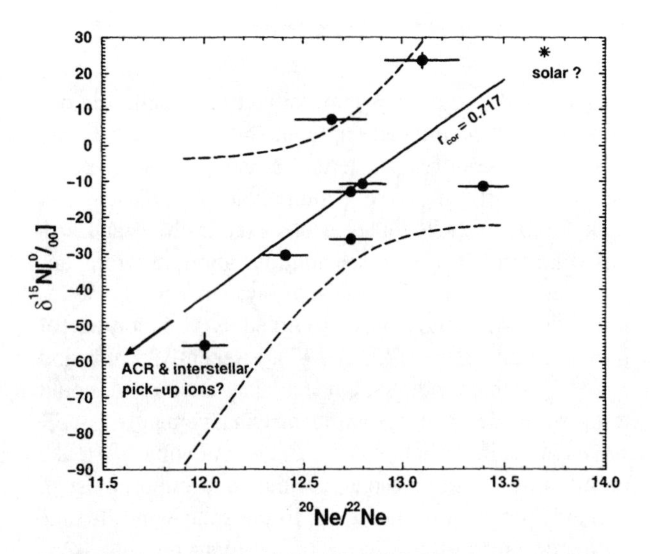

FIGURE 2. Correlation of $\delta^{15}N$ with the $^{20}Ne/^{22}Ne$ abundance ratio in recently irradiated lunar ilmenite grains. Adapted from (9).

$^{20}Ne/^{22}Ne$ in a recently (<100 My) irradiated lunar ilmenite. They conclude that the impulsive flare acceleration process proposed by (10) which would predict an anticorrelation among $\delta^{15}N$ and $^{20}Ne/^{22}Ne$ is contradicted by this observation. However, the conclusion of Mathew et al. is incorrect if the particles in question do not originate from impulsive flares but from interstellar gas as will be discussed shortly.

An alternative interpretation appears to be consistent with the observed isotopic signatures as is illustrated in Figure 2. With the chemical evolution of the galaxy between the formation of the solar system 4.6 Gy ago and today, the metallicity of the galaxy (i.e., the O/H abundance ratio in interstellar matter) has increased. ^{20}Ne is essentially a primary product of nucleosynthesis. It is synthesized in supernova explosions by addition of He nuclei. ^{22}Ne is secondary, and is produced in less massive stars with longer lifetimes and with some delay relative to the production of ^{20}Ne. Therefore, in the present-day local interstellar medium (LISM), one expects to find an enhanced amount of ^{20}Ne, but even more of ^{22}Ne and, correspondingly, a $^{20}Ne/^{22}Ne$-ratio which is lower than solar. The situation is less clear for the pair ^{14}N and ^{15}N. Wielen and Wilson (13) find that $^{14}N/^{15}N \approx 400$ in the local interstellar medium (the solar value is 200 ± 55 (11)!) indicating that ^{15}N is "more primary" than ^{14}N. On the other hand they also find a marginally significant positive galactic radial gradient of $^{14}N/^{15}N$ which points towards ^{14}N being primary, although the galactic center, which has been most thoroughly processed has the highest $^{14}N/^{15}N$ ratio (Figure 3). The recent finding of (12), that the Large Magellanic Cloud, which has a comparatively low metallicity, has a rather low $^{14}N/^{15}N$ ratio of approximately 100 fits into the picture of a primary ^{15}N. Considering this evidence, we favor ^{15}N as the primary nitrogen isotope and claim that the correlation between $^{15}N/^{14}N$ and $^{20}Ne/^{22}Ne$ in recently irradiated ilmenite grains is the imprint of interstellar matter with a variable degree of nucleosynthetic evolution. Thus the observation of such a correlation in lunar soils, apart from eliminating solar impulsive flares as a source, supports our hypothesis that the "SEP" component in lunar ilmenites is of non-solar, possibly interstellar origin.

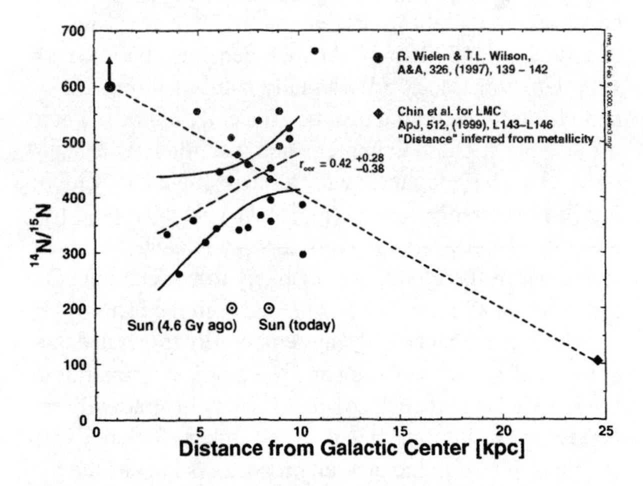

FIGURE 3. Galactic radial dependence of $^{14}N/^{15}N$ (13). The recently determined $^{14}N/^{15}N$ of the Large Magellanic Cloud (12) has been inserted according to its metallicity and following a suggestion of André Maeder (14).

ABSOLUTE ABUNDANCES

As stated earlier, it is impossible to reconcile the observed "SEP" abundances in lunar soils with present-day observed fluxes of suprathermal or energetic solar particles. Similarly, they are incompatible with present-day pick-up ion or ACR fluxes as observed in the near-Earth environment. However, speculations about previous encounters of the Sun and the solar system with dense interstellar clouds have been promoted in the literature for half a century: e.g. (15, 16, 17). Recently, Zank and Frisch (17) have pointed out that such encounters could temporarily lead to a significant enhancement of ACR fluxes in the inner heliosphere. In the following, we make an attempt to estimate the possible contribution of such enhancements and relate them to the solar wind fluence collected during the lifetime of the lunar regolith. Average neutral gas densities in the interstellar medium of the galactic disk are 1 particle per cm^3, i.e. a factor of ten above the ambient LISM. Correspondingly, the present-day influx of neutral gas into the inner heliosphere, the pick-up ion flux and the ACR flux cannot be considered representative for the lifetime of the lunar regolith. An estimate for the flux of pick-up ions averaged over the entire history of the solar system is readily obtained from the following argument: If the average density of interstellar gas is ten times the density of the contemporary LISM, the integrated flux average of pick-up ions and the flow at any place in the solar system and the flow into the lunar regolith must be ten times the contemporary flow, assuming that the ionization efficiency and incorporation of the pick-up ions into the solar wind stream remain unchanged. This scenario changes drastically when the mean free path of interstellar particles becomes comparable to the parameter, $2GM_\odot/v_\infty^2$, i.e., at densities $n_\infty > 300 cm^{-3}$. At such densities the ionizing solar UV will also be substantially reduced within 1 AU, and Ne and N will no longer be efficiently ionized. Thus, if the lunar regolith is impregnated with interstellar matter during the encounter with a dense molecular cloud, it will presumably contain (molecular) nitrogen (and hydrogen). Neutral noble gases such as He and Ne, even if present in the cloud, are not likely to be retained. On the other hand molecular nitrogen, e.g. in the form of N_2 or NH_3 has a chance of being temporarily trapped, especially if the solar radiation and the ambient temperature are reduced. Later such compounds may be gradually released and re-implanted with much higher efficiency into the lunar soil with the pick-up process. As long as the solar wind ionizes and blows beyond 1 AU, it is expected to become increasingly turbulent due to mass loading with the increasing inflow of neutrals. A significant fraction of the energy transferred from the solar wind into the inflowing neutral gas will lead to instabilities, to growth of MHD waves, and ultimately contribute to the efficient acceleration of freshly produced pick-up ions. For lack of better insight we thus assume that the fluxes of suprathermal ions of interstellar origin increase according to n_∞^{1+S}, taking at least a linear correlation (S=0) between density of the ISM and pick-up flux for granted and expecting some enhancement ($S > 0$) due to increased turbulence.

To obtain a rough estimate of the density distribution of clouds encountered during the journey of the solar system along the galactic plane the following assumptions are made. It is assumed that sizes of clouds follow a power law distribution according to (18)

$$f(L) = f(L_0)\left(\frac{L}{L_0}\right)^{-D}; \qquad (1)$$

reasonable values for D range from 1.2 to 2.7 (18), to simplify matters we assume D=2. Another ingredient of importance is the size-density relation

$$\rho(L) = \rho(L_0)\left(\frac{L}{L_0}\right)^{-R}. \qquad (2)$$

Here we assume R=1, i.e., with decreasing size the density of clouds increases. For instance, the combination D=2 and R=1 describes the steady state situation of fragmenting and collapsing clouds where mass is conserved in each generation of new fragments.

Now consider a given volume of size L_0^3 consisting of one cloud at the lowest density ρ_0 (i.e. $f(L_0) = 1$). This volume also contains clouds of smaller sizes according to equation (1). The proportionality factor between density and size in equation (2) can now be extracted from the known average density in the given volume. The total mass contained in the volume L_0^3 is

$$M = -\int_{L_1}^{L_0} \frac{df(L)}{dL} L^3 \rho(L) dL, \qquad (3)$$

from this we derive

$$<\rho> = \rho_0 \frac{D}{3-D-R}\left[1-\left(\frac{L_1}{L_0}\right)^{3-D-R}\right], \qquad (4)$$

which simplifies for the case R=1, D=2 to

$$<\rho> = 2\rho_0 \ln\left(\frac{L_0}{L_1}\right). \qquad (5)$$

With one big cloud of size $L_0 = 10^3$pc, and the smallest clouds to be of size $L_1 = 0.1$pc, and assuming that the average density in the entire volume is $<\rho> = 1 cm^{-3}$, one finds that the density of the largest cloud is $\rho_0 = 0.05 cm^{-3}$. The average flux of interstellar pick-up ions can then be related to the present-day flux using the rule previously discussed

$$\phi(L) = \phi_{today}\left(\frac{\rho(L)}{\rho_{today}}\right)^{1+S}. \quad (6)$$

Certainly, the Sun will cross small clouds with high densities in shorter times than large clouds. If the speed is independent of the cloud properties, this is taken into account assuming residence probabilities which are proportional to the volume occupied by clouds of different sizes.

$$dp = (3-D)dL\frac{L^{2-D}}{L_0^{3-D}}\frac{1}{\left[1-\left(\frac{L_1}{L_0}\right)^{3-D}\right]}. \quad (7)$$

The average pick-up flux then relates to the present-day flux according to

$$<\phi>=\phi_{today}\cdot\left(\frac{\rho_0}{\rho_{today}}\right)^{1+S}\cdot\frac{3-D}{3-D-R(1+S))}\cdot\frac{\left[1-\left(\frac{L_1}{L_0}\right)^{3-D-R(1+S)}\right]}{\left[1-\left(\frac{L_1}{L_0}\right)^{3-D}\right]}. \quad (8)$$

Using again D=2 and R=1, one obtains

$$<\phi>\approx\frac{\phi_{today}}{S}\left(\frac{\rho_0}{\rho_{today}}\right)^{1+S}\left(\frac{L_0}{L_1}\right)^{S}. \quad (9)$$

For S=1 (enhanced turbulent acceleration) one finds

$$<\phi>\approx 2500\cdot\phi_{today}. \quad (10)$$

CONCLUSIONS

The above estimates crucially depend on the link between ambient interstellar gas density and pick-up ion flux. They are less sensitive to the assumption on the relation between cloud sizes and densities and on the fractal dimension of cloud size distribution and on the absolute sizes of clouds considered. In any case, the result confirms the conclusion of (17) that during encounters of the solar system with dense clouds, an enhanced pick-up and ACR ion flux in the near-Earth environment is likely. The putative "SEP" component in the lunar regolith with the mysterious compositional features should cautiously be renamed "HEP" (Heliospheric Energetic Particles) and re-examined for compositional signatures which exclude a solar origin. A prime candidate for identification is deuterium. Another positive aspect of such an investigation, apart from contributing to the solution of the "SEP-mysteries", is the opening of a new and attractive possibility to study the past of the solar system: Deciphering the archive of volatiles in the lunar regolith could not only help us to learn about the past of solar activity but also enable us to read the travel-diary of the solar system as recorded during its galactic journey.

ACKNOWLEDGEMENTS

We wish to acknowledge helpful discussions with O. Eugster, A. Maeder, and R. Wieler. This work was supported by the Schweizerischer Nationalfonds.

REFERENCES

1. Eugster, O., Grögler, N., Eberhardt, P., and Geiss, J., *Proc. Lunar Planet. Sci. Conf.* **10**, (1979) 10:1351–1379.
2. Wieler, R., and Baur, H., *Astrophys. J.* **453**, (1995) 987–997.
3. Wieler, R., *Space Sci. Rev.* **85**, (1998) 303–314.
4. Geiss, J., Bühler, F., Cerutti, H., Eberhardt, P., and Filleux, C., *Apollo 16 Prelim. Sci. Rep. NASA SP* **315** (14), (1972) 14.1–14.10.
5. Kallenbach, R., *et al.*, *J. Geophys. Res.* **102**, (1997) 26895–26904.
6. Wimmer-Schweingruber, R. F., Bochsler, P., Kern, O., Gloeckler, G., and Hamilton, D. C., *J. Geophys. Res.* **103**, (1998) 20621 – 20630.
7. Frick, U., Becker, R. H., and Pepin, R., *Proc. Lunar Planet. Sci. Conf.* **18**, (1987) 87–120.
8. Kerridge, J. F., *Rev. Geophys.* **31**, (1993) 423–437.
9. Mathew, K., Kerridge, J. F., and Marti, K., *Geophys. Res. Lett.* **25** (23), (1998) 4293–4296.
10. Bochsler, P., and Kallenbach, R., *Meteoritics* **29**, (1994) 652–658.
11. Kallenbach, R., Geiss, J., Ipavich, F. M., Gloeckler, G., Bochsler, P., Gliem, F., Hefti, S., Hilchenbach, M., and Hovestadt, D., *Astrophys. J.* **507**, (1998) L185–L188.
12. Chin, Y.-N., Henkel, C., Langer, N., and Mauersberger, R., *Astrophys. J.* **512**, (1999) L143–L146.
13. Wielen, R., and Wilson, T., *Astron. Astrophys.* **326**, (1997) 139–142.
14. Maeder, A., *Personal communication*.
15. Bondi, H., and Hoyle, F., *Mon. Not. R. Soc. Astr.* **104**, (1944) 273–282.
16. Talbot, R., and Newman, M., *Astrophys. J. Suppl.* **34**, (1977) 295–308.
17. Zank, G. P., and Frisch, P. C., *Astrophys. J.* **518** (2), (1999) 965–973.
18. Henriksen, R., *In: Fragmentation of molecular clouds and star formation (E. Falgarone, F. Boulanger, and G. Duvert eds.), Kluwer Academic Publisher* **IAU Symposium No. 147**, (1991) 83–92.

Hybrid Simulations of Preferential Heating of Heavy Ions in the Solar Wind

Paulett C. Liewer, Marco Velli* and Bruce E. Goldstein

Jet Propulsion Laboratory, California Institute of Technology, Pasadena, CA 91109

Abstract. We present results from the first fully self-consistent 1D hybrid (kinetic ions/fluid electrons) simulations of the preferential heating of alphas and heavier minor ions by a flat spectrum of Alfvén-ion cyclotron waves in a collisionless plasma. We find that the simulations reproduce the observed solar wind scaling $T \propto M$ for alphas and heavier minor ions when the alphas and the minor ions have equal charge to mass ratios, q/M, and equal initial thermal velocities, $V_{th}=(T/M)^{1/2}$. This scaling is interpreted as a result of the basic physics: the time evolution of the Vlasov/Maxwell system without collisions depends only on the ratio q/M and not q or M separately. Because this result follows from the basic nature of the physical model, the $T \propto M$ scaling would be obtained for any spectrum of waves. For minor ions with q/M different from the alphas but equal initial thermal velocities, the final thermal velocity is seen to vary by ±50% from that of the alphas in the simulations presented here.

INTRODUCTION

It has long been observed that, in the solar wind, the alpha particles and heavier minor ions have higher temperatures than the protons (1,2). In coronal hole associated solar wind, the heavy ion temperatures are found to be proportional to the heavy ion mass, $T \propto M$, resulting in equal thermal velocities (3,4). Recently von Steiger et al. (5) extended this result to more ion species using SWICS/Ulysses data. Wave-particle interactions between the ions and Alfvén/ion cyclotron waves are generally considered to be the cause of the preferential heavy ion heating.

Quasilinear models (6 and references therein) have successfully shown that the alphas can be heated and accelerated relative to the protons. More recently, a non-linear hybrid simulation model which includes the effects of solar wind expansion on the evolution of the wave spectrum and frequencies has also reproduced preferential heating and acceleration of the alphas relative to the protons (7).

This paper focuses on the heating of multiple heavy ion species. Quasilinear models of the heating of minor ions have not been able to reproduce the $T \propto M$ scaling for alphas and heavier ions (8). Based on an analysis of the linear dispersion relation for a multi-ion plasma, Gomberoff and Astudillo (9) have suggested a scenario that could lead to the $T \propto M$ scaling, but a non-linear analysis is needed to verify this suggestion.

Here we present results of self-consistent 1D hybrid (kinetic ions/fluid electrons) simulations of the interaction of alpha particles and heavy ions with a spectrum of Alfvén-ion cyclotron waves. In these simulations, the heavy ions, alpha particles and protons, are all treated self-consistently and thus the linear and non-linear effects of the heavy ions on the waves are included. These effects include, for example, the introduction of the frequency gap at the alpha cyclotron frequency by the 4% alpha population (linear effect) and the damping of the waves by resonant ions of all 3 species (non-linear effect). We believe these are the first simulations of minor ion heating by Alfvén waves in which the alphas and other minor ions, as well as the protons, are treated self-consistently with the wave fields. The simulations are initialized with a spectrum of circularly polarized Alfvén-ion cyclotron waves. For the results presented here, the initial spectrum of waves was flat out to wave numbers $kV_A=1.5\Omega_P$ where Ω_P (= eB/M_Pc) is the proton gyro- frequency. The waves heat and accelerate the ions and the ions damp the waves.

We find that the simulations reproduce the observed scaling $T \propto M$ for alphas and heavier minor ions when the minor ions and the alphas have the same q/M and the same initial *thermal velocities* $V_{th}=(T/M)^{1/2}$. This result is explained as a consequence of the dependence of the Lorenz force solely on the ratio of q/M and not q or M independently and would be obtained for any wave spectrum. For minor ions with q/M different from the alphas but with the same initial

* Permanent address: Dipartimento di Astronomia e Scienza dello Spazio, Universitá di Firenze, 50125 Firenze, Italy

CP528, *Acceleration and Transport of Energetic Particles Observed in the Heliosphere: ACE 2000 Symposium*, edited by Richard A. Mewaldt, et al.
© 2000 American Institute of Physics 1-56396-951-3/00/$17.00

thermal velocities, we find that the final thermal velocity may vary by ±50% from that of the alphas in our simulations. For minor ions with same q/M and equal initial *temperatures*, the heavier ions have a higher final temperature, but the $T \propto M$ scaling is not strictly observed.

THE MODEL

The plasma simulation model used is the 1D hybrid code of Winske and Leroy (10), modified to include multiple ion species. All three ion species – protons, alphas and the minor ions– are treated as particles and the code updates the ions at each time cycle using

$$\frac{d\vec{x}}{dt} = \vec{v} \quad \text{and} \quad \frac{d\vec{v}}{dt} = \frac{q}{M}\left(\vec{E} + \vec{v} \times \vec{B}\right) \qquad (1)$$

where the electric and magnetic fields are computed self-consistently from the particle positions and velocities (10). Thus all ion species are treated in a fully self-consistent manner. The electrons are treated as a neutralizing fluid. This model has also been extended to include the effects of solar wind expansion on ion cyclotron heating (7), but that capability has not yet been used in this study.

The code is initialized with a spectrum of outward propagating left-hand circularly polarized Alfvén-ion cyclotron waves. For the results presented here, the initial spectrum of waves was flat out to wave numbers $kV_A=1.5\Omega_P$ where $\Omega_P(= eB/M_Pc)$ is the proton gyro-frequency. Various fluctuations levels were used, $|\delta B/B|^2=0.16 - 0.55$. The alpha particles can resonate with and damp waves with wave numbers $kV_A \sim \Omega_{He}=0.5\Omega_P$ and higher and likewise for heavier minor ions with $q_{MI}/M_{MI} \sim q_{He}/M_{He} < q_p/M_p = 1$ (in our units). Only wave numbers, not wave frequencies, are specified in hybrid simulation codes; the time evolution of the waves is determined by the self-consistent time integration of the particle and field equations. Thus the alpha minor ion effects on the linear dispersion relation are correctly treated in our simulation model. The longest wave length modes, those with $kV_A \ll \Omega_P$, satisfy $\omega = kV_A$. In a pure proton plasma, as k increases the left-hand circularly polarized branch approaches $\omega = \Omega_P$ asymptotically. In cold plasmas with an alpha density typical of the solar wind (~4%), the low frequency LHCP branch asymptotes to $\omega=\Omega_{He}$ and a gap appears between this low frequency branch and the proton cyclotron frequency branch at $\omega \sim \Omega_P$. The gap can disappear as the alphas are heated and accelerated (see, e.g. Ref. 9).

SIMULATION RESULTS

In the results presented here, each simulation run had three ions species: protons (p), alphas (He) and one minor ion specie (MI); separate runs were made to determine the dependence of the heating on the particular minor ion. The alphas contributed 4% of the number density, the minor ion 0.1% with protons making up the remaining 95.9%. The parameters for these runs were as follows: time step $\Delta t\ \Omega_P=0.001$, $\omega_{pi}/\Omega_P = 100$, grid size $L_x = 128\ V_A/\Omega_P$, number of grid cells $N_x = 256$ with 40,240 particles for each for the three species. The protons were initialized with $\beta_P = 0.05$. Results are presented here for two initializations of the alpha and heavy ion temperatures. In the first set of runs, the protons, alphas and minor ions were initialized with equal thermal velocities $V_{th}=(T/M)^{1/2}$; runs were made for cases with various minor ion species (Carbon, Oxygen, Neon, Magnesium and Iron) in various charge states. The initial level of turbulence was $\delta B/B=0.4$.

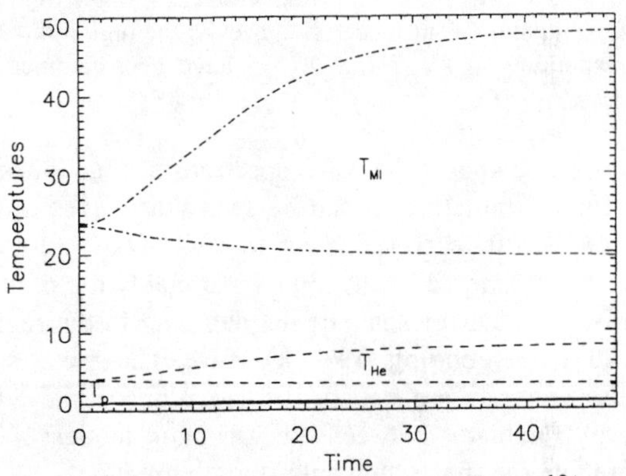

Figure 1. Proton (__), He^{++} (---) and Mg^{12+} (._._) perpendicular and parallel temperatures as a function of time in the simulation. The upper of the two curves for each species is the perpendicular temperature and the lower the parallel temperature. Note that the Mg^{12+} temperature is 6 times the He^{++} temperature for all times in the simulation, giving $T \propto M$. Very little proton heating is seen (~10%).

Figure 1 shows the time history of the perpendicular and parallel temperatures for a case with Mg^{12+}. To obtain the V_{th} for each species, each species mean velocity is first calculated at each grid point. The thermal velocity is then the square root of the sum over all ions of a species of the squares of each ion's relative velocity, as determined by interpolation of the mean velocity to the particle's position. In the simulations, most of the heating of the ions occurred by $t\Omega_P \approx 30$. Very little proton heating (~10%) occurred. In the figure, it can be seen that the alphas are heated more than the protons and the minor ions are heated more than the alphas; the heating is all perpendicular with a slight cooling in the parallel direction as expected from cyclotron damping. The alphas and Mg^{12+} ions can resonate with waves $kV_A \geq \Omega_{He} = \Omega_{Mg12+} = 0.5\Omega_p$, and thus have access to more of the spectral energy than the protons. For a case with the same parameters but no alphas or minor ions, the protons are heated somewhat more than in this case (15% increase in temperature vs. the 10% seen in the Figure 1 case).

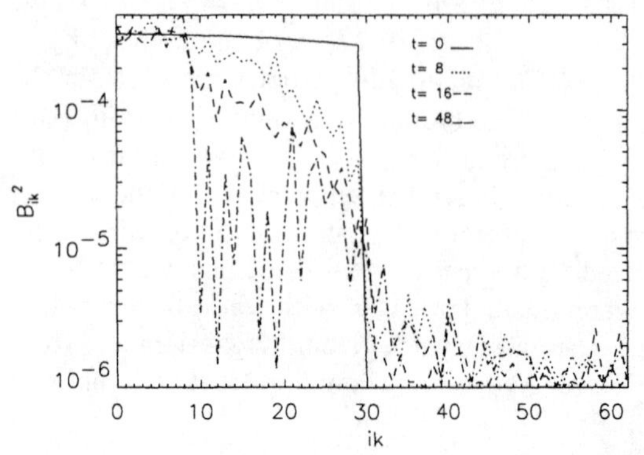

Figure 2. Power spectrum of the waves at 4 times from the same simulation as in Figure 1. Waves in the region ik > 10, corresponding to $kV_A > \Omega_{He} = \Omega_{Mg12+}$, have been damped by the ions.

Figure 2 shows the wave spectrum at four times in the same simulation. It can be seen that waves in the region ik>10, corresponding to $kV_A > \Omega_{He} = \Omega_{Mg12+}$, have been damped by the ions. Presumably the protons are responsible for some of the damping in the region ik>20, corresponding to $k = kV_A > \Omega_P$. The spikes seen in the spectrum are statistical fluctuations as the wave energy fluctuates between the various k modes; time-averaging the spectrum would smooth these out.

Figure 3 plots the thermal velocities of the heavy ions relative to the thermal velocity of the alphas from the various simulations with equal heavy ion initial temperatures at $t\Omega_P = 0$. Ions with the same q/M as alphas (O^{8+}, Ne^{10+}, $Mg12^+$, plotted as ◊) have the same thermal velocities for all times (all the points fall on top of each other). The velocities of ions with different q/M are plotted at $t\Omega_P = 50$ (Δ) and 100 (❑). By $t\Omega_P = 100$, the rate of change of the temperatures is small. The thermal velocities observed for ions of different q/M vary by about ±50% from that of the alphas and other q/M=0.5 ions. Thus we are able to reproduce the observation $T \propto M$ exactly, but only for alphas and heavy ions which have the same q/M as the alphas and the same initial thermal velocities.

The variations in the thermal velocity with q/M are not presently understood. Numerical experiments have shown that the results are not dependent on the spatial resolution, and thus k-space density, of the simulations. We plan to investigate how these deviations depend on the spectrum of the waves and other parameters.

The result that $T \propto M$ for heavy ions that start with the same thermal velocities and have the same q/M as the alphas can be understood from the basic physical system. In a collisionless plasma, the particle distribution functions evolve under the influence of the Lorenz force that depends only on the *ratio* of q/M, and not on either separately:

$$\frac{\partial f}{\partial t} + \vec{v}\frac{\partial f}{\partial \vec{x}} + \frac{q}{M}\left(\vec{E} + \vec{v} \times \vec{B}\right)\frac{\partial f}{\partial \vec{v}} = 0 \qquad (2)$$

If the alphas and heavy ions have the same f(x,v,t) initially and the same q/M, the distribution functions, and hence the thermal velocities, will remain the same for all times. The hybrid code uses instead equation (1), but the physical argument is the same. Moreover, because the scaling $T \propto M$ for ions with equal initial thermal velocities and q/M is a result of the nature of the physical system, this scaling would be obtained for any wave spectrum. Note that including the Sun's gravitational force would not add any dependence on mass alone and the physical argument is unchanged.

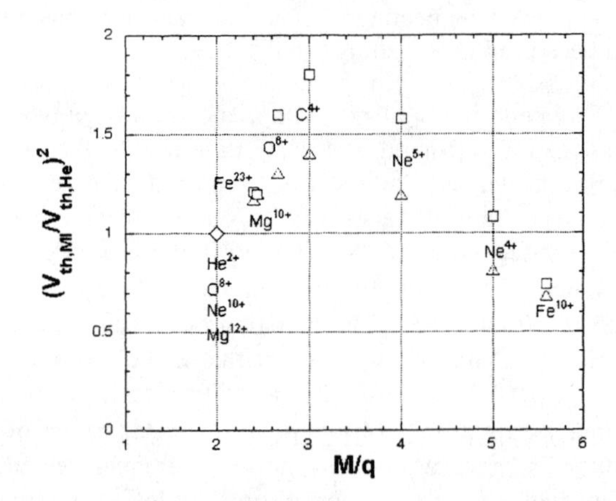

Figure 3. Square of thermal velocities of the heavier minor ions relative to the thermal velocity of the alphas from the various simulations with equal heavy ion initial temperatures. Minor ions with the q/M of alphas (◊) all fall on top of each other for all times; for these ions, we reproduce the $T \propto M$ scaling. The thermal velocities of minor ions of other q/M's (Δ for $t\Omega_P$=50; ❑ for $t\Omega_P$=100) vary by about ±50%.

In the second set of simulation runs, the three species (proton, alpha and the minor ion), all with $q_{MI}/M_{MI} = q_{He}/M_{He} = 0.5$) were initialized with equal temperatures ($T_P = T_{He} = T_{MI}$). Cases were run with two levels of turbulence ($|\delta B/B|^2 = 0.16$ and 0.55) and two minor ions [O^{8+} (M_{MI}=16) and Mg^{12+} (M_{MI}=24)]. Results are shown in Figure 4 again at time $t\Omega_P = 100$.

It can be seen that the heavy ion temperature increases with mass, but the scaling $T \propto M$ is not strictly obeyed. However, as the fluctuation level increases, the initial condition becomes less important and the scaling is approached. We suspect that if there was a source of wave energy to continue the heating, then the $T \propto M$ would again be obtained.

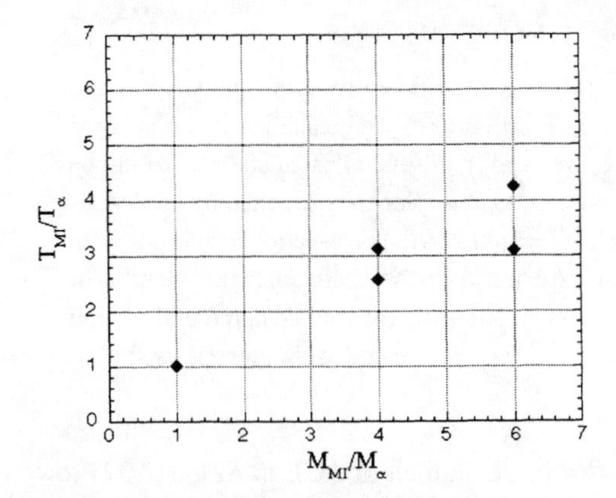

FIGURE 4. Temperature versus mass from simulations with equal initial temperatures and heavier ions with $q_{MI}/M_{MI}=q_{He}/M_{He}=0.5$. For each mass, upper (lower) point is for initial fluctuation level $|\delta B/B|^2= 0.55$ (0.16).

DISCUSSION

We have presented simulation results showing that a spectrum of waves heats alpha particles and heavier minor ions to temperatures $T \propto M$ for minor ions with the same q/M as the alphas and the same initial thermal velocities. This result stems from the basic nature of the physical model governing the evolution of the system: the Lorenz forces depends only on the ratio q/M. Thus this result would be obtained for any level of turbulence and any wave spectrum. This result only holds in collisionless plasmas. The Coulomb collision frequency depends on q^2/M, and the temporal evolution no longer depends only on the ratio q/M.

How does this relate to the observations (1-3,5)? The solar wind observations of von Steiger et al. (5) showed that $T \propto M$ in fast solar wind where collisions are not important and there is a large flux of Alfvén waves. The M/q of the ions in Ref. 5 was not identical to that of the alphas, but the range was not very great: M/q = 2 for O^{8+} to M/q=3 for C^{4+}. Deviations from $T \propto M$ of the size seen in our simulations (Figure 3) were within the error bars on the temperatures in Ref. 5.

We have not addressed here the question of how the wave power reaches the high frequencies used in the simulation spectrum. However, in related work using a hybrid code with the effects of solar wind expansion included (7), we have shown that a spectrum of waves below the alpha and proton cyclotron frequencies becomes resonant first with the alphas as the solar wind expands and the mode frequencies decrease ($\sim 1/R$) less rapidly than the cyclotron frequencies ($\sim 1/R^2$). We plan to continue these studies with minor ions included. The minor ion with the lowest q/M will come into resonance first as the solar wind expands (cf. Ref. 9).

ACKNOWLEDGMENTS

We would like to thank M. Neugebauer, R. von Steiger and T. Zurbuchen for several useful discussions. This research was performed at the Jet Propulsion Laboratory under a contract between the California Institute of Technology and NASA.

REFERENCES

1. Neugebauer, M., *Fund. of Cosmic Physics* 7,131-1099 (1981).

2. Zurbuchen, T. H., Fisk, L. A., Schwadron, N. A., and Gloeckler, G., this proceedings.

3. Schmidt, W. K. H., Rosenbauer, H., Shelley, E. G., Sharp, R. D., Johnson, R. G., and Geiss, J., *Geophys. Res. Lett.* 7, 697-700 (1980).

4. Ogilvie, K. W., Bochsler, P., Geiss, J., and Coplan, M. *J. Geophys. Res.* A **85**, 6069-6074 (1980).

5. Von Steiger, R., Geiss, J., Gloeckler, G., and Galvin, A. B., *Space Science Reviews* **72**, 71-76 (1995).

6. Marsch, E., Goertz, C. K., and Richter, K., *J. Geophys. Res.* A **87**, 5030-5044 (1982).

7. Liewer, P., Velli, M. and Goldstein, B., in *Solar Wind Nine*, edited by S. Habbal, R. Esser, J. V. Hollweg and P. A. Isenberg (AIP Conference Proceedings 471) 449 (1999).

8. Isenberg, P. A., and Hollweg, J. V., *J. Geophys. Res.* A **88**, 3923-3935 (1983).

9. Gomberoff, L., and Astudillo, H. F., in Solar *Wind Nine*, edited by S. Habbal, R. Esser, J. V. Hollweg and P. A. Isenberg (AIP Conference Proceedings 471) 461 (1999).

10. Winske, D., and Leroy, M. M., in *Computer Simulation of Space Plasmas – Selected Lectures at the First ISSS*, edited by H. Matsumoto and T. Sata, D. Reidel, Norwell, MA (1985).

Coherent frequency variations in electron fluxes at 1 and 5 AU in the inner heliosphere

D. J. Thomson, L. J. Lanzerotti and C. G. Maclennan

Bell Laboratories, Lucent Technologies, Murray Hill, New Jersey 07974

Abstract. Several years ago we reported evidence for distinct modal structure in particle flux data from Ulysses and other interplanetary spacecraft, based on detailed time series analyses (1). Furthermore, we concluded that numerous discrete frequencies could be identified with known solar p-mode frequencies and with several calculated solar g-mode frequencies. This paper continues to explore the nature of the modal structure of the interplanetary medium by investigating the statistics of the coherence in electron fluxes measured at two separate locations in the near-ecliptic heliosphere, at 1 AU and at 5 AU, on the ACE and the Ulysses spacecraft, respectively. We demonstrate that there are highly significant coincident frequency peaks at the two locations in data taken at different times, and that, for data acquired at the same time at the two locations, there are high levels of coherence at multiple frequency pairs across a wide spectral band.

INTRODUCTION

Thomson et al. (1) (hereafter TML) reported that careful time series analysis of interplanetary particle (ion and electron) fluxes in the heliosphere showed evidence for distinct frequency modes. These authors asserted that the source of these modes might be solar acoustic (p-mode) and gravitational (g-mode) oscillations. This conclusion elicited a spirited discussion in the literature regarding both the solar and the interplanetary implications *e.g.* (2, 3, 4, 5, 6, 7).

One of the implications of the existence of long period modes in the heliosphere (from whatever source) is that particles of very dissimilar kinematic parameters should demonstrate responses to the same frequencies. Long period (>2 day) fluctuations in galactic cosmic ray fluxes and interplanetary magnetic fields have been reported in (8) and (9), respectively. Similarly, electron fluxes (~0.5 MeV) and galactic cosmic ray protons measured on the Ulysses (ULS) spacecraft have been shown in (10) to have coherent oscillations at a long period of ~2.3 days. Moreover, it was shown in (11) that an oscillation at the same frequency was observable in central England temperatures, and was stable for the length of the record, more than 200 years. Other tests of the conclusion of (1) regarding the existence of discrete frequencies in the interplanetary medium can be devised, such as those reported herein for frequency comparisons at widely separated spatial locations.

In this paper we report a statistical study of low energy (~40–100 keV) electron fluxes as measured on the ACE and the ULS spacecraft in the heliosphere in 1997 to 1999. From the launch of ACE in August 1997 toward the sunward Lagrangian point along the Earth-Sun line to mid-1999, ULS was within about 30° of the ecliptic plane and at a helioradius of ~5 AU from the Sun. Ulysses was just completing its first solar polar (Solar Minimum) orbit of the Sun, and was beginning its Solar Maximum orbit.

We report here two separate findings: (1) a comparison of power spectra from data series acquired at different times on the two spacecraft which shows coincident frequency peaks at high confidence levels, and (2) a dual coherence analysis of electron data taken at the same time on both spacecraft which shows multiple frequency coherence across a wide spectral band. This finding of a correlation at different frequencies proves that the two data sets are nonstationary. There are no previous studies that we are aware of that have examined correlations between separated locations at these frequencies.

ANALYSIS AND SPECTRAL COMPARISONS

The electron data analyzed here are acquired by the deflected electron (DE) telescopes of the EPAM and the HI-SCALE instruments on ACE and Ulysses, respectively. The DE telescopes are virtually identical as the EPAM instrument is the back-up HI-SCALE instrument from ULS (12). The 38–53 keV and 53–103 keV (DE-1 and DE-2) electron energy channels from each instrument are used for the analysis. (For simplicity in the following discussions, the spacecraft designations will be used in lieu of the instrument names.)

CP528, *Acceleration and Transport of Energetic Particles Observed in the Heliosphere: ACE 2000 Symposium*, edited by Richard A. Mewaldt, et al.
© 2000 American Institute of Physics 1-56396-951-3/00/$17.00

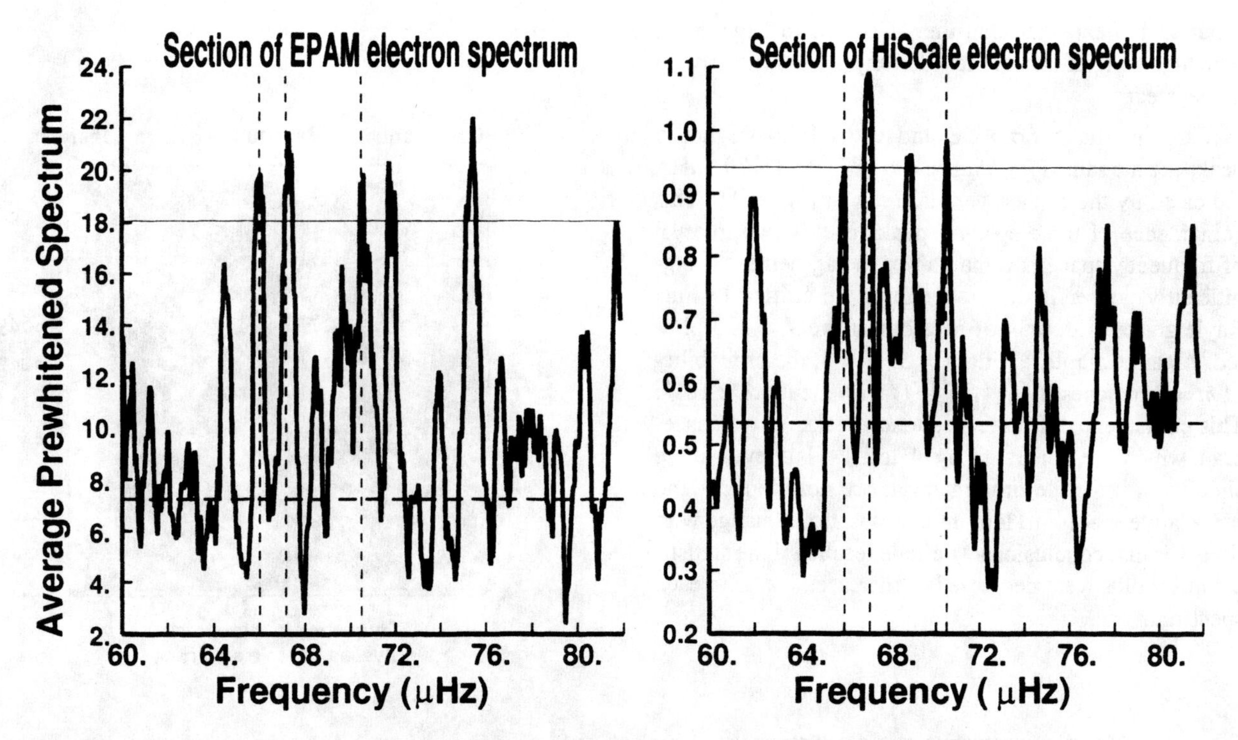

FIGURE 1. Averages of prewhitened spectra of ACE EPAM and Ulysses HiScale electron channels, 38–53 and 53–103 keV. The dashed vertical lines mark 3 peaks which are the same in both spectra.

The data were prepared in the same manner as reported in TML: we used an initial linear interpolation across the few short data gaps, logarithms of electron fluxes, followed by a refined Wiener interpolation. The power spectra were estimated using a multitaper method (13, 14, 15) (see also (16) for a review and commentary). The data tapers (data windows) are discrete prolate spheroidal (Slepian) sequences. The total number of data samples, including the interpolations, is denoted N. The bandwidth W of the analysis is chosen as a compromise between spectral resolution where W is small, and statistical stability where W is large. The Fourier transforms of these prolate spheroidal sequences have the greatest energy concentration possible for a sequence of length N and bandwidth W so they are ideal data windows (16).

For purely non-deterministic random data with a spectrum that does not vary too rapidly in the frequency, f, band $(f-W, f+W)$, the probability distribution of this estimate is chi-squared, χ^2, with $2K$ degrees-of-freedom. In this study we use a time-bandwidth product of $NW = 5$ and $K = 8$ windows, resulting in power spectra with 16 degrees-of-freedom. A fast Fourier transform algorithm was used to compute the spectra.

Shown in Figure 1 are sections of power spectra, 60–82 μHz (between ~3.5 and 4.5 hour periods), for ACE and ULS electron fluxes, respectively. Each spectrum is the average of the pre-whitened spectra for the DE-1 and DE-2 (38–53 and 53-103 keV) channels individually on each spacecraft, and thus each spectrum has 32 degrees-of-freedom, Δ. The spectra for the ACE data are for the time interval August 30, 1997, to March 26, 1998; the spectra for the ULS data are for the time interval April 19, 1997, to November 13, 1997, overlapping the ACE data series by about 2.5 months. The dashed horizontal line in each figure is the spectrum mean, obtained by fitting the lower 5% points of the spectrum estimates, and scaling by $\frac{\Delta}{Q_{\chi^2}(0.05,\Delta)} = 1.74$ where $Q_{\chi^2}(0.05,\Delta)$ is the χ^2 quantile at the 5% point, and here $\Delta = 32$. When a spectrum contains line components this is a much more robust estimate of the base level than a simple average (16). Since $\Delta = 32$, the 99% points are a factor of 1.67 above these lines. The spectral values for ACE in Figure 1 are larger than those for ULS because the particle fluxes are higher at ACE, which is closer to the Sun. The ULS data are also more "noisy" than the ACE data because of the ULS RTG power source which provides more background counts in some energy channels.

Five frequency peaks at ACE (A) have a significance level $> 6\sigma$ (denoted by light horizontal line) above the spectrum mean, and three are $> 7\sigma$. At ULS, four frequency peaks (U) have $> 3\sigma$ (light horizontal line) significance. Because of the larger count rates at ACE compared to ULS, and therefore more counts above any background, it is expected that the ACE spectra have peaks with higher significance, provided any peaks were present in the spectra.

Three of the $> 6\sigma$ ACE and $> 3\sigma$ ULS frequency peaks are the same ($f \sim 66\mu$Hz, $\sim 67\mu$Hz, and $\sim 70.5\mu$Hz, indicated by the dashed vertical lines in Figure 1). This coincidence of three spectral peaks in this one interval of frequency space is remarkable: the agreement is significantly poorer if the frequencies are shifted by one Rayleigh resolution, ~ 56 nHz out of the $J = 395$ plotted. As an example, taking $A = 5$, $U = 4$, the probability of $m = 3$ matches is $\binom{A}{m}\binom{J-A}{U-m}/\binom{J}{U}$ or about 9.8×10^{-6}. This coincidence of three frequency peaks at two spacecraft which are separated by 4 to 5 AU is even more significant since the time intervals are not identical and the relative spacecraft longitudinal locations change with time. Similar conclusions would be reached if the statistical thresholds were decreased or increased by 1σ in each spectrum.

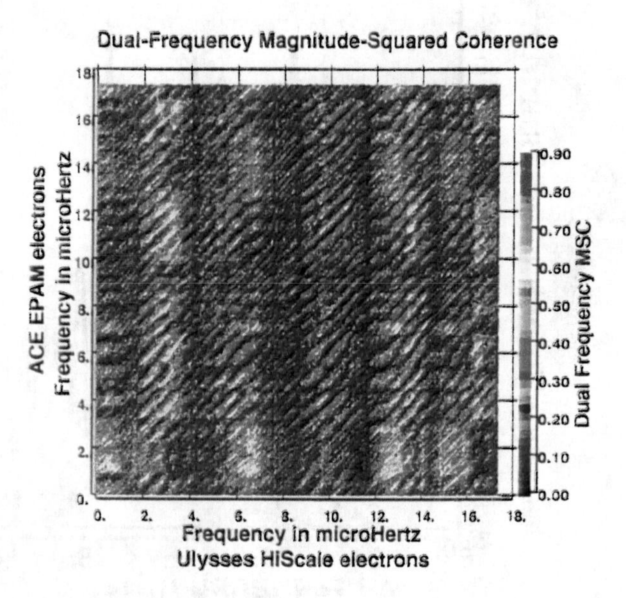

FIGURE 2. Dual-frequency magnitude-squared coherence between ACE and ULS electron fluxes

DUAL-FREQUENCY SPECTRA

The dual-frequency power spectrum between ACE and ULS electron fluxes (53–103 keV energy channel on both spacecraft) was determined for particle time series acquired between March 14, 1998, and July 30, 1998. The dual-frequency spectrum $\gamma(f_1, f_2)$ is estimated by standard procedures; see, for example, §XIV of (13). As in the previous section, multitapers of prolate spheroidal sequences were used on the data. As with ordinary cross-spectra, it is most useful to work with dual-frequency magnitude-squared coherences $\widehat{C}_{pq}(f_1, f_2)$. Shown color-coded in Figure 2 is the dual-frequency magnitude-squared coherence between the ACE and ULS electron fluxes in the frequency band 0–18 μHz (periods $\gtrsim$ 16 hours).

There is high coherence between the electron fluxes at the two spacecraft at many frequency pairs in this band. The high coherence tends to occur on diagonals across the plot, approximately parallel to the principal diagonal. Such enhancements in coherence would not occur at significant levels in stationary time series without periodic components. In a stationary time series, high coherence levels, if there were coherence between two series, would occur only along the $f_1 = f_2$ diagonal where the processes would exhibit the ordinary coherence (see Figure 3 below). If the interplanetary medium were simply an approximately stationary, turbulent process, there would be no known reason for the appearance of the many diagonals, nor the column appearances of deep nulls in the coherence along the ULS electron frequency axis.

Alternatively, if one considers a source process with *many* stable frequencies (such as might originate in the solar core and that are only observable after being propagated through the convection zone), the result is reasonably explained. Modes that have different frequencies with similar spatial (solar latitude and longitude) structure may be given similar modulations by the convection process and one would then see frequency-translated versions of this process in the interplanetary medium. Because the two spacecraft are in different orbits, they will not see all the same modes, or the same convection modulation, so the dark lines might be simply explained by a mode having a large amplitude at one spacecraft and not at the other. Bright regions occur when modes of similar type are present at both.

Shown in Figure 3 is a line plot of the magnitude-squared coherence (left ordinate of the figure) taken along the principal diagonal axis of Figure 2. This is the result that would be obtained with conventional coherence analysis between two time series. A visual comparison of the color-coded coherence along the principal diagonal of Figure 2 and the line plot of Figure 3 shows qualitatively that the two presentations are the same. A vertical axis is provided on the right of Figure 3 for the cumulative prob-

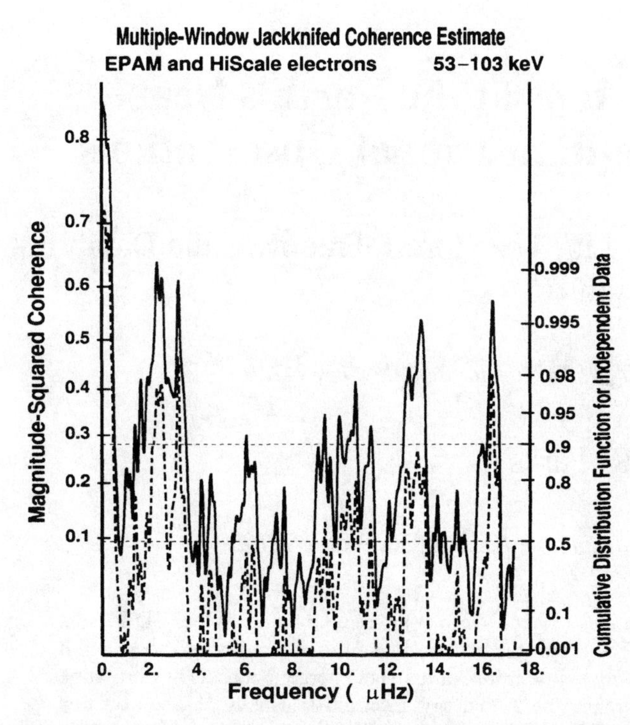

FIGURE 3. Magnitude-squared coherence along the diagonal of Figure 2

ability distribution function (assuming the two electron flux time series are independent data). There are numerous high coherence peaks above 90% probability, and at least 4 above 99.5% probability. The lower dashed line in Figure 3 is one standard deviation below the estimate (determined using the jackknife procedure; see (17, 18)).

CONCLUSIONS

The spectra of Figure 1 show highly statistically significant peaks at similar frequencies in electron flux data sets that were acquired at different times and different helioradii locations. Figures 2 and 3 show definitively that the electron fluxes acquired over a nearly five month interval in 1998 at two locations in the inner heliosphere, at 1 AU and ~ 5 AU, have significant coherence.

Taking the results of Figures 1–3 together, it is clear that the electron flux time series are non-stationary, and that they are being modulated by a set of low frequency modes that are present in the heliosphere, at least inside 5 AU. TML used some spacecraft data beyond 5 AU in their analysis, so that it can be concluded that such frequency modes are prevalent more widely than just inside the orbit of Jupiter. It is not unreasonable to conclude that a logical driving source for the frequency modes is the Sun. The solar source for the modes is unknown, but TML have argued that in this frequency range, it could be long period solar gravitational modes.

REFERENCES

1. Thomson, D. J., Maclennan, C. G., and Lanzerotti, L. J., *Nature* **376**, (1995) 139–144.
2. Roberts, D., Ogilvie, K., and Goldstein, M., *Nature* **381**, (1996) 31–32.
3. Thomson, D., Maclennan, C., and Lanzerotti, L., *Nature* **381**, (1996) 32.
4. Kumar, P., Quataert, E., and Bahcall, J., *Astrophys. J. Lett.* **458**, (1996) L83–L85.
5. Riley, P., and Sonett, C., *Geophys. Res. Lett.* **23**, (1996) 1541–1544.
6. Hoogeveen, G., and Riley, P., *Solar Physics* **179**, (1998) 167–177.
7. Denison, D., and Walden, A., *Astrophys. J.* **514**, (1999) 972–978.
8. Blake, J., Looper, M., Keppler, E., Heber, B., Kunow, H., and Quenby, J., *Geophys. Res. Lett.* **24**, (1997) 671–674.
9. Neugebauer, M., Goldstein, B., McComas, D., Suess, S., and Balogh, A., *J. Geophys. Res.* **100**, (1995) 23,389–23,395.
10. Thomson, D., Maclennan, C., Lanzerotti, L., Heber, B., Kunow, H., and Gold, R. E., *submitted* **xxx**, (2000) xxx–xxx.
11. Thomson, D., Lanzerotti, L., and Maclennan, C., in *Proc. of the SOHO 6/GONG 98 Workshop, 'Structure and Dynamics of the Sun and Sun-like Stars'*, vol. ESA SP-418 Vol. 2, 1998 pp. 967–971.
12. Lanzerotti, L., Gold, R., Anderson, K., Armstrong, T., Lin, R., Krimigis, S., M.Pick, Roelof, E., Sarris, E., Simnett, G., and Frain, W., *Astron. and Astrophysics Suppl.* **92**, (1992) 349–363.
13. Thomson, D. J., *Proc. IEEE* **70**, (1982) 1055–1096.
14. Thomson, D. J., *Phil. Trans. R. Soc. Lond.* **A 330**, (1990) 601–616.
15. Thomson, D. J., *Phil. Trans. R. Soc. Lond.* **A 332**, (1990) 539–597.
16. Thomson, D. J., in *Nonlinear and Nonstationary Signal Processing*, edited by W. Fitzgerald, R. Smith, A. Walden, and P. Young, Cambridge Univ. Press, 2000 .
17. Thomson, D. J., and Chave, A. D., in *Advances in Spectrum Analysis and Array Processing*, edited by S. Haykin, vol. 1, Prentice-Hall, 1991 pp. 58–113.
18. Efron, B., *The Jackknife, the Bootstrap, and Other Resampling Plans*, SIAM, Philadelphia, 1982.

Acceleration of Energetic Ions at the Earth's Near Perpendicular Shock: Three-dimensional Observations

Karim Meziane, Arthur J. Hull, Robert P. Lin, Theodore J. Freeman, and Davin E. Larson

Space Sciences Laboratory, University of California, Berkeley, CA 94720

George K. Parks

Geophysics Program, University of Washington, Box 351650, Seattle 98195

Abstract. We discuss in detail the observation of energetic ions (30 *keV* - 2 *MeV*) measured by the WIND-3D Plasma and Energetic Particle experiment at the Earth's bow shock. The events are observed near the perpendicular shock, and it seems that their occurrence requires the presence of a relatively high ion flux of ambient energetic ions as in Corotating Interaction Region or Solar Energetic Particle events. Upstream, the ion energy spectrum is usually peaked at a few hundred *keV*. The ions propagate upstream with an angular distribution peaked at 150° pitch-angle. However, the angular distribution is non-gyrotropic with a range of gyrophase ≤ 180°. In the downstream side of the shock, the angular distribution of energetic particles is peaked at ~ 90° and is also non-gyrotropic. We show that the energy-spectrum is consistent with an adiabatic-like reflection at the shock of the incoming Corotating Interaction Region ions. However, the pitch angle distribution is satisfied only within a certain range of gyrophase angle. A similar comparison is made for energetic ions observed downstream.

INTRODUCTION

Observations have established that fast mode shock waves are associated with enhancements of energetic particles. Ions from ~1 *keV* up to several *MeV* are observed ahead of interplanetary shocks [1] and the Earth's bow shock [2]. The characteristics of the produced particles strongly depend on whether θ_{Bn} the angle between the magnetic field and the local shock surface normal, is smaller or larger than 45°. Ion spikes up to ~ 1 *MeV* observed at nearly perpendicular interplanetary shocks have been reported using Explorer data [3]. It has been suggested that the Shock Drift Acceleration (SDA) mechanism is responsible for the production of these spikes [4]. To explain the observed count rates, relatively high fluxes of a pre-accelerated population is necessary to overrun the shock velocity which may be very high when θ_{Bn} approaches 90°. The observation of energetic particles upstream of the nearly perpendicular shock is not well documented in general; few events have been reported so far [5, 6]. Also to date, no observational diagnostic on the SDA mechanism has been developed. In the present study we provide a detailed analysis of ion spikes observed upstream and downstream at the nearly perpendicular shock. The observations are then compared with the predictions of the SDA mechanism.

OBSERVATIONS

We use data from the WIND-3DP and Energetic Particles experiment [7]. The high energy ions are detected by three pairs of Solid State Telescopes (SST) providing full 4π steradians coverage. One end of each telescope (SST-FOIL) stops protons up to ~400 *keV*, leaving the electrons unaffected. While the opposite end (SST-OPEN) leaves the ions unaffected and sweeps away the electrons. Given this, some information on the ion species may be obtained by comparing the response of both SSTs. The plasma data and the magnetic field data are provided by the WIND 3DP PESA-Low analyser and the MFI experiment [8],

CP528, *Acceleration and Transport of Energetic Particles Observed in the Heliosphere: ACE 2000 Symposium*, edited by Richard A. Mewaldt, et al.
© 2000 American Institute of Physics 1-56396-951-3/00/$17.00

respectively. Electrostatic wave data from the TNR instrument [9] is also used.

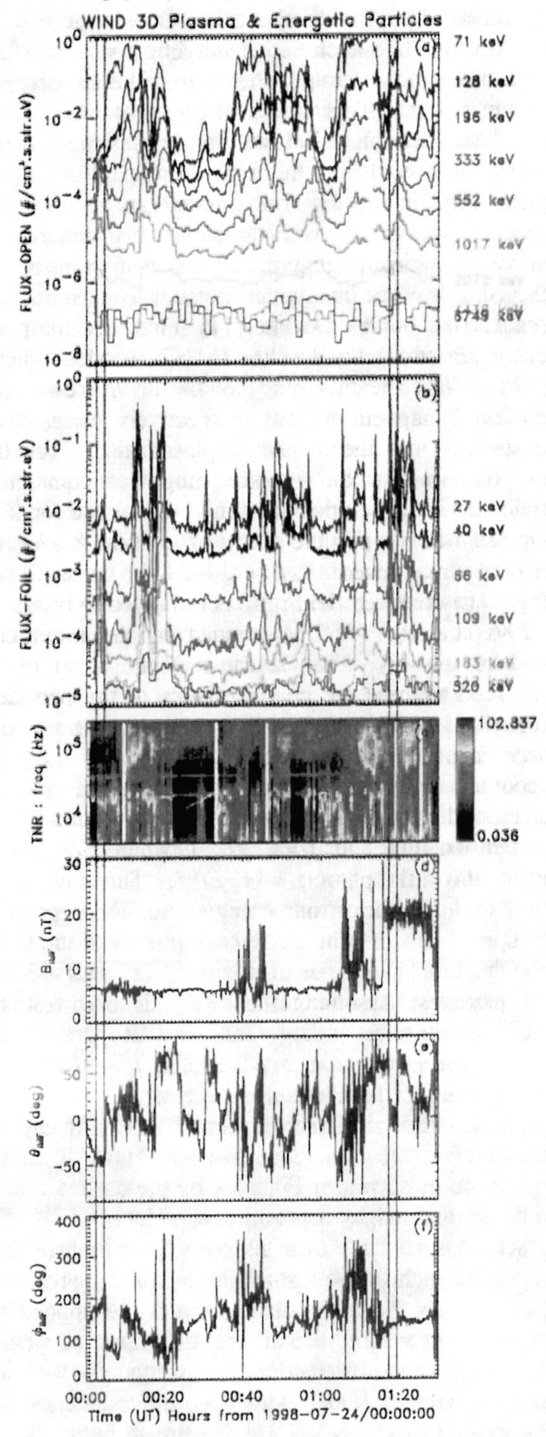

Figure 1. The two top panels show the fluxes registered by the SST-OPEN (a) and the SST-FOIL (b). The electrostatic wave measurements are given in panel (c); the interplanetary magnetic field magnitude, the polar and azimuthal angles are respectively given panels (d), (e) and (f).

Figure 1 shows the observations made on 1998 July 24, between 00:00 and 01:30 UT. We note that WIND detected the onset of a Corotating Interaction Region (CIR) the day before at ~04:00 UT. WIND was travelling inbound, and as indicated by panel *(d)*, had several partial shock crossings in the dusk side of the $(Y\text{-}Z)_{GSE}$ plane. At 01:15:30 UT, WIND crossed a supercritical quasi-perpendicular shock and entered the magnetosheath. Panels *(a)* and *(b)* show several flux enhancements. We will focus on the two intervals indicated by vertical bars.

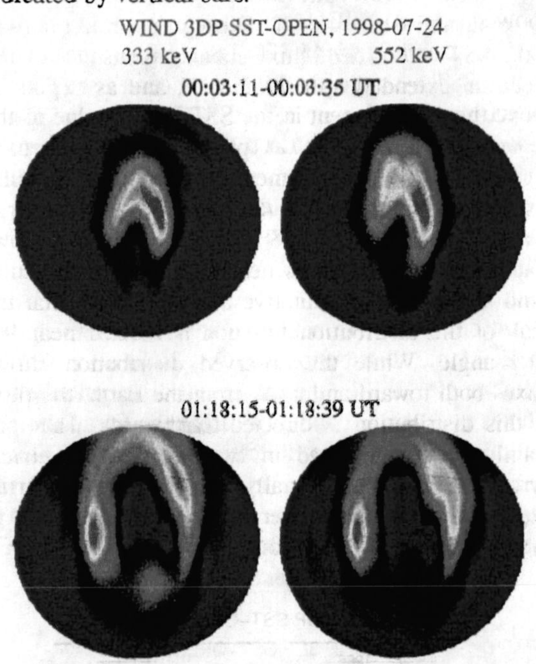

Figure 2. Three dimensional proton angular distributions sampled in the solar wind frame. The Hammer-Aitoff projection is used to display 4π steradians coverage. The **B** vector is located at the center and the asterisk symbol indicates the solar wind direction. The two top frames show the 333 and 552 *keV* ion angular distributions measured upstream; the bottom panels show the downstream measurements.

Upstream event: At 00:01:34 UT the Interplanetary Magnetic Field line crossing WIND is tangent to the bow shock as indicated by the appearance of an intense burst of narrow band wave activity at the plasma frequency (Figure 1c). This electrostatic burst is followed by an ion burst of 196 *keV* up to ~ 2 *MeV* detected by the SST-OPEN. The 71 *keV* and 128 *keV* channels also detected a small flux enhancement. The 196 *keV*-2 *MeV* bursts are simultaneously accompanied by a burst in all energy channels up to ~ 520 *keV* observed by the FOIL telescope. Protons of ~ 500 *keV* energy penetrate the FOIL detector, and the FOIL fluxes are consistent with the observed count rates. The three-dimensional angular distribution for 333 and 552 *keV* protons registered at the maximum of the burst, shown in Figure 2 (top frames), indicate that the distribution is non-gyrotropic, and the gyrophase range is limited to ≤180°. The proton flux pitch-angle dependence for E = 552 keV is shown in Figure 3. There is a clear peak in this distribution at 150° indicating that most of the protons at this energy travel away from the shock. The complete proton energy spectrum is shown on Figure 4 (open circles).

Downstream event: Just downstream of the Earth's bow shock the IMF direction is quasi-steady between 01:16:42 and 01:19:45 UT. During this time interval, both SSTs recorded flux enhancements. The ion spectrum extended up to ~ 2 MeV, and as explained above the enhancement in the SST-FOIL is due to the penetration of the ~500 keV protons into the detector. Figure 2 (bottom frames) shows the angular distribution for the 333 keV and 552 keV energy channels at 01:18:15-18:39 UT. Figure 2 also shows that the IMF direction is nearly parallel to the solar wind direction; a quantitative analysis shows that the peak of the distribution function is located near 90° pitch-angle. While the observed distribution shows fluxes both toward and away from the Earth, the peak in this distribution is directed Earthward. The ions mainly appear bunched in two nearly symmetrical gyrophase domains. Finally, the energy spectrum average over the time interval 01:18:15-18:39 UT is shown in Figure 4 (asterisks).

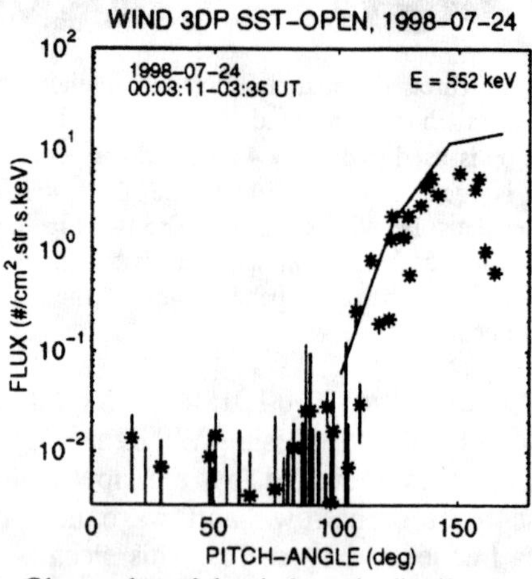

Figure 3. Observed particle pitch-angle distribution for E = 552 keV protons (asterisks) for the upstream event. The solid line corresponds to the SDA prediction for E_S = 130 keV.

SHOCK DRIFT ACCELERATION

The 332-552 keV ion count rates associated with the events described above are not consistent with the energetic particle fluxes observed during quiet times at ~1 AU. The CIR particles play a crucial role here because they provide the necessary pre-accelerated population. The upstream event is consistent with a reflection at the shock. The energetic ions observed downstream mostly propagate Earthward and therefore are likely transmitted from the upstream side after being accelerated. Below, we examine these possibilities by using the expectations of the SDA mechanism. In this process, ions gain energy by successive encounters with the shock; the ions escape the shock when magnetic moment conservation is reached [10]. The analytical treatment of SDA has been carried out by Decker [11]. First, the jump in field B_d/B_u strength across the shock is needed, where B_d and B_u are the magnitude of the downstream and upstream magnetic field, respectively. For this purpose, we use the measured plasma and magnetic field as input to the Rankine-Hugoniot equations. During the selected upstream time interval and for θ_{Bn} ~ 75°-89°, the jump in the field B_d/B_u = 3.6-3.8. During the downstream event, B_d/B_u = 2.7-2.8 for the same θ_{Bn} range. The observed field ratio is estimated to be B_d/B_u = 2.86, and θ_{Bn} ~86° determined by the minimum variance analysis of the shock structure at 01:15:30 UT. To determine the characteristics of the particles accelerated by the shock, we assume that the phase space density is conserved during the particle encounter with the shock; the differential flux corresponding to the accelerated ions j_f is related to the incident ion flux, j_i as: $j_f = (E_f/E_i)\, j_i$, where E_f/E_i is the final to the initial particle energy ratio. The expression E_f/E_i is obtained for a single encounter with the shock in the case of the reflection or transmission downstream [11]. The incident flux j_i is taken from the CIR particles. Assuming that the CIR particles are nearly gyrotropic, we may assume $j_i = J(E_i)H(\mu_i)$, where J and H functions are the energy-spectrum and the pitch-angle distribution respectively, μ_i being the cosine of initial particle pitch-angle. The CIR energy spectrum, as shown in Figure 4 by the dashed line, is well represented by a power law: $J(E_i) = 4E_i^{-2.1}$. The observed pitch angle distribution is fitted in three lines segments each of the form $H(\mu_i) = a_k\mu_i + b_k$, where k = $1,2,3$. Using the above analysis and the appropriate expression for E_f/E_i, the differential fluxes as well as the pitch-angle distribution are computed for each event. The flux j_f are plotted as the solid curves in Figure 4. The 196 keV-2 MeV upstream bursts flux is well fitted for θ_{Bn} = 83.4°, equivalent to E_S = 130 keV, where $E_S = (½)m_pV_S^2$, and V_S is the parallel shock velocity. By tracing ion trajectories from the spacecraft back to the shock, we have found that a source region located at the shock for the 196 keV-1 MeV bursts is consistent with the above θ_{Bn} value; whereas the 128 keV protons are associated with lower θ_{Bn}. The downstream proton energy spectrum is modeled using $\theta_{Bn} = 81^\circ$ ($E_S = 43\ keV$). The upstream proton pitch-angle distribution for $E = 552\ keV$ is compared to the SDA model prediction (Figure 3). The observed range of pitch-angles agrees very well with the SDA expectation. The flux level, however, associated with the nearly field-aligned protons is inconsistent this theory. We have found that the good agreement is satisfied for particles having a gyrophase angle within a ~ 160° range. It is important to note that the model assumes that the pre-accelerated particles are nearly gyrotropic to allow analytical calculation of SDA predictions. The observation of the CIR protons show that the hypothesis is not quite true.

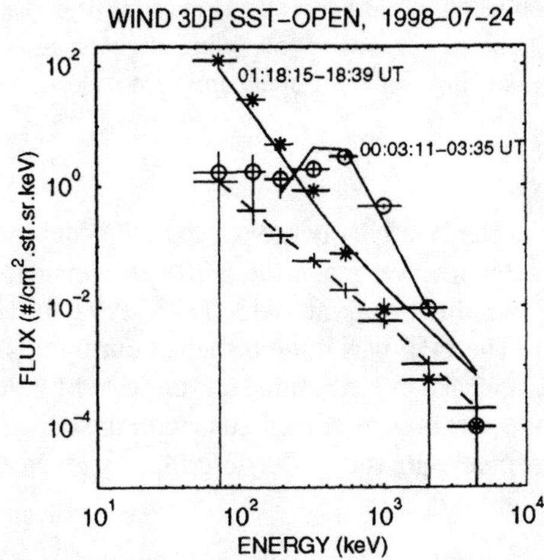

Figure 4. The energy spectrum for the protons measured upstream at 00:03:11-03:35 UT (open circles), and downstream at 01:18:15-18:39 UT (asterisks). The dashed line is the best linear fit of the CIR protons. The solid curves are the spectrum predicted by the adiabatic reflection of CIR protons.

CONCLUSION

We have analyzed the properties of energetic proton bursts observed upstream and downstream at a nearly perpendicular shock. These bursts are usually observed in the presence of Corotating Interaction Region energetic particles. The proton energy spectrum is consistent with a diffusion free Shock Drift Acceleration mechanism in which Corotating Interaction Region particles constitute the pre-accelerated population. The results are in agreement with recent studies [6]. The determination of the shock parallel velocity from the particle properties provides a good estimate of the drift distance along the shock surface. This distance is of order of the size of the bow shock.

REFERENCES

1. Tsurutani, B. T., and Lin, R. P., *J.Geophys. Res.* **90**, 1-11 (1985)

2. Skoug, R. M., Winglee, R. M., McCarthy, M. P., Parks, G. K., Lin, R. P., Anderson, K. A., Carlson, C. W., Ergun, R. E., Larson, D. E., McFadden, J., Reme, H., Bosqued, J., d'Uston, C., Sanderson, T. R., Wenzel, K.-P., Lepping, R., P., and Szabo, A., *Geophys. Res. Lett.* **23**, 1223-1226 (1996).

3. Armstrong, T. P., Krimigis, and Behannon, K. W., *J. Geophys. Res.* 75, 5980-5988 (1970).

4. Sarris, E. T., and Van Allen, E. T., *J. Geophys. Res.* **79**, 4157-4173 (1974).

5. Anagnostopoulos, G. C., Sarris, E. T., Krimigis, S. M., *J. Geophys. Res.* **93**, 5541-5546 (1988).

6. Meziane, K., Lin, R. P., Larson, D. E., Parks, G. K., Larson, D. E., Bale, S. D., Mason, G. M., Dwyer, J. R., and Lepping, R. P., *Geophys. Res. Lett.* **26**,2925-2928 (1999).

7. Lin, R. P., Anderson, K. A., Ashford, S., Carlson, C. W., Curtis, D., Ergun, R. E., Laron, D. E., McFadden, J., McCarthy, Parks, G. K., Rème, H., Bosqued, J. M., Coutelier, J., Cotin, F., d'Uston, C., Wenzel, K.-P., Sanderson, T. R., Henrion, J., Ronnet, J. C., and Paschmann, G., *Space Sci. Rev.* **71**, 125-153 (1995).

8. Lepping, R. P., Acuna, M. H., Burlaga, L. E., Farrell, W. M., Slavin, J. A. Schatten, K. H., Mariani, F., Ness, N. F., Neubauer, F. M., Whang, Y. C., Byrnes, J. B., Kennon, R. S., Panetta, P. V., Scheifele, J., Worley, E. M., *Space Sci. Rev.* **71**, 207-229 (1995).

9. Bougeret, J. L., Kaiser, M. L., Kellogg, P. J., Manning, R., Goetz, K., Monson, S. J., Monge, N., Friel, L., Meetre, C. A., Perche, C., Sitruk, L., Hoang, S., *Space Sci. Rev.* **71**, 231-263 (1995).

10. Pesses, M. E., and Decker, R. B., *J. Geophys. Res.* **91**, 4143-4148 (1986)

11. Decker, R. B., *J. Geophys. Res.* **88**, 9959-9973 (1983).

Characteristics of MeV upstream ion bursts observed by Wind

T. J. Freeman[1], D. E. Larson[1], R. P. Lin[1], K. Meziane[1], G. K. Parks[2]

[1]*Space Sciences Laboratory, University of California, Berkeley*
[2]*Geophysics Program, University of Washington, Seattle*

Abstract. The 3D Plasma and Energetic Particle instrument on the Wind spacecraft has recorded more than 100 upstream ion events which contain heavy ions of ≈2 MeV. The detailed in situ observations made by Wind can shed light on a process that not only produces high energy ions upstream of the Earth's bow shock, but which scales geometrically to larger astrophysical systems, and can therefore provide insight into the production of cosmic rays. Three-dimensional distribution functions of these bursts show these events to be highly nongyrotropic, from ≈100 keV to ≈2 MeV. In a typical event, nearly all of the flux is gyrophase-bunched, with an angular width of ≈60 degrees. These nongyrotropic bursts are explained by a model of Fermi acceleration in which suprathermal heavy ions of solar wind origin, such as O^{6+}, are energized through repeated reflections between large-scale magnetic field rotations in the upstream region and the bow shock.

INTRODUCTION

For many years, energetic ions which originate at the Earth's bow shock have been observed in the solar wind (e.g., Asbridge *et al.*, 1968; Lin *et al.*, 1974). With the launch of Wind in 1994, we have entered a new era in the study of upstream ion events. The 3D Plasma and Energetic Particle (3DP) instrument (Lin *et al.*, 1995) has made it possible to study these events in unprecedented detail. These events have been shown to contain ions up to ≈2 MeV (Skoug *et al.*, 1996), and in this paper we show that these bursts are highly nongyrotropic. Using data from the Suprathermal and Energetic Particle experiment, Desai *et al.* (2000) have shown these bursts to be of solar wind composition and dominated by heavy ions above 1 MeV. In this paper, we demonstrate how these observations can be explained by a model of Fermi acceleration in which suprathermal solar wind ions gain energy as they are reflected between interplanetary magnetic field (IMF) rotations and the bow shock (Freeman and Parks, 2000).

WIND OBSERVATIONS

A typical ≈2 MeV event is shown in Figure 1. This 4.25 hour interval on October 3, 1996 was extremely active. Wind was in the solar wind at this time, approaching the shock. Omnidirectional ion fluxes from the solid state telescopes (SSTs) are shown in the top panel. The central energy for each channel is given. The flux is the actual value in the lowest channel, and is divided by a factor of 2 for each successive channel to separate the data. Several bursts in this interval reach the ≈2 MeV channel, especially those which peak at 1813, 1823, 1902, 1921, and 2105 UT. The IMF was quite turbulent during this interval, and several large rotations are indicated by vertical lines that occur in near coincidence with the bursts. This is in agreement with the model described by Freeman and Parks (2000).

NONGYROTROPIC DISTRIBUTIONS

With the the 3DP instrument, for the first time we have the ability to measure fully three-dimensional distribution functions of these ions. A typical example is shown in Figure 2, for the first ion burst shown in Figure 1, which occurs at 1812:48–1817:38 UT. The distributions have been rotated so that the IMF direction is in the center of each plot. Each energy channel has been normalized on its own linear scale. Almost all of the flux is seen to be concentrated in a particular range of pitch angle and gyrophase. This is a robust feature of these bursts, extending from tens of keV to ≈2 MeV. This 5 min interval has been selected to improve statistics, but the same features are visible in 48 s distributions and are seen in almost every ≈2 MeV event observed by Wind.

The two explanations that have been offered for many years to explain the origin of upstream energetic ions are diffusive Fermi acceleration (e.g. Terasawa, 1979; Eichler, 1981; Lee, 1982; Ellison, 1985) and magnetospheric leakage (e.g. Anagnostopoulos *et al.*, 1986; Sarris *et al.*, 1987; Baker *et al.*, 1988; Sibeck *el al.*, 1988). In either

CP528, *Acceleration and Transport of Energetic Particles Observed in the Heliosphere: ACE 2000 Symposium*, edited by Richard A. Mewaldt, et al.
© 2000 American Institute of Physics 1-56396-951-3/00/$17.00

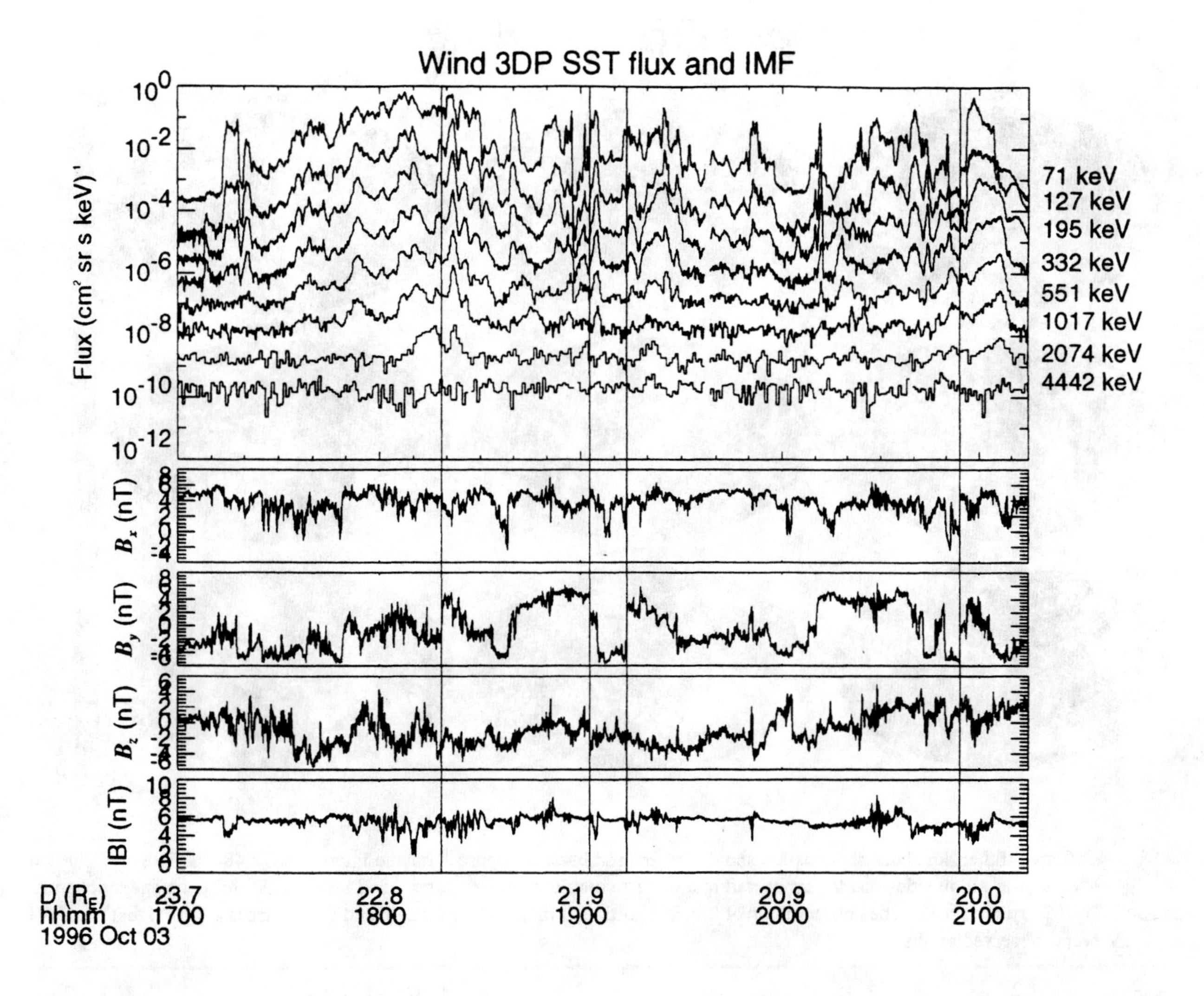

FIGURE 1. Wind 3DP observations of energetic upstream ions on October 3, 1996. The top panel shows omnidirectional ion fluxes measured by the SSTs. The flux is the actual value for the 71 keV energy channel, with successive channels divided by factors of 2 to separate the data. The Interplanetary Magnetic Field data is also given. Several large IMF rotations are indicated by vertical lines, which are in near coincidence with bursts in the ≈ 2 MeV channel.

case, the persistence of these nongyrotropic distributions is difficult to explain. A new type of Fermi model, in which heavy ions are accelerated between the shock and IMF rotations upstream of the shock (Freeman and Parks, 2000) does predict such nongyrotropic features. In this model, many single particle trajectories are calculated using measured IMF data and a three-dimensional model of the shock. The model shows that suprathermal O^{6+} ions that originate at the quasi-parallel shock and are then subsequently energized at the quasi-perpendicular shock can account for the bursts of ≈ 2 MeV ions observed by Wind. Because the full Lorentz-force motion of the particles is calculated, nongyrotropic features of the simulated events match the observed anisotropic distributions closely. This occurs because particle reflection at the shock and at IMF rotations is gyrophase-selective.

FERMI MODEL

A single particle trajectory is shown in Figure 3. Projections of the particle motion are shown in the x_A–y_A plane and the x_A–z_A plane (the coordinates are "aligned" coordinates, which have been rotated from the GSE coordinate directions by $\approx 3°$ to align the $-x_A$ direction with the solar wind flow direction). In this example calculation, an O^{6+} ion of energy 1.8 MeV starts at the spacecraft location at the time of the first burst, and is traced backward in time, using the measured IMF data and a model

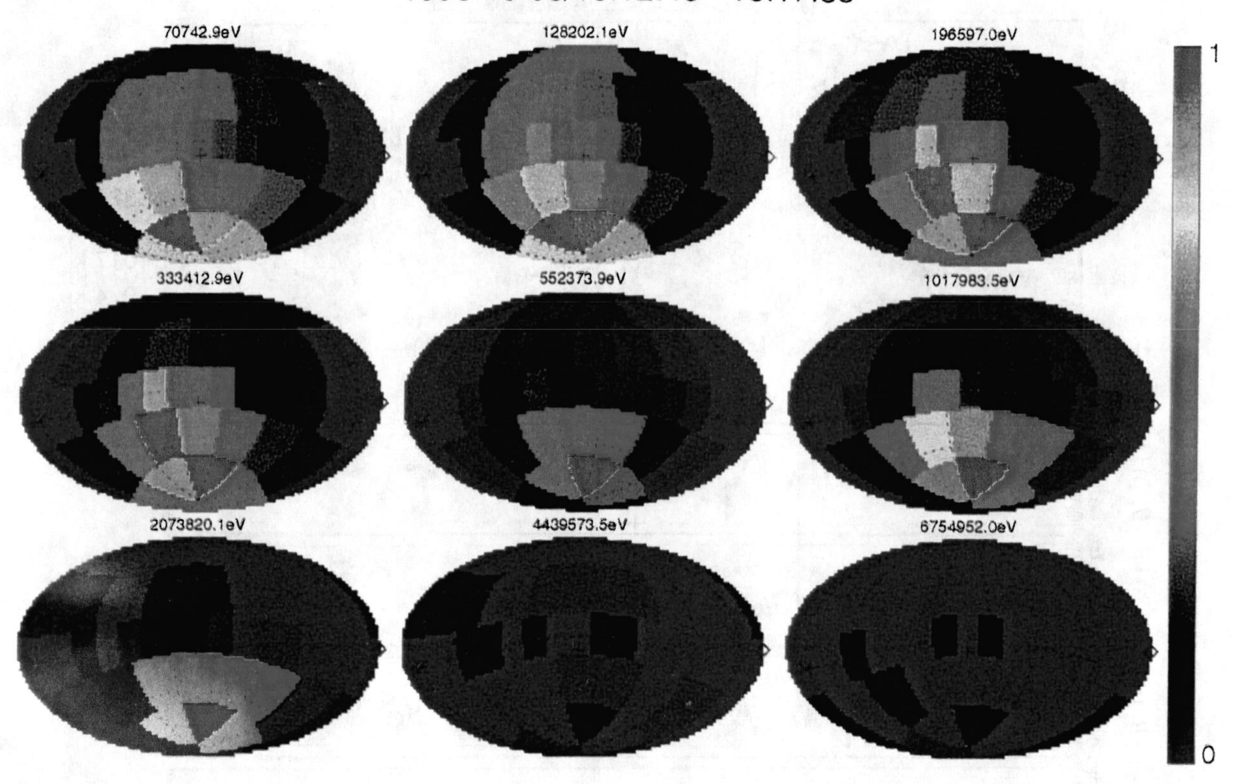

FIGURE 2. Three-dimensional distributions are shown for the first burst in Figure 1 for the interval 1812:48–1817:38 UT. The ion [illegible] measure flux in all directions as the spacecraft rotates. The distributions are normalized individually for each energy channel and are shown on a linear scale. The concentration of flux in a particular range of pitch angle and gyrophase is a typical feature of the ≈2 MeV events observed by Wind.

of the shock position that is in part determined by the upstream solar wind conditions and the point at which the spacecraft crossed the shock (Freeman and Parks, 2000). The particle is found to originate at the shock with an energy of 65 keV, on the opposite side of the shock from the spacecraft.

A statistical analysis of one million such particle trajectories (not shown) shows that this calculation is typical. The particles are drawn from the nongyrotropic distribution in Figure 2, and almost all of the trajectories are traced back across the top of the shock to the other side. The first reflection of the particle at the shock almost always occurs at the quasi-parallel shock, while later interactions are at the quasi-perpendicular shock.

The nongyrotropic distribution of energetic upstream ion bursts provides an important clue about the origin of these particles. The results of this backward particle tracing depend sensitively on the three-dimensional distribution. The reflection of particles from the bow shock and from IMF rotations are gyrophase-selective processes, so the nongyrotropic nature of the bursts is entirely consistent with the model of Freeman and Parks (2000).

CONCLUSION

The capability of the 3DP instrument to measure complete three-dimensional distributions of energetic upstream ions has given us a new test for models of the origin of these ions, whether Fermi acceleration of solar wind ions or magnetospheric leakage. In this paper we have used a measured nongyrotropic distribution as an initial condition for the calculation of particle orbits, starting at the spacecraft location and following the particles backward in time to their points of origin at the shock. The results are consistent with an origin of suprathermal heavy ions such as O^{6+} at the quasi-parallel shock.

If these bursts are instead due to the leakage of energetic ions from the Earth's magnetosphere, then the

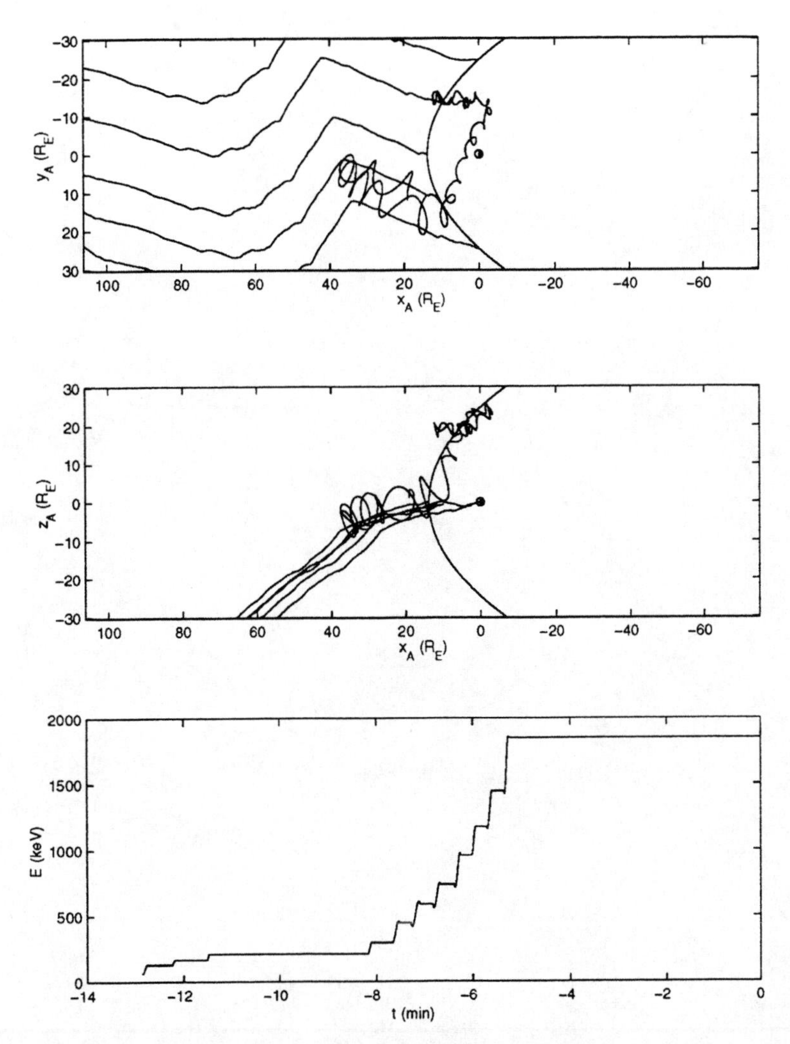

FIGURE 3. The top two panels show projections into the x_A–y_A plane and the x_A–z_A plane of an $^{6+}$ ion trajectory. IMF lines and the location of the shock are also shown. The bottom panel shows the kinetic energy of the particle in the solar wind frame of reference. The particle starts at the location of the spacecraft at the time of the first burst in Figure 1, and is followed backwards in time. This trajectory is typical of 1 million calculated trajectories in that the particle is found to originate on the other side of the shock where the IMF connects to the shock in a quasi-parallel geometry, at an energy of 65 keV.

nongyrotropic distributions must be explained in some other way. Likewise, the coincidence between IMF rotations and upstream ions must be explained. We leave these as challenges to the proponents of the magnetospheric leakage model, and state simply that the Wind 3DP observations are consistent with the Fermi acceleration model of Freeman and Parks (2000).

REFERENCES

Asbridge, J. R. *et al.*, *J. Geophys. Res.*, **73**, 5777, 1968.
Anagnostopoulos, G. C. *et al.*, *J. Geophys. Res*, **91**, 3020, 1986.
Baker, D. N. *et al.*, *J. Geophys. Res.*, **93**, 14,317, 1988.
Desai, M. I., *et al.*, *J. Geophys. Res.*, **105**, 61, 2000.
Eichler, D., *Astrophys. J.*, **244**, 711, 1981.
Ellison, D. C., *J. Geophys. Res.*, **90**, 29, 1985.
Freeman, T. J., and G. K. Parks, *J. Geophys. Res.*, in press, 2000.
Lee, M. A. *et al.*, *J. Geophys. Res.*, **87**, 5080, 1982.
Sarris, E. T. *et al.*, *J. Geophys. Res.*, **92**, 12,083, 1987.
Sibeck, D. G., *et al.*, *J. Geophys. Res.*, **93**, 14,328, 1988.
Skoug R. M. *et al.*, *Geophys. Res. Lett.*, **23**, 1223, 1996.
Terasawa, T., *Planet. Space Sci.*, **27**, 365, 1979.
Lin, R. P. *et al.*, *J. Geophys. Res.*, **79**, 489, 1974.
Lin, R. P. *et al.*, *Space Sci. Rev.*, **71**, 125, 1995.

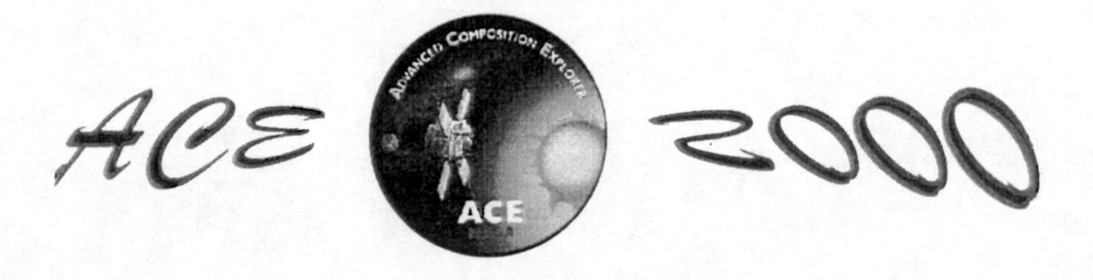

Anomalous and Galactic Cosmic Rays in the Heliosphere

Observations of Anomalous Cosmic Rays at 1 AU

R. A. Leske[1], R. A. Mewaldt[1], E. R. Christian[2], C. M. S. Cohen[1], A. C. Cummings[1], P. L. Slocum[3], E. C. Stone[1], T. T. von Rosenvinge[2], and M. E. Wiedenbeck[3]

[1]*California Institute of Technology, Pasadena, CA 91125 USA*
[2]*NASA/Goddard Space Flight Center, Greenbelt, MD 20771 USA*
[3]*Jet Propulsion Laboratory, Pasadena, CA 91109 USA*

Abstract. Anomalous cosmic rays (ACRs) provide a sensitive probe of the access of energetic particles to the inner heliosphere, varying in intensity by more than two orders of magnitude during the course of the solar cycle. New data which are becoming available from the Advanced Composition Explorer (ACE) can provide a detailed record of ACR intensity and spectral changes on short ($\sim$ 1 day) time scales during the approach to solar maximum, which will help address issues of ACR modulation and transport. The elemental and isotopic composition of ACRs provides important information on the source or sources of these particles, while their ionic charge state composition and its energy dependence serves as a diagnostic of their acceleration time scale. We review measurements of the ACR elemental, isotopic, and charge state composition and spectra as determined at 1 AU by SAMPEX, ACE, Wind, and other spacecraft. These results are important input to models of the acceleration, modulation, and transport of ACRs.

INTRODUCTION

Anomalous cosmic rays (ACRs) were discovered some 25 years ago as unexpected intensity enhancements in low energy quiet time He, N, and O spectra (1, 2, 3), with a composition quite unlike that previously seen in other samples of energetic particles. Although the correct interpretation of these particles was suggested very soon after their discovery, verification of the model has taken much longer. In the standard scenario (4, 5), most ACRs begin in the interstellar medium (ISM) as neutral atoms that flow relatively unimpeded into the heliosphere, where they become ionized by solar UV or by charge exchange with the solar wind as they approach the Sun. Once singly ionized, they are convected into the outer heliosphere by the solar wind as "pickup" ions and carried to the solar wind termination shock, where they are accelerated to energies of several to tens of MeV/nucleon and become the ACRs observed throughout the heliosphere.

Many predictions of this model have now been confirmed experimentally. Other elements likely to be primarily neutral in the ISM due to their high first ionization potential (FIP), such as H (6), Ne (7), and Ar (8), are now known to have an ACR component; neutral interstellar He has been found flowing into the heliosphere (9); pickup ions have been detected (10) with a composition consistent with an origin as interstellar neutrals (11); and low energy ACRs have been shown to be predominantly singly ionized (12, 13). The distribution of ACRs within the heliosphere, both radially and latitudinally (e.g., (14, 15)), the reversal of the latitudinal gradients when the solar magnetic field changes polarity (16), and the continually unfolding spectra seen by Voyager 1 as it approaches the termination shock (17) also agree with the general expectations of the standard model.

Recently, some low-FIP elements not expected to be neutral in the ISM have been found to exhibit ACR-like enhancements (18, 19, 20). Also, above $\sim$ 20 MeV/nucleon, ACRs have been shown to be mainly multiply ionized instead of singly charged (21, 22). Such findings may lead to some extensions to the standard scenario and have prompted revisions in models of the acceleration and transport of ACRs in the heliosphere (23, 24, 25).

In this article, we review measurements of ACRs made during the past solar minimum with new spacecraft at 1 AU, in particular the Solar, Anomalous, and Magnetospheric Particle Explorer (SAMPEX) in a polar orbit about the Earth, Wind in the nearby interplanetary medium, and the Advanced Composition Explorer (ACE) in orbit about the upstream Sun-Earth Lagrange point L_1. We discuss ACR temporal variations and their elemental, isotopic, and especially charge state spectra.

TIME VARIABILITY

Studies of ACRs at 1 AU are essentially limited to solar minimum periods. As shown in Figure 1, the intensity

CP528, *Acceleration and Transport of Energetic Particles Observed in the Heliosphere: ACE 2000 Symposium*, edited by Richard A. Mewaldt, et al.
© 2000 American Institute of Physics 1-56396-951-3/00/$17.00

of ACRs varies by a factor of ~ 100 over the course of the solar cycle, with ACRs becoming undetectable above the galactic cosmic ray (GCR) or solar particle backgrounds at 1 AU during solar maximum. These intensity variations tend to track those seen in the modulation of GCRs, as reflected in the variations in the neutron monitor count rate, although the ACR intensity recovered more quickly than the GCRs at the onset of the last solar minimum (26).

The data in Figure 1 are averaged over periods of a solar rotation or longer. Using the Solar Isotope Spectrometer (SIS) on ACE, with its large (~ 40 cm^2sr) collecting power (27), it is possible to obtain more detailed records of ACR time variability and even track day-to-day variations in ACR intensity, as illustrated in Figure 2. These data were selected for solar quiet days (when the 3.4–7.3 MeV/nucleon He intensity was $< 10^{-4}$ (cm^2·sr·s·MeV/nuc$)^{-1}$) and have had the GCR background subtracted. This background correction was $\lesssim 5\%$ early in the period, and grew to ~ 30% by the end of the period. The ACR variability is considerable, with factors of ~ 5 intensity fluctuations sometimes observed during half a solar rotation (e.g., near day 200 of 1999), possibly due to modulation by solar wind structures (28). Even on these short time scales, the correlation between the ACR intensity and the Climax neutron monitor count rate is evident. These new ACE data, publicly available from the ACE Science Center (*http://www.srl.caltech.edu/ACE/ASC*), will provide new challenges and tests for modulation models.

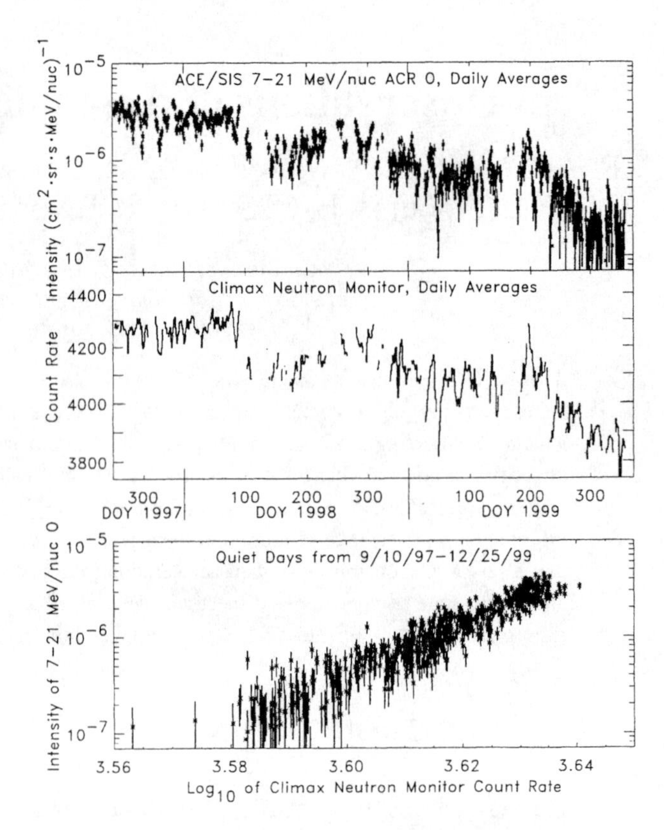

FIGURE 2. Time dependence of the daily average quiet time fluxes of 7–21 MeV/nucleon O from SIS (*top*) and the count rate of the Climax neutron monitor (*middle*). The two quantities are strongly correlated, as indicated in the cross plot (*bottom*).

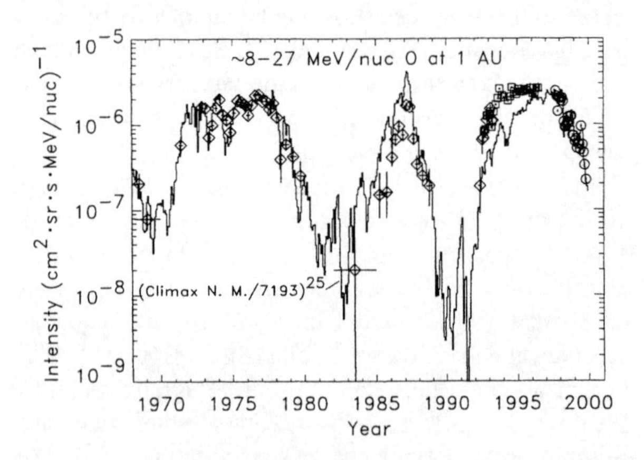

FIGURE 1. Quiet time intensities of ~ 8–27 MeV/nucleon ACR O at 1 AU over the past 3 solar cycles (*data points*), compared with the Bartels rotation averaged count rate of the Climax neutron monitor scaled as indicated (*histogram*). Recent ACR data are from SAMPEX (*squares*) and from SIS on ACE (*circles*); earlier data (*diamonds*) are from OGO-5 (29, 30) and IMP 6 (7) for 1968-1971, and IMP 7 and 8 (26) for 1972-1992.

The ACR intensity is falling dramatically as the maximum of solar cycle 23 approaches, and the amount of modulation is energy dependent, as shown in Figure 3. Since the launch of ACE, the intensity of ACR O at 1 AU at an energy of 10 MeV/nucleon has fallen by a factor of ~ 20. ACR O is still present, however, as indicated by the low energy turn up (although the very lowest energy point may have some residual contamination from solar particles during the latest period). GCR O, on the other hand, has so far decreased by only a factor of ~ 2 at 100 MeV/nucleon. This is due largely to the fact that 100 MeV/nucleon is below the peak of the GCR intensity, and adiabatically cooled particles from higher energies fill in for the particles lost from lower energies. The ACRs at ~ 10 MeV/nucleon, on the other hand, are well above the peak of the steeply falling ACR spectrum, and fewer higher energy particles are available to compensate for the low energies losses (28). Also, the velocity of 100 MeV/nucleon GCRs is more than 3 times that of 8 MeV/nucleon ACRs, even though the rigidity of these fully ionized nuclei is less than half that of the singly ionized ACRs. If the diffusion coefficient scales as the product of velocity and a weak function of rigidity (see, e.g., (31)), the GCRs have a larger diffusion coefficent which also results in less solar modulation than for ACRs.

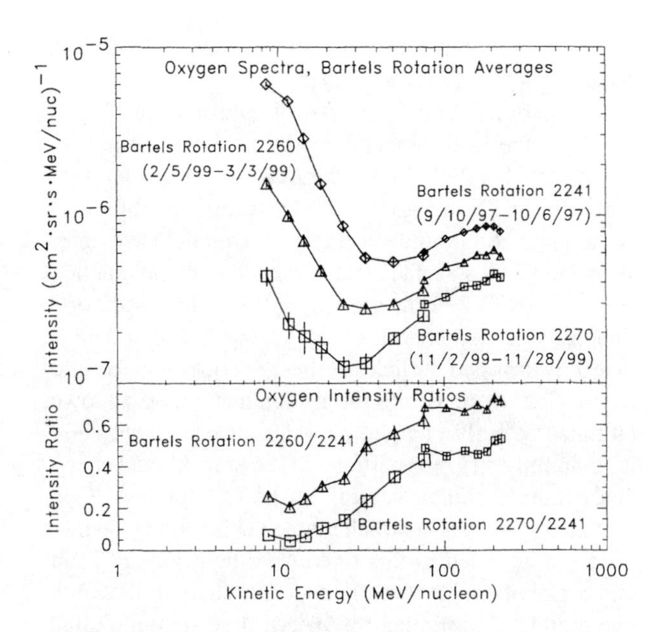

FIGURE 3. Temporal evolution of O spectra during the approach to solar maximum is illustrated for 3 selected Bartels rotations (*top*), using data from SIS (*larger symbols*) and CRIS (*smaller symbols* above $\sim$ 80 MeV/nucleon) on ACE. Modulation differences with energy are more clearly illustrated by the intensity ratios (*bottom*).

ELEMENTAL AND ISOTOPIC SPECTRA

Elemental spectra of ACRs during solar minimum are illustrated in Figure 4, combining data from SIS, higher energy data from the Cosmic Ray Isotope Spectrometer (CRIS) on ACE (32) and lower energy data from the Low-Energy Matrix Telescope (LEMT) on Wind (19). The ACE data were accumulated during quiet days from 8/27/97 through 3/23/98, when ACR fluxes were high and relatively constant (see Figure 2). The Wind data set used here runs from November 1994 to April 1998, with most of the quiet time periods occurring in 1996 and early 1997 when the ACR fluxes were similar to those during the ACE analysis period (19). The agreement between measurements from all three instruments represented in Figure 4 is remarkably good in general. These combined spectra represent the best ACR spectra at 1 AU obtained to date (compared, e.g., with (33)).

The ACR low energy intensity enhancements are clearly exhibited for N, O, Ne, and Ar in Figure 4, but for other elements the turn-up, if present, is much smaller and starts at lower energies. In particular, the ACR C abundance is somewhat controversial. Other observations at 1 AU from SAMPEX (26) and Geotail (34) have found a rather large amount of ACR-like C present, with C/O $\sim$ 0.1, but further observations from SAMPEX (35, 36) indicate that most of this C is not singly ionized. The Wind results (19) suggest that the C turn-up is greatly reduced when tighter quiet time cuts are imposed, but large instrumental background corrections were necessary for C, and the resulting spectrum is inconsistent at higher energies by a factor of 2 with results from SIS (which may also have some background at the lowest energies). The question of just how much ACR C is present at 1 AU under strict quiet time conditions appears to be unresolved.

Turn-ups at energies < 5 MeV/nucleon in the spectra of the low-FIP elements Mg, Si, and S from Wind (19) seem to be larger relative to ACR O, Ne, or Ar than those found in the outer heliosphere by Voyager, suggesting that other sources besides interstellar neutrals may contribute to these species at 1 AU (20). Other than the possibility of contamination from particles accelerated at corotating interaction regions, potential contributors to this population of low-FIP elements include an "inner source" from adsorbed, neutralized, and desorbed solar wind on interplanetary dust grains (37) or grain destruction products (38), or solar wind reaccelerated at the termination shock (39).

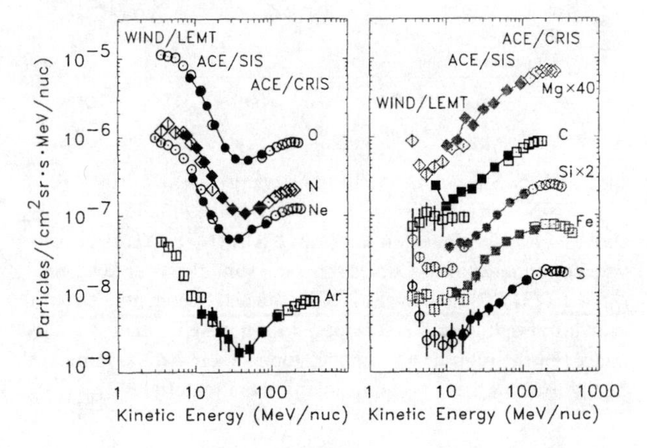

FIGURE 4. Elemental spectra taken from (40), using data from the SIS (*filled symbols*) and CRIS (*high energy open symbols*) instruments on ACE (41) and from the LEMT instrument (*low energy open symbols*) on Wind (19). The statistical uncertainties shown are often much smaller than the plotted symbols.

Besides elemental spectra, recent ACR results from ACE include isotopic spectra of ACR N, O, and Ne, as shown in Figure 5 (40). The isotopic abundances are quite different for ACRs and GCRs. In GCRs, relatively large quantities of ^{15}N, ^{18}O, and ^{21}Ne are produced by cosmic ray spallation during transport through the Galaxy, so the observed abundances are not directly representative of the GCR source. Unlike GCRs, the ACRs have passed through a negligible amount of material, and no secondary spallation products should be present.

In addition to large ACR turn-ups of the dominant isotopes ^{14}N, ^{16}O, and ^{20}Ne, the rarer species ^{18}O and ^{22}Ne also show clear low-energy ACR enhancements (40). The

ACR $^{18}O/^{16}O$ and $^{22}Ne/^{20}Ne$ ratios are consistent with those found in solar system material (42), with the Ne isotopic composition more closely resembling that found in the solar wind (43) than that present in other meteoritic components (see, e.g., (44)). The ACR $^{22}Ne/^{20}Ne$ ratio, which should represent the ratio in the local ISM, is a factor of ~ 5 below that deduced for the GCR source (see, e.g., (45, 46)), indicating that GCRs cannot be only an accelerated sample of the local ISM (40, 47). The lack of low energy enhancements in the spectra of ^{15}N and ^{21}Ne suggest that the ACR $^{15}N/^{14}N$ and $^{21}Ne/^{20}Ne$ ratios are no more than 5 and ~ 10 times, respectively, those found in standard solar system abundances (42).

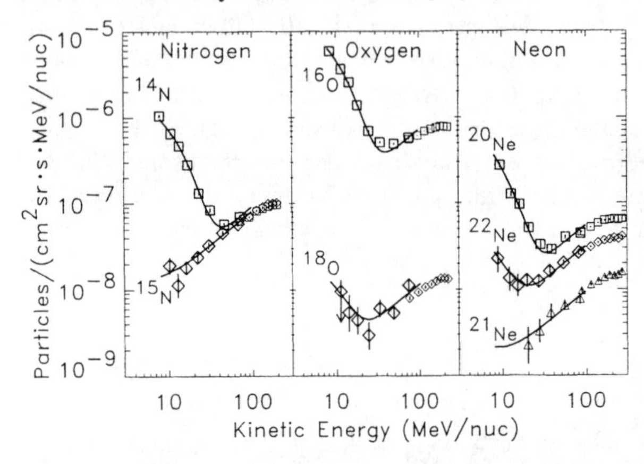

FIGURE 5. SIS quiet time isotopic energy spectra from 8/97-3/98 for N (*left*), O (*center*), and Ne (*right*); smaller symbols above ~ 80 MeV/nucleon are from CRIS (N. E. Yanasak, private communication). Curves show the sum of power law spectra ($\propto (E/M)^{0.8}$) fit to the SIS GCR data and exponential spectra drawn through the ACR isotopes, with the ACR relative abundances of the rarer heavy isotopes for each element assumed to be the same as those found in solar system material (42).

CHARGE STATE COMPOSITION

Although of great interest for determining the isotopic composition of the neutral component of the ISM and possibly for helping to constrain the estimation of mass-dependent acceleration efficiencies, the presence of heavier isotopes of ACR O and Ne has little effect on the overall shape of the elemental spectra, as their abundances are low and their rigidities differ by only 10% from those of the dominant isotopes. A recent finding which has a much more critical impact on understanding ACR acceleration and transport has been the determination of their charge state composition as a function of energy (12, 13, 21). ACE and Wind cannot make such measurements, as the only method available at present to directly determine charge states at ACR energies requires the use of the Earth's magnetic field as a particle rigidity filter. This in turn requires a spacecraft in a high inclination or polar Earth orbit, such as SAMPEX.

The geomagnetic filter effect is illustrated in Figure 6, which shows the kinetic energy vs. invariant latitude for O particles detected by the Mass Spectrometer Telescope (MAST) on SAMPEX. (The invariant latitude, Λ, is the magnetic latitude at which a given field line intersects the Earth's surface and is related to the magnetic L shell by $\cos^2 \Lambda = 1/L$; see, e.g., (48)). The abrupt drop in density at high energies below $\sim 60°$ is due to the fact that fully stripped particles at these energies have a rigidity too low to penetrate the geomagnetic field to lower latitudes. Singly charged particles have a higher rigidity than fully stripped particles at the same kinetic energy and exhibit a similar "cutoff" at lower latitudes. (Geographic longitudes at which trapped O is found (49) have not been included in this figure). Although GCR O nuclei are abundant at high latitudes throughout the entire energy interval sampled by MAST, they are quite effectively excluded at latitudes below the geomagnetic cutoff for fully stripped ($Q = 8$) oxygen, leaving only the ACRs extending to these lower latitudes.

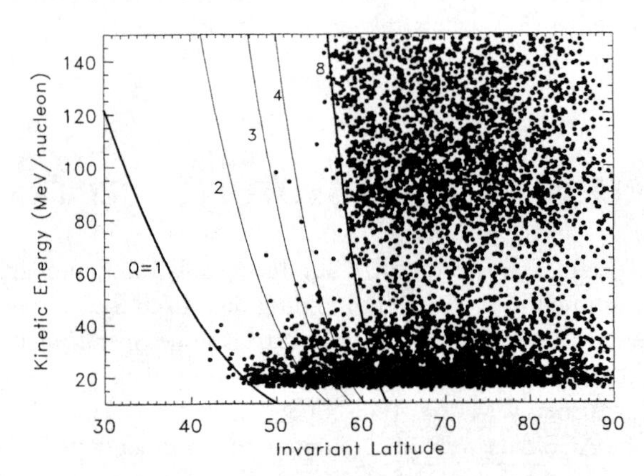

FIGURE 6. Kinetic energy vs. invariant latitude for quiet time O particles measured by MAST/SAMPEX, as in (21). The low density region from ~ 45–75 MeV/nucleon is due to a detector failure part way through the mission, while the decreasing density towards higher latitudes ($\gtrsim 80°$) reflects the lesser amount of time spent at such latitudes in SAMPEX's 82° inclination orbit. Curves show the calculated approximate locations of the geomagnetic cutoffs for O with the charge states (Q) indicated, based on an empirically derived cutoff-rigidity relation (50).

Energy spectra obtained using only those particles which penetrate more than several degrees below the cutoff for fully stripped nuclei are shown in Figure 7. The geomagnetic filter approach allows the pure ACR component to be revealed even in the presence of an interplanetary background of GCRs with intensity $\gtrsim 100$ times greater. Combined with MAST's ability to resolve isotopes, this powerful background suppression has also been used to study the isotopic composition of pure ACRs

(47), complementing the results from ACE (Figure 5) which have better statistical accuracy but require GCR background corrections.

Separated from the GCR background in Figure 7, ACRs are seen to extend to energies of at least ~ 100 MeV/nucleon with smoothly falling spectra. The diffusive shock-drift ACR acceleration model (31) predicts that the "maximum" energy gain ΔE (beyond which the spectrum is expected to fall off rapidly) for a particle of charge Qe at a quasi-perpendicular shock is $\Delta E \simeq Qe\Delta\phi$, where $\Delta\phi$ is the change in potential in drifting along the shock. Jokipii (31) calculates that $\Delta\phi \simeq 240$ MV between the heliospheric equator and pole for the termination shock, which implies that the spectrum of singly charged ^{16}O should steepen significantly beyond ~ 15 MeV/nucleon. Conversely, as pointed out by Mewaldt et al. (35), ^{16}O with a higher charge state would be more easily accelerated to the observed ~ 100 MeV/nucleon.

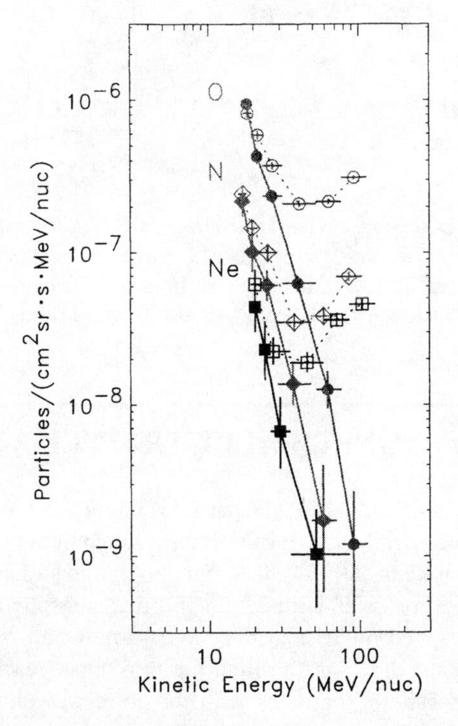

FIGURE 7. MAST/SAMPEX energy spectra of pure ACR N, O, and Ne (*filled symbols; solid lines*) obtained using the geomagnetic filter approach, compared with interplanetary spectra at $\Lambda > 65°$ (*open symbols; dotted lines*) measured by MAST during the same time period (7/6/92-1/7/95), as in (35).

Only an upper limit on the charge state of any individual particle can be determined using the geomagnetic filter technique, not the actual charge state itself. However, the distribution in latitude of a collection of particles does reveal the charge state composition of the sample. For example, none of the events in Figure 6 around 30–40 MeV/nucleon between the $Q = 2$ and $Q = 3$ cutoffs could have $Q > 2$, or else they could not be found at such low latitudes, but any particular event might have either $Q = 1$ or $Q = 2$. However, the significant drop in density below the $Q = 2$ cutoff shows that most of these events must in fact be doubly ionized, and the relative densities on either side of the $Q = 2$ cutoff indicate the relative proportions of singly and doubly ionized particles at this energy.

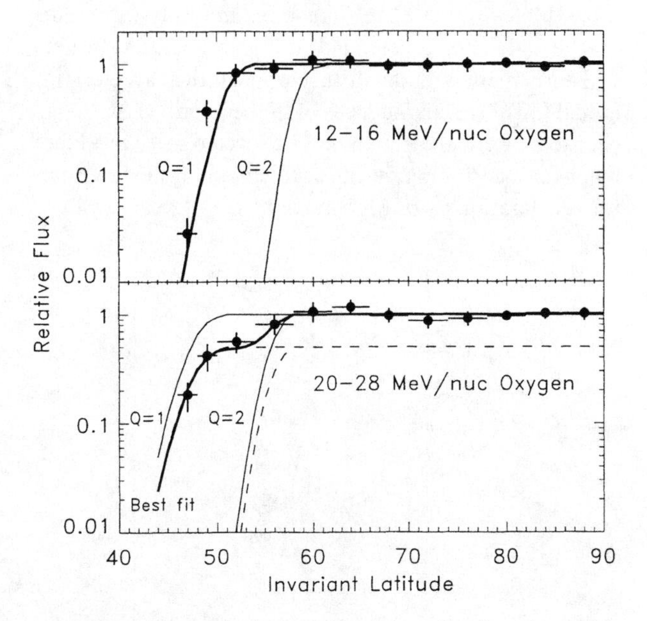

FIGURE 8. Relative flux profiles (normalized to 1 at high latitudes) as a function of invariant latitude for O at 12–16 MeV/nucleon (*top*) and 20–28 MeV/nucleon (*bottom*) from HILT/SAMPEX (13, 22), using quiet time periods between October 1992 and January 1995. Curves show the calculated profiles for particles with $Q = 1$ and $Q = 2$, and the weighted sum of these profiles (equal to the sum of the dashed curves) which best fits the high energy data.

The process of obtaining charge state distributions using the geomagnetic filter is illustrated more quantitatively in Figure 8, which shows how the O flux varies with latitude at two different energies measured by the Heavy Ion Large Telescope (HILT) on SAMPEX (13, 22). Although variations in geomagnetic activity or the arrival directions of the particles may produce some blurring of the cutoff locations, the expected difference in cutoff latitude for O with $Q = 1$ and $Q = 2$ (Figures 6 and 8) is so large that these effects are relatively unimportant. (Such effects have a greater impact when using this same technique to study solar energetic particle charge states as in (51), where the charge states are much higher and the cutoffs for different Q more closely spaced.) Below 16 MeV/nucleon, the flux profile with respect to latitude agrees with that expected for $Q = 1$ particles (13). However, at the higher energies, there is a pronounced dip in the profile at low latitudes starting near the expected position of the $Q = 2$ cutoff. The data are well fit by the sum of two step profiles of roughly equal height, one extending to the $Q = 1$ cutoff, the other only to the $Q = 2$ cutoff latitude, indicating a mixture of equal amounts of both charge states at these energies.

Similar analyses of MAST data (21, 52) yield the O charge state spectra shown in Figure 9. Singly charged O is seen to have a very steep spectrum, possibly with a break at $\sim$ 20–30 MeV/nucleon, and only the higher charge states are accelerated to higher energies, in general agreement with the diffusive shock-drift acceleration model (53). The abundances of higher charged ACRs appear to be consistent with those expected to be produced by electron stripping during acceleration, if the timescale for acceleration to 10 MeV/nucleon is $\sim$ 1 year (21).

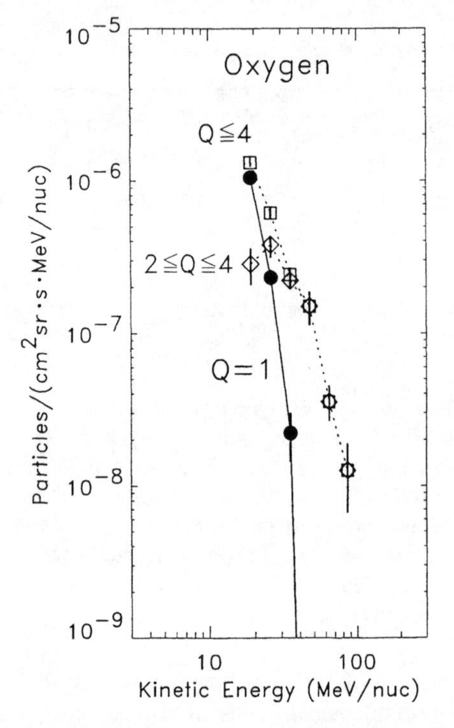

FIGURE 9. Energy spectra of ACR O^{+} (*filled circles*) compared with the spectra of $O^{+2}+O^{+3}+O^{+4}$ (*open diamonds*) and their sum, $O^{+}+O^{+2}+O^{+3}+O^{+4}$ (*open squares*), from (52). Differences in the $Q \leq 4$ spectrum shown here and the filtered O spectrum in Figure 7 arise from the assumption made in the earlier work (35) that ACRs are singly ionized at all energies, which would significantly increase the size of the latitude interval over which they could be detected (Figure 6) and hence the effective collection time.

Carrying out the same analysis for N and Ne in addition to O (22, 52), it is found that in each case the fraction of ACRs that is singly ionized decreases rapidly with increasing energy. The energy dependence of the singly ionized fraction seems better organized by total energy as in Figure 10 than by energy per nucleon (33). Above energies of $25.5 \pm 2.5, 22 \pm 2$, and 17.5 ± 2.0 MeV/nucleon for N, O, and Ne, respectively, more than half the ACRs are *not* singly ionized and multiply charged ACRs dominate (33). This corresponds to a common total energy of $\sim$ 350 MeV for all three species, which again supports the diffusive shock-drift acceleration picture (23). Attempts to model and understand this transition to higher charge states in more detail are ongoing (23, 24).

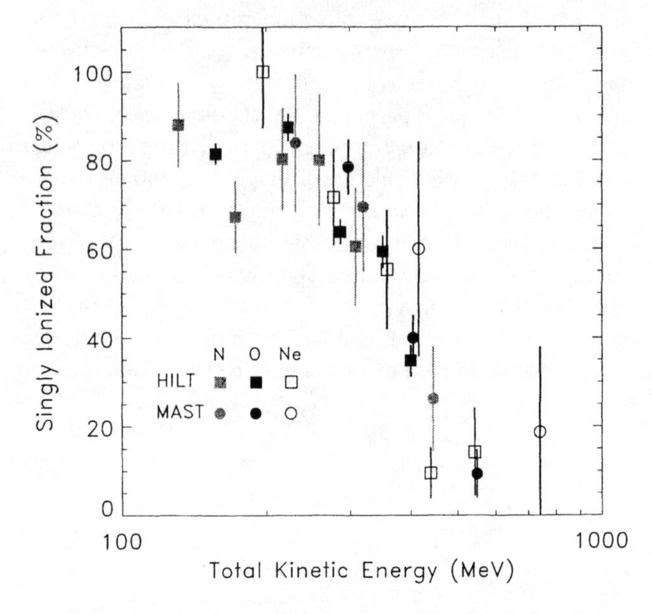

FIGURE 10. Singly ionized fraction of ACR N (*filled gray symbols*), O (*filled black symbols*), and Ne (*open symbols*) as a function of total energy, combining data from HILT (*squares*; (22)) and MAST (*circles*; (52)) on SAMPEX. Adapted from (33).

GEOMAGNETICALLY TRAPPED ACRS

One population of ACR particles unique to 1 AU (or, more specifically, low Earth orbit) is found in the Earth's radiation belts (49, 54, 55). Singly charged ACRs can penetrate to lower latitudes in the magnetosphere than fully stripped nuclei and may lose some or all of their electrons if they pass through the thin upper reaches of the atmosphere. Particles with the correct combination of energy and pitch angle relative to the magnetic field entering at latitudes near the $Q = 1$ cutoff will be stably trapped in the geomagnetic field after this abrupt change in their rigidity (56) and form a radiation belt around $L = 2$ (which would appear near $\Lambda = 45°$ straddling the $Q = 1$ cutoff line in Figure 6).

The trapped ACRs are potentially attractive for ACR studies because they are concentrated in the radiation belt to intensities $\sim$ 100 times greater than normally found in interplanetary space at 1 AU (57). Like the geomagnetically filtered ACRs described above, this is a pure sample of ACRs, free of GCR or solar particle backgrounds. Their elemental abundances are depleted in C, N, and perhaps Ne relative to O (57). This may be due to differences between the elements in their electron stripping cross sections or the fact that ions with more electrons to lose can undergo a greater change in rigidity during stripping and should be more likely to be trapped (57). Their composition is also influenced by the fact that only the singly-ionized fraction should be able to penetrate to $L = 2$ and become trapped. Indeed, the trapped ACR energy spectra are observed to be very similar to those of the singly-ionized components and are much softer than those of the overall filtered ACRs (57). Their isotopic composition is similar to that found in interplanetary ACRs but with higher statistical accuracy (47). Finally, the trapped ACR intensity seems to track the variations seen in interplanetary ACRs (54, 57), even reflecting the recent decreases shown in Figure 2 as solar maximum approaches (58). It will be interesting to see if ACRs are still detectable in this distilled sample long after they have faded beneath the GCR background in interplanetary space.

SUMMARY

The presently declining ACR fluxes at 1 AU mark the close of a solar cycle minimum during which tremendous progress has been made in both measuring and understanding these particles. The elemental spectra have been determined more precisely than ever before and have revealed new species with ACR-like enhancements, possibly with different origins and histories than the "classic" ACRs. Measurements of their isotopic composition are beginning to be made, which may provide insight into the nucleosynthetic history of the ISM. The GCR background has been stripped away to uncover previously unsuspected ACRs out to energies as high as 100 MeV/nucleon, indicating that the termination shock is capable of accelerating particles to a total energy as great as 1.6 GeV. Clues as to how such high energies are achieved were provided by the revelation that high energy ACRs are multiply charged, which can also help to constrain the acceleration time scale, while low energy ACRs have been confirmed to be singly ionized as predicted.

Measurements utilizing the large collecting power of instruments on Wind and ACE, or taking advantage of the geomagnetic filter or the concentrated collection of trapped ACRs with SAMPEX, may allow ACR intensities to be tracked longer into solar maximum than in previous cycles, furthering studies of ACR modulation under extreme conditions. While theorists incorporate the recent results into their latest models to improve our understanding of ACRs, we eagerly anticipate the return of these particles to 1 AU around 2003 and the discoveries yet to come during the next solar minimum.

ACKNOWLEDGMENTS

This research was supported by NASA at the California Institute of Technology (under grant NAG5-6912), the Jet Propulsion Laboratory, and the Goddard Space Flight Center. We thank R. S. Selesnick, N. E. Yanasak, and B. Klecker for providing data used in some of the figures. Climax neutron monitor data were obtained from the web, courtesy of the University of Chicago, National Science Foundation grant ATM-9613963.

REFERENCES

1. Garcia-Munoz, M., Mason, G. M., and Simpson, J. A., *Astrophys. J. Letters* **182**, L81-L84 (1973).
2. Hovestadt, D., Vollmer, O., Gloeckler, G., and Fan, C. Y., *Phys. Rev. Lett.* **31**, L650-L653 (1973).
3. McDonald, F. B., Teegarden, B. J., Trainor, J. H., and Webber, W. R., *Astrophys. J. Letters* **187**, L105-L108 (1974).
4. Fisk, L. A., Kozlovsky, B., and Ramaty, R., *Astrophys. J. Letters* **190**, L35-L37 (1974).
5. Pesses, M. E., Jokipii, J. R., and Eichler, D., *Astrophys. J. Letters* **246**, L85-L88 (1981).
6. Christian, E. R., Cummings, A. C., and Stone, E. C., *Astrophys. J. Letters* **334**, L77-L80 (1988).
7. von Rosenvinge, T. T. and McDonald, F. B., *Proc. 14th Internat. Cosmic Ray Conf. (Munich)* **2**, 792-797 (1975).
8. Cummings, A. C. and Stone, E. C., *Proc. 20th Internat. Cosmic Ray Conf. (Moscow)* **3**, 413-416 (1987).
9. Witte, M., Rosenbauer, H., Banaszkiewicz, M., and Fahr, H., *Adv. Space Res.* **13**, 121-130 (1993).
10. Möbius, E., Hovestadt, D., Klecker, B., Scholer, M., Gloeckler, G., and Ipavich, F. M., *Nature* **318**, 426-429 (1985).
11. Geiss, J., Gloeckler, G., Mall, U., von Steiger, R., Galvin, A. B., and Ogilvie, K. W., *Astron. Astrophys.* **282**, 924-933 (1994).
12. Adams, J. H. Jr. et al., *Astrophys. J. Letters* **375**, L45-L48 (1991).
13. Klecker, B. et al., *Astrophys. J. Letters* **442**, L69-L72 (1995).
14. Cummings, A. C. et al., *Geophys. Res. Letters* **22**, 341-344 (1995).
15. Trattner, K. J., Marsden, R. G., Bothmer, V., Sanderson, T. R., Wenzel, K.-P., Klecker, B., and Hovestadt, D., *Astron. Astrophys.* **316**, 519-527 (1996).
16. Stone, E. C., *Proc. 20th Internat. Cosmic Ray Conf. (Moscow)* **7**, 105-114 (1987).
17. Stone, E. C., Cummings, A. C., Hamilton, D. C., Hill, M. E., and Krimigis, S. M., *Proc. 26th Internat. Cosmic Ray Conf. (Salt Lake City)* **7**, 551-554 (1999).

18. Takashima, T. et al., *Astrophys. J. Letters* **477**, L111-L113 (1997).

19. Reames, D. V., *Astrophys. J.* **518**, 473-479 (1999).

20. Cummings, A. C., Stone, E. C., and Steenberg, C. D., *Proc. 26th Internat. Cosmic Ray Conf. (Salt Lake City)* **7**, 531-534 (1999).

21. Mewaldt, R. A., Selesnick, R. S., Cummings, J. R., Stone, E. C., and von Rosenvinge, T. T., *Astrophys. J. Letters* **466**, L43-L46 (1996).

22. Klecker, B., Oetliker, M., Blake, J. B., Hovestadt, D., Mason, G. M., Mazur, J. E., and McNab, M. C., *Proc. 25th Internat. Cosmic Ray Conf. (Durban)* **2**, 273-276 (1997).

23. Jokipii, J. R., in *Acceleration and Transport of Energetic Particles Observed in the Heliosphere*, AIP Conf. Proc. this volume, New York: AIP Press (2000).

24. Barghouty, A. F., Jokipii, J. R., and Mewaldt, R. A., in *Acceleration and Transport of Energetic Particles Observed in the Heliosphere*, AIP Conf. Proc. this volume, New York: AIP Press (2000).

25. Steenberg, C. D., in *Acceleration and Transport of Energetic Particles Observed in the Heliosphere*, AIP Conf. Proc. this volume, New York: AIP Press (2000).

26. Mewaldt, R. A. et al., *Geophys. Res. Letters* **20**, 2263-2266 (1993).

27. Stone, E. C. et al., *Space Sci. Rev.* **86**, 357-408 (1998).

28. von Rosenvinge, T. T. and Paizis, C., *Proc. 17th Internat. Cosmic Ray Conf. (Paris)* **10**, 69-72 (1981).

29. Teegarden, B. J., McDonald, F. B., and Balasubrahmanyan, V. K., *Acta Physica Hungarica* **29**, Suppl. 1, 345 (1970).

30. Mogro-Campero, A., Schofield, N., and Simpson, J. A., *Proc. 13th Internat. Cosmic Ray Conf. (Denver)* **1**, 140-145 (1973).

31. Jokipii, J. R., in *Physics of the Outer Heliosphere*, S. Grzedzielski and D. E. Page, eds., Oxford: Pergamon Press (1990).

32. Stone, E. C. et al., *Space Sci. Rev.* **86**, 285-356 (1998).

33. Klecker, B. et al., *Space Sci. Rev.* **83**, 259-308 (1998).

34. Hasebe, N. et al., *Geophys. Res. Letters* **21**, 3027-3030 (1994).

35. Mewaldt, R. A., Cummings, J. R., Leske, R. A., Selesnick, R. S., Stone, E. C., and von Rosenvinge, T. T., *Geophys. Res. Letters* **23**, 617-620 (1996).

36. Oetliker, M., Klecker, B., Mason, G. M., McNab, M. C., and Blake, J. B., *Proc. 25th Internat. Cosmic Ray Conf. (Durban)* **2**, 277-280 (1997).

37. Fahr, H. J., Ripken, H.W., and Lay, G., *Astron. Astrophys.* **102**, 359-370 (1981).

38. Geiss, J., Gloeckler, G., and von Steiger, R., *Space Sci. Rev.* **78**, 43-52 (1996).

39. Mewaldt, R. A., *Proc. 26th Internat. Cosmic Ray Conf. (Salt Lake City)* **7**, 547-550 (1999).

40. Leske, R. A., in *Summary-Rapporteur Vol. of the 26th Internat. Cosmic Ray Conf.*, B. Dingus, ed., New York: AIP Press (2000), in press.

41. Christian, E. R. et al., *Proc. 26th Internat. Cosmic Ray Conf. (Salt Lake City)* **7**, 559-560 (1999).

42. Anders, E. and Grevesse, N., *Geochim. Cosmochim. Acta* **53**, 197-214 (1989).

43. Geiss, J., Buehler, F., Cerruti, H., Eberhardt, P., and Filleux, Ch., *Apollo 16 Prelim. Sci. Report, NASA SP-315* **231**, 14-1–14-10 (1972).

44. Podosek, F. A., *Ann. Rev. Astron. Astrophys.* **16**, 293-334 (1978).

45. Connell, J. J. and Simpson, J. A., *Proc. 23rd Internat. Cosmic Ray Conf. (Calgary)* **1**, 559-562 (1993).

46. Lukasiak, A., Ferrando, P., McDonald, F. B., and Webber, W. R., *Astrophys. J.* **426**, 366-372 (1994).

47. Leske, R. A., Mewaldt, R. A., Cummings, A. C., Cummings, J. R., Stone, E. C., and von Rosenvinge, T. T., *Space Sci. Rev.* **78**, 149-154 (1996).

48. Roederer, J. G., *Dynamics of Geomagnetically Trapped Radiation*, New York: Springer (1970).

49. Cummings, J. R., Cummings, A. C., Mewaldt, R. A., Selesnick, R. S., Stone, E. C., and von Rosenvinge, T. T., *Geophys. Res. Letters* **20**, 2003-2006 (1993).

50. Leske, R. A., Cummings, J. R., Mewaldt, R. A., Stone, E. C., and von Rosenvinge, T. T., in *High Energy Solar Physics*, AIP Conf. Proc. 374, R. Ramaty, N. Mandzhavidze, and X.-M. Hua, eds., New York: AIP Press, 86-95 (1996).

51. Leske, R. A., Cummings, J. R., Mewaldt, R. A., Stone, E. C., and von Rosenvinge, T. T., *Astrophys. J. Letters* **452**, L149-L152 (1995).

52. Selesnick, R. S., Mewaldt, R. A., and Cummings, J. R., *Proc. 25th Internat. Cosmic Ray Conf. (Durban)* **2**, 269-272 (1997).

53. Jokipii, J. R., *Astrophys. J. Letters* **466**, L47-L50 (1996).

54. Grigorov, N. L. et al., *Geophys. Res. Letters* **18**, 1959-1962 (1991).

55. Mewaldt, R. A., Selesnick, R. S., and Cummings, J. R., in *Radiation Belts: Models and Standards*, Geophys. Monograph 97, J. F. Lemaire, D. Heynderickx, and D. N. Baker, eds., Washington, D.C.: AGU Press, 35-41 (1996).

56. Blake, J. B. and Friesen, L. M., *Proc. 15th Internat. Cosmic Ray Conf. (Plovdiv)* **2**, 341-346 (1977).

57. Selesnick, R. S., Cummings, A. C., Cummings, J. R., Mewaldt, R. A., Stone, E. C., and von Rosenvinge, T. T., *J. Geophys. Res.* **100**, 9503-9518 (1995).

58. Selesnick, R. S. et al., *Geophys. Res. Letters*, submitted.

The Morphology of Anomalous Cosmic Rays in the Outer Heliosphere

C.D. Steenberg

California Institute of Technology, Mail code 220-47, Pasadena, CA, 91125, USA

Abstract. The well established cosmic ray transport equation describes the physics of cosmic ray modulation in the heliosphere and is thought to be complete. Its solution for anomalous cosmic rays has special characteristics due to the local acceleration of these particles at the solar wind termination shock. Some of these characteristics are demonstrated here, namely effects by the strength of the shock, shock drift, the cutoff in the spectrum at the shock, species scaling, drift, and ionization.

INTRODUCTION

In a test particle approach the distribution of cosmic ray particles in the heliosphere can be described by Parker's transport equation (TPE) [18]. In a recent overview of the state of modulation theory [5] it is argued that the physics contained in this equation is sufficiently complete to describe cosmic ray modulation processes. The problem that remains is to develop a detailed description of the time-dependent characteristics of the solar wind plasma causing the modulation of cosmic rays. This description needs to be put into the TPE to produce a realistic modulation model. It is expected that such a realistic model would consist of a system of equations, including an MHD description of the solar wind plasma.

The characteristics of the modulation of cosmic rays is determined to a large extent by the "source" or boundary spectrum that is modulated. The source of anomalous cosmic rays (ACRs) is widely believed to be the the acceleration of so-called pick-up ions [6] at the solar wind termination shock (SWTS) as proposed by [16]. In this paper solutions of Parker's equation will be used to demonstrate the unique features of ACRs and their dependence on the location, strength, diffusion parameters at the SWTS, ionization, and drift effects.

A set of modulation parameters with simple spatial and energy dependence is chosen to demonstrate these features as clearly as possible. More realistic parameters would yield results that are quantitatively different, but still qualitatively similar. This paper is an extension of the work published in [17, 20].

TRANSPORT AND ACCELERATION OF COSMIC RAYS

The transport equation contains the modulation effects of convection, diffusion, drifts in the HMF, adiabatic energy losses, and acceleration at the SWTS. In terms of the omnidirectional distribution function f, the TPE can be written as

$$\frac{\partial f}{\partial t} = \nabla \cdot (\mathbf{K} \cdot \nabla f - \mathbf{V} f) + \frac{1}{3}(\nabla \cdot \mathbf{V}) \frac{\partial f}{\partial \ln P} + Q \qquad (1)$$

where $P = pc/Z$ is particle rigidity in terms of momentum, p, the speed of light, c, and ionic charge Z. Diffusion and drift effects are described by the diffusion tensor, $\mathbf{K}$, $\mathbf{V}$ is the solar wind vector, and Q represents sources/sinks of particles. The diffusion tensor contains anti-symmetrical terms quantifying drift (κ_T) and diffusion parallel and perpendicular to the mean magnetic field ($\kappa_{\|}, \kappa_{\perp 1}, \kappa_{\perp 2}$).

The diffusive shock (first order Fermi) acceleration mechanism produces spectra at the SWTS (e.g. [3]) of the form

$$f(r_s, P) = qP^{-q} \int_0^P \frac{Q(P')}{V_r} P'^{q-1} dP' \qquad (2)$$

In the special case of a mono-energetic source spectrum $Q(P,r) = Q_0 \delta(P - P_i)\delta(r - r_s)$, where δ is the Dirac-delta function, injected at rigidity P_i, into a steady, plane SWTS with compression ratio s, positioned at a distance r_s from the sun, this reduces to

$$f(r_s, P) \propto P^{-q} \qquad (3)$$

CP528, *Acceleration and Transport of Energetic Particles Observed in the Heliosphere: ACE 2000 Symposium*, edited by Richard A. Mewaldt, et al.
© 2000 American Institute of Physics 1-56396-951-3/00/$17.00

with spectral index $q = 3s/(s-1)$. In terms of intensity, $j_T \propto P^2 f$, and kinetic energy, T, this accelerated spectrum reduces to

$$j_T \propto T^{-(q-2)/2} \quad \text{or} \quad j_T \propto T^{-(q-2)} \tag{4}$$

in the non-relativistic and relativistic limits respectively.

This power-law accelerated spectrum will extend to a maximum energy limited by the available acceleration time, the finite dimensions of the shock, adiabatic energy losses (e.g. [3, 9]), and also ionization [11]. Considering the short acceleration time needed to extend the power law to ACR energies [10], compared to the long-lived nature of the SWTS, and the length of a solar cycle, the acceleration time is not thought to be responsible for limiting the maximum energy gain by ACRs at the quasi-perpendicular SWTS.

Instead, particles are expected [3] to escape from a spherical SWTS without further acceleration when the diffusive length scale becomes comparable to the radius of curvature of the shock

$$\frac{\kappa_{rr}}{V} \sim r_s. \tag{5}$$

Adiabatic deceleration will decrease this maximum energy, as it competes with the diffusive shock acceleration mechanism.

The maximum energy a particle can attain by acceleration at the SWTS can also be estimated in an interesting special case as follows: by noting [8] that in absence of diffusion, for a purely perpendicular shock, and for particles that can not escape from the shock, the energy gain due to first order Fermi acceleration is equivalent to an energy gain of a charged particle in a potential field $\mathbf{E} = -\mathbf{V} \times \mathbf{B}$. In this case a positively charged particle gains energy as it drifts along the shock from the equator to the pole (pole to equator) in the positive (negative) drift cycle. The maximum energy gain is then limited by the potential difference between equator and pole. This upper limit was estimated to be a total energy of $\sim$300 MeV by [11] and references therein.

THE POWER LAW SPECTRUM AND CUTOFF

It is instructive to to solve (1) for a spherically symmetric heliosphere to illustrate the power law and its cutoff at the SWTS.

Using the numerical model of [22], a baseline solution of (1) is obtained for 140 radial intervals between a reflecting inner boundary at $r_i = 0.05$ AU and outer free escape boundary $r_b = 120$ AU, with a discontinuous, strong shock at $r_s = 90$ AU [23]. A diffusion mean free path perpendicular to the SWTS of $\lambda_{rr} = 3\kappa_{rr}/v = \lambda_{rr0} P/P_0 V/V_0$ AU is used, with $P_0 = 1$ GV, and v the particle speed in the rest frame of the sun. The solar wind speed is $V = V_0$, with $V_0 = 400$ km/s inside the shock, and $V = V_0 (r_s/r)^2/s$ outside the shock. A mono-energetic spectrum of particles are injected at a rigidity P, with the rigidity domain divided into 400 logarithmically spaced intervals. The model is run for a total of 4.8 years, after which the solution has reached a stationary state.

For a baseline solution ACR O^+, with atomic mass to charge ratio $A/Z = 16$, is modulated by setting $\lambda_{rr0} = 0.5$, with $P_i = 0.2$ GV (72 keV/nuc), and accelerated by a strong ($s = 4$) shock. The resulting spectrum at the SWTS is shown by the solid line in Figure 1 (top). This spectrum has the expected form of a power law T^{-1} at low energies, with a cutoff at 5.3 MeV/nuc, using a $1/e$ deviation from the power law as a cutoff criterion.

Also shown in Figure 1 are two parameter variations relative to the baseline: (i) Changing only the compression ratio of the shock to $s = 2.5$, results in a shock spectrum with a power law of $T^{-1.5}$ (dashed line), as expected from (4). Note that the cutoff still occurs at 5.3 MeV/nuc. (ii) Increasing the mean free path to $\lambda_{rr0} = 1.0$ results in a shock spectrum (dash-dotted line) identical to the the baseline, except that the cutoff energy is reduced by a factor of two, to 2.65 MeV/nuc.

In the bottom panel of Figure 1 the modulated spectra at 60 AU show that the spectral peak for case (i) is moved to slightly lower energies, due to stronger modulation which stems from the steeper spectrum at the shock. In case (ii) the spectral peak is shifted to lower energies by a factor of approximately two, with a lower level of modulation than in the baseline solution.

It was found that for a wide range of parameters [17, 20] that the cutoff in the shock spectrum occurs where

$$\frac{V r_s}{\kappa_{rr}} = C_c \tag{6}$$

where $C_c \approx 4.5$ for the above solutions, using the $1/e$ criterion to determine the cutoff energy. This number is higher than the lower limit of $C_c = 1$, in (5), which may be due to the effects of adiabatic energy losses not being taken into account in that estimate, or may simply be dependent on the criterion used to determine the cutoff energy. In comparison [17, 20] obtained a value of $C \approx 10$, by essentially taking the smallest detectable deviation from a power law to indicate a cutoff.

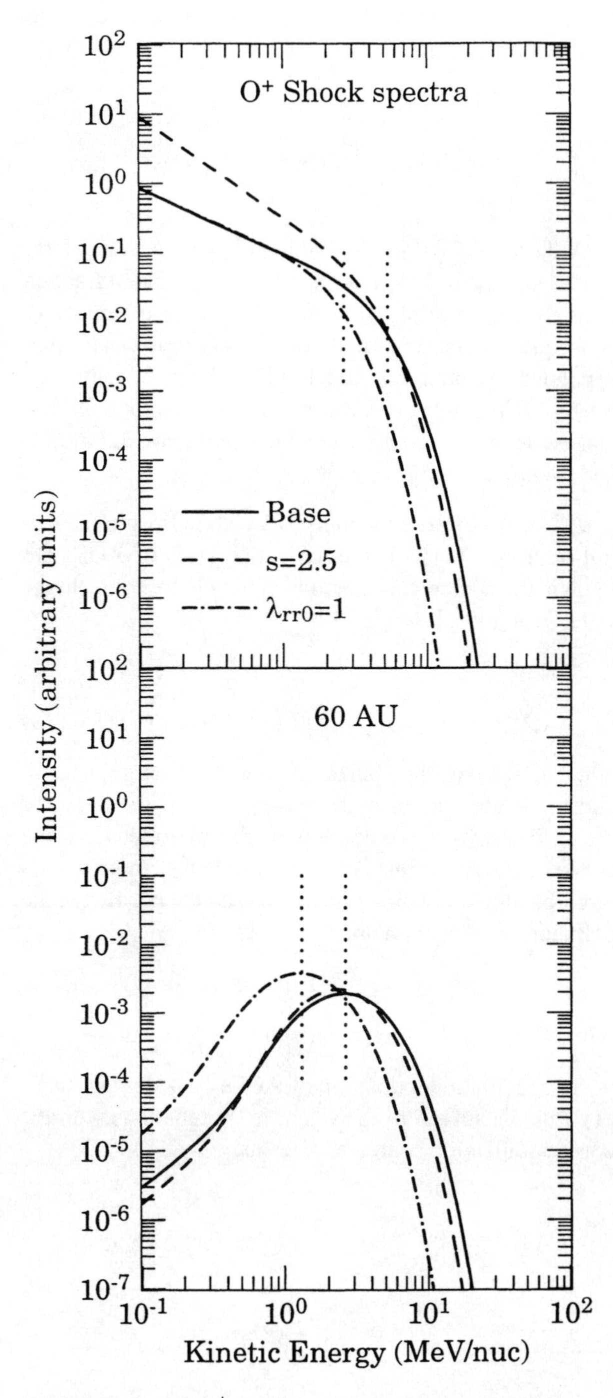

FIGURE 1. ACR O^+ shock spectra (top) and at 60 AU (bottom) for (i) baseline solution with $s = 4$, (ii) $s = 2.5$, (iii) solution with $\lambda_{rr0} = 1$. Vertical lines (dots) indicate the cutoff energies of the shock spectra (top), and the peak energies of the modulated spectra (bottom). Both the cutoff energy, and peak energy of the modulated spectrum in the $\lambda_{rr0} = 1$ case are lower by a factor of two. The normalization was chosen arbitrarily to emphasize the effects of the parameter variations.

More important than its value is the fact that C_c is a constant, and that the cutoff energy can easily be determined from (6), as long as the same criterion for the cutoff is applied consistently.

SPECIES SCALING

Modulated ACR spectra for different species (elements) have very similar shapes, but have different abundances and appear to be shifted in energy per nucleon ([23] and references therein). The difference in abundances are controlled by the seed particle abundance injected into the shock, while the so-called *species scaling* in energy is caused by the species dependence of diffusion coefficients.

The species scaling of modulation can be derived [4, 21] as follows: If the effective radial diffusion coefficient for more than one species of speed βc and rigidity P has an energy dependence of βP^γ, the (non-relativistic) energy per nucleon at which spectral features occur, scale as $T_2/T_1 \approx \alpha_1/\alpha_2$ in the relativistic limit, with $\alpha = A/Z$. In the non-relativistic limit, which is applicable to ACRs, the species scaling factor is

$$T_2/T_1 \approx \left(\frac{\alpha_1}{\alpha_2}\right)^{\frac{2\gamma}{\gamma+1}} \tag{7}$$

which only depends on the power γ. In this derivation it is assumed that the shock spectrum (or local interstellar spectrum in the case of GCRs) that is modulated, has the same form for both species, but scaled in energy according to the species scaling factor above. This is the case for ACR spectra accelerated at the SWTS with a power law up to s characteristic cutoff energy. This cutoff is controlled by the magnitude of the radial diffusion coefficient in (6), so that the cutoff energies will scale with species in the same way as modulation.

Figure 2 demonstrates the species scaling between ACR O^+, He^+, and H^+ by solving the TPE with the baseline parameters for these species. In the top panel vertical lines show the cutoffs in the three shock spectra at 5.3, 21.2, and 84.8 MeV/nuc, as expected from (7) for the $\gamma = 1$ used in all three cases. Similar lines are also drawn in the bottom panels to show that modulated peak energies at about 2.6, 5.3, and 21 MeV/nuc for each species, respectively.

Because the species scaling factor is a unique function of the rigidity dependence γ in the non-relativistic limit, the observed species scaling factor may be used to determine this parameter. For Voyager 1 and 2 ACR O,

He, and H, [21] found that the species scaling factor at the spectral peaks was consistent with a value of $\gamma \approx 2$. Conversely, observed 1998 modulated spectra of several different species was fitted [19] by solutions of the TPE by using a value of $\gamma = 1.6$ for rigidities up to 0.4 GV.

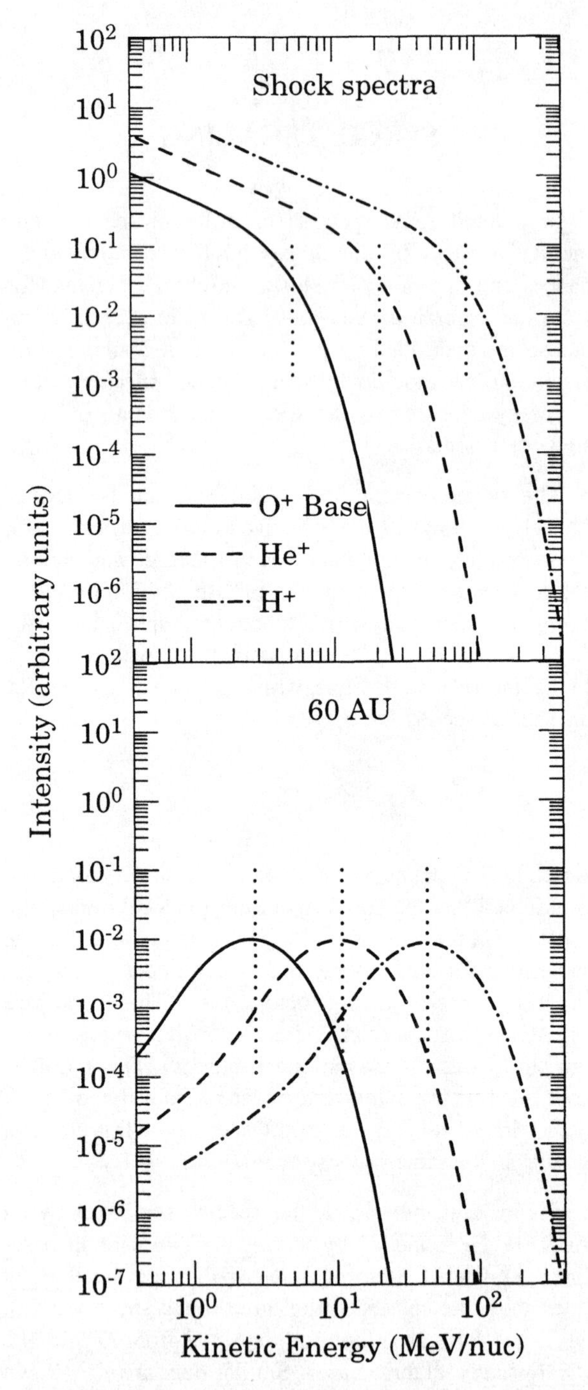

FIGURE 2. ACR O^+ shock spectra (top) and at 60 AU (bottom) for (i) baseline O^+ solution, (ii) He^+, (iii) H^+. Vertical lines (dots) indicate the cutoff energies of the shock spectra (top), and the peak energies of the modulated spectra (bottom). Both the cutoff energy, and peak energies for He^+ and H^+ are higher by factors of 4 and 16 respectively. The normalization was chosen arbitrarily so that the modulated intensities at 60 AU are equal at the spectral peaks.

IONIZATION OF ACRS

Although ACRs are predominantly singly charged, recent observations [15, 14] show that a significant fraction may also be multiply charged. Ionization effects were consequently incorporated in a axisymmetrical time-dependent modulation model [11] and subsequently applied to SAMPEX observations [2]. The method used in that model was also applied to the Steenkamp [22] model, and is summarized below.

ACRs are ionized by collisions with heliospheric neutral hydrogen [11]. If the ionization rates of ACRs are known, the source/sink function Q_i^q in (1) for each charge state, q, is given by

$$Q_i^q(\mathbf{r},P,t) = n_H\left[R^{q-1}(T)f^{q-1} - R^q(T)f^q\right] \quad (8)$$

where $R^q(T)$ is the ionization rate for charge state q, and n_H is the neutral hydrogen density. If the speed of the neutral hydrogen atoms is negligible relative to the speed of the ACR ions, and the particle distribution f^q is isotropic, the ionization rate is given by the product of the collision cross section and the ion speed,

$$R^{q-1}(T) = \sigma^q(T)v = \sigma^q(T)\beta c \quad (9)$$

with σ^q the ion-hydrogen ionization cross section.

To model the modulation of ACRs of N charge states, (1) must be solved for a system of N equations with the correct ionization source/sink terms:

$$\begin{aligned} \frac{\partial f^1}{\partial t} &= \ldots - Q_i^1(\mathbf{r},P,t) \\ \frac{\partial f^2}{\partial t} &= \ldots - Q_i^2(\mathbf{r},P,t) \\ &\vdots \\ \frac{\partial f^N}{\partial t} &= \ldots - Q_i^N(\mathbf{r},P,t) \end{aligned} \quad (10)$$

Each charge state undergoes modulation due to convection, diffusion, drifts, adiabatic energy losses, acceleration at the shock, with an additional sink term to account for ionization losses. For the fully ionized state, $Q_i^N(\mathbf{r},P,t) = 0$.

Assuming that only singly charged O^+ is injected at the shock, we repeat the baseline solution for ACR O and calculate the spectra and relative abundances of all 8 charge states, with the results shown in Figures 4 and 5. A neutral hydrogen density of $n_H = 0.115$, an estimate for the value at the SWTS [7], was used throughout the heliosphere. Ionization cross sections, σ^q, from [1] for O, shown in Figure 5, have maxima in the energy range of ~ 100 to ~ 200 keV/nuc for charge states 1 to 5, and ~ 1 MeV/nuc for charge states 6 and 7. An injection energy of 180 keV/nuc is used, instead of 72 keV/nuc in the baseline solution, for stability reasons, and is sufficiently low to allow a realistic demonstration.

Firstly, note that the O^+ spectra are only slightly depleted relative to the no-ionization baseline solution, with the depletion increasing with energy. At low energies, the T^{-1} power law is also still clearly present at the shock. This indicates that the source of particles due to acceleration is large relative to the sink due to ionization in (8). Secondly, there is also a clear trend that the abundance of a particular charge state O^{N+} relative to $O^{(N-1)+}$ decreases with increasing charge, N: At 60 AU, O^{2+} makes up 5% to 10% of the total in the energy range between 1 and 10 MeV/nuc, where the peak intensity occurs. In this energy range, O^{3+} contributes 0.1% to 0.5% of the total.

At the shock, the spectra of multiply charged ACRs do not have the T^{-1} power law at low energies. This indicates that the majority of these particles are the products of ionization of lower charge states, rather than being ionized mainly at low energies and then accelerated. As expected from (6), the multiply charged ACRs are accel-

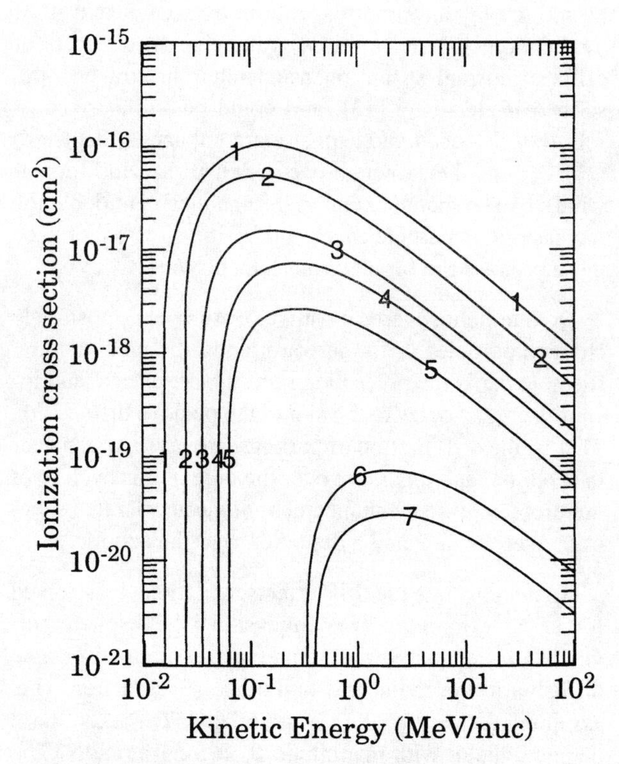

FIGURE 3. Ionization cross sections for charge states 1 to 7 of oxygen, from [1].

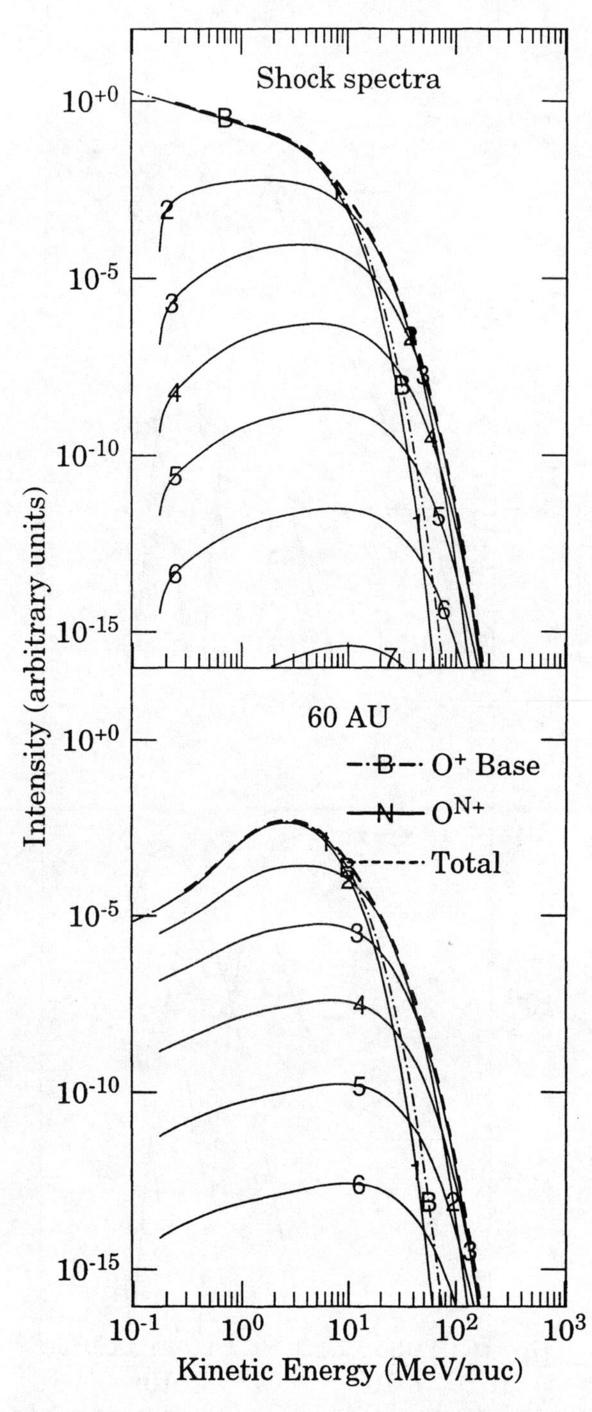

FIGURE 4. ACR $O^+ \ldots O^{8+}$ shock spectra (top) and at 60 AU (bottom). The number on each curve is the charge state, the dashed line is the total of all 8 charge states. The line marked B is the baseline solution.

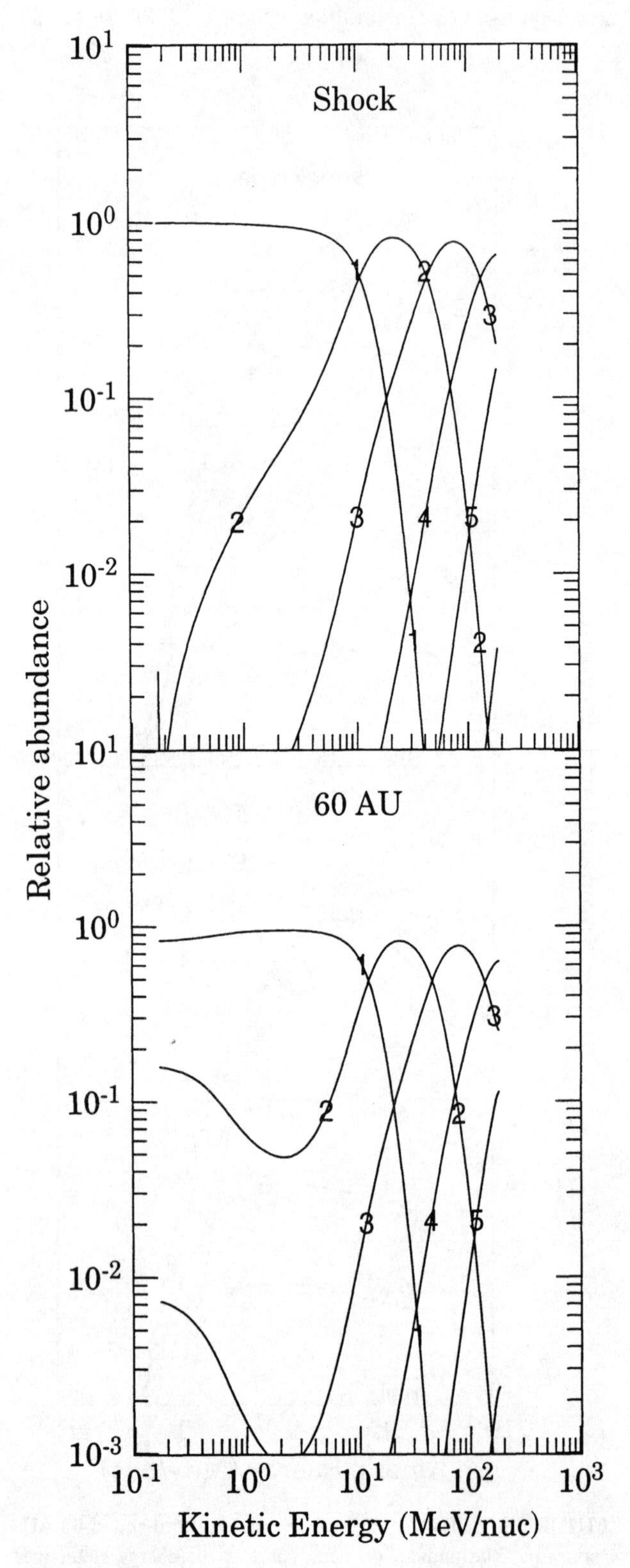

FIGURE 5. ACR $O^{+}\ldots O^{8+}$ relative abundances at the shock (top) and at 60 AU (bottom).

erated more efficiently at the SWTS, reaching higher energies (in energy per nucleon) than singly charged particles. In Figure 5 we show that at the shock, O^{2+} becomes more abundant than O^{+} at 10 MeV/nuc (11 MeV/nuc at 60 AU), while O^{3+} in turn becomes more abundant at 40 MeV/nuc (44 MeV/nuc at 60 AU). These 'crossovers' are a natural consequence of charge dependence of the acceleration cutoff.

Although ionization is a relatively unimportant modulation mechanism compared to convection, diffusion, adiabatic energy losses and acceleration, it is a significant source of particles for the acceleration mechanism at the shock. The sum of the accelerated spectra of all charge state can be a significantly broader spectrum compared to a similar spectrum calculated for singly charged ions without taking ionization into account. This is expected to be particularly important for interpreting observations of ACRs with larger ionization cross-sections. The contribution of multiply charged ACRs should be more easily detectable for species where the GCRs are significantly less abundant than ACRs.

DRIFT EFFECTS

During the solar minimum period, drift effects are expected to play an important role in both GCR and ACR modulation, [5] and references therein. The effects of drifts at current sheets on accelerated spectra was described in detail by [13], and could be summarized as follows: a) Accelerated spectra are enhanced (depleted) near current sheets where particles drift upwind (downwind), b) The spatial extent of the enhancement/depletion increases with particle energy and c) the effects should be more pronounced for electrons than protons.

In the heliosphere, positive drift cycle, positively charged particles drifts outward along the wavy neutral sheet, towards the poles along the shock surface, and inward from the poles to the sun in the positive drift cycle,. That is, the drift motion of particles are *downstream* near the ecliptic, and *upstream* over the pole in this cycle. We therefore expect an enhancement of the accelerated spectrum over the pole and a depletion near the ecliptic.

To demonstrate the drift effects, equation (1) is solved for an axisymmetric two-dimensional heliosphere, divided into 30 equally spaced intervals, assuming isotropic diffusion in the radial and latitudinal (θ) directions (i.e. assuming $\lambda_{\theta\theta} = \lambda_{rr}$), and a modified [12] Parker spiral magnetic field, with magnitude $B_e = 5$ nT at earth. The diffusion parameters and the solar wind were multiplied by $l(\theta) = 1.5 + 0.5\cos 2\theta$, thereby increasing their magnitude by a factor of 2 over the poles relative to the ecliptic. A simulated wavy neutral sheet with a tilt angle of 10° was used to approximate solar minimum conditions.

A comparison of the drift solution with the baseline solution is shown in Figure 6. At the shock, the ecliptic spectrum is depleted by 20% relative to the no-drift baseline at the cutoff energy of 5.2 MeV/nuc, while the spectrum at the pole is enhanced by a factor of 4 at this energy, taking the latitudinal gradient into account. In both cases, the modification does neither destroy the power law shape of the spectra at low energies, nor is the cutoff energy significantly modified. For our choice of parameters, the modulated ACR spectra at 60 AU are not substantially modified by drift effects.

SUMMARY

One-and two-dimensional solutions of the cosmic ray transport equation was used to demonstrate important characteristics of ACR spectra accelerated at the SWTS, which can be summarized as follows:

- Modulated ACR spectra have a peak intensity that is determined by the cutoff in the spectrum at the SWTS.
- ACR spectra at the SWTS have a characteristic power law form, with a spectral index determined by the compression ratio of the shock.
- This power law extends to a certain cutoff energy, which occurs at an energy where $Vr_s = C_c\kappa_{rr}$. Defining the cutoff to be where the spectrum has fallen to $1/e$ of its power law extrapolation, it is found that $C_c \sim 4.5$.
- The species dependence of the diffusion and drift terms in the diffusion tensor and the dependence of the cutoff in the shock spectra on κ_{rr} give rise to species dependent modulation, or species scaling, of ACRs.
- The power law spectrum may be modified by drift effects, which either enhances of suppresses acceleration, depending on the direction of the particle drift motion relative to the direction of convection by the solar wind.
- ACRs are ionized by collisions with neutral hydrogen atoms in the heliosphere, giving rise to multiply charged ions.

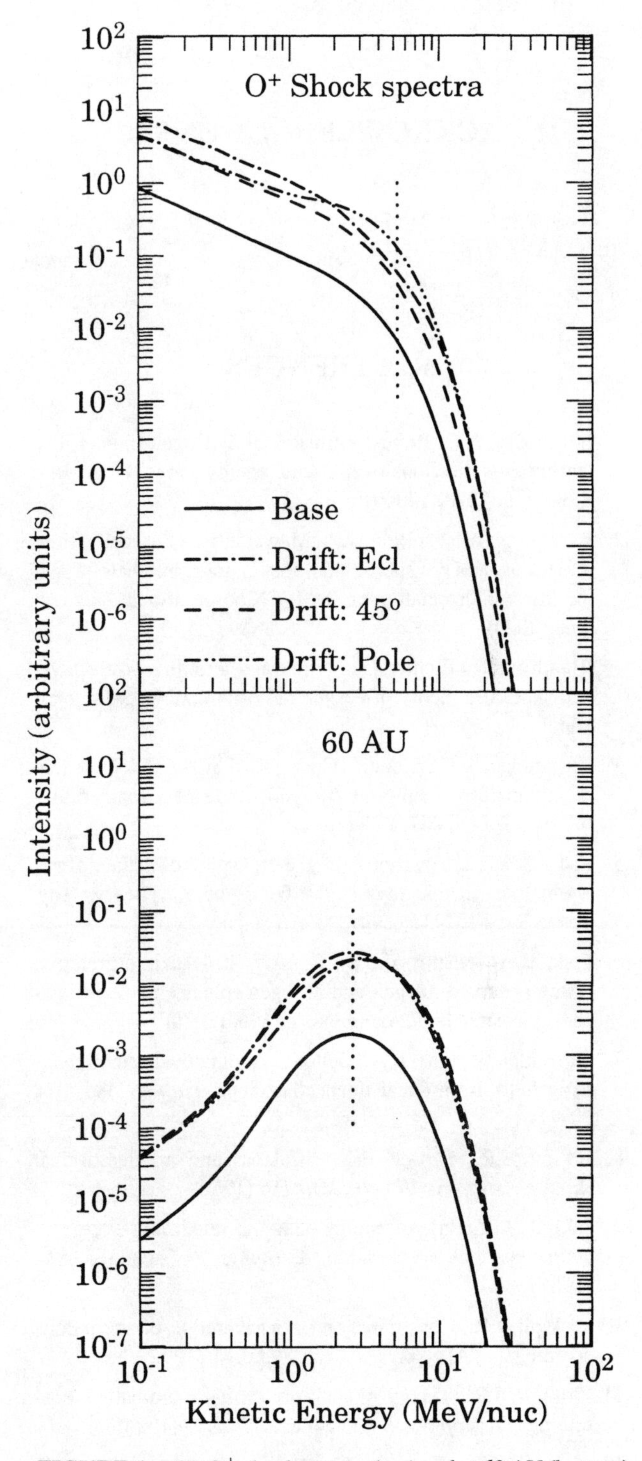

FIGURE 6. ACR O^+ shock spectra (top) and at 60 AU (bottom) comparing the no-drift baseline spectra (solid lines) with drift-modified spectra in the ecliptic (dashes), at 45° latitude (dash-dot), and at the pole (dash-dot-dot). Vertical lines (dots) show the cutoff at 5.2 MeV/nuc (top), and the spectral peaks at ~2.6 MeV/nuc. The drift spectra are offset by a factor of 10 relative the the baseline spectra.

- Multiply charged ACRs give rise to modulated spectra that are broader in energy than those consisting only of singly charged ACRs.

ACKNOWLEDGEMENT

This work was supported by NASA at Caltech (contract NAS7-918).

REFERENCES

1. Barghouty, A.F., Robust estimates of hydrogen-impact ionization cross sections over a wide energy range, *Phys. Rev. A*, **61**, nr. 052702 (2000)
2. Barghouty, A.F., Jokipii, J.R., Mewaldt, R.A., Acceleration, ionization, and transport of multiply charged anomalous cosmic rays: interpretation of SAMPEX observations, *This volume* (2000)
3. Blandford, R., Eichler, D., Particle acceleration at astrophysical shocks: a theory of cosmic ray origin, *Physics Reports*, **154**(1), 1 (1987)
4. Cummings, A.C., Stone, E.C., Webber, W.R., Evidence that the anomalous cosmic-ray component is singly ionized, *Astrophys. J. Lett*, **287**, L99 (1984)
5. Fisk, L.A., An overview of the transport of galactic and anomalous cosmic rays in the heliosphere: Theory, Adv. Space Res., **23**, 415 (1999)
6. Fisk, L.A., Kozlovsky, B., Ramaty, R., An interpretation of the observed oxygen and nitrogen enhancements in low-energy cosmic rays, *Astrophys. J.*, **190**, L35 (1974)
7. Gloeckler, G., Fisk, L.A., Geiss, J., Anomalously small magnetic field in the local interstellar cloud, *Nature*, **386**, 374 (1997)
8. Jokipii, J.R., Particle drift diffusion, and acceleration at shocks, *Astrophys. J. Lett.*, **255**, 716 (1982)
9. Jokipii, J.R., Rate of energy gain and maximum energy in diffusive shock acceleration, *Astrophys. J. Lett.*, **313**, 842 (1987)
10. Jokipii, J.R., Constraints on the acceleration of anomalous cosmic rays, *Astrophys. J. Lett.*, **393**, L41 (1992)
11. Jokipii, J.R., Theory of multiply charged anomalous cosmic rays, *Astrophys. J. Lett.*, **466**, L47 - L50 (1996)
12. Jokipii, J.R., Kóta, J., The polar heliospheric magnetic field, *Geophys. Res. Lett.*, **16**, 1 (1989)
13. Kóta, J., Jokipii, J.R., Diffusive shock acceleration in the presence of current sheets, *Astrophys. J.*, **429**, 385 (1994)
14. Klecker, B., Mewaldt, R., *et al.*, A search for minor ions in anomalous cosmic rays, *Space Sc. Rev.*, **83**, 259 (1998)
15. Mewaldt, R.A., Selesnick, R.S., Cummings, J.R., Stone, E.C., von Rosenvinge, T.T., Evidence for multiply charged anomalous cosmic rays, *Astrophys. J. Lett.*, **466**, L43 (1996)
16. Pesses, M.E., Jokipii, J.R., Eichler, D., Cosmic ray drift, shock-wave acceleration, and the anomalous component of cosmic rays, *Astrophys. J. Lett.*, **246**, L85 (1981)
17. Moraal, H., Steenberg, C.D., Basic properties of anomalous cosmic ray spectra, *Proc. 26th Int. Cosmic Ray Conf. (Salt Lake)*, **7**, 543 (1999)
18. E.N. Parker, The passage of energetic particles through interplanetary space, *Planet. Space Sci.*, **13**, 9 (1965)
19. Steenberg, C.D., Cummings, A.C., Stone, E.C, Drift calculations on the modulation of anomalous cosmic rays during the 1998 solar minimum period, *Proc. 26th Int. Cosmic Ray Conf. (Salt Lake)*, **7**, 593 (1999)
20. Steenberg, C.D., Moraal, H., Form of the anomalous cosmic ray spectrum at the solar wind termination shock, *J. Geophys. Res.*, **104**(A11), 24879 (1999)
21. Steenberg, C.D., Moraal, H., McDonald F.B., Species scaling of modulated anomalous cosmic ray spectra, *Proc. 25th Int. Cosmic Ray Conf. (Durban)*, **2**, 233 (1997)
22. Steenkamp, R., Shock acceleration as source of the anomalous component of cosmic rays in the heliosphere, *Ph.D. Thesis*, Potchefstroom University for CHE, South Africa (1995)
23. Stone, E.C., Cummings, A.C., Webber, W.R, The distance to the termination shock in 1993 and 1994 from observations of anomalous cosmic rays, *J. Geophys. Res.*, **101**(A5), 11017 (1996)

Acceleration, Transport and Fractionation of Anomalous Cosmic Rays

J. R. Jokipii

University of Arizona, Tucson, AZ, 85721

Abstract. The effects acceleration and transport on the charge and elemental composition of anomalous cosmic rays is discussed in the context of the model of acceleration at the solar-wind termination shock. Since the transport coefficients depend on the mass and charge of the particles, changes of composition are expected, both in the acceleration and the transport process. These effects are shown for different species. Special attention will be given to the production of multiply-charged ionic species from the originally singly-charged species, as a result of the acceleration at the termination shock and subsequent propagation. Good agreement is found with these observations, suggesting that the models are capturing much of the basic physics. In particular, the energy where the singly-charged anomalous cosmic rays give way to multiply-charged particles is very sharp and at very nearly the same energy for all species observed, and also in the model.

INTRODUCTION

The anomalous component of the quiet time cosmic-ray flux (hereinafter ACR) manifests itself as a bump, or enhancement in the observed flux at kinetic energies of the order of two hundred MeV, or some 10-20 Mev/nucleon for Oxygen. It was discovered in the early 1970's (1,2,3). It is composed of fluxes of helium, nitrogen, oxygen, neon, protons and low levels of carbon, and other species (for a recent review, see reference 4). The observed radial intensity gradient is positive out to the maximum distance reached by current spacecraft, indicating that this component probably originates in the interaction of the solar wind with the interstellar medium. The observed spectra of ACR oxygen, helium and protons are shown on the left in Figure 1.

The ACR were recognized early on as being special, in that their characteristics seemed quite different from the well-known galactic cosmic rays: hence the name anomalous. A number of observational and theoretical advances over the past 25 years have increased considerably our knowledge of these particles – to the extent that, perhaps, the name "anomalous" is anachronistic and misleading. This field of research has been characterized by an exemplary cooperative effort by theorists and observers to extend our knowledge. Arguably, we now know more about the ACR than the galactic cosmic rays.

Fisk, Koslovsky and Ramaty (6) suggested that the anomalous component was the result of heliospheric acceleration (by some unspecified mechanism) of freshly-ionized interstellar neutral particles which stream into the solar system. The observed behavior of the particles suggested that they were indeed, mainly singly-charged. Once ionized, the then singly-charged particles (termed pickup ions) are then swept out of the inner solar system by the solar wind and subsequently accelerated. This accounts naturally for the presence of oxygen and near-absence of carbon. The charge state of anomalous oxygen at Earth, at an energy of $\approx$ 10 MeV/Nucleon, at 1 A.U., has been observationally determined by Adams, et al, (7) to be 0.9(+0.3,-0.2), consistent with the interstellar neutral origin hypothesis.

Pesses, Jokipii and Eichler (8) pointed out that many features of the anomalous component could be explained if the acceleration of the freshly-ionized particles occurs at the termination shock of the solar wind, by the mechanism of diffusive shock acceleration. Jokipii (9) presented results from a quantitative two-dimensional numerical simulation, in which the full transport equation was solved. It was found that the basic observed features of the spectrum and spatial gradients could be explained very naturally in terms of this model. Because of the small charge and consequent high magnetic rigidity of the ACR, gradient and curvature drifts of the particles, both along the face of the shock and in the solar wind play a major role. Steenberg and Moraal (10,11) presented results from a simulation neglecting drifts and with drifts. In the past few years the suite of species seen or claimed has increased, and the first observations of a transition to multiply-charged ACR at higher energies were reported (12). The theory then showed (13) that the characteristics

CP528, *Acceleration and Transport of Energetic Particles Observed in the Heliosphere: ACE 2000 Symposium,* edited by Richard A. Mewaldt, et al.
© 2000 American Institute of Physics 1-56396-951-3/00/$17.00

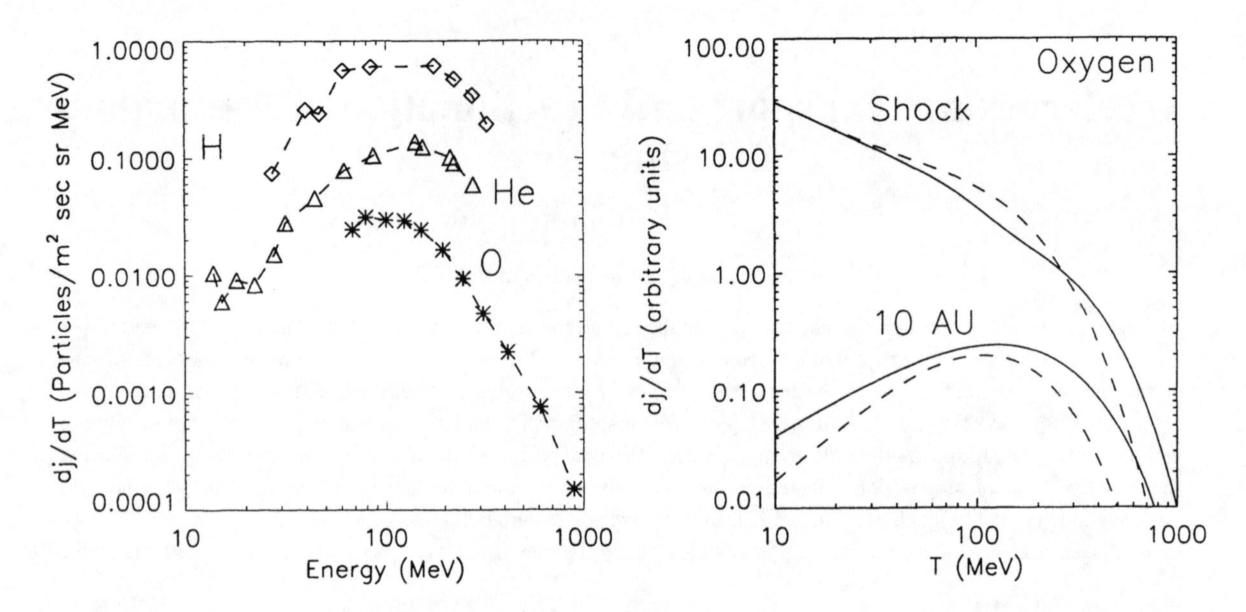

FIGURE 1. Left panel: Energy spectra of anomalous oxygen, helium and hydrogen observed near 21 A.U. in 1985, prepared from data in reference 5. Right panel: Plot of the computed energy spectra of anomalous Oxygen near the heliospheric equator, for two radii, and for the two signs of the heliospheric magnetic field. The solid lines correspond to the case where the northern heliospheric magnetic field is directed outward $A > 0$, corresponding to the present sunspot minimum. The dashed lines are the same for $A < 0$, corresponding to the 1986 sunspot minimum.

of the transition to multiply-charged ACR was a natural consequence of the models as well.

THE STAGE

The transport of charged particles in the solar wind is governed by the ambient electric and magnetic fields, because particle-particle collisions are totally negligible. Fast charged particles are subject to four distinct transport effects. Because of the magnetic field, they are convected with the fluid flow at velocity $\mathbf{V}_w$. In addition, because the magnetic field varies systematically over large scales, the curvature and gradient drifts (velocity $\mathbf{V}_d$) are coherent over large distances. The expansion or compression of the wind causes associated energy changes. Finally, there is an anisotropic random walk or spatial diffusion (diffusion tensor κ_{ij}) which is caused by scattering due to random magnetic irregularities carried with the flow, which also maintains a nearly-isotropic angular distribution in the fluid frame. The magnitude of the diffusion coefficient depends on the particle charge and mass, as well as it's velocity, so the composition of various species will in general change with energy, time and position.

The resulting superposition of coherent and random effects were combined first by Parker (14) to obtain the generally accepted transport equation, which may be written for the quasi-isotropic distribution function $f(\mathbf{r},p,t)$ of cosmic rays of momentum p at position $\mathbf{r}$ and time t:

$$\frac{\partial f}{\partial t} = \frac{\partial}{\partial x_i}\left[\kappa_{ij}\frac{\partial f}{\partial x_j}\right] - V_{w,i}\frac{\partial f}{\partial x_i} - V_{d,i}\frac{\partial f}{\partial x_i} + \frac{1}{3}\frac{V_{w,i}}{\partial x_i}\left[\frac{\partial f}{\partial \ell np}\right] + Q(\mathbf{r},p,t) \quad (1)$$

The guiding-center drift velocity is given in terms of the local magnetic field $\mathbf{B}$ and the particle charge q and speed w by $\mathbf{V}_d = (pcw/3q)\,\nabla \times (\mathbf{B}/B^2)$, where c is the speed of light. This transport equation is remarkably general, and is a good approximation if there is enough scattering by the magnetic irregularities to keep the distribution function nearly isotropic, and if the particles have random speeds substantially larger than the background fluid convection speed. In particular, since the fluid velocity need not be a continuous function of position, **all** of the standard theory of diffusive shock acceleration is contained in this equation. One may place step function changes in the definition of the flow velocity at appropriate places and ensure that the discontinuity is treated properly.

The general configuration of the inner heliosphere is well understood, and extrapolation to the termination shock probably does not introduce significant uncertainties. The configuration beyond the shock is uncertain, and

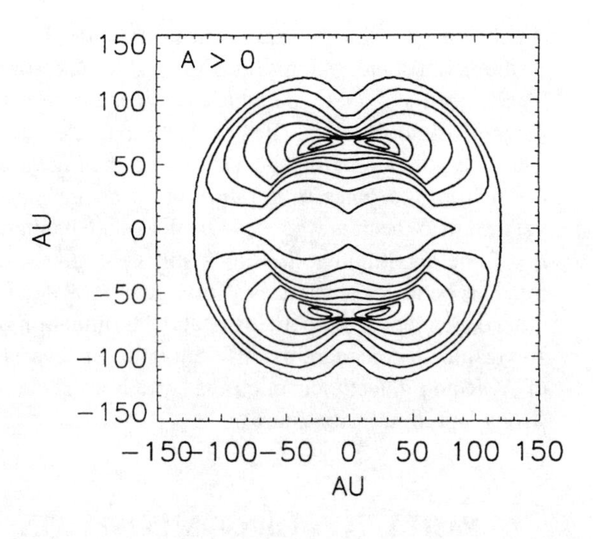

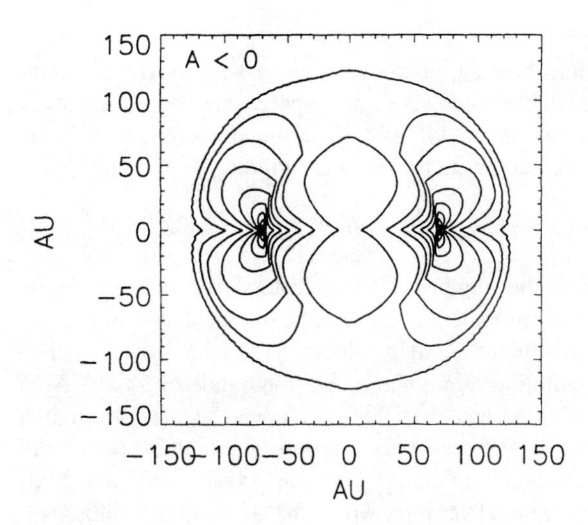

FIGURE 2. Equal intensity contours computed for 8 MeV/nucleon anomalous oxygen, as a function of position in a meridional plane, as computed for the model parameters discussed in the text. The left panel is for $A > 0$ and the right is for $A < 0$. In this case, the shock was at a radius of 70 AU and the outer boundary at 120 AU.

we adopt a simple configuration which contains the basic features expected. Fortunately, the conclusions obtained from the models, and the comparison with data, up till now, have not been very sensitive to our lack of knowledge of this region.

The solar wind velocity is taken to be radial out to a spherical termination shock at a radius R_{ts} at which it drops by a factor of r_{sh} (which is 4 for a strong shock), and then decreases as $1/r^2$ (small Mach number flow) out to an outer boundary R_b, where the energetic particles are presumed to escape. Typically, R_{ts} is taken to be some 70–100 AU and R_b some 30%–50% larger.

During several years around each solar sunspot minimum, the interplanetary magnetic field is generally organized into two hemispheres separated by a thin, roughly equatorial current sheet, across which the field reverses direction. In each hemisphere the field is generally assumed to be a classical Archimedean spiral, with the sense of the field being outward in one hemisphere and inward in the other. The field direction alternates with each 11-year sunspot cycle, so that during the 1996 sunspot minimum, the northern field was directed outward from the sun (conventionally denoted as $A > 0$), but in 1986 the northern field pointed inward ($A < 0$). This field is assumed to continue beyond the termination shock, with the spiral angle reflecting the local V_w.

There is now evidence that the **polar** magnetic field differs considerably from this spiral (15,16), so the polar field is modified in our simulations. The magnetic-field structure for the years near sunspot maximum is not simple, so the following discussion is most-relevant during the several years around sunspot minimum.

ACCELERATION

As mentioned above, the phenomenon of diffusive shock acceleration is contained within the Parker cosmic-ray transport equation (1), if one allows the fluid flow to have a compressive discontinuity corresponding to the shock wave. The acceleration can occur for any angle between the shock normal and the magnetic field and is in general due to both compression and drift. It may be demonstrated that the acceleration process at quasi-perpendicular shocks is more closely related to drift in the $\mathbf{V_w} \times \mathbf{B}$ electric field than to compressive Fermi acceleration. Jokipii (17, 18, 19) showed that if the scattering frequency is significantly less than the gyro-frequency, the energy ΔT gained by a particle having electric charge Ze, at a planar, quasi-perpendicular shock, is approximately Ze times the electrostatic potential $\Delta\phi$ gained in drifting along the shock face: $\Delta T \approx Ze\Delta\phi$. Note that the scattering must be sufficient to maintain near isotropy. For billiard-ball scattering this requires $\lambda_{\parallel}/r_g \ll w/V_w$. It follows, then, that there is in general a characteristic energy $T_c \approx Ze(\Delta\phi)_{max}$ above which the spectrum begins to decrease rapidly. Below this energy the spectrum is often a power law. Particles will "drift off" of the shock before gaining more energy. The cutoff need not be sharp, since some particles will be scattered back and forth, receiving more or less energy than T_c. In any real situation, of course, boundary conditions and other physical phenomena such as changes in flow speed, etce, occureing away from the shock will also affect the acceleration.

We may write for the solar wind (18)

$$Ze(\Delta\phi)_{max} = ZeB_r r^2 \Omega_\odot / c \approx 240Z \text{ MeV} \qquad (2)$$

where the numerical value results from using a radial magnetic field of 3.5γ at a radius of 1 AU and a solar rotational angular velocity $\Omega_\odot = 2.9 \times 10^{-6}$ (corresponding to a magnetic field magnitude of 5γ at 1 AU). Singly-charged particles accelerated at the termination shock would then have a spectrum which exhibits a decrease above an energy between 200 and 300 MeV. Multiply charged particles would, because of their higher values of Z, have a characteristic energy which is a multiple of this.

SIMULATION RESULTS

Our basic numerical model (including two spatial dimensions, heliocentric radius and polar angle) follows the acceleration of low-energy, singly-charged particles injected into the solar wind *a la* Fisk Kozlovsky and Ramaty (6), and then accelerated at the termination shock. The injection energy used is typically of the order of 100 keV, because of the limited dynamical range available in numerical solutions. The method of solution is then to follow these particles in time as they are accelerated at the shock and propagate throughout the heliosphere, until the distribution reaches a steady state. The characteristic time to approach a steady state is found to be 2–3 years or less, except for the highest-energy ACR.

The theoretical spectra from a simulation, in which the injection of low-energy particles was spatially uniform over the shock, are shown in the right panel of Figure 1. Here nominal values for the parameters were used. V_w = 400 km/sec, $\kappa_{\|} = 1.5 \times 10^{22} P^{.5} \beta$ cm^2/sec, $\kappa_\perp = .1\ \kappa_{\|}$, R_{ts} = 70 AU, and R_b = 130 AU (here P is the particle rigidity in GV, and β is the ratio of the particle speed to the speed of light). A modified polar field (15) was used. Note that the relative normalization (amplitude) of the two spectra are not significant and may be changed by small variation of the parameters. The computed energy spectra of anomalous protons, helium, neon, etc are in good agreement with the observed spectra.

Figure 2 illustrates typical contour plots of the intensity of the modeled anomalous oxygen as a function of heliocentric radius and polar angle, in a solar meridional plane. We see that the intensity increases with radius, much as does that of the galactic cosmic rays, out to the termination shock (at a radius of 60 AU). Beyond the shock, the intensity decreases out to the outer boundary of the heliosphere. Along the shock, the **maximum** intensity occurs at a latitude which shifts as the sign of the magnetic field changes. If $A > 0$, the particles drift toward the pole along the shock face and then inward and down from the poles to the current sheet, the intensity maximum is near the poles. On the other hand, when $A < 0$ the maximum at the shock shifts toward the equator. This behavior introduces a strong temporal and latitudinal dependence of the intensity at the termination shock and results in considerably different latitudinal variations of ACR and galactic cosmic rays, which do not have as strong latitudinal variations.

VARIATION OF COMPOSITION

The composition of the ACR is determined by a large number of distinct processes ranging from the initial composition of the neutrals coming into the heliosphere to the final process of acceleration at the termination shock. In between is the still poorly-understood initial acceleration of the pickup ions to energies where the Parker equation (1) is applicable. I will address here only the final process whereby the particles are accelerated to ACR energies and then propagate through the heliosphere. In this process the particles may be stripped of more electrons, changing their charge state. This will be discussed separately below.

It is apparent that the process of acceleration and transport will depend on the charge and mass. Illustrated in Figure 3 are the results of comparing the intensity of ACR Oxygen and Helium. In these calculations the injection at low energy was taken to be the same for both, so any effects of fractionation in the pickup process or pre-acceleration must be added. These give the effect of the diffusive acceleration at the termination shock and the associated spatial transport.

The computation shows that the composition varies considerably with space and with energy, and with the sign of the interplanetary magnetic field. Clearly any comparison with observations must take this into account.

THE CHARGE STATE

The charge state of anomalous cosmic-ray oxygen at energies of ≈ 10 MeV/n has been shown to be essentially unity (7). This supports the idea that the ACR are the result of local acceleration of freshly-ionized interstellar neutral atoms (pickup ions), which have entered the heliosphere because of the motion of the heliosphere through the interstellar gas (6). However, because the process of

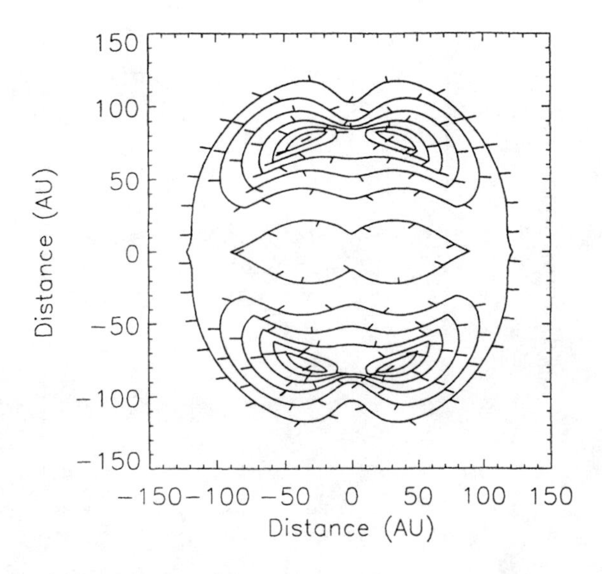

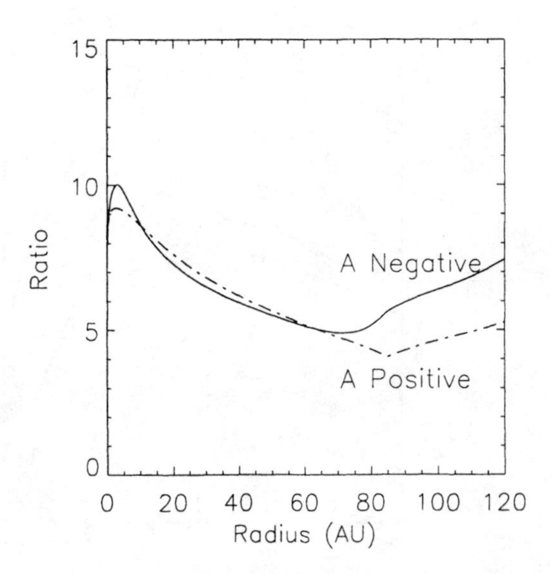

FIGURE 3. Illustration of the spatial variation of the ratio of singly-charged ACR Oxygen to Helium in a model heliosphere. The left panel contains a contour plot of the ratio for the case A>0 (northern heliospheric magnetic pointed outward, as in 1996). The tick marks point downhill (to lower values). There are 8 equally-spaced contours. The maximum and minimum values of the ratio are .21 and 0, respectfully. The right panel shows the ratio along a radius at 10 degrees latitude, for both signs of the magnetic field.

accelerating the singly-charged ions to the observed energies takes a significant amount of time, some of the ions will be ionized further to produce multiply-charged ions which can then be accelerated further and which will contribute to the ACR flux.

Adams, et al (7) used the observed average charge (0.9, +.3,-.2) of anomalous oxygen, at an energy of 10 Mev/nucleon, to argue that the ACR could not have traveled more than some .2 parsec since acceleration. They pointed out that this implied a local source for the ACR.

Jokipii (20) subsequently used this same observation to put an upper limit of some 4-6 years on the acceleration *time* of the $\approx$ 10 MeV/nucleon anomalous oxygen. He concluded that diffusive shock acceleration at the quasi-perpendicular termination shock of the solar wind could readily satisfy this constraint, but that other proposed acceleration mechanisms had difficulties. However, anomalous oxygen has now been observed up to at least 100 MeV/nuc, and acceleration to these energies will take even longer, so multiply-charged ACR must become significant at some energy, as pointed out by Mewaldt, et al, (12), who reported observations of multiply-charged ACR.

It is expected that, in general, multiply-charged particles will be accelerated faster and to higher energies than singly-charged particles. This may be understood in two ways. First, on general grounds, particles with higher charges interact more strongly with electric and magnetic fields so one expects more acceleration. Second, and more specifically, the acceleration at the termination shock is closely related to the electrostatic potential in the $\mathbf{U} \times \mathbf{B}$ electric field, which is higher for higher charges (17). Hence, at high energies, the intensity of multiply-charged ACR should be higher at the shock than that of singly-charged particles. The situation is complicated by the fact that the higher charge also cause the diffusion and drift to be slower, so that the multiply-charged anomalous cosmic rays cannot penetrate as readily to the inner heliosphere. Hence, the spectrum in the inner heliosphere is the result of several competing effects, and a global numerical simulation is needed to attempt to determine the importance of the various effects.

In an extension of previous work on singly-charged ACR (9), a probability of further ionization is introduced, and the transport and further acceleration of the newly created, multiply-charged ions are followed in exactly the same manner as are the singly-charged ions. The multiply-charged ions are accelerated more rapidly at the shock, but their intensity is decreased more in their transport into the inner heliosphere. The simulation (13) tracks these competing effects.

For the analysis of multiply-charged species, our original model was extended to keep track of more than one charge species. The major difference between the various species in the model is that it is assumed that the injected particles at the lowest energy considered (typically of the order of 100 keV) are only singly charged. The singly-charged particles, as they evolve in energy and

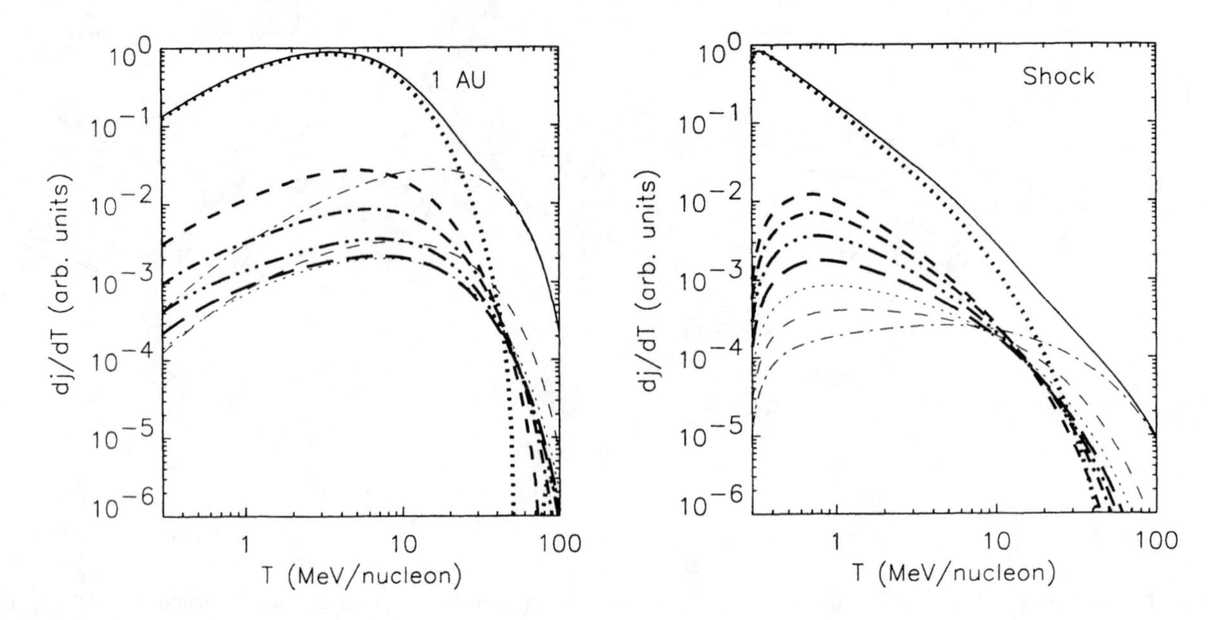

FIGURE 4. The computed spectra (13) of the 8 charge states of anomalous oxygen, and their sum, as a function of energy, at two different radii, near the heliospheric equator, for the parameters given in the text. The left curves refer to 1 AU and the right to the termination shock. The individual charge states from 1 to 8 and the sum are distinguished in each plot by different line styles and thicknesses. The lines are quite ordered in the right plot and are readily identified there. The 1 AU curves are more complex, but the line styles are consistent with the curves at the shock. In the right plot, the curve which is highest at low energy ($\approx$ 1 Mev) is charge state 1, the second highest is charge state 2. The highest, solid, curve is the sum of all 8 charge states.

location, have in addition a loss probability due to further ionization. The higher charge states therefore have a distributed, energy and time-dependent source which is due to the possibility that the ionization of smaller charge states produce higher-charge particles at the given rate. All charge states except the highest, fully stripped, state, also have a loss rate due to ionization. In addition to the loss and gain due to ionization, the spatial transport coefficients and drift velocity of the higher charge states are reduced to reflect their higher charge.

Some results of this model simulation are illustrated in Figure 4. The left panel shows the spectra of the 8 charge states of Oxygen and their sum at 1 AU and the right panel the same at the termination shock. Note that the overall normalization is arbitrary, reflecting the unknown normalization of the injected particles at charge 1. However, the *relative* normalizations of the different spectra are computed by the simulation. At $T \approx 20$ MeV/n, the sum spectrum shows a characteristic bump, reflecting the transition to a dominance of the higher charge states at higher energies. From this it is clear that the higher charge states begin to be important at some 20-30 MeV/n. Note the abrupt transition from essentially pure O^+ below some 20 MeV/n to essentially all multiply charged oxygen beyond some 40 MeV/n. In the present simulations, this abrupt change in the charge state occurs at a characteristic energy which is not very sensitive to the cross-section used, but does depend somewhat more on the transport parameters, the wind velocity and the distance to the termination shock.

It is interesting that the energy of transition corresponds to a total energy of about 300 MeV, which is close to the potential energy change from pole to equator of the heliosphere. As has been shown elsewhere (18, 19), this corresponds to a characteristic energy for acceleration of singly-charged ions at a quasi-perpendicular termination shock. It was suggested that the basic acceleration process can accelerate singly-charged ACR to approximately this energy quite readily, but that there is a rapid decrease in the singly-charged ACR beyond this energy. However, multiply-charged particles can pick up $\approx Z$ times this potential (18), and hence become dominant at higher energies. This effect can be seen in Figure 1 where, near the shock, the progressively higher charge states extend to progressively higher energies. This appears to be the major effect responsible for the rapid transition. The very high intensity of fully-stripped oxygen at energies near 100 MeV/nucleon in the 1 AU spectra (more than 10 times as intense as charge 7) is probably a consequence of the crude cross sections used in this work.

Since the earlier work, data concerning other species has become available and the work of Nasser Barghouty

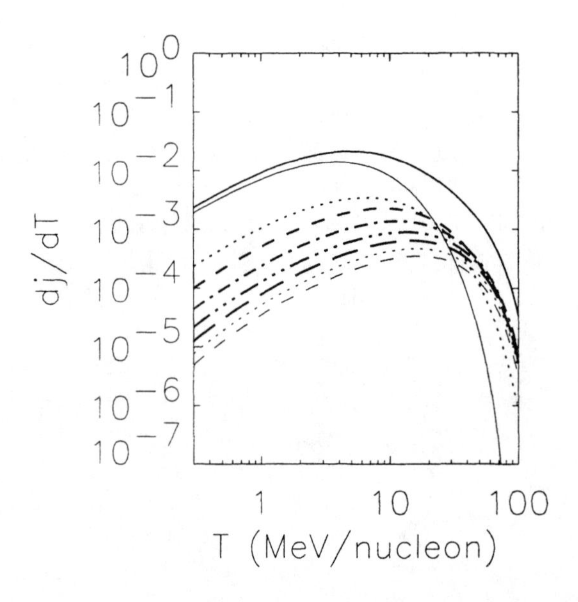

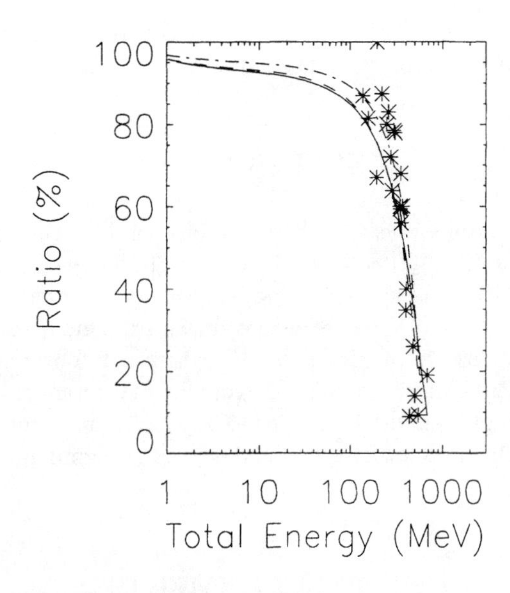

FIGURE 5. Illustration of results obtained with the new cross-sections. The left panel shows the spectra of the 8 species for 1 AU, to be compared with the results in Figure 4 for the old cross-sections. The right panel shows the ratio of singly charged to total as a function of total energy for the three species Neon, Oxygen and Nitrogen, for which data is available. The stars show the data points for the three species overlaid, showing how well the data are organized by plotting as a functio of total energy.

has led to more accurate cross sections for the stripping of electrons. These are discussed in Barghouty, et al, in this book. I present some of the results here, to place them in the context of the above.

The new cross-sections are available for Neon, Nitrogen and Carbon, as well as Oxygen. They differ considerably from those used previously (13), which were based on extrapolations from measurements, and are expected to be considerably more accurate. They are based on computations.

Shown in Figure 5 are sample results from models run with the new cross-sections. Comparison of the left panel with the left panel of Figure 4, carried out with th old, more approximate cross-sections, shows considerable difference in the levels of the higher-charge components between the old and the improved cross-sections. The panel on the right in Figure 5 shows the computed ratio of the singly-charged species to the sum of charges 1, 2 and 3 (three lines), for the three species Neon, Oxygen and Nitrogen (top to bottom) , together with the data (stars) for all three species. The transition between singly and multiply-charged species clearly occurs at nearly the same total kinetic energy for all three elements, and is well-modelled by the simulation. Plotting the points as a function of energy per nucleon is clearly not as good.

In these computations the wind and transport parameters used are: solar wind velocity: 350 (equator) - 700 km/s (poles), Termination shock radius: 90 AU ,Termination shock strength: 2.5, injection Energy: <100 keV/nucl, neutral hydrogen density: $0.1cm^{-3}$, heliospheric boundary: $150AU$, A > 0, $\kappa_{\|} = 1.5 \times 10^2 2R{\cdot}5\beta|B_{Earth}|/|\vec{B}(r,\theta)|$, $\kappa_{\perp} = .03\times\ \kappa_{\|}$. The rotation rate of the sun was taken to vary with latitude as the observed differential rotation:

$$\Omega_{\odot} = 2.9\times10^{-6} + [.464 sin^2(\lambda) + .328 sin^4(\lambda)]\times10^{-6} \quad (3)$$

where λ is the heliographic latitude.

The excellent agreement between the data and the observations suggests that much of the essential physics is contained in the models. It is remarkable that the transition from singly to multiply charged ACR is so sharp and so well in agreement with observation. We have carried out a comparison of the present model results with those using the old, less accurate cross-sections, and, although the spectra differ considerably, the transition is very nearly independent of the cross-sections. Indeed, when plotted as in the right panel of Figure 5, the curves for oxygen are virtually indistinguishable. From this, we conclude that the transition energy depends almost entirely on the cutoff of singly-charged particles and the further acceleration of multiply-charged particles, perhaps for the reasons discussed in equation (2) and related discussion. On the other hand, the cutoff is also due in part to the transport parameters and the loss out of the system through the interface to the galaxy.

DISCUSSION

The results presented here demonstrate the remarkable variety of effects which result from the acceleration of different species at the heliospheric termination shock. Many of the conclusions drawn from the model calculations are in very good agreement with observations. Particularly striking is the rapidity of the transition from singly to multiply charged ACR which is apparently caused by transport effects at and beyond the termination shock.

ACKNOWLEDGEMENTS.

This work was supported, in part by NASA under grants NAG5-7793 and NAG5-6620, and by the National Science Foundation under Grant ATM 9616547.

REFERENCES

1. Garcia-Munoz, M. , Mason, G. M. and Simpson, J. A. , *Astrophys. J. (Letters), 182*, L81 (1973)
2. Hovestadt, D., Vollmer, O. , Gloeckler, G. and Fan, C. Y., *Phys. Rev. Letters, 31*, 650 (1973)
3. McDonald, F. B., Teegarden, B. J., Trainor, J. and Webber, W. B., Astrophys. J. (Letters), 187, L105 (1974)
4. Klecker, B., *Space. Sci. Rev.*, 72, 419 (1995)
5. Cummings, A. C., and Stone, E. C., Energy spectra of anomalous cosmic-ray oxygen during 1977-1987m *Proc. 20th Int. Cosmic Ray Conf.*, Moscow, 1987 , 3, 421.
6. Fisk, L. A., Kozlovsky, B. and Ramaty, R., *Astrophys. J. Lett.*, 190, L35 (1974)
7. Adams, J. H., Jr., *et al.*, *Astrophys. J. Lett.*, *375*, L45 (1991)
8. Pesses, M. E., Jokipii, J. R. and Eichler, D., *Astrophys. J. Lett.*, 246, L85 (1981)
9. Jokipii, J. R., *J. Geophys. Res.*, 91, 2929 (1986)
10. Steenberg, C. D., and Moraal, H., *Astrophys. J, 463*, 776 (1996)
11. Steenberg, C. D., and Moraal, H., *J. Geophys. Res, 104*, 24879 (1999)
12. Mewaldt, R. A., et al, *Astrophys. J.* 466, L43 (1996)
13. Jokipii, J. R., *Astrophys. J.*, 466, L47 (1996)
14. Parker, E. N. , Planet. Space Sci., 13, 9 (1965)
15. Jokipii, J. R., and Kóta, J., *Geophys. Res. Lett., 16*, 1 (1989)
16. Jokipii, J. R., Kóta, J. ,Giacalone, J. , Horbury, T. S. and Smith, E. J., *Geophys. Res. Lett.*, *22*, 3385 (1995)
17. Jokipii, J. R., *Astrophys. J.*, 255, 716 (1982)
18. Jokipii, J. R., *Physics of the Outer Heliosphere*, COSPAR Colloquia Series Volume 1, 1990, 169-178.
19. Jokipii, J. R., *Astrophys. J.*, 313, 842 (1987)
20. Jokipii, J. R., *Astrophys. J. Lett*, 393, L41 (1992)

The Injection Problem for Anomalous Cosmic Rays

G.P. Zank, W.K.M. Rice, J.A. le Roux, J. Y. Lu and H.R. Müller

Bartol Research Institute
The University of Delaware
Newark, DE 19716

Abstract. A critical examination of the injection problem for anomalous cosmic rays is presented together with a discussion of possible mechanisms.

1. INTRODUCTION

That interstellar pickup ions are related to the anomalous cosmic ray (ACR) component is now well established [Fisk et al., 1974]. However, pickup H^+ has typical energies of 1-3 keV, sufficiently energetic to form a suprathermal, energetically important plasma component in the outer heliosphere, but well below ACR energies which extend to several 100 MeV/nuc. How some fraction of the interstellar pickup ion population becomes preferentially energized up to such high energies remains a central question in the physics of cosmic ray acceleration. It is generally accepted that for sufficiently energetic pickup ions, first-order Fermi acceleration, otherwise known as diffusive shock acceleration, at the heliospheric termination shock can account for the accelerated ACR spectrum [Pesses et al., 1981; Jokipii, 1986]. However, the precise mechanism by which some pickup ions are selected and possibly pre-energized up to energies sufficiently large that they can be Fermi accelerated remains a puzzle. This is the so-called "injection problem" for ACRs.

2. OBSERVATIONS

Although we shall concentrate on observations related to pickup ions and ACRs below, a classic paper by Gosling et al. [1981] allows an important point to be made. By combining data from two instruments, Gosling et al. [1981] were able to determine the ion velocity distribution function from solar wind energies (~1 keV) up to 1.6 MeV during the post-shock phase of an energetic storm particle event. The basic conclusions from this study were (1) that the energized ions emerged smoothly from the solar wind ion population and are nearly isotropic in the solar wind frame; (2) that the energized ion distribution possessed initially a power law in energy ($E^{-2.4}$) in the solar wind frame up to ~40 keV, but no single power law exponent could describe the data for higher energies; and (3) that the injection efficiency was such that ~1% of the solar wind population is accelerated to suprathermal energies. Gosling et al. further noted that such events were common and that 40 keV post-shock ions are accelerated directly out of the solar wind population.

Consider now the energization of interstellar pickup ions at an interplanetary shock. Observations of interstellar pickup ions accelerated by a forward shock have been made by Ulysses at 4.5 AU [Gloeckler et al., 1994]. The observed shock was weakly perpendicular with a compression ratio of ~2.4. Gloeckler et al. [1994] defined an "injection efficiency" as the ratio of the number of ions in the tail of the interstellar pickup ion distribution to the total number of pickup ions. Gloeckler et al. find (i) that injection efficiencies for pickup ions are much higher than those of solar wind ions (although the efficiency for solar wind protons is similar to that found by Gosling et al., [1981]); (ii) that injection efficiency for interstellar pickup ions scales inversely with mass e.g., pickup H^+ is injected more efficiently than He^+; (iii) that both pickup H^+ and He^+ have identical power law spectral slopes above the pickup ion cut-off speed, and the power law indices are considerably harder than expected of diffusive shock acceleration. Most importantly, like Gosling et al. [1981], the accelerated pickup ions emerge directly and smoothly out of the "thermal pickup" ion distribution. Thus, although similar in many respects to the observations for solar wind protons, low energy pickup ions accelerated at an interplanetary shock possess important differences.

These differences have been highlighted by a further set of intriguing observations reported by Fränz et al. [1999]. These authors investigate the elemental composition of 0.1-4.0 MeV/nuc. ions in corotating

CP528, *Acceleration and Transport of Energetic Particles Observed in the Heliosphere: ACE 2000 Symposium*,
edited by Richard A. Mewaldt, et al.
© 2000 American Institute of Physics 1-56396-951-3/00/$17.00

interaction regions (CIRs) from 1 to 5 AU. By comparing the abundances of energetic minor ions, Fränz et al. [1999] find that, with the exception of He (and Ne at 1 AU), energetic particle abundances are in rough agreement with the respective solar wind thermal abundances. He is overabundant by a factor of 2.5 at the CIR reverse shock at distances of 4-5 AU, but no He enhancement is seen at the forward shock. Fränz et al. [1999] suggest that the energized He ions are in fact accelerated pickup ions since neutral interstellar He penetrates to some 0.3 AU. Such an interpretation implies that the special character of the pickup ion distribution (shell-like, suprathermal) allows it to be more easily injected and subsequently accelerated at interplanetary shocks than their thermal solar wind counterparts.

In summary, the above three sets of observations suggest that (i) accelerated ions emerge directly from the thermal pool; (ii) the injection efficiency appears to depend on the nature of the underlying particle distribution function, and (iii) the observed pickup ion injection efficiency appears to correlate inversely with mass.

Consider now observations related to the injection of ACRs at the termination shock. Cummings and Stone [1996] used ACR energy spectra measured by Voyager 1 and 2 during 1994 to determine an injection efficiency for ACRs, now defined as the flux of ACRs relative to the flux of pickup ions at an assumed termination shock location. The observed modulated ACR spectra for various ACR species are used by Cummings and Stone [1996] to infer a source spectrum at the termination shock, from which they can infer the injection efficiency. This approach depends on several uncertainties in the modeling, but, nonetheless, a basic conclusion is that, even if the estimated injection efficiencies are possibly inaccurate, heavier ions are injected preferentially. Furthermore, besides the ACR "injection efficiencies" of Cummings and Stone [1996] being in the opposite sense of those presented by Gloeckler et al. [1994], the "injection efficiencies" of the latter authors are considerably higher than those found for ACRs (recognizing the distinct sense of the terms).

Finally, a further important result related to the injection and acceleration of ACRs at the termination shock is the diffusive shock acceleration time scale for ions. Jokipii [1992] argued that to ensure that most ACRs remain singly ionized, the characteristic acceleration time at the termination shock was limited to ~4.6 years for 10 MeV/nuc. ions. However, the diffusive acceleration time for a parallel shock far exceeds the 4.6 year acceleration time constraint, even in the Bohm limit. Instead, only by assuming weak scattering in the hard-sphere scattering model at a quasi-perpendicular termination shock could Jokipii [1992] reduce the diffusive shock acceleration time scale to meet the acceleration time constraint. The requirement that scattering be weak (in the hard sphere sense) requires that pickup ions must already be energetic for them to be Fermi accelerated. If η_c is a measure of the scattering strength (defined precisely below), so that $\eta_c >> 1$ corresponds to weak scattering, then for a particle to be scattered multiple times as a field line convects through a shock, the particle velocity must satisfy $v >> V_{sh}\eta_c$, where V_{sh} is the shock speed [Jokipii, 1987; Webb et al., 1995]. For the parameters discussed by Jokipii [1992], this requires that pickup ions have initial energies close to ~1 MeV [Zank et al., 1996]. As a result, this has led to suggestions that some pre-energization of pickup ions is necessary [Kucharek and Scholer, 1995; Zank et al., 1996; Lee et al., 1996; Giacalone et al., 1997]. It has also been suggested that very high levels of magnetic turbulence are present at the termination shock [Giacalone et al., 1994; Ellison et al., 1999], but, for hard sphere scattering, this would contradict the requirement that $\eta_c >> 1$.

A successful resolution of the "injection problem" at the quasi-perpendicular termination shock must relate the above sets of observations consistently.

3. INJECTION MECHANISMS

From the time scale argument described above, pickup ions appear to require some pre-energization in order to boost them to energies sufficiently large that they can then be accelerated by a diffusive shock acceleration mechanism.

Four ideas have been advanced to address the injection problem. The first, due to Lee et al. [1996] and Zank et al. [1996], is the Multiply Reflected Ion (MRI) acceleration or shock surfing mechanism. MRI acceleration can explain the observations of Gloeckler et al. [1994] and rapidly provide a pre-energized pickup ion component for a second-stage diffusive shock acceleration process. However, it appears to provide the wrong dependence of ACR injection efficiency with ion mass in the sense described by Cummings and Stone [1996].

A second approach is to use interplanetary shocks to accelerate ions up to energies appropriate to diffusive acceleration at the termination shock [Giacalone et al., 1997]. However, an injection mechanism is still needed presumably for an interplanetary shock, and it would be surprising indeed were the largest shock in the heliosphere incapable of injecting particles when interplanetary shocks are expected to do so. Giacalone et al. [1997] suggest that interplanetary shocks might have a reduced injection energy threshold since they propagate, but, of course, the termination shock is in constant motion due to variations in solar wind ram pressure, both on small scales due to shocks [Story and Zank, 1997] and on large scales associated with solar cycle variation [Zank, 1999]. The interplanetary shock injection mechanism described by Giacalone et al. [1997] cannot address the Gloeckler et al. [1994] and Fränz et al. [1999] observations directly. It is also unclear whether interplanetary shocks do in fact inject and accelerate ions throughout the heliosphere [Rice et al., 2000].

A third approach is the stochastic acceleration of interstellar pickup ions by solar wind turbulence [Isenberg, 1987; Bogdan et al., 1991; Chalov et al., 1995; le Roux and Ptuskin, 1998]. Chalov et al. [1997] also included energization by pickup ion interaction with multiple weak shocks ("supersonic turbulence").

Finally, Ellison et al. [1999] suggest that all ions, regardless of energy, are diffusive and that their transport is described adequately by the hard sphere scattering model. All ions therefore experience diffusive shock acceleration, regardless of energy and one does not then have an injection problem. The basic assumptions on which the Ellison et al. [1999] Monte Carlo model are based are very difficult to accept for low energy ions and, in particular, the transport of low energy ions is likely to be dominated by effects other than simple hard sphere scattering (magnetic field line wandering, trapping, shock structure, etc., and it should be recognized that the hard sphere diffusion coefficients were derived under the assumption that particle velocities satisfy $v >> u$, the solar wind speed). Finally, the fluxes derived by Ellison et al. [1999] for ACR He^+ and O^+ are at least a factor of ~4 too small.

3.1 Stochastic pre-acceleration

In a steady, spherically symmetric expanding solar wind, an isotropic pickup ion distribution evolves as

$$\frac{\partial f}{\partial t}+u\frac{\partial f}{\partial r}-\frac{2u}{r}\frac{v}{3}\frac{\partial f}{\partial v}=\frac{1}{v^2}\frac{\partial}{\partial v}\left(v^2 D_{pp}\frac{\partial f}{\partial v}\right)+S, \quad (1.1)$$

where $f(r,v,t)$ is the isotropic pickup ion distribution in the solar wind at heliocentric distance r with velocity v at time t. S denotes the source term and D_{pp} the momentum diffusion coefficient for the stochastic acceleration of pickup ions.

Equation (1.1) in the absence of D_{pp} yields the well-known Vasyliunas and Siscoe [1976] solution for the radial evolution and transport of pickup ions in the expanding solar wind. The characteristic sharp cutoff in the pickup ion distribution (in the solar wind rest frame) observed in the pickup ion distribution [Gloeckler et al., 1993] indicates that adiabatic cooling dominates energy diffusion. Nonetheless, Isenberg [1987] has generalized the Vasyliunas and Siscoe [1976] solution by retaining the energy diffusion term D_{pp} in (1.1). The energy diffusion coefficient is derived from the quasi-linear theory of resonant wave-particle interactions. Subject to the assumptions that (i) the wave field is unpolarized; (ii) has zero cross-helicity; (iii) the turbulence spectrum has a power law form ($k^{-\gamma}$, k the wavenumber); (iv) the wave power decays with heliocentric distance as r^{-3} (i.e., WKB theory), and finally, (v) the IMF is purely azimuthal, the energy diffusion term in (1.1) may be expressed as $D_{pp} = Cv^{\gamma-1}r^{-1}$ [Isenberg, 1987]. Remarkably, Isenberg was able to solve (1.1) with D_{pp} above. Initially, ($r \sim 1$ AU) the distribution remains close to the shell distribution. With the adiabatic cooling of the pickup ions with increasing heliocentric distance, the characteristic "flat" distribution at energies lower than $v/u = 1$ develops. A sharp transition to higher energies results from the energy diffusion term in (1.1) and the slope and maximum energy are determined by the assumed wave intensity.

Both Bogdan et al. [1991] and le Roux and Ptuskin [1998] have extended the original calculation of Isenberg considerably by including explicitly the effects of wave growth and damping driven by the pickup ions themselves. We follow the approach of the latter authors. le Roux and Ptuskin assume that the spectral wave energy density W_k satisfies a WKB-like transport equation in wave number space and wave-wave interactions are described via diffusion in wave number space. Pickup ions both damp and act as a source of wave energy density. The pickup ions and waves are coupled through momentum diffusion and wave damping. le Roux and Ptuskin [1998] find that (i) pickup H is responsible for wave damping at large k, which in turn suppresses the acceleration of H; and (ii) pickup He and O are unaffected by wave damping. This conspires to produce an under abundance in H relative to the heavier ions, which provides a possible explanation for the observed under abundance in ACR H^+. However, the acceleration of pickup ions is rather weak with the particles reaching maximum energies of about 40 keV/nuc. in the solar wind frame.

3.2 Acceleration at interplanetary shocks

Voyager 2 observations in the outer heliosphere (Lazarus et al. [1999]) show significant delays between the arrival of interplanetary shock fronts and the peaks in the 0.52 – 1.45 MeV energetic particle fluxes. At 5 AU the energetic particle peaks and the shock fronts are virtually coincident. At 47 AU, however, the peaks in the energetic particle flux lag the shock fronts by about 7 days. Rice et al. [2000] modeled this by assuming that the injection of particles into the diffusive shock acceleration process stopped once the shock compression ratio dropped below a certain value (< 1.5). At 5 AU the shock is strong and particle injection occurs. By 47 AU, the shock is much weaker and the peaks in the energetic particle flux lag the shock front by about 2 AU. With a solar wind speed of 400 km s^{-1}, this corresponds to a delay of about 8 days.

These results suggest that some form of injection exists at interplanetary shocks. If not (i.e. if particles are pre-accelerated by some mechanism that is independent of shock strength) then diffusive shock acceleration should continue and the energetic particle peaks should remain coincident with the shock fronts. The observations and results of Rice et al. [2000] suggest that injection efficiency depends on shock strength and that a shock may become too weak to inject particles efficiently into the acceleration process.

3.3 Pre-energization by interplanetary shocks

Giacalone et al. [1997] suggest that ACRs are produced by the acceleration of ions at the termination shock after pre-energization at interplanetary shock waves. They considered energetic particles at 10 AU, normalized using Voyager 2 data from ~ 40 AU, as a source spectrum. Using a transport model that included drifts, convection, energy losses through expansion, and diffusion, Giacalone et al.showed that the modulated ACR spectra measured by Voyager 1,2 could be explained by the acceleration of these particles at the termination shock. Assuming an injection energy of 50 keV, the peak energies in their modulated ACR spectra show the correct radial dependence and the fluxes are consistent with those observed. As suggested in Section 2 above, however, the injection energy may be considerably higher than 50 keV.

Rice et al. [2000 – in preparation] use an analytic model to further study this possibility. They use a source function similar to that used by Giacalone et al. [1997], normalized to the energetic particle fluxes measured by Voyager 2 at ~ 40 AU, and assume [Decker et al., 1992] an r^{-2} dependence to obtain the energetic particle flux at a termination shock located at 80 AU. During solar quiet periods shock waves are generally confined to within 30° of the solar equator, and hence the average flux over the entire termination shock should be ~ 1/2 the flux determined from Voyager 2 measurements. The resulting spectrum provides the energetic particle source at the termination shock. The steady state convection-diffusion equation was then used with various injection energies to calculate the accelerated ACR spectra. With a diffusion coefficient model used by Cummings and Stone [1996], the accelerated ACR spectra were modulated back to 57 AU. Figure 1 shows the source spectrum (thick solid line) at 80 AU, the accelerated spectra for injection energies of 50 keV (thin solid line) and 500 keV (dashed line), and the resulting modulated ACR fluxes at 57 AU. For an injection energy of 50 keV, the ACR flux at 57 AU is very similar to that measured by the Voyager satellites and consequently similar to the result obtained by Giacalone et al. [1997]. With an injection energy of 500 keV the modulated ACR flux at 57 AU is ~ 5-10 times too small.

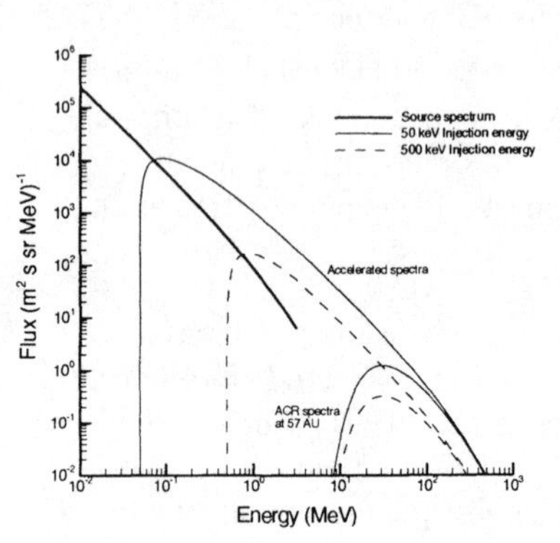

Figure 1. Source spectrum (thick solid line) at 80 AU together with the accelerated spectra (50 keV and 500 keV injection energies) and the modulated ACR spectra at 57 AU.

The suggestion by Ellison et al. [1999] that pickup ions can be directly accelerated at the termination shock would imply a very low injection energy. Since an injection energy of 50 keV seems to produce modulated ACR fluxes similar to those observed, an even lower injection energy should produce modulated ACR fluxes that are greater than observed. An injection energy of ~ 50 keV, would suggest that all ACRs could be produced by particles pre-energized at interplanetary shocks. An injection energy significantly greater than 50 keV, would imply that ACRs are produced by a combination of pre-accelerated particles and particles injected at the termination shock itself.

3.4 MRI acceleration

A fairly well developed theory exists for high plasma beta $\beta_p \sim O(1)$ perpendicular shocks [e.g., Leroy, 1983], and a self-sustaining mechanism of ion reflection at the electrostatic cross shock potential is thought to be the primary dissipation mechanism. Zank et al. [1996] and Lee et al. [1996] observed that part of the pickup ion distribution function in the shock frame has a very small normal velocity component at the shock interface, which prevents their overcoming the electrostatic cross shock potential ϕ. From the shock potential [Leroy, 1983], one can estimate the fraction of incident solar wind or pick-up ions that are reflected. For a pickup ion shell ahead of the shock, the fraction of the distribution R_{ref} that is incapable of surmounting the cross shock potential barrier is found to be [Zank et al., 1996]

$$R_{ref} = \left[\frac{Zm}{M} \frac{\eta}{2M_{A1}^2}(r-1) \right]^{1/2}, \qquad (1.2)$$

where m refers to the proton mass, M and Z to the mass and charge of the particle of interest (pick-up H^+, He^+, etc.) and the parameter η has been introduced to approximate the contribution to ϕ of the deflected bulk velocity u_y which results from ion reflection at the shock ramp. These reflected ions are capable of being accelerated to large energies. Several points can then be inferred if we interpret reflection efficiency as injection efficiency. (i) Heavier pickup ion species, i.e., with $M > m$, are less efficiently injected, and (ii) injection efficiency increases with increasing particle charge state. Thus, pick-up H^+ should be injected twice as efficiently as pick-up He^+. This appears to be what is observed by the Ulysses spacecraft at a forward shock [Gloeckler et al., 1994].

Many of the reflected pickup ions can experience multiple reflections before being transmitted downstream. The downstream or transmitted pickup ion distribution may be computed [Zank et al., 1996], resulting in a power law in energy that is extremely hard. The spectra obtained from MRI acceleration are very different from the spectra expected of a first-order Fermi shock acceleration mechanism. For Fermi acceleration, the non-relativistic energy spectrum is $\propto v^{-q}$, where $q = 3r/(r-1)$. The hardest spectra, $q = 4$ are therefore associated with the strongest shocks whereas, even for weak supercritical shocks, MRI acceleration yields spectra much flatter (harder) than this.

Several effects conspire to prevent a reflected pickup ion gaining unlimited energy from the motional electric field. The most important is that, as the reflected pickup ions acquire a large velocity v_y in the transverse direction associated with the motional electric field, the Lorentz force must eventually exceed the force exerted by the electrostatic potential so allowing the reflected ion to escape downstream. Evidently, the gradient of the electrostatic shock potential is crucial in determining the maximum pickup ion energy gain from MRI acceleration. The balancing of the particle Lorentz force against the electrostatic potential gradient shows [Zank et al., 1996; Lee et al., 1996] that the maximum energy gain is proportional to the ratio of an ion gyroradius (whose velocity is that of the solar wind) to the smallest characteristic electrostatic shock potential length scale. v_y can become very large as the shock ramp thickness $L_{ramp} \to 0$. If the length scale L_{ramp} is that of the thermal solar wind ion gyroradius, the initially very low velocity pickup ions will be accelerated up to no more than the ambient solar wind speed. However, our current understanding of the micro-structure of quasi-perpendicular shocks is that fine structure in the shock potential can be on the order of electron inertial scales, so yielding pickup ion energies up to 0.5 MeV at even weak interplanetary shocks.

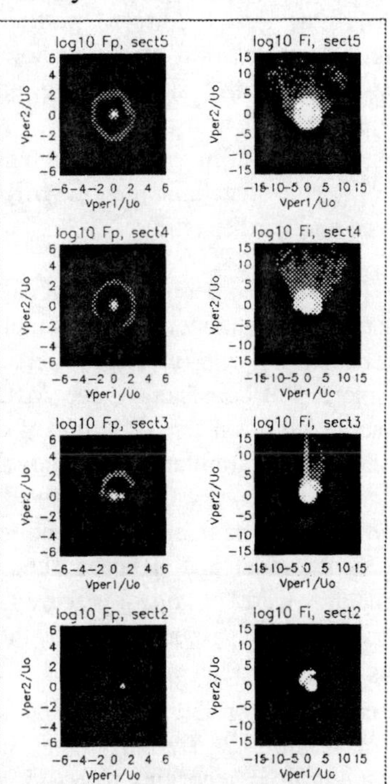

Figure 2. The projected velocity space distribution of protons (left column) and pickup H^+ for spatial sections ranging from upstream to downstream. (Lipatov and Zank, 1999)

Evidently, for the MRI or shock surfing mechanism to be effective, it is important that sharp gradients exist within the shock, either the ramp itself or other fine scale structure. From the theoretical model of Kennel and Sagdeev [1967], it is expected that ramp thickness scales with the electron inertial length. Observations of the Earth's bow shock provide such examples [Russell and Greenstadt, 1979; Scudder et al., 1986; Balikhin et al., 1995; Newbury and Russell, 1996] and some support comes from simulations [Tokar et al., 1986]. Furthermore, observations by Newbury et al. [1998] of some 20 supercritical, quasi-perpendicular bow shock crossings show large amplitude fine scale structure in the ramp itself. The spatial scale of the sub-ramp fluctuations is much smaller than the ion inertial length scale, corresponding instead to electron inertial scales. The importance of internal sub-structure was emphasized by Zilbersher and Gedalin [1997] who included micro-structure in MRI acceleration simulations to obtain large energies at a shock which scaled with the ion inertial length.

Lipatov et al. [1998] used direct particle simulations to explore the acceleration of pickup ions at collisionless quasi-perpendicular shocks with an assumed fixed profile for the transition layer. In particular, they found that, even with turbulent magnetic fields, the basic mechanism of MRI acceleration is preserved and hard spectra result.

Lipatov and Zank [1999] used a one-dimensional, (1+2/2)D hybrid kinetic electromagnetic code which included anomalous resistivity and electron inertia terms to investigate more self-consistently the MRI mechanism and its backreaction on the structure of a quasi-perpendicular shock. The simulation was novel in that it allowed the ramp to be very finely resolved. The basic tenets of the MRI mechanism were found to hold in the self-consistent model and detailed shock and phase space structure were obtained. Figure 2 shows the velocity components perpendicular to the magnetic field, $v_{\perp 2}$ vs $v_{\perp 1}$, for the protons and H^+ pickup ions at different locations relative to the shock ramp. The panels of Figure 2 are arranged in ascending order from the bottom according to position as follows: far upstream (bottom panel), on the shock front, just downstream of the shock, and, the top panel, far downstream. The left column illustrates the projected proton distribution at various distances from the shock ramp. The proton distribution function has a supersonic core ahead of the ramp and a subsonic core downstream of the shock front. Before the ramp, we see reflected protons while downstream, the transmitted protons form a halo due to phase mixing. The right column shows the projected pickup ion distribution. The bottom right panel illustrates a very typical distribution which results from ion reflection at a perpendicular shock. The second panel from the bottom shows the pickup ion distribution at the shock ramp and a strong transverse acceleration of pickup ions along the shock front is evident with the formation of an extended "tongue" along $v_{\perp 2}$. Finally, phase mixing occurs downstream. The energy spectrum is very similar to that found by Zank et al. [1996].

3.5 A synthesis

Although pickup ions are energized by the MRI mechanism, not all the MRI accelerated ions are sufficiently energetic to be further accelerated by a second-stage diffusive shock acceleration process. For diffusion theory to be applicable at a perpendicular shock, particles downstream of the shock must be capable of diffusing upstream. The probability that a downstream particle will escape downstream and not return is $P_{esc} = 2V_{n2}/V_{D2}$ [Webb et al., 1995], where $V_{n2(1)}$ is the downstream (upstream) flow speed normal to the stationary shock and V_{D2} is the effective downstream guiding center speed normal to the shock, $V_D = (v/2)\sqrt{\kappa_\perp/\kappa_\parallel}$. Here, $\kappa_{\parallel,\perp}$ are the parallel and perpendicular diffusion coefficients. For a particle to be diffusive in the context of particle acceleration at a perpendicular shock, P_{esc} must be small, i.e., [Webb et al., 1995]

$$v >> \frac{4V_{n1}}{r}\left(\frac{\kappa_\parallel}{\kappa_\perp}\right)^{1/2}, \qquad (1.3)$$

which is related to a result of Jokipii [1987]. Zank et al. [2000] considered the implications of (1.3) in the context of MRI acceleration as a pre-acceleration mechanism.

Hard-sphere scattering [Axford, 1965] is used frequently to describe the transport of diffusive particles, for which

$$\kappa_\parallel / \kappa_\perp = 1+\eta_c^2, \quad \eta_c = \frac{3\kappa_\parallel}{v r_g} \equiv \frac{\lambda_\parallel}{r_g}. \qquad (1.4)$$

$\lambda_\parallel$ is the parallel mean free path for a particle of speed v and $r_g = pc/(QB)$ is the particle gyroradius (p is particle momentum, Q the charge, c the speed of light, and B the magnetic field strength). In the hard sphere scattering model, η_c is a measure of the strength of scattering: η_c small implies strong scattering whereas η_c large corresponds to weak scattering. For resonant scattering, a quasi-linear model for the parallel diffusion coefficient shows that [Zank et al., 1998; equation 4] $\lambda_\parallel \propto R^{1/3}$, where $R \equiv pc/Q$ is the particle rigidity. It follows then that $\eta_c \propto R^{-2/3}$, or

$$\eta_c = \eta_{cp}\left(\frac{m}{M}\right)^{2/3}, \eta_{cp} \equiv \left(\frac{\lambda_\parallel}{r_g}\right)_{proton} . \qquad (1.5)$$

A particle may be regarded as diffusive if

$$v_{inj}=\alpha\frac{4V_n l}{r}\left(1+\eta_c^2\right)^{1/2}, \quad (1.6)$$

for some parameter α. The important implication of (1.6) is that even if scattering is weak for protons, i.e., $\eta_{cp} >> 1$, the inverse dependence of η_c on M implies that η_c can be much smaller for heavy ions. Thus, heavy ions are accelerated diffusively at a much lower threshold velocity than lighter ions. Consequently, a larger fraction of heavy MRI accelerated ions will enter a second-stage diffusive shock acceleration process than light MRI accelerated ions.

It remains to determine whether the two competing mass dependence effects (1.2) and (1.6) conspire to satisfy the Cummings and Stone [1996] ACR injection criterion. Assume that the source MRI accelerated spectrum is approximately v^{-4}. For parameters, consistent with values used by Cummings and Stone [1996], we may plot the ratio of the number density of ions injected into the second-stage diffusive shock acceleration process n_{inj} (i.e., the number density of pickup ions that will become part of the ACR component) to the total number density of pickup ions of mass M for the pickup ions H^+, He^+, C^+, N^+, O^+, and Ne^+ (Figure 3). The injection energies needed to render an ion diffusive range from 337 keV for pickup H^+ to about 185 keV for C^+, N^+, O^+, and Ne^+.

Since $R_{ref}(M) = R_{ref}\left(H^+\right)\left(m/M\right)^{1/2}$, we plot n_{inj}/n as a function of $R_{ref}(H^+)$. For weak scattering with $\eta_{cp} = 10$, Figure 3 illustrates that the efficiency with which pickup H^+ is injected into ACR H^+ is considerably lower than that of the heavier ions. The injection efficiency curves for C^+, N^+, O^+, and Ne^+ are clustered closely in Figure 3 and are ordered by mass. Although much more efficiently injected than H^+, pickup He^+ is intermediate to H^+ and C^+, N^+, O^+, and Ne^+. The very distinct injection efficiencies for H^+, He^+, and collectively C^+, N^+, O^+, and Ne^+ are exactly consistent with the relative efficiencies derived by Cummings and Stone [1996] (see in particular their Figure 5), including the clustering of C^+, N^+, O^+, and Ne^+.

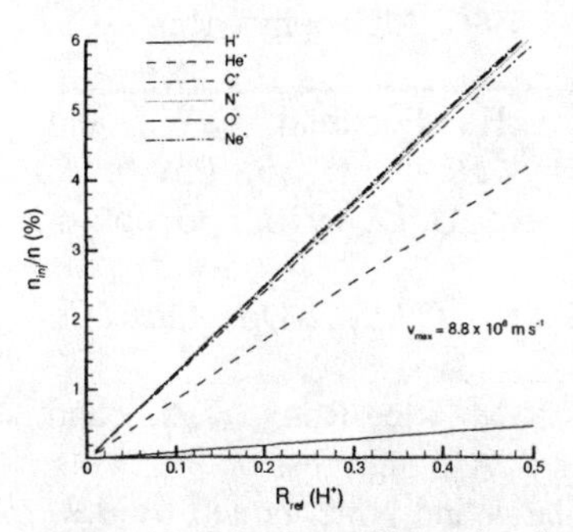

Figure 3. Injection efficiencies for H^+, He^+, C^+, N^+, O^+, and Ne^+, plotted against the reflection efficiency for H^+ ($R_{ref}(H^+)$). [Zank et al., 2000]

Figure 3 shows that the injection efficiency of heavy pickup ions exhibits much less variance than exists between H^+ and the heavy ions. In this sense, the injection efficiency of pickup H^+ is anomalous with respect to heavy ions, provided the scattering is weak. However, in the limit of strong scattering, the injection threshold for ions to be viewed as diffusive is virtually identical for all ion species. Consequently, no difference exists between the injection efficiencies of different mass ion species in the case of strong scattering. This is an important point since it relates for the first time the differential injection efficiency of pickup ions of different masses to the particle scattering strength. Zank et al. [2000] consider too models of the perpendicular diffusion coefficient based on magnetic field line wandering, arriving at conclusions very similar to those based on hard sphere scattering.

The combined pickup ion, MRI accelerated and ACR spectrum at the termination shock, assumed to be located at 80 AU, is illustrated in Figure 4. In computing the ACR flux, we assumed that $R_{ref}\left(H^+\right) = 0.01$ (i.e., 1% reflection efficiency). Our reason for choosing such a low value is prompted by the time-dependent computations of le Roux et al. [2000]. These authors have constructed a self-consistent model of MRI and diffusive shock acceleration at the termination shock in which the pressure contributed by the energized reflected ions and anomalous cosmic rays reacts back on the flow and, most importantly, on the structure of the shock itself. le Roux et al. find that the termination shock can fluctuate between a supercritical (with a cross-shock electrostatic potential) and a subcritical (i.e., without a cross-shock electrostatic potential) state. Since MRI acceleration can occur only at a supercritical termination shock, the mechanism is highly time dependent. Thus, even though individual episodes of ion reflection and MRI acceleration can be very efficient, the averaged value of $R_{ref}\left(H^+\right)$ can be much lower. As discussed by Cummings and Stone [1996] and le Roux et al. [2000], the flux of ACRs illustrated in Figure 4 is in accord

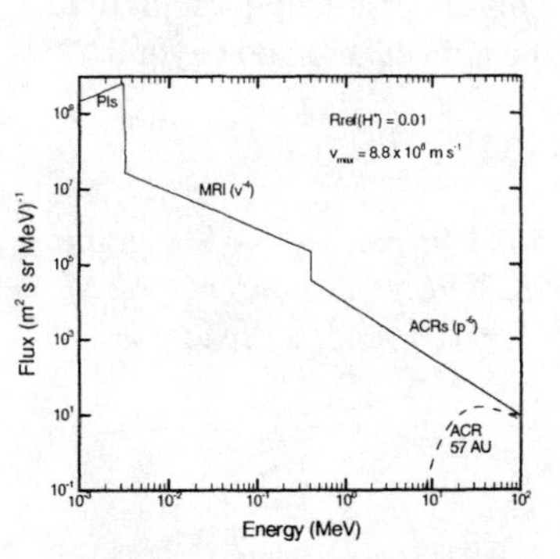

Figure 4. Fluxes of pickup H^+, MRI accelerated H^+, and H^+ ACRs at the termination shock, together with the resulting modulated ACR flux at 57 AU.

with the expected ACR source spectrum used to model the observed modulated ACR flux within the heliosphere. In computing the modulated spectrum, the same basic model parameters as Cummings and Stone [1996] were used.

4. CONCLUDING REMARKS

A review of the injection question for particles accelerated at interplanetary shocks waves has been presented. We have concentrated to some extent on the heliospheric termination shock because of its perceived importance in accelerating anomalous cosmic rays. The various injection mechanisms that have been proposed were discussed. With the improved resolution and coverage offered by new spacecraft such as ACE, WIND and Ulysses, the outstanding issues related to the injection of particles into shock acceleration mechanisms are now becoming amenable to analysis and theory and we may expect substantial progress in our understanding of particle acceleration throughout the heliosphere and beyond.

ACKNOWLEDGMENTS

This work was supported in part by NASA grants NAG5-6469 and NAG5-7796, NSF grant ATM-9713223, JPL contract 959167 and a Space Grant College Award.

REFERENCES

Adams, J.H., and Leising, M.D., in Proc. International Cosmic-Ray Conference Ser. 22, Vol. 3 (Dublin : Dublin Inst. Adv. Studies), 1991, pp. 304

Balikhin, M.A., Krasnosselskikh, V., & Gedalin, M. 1995, Adv. Space Res., 15, 247

Bogdan, T. J., et al., 1991, J. Geophys. Res., 96, 161

Chalov, S., Fahr, H.J., & Izmodenov, V., 1995, Astron. Astrophys., 304, 609

Chalov, S., Fahr, H.J., & Izmodenov, V., 1997, Astron. Astrophys., 320, 659

Cummings, A.C., & Stone, E.C. 1996, Space Sci. Rev., 78, 117

Decker, R. B., 1988, Space Sci. Rev., 48, 195

Ellison, D.C., Jones, F.C., & Baring, M.G. 1999, ApJ, 512, 403

Fisk, L.A., Koslovsky, B., and Ramaty, R., 1974,ApJ., 190, L35

Fränz et al.,1999, Geophys. Res. Lett., 26, 17

Giacalone, J., et al. 1997, ApJ, 486, 471

Giacalone, J., Jokipii, J.R., & Kota, J. 1994, J. Geophys. Res., 99, 19351

Gloeckler, G., et al. 1994, J. Geophys. Res., 99, 17637

Gloeckler, G., et al. 1993, Science, 261, 70

Gosling, J.T., et al. 1981, J. Geophys. Res., 86, 547

Isenberg, P.A., 1987, J. Geophys. Res., 92, 1067

Jokipii, J.R. 1992, ApJ, 393, L41

Jokipii, J.R. 1987, ApJ, 313, 842

Jokipii, J.R. 1986, J. Geophys. Res., 91, 2929

Kennel, C.F., and R.Z. Sagdeev, 1967

Kucharek, H., & Scholer, M. 1995, J. Geophys. Res., 100, 1745

Lazarus, A., Richardson, J.D., and Decker, R., 1999, ISSI Proceedings, in press

Lee, M.A., Shapiro, V.D., & Sagdeev, R.Z. 1996, J. Geophys. Res., 101, 4777

Le Roux, J.A., Fichtner, H., Zank, G.P., & Ptuskin, V.S., 2000, J. Geophys. Res., submitted

Le Roux, J.A., & Ptuskin, V.S., 1998, J. Geophys. Res., 103, 4799

Leroy, M. M., 1983, Phys Fluids, 26, 2742

Lipatov, A.S., Zank, G.P., & Pauls, H.L. 1998, J. Geophys. Res., 103, 26979

Lipatov, A.S., & Zank, G.P. 1999, Phys. Rev. Lett., 82, 3609

Newbury, J.A., Russell, C.T., & Gedalin, M. 1998, J. Geophys. Res., 103, 29581

Newbury, J.A., & Russell, C.T. 1996, Geophys. Res. Lett., 23, 781

Pesses, M.E., Jokipii, J.R., and Eichler, D., 1981, ApJ., 246, L85

Rice, W.K.M., G.P. Zank, J.D. Richardson, and R.B. Decker, 2000, Geophys. Res. Lett., 27, 509

Russell, C.T., & Greenstadt, E.W. 1979, Space Sci. Rev., 23(1), 3

Scudder, J.D., et al. 1986, J. Geophys. Res., 91, 1053

Story, T.R., & Zank, G.P. 1997, J. Geophys. Res., 102, 17381

Tokar, R.L., Aldrich, C.H., Forslund, D.W., and Quest, K.B., 1986, Phys. Rev. Lett., 56, 1059

Vasyliunas, V. M., & Siscoe, G. L., 1976, J. Geophys. Res., 81, 1247

Webb, G.M., Zank, G.P., Ko, C.M., & Donohue, D.J. 1995, ApJ, 453, 178

Zank, G.P., Rice, W.K.M., le Roux, J.A., and Matthaeus, W.H., 2000, ApJ., in press

Zank, G.P. 1999, in Solar Wind Nine, edited by S.R. Habbal, R. Esser, J.V. Hollweg, and P.A. Isenberg, The American Inst. Phys., 783

Zank, G.P., Matthaeus, W.H., Bieber, J.W., & Moraal, H. 1998, J. Geophys. Res., 103, 2085

Zank, G.P., Pauls, H.L., Cairns, I.H., & Webb, G.M. 1996, J. Geophys. Res., 101, 457

Zilbersher, D., & Gedalin, M. 1997, Planet. Space Sci., 45(6), 693

Imaging the Global Distribution of Anomalous Cosmic Rays

K. C. Hsieh[1], A. Czechowski[2] and M. Hilchenbach[3]

[1]*Department of Physics, University of Arizona, Tucson, AZ 85721, USA*
[2]*Space Research Centre, Polish Academy of Sciences, 00716 Warsaw, POLAND*
[3]*Max-Planck-Institut für Aeronomie, 37191 Katlenburg-Lindau, GERMANY*

Abstract. Since the discovery of the anomalous component of cosmic rays (ACR), more detailed information on the composition and intensity variations over solar cycles have been accumulated and studied. As the two Voyagers approach the solar-wind termination shock, anticipation of a decisive test on the theory of transport and acceleration of ACR heightens. The fact that all existing ACR measurements have been *in situ*, *i.e.* by directly detecting the modulated particles along specific trajectories of ACR-measuring spacecraft; that ACR is an important link in our understanding of our plasma environment; and that the opportunity to explore the outer edges of the heliosphere will not be readily available; all suggest that the means to survey the global distribution of ACR should be examined and implemented. Since the use of detecting energetic neutral atoms (ENA) as a means to study ACR at and beyond the termination shock was first proposed, the idea has gained acceptance and the techniques for detecting ENA have matured. It is believed that the first signal of energetic neutral H atoms of ACR origin has been detected. Where do we go from here? What can we expect to accomplish in the near future?

INTRODUCTION

The anomalous cosmic rays (ACR), like all distinct populations of energetic charged particles in space plasma, carries with it a story that ties to other particle populations and the magnetic field structures in the regions of its dwelling. From ACR's distributions in space, time and energy, one could discover its parentage, the *original particle population* from which ACR come, and its migratory pattern, which includes *acceleration mechanisms* from which it efficiently gains energy and *propagation mode* by which it populates the various regions of space. We want to piece together a reasonable ACR story confirmed by observation, so that it may accurately describe the real ACR population and its environment. Our present story is this: ACR differs from the galactic cosmic rays (GCR) in that ACR has its origin within the immediate solar neighborhood, *i.e.* interplanetary space, heliosphere and the local interstellar space. The ACR detected in interplanetary space and the inner heliosphere are believed to be those pickup ions accelerated at the solar-wind termination shock and diffused back into the inner heliosphere [1,2]. The pickup ions are in turn portions of the neutral gas of the local interstellar medium (LISM) that penetrated the interplanetary space and then ionized by either solar photons or particles and then swept out by the solar wind to the termination shock [3]. This story guides us to learn more about the interrelations among the particle populations, acceleration mechanisms, and propagation modes of charged particles within our immediate neighborhood extending from the Sun to the local interstellar space. More detailed renditions of the ACR story are found within this volume.

Here, we review and update the gathering of thus far missing information on ACR., *i.e.* the global distribution of ACR. We shall, therefore, review: how distributions of ACR in the vast remote regions of space - not easily accessible to the existing *in situ* ACR-detection techniques - can be observed, what information can be deduced from the detection of energetic neutral atoms (ENA) of ACR origin, what have we seen so far; and then look forward to what awaits us in the coming years. Page limit prevents us to cite some significant earlier works, which, however, can be traced from the more recent references.

CP528, *Acceleration and Transport of Energetic Particles Observed in the Heliosphere: ACE 2000 Symposium*, edited by Richard A. Mewaldt, et al.
© 2000 American Institute of Physics 1-56396-951-3/00/$17.00

REMOTE SENSING OF DISTANT ENERGETIC IONS *VIA* ENA

Energetic ions in space plasma spiral along magnetic field lines and yet do not radiate any detectable electromagnetic waves. These two features – confinement by magnetic field and not broadcasting signals - limit ACR observation to *in situ* measurements, *i.e.* the measurements that can only be done on portions of ACR at the location, (x, t), of a given instrument along the trajectory of a specific spacecraft. As such, *in situ* measurements are local and instant, hence, incapable of yielding a global view that could separate temporal-dependence from spatial-dependence of the observed variations in the particle distributions. Such separation is crucial to the understanding of the dynamics of the population. So far, we have spacecraft anchored at 1 AU, including ACE in whose honor this volume is dedicated, and those venturing out to the outer heliosphere and eventually into interstellar space, namely, Voyagers 1 and 2 and Pioneers 10 and 11. Measurements performed on these spacecraft are all *in situ*, thus, unable to provide a global view for a more complete ACR story. This constraint is about to be or has begun to be removed and here is how.

A global view of an ion population distributed over a vast region of space, *e.g.* ACR over the entire heliosphere or energetic ions in magnetospheres, can only be captured by "viewing" a large portion of the object from a distance, *i.e.* by remote sensing, as in photography. Photography does not apply to space plasma for lack of observable light signals. Luckily, there are actually messengers, not just signals, that make remote sensing of energetic ion populations, especially those not easily accessible to space missions, possible. It was recognized about four decades ago that space plasma radiate in energetic neutral atoms [4]. The mechanism is charge exchange between an energetic +1 ion and a nearby neutral atom. Because charge exchange occurs at inter-nuclear distances of atomic scale, the momentum transfer is negligible. This energetic neutral atom (ENA) is liberated from the magnetic field and follows a ballistic trajectory, determined by its momentum at the moment of neutralization, to remote regions of space unreachable by the original ion. Thus, ENA are direct samples of remote ion populations, even more intimate than photography. Charge exchange requires an ambient neutral gas to neutralize the ions. We are fortunate that, for the magnetospheric plasma we have the geocorona [5]; and for the heliospheric plasma the interstellar neutral gas [6]. Figure 1 shows the remote sensing of space plasma in ENA.

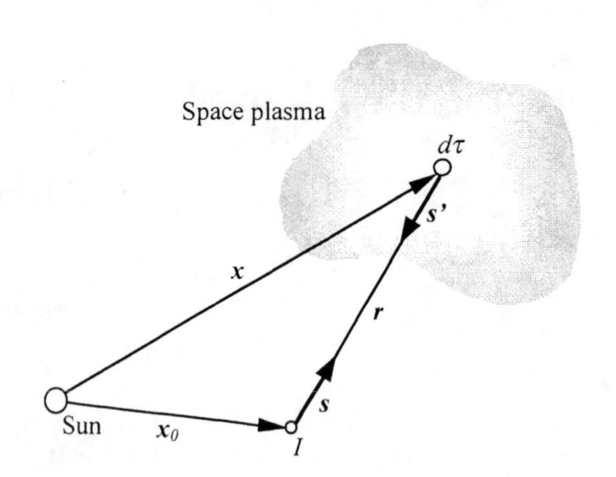

FIGURE 1. Geometry of ENA observation in the helio-centric system. An ENA detector, I, at $\boldsymbol{x_0}$ receive ENA produced in a volume element, $d\tau$ at $\boldsymbol{x}$, to come from the direction $\boldsymbol{s}$. The net ENA intensity reaching I from that direction is the sum of the ENA produced in the line of sight.

The intensity of ENA of energy E of the ith species at point, I, from the direction $\boldsymbol{s}$, is the column density of ENA along the line of sight, as expressed by

$$j_i(\boldsymbol{s},E;t) = \int r^{-2} j_{i,+1}(\boldsymbol{x},\boldsymbol{s}',E;t') \sum_k \sigma_{ik}(E) n_k(\boldsymbol{x}) \cdot \exp[-D(\boldsymbol{s}',E)] d\tau \qquad (1)$$

where $j_{i,+1}$ is the parent ion intensity, σ_{ik} the cross section for charge exchange with k^{th} species of neutrals, whose number density is n_k. The time of detection, t at I, and the time of neutralization, t' at $\boldsymbol{x}$, differ by the travel time, r/v, where v is the velocity of the ENA E. The exponential is the extinction of the ENA intensity between birth at $d\tau$ to detection at I.

$$D(\boldsymbol{s}',E) = \int [\frac{\beta(\boldsymbol{x})}{v} + \sum_j \sigma_{ij}(E) n_j(\boldsymbol{x})] dl \qquad (2)$$

where $\beta(\boldsymbol{x})$ is the photoionization rate, σ_j the cross section for impact ionization by the jth species such as electrons, protons, and neutral atoms, and n_j the number density of jth species of ionizing particles.

The ENA spectrum, $j_i(E)$ in Eq. (1), is that of the original ion spectrum, $j_{i,+1}(E)$, modified by the sum of the charge exchange cross sections weighted by the number density of the respective neutral atoms in the ambient gas. Typically, space plasma exhibits ion spectra that decrease with increasing E. Since the charge-exchange cross sections above 10 keV amu^{-1} also show similar trend in E-dependence, we expect $j_i(E)$ to decrease with increasing E more rapidly than $j_{i,+1}(E)$ does. It is then only practical to concentrate on the detection of ENA of energies < 1 MeV amu^{-1}.

An instrument at I capable of detecting ENA with sufficient angular resolution can obtain a sky map in ENA. This is "imaging" space plasma in ENA. This is applicable to all appropriate ion populations, but this article shall concentrate on ENA of ACR origin.

ACRENA: ENA OF ACR ORIGIN

The production of ENA is assured when a plasma cohabitate with a neutral gas. If ACR is accelerated at the termination shock and propagates according to the conditions at and beyond the termination shock, then we expect ACR to generate ENA (ACRENA) through charge exchange with the neutral atoms of the local interstellar medium (LISM) in that very region, as ACR of energies < 1 MeV per amu^{-1} are kept out of the inner heliosphere by solar modulation. Figure 2 shows the genealogy of ACRENA.

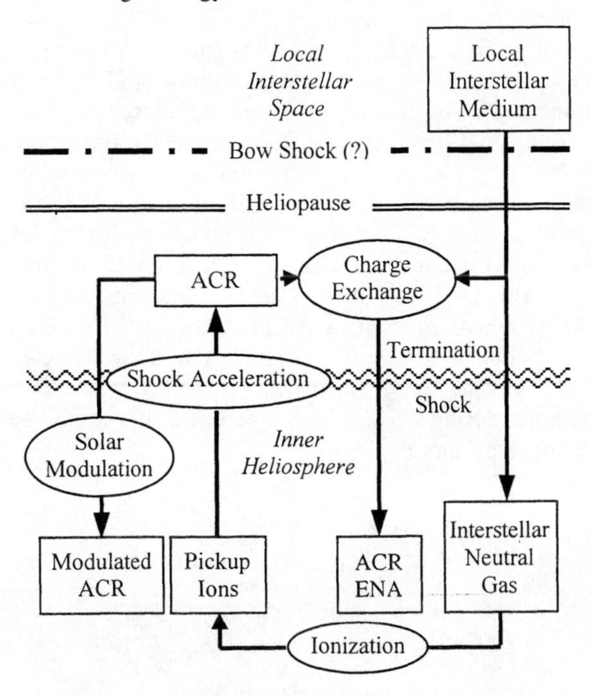

FIGURE 2. The immediate populations of particles and processes involved in the birth of ACRENA. Processes are in ovals and populations in rectangles. ACRENA and ACR take different paths to reach the inner heliosphere.

Based on the production scheme shown in Figure 2 and using the known cross sections, an ACR spectrum based on shock acceleration, the best values at the time for the densities of H and He in LISM, and other parameters in Eqs. (1) and (2), the spectrum of energetic H atoms of ACR origin was estimated in anticipation of their eventual detection [7].

WHAT CAN WE LEARN FROM ACRENA

The detection of ACRENA is of direct interest to the study of cosmic-ray propagation in the immediate neighborhood of the Sun, *i.e.* in the inner heliosphere bounded by the termination shock and in the heliosheath just outside the shock. The detection of ACRENA can also give an indirect way of finding the direction of the local interstellar magnetic field.

Since ACR of E < 1 MeV amu^{-1} are effectively kept out of the inner heliosphere by the solar wind, as our present understanding of solar modulation predicts, ACRENA of E < 1 MeV amu^{-1} present in the inner heliosphere, *e.g.*, at 1 AU, can only be samples of ACR in the outer heliosphere, at and beyond the termination shock. The distribution of ACR in (E, θ, ϕ) can be obtained by an ENA imager with sufficient energy and directional resolution. This is "imaging the global distribution of ACR". The detection of ACRENA intensity from a given vantage point will be unable to give the location of the termination shock, since the detected intensity is a column density as indicated by Eq. (1). Having ENA imagers on multiple spacecraft, 3-D tomographic imaging of the termination shock and heliosheath may be the solution.

Since the ACRENA detected in the inner heliosphere are ACR from at or beyond the termination shock, they are ACR *without* modulation, in direct contrast to the ACR of higher energies detected in the same inner heliosphere. By comparing the information gathered from the two populations of the same origin but transported through distinctly different paths, it is possible to separate the effects of solar modulation from those of acceleration at the termination shock. The two-prong attack on the central problem of particle transport in the solar neighborhood is outlined in Figure 3. Without the benefit of remote sensing *via* ENA, ACR spectrum beyond the termination shock, $j_{i,+1}(E)$ in Eq. (1), can only be estimated by de-modulating the ACR spectrum detected by *in situ* instruments in the inner heliosphere with the best information on the location of the termination shock and the heliospheric magnetic field [*e. g.*, 8]. This is Approach 1 shown in Figure 3. The detection of ACRENA opens up Approach 2. By iterative comparison between the detected ACRENA intensity, $j_i(E)$ in Eq. (1), with models of conditions in the heliosheath, the likely habitat of ACRENA, and of shock acceleration, one could produce an ACR spectrum, $j_{i,+1}(E)$, independent of modulation. By direct comparison between the two independently derived "best" spectra, we can expect to reach a more accurate description of both modulation and shock acceleration than any single approach could. This we believe is the best contribution of remote sensing of ACR via ENA to the understanding of particle transport in the solar neighborhood.

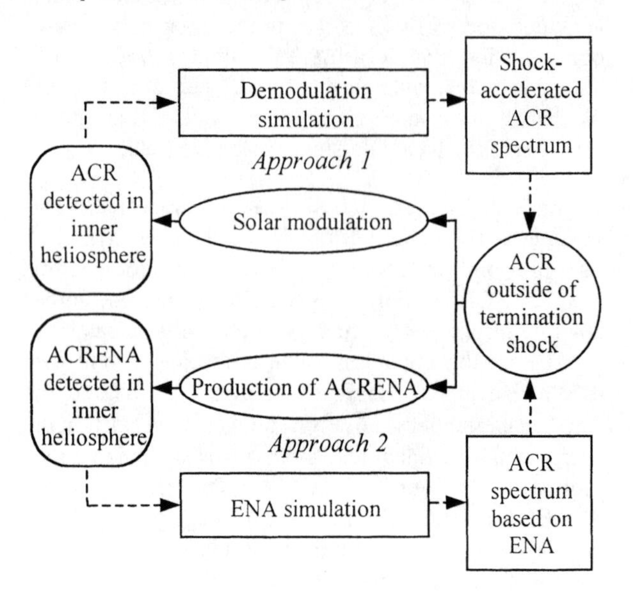

FIGURE 3. Schema for combining *in situ* detection of ACR (Approach 1) with remote sensing of ACR through ACRENA (Approach 2) from the inner heliosphere to arrive at the ACR spectrum beyond the termination shock. The combination uncouples solar modulation from shock acceleration. The circle and ovals connected by solid lines are natural phenomena; and the rectangles connected by dashed lines represent computational efforts.

WHAT HAVE WE SEEN SO FAR

ENA of planetary magnetospheric origin have been detected since the 1980s [9-13], the first instrument designed with the intent to detect heliospheric ENA is the High-energy Superthermal Time-of-Flight sensor (HSTOF) of the Charge, Element, and Isotope Analysis System (CELIAS) on SOHO [14], launched on 1995 December 2. HSTOF uses TOF and residual energy to identify the mass and energy of particles. A parallel-plate E/Q filter repels ions of $E < 100$ keV/e. Its geometrical factor of 0.22 cm^2-sr is framed by ±2° in the ecliptic and ±17° out of the ecliptic, about its bore-sight, which is set at 37° west of the Sun-SOHO line. As SOHO moves about Earth's L1 point and the Earth around the Sun, HSTOF scans the ecliptic in a ±17° band once a year.

In 1996 and 1997, the last solar minimum, 55 – 80 keV H atoms detected during quiet interplanetary conditions showed a higher flux, when HSTOF was pointing in the heliospheric anti-apex direction with respect to LISM, as shown in Figure 4 [15]. The tumbling of SOHO in 1998 and the increased solar activity in 1999 prevented the recurrent observation. Hilchenbach *et al.* [15] interpreted the detected H atoms as energetic H converted from ACR protons, as anticipated [7]. The apparent anisotropy coincides with the prediction of Czechowski, Grzedzielski and Mostafa [16] based on the simulations of the ACRENA production, when the transport of ACR in the heliosheath is considered. The first detected heliospheric ENA is, therefore, perhaps the first remote sensing of ACR in the heliosheath, outside the termination shock.

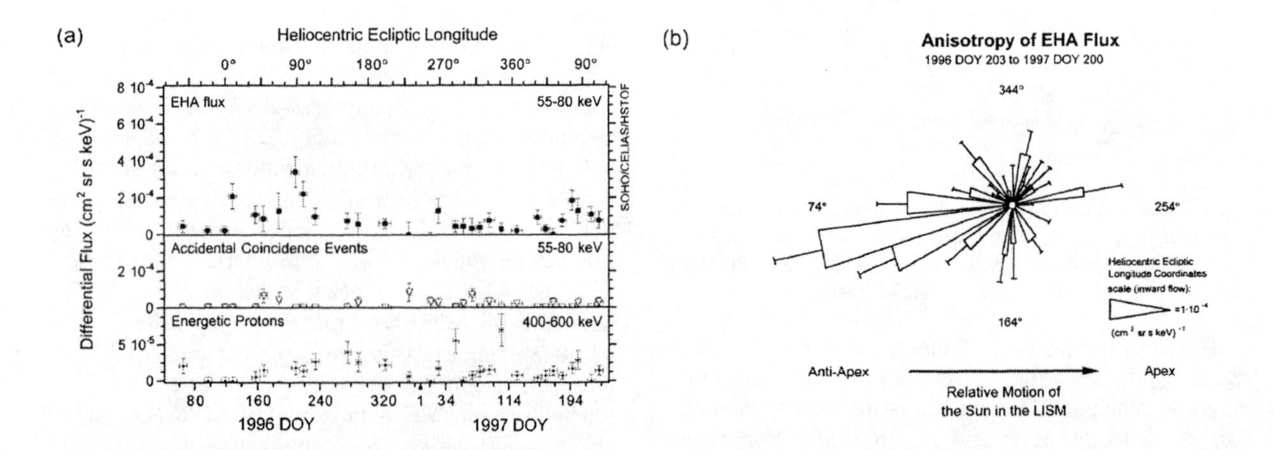

FIGURE 4. a. Differential intensity of quiet-time 55-80 keV energetic H atoms observed by HSTOF of CELIAS on SOHO. b. Same data shown in polar plot. The anti-apex direction is 74° in heliocentric ecliptic longitude. [15]

From the termination shock, low-energy ACR can not diffuse too far upstream before being convected downstream with the solar plasma into the heliotail. The larger charge-exchange cross sections at lower energies and the larger column length in the tail region yield a maximum ENA flux from the heliotail. Figure 5 compares the SOHO data with simulated H intensity from 63 keV ACR protons based on two models [17]. The models differ in the shapes of the heliopause and the termination shock, which in turn differ in plasma flow, diffusion coefficient, and charge-exchange loss rate. The model, based on Parker's analytical solutions for the heliospheric flow, has a spherical termination shock and a cylindrical tail parallel to the Sun-apex axis [18]. A 0.1 H cm^{-3} density is used for inside the heliopause. The model based on Kausch [19] has a non-spherical termination shock and a slightly flare out heliopause. It uses 0.1 H cm^{-3} in LISM. In both cases, charge exchange with He in LISM contributes ~30% of the H flux. The variation in the ACR proton spectrum over the shock surface, reflecting the variation of shock strength and the upstream flow speed, is incorporated. The ACR intensity at the shock is matched to the spectrum at the shock [8]. For both models, the H flux is predominantly from the heliotail direction. The poor statistics of the SOHO data does not make a clear choice between the models.

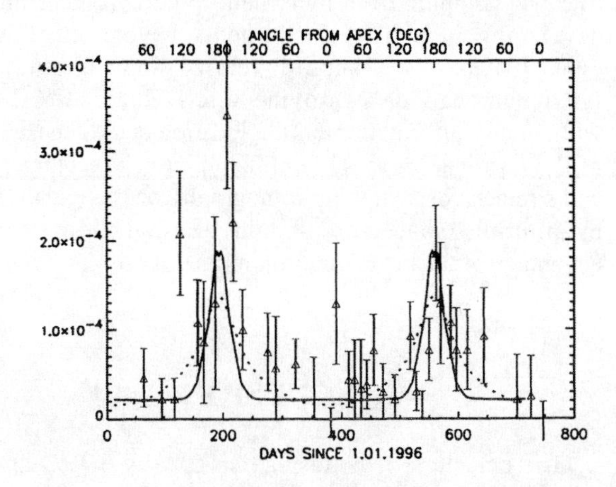

FIGURE 5. The H differential intensity displayed in Figure 5 is superposed on the simulated H intensity from 63 keV ACR protons as functions of direction in the ecliptic plane relative to the apex of the heliosphere. The models of the heliosphere used are that of Kausch (solid line) and Parker (dotted line) [17].

In the simulations just described, local interstellar magnetic field is not included. In modeling the heliosphere, Czechowski and Grzedzielski [20] and Ratkiewicz et al. [21] showed independently that the direction and shape of the heliotail depends on the direction and strength of the local interstellar magnetic field. It is tempting to ascribe the apparent shift in the peak in the H intensity (Figure 4) to a particular direction of the local interstellar magnetic field [22].

The instrument HSTOF of CELIAS on SOHO gave us the first detection of heliospheric ENA, and possibly that of ACRENA. Although the quality of the data is insufficient for selecting between models or determining the local interstellar magnetic field, the interplay between observation and modeling in studying ACR and the heliosphere through ENA, *i.e. Approach 2* in Figure 3, has begun.

WHAT AWAITS US IN THE COMING YEARS

The work just reviewed is the humble beginning of the noble quest of imaging the global distribution of ACR. Here we forecast some of the things to come and outline the tasks yet to be done

The solar minimum of 1996-1997 gave us a favorable time for observing ACR in the heliosheath, the yet unexplored outer heliosphere. The detection of energetic H of the larger family of ACRENA urges us to look further. Using HSTOF of CELIAS on SOHO to search for energetic neutral He, N and O is underway. At the same time, modeling effort widens to include other participants, such as dust, and also interstellar magnetic field.

While HSTOF of CELIAS on SOHO continues to monitor heliospheric ENA intensity at Earth's L1 point, Cassini carrying INCA (Ion & Neutral Camera) of MIMI (Magnetospheric Imaging Instruments) is en route to Saturn *via* Jupiter. Imager INCA is dedicated to imaging Saturn's magnetosphere in ENA [23]. Its 2.6 cm^2 sr geometrical factor has ±45° in azimuth and ±60° in elevation, with the capability of scanning 360° in azimuth. Its angular resolution depends on statistics and varies with the energy and mass of the ENA. Its energy threshold for H is 10 keV and for O is 25 keV. Its mass resolution is derived from the pulse height of its microchannel-plate signal caused by the incident ENA. Statistically, it can separate O from H, but leaving He more difficult to identify. Although INCA is dedicated to the Saturn mission, it will search for heliospheric ENA during Cassini's interplanetary cruise. With its larger geometrical factor, ~10 times that of HSTOF, and its lower energy threshold, INCA has great potential to actually image the entire heliosphere in ENA. An image of Earth's ring current in 10-100 keV neutral H with 8° × 8° resolution taken during its last Earth fly-by on 18 August 1999 demonstrated INCA's ENA imaging capability [24].

In view of the ENA capability of INCA on Cassini, Figure 6 shows a 2-D image of the heliosheath in 63 keV ACREA H atoms as might appear to an observer in the inner interplanetary space. This image is from the same model that produced the dotted-line intensity profile in Figure 5. Simulations indicate INCA should be able to choose between models, thus setting values on heliospheric and LISM parameters for particle propagation and acceleration, and also the direction of the local interstellar magnetic field. Great strides in the study of ACR *via* ENA are to be made by INCA.

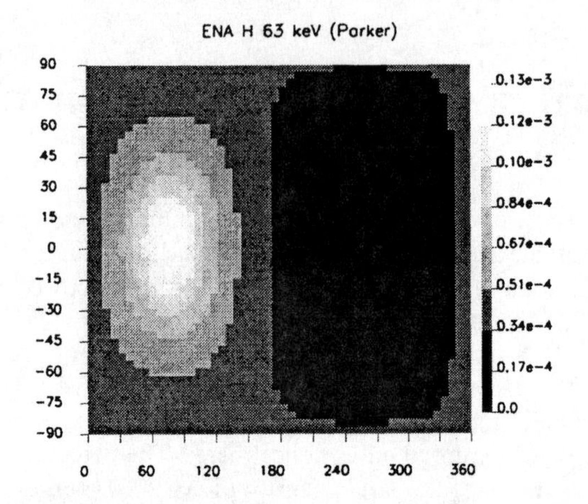

FIGURE 6. The intensity of 63 keV ACRENA in different portions of the heliosheath based on Parker's model of the termination shock and heliopause [18]. The coordinates are ecliptic. The Sun is located at (180°, 0°), and the heliospheric apex (254°, -6°). The intensity scale in units of $(cm^2\ sr\ s\ keV)^{-1}$ is shown on the right side of the figure. A minimum in ACRENA intensity is seen around the apex, while the maximum centers around the anti-apex direction.

While HSTOF had the fortune of the solar minimum to capture a few ACRENA, INCA with its more advanced ENA capability will have to face the challenge of an increasingly active Sun. The more intense ENA fluxes generated by energetic ions of either solar or interplanetary origin will interfere with INCA's observation on ACR in the outer heliosphere. This problem was anticipated [7], but it is not a problem when studying interplanetary shocks and co-rotating interaction regions (CIR) [25]! To check whether the ENA detected by HSTOF could be due to those generated in CIR (CIRENA), Hsieh et al. [26] used a 3-D CIR model based on the tilted dipole model of the Sun [27,28] using a MHD code [29] simplified to having a current sheet that remains stable for several solar rotations – a condition more suitable for solar minimum, and CIR-accelerated 100 keV proton intensity is taken at 10 $(cm^2\ sr\ s\ keV)^{-1}$, consistent with Ulysses observation [30]. After comparison, the ACR interpretation of the SOHO data is still favored.

In anticipation of INCA's observation, Figure 7 shows a 2-D map in 25 keV H from the same CIRENA model just described [courtesy of J. Kóta]. The CIRENA concentrate in lower and mid-latitudes, while leaving the polar regions clear for ACR observation. The mission IMAGE (Imager for Magnetopause-to-Aurora Global Exploration) will provide another opportunity to image the global distribution of ACR and the heliosphere, if launched successfully in March 2000. IMAGE has three ENA instruments covering a range from 10 eV to 500 keV.

The imminent observational opportunities for our quest of imaging the global distribution of ACR are most promising. To accomplish the eventual goal of gaining reliable knowledge of the pre-modulated ACR spectrum at and beyond the termination shock would require efforts in two directions:

1. Developing computational tools to arrive at $j_{i,+1}(\mathbf{x}, \mathbf{s}', E; t')$ in Eq. (1) from observed $j_i(\mathbf{s}, E; t)$. This is necessary, in terms of the strategy outlined in Figure 4, for Approach 2 to reach the same level of performance as Approach 1. Only then the iterative interaction between the two approaches can be fruitful.

2. Developing tomographic capability to achieve true 3-D mapping from inner heliosphere. Aside from developing new ENA instruments beyond existing ones [31,32], we should realize that a single instrument can only provide a 2-D map of ACR without any information on the location and dimension of the termination shock and heliopause. Techniques and strategy for achieving tomography or triangulation by multiple spacecraft should be studied to take advantage of the next solar minimum.

CONCLUSION

The detection of heliospheric ENA by SOHO and the modeling efforts accompanying it offer a glimpse of the potential benefit of using ACRENA to study ACR. It took four decades since Parker's study of the heliosphere to the first comparison with observation. Innovative instrumentation and mission planning can the explore much of the outer heliosphere and solve the central problem of charged particle acceleration and propagation throughout the solar neighborhood without leaving interplanetary space or the inner heliosphere. We hope the pace will be hastened in the wake the progress made in the last decade of the 20th Century. As a reminder and a challenge, Figure 8 shows the network of particle populations in our solar neighborhood.

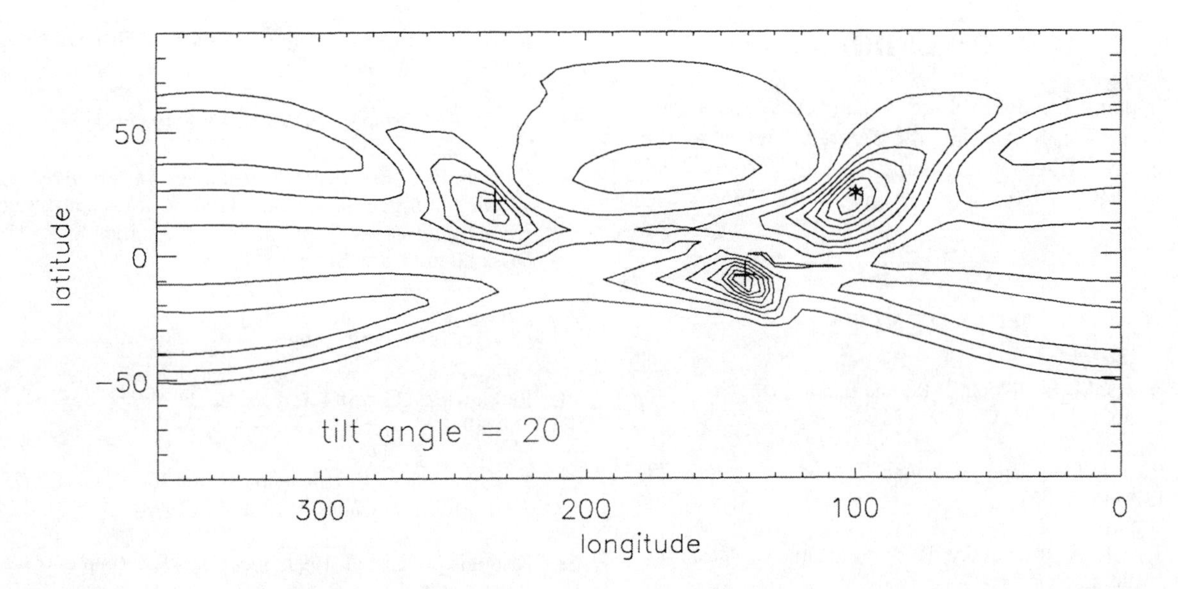

FIGURE 7. Intensity contour of 25 keV H from a stable CIR simulated for the location of Cassini around its Jupiter swing-by. Contour lines correspond to 10% increments between the extremes, over a range of three orders of magnitude. The direction of the Sun is at (180°, 0°). Maximum intensity is indicated by * symbol, and additional intensity enhancements are indicated by + symbols. The precise location of intensity maxima exhibits a 27-day variability due to the rotation of the CIR. [From J. Kóta.]

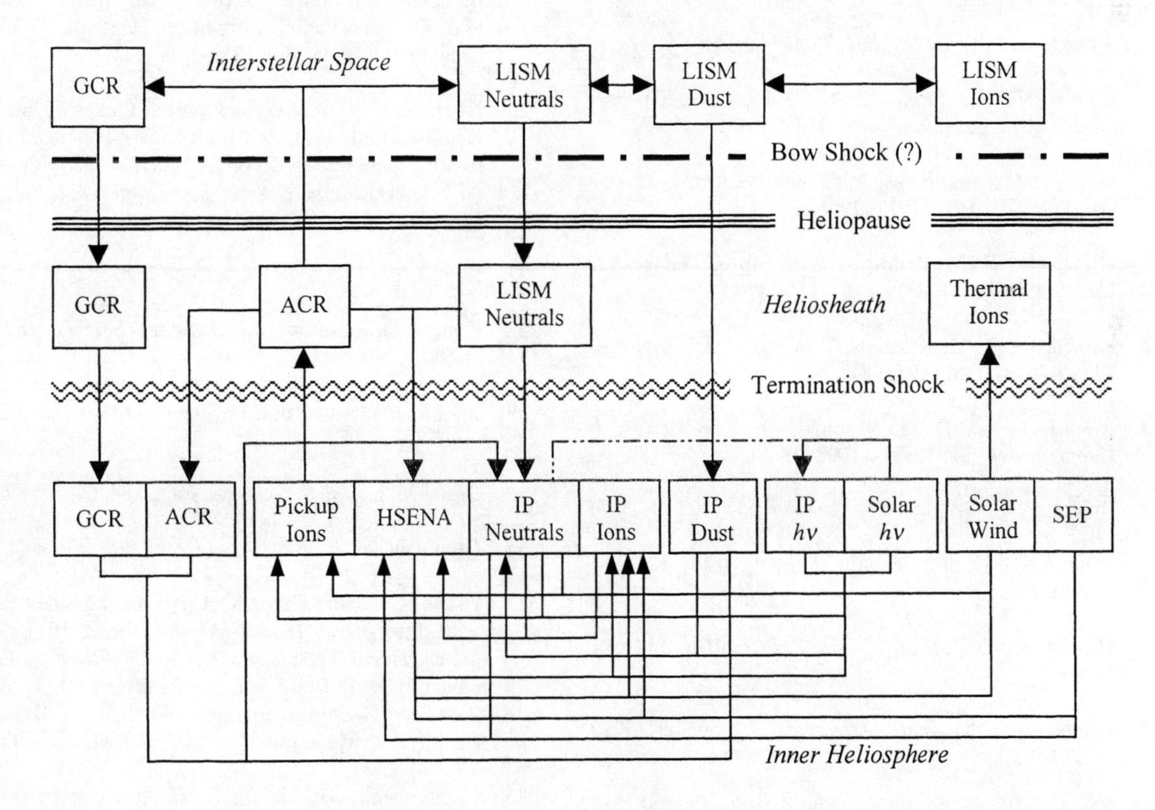

FIGURE 8. The interrelations of the various populations of particles in the solar neighborhood. ACRENA is a member of HSENA (heliospheric ENA), which includes ENA of inner heliospheric origin. Some of HSENA are ionized to become IP ions [33]. Innovative investigation of all these populations within the inner heliosphere can teach us much about the entire solar neighborhood. Not represented here is the magnetic field that permeates the whole neighborhood and inseparable from the story of the particle populations. (IP stands for interplanetary, *hν* for photons, and SEP for solar energetic particles.)

ACKNOWLEDGEMENT

We thank J. Kota for Figure 7, and A. W. Shaw and A. Levine for formatting the figures. Work at the University of Arizona is supported in part by NSF grant ATM9727080 and NASA grant NAG57966.

REFERENCES

1. Pesses, M. E., Jokipii, J. R., and Eichler, D., *Astrophys. J.* 246, L85 (1981).

2. Jokipii, J. R., and Giacalone, J., *Space Sci. Rev.* **78**, 137-148 (1996).

3. Fisk, L. A., Koslovsky, B., and Ramaty, R., *Astrophys. J.* **190**, L35 (1974).

4. Dessler, A. J., W. W. Hanson, E. N. Parker, *J. Geophysi. Res.* **66**, 3631 (1961).

5. Rairden, R. L., L. A. Frank, and J. D. Craven, *J. Geophysi. Res.* **91**, 13631 (1986).

6. Axford, W. I., *Space Sci. Rev.* **78**, 9-14 (1996).

7. Hsieh, K. C., Shih, K. L., Jokipii J. R., and Grzedzielski, S., *Astrophys. J.* **393**, 756-763 (1992).

8. Stone, E. C., Cummings, A. C., and Webber, W. R., *J. Geophys. Res.* **101**, 11017 (1996).

9. Kirsch, E., S. M. Krimigis, W.-H. Ip, G. Gloeckler, *Nature* **292**, 718 (1981).

10. Kirsch, E., S. M. Krimigis, J. W. Kohl, E. P. Keath, *Phys. Rev. Lett.* **8**, 169 (1981).

11. Roelof, E. C., D. G. Mitchell, D. J. Williams, *J. Geophys. Res.*, **90**, 10991-11008 (1985).

12. Roelof, E. C., *Geophys. Res. Lett.* **14**, 652 (1987).

13. Wilken, B., *et al.*, *Geophys. Res. Lett.* **24**, 111-114 (1997).

14. Hovestadt, D., *et al.*, *Solar Physics* **162**, 441-481 (1995).

15. Hilchenbach, M. *et al.*, *Astrophys. J.* **503**, 916-922 (1998).

16. Czechowski, A., Grzedzielski, S., and Mostafa, I., *Astron. & Astrophys.* **297**, 892-898 (1995).

17. Czechowski, A. *et al.* "Low Energy ACR Beyond the Termination Shock as a Source of Energetic Neutrals: Models and Observations," in *Proceedings of the 26th International Cosmic Ray Conference*, Salt Lake City, 1999, **7**, pp.589-592.

18. Parker, E.N., *Planet. Space Sci.* **13**, 9-49 (1965).

19. Kausch, T., "*Ein hydrodynamisches Mehrkomponentmodell zur Beschreibung der Wechselwirkung zwischen Heliosphaere und interstellaren Medium*," Ph.D. Dissertation, Univ. Bonn (1998).

20. Czechowski, A. and Grzedzielski, S., *Geophys. Res. Lett.* **25**, 1855-1858 (1998).

21. Ratkiewicz, R. *et al.*, *Astron. & Astrophys.* **335**, 363-369 (1998).

22. R. Ratkiewsicz, M. Hilchenbach, and K. C. Hsieh, *EOS Trans. AGU, Supplement*, **79**, F707 (1998).

23. Krimigis, S. K. et al. 1999, *Space Sci. Rev.* (in press).

24. Roelof, E. C., Mitchell, D. G., and Krimigis, S. M., *EOS Trans. AGU, Supplement*, **80**, F873 (1999).

25. Roelof, E. C., "Imaging Heliospheric Shocks using Energetic Neutral Atoms," in *Solar Wind Seven*, edited by E. Marsch and R. Schwenn, edited by E. Marsch, and R. Schwenn, Pergamon, Oxford, COSPAR Colloquium #3, 1992, pp.385-390.

26. Hsieh, K. C. *et al.*, "Estimating the Fluxes of Energetic Neutral Atoms Produced from Ions Accelerated in Corotating Interaction Regions," in *Proceedings of the 26th International Cosmic Ray Conference*, Salt Lake City, 1999, **6**, pp.492-495.

27. Kóta, J. and Jokipii, J. R., *Science* **268**, 1024 (1995).

28. Kóta, J. and Jokipii, J. R., *Space Sci. Rev.* **83**, 137 (1998).

29. Jokipii, J. R. and Kóta, J., *Space Sci. Rev.* **72**, 379-384 (1995).

30. Desai, M. I. *et al. J. Geophys. Res.*, **104**, 6705 (1999).

31. Gruntman, M. A., *Rev. Sci. Instr.* **68**, 3617-3656 (1997).

32. Hsieh, K. C. and Curtis, C. C., "Imaging Space Plasma with Energetic Neutral Atoms above 10 keV" in *Measurement Techniques for Space Plasmas: Fields*, edited by R. F. Pfaff, J. E. Borovsky and D. T. Young, American Geophysical Union, Washington, DC, 1998, Geophysical Monograph Series, # 103, pp. 235-249.

33. Hilchenbach, M., Hsieh, K. C. and Czechowski, A. "Heliospheric energetic Hydrogen Atoms as a Source of Interplanetary Energetic Protons," in *Proceedings of Solar Wind 9*, edited by S. R. Habbal, R. Esser, J. V. Hollweg, and P. A. Isenberg, American Institute of Physics, CP 47,1999, pp. 779–782.

Observations of Pick-up Ions in the Outer Heliosphere by Voyagers 1 and 2

S.M. Krimigis[1], R.B. Decker[1], D.C. Hamilton[2], and G. Gloeckler[2]

[1] *Applied Physics Laboratory Johns Hopkins University*
[2] *University of Maryland College Park*

Abstract. Observations from the Low Energy Charged Particle (LECP) instrument on the Voyager 1 (V1) spacecraft at ~70 AU during the last solar minimum revealed an excess of counts from the sunward direction in the nominal proton energy range ~40 to ~140 keV. This, above background, response was also seen by Voyager 2 at ~55 AU and was present during the previous solar minimum at V1 at ~28 AU. The angular distribution is inconsistent with these counts being due to hot protons convected into the sunward-viewing sector only. We have examined possible sources for the observed counts and have considered that they may be due to singly-ionized heavy (A≥16) ions picked up by the solar wind and convected into the sunward sector of the detector. The spectrum is steep ($dj/dE \propto E^{-6}$), and exhibits a cutoff at ~25 keV/nuc. This observation suggests that pickup ions may be present beyond 5 AU, and could be the pre-accelerated seed population for anomalous cosmic rays (ACR).

INTRODUCTION

Anomalous cosmic rays (ACR) have been observed throughout the solar system, from points inside the orbit of earth to the position of V1 at 72 AU (1,2), and are thought to originate from neutral interstellar atoms entering the solar system, becoming photoionized, sensing the solar wind electric field, and eventually becoming accelerated to energies up to 100's of MeV at the solar wind termination shock (3). The seed population is thought to be pickup ions observed principally by Ulysses (4).

The vacuum of observations of pickup ions beyond 5 AU is principally due to lack of instruments that possess the appropriate resolution in energy and species to make such observations possible on those spacecraft that have reached the outer solar system such as Pioneer and Voyager. It is possible however, as described below, that identification of such ions at times of solar minimum may be made by appropriate analysis of observations that address the low end of the energy spectrum, namely in the range of total energy from ~30 to ~500 keV.

The data presented below suggest that newly ionized oxygen ions were observed near solar minimum in the 1994-1998 time frame, and previously during the 1986 solar minimum by the LECP operating on the V1 and V2 spacecraft located at ~70 AU and ~55 AU, respectively during the latest period. The LECP sensor of interest to the present study is detector α (in the LEMPA subsystem of the LECP) that is a solid state detector whose thickness is 96.5 μ on V1 and 89.1 μ in V2, has a field of view of 45° with an overall geometry factor of 0.04 cm^2-sr and 0.113 cm^2-sr in V1 and V2, respectively. The energy thresholds for each of the single parameter, total energy channels for the 2 spacecraft are listed in (5). It is noted from the reference that the lowest threshold at V1 for protons is ~40-53 keV, while that for oxygens is ~100-135keV.

OBSERVATIONS

The 10-day averages for ~0.5 MeV protons in Figure 1 show that minimum intensities were observed during the 1986-1988 interval and again from 1995 through the recent upturn in activity in the latter part of 1998. Also shown is the plot of the sunward-viewing sector-1 (third from top) from detector α referred to above, corrected for

CP528, *Acceleration and Transport of Energetic Particles Observed in the Heliosphere: ACE 2000 Symposium*, edited by Richard A. Mewaldt, et al.
© 2000 American Institute of Physics 1-56396-951-3/00/$17.00

background, and also sector-7, at ~90° to sector-1. The reason for including this detector channel (PL01) will be explained in the next figure. It is evident that intensities at this low energy are minimal during the same periods as those for the ~0.5 MeV protons.

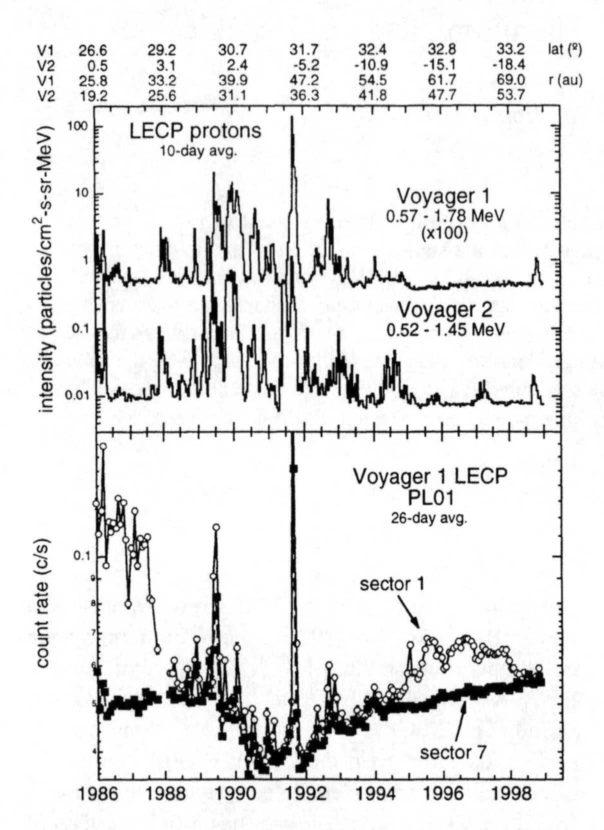

FIGURE 1. Overview of low energy proton intensities in the outer heliosphere from two-parameter measurements on V1 and V2, 1986-1998 (upper panel). Intensities of low energy ions (lower panel), corrected for background are shown for two viewing directions.

Since detector α makes a 1-parameter, total energy measurement only, it is not possible to readily identify nuclear or ionic species. Further, during times of low foreground activity, penetrating cosmic rays contribute a background which is different for each of the detector's 8 energy channels. There exists, however, angular information that permits separate observations of each count rate as a function of direction. The sectoring scheme and view directions for each of the 8 sectors is shown in Figure 2; the pie diagram also shows that sector 8 is permanently blocked.

As one can see in the bottom panel of Figure 2, 6 out of 7 sectors plotted have count rates (left axis) that are within one sigma of each other, suggesting that these directions represent the instrument background, mostly due to penetrating cosmic rays. Sector 1, however, shows a large increase, apparently due to foreground counts. Note that the edge of sector 1 is in the direction of the sun, and that is the sector which is most likely to be viewing any energetic component associated with the sun or the solar wind.

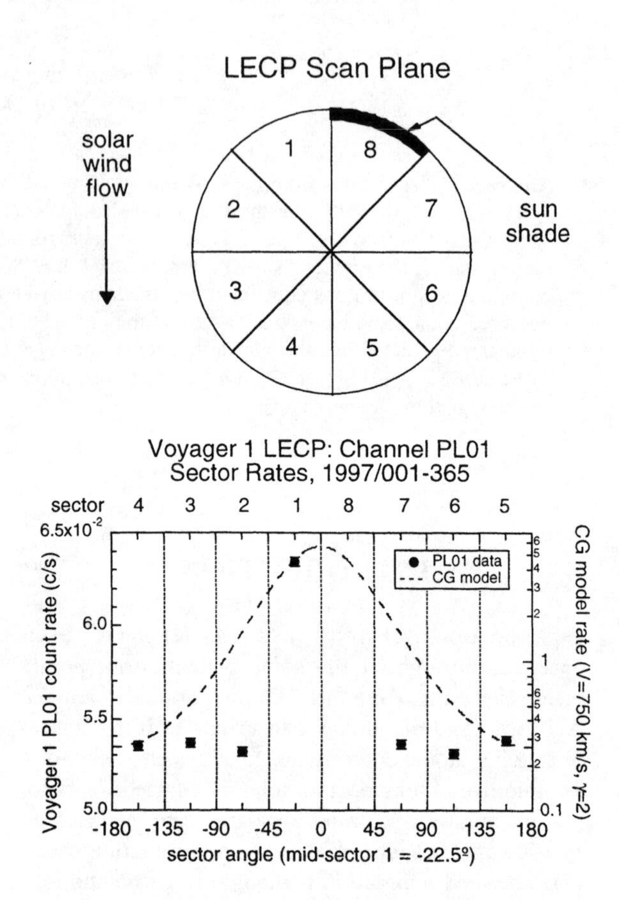

FIGURE 2. Top: Sectoring scheme near the ecliptic plane for the low energy detector α. Bottom: The response of each sector (symbols), averaged for 1997, shows only sector-1 to be clearly above background. The distribution cannot be explained by Compton-Getting effect (dotted line, right axis) for a high-speed solar wind 750 km/s.

We note that the detector is sensitive to x-rays (6) and the post-Jupiter efficiency is estimated at ~2%/keV for the lowest energy channel. The low efficiency, combined with the distance of V1, make it most unlikely that the response of the detector can be explained by solar x-rays. An inspection of solar x-ray emission records from 1994-1998 show that even the hardest solar x-ray event could not have produced a measurable response in detector α. Other, potential, spurious effects such as cross-talk, pulse pileup, etc. have been investigated and found to be insignificant.

Also shown in Figure 2, is the expected angular response (count rate along right axis) if the assumption is made that the detector is counting protons which would be convected past the spacecraft and therefore subject to the Compton-Getting effect (e.g., 7). To estimate the Compton-Getting response, we used a solar wind flow speed of 750 km/s and an ambient energetic proton intensity $dj/dE \propto E^{-\gamma}$, with spectral slope γ=2. It is evident from the calculated profile that such an assumption does not at all fit the data.

Figure 3 shows the responses of the sunward sector of the first three channels of LEMPA, translated into fluxes under the assumption that they are due either to protons (right line) or oxygens (left line). Under both assumptions the spectra are steep ($dj/dE \propto E^{-6}$), and when extrapolated to higher energies predict fluxes in the general vicinity of the PHA spectra (2) at energies ≥300 keV/nuc.

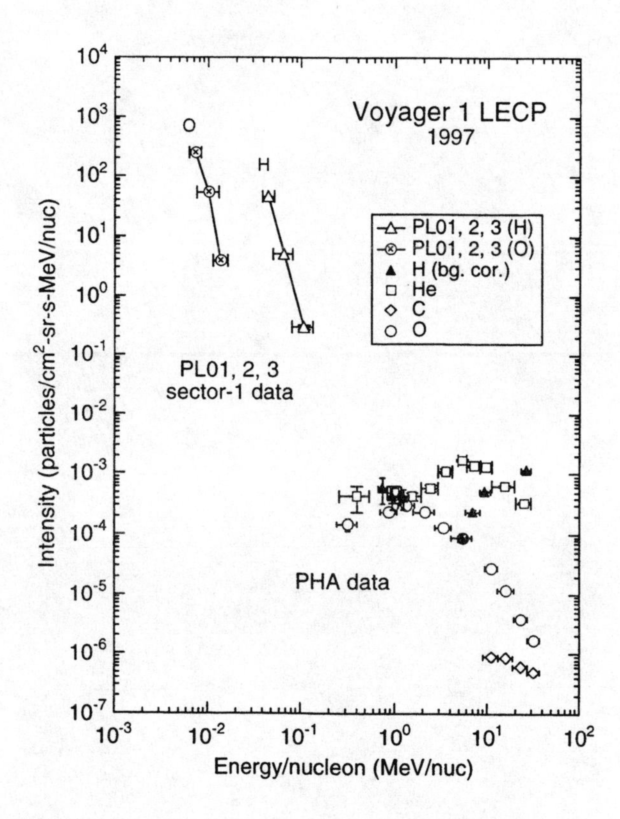

FIGURE 3. The observed spectrum under the proton (H) or oxygen (O) assumptions, plotted on the same scale as the two-parameter measurements of Hamilton et al. (2).

Discussion

Two possibilities are considered: (a) the ions are accelerated protons; and (b) that the ions are oxygens. The former possibility arises from the likelihood that CIR-associated increases may well be operating in the vicinity of V1 or that such accelerated protons may have been generated in the vicinity of the heliospheric current sheet and then transported to the high latitudes where V1 is located. Inspection of Figure 1, however, does not reveal the presence of CIRs, in the way that they are normally manifested as increases in ~0.5 MeV protons. Secondly, if these were indeed CIR protons their spectrum would likely resemble that observed previously in an earlier part of the solar cycle (8), but that does not appear to be the case. The earlier CIR intensity energy spectra are very hard, with exponents ~1.5 in this energy range (~30-140 keV) while the observed intensity spectrum in this case is exceptionally soft with an exponent of ~6.

The possibility that these are pickup oxygens seems to be much more plausible. Interstellar oxygen atoms entering the heliosphere and becoming ionized either through solar UV or charge exchange would sense the solar wind electric field and then acquire velocities that range up to twice that of the solar wind (4). A solar wind velocity of 550 km/s would provide enough energy to the pickup ions to bring them into the lowest energy oxygen threshold of the V1 detector at ~100 keV, i.e. 6.2 keV/nuc. At twice that speed, where the cutoff energy for the oxygens would occur, the energy for such particles would be ~25 keV/nuc (i.e. close to where the present cutoff in the oxygen spectrum in Figure 3 is observed).

We note that velocities substantially >500 km/s are likely to obtain at the location of V1, although no direct measurement of the solar wind is available from this spacecraft. The results from the high latitude pass from Ulysses (9), however, when projected to the location of V1, also at north latitudes of ~33° suggest that velocities as high as 750 km/s obtained in the northern polar coronal hole in late 1996 and quite likely throughout 1997. Further, if these ions were oxygen they would indeed be confined in the solar sector of the LECP detector and would not be affected by Compton-Getting to the extent that higher energy protons would have been. Thus, the data is consistent with the angular distribution shown in Figure 2. It is important to note here that if the ions were pickup helium instead of oxygen, their energy would not have been sufficiently high at these solar wind velocities to bring these particles into the energy range of the LECP detector. The ions could have been Fe at high solar wind speeds (10), but a variety of considerations, including observations at V2 where solar wind speed is measured, makes that interpretation unlikely. Preliminary calculations show that the densities are comparable to those predicted (11) for pickup oxygen at these distances.

ACKNOWLEDGMENTS

This work has been supported by NASA grant NAG5-4365 to the Johns Hopkins University Applied Physics Laboratory and by subcontract at the University of Maryland.

REFERENCES

1. Cummings, A.C. and Stone, E.C., Anomalous cosmic rays and solar modulation, *Space Sci. Rev.*, **83**, 51-62 (1998).

2. Hamilton, D.C. et al, Changes in ACR spectra in the outer heliosphere between the 1987 and 1997 solar minima, *EOS*, **79**, F705, (1998).

3. Fisk, L.A., Kozlovsky, A.B., and Ramaty, R., An interpretation of the observed oxygen and nitrogen enhancements of low energy cosmic rays, *Astrophys. J.*, **190**, L35 (1974).

4. Gloecker, G. and Geiss, J., Interstellar and inner solar pickup ions observed with SWICS on Ulysses, *Space Sci. Rev.*, (1998).

5. Krimigis, S.M. et al, Characteristics of hot plasma in the Jovian magnetosphere; Results from the Voyager spacecraft, *J. Geophys. Res.*, **86,** 8227-8257, (1981).

6. Ipavich, F.M., The Compton-Getting effect for low energy particles, *Geophys. Res. Lett.*, **1,** 149, (1974).

7. Kirsch, E., Krimigis, S.M., Kohl, J.W., and Keath, E.P., Upper limits for x-ray and energetic neutral particle emission from Jupiter, *Geophys. Res. Lett.*, **8,** 169, (1981).

8. Decker, R.B., Krimigis, S.M., McNutt, R.L., and Kane, M.M., Spatial gradients, energy spectra, and anisotropies of ions $\geq$ 30keV and CIR shocks from 1 to 50AU, Proceedings 24th ICRC, *1995,* 4, pp. 421-425.

9. McComas, D.J. et al, Ulysses return to the slow solar wind, *Geophys. Res. Lett.*, **25,** 1-4, (1998).

10. Mitchell, D.G. and Roelof, E.C., Thermal ions in high speed solar wind streams: detection by the IMP 7/8 energetic particle experiments, *Geophys. Res. Lett.*, **1,** 661, (1980).

11. Le Roux, J.A. and Ptuskin, V.S., Self-consistent stochastic preacceleration of interstellar pickup ions in the solar wind including the effects of wave coupling and damping, *J. Geophys. Res.* **103**, 4799, (1998).

The Transition from Singly to Multiply-Charged Anomalous Cosmic Rays: Simulation and Interpretation of SAMPEX Observations

A.F. Barghouty[1], J.R. Jokipii[2], and R.A. Mewaldt[1]

[1] *California Institute of Technology, Pasadena, CA 91125, USA*
[2] *University of Arizona, Tuscon, AZ, 85721, USA*

Abstract. Multiply-charged anomalous cosmic rays (ACRs) can arise when singly-charged ACR ions are stripped of one or more of their electrons during their acceleration via, e.g., the process of diffusive shock-drift acceleration at the solar-wind termination shock. Recent measurements of the charge states of ACR neon, oxygen, and nitrogen by SAMPEX at 1 AU have shown that above ≈ 25 MeV/nucleon these ions are multiply charged. In addition, SAMPEX observations have also established that the transition from mostly singly-charged to mostly multiply-charged ACRs (defined as the 50% point) occurs at a total kinetic energy of ≈ 350 MeV. Preliminary simulations for ACR oxygen based on a theory of multiply-charged ACRs were able to show a transition energy at ≈ 300 MeV. However, the simulated intensity distribution among the various charge states was inconsistent with observations. This paper reexamines the predictions of the theory in light of new SAMPEX ACR observations and recently developed and refined estimates of hydrogen-impact ionization cross sections. Based on simulations for multi-species ACR ions, we find that the transition energy is only weakly dependent on characteristic transport parameters, and that the new ionization rates distribute the intensity among the charge states in a manner consistent with observations. The calculated transition energy is in excellent agreement with the measured value.

INTRODUCTION

Unlike galactic cosmic rays (GCRs) or solar energetic particles ions (SEPs) observed at 1 AU, anomalous cosmic rays (ACRs) are expected to be predominantly singly-charged ions. According to the theory (1), ACRs are believed to originate as interstellar neutrals that penetrate the heliosphere before getting ionized —either by solar radiation or by charge-exchange collisions with solar-wind protons— to become singly-charged pickup ions. These pickup ions are then convected by the solar wind to the solar-wind termination shock (SWTS) where they are accelerated up to tens of MeV/nucleon via, e.g., the process of diffusive shock-drift acceleration (2).

However, recent observational evidence from SAMPEX (3, 4) have shown that ACR neon, oxygen, and nitrogen above ≈ 25 MeV/nucleon are multiply charged, with ionic charge states of 2, 3, and higher. At energies lower than ≈ 20 MeV/nucleon most of the observed ACRs are singly charged. SAMPEX observations (4) have further established that the transition from mostly singly-charged to mostly multiply-charged ACRs (defined as the 50% point) occurs at a total kinetic energy of ≈ 350 MeV.

The predominance of multiply-charged ACRs at high energy has been interpreted as evidence that some ACR ions are stripped of one or more of their electrons during their acceleration at the SWTS (3, 6).

According to the theory of diffusive shock-drift acceleration (5, 6) the amount of energy an ACR ion gains in drifting along the SWTS (from the equator towards the poles during the 1990s) is proportional to its ionic charge q; energy gain $\approx 240 \times q$ MeV. As a result, multiply-charged ACRs are accelerated to higher energy per nucleon than the more abundant singly-charged ions. Thus, at energies > 240 MeV, the theory predicts that multiply-charged ACR ions will dominate.

Preliminary simulations based on a theory of multiply-charged ACRs (6) applied to ACR oxygen and using an older set of ionization cross sections (7), did show the presence of a transition energy around 300 MeV, above which the ACR oxygen is mostly multiply charged. However, the simulated intensity distribution among the charge states was skewed towards fully or almost fully stripped oxygen ions, in disagreement with SAMPEX observations (3).

The recent development (8, 9) of a refined set of ionization cross sections warrants a reexamination of the predictions of the theory. This paper focuses on the simula-

CP528, *Acceleration and Transport of Energetic Particles Observed in the Heliosphere: ACE 2000 Symposium*, edited by Richard A. Mewaldt, et al.
© 2000 American Institute of Physics 1-56396-951-3/00/$17.00

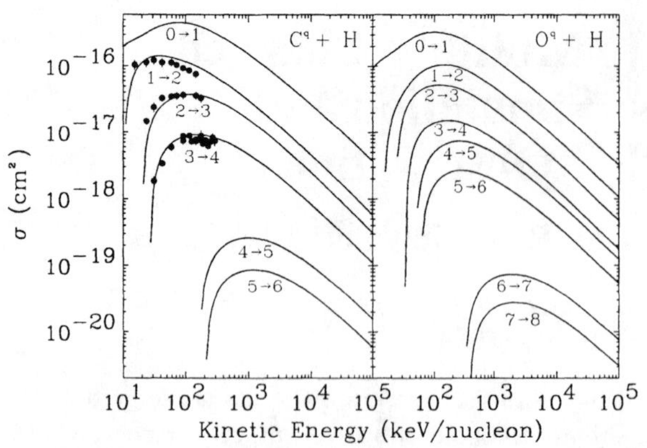

FIGURE 1. Estimated hydrogen-impact ionization cross sections for carbon and oxygen. Curves depict the cross sections $\sigma_{0\to1}$ (i.e., the differential cross section for the reaction $X^0 + H \to X^1 + H^* + e'$), $\sigma_{1\to2}$, etc., and terminate as the kinetic energy approaches the Born energy. [See (9) for reference on the experimental data for carbon.]

tion —using the new cross sections— and comparison to SAMPEX observations of ACR N, O, and Ne spectra at 1 AU, in particular, the observed transition from singly to multiply-charged ions.

IONIZATION CROSS SECTIONS

A new, theoretically-derived set of hydrogen-impact ionization cross sections applicable over a wide energy range has recently been developed (8, 9). This new cross-section set is derived using a formalism that is valid in the first Born approximation using measured (or estimated) electron-impact ionization cross sections (e.g., (10, 11, 12)). The formalism is based on a simple and elegant functional connecting the two sets of cross sections known as the Bates-Griffing relation. The Bates-Griffing relation is reformulated and a correction factor due to multiple transitions, i.e., sum over electrons' shells, is introduced. Fig. 1 shows a sample calculation using this new formalism for carbon and oxygen, where available data for carbon are also shown. Multi-electron removal is ignored in this calculation.

The previous simulations of multiply-charged ACRs (6) relied on ionization cross-section estimates based on oxygen-oxygen collisions (7). These estimates grossly overestimate the cross sections for multi-electron removal when applied to ion-hydrogen collisions. While data remain scarce for proton or hydrogen-impact ionization cross sections, multi-electron removal cross sections have been shown to be insignificant compared to single-electron removal in ion-electron collisions at energies relevant to ACR studies (e.g., (13, 14)).

For hydrogen-impact collisions at such energies, multi-electron removal is also expected to be insignificant compared to single-electron removal. For example, the cross section for the two-electron removal process $p+He \to p + He^{+2}+2e$ at 1 MeV/nucleon is found both experimentally (15) and theoretically (16) to be about 2.5 orders of magnitude smaller than that for the single-electron removal process $p+He \to p+He^{+1}+1e$.

SIMULATIONS

The simulations presented here are produced using the acceleration and transport model described in (6). The model solves the time-dependent Parker transport equation in 2 dimensions with drift terms:

$$\begin{aligned}\frac{\partial f_q}{\partial t} &= \frac{\partial}{\partial x_i}\left(\kappa_{ij}\frac{\partial f_q}{\partial x_j}\right) - V_{w,i}\frac{\partial f_q}{\partial x_i} \\ &\quad - V_{d,i}\frac{\partial f_q}{\partial x_i} + \frac{1}{3}\frac{\partial V_{w,i}}{\partial x_i}\frac{\partial f_q}{\partial \ln p} \\ &\quad + \text{sources}_q - \text{sinks}_q\ ,\end{aligned}$$

where $f_q(\vec{r},p,t)$ is the distribution function of the ACR ion with charge q. From left to right, the RHS terms in the equation depict spatial diffusion, convection, drift, energy gains and losses, and particle sources and sinks due to ionization.

The salient transport parameters used for this study are tabulated in Table 1. In addition, the simulations are presented with and without taking differential solar rotation into account. The reader is referred to Ref. (17) in this volume for further discussion of the theory and simulations. Below we discuss the simulations as they pertain to the recent SAMPEX ACR observations (4).

DISCUSSION

Fig. 2 shows the calculated ACR oxygen spectra at 1 AU. Comparing Fig. 2 with Fig. 1 of Ref. (6) there are significant differences in the charge state distribution. In Fig. 1 of Ref. (6), using the older set of ionization cross sections (7), the calculated spectrum at higher energies is dominated by charges 7 and 8, inconsistent with SAMPEX observations (3). In contrast, the new simulations show a summed spectrum that is dominated by charges 2 and 3 at higher energies. Although not shown in Fig. 2 for sake of clarity, the calculated contributions of charges

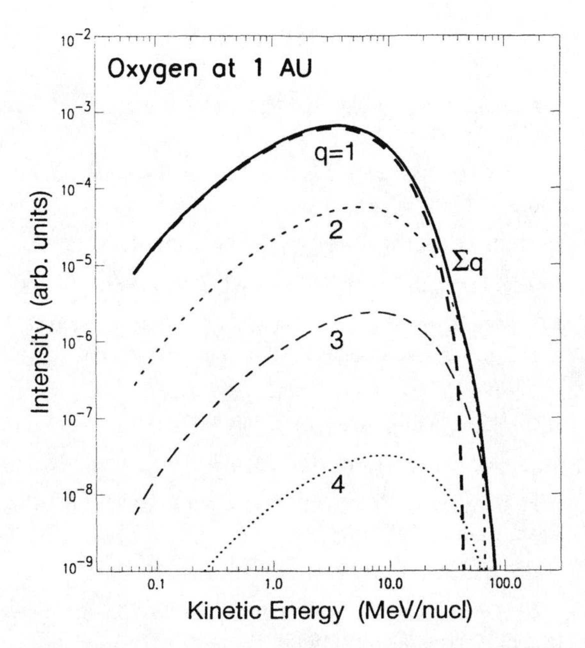

FIGURE 2. Simulated ACR oxygen spectra (in arbitrary units) at 1 AU. The top curve depicts the summed spectrum of charge states 1 through 8; while the second curve from the top depicts the spectrum for charge 1; third from the top charge 2, etc. The spectra are simulated taking differential solar rotation into account.

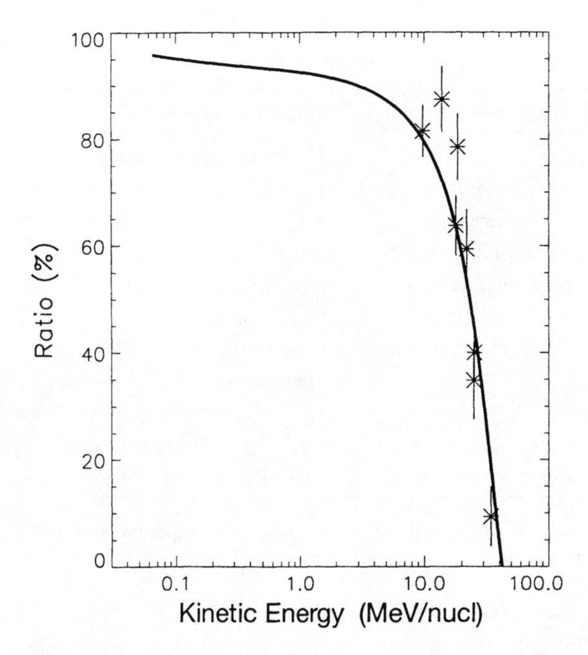

FIGURE 3. Calculated ratio of the singly-charged ACR oxygen to charges 1 through 3 as a function of energy. The 50% point is at $\approx$ 24 MeV/nucleon. Data points are from Ref. (4).

7 and 8 to the total spectrum at both high and low energies are, consistent with observations, negligibly small. Simulated spectra for nitrogen and neon show very similar trends.

Because of the difficulty in resolving charge states higher than 3, Ref. (4) defines the transition energy as the energy that corresponds to the 50% point of the relative abundance of charge 1 to the sum of charges 1 through 3. Below that energy point the spectrum is dominated by singly-charged ACR ions and above it by multiply-charged ions. Using this definition of the transition energy, Fig. 3 shows the calculated ratio as a function of energy and compares it to the measured ratio (4). The measured transition energy for ACR oxygen is 22 ± 2 MeV/nucleon. In excellent agreement with observations, from Fig. 3, the calculated transition energy is ≈ 24 MeV/nucleon. Again, although not shown, the same trends are seen for the calculated transition energies for nitrogen and neon.

In Fig. 4 we show the calculated transition energies for nitrogen, oxygen, neon as well as argon (although no measurement currently exists for argon) as a function of nuclear charge of the element. Multiply-charged ions can gain energy roughly in proportion to their ionic charge $\times$ the SWTS electrostatic potential while being accelerated at the shock (18).

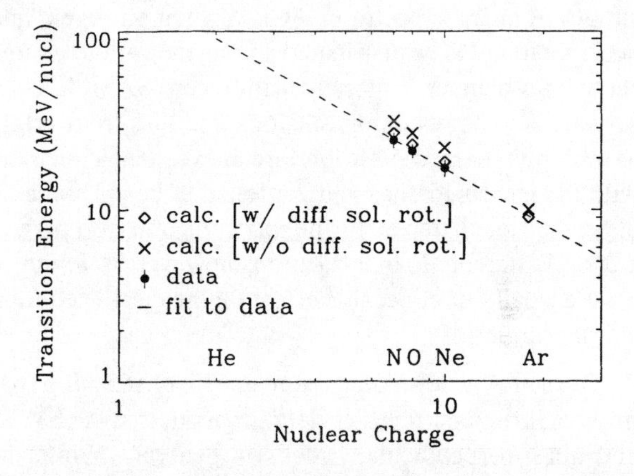

FIGURE 4. Calculated and measured 50%-point transition energy as a function of nuclear charge demonstrating the effect of differential solar rotation. Data points are from Ref. (4).

Calculated transition energies with and without the effect of differential solar rotation are shown in Fig. 4 (simulated spectra are qualitatively unchanged). Although it is not easy to surmise directly from the simulation model, this effect can decrease the electrostatic potential between the equator and the poles so as to give a somewhat lower transition energies as evident from Fig. 4. For argon, the model predicts a transition energy of ≈ 9 MeV/nucleon.

Table 1. Simulation parameters:

Transport parameter	Value/Description
Solar wind velocity	350 (equator) - 700 km/s (poles)
SWTS radius	100 AU
SWTS strength	2.5
Heliospheric boundary	160 AU
Neutral-H density	0.16 cm^{-3}
Injection energy	60 keV/nucleon
Heliospheric B-field	Parker's+polar modification +flat current sheet +$qA > 0$ conditions
Paral. diffusion coeff.*	$\propto R^{1/2}\,\beta/B(\vec{r})$
Perp. diffusion coeff.	0.03×Paral.

* R is ion's rigidity, β is its speed divided by the speed of light, and B is the strength of the magnetic field at the heliospheric point $\vec{r}$

CONCLUSIONS

Using the new ionization cross-section estimates the simulated energy spectra of ACR oxygen, nitrogen, and neon are found to be distributed among the various charge states in a manner consistent with recent SAMPEX observations. The resulting transition energies from singly to multiply-charged ACR ions are in excellent agreement with the measured ones and are found to be only weakly dependent on characteristic heliospheric transport parameters. Differential solar rotation, however, is found to have a small but noticeable effect on the calculated transition energies.

The achieved level of agreement with observations has implications to particle acceleration models at the SWTS and quasi-perpendicular shocks in general. While the reader, again, is referred to Ref. (17) in this volume for further discussion of such implications, we conclude here that the agreement achieved in this study clearly supports the diffusive shock-drift acceleration model of ACRs. Also, the agreement underscores the critical role precise ionization cross sections play in this and other models of charged particle transport in the heliosphere.

ACKNOWLEDGMENTS

Work is supported by NSF grant no. 9810653 and NASA-JOVE NAG8-1208 (A.F.B.) and by NASA grants NAS5-30704 and NAG5-6912 at Caltech.

REFERENCES

1. L.A. Fisk, B. Koslovsky, and R. Ramaty, Astrophys. J. Lett. **190**, L35 (1974).
2. M.E. Pesses, J.R. Jokipii, and D. Eichler, Astrophys. J. Lett. **246**, L85 (1981).
3. R.A. Mewaldt et al., Astrophys. J. Lett. **466**, L43 (1996).
4. B. Klecker et al., Space Sci. Rev. **83**, 259 (1998).
5. J.R. Jokipii, in *Physics of the Outer Heliosphere*, eds. S. Grzedzielski and D.E. Page (Oxford: Pergamon), p. 169 (1990).
6. J.R. Jokipii, Astrophys. J. Lett. **466**, L47 (1996).
7. W. Spjeldvik, Space Sci. Rev. **23**, 499 (1979).
8. A.F. Barghouty, Proc. 26th Int. Cosmic-Ray Conf. (Salt Lake, Utah) **7**, 555 (1999).
9. A.F. Barghouty, Phys. Rev. A**61**, 052702 (2000).
10. M. Arnaud and R. Rothenflug, Astron. & Astrophys. (Suppl. Series) **60**, 425 (1985).
11. M. Arnuad and J. Raymond, Astrophys. J. **398**, 394 (1992).
12. V.P. Shevelko et al. 1983, Monthly Notices Royal Astron. Soc. **203**, 45P (1983).
13. E. Krishnakumar and S.K. Srivastava, Int'l. J. Mass Spect. & Ion Proc. **113**, 1 (1992).
14. H. Deutch et al., Int'l. J. Mass Spect. **192**, 1 (1999).
15. M.B. Shah and H.B. Gilbody, J. Phys. B **18**, 899 (1985).
16. M.L. McKenzie and R.E. Olson, Phys. Rev. A **35**, 2863 (1987).
17. J.R. Jokipii, *this volume*, (2000).
18. R.A. Mewaldt et al., Geophys. Res. Lett. **23**, 617 (1996).

A Self-consistent, Unified Model Of Pickup Ion And Anomalous Cosmic Ray Transport

Jakobus A. le Roux[1], Horst Fichtner[2], Gary P. Zank[1], and Vladimir S. Ptuskin[3]

[1]*Bartol Research Institute, University of Delaware, Newark, DE, 19716*
[2]*Institut für Theoretische Physik, Lehrstuhl IV: Weltraum- und Astrophysik, Ruhr-Universität Bochum, Bochum, Germany*
[3]*Institute for Terrestrial Magnetism, Ionosphere, and Radiowave Propagation (IZMIRAN), Troitsk, Moscow District, Russia*

Abstract. A self-consistent model is presented that gives a unified, fully theoretical description of pickup ion (PUI) and anomalous cosmic ray (ACR) proton transport in the near ecliptic regions of the upwind heliosphere. Unification is achieved through multiply reflected ion (MRI) acceleration theory which is used for the injection of pickup protons into the process of diffusive shock acceleration at the quasi-perpendicular, supercritical termination shock (TS). The results show that MRI accelerated pickup protons strongly mediate the TS while anomalous protons also significantly modify it. Although this lowers the cutoff speed of the MRI spectrum, a sufficient population of anomalous protons is formed at the TS to give a favorable comparison of modulated spectra upstream with observations at large heliocentric distances.

INTRODUCTION

An interesting outstanding problem in heliospheric physics is whether and to what degree the TS as the generator of ACRs is mediated by these particles. Unfortunately, models based on standard non-linear diffusive shock acceleration theory are ill-suited to describe the highly anisotropic nature of the reflection, initial acceleration and injection of PUIs into diffusive shock acceleration in a realistic way so that injection is treated as a free parameter constrained by indirect means [1]. Self-consistent hybrid calculations, on the other hand, are very computer intensive. To overcome these limitations, the pickup proton source spectrum for non-linear diffusive shock acceleration at the TS is determined self-consistently from MRI theory [2] in a time-dependent model. The model is used to solve a set of 3-fluid equations for the solar wind plus pickup protons, the MRI-accelerated pickup protons, and ACR protons to determine the TS structure and position, and to solve the PUI and ACR transport equations to determine the particle spectra self-consistently at the TS.

INTRODUCTION

The PUI distribution function f_{PUI} is found from the analytical solution to the standard PUI transport equation given by

$$f_{PUI}(v) = \int_0^r \frac{Q(r_2)\delta\left(v - v_f\right) dr_2}{\left[r_2^2 u(r_2)\right]^{2/3}\left[r^2 u(r)\right]^{1/3}}, \tag{1}$$

where r is heliocentric distance, u is solar wind speed, $v_f = u(r_2)[r_2^2 u(r_2)/r^2/u(r)]^{1/3}$, and $Q(r_2) = n_H(r_2)\nu_i(r_2)r^2/4/\pi/u^2(r_2)$ with n_H the interstellar H density, and ν_i the ionization frequency of interstellar H due to predominantly charge exchange with solar wind protons. Equation (1) is a more general solution valid for a variable solar wind speed compared to the standard one valid for a constant upstream solar wind speed [3].

On the basis of MRI theory the cross-shock potential for pickup protons at a supercritical perpendicular shock [4] is estimated as

$$e\phi(x) = \eta(s-1)mu_1^2 / M_{A1}^2, \tag{2}$$

CP528, *Acceleration and Transport of Energetic Particles Observed in the Heliosphere: ACE 2000 Symposium*, edited by Richard A. Mewaldt, et al.
© 2000 American Institute of Physics 1-56396-951-3/00/$17.00

where e is the charge, ϕ is the potential, s is the compression ratio of the shock, m is the proton mass, u_1 is the upstream flow speed, M_{A1} is the Alfvénic Mach number upstream, and $\eta = 3$ initially. By using equation (2), one finds that pickup protons will be reflected at the shock when $v_x \leq V_{spec}$ [2] where

$$V_{spec} = \left[2\eta(s-1)u_1^2 / M_{A1}^2\right]^{1/2}. \tag{3}$$

Viewing the upstream pickup proton distribution at the TS as consisting of a series of thin nested isotropic shells in velocity space due to adiabatic cooling or heating, the fraction of PUIs reflected by the shock potential in a given shell with particle speed v is given by

$$R_{refl}^{shell}(v) = 0.5\left[(v - v_{\lim})/v\right]H(v - v_{\lim}), \tag{4}$$

where $v_{\lim} = u_1 - V_{spec} - v_{sh}$, v_{sh} is the TS speed, and the Heaviside function H ensures that $R_{refl}^{shell}(v) = 0$ for those shells unaffected by the shock potential. Assuming that the MRI acceleration process causes an isotropic, power law MRI distribution for PUIs to be attached to each PUI shell, the resultant MRI spectrum is determined by

$$f_{MRI}(v) = (3-a)v^{-a} \times \int_0^v \frac{w^2 f_{PUI}(w) R_{refl}^{shell}(w)}{v_{\max}^{3-a} - w^{3-a}} dw, \tag{5}$$

where a is the power law index, and $v_{\max}$ is the maximum speed pickup protons reach along the TS front due to MRI acceleration. This speed is calculated according to [2]

$$v_{\max} = \eta(m/m_e)^{1/2} V_{A1}(s-1)/x, \tag{6}$$

where m_e is the electron mass, V_{A1} is the Alfvén speed upstream, and x is the assumed ratio of the TS (sub)ramp width to an electron inertial length . The pressure of the MRI accelerated pickup protons is determined according to the standard expression

$$P_{MRI} = \int_{v_{\lim}}^{v_{\max}} \frac{4\pi}{3} p^3 v f_{MRI}(p) dp. \tag{7}$$

To determine the collective behavior of the solar wind plus pickup proton fluid (TS structure and position) under the dynamic influence of MRI accelerated pickup and ACR protons as low mass density fluids, a set of 3-fluid equations is solved numerically. The total mass density equation for the 3 fluids is

$$\frac{\partial \rho}{\partial t} + \frac{1}{r^2}\frac{\partial}{\partial r}\left(r^2 \rho u\right) = mQ_{ph}, \tag{8}$$

where ρ is the total mass density of the solar wind plus pickup protons, and Q_{ph} the pickup proton production rate as interstellar H is photo-ionized by solar radiation. The total momentum density equation is

$$\frac{\partial \rho u}{\partial t} + \frac{1}{r^2}\frac{\partial}{\partial r}\left(r^2 \rho u^2\right) = -\frac{\partial P}{\partial r} - \frac{3}{2}\frac{P_{MRI}}{\Delta r} - \frac{\partial P_{ACR}}{\partial r} - muQ_{ce} \tag{9}$$

where P is the combined pressure of the solar wind plus pickup protons, P_{MRI} is the pressure of MRI accelerated pickup protons, P_{ACR} is the pressure of ACR protons, and Q_{ce} is the production rate of pickup protons due to charge exchange of neutral H with the solar wind. The total energy density equation is

$$\frac{\partial e}{\partial t} + \frac{1}{r^2}\frac{\partial}{\partial r}\left(r^2 u[e+P]\right) = -u\frac{\partial P_{ACR}}{\partial r} - u_1\frac{3P_{MRI}}{2\Delta r} - \frac{1}{2}m\left[u^2 H(r_{sh} - r) + v_{rms}^2 H(r - r_{sh})\right]Q_{ce} \tag{10}$$

where $e = 0.5\rho u^2 + P/(\gamma-1)$, $\gamma = 5/3$, v_{rms} is the root-mean-square average speed of solar wind thermal protons, and r_{sh} is the heliocentric distance to the TS. For more details see [5].

The 3-fluid equations are closed by solving numerically the standard transport equation for ACR protons

$$\frac{\partial f}{\partial t} + u\frac{\partial f}{\partial r} - \frac{1}{r^2}\frac{\partial}{\partial r}\left(r^2 \kappa_{rr}\frac{\partial f}{\partial r}\right) - \frac{1}{3r^2}\frac{\partial}{\partial r}\left(r^2 u\right)p\frac{\partial f}{\partial p} = S(r_{sh}, p, t), \tag{11}$$

where $f(r, p, t)$ is the isotropic part of the near-isotropic ACR distribution function, κ_{rr} is the spatial diffusion coefficient in the radial direction, and S is a source function based on the combined spectrum of unreflected pickup protons (equation (1) and MRI accelerated pickup protons (equation (5)) of which a portion of the MRI spectrum is injected into the process of diffusive shock acceleration at the TS for particle speeds $v \geq v_{inj} = 4\alpha u_2(\kappa_{\parallel}/\kappa_{\perp})^{0.5}$ where $\alpha = 2$, u_2

is the downstream solar wind speed, $\kappa_{\parallel}$ is the parallel, and $\kappa_{\perp}$ is the perpendicular diffusion coefficient, respectively [6]. κ_{rr} is determined theoretically on the basis of quasi-linear theory for parallel diffusion, a theory for the random walk of field lines applied to perpendicular diffusion, and a transport theory for MHD turbulence in the solar wind [7, 8].

RESULTS

In Figure 1 it is shown how the quasi-perpendicular TS structure and position (TS initially at ~96 AU) is affected by MRI accelerated pickup protons alone (dashed curve in left and right panels), and by both MRI particles and ACR protons (solid curve in left and right panels). The simulation was done by assuming that $f_{MRI} \propto v^{-2}$ [9] and that $x = 2$ [10]. For $n_H = 0.115$ part. cm^{-3} we find that MRI accelerated pickup protons combined effect of MRI and diffusive shock acceleration of pickup protons reduces the upstream solar wind speed by ~38% and the TS compression ratio to $s \approx 1.75$.

The fraction of reflected pickup protons $R_{refl} \approx 14\%$ without ACR effects. This is considerably higher by a factor of ~ 4-5 compared to the case where the solar wind speed is constant upstream of the TS because of (1) a linearly decreasing solar wind speed due to charge-exchange of interstellar H with solar wind protons which reduces the effect of adiabatic cooling on the PUIs, (2) adiabatic heating in the MRI-induced TS foot, and (3) the lower convection speed of the PUI distribution toward the TS (see equation (4)). When ACR protons are included, there is an increase to $R_{refl} \approx 19\%$ due to additional adiabatic heating in the TS precursor and a further reduction in the solar wind convection speed.

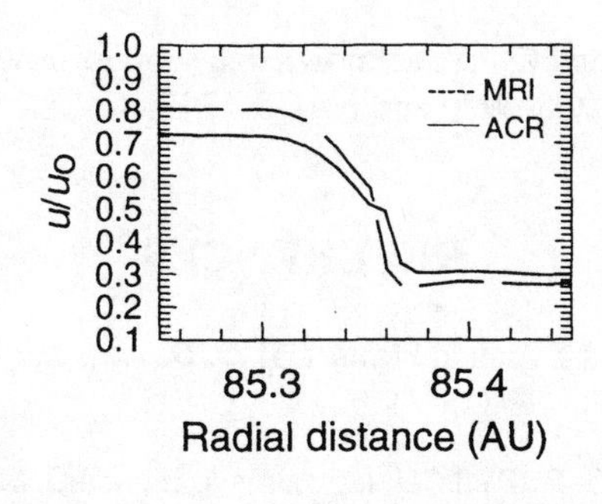

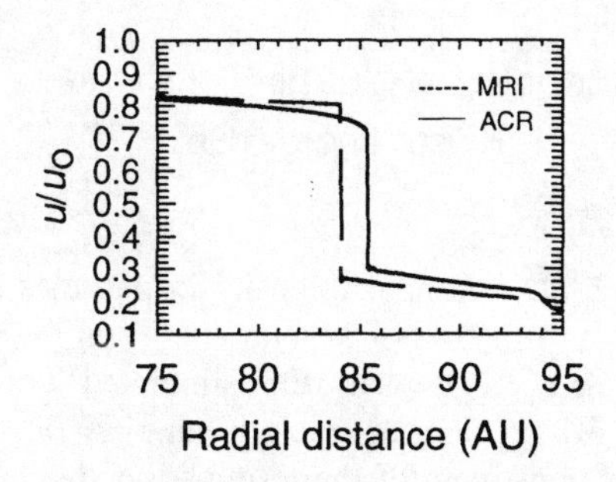

FIGURE 1. Solar wind speed normalized to 400 km/s as a function of heliocentric distance in AU. The dotted curves denote the case where MRI-accelerated pickup protons mediate the TS while the solid curves also include the modifying effects of diffusively TS accelerated ACR protons. The left panel provides a close-up view of TS.

strongly mediates the TS with the compression ratio being reduced from the initial value $s = 3.1$ to $\sim s = 1.9$ (see dotted curve in see left panel for a close-up view) which constitutes a ~30% reduction in the upstream solar wind speed. This MRI-induced foot has a length scale of ~7×10^{-3} AU corresponding to the gyroradius of the MRI particles with the highest speed. Viewed on a larger scale (dashed curve in right panel), appropriate for detecting TS modification by ACR protons, the TS modification by MRI particles becomes almost invisible. A comparison of the curves in the right panel reveals that the TS is significantly modified by ACR protons on a much longer (e-folding) length scale of ~2.2 AU compared to MRI particles. The TS compression ratio in this case appears to be $s \approx 2.3$, which is the value that the more energetic ACR protons see during the diffusive shock acceleration process. The close-up view of the TS structure (solid curve in left panel) shows that the

In Figure 2 simulated particle spectra are shown at various heliocentric distances throughout the heliosphere. The spectrum downstream of the TS (top solid curve) features the combination of unreflected pickup protons (kinetic energies E_k < 1-2 keV), MRI-accelerated pickup protons (1-2 < E_k < 9-10 keV), and ACR protons (E_k > 20-30 keV). The dashed curve on the left denotes the PUI distribution upstream while the dashed curve on the right represents the MRI spectrum upstream. The peaks in the unreflected PUI spectra near 1 keV at their cutoffs are caused by the decrease in the solar wind speed upstream of the TS (less adiabatic cooling). The peak shifts to somewhat higher energies from upstream to downstream as the PUIs are adiabatically heated while being convected through the TS. A similar shift in energy is noticeable in the in MRI spectra. Such a shift in energy is useful in that it makes it easier for the MRI particles to cross the threshold of diffusive shock acceleration. An elbow-like transition region connects the unreflected portion of the PUI spectrum with the MRI-accelerated part of the PUI spectrum. The size of the elbow depends on R_{refl} as well as on the power law index of the MRI spectrum and v_{max} (see equation (5)). A large elbow connects the MRI spectrum with the ACR spectrum formed by the process of diffusive shock acceleration. This weak connection results from the fact that $v_{max} < v_{inj}$. Initially, $v_{max} > v_{inj}$ but modification of the TS by ACRs reduces v_{max} (see equation (6)). Although $v_{max} < v_{inj}$ the MRI and ACR

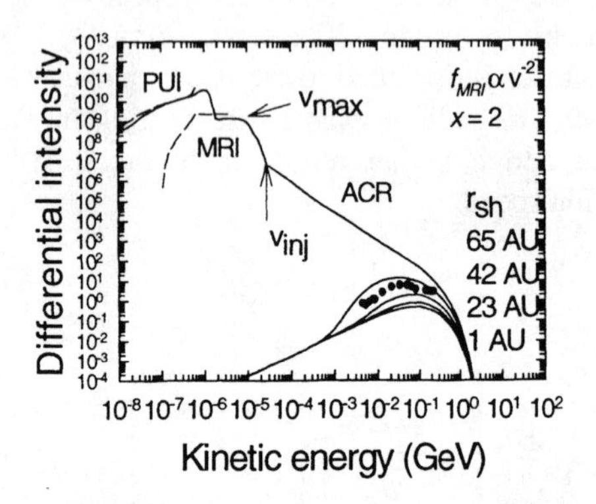

FIGURE 2. Differential intensity in particles m^{-2} s^{-1} sr^{-1} MeV^{-1} as function of kinetic energy in GeV. The curve labeled r_{sh} depicts the simulated combined pickup, MRI-accelerated pickup, and ACR proton spectrum downstream of the termination shock, while the modulated ACR proton spectra upstream are shown at heliocentric distances 1, 23 42, and 65 AU. The dashed curve on the left (right) is the PUI (MRI) spectrum upstream. The filled circles denote a reproduction of Voyager 1 data observed at 52 AU during 1997 [11].

spectra stay connected because both v_{max} and v_{inj} fluctuates sufficiently in time. Consequently, enough MRI accelerated pickup protons are injected into the process of diffusive shock acceleration so that the modulated ACR proton spectra upstream compare favorably with observational data (filled circles in Figure 2). Other test runs underscored the robustness of the result discussed in this paper because they yielded comparable results when the power law index of the MRI spectrum was varied between a = 2-4 and when the (sub)shock width was changed from x = 1-4.

CONCLUSIONS

A self-consistent model is presented that gives a unified description of PUI and ACR proton transport in the near ecliptic regions of the upwind heliosphere. The model uses a 3-fluid approach for determining the large-scale solar wind and TS structure, and PUI and ACR transport theory for calculating the PUI and ACR proton spectra. Unification is achieved through MRI theory which is used for the injection of pickup protons into the process of diffusive shock acceleration at the quasi-perpendicular, supercritical TS. The results show that MRI accelerated pickup protons strongly mediate the TS while ACR protons also significantly modify it. Although the TS modification by ACR protons lowers the cutoff speed of the MRI spectrum to below the injection speed for diffusive shock acceleration, a sufficient population of ACR protons is formed at the TS to give a favorable comparison of modulated spectra upstream with observations at large heliocentric distances.

ACKNOWLEDGMENTS

J. A.l.R. and G.P.Z. acknowledge support from NASA grants NAG5-6969 and NAG5-7796.

REFERENCES

1. le Roux, J. A., and Fichtner, H., *J. Geophys Res.* **102**, 17365-17380 (1997).
2. Zank G. P., Pauls, H. L., Cairns, I. H., and G. M. Webb, *J. Geophys Res.* **101**, 457-477 (1996).
3. Vasyliunas, V. M., and Siscoe, G. L., *J. Geophys. Res.* **81**, 1247-1252 (1976).
4. Leroy, M. M., *Phys. Fluids* **101**, 2742-2753 (1983).
5. le Roux, J. A., Fichtner, H., Zank, G. P., and Ptuskin, V. S., *J. Geophys. Res.*, in press (2000).
6. Webb, G. M., Zank, G. P., Ko, C. M., and Donohue, D. J., *Astrophys. J.* **453**, 178-206 (1995).
7. Zank, G. P., Matthaeus, W. H., Bieber, J. W., and Moraal, H., *J. Geophys. Res.* **103**, 2085-2097 (1998).
8. Zank G. P., Matthaeus, W. H., and Smith, C. W., *J. Geophys Res.* **101**, 17093-17107 (1996).
9. Lipatov, A. S., Zank, G. P., and Pauls, H. L., *J. Geophys Res.* **103**, 29679-29696 (1998).
10. Newbury, J. A., Russell, C. T., and Gedalin, M., *J. Geophys. Res.* **103**, 29581-29593 (1998).
11. McDonald, F. B., *Space Sci. Rev.* **83**, 33-50, 1998.

Energetic particles beyond the heliospheric shock: Anomalous Cosmic Rays (ACRs), Pick-Up Ions (PUIs) and the associated Energetic Neutral Atoms (ENAs)

Horst Fichtner[1], Andrzej Czechowski[2], Hans J. Fahr[3] and Günter Lay[3]

[1]*Institut für Theoretische Physik IV: Weltraum- und Astrophysik, Ruhr-Universität Bochum, 44780 Bochum, Germany*
[2]*Space Research Centre, Polish Academy of Sciences, Bartycka 18A, 00-719 Warsaw, Poland*
[3]*Institut für Astrophysik und Extraterrestrische Forschung, Universität Bonn, Auf dem Hügel 71, 53121 Bonn, Germany*

Abstract. The Voyager 1 spacecraft is expected to encounter the heliospheric termination shock within the next decade. Besides the ongoing discussion how to possibly predict the time of this encounter, there is a growing interest into a more detailed description of the region beyond the heliospheric shock, i.e. the heliosheath. Refinements of the so far rather crude models will facilitate interpretation of forthcoming data. We report on results obtained with our model of the transport of ACRs in the heliosheath. In improvement of earlier approaches it is based on a solar wind background flow computed with a self-consistent large-scale model of the heliosphere. Besides these downstream ACR spectra, which will become accessible for in-situ observation as soon as the Voyager spacecraft will have crossed the heliospheric shock, we study the potential of observations of the flux of ENAs to remotely explore the structure of the heliosheath. In particular, as part of a comparison of the various ENA sources, we will address the significance of the contribution of those ENAs resulting from a de-charging of PUIs.

MOTIVATION

After almost three decades of in-situ exploration of the outer heliosphere, beginning with Pioneer 10's visit to Jupiter in 1972, there now exists increasing interest in observations of the so-called *heliospheric interface*, i.e. the region beyond the solar wind termination shock, also called the heliospheric shock.

This is because the majority of scientists working in the field believes that the nowadays outermost deep space probe Voyager 1, which is currently (Feb 2000) located at about 76 AU, must be relatively close to the heliospheric shock. In contrast to the situation about a decade ago, when estimates of the shock location could not be reasonably derived from data but were rather simple guesses reflecting more or less the personal expectation, today the cosmic ray observations from the deep space probes provide a much improved data base.

The spread in the predictions became significantly smaller over the last decade (see Fig. 1), and different methods now result in a shock location of 85±10 AU. The uncertainty is not only due to the estimation method but also to a motion of the shock (see, e.g., 14,18,20). The value of 85 AU is, interestingly, very close to one of the first estimates made about three decades ago (2).

Comparing this distance estimate with the trajectory of Voyager 1 results in the finding that the spacecraft could encounter the shock in 2003 ± 3. Therefore, one can expect this first encounter to happen soon and should be prepared with respect to an interpretation of forthcoming data. Or, in other words: what are the deep space probes expected to see in the region beyond the termination shock, i.e. in the heliospheric interface?

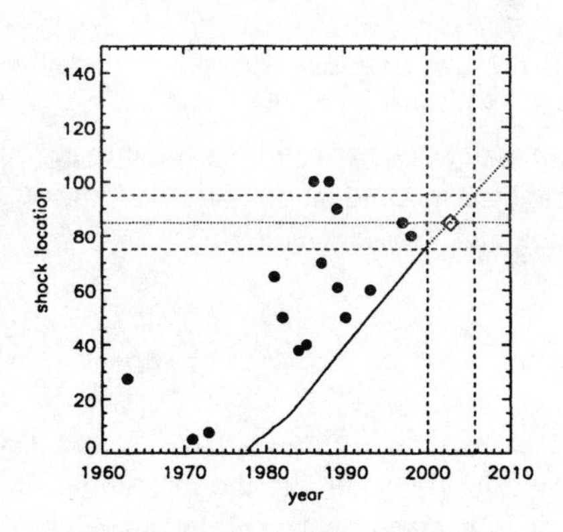

FIGURE 1. Some predictions (filled dots) of the location of the heliospheric shock made during the last three decades. The solid line indicates the heliocentric distance of Voyager 1. The diamond marks a hypothetical shock encounter at 85 AU, the dashed lines indicate the ranges of uncertainty.

CP528, *Acceleration and Transport of Energetic Particles Observed in the Heliosphere: ACE 2000 Symposium*, edited by Richard A. Mewaldt, et al.
© 2000 American Institute of Physics 1-56396-951-3/00/$17.00

ENERGETIC PARTICLE POPULATIONS BEYOND THE HELIOSPHERIC SHOCK

The model

At first sight, there are at least three populations of energetic particles of interest as agents to infer information about the heliospheric interface. Galactic cosmic rays (GCRs), coming from interstellar space, traverse the whole interface. Anomalous cosmic rays (ACRs), supposedly originate at the heliospheric shock. And, energetic neutral atoms (ENAs), which result from charge exchange of both pick-up ions (PUIs) and ACRs with neutrals from the local interstellar medium (LISM) upstream and downstream of the shock.

Any reasonable model of the region beyond the heliospheric shock has to take into account, at least approximately, its overall geometrical structure. This structure, as resulting from contemporary modelling, is sketched in Fig. 2.

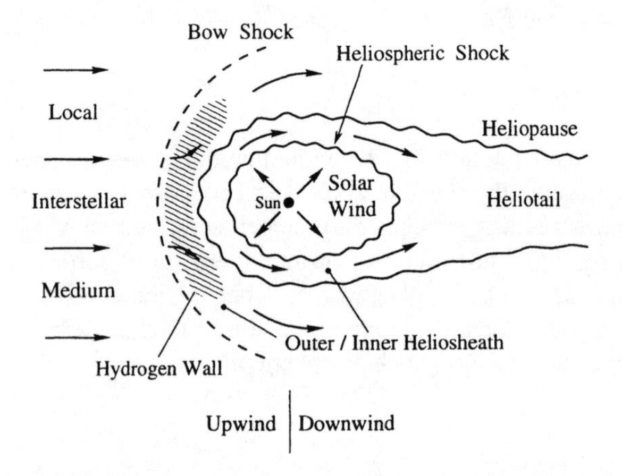

FIGURE 2. The large-scale structure of the heliosphere.

This overall structure can be taken into account in the cosmic ray transport equation (particle momentum is denoted by p, and S is a source term):

$$\frac{\partial f}{\partial t} = \nabla \cdot (\kappa \nabla f) - \vec{u} \cdot \nabla f + \frac{1}{3} (\nabla \cdot \vec{u}) \frac{\partial f}{\partial ln p} + S(\vec{r}, p, t)$$

via the velocity field $\vec{u}$. Inside the heliopause, i.e. the surface separating the solar and the interstellar plasma, the velocity field is given by the solar wind velocity $\vec{u} = \vec{u}_{sw}$, outside by the velocity of the LISM $\vec{u} = \vec{u}_{LISM}$. We have used the velocity field $\vec{u}$ that was obtained by (13), see also (8), for a configuration similar to that shown in Fig. 2, which resulted from a self-consistent 5-fluid computation taking into account the solar and interstellar wind, PUI-, ACR-, GCR protons as well as a neutral hydrogen background. Within the framework of such modelling the heliospheric interface is a *two-shock* or *Baranov interface* as opposed to a *one-shock* or *Parker interface* resulting from the classical Parker model (17).

To solve the transport equation for the region beyond the heliospheric shock, we assumed a simple power law shock spectrum (for a different view see (6)) as derived by (19) from Voyager data $j(E_k) = j(100keV) \left(\frac{E_k}{100keV}\right)^{-1.41}$, where $j(E_k)$ is the differential flux at kinetic energy E_k, $j(100keV) = 7 \cdot 10^4/(m^2 ssrMeV)$ and a scalar diffusion coefficient

$$\kappa(R) = \begin{cases} \kappa_{helio}(R) & \text{inside the heliopause} \\ 100\,\kappa_{helio}(R) & \text{outside the heliopause} \end{cases}$$

where $\kappa_{helio}(R)$ was taken from (16); R denotes particle rigidity. The isotropic, scalar nature of the diffusion should mainly be considered as a necessary assumption enabling us to limit the model computations to cases with axisymmetry in configuration space. Both the roles of anisotropic diffusion as well as of drifts in the heliospheric interface, which are neglected here but were recently studied by (10), are issues to be addressed in future work. The increase of κ at the heliopause is motivated by supposedly generally lower turbulence levels in the LISM.

Since the assumed velocity field is axisymmetric with respect to the upwind-downwind axis, the transport equation has to be solved in a three-dimensional phase space with ccordinates r (heliocentric distance), ϑ (off-upwind axis angle) and p (particle momentum), a task that we performed employing the VLUGR3 code (1).

ACR spectra beyond the shock

To facilitate the discussion we select the upwind and the downwind direction to show the spectral evolution with heliocentric distance. Figs. 3 and 4 give the ACR proton spectra from the shock outwards (upwind: r_{sh} = 89 AU; downwind: r_{sh} = 187 AU) for the 5-fluid model (8, 13). The differential flux is in arbitrary units.

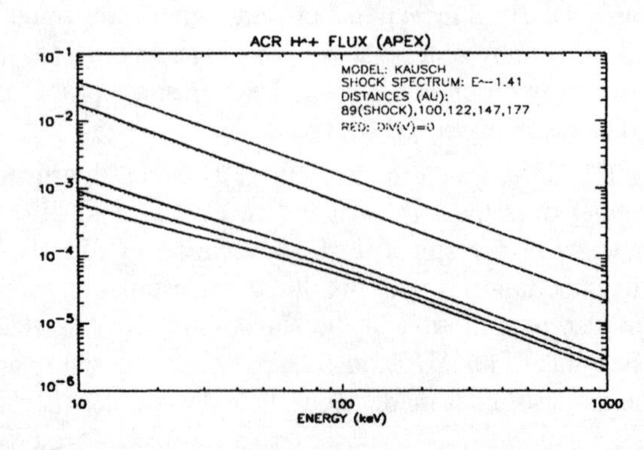

FIGURE 3. Upwind ACR spectra resulting from the 5-fluid model (8, 13).

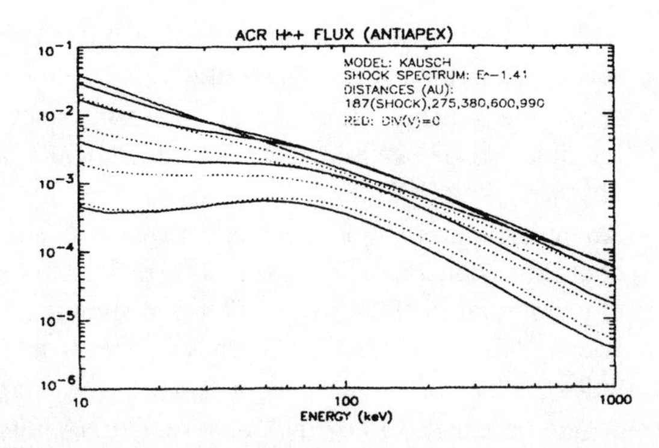

FIGURE 4. Downwind ACR spectra resulting from the 5-fluid model (8, 13). The dotted lines check on the influence of the velocity field $\vec{u}$: the transport equation was solved with the additional assumption that $div(\vec{u}) = 0$, for details see (5).

An inspection of the spectra reveals that the evolution of their shapes with distance is predominantly due to charge exchange. The overall flux decrease with increasing heliocentric distance is simply a geometrical effect.

Obviously, the latter is the dominant effect in the upwind direction (apex): the spectral gradient does not change, only for rather large distances of $r >$ 100-120 AU the charge exchange effect becomes weakly noticable. It is, in fact, impossible to distinguish between the 5-fluid and, for example, the Parker model (17) on the basis of the spectral evolution in the upwind direction, see (5).

The situation is quite different in the downwind direction (anti-apex) where the huge so-called heliotail, i.e. the subsonic solar wind region between the heliospheric shock and the heliopause in the downwind heliosphere (see Fig. 2), serves as storage volume for ACRs due to their relatively low diffusion there. As a consequence of the higher ACR number density in the heliotail, the charge exchange process operates significantly in a much extended distance range as compared to the upwind interface. Therefore, the corresponding spectra show, with strongly decreasing flux levels at low energies, a clear signature of the process.

While with these results we have, at least in some approximation, indeed fulfilled the task to compute the ACR spectra beyond the termination shock, the results are somewhat disappointing: it appears that we will, on the basis of the forthcoming ACR observations by the two Voyagers which are relatively close to the upwind direction, not be in the position to discriminate between different heliospheric interfaces because the spectra do not exhibit sufficient details there.

However, the finding that charge exchange is, if noticable at all, the dominant process, points to another particle population, which should be a suitable agent for obtaining better information about the large-scale heliospheric structure: given that the charge exchange operates so much more efficient in the downwind heliosphere, which will remain unaccessible for *in-situ* observations for the next few decades at least, one should expect the flux of the resulting ENAs from there to be significantly higher than that from the upwind direction.

The ENA flux for different heliospheric interface models

The expected flux asymmetry, a high downwind-to-upwind ratio of ENAs, has indeed been observed with the Solar Heliospheric Observatory (SOHO), see Fig. 5 and references (12) and (4). In the ENA energy range from 55-80 keV the flux was clearly enhanced when the CELIAS/HSTOF instrument on SOHO was looking into the downwind direction (around day 200 and day 560).

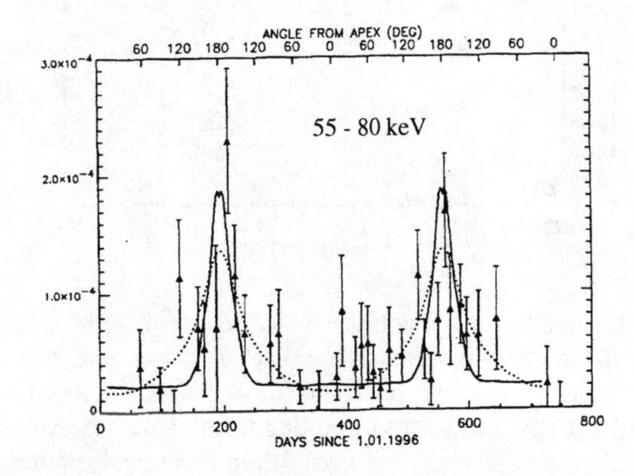

FIGURE 5. Modelling of the ENA flux observed with CELIAS/HSTOF aboard SOHO employing the 5-fluid model (solid line) and the Parker model (dotted line).

The dotted line gives the expectation from the one-shock Parker model, the solid line that from the two-shock 5-fluid model. Clearly, a comparison with the ENA fluxes allows a better discrimination between the two models and, thus, a check on the large-scale heliospheric structure. While this result appears satisfactory, the energy range of observed ENA fluxes is rather low, and one could ask: are these particles really originating from a charge exchange with ACRs?

ACR-ENAS OR PUI-ENAS?

Not only is the range from 55-80 keV rather low for ACRs, but also, in view of the observed PUI pre-acceleration and related modelling, see, e.g., (3, 9, 11) and (15), it appears quite plausible that PUIs do contribute appreciably to the observed ENA flux as well. The various modellers cited expect a significant PUI pre-acceleration well into the 50-100 keV regime and above.

This pre-acceleration causes the PUI flux at low energies to be higher than the ACR flux, also beyond the heliospheric shock. Consequently, there should be more ENAs resulting from charge exchange with PUIs in the energy range of interest. In order to check on this, we have first computed the PUI spectra upstream ('pre-shock') and downstream ('post-shock') of the heliospheric shock (5). Those spectra can then be used to compute the ENA flux. Fig. 6 gives a comparison of PUI-ENA with ACR-ENA and other ENA spectra, resulting from particle populations which, supposedly, contribute to the overall ENA flux (7).

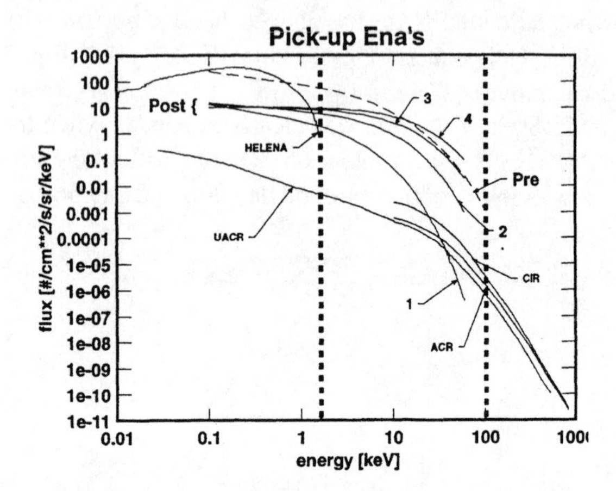

FIGURE 6. Comparison of the (pre- and post-shock) PUI-ENA flux with those resulting from charge exchange with other particle populations, namely [unmodulated] ACRs ([U]ACRs), heliospheric ENAs (HELENA) resulting from charge exchange with solar wind protons, and ENAs from de-charged protons near corotating interaction regions (CIR). The numbers label the PUI-ENA spectra according to the assumed compression ratio $s = 1,2,3,4$ of the heliospheric shock. The vertical lines mark the energy range dominated by PUI-ENAs, adopted from (7).

Evidently the PUI-ENA flux is dominating in the energy interval 1-100 keV and, therefore, SOHO is probably observing PUI-ENAs. This finding has no impact on the general idea of probing the region beyond the termination shock: PUI-ENAs do allow one to study the heliospheric interface as well, of course.

SUMMARY

The following statements summarize our findings:

- If the heliospheric shock is located around 85 AU, Voyager 1 will soon start seeing downstream ACR spectra.
- A first computation of ACR spectra in a 'realistic' heliosheath, i.e. one taking into account the expected large-scale upwind-downwind asymmetry of the heliospheric interface, shows that there is almost no change in the spectral gradients with heliocentric distance in the upwind direction. In downwind direction the spectra do exhibit an evolution in shape with distance, but they are unaccessible to *in-situ* observations in the foreseeable future.
- A remote sensing of the heliospheric interface is possible via the asymmetric ENA fluxes which are observed with SOHO.
- An investigation of PUI spectra reveals that they contribute considerably to the ENA flux in the energy interval 1-100 keV, so that a distinction between PUI-ENAs and ACR-ENAs appears advisable.
- A comparison of PUI-ENA and ACR-ENA fluxes suggests that the ENAs observed with SOHO are of PUI origin.

REFERENCES

1. Blom, J.G. and Verwer, J.G., Appl. Numer. Math. 16, 129 (1994).
2. Blum, P.W. and Fahr, H.J., Nature 223, 936 (1969).
3. Chalov, S.V., Fahr, H.J. and Izmodenov, V.V., A&A 304, 609 (1995).
4. Czechowski, A., Fichtner, H., Grzedzielski, S., Hilchenbach, M., Hsieh, K.C., Jokipii, J.R., Kausch, T., Kòta, J. and Shaw, A., Proc. 26th ICRC, Salt Lake City, USA, 7, 589 (1999).
5. Czechowski, A., Fahr, H.J., Fichtner, H. and Kausch, T., Proc. 26th ICRC, Salt Lake City, USA, 7, 464 (1999).
6. Dworsky, A. and Fahr, H.J., A&A 353. L1 (2000).
7. Fahr, H.J. and Lay, G., A&A, in press (2000).
8. Fahr, H.J., Kausch, T. and Scherer, H., A&A, in press (2000).
9. Fichtner, H., le Roux, J.A., Mall, U. and Ruciński, D., A&A 314, 650 (1996).
10. Florinski, V. and Jokipii, J.R., ApJ 523, L185 (1999).
11. Giacalone, J., Jokipii, J.R., Decker, R.B. and Krimigis, S.M., Scholer, M. and Kucharek, H., ApJ 486, 471 (1997).
12. Hilchenbach, M., et al., ApJ 503, 916 (1998).
13. Kausch, T., Ph.D. thesis, Univ. of Bonn, Germany (1998).
14. leRoux, J.A. and Fichtner, H., JGR 104, 4709 (1999).
15. le Roux, J.A. and Ptuskin, V.S., JGR 103, 4799 (1998).
16. leRoux, J.A., Potgieter, M.S. and Ptuskin, V.S., JGR 101, 4791 (1996).
17. Parker, E.N., *Interplanetary dynamical processes*, Interscience Publishers New York (1963).
18. Stone, E.C. and Cummings, A.C., Proc. 26th ICRC, Salt Lake City, Vol. 7, 500 (1999).
19. Stone, E.C., Cummings, A.C. and Webber, W.R., JGR 101, 11017 (1996).
20. Whang, Y.C., Burlaga, L.F., and Ness, N.F., JGR 100, 17015 (1995).

Evolution of cosmic ray fluxes during the rising phase of solar cycle 23: ULYSSES EPAC and COSPIN/KET observations.

B. Heber,[1] J. B. Blake,[2] M. Fränz,[3] E.Keppler,[1] H. Kunow,[4]

[1] *Max-Planck-Institut für Aeronomie, Katlenburg-Lindau, Germany* [2] *The Aerospace Corporation, Los Angeles, CA 90009, USA* [3] *Queen Mary & Westfield College, Astronomy Unit, London E1 4NS, UK* [4] *Institut für Experimentelle und Angewandte Physik, Universität Kiel, Kiel, Germany*

Abstract. Galactic cosmic rays are entering the heliosphere from the interstellar medium, while anomalous cosmic rays are believed to be pickup ions accelerated at the heliospheric termination shock. Both particle species are modulated by the solar wind and the heliospheric magnetic field. Since 1997 solar activity increased and as a consequence the flux of galactic and anomalous cosmic ray decreased. In this paper we will discuss the variation of low energy anomalous cosmic rays as measured by the Ulysses Energetic Particle Composition Experiment (EPAC) and the Kiel Electron Telescope (KET) on board Ulysses. Specifically we are addressing the question: Are there differences in the modulation of galactic and anomalous cosmic rays and what are possible implication for the modulation of cosmic rays in the heliosphere?

INTRODUCTION

Galactic cosmic rays entering our heliosphere encountered an outward-flowing solar wind carrying a turbulent magnetic field. Since the discovery of anomalous cosmic rays (ACRs) there has been an enormous progress in delineating their observed properties and in deducing the processes by which they originate (1, 2). Today the bulk of anomalous cosmic rays are believed to result from interstellar neutrals that have been swept into the heliosphere and than ionized to become PickUp Ions (3). These PUIs are convected out and become accelerated at the heliospheric termination shock (4). The interaction between GCRs and ACRs and the solar wind reduces the particle intensities with decreasing distance to the Sun. Studying the modulation processes is important a) to determine the local interstellar and source spectra for GCRs and ACRs, respectively; b.) to explore the structure and dynamics of the outer heliosphere; and c) to understand the transport of energetic particles in the heliosphere.

OBSERVATIONS

Ulysses, launched in October 1990 in the declining phase of solar cycle 22, encounters in February 1992 the planet Jupiter, and using a gravity assist began its journey out of the ecliptic plane. It crossed in April 1998 the ecliptic plane and completed its first out of ecliptic orbit.

Fig. 1 shows Ulysses heliographic latitude as function of its radial distance during the first and second or-

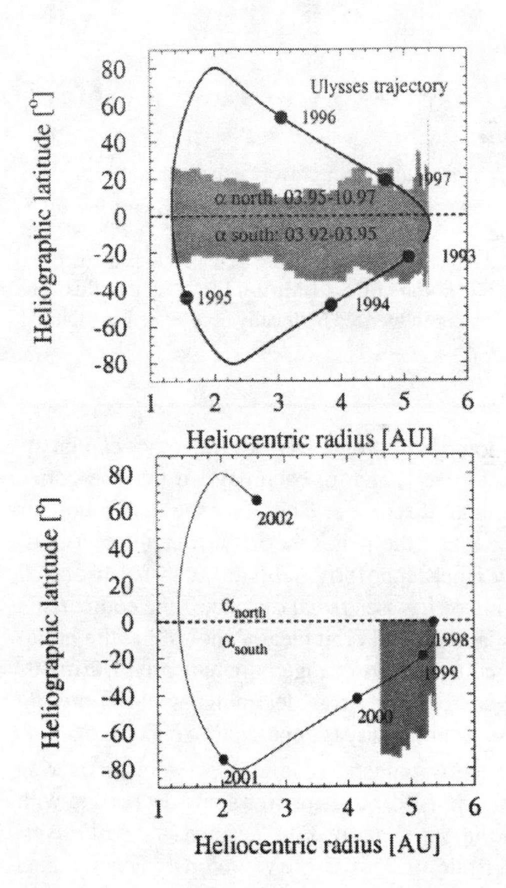

FIGURE 1. Ulysses trajectory from beginning of 1993 to the end of 1997, and from beginning of 1998 to end of 2001. Solid circles mark the start of each year. The histogram shows the evolution of the tilt angle α.

CP528, *Acceleration and Transport of Energetic Particles Observed in the Heliosphere: ACE 2000 Symposium*, edited by Richard A. Mewaldt, et al.
© 2000 American Institute of Physics 1-56396-951-3/00/$17.00

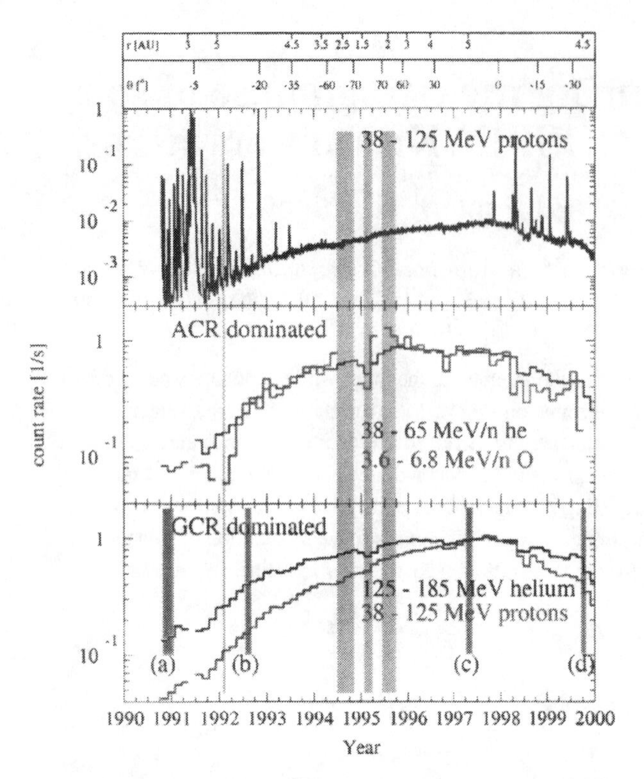

FIGURE 2. Daily and 52-day averaged count rates of 38 - 125 MeV protons, 35 - 65 MeV/n helium, and 3.6 - 6.8 MeV oxygen, and 38 - 125 MeV protons, and 125 - 185 MeV/n helium. Only times were taken into account, when the 38 - 125 MeV KET proton or 0.5 - 1 MeV/n EPAC helium flux (not shown here) is not influenced by locally accelerated particles.

bit. Each point (r,θ) on the Ulysses trajectory can be attributed to a time t, and for each day we find the corresponding radial distance r. Therefore the maximum latitudinal extent of the heliospheric current sheet, as calculated by Hoeksema (priv. comm.), can be displayed as a function of Ulysses radial distance. The comparison of the first and second orbit clearly shows that the heliospheric conditions have changed dramatically. During its first orbit solar activity was declining, while at present (Jan. 2000) solar activity is approaching maximum.

Consequently galactic cosmic rays were recovering from 1990 to fall 1997 and are now again decreasing with time. The upper panel of Fig. 2 shows the daily averaged count rate of 38-125 MeV protons. This channel has been used to define "quiet" time periods when the measurements are not influenced by solar or interplanetary particle events. The middle panel displays 52-day quiet time count rates of 35-65 MeV/n helium and 2.1-6.8 MeV/n oxygen. Both channels are dominated by the anomalous component. The "quiet" time profile of predominantly 38-125 MeV protons and 125-185 MeV/n helium is shown in the lowest panel. Marked by shading are time periods, when the spacecraft was above 70°in 1994 and 1995, and when it crossed the heliographic equator in 1995. In the lowest panel (a), (b), (c) and (d) mark time periods in 1990, 1992, 1997, and 1999, when we determined 78 day averaged proton, helium and oxygen spectra.

Table 1. Characteristics of channels used, by assuming that ACRs are singly ionized and GCRs are fully stripped.

	E_l MeV	P_l GV	$(\beta\cdot P)_l$ GV	E_u MeV	P_u GV	$(\beta\cdot P)_u$ GV
O	3.6	1.3	0.11	5.1	1.6	0.16
O	5.1	1.6	0.16	6.8	1.8	0.22
He	35	1.0	0.27	65	1.4	0.50
He	65	1.4	0.50	85	1.6	0.65
He	125	1.0	0.47	185	1.2	0.68

DATA ANALYSIS

Table 1 summarizes the characteristics of the channels used in this analysis. The channels are chosen such that the rigidity P or $\beta\cdot P$ of anomalous oxygen, and helium as well as galactic helium are approximately the same. β denotes the particle velocity in units of the speed of light. The EPAC and KET instruments are described in (5) and (6), respectively.

The proton, helium, and oxygen spectra are displayed in Fig. 3 and represent the change of the CR flux during the declining and rising phase of the last and current solar cycle. The lines superimposed are approximations of a power law with spectral index $\gamma = 1$. Such a power law is expected for galactic cosmic rays, if drift effects are negligible (Moraal, 1993) (7):

Protons: The proton flux increased from end of 1990 to mid 1997 by approximately 1.5 orders of magnitude, in the range of 40 - 100 MeV independently of energy. A power law with spectral index of 1 is always a good approximation to the data set, indicating no major contribution of ACR hydrogen in the inner heliosphere.

Helium: In contrast to the proton spectrum the helium spectrum changes from a GCR dominated spectrum in 1990 to an ACR dominated spectrum in 1997. With increasing solar activity a decrease of the ACR component is observed.

Oxygen: The oxygen spectrum in 1990, and 1992 is dominated by a solar or interplanetary component.

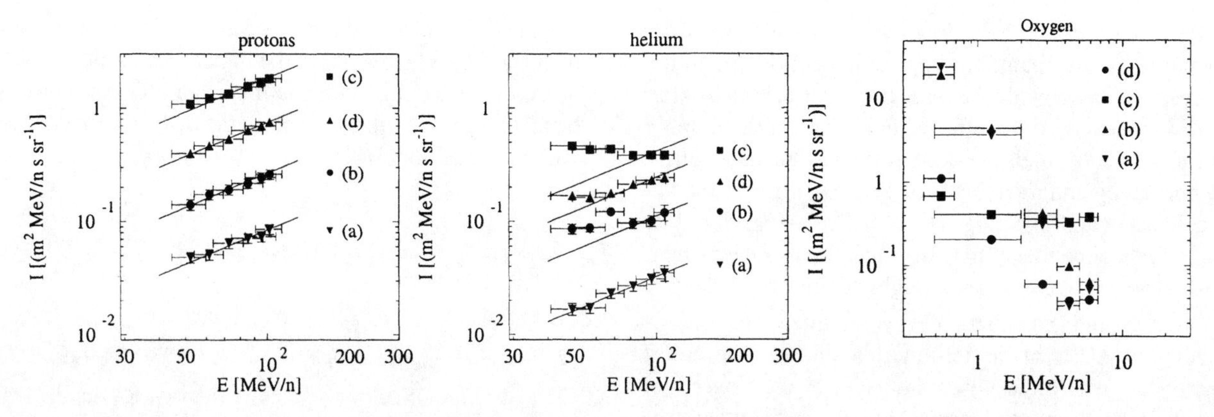

FIGURE 3. 78-day averaged proton, helium, and oxygen spectra from day 300 in 1990 (a), 200 in 1992 (b), 100 in 1997 (c), and 260 in 1999 (d). Note that for the spectra no quiet time criteria have been applied.

At solar minimum this component is reduced by an order of magnitude , so that we measure a ACR dominated oxygen spectrum in 1997. With increasing solar activity the local component is increasing again.

To compare the temporal variation of ACRs with GCRs, the count rate measured on board Ulysses have to be corrected for the special variation of the spacecraft. We used the method described in (8, 9), to determine the heliographic equator equivalent count rate. The oxygen channel has been corrected by using a mean latitudinal gradient (10), because the instrument has been switched to a different mode during the rapid pole to pole passage (11). Note no correction for radial gradients have been applied.

The helium channel measures galactic as well as anomalous helium: The ACR helium time profile in the energy ranges from 35-65 MeV/n and 65-85 MeV/n can be calculated when we assume that the 85-125 MeV/n count rate $\overline{C}_3$ is dominated by GCRs. Then the GCR contribution at lower energies $\overline{C}_i$ can be approximated by $\overline{C}_i(t) = \alpha \cdot \overline{C}_3(t)$, with α determined in 1990, when no obvious ACR component was seen. The ACR count rate C_i is then given by $C_i = Cm_i - \overline{C}_i$, with Cm_i the measured count rate.

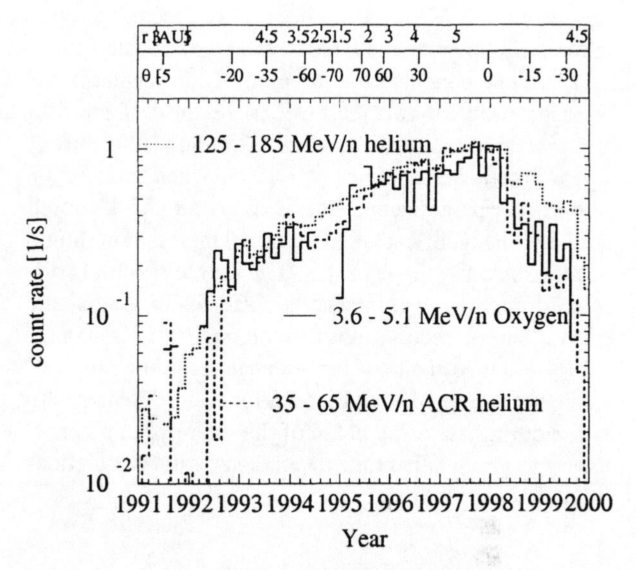

FIGURE 4. Quiet time 52-day averaged count rates of 35 - 65 MeV/n, 125 - 185 MeV/n helium, and 3.6 - 5.1 MeV/n oxygen

DISCUSSION

Fig. 4 displays the heliographic equator equivalent count rates of ~1.2 GV particles, namely 3.6-5.1 MeV/n ACR oxygen, 35-65 MeV/n ACR, and 125-185 MeV/n GCR helium. The time profiles of both ACR components are in good agreement with each other. Slight differences might be due to different radial gradients. The GCR flux varies only to a smaller amount. The same mismatch between the temporal variation of ACR and GCR persists, when comparing the two helium channels with $\beta \cdot P \sim 0.6$ (not shown here). Since Ulysses moved only 0.5 AU inward in 1997 and 1998 it seems unlikely that different radial gradients for ACRs and GCRs only, could cause different modulation for ACRs and GCRs. Therefore neither P nor $\beta \cdot P$ is a good parameter to order the temporal variation of ACRs and GCRs.

It is well known, that ACRs are much more sensitive to changing modulation conditions in the heliosphere. The radial and latitudinal gradients exhibit the one of GCRs (12). Their response to Corotating Interaction Regions exhibit the one of GCRs (13, and references therein).

As can be seen in Fig. 4 the intensity of all channels

starts to decrease in fall 1997. GCR modulation is determined by the changing propagation conditions in the heliosphere only, while the ACR time profile is also determined by several other effects: For example the number of Pick Up Ions, their pre-acceleration in the heliosphere by corotating and traveling shock waves, the efficency of accelerating and injecting these PUIs as ACRs could be solar cycle dependent (14). In principle also the properties of the termination shock might be time dependent.

Moraal and Steenberg (15) investigate the energy spectra of ACRs at the termination shock. The cutoff energy T_c, where the power law spectra rolls over into an exponential one, is depending on the radial mean free path λ_{rr} inside the shock. For typical values of λ_{rr} they found $T_c = 37$ MeV. While the spectral index is depending on the shock strength, this cutoff energy is nearly independent of it. If no drift effect would occur, the peak in the ACR spectra would be at roughly the same energy as T_c. They conclude that as modulation increases the intensity drop for two reasons: (i) because of the normal increased modulation as for GCRs and (ii) a shift of T_c towards lower energies. In 1997, when modulation sets in, the parameters at the termination shock should not have changed, so that only normal modulation should occur. Since the shape of the ACR source spectra is different from the Local Interstellar GCR Spectra, adiabatic cooling should become more important for ACRs than for GCRs. To investigate, if indeed changes in the propagation parameters only can account for the different modulation during the rising phase of the solar cycle, it is necessary to use detailed time dependent model calculations of GCR's and ACR's.

SUMMARY

We investigate the time profiles of 3.6 to 6.8 MeV/n oxygen, 35 to 85 MeV/n ACR helium and 125 to 185 MeV/n GCR helium using Ulysses EPAC and KET measurements. To correct the measured profiles for the spatial variation of the spacecraft, we used the fast latitude scan, to determine a heliographic equator equivalent time profile. The heliographic equator equivalent variation of ACR oxygen and helium at a rigidity of $P \sim 1.2$ GV are in good agreement with each other. In contrast to ACR helium GCR helium at the same P or $\beta \cdot P$ varies only to a smaller amount. Herein β is the speed of the particle in units of the speed of light. We conclude therefore, that neither P nor $\beta \cdot P$ can order our data set. The ACR spectra are expected to have a cutoff energy at $T_c = 37$ MeV, where the power law spectra roll over into an exponential spectra. Since the spectral shape of the local interstellar spectrum is different from the ACR source spectrum, adiabatic energy losses should be more important for ACRs than for GCRs. However, time dependent model calculations are necessary to decide if further processes have to be taken into account, to explain the different variation of GCRs and ACRs (16).

REFERENCES

1. Klecker, B., *Spac. Sci. Rev.* **72**, (1995) 419–430.
2. Simpson, J., *Adv. in Spac. Res.* **16**, (1995) 135–149.
3. Fisk, L. A., Koslovsky, B., and Ramaty, R., *Astrophys. J. (Lett.)* **190**, (1974) L35.
4. Pesses, M., Jokipii, J., and Eichler, D., *Astrophys. J. Lett.* **246**, (1981) L85–L88.
5. Keppler, E., Blake, J., Hovestadt, D., Korth, A., Quenby, J., Umlauft, G., and Wock, J., *Astron. Astrophys., Suppl.* **92** (2), (1992) 317–331.
6. Simpson, J., Anglin, J., Barlogh, A., Bercovitch, M., Bouman, J., Budzinski, E., Burrows, J., Carvell, R., Connell, J., Ducros, R., Ferrando, P., Firth, J., Garcia-Munoz, M., Henrion, J., Hynds, R., Iwers, B., Jacquet, R., Kunow, H., Lentz, G., Marsden, R., McKibben, R., Müller-Mellin, R., Page, D., Perkins, M., Raviart, A., Sanderson, T., Sierks, H., Treguer, L., Tuzzolino, A., Wenzel, K.-P., and Wibberenz, G., *Astron. Astrophys. Suppl.* **92** (2), (1992) 365–399.
7. Moraal, H., *Nuclear Phys. B*, **33A**, (1993) 161–178.
8. Heber, B., Dröge, W., Ferrando, P., Haasbroek, L., Kunow, H., Müller-Mellin, R., C.Paizis, Potgieter, M., Raviart, A., and Wibberenz, G., *Astr. and Astrophys.* **316**, (1996) 538–546.
9. Heber, B., Ferrando, P., Raviart, A., Wibberenz, G., Müller-Mellin, R., Kunow, H., Sierks, H., Bothmer, V., Posner, A., Paizis, C., and Potgieter, M. S., *Geophys. Res. Let.* **26** (14), (1999) 2133–2136.
10. Heber, B., Keppler, E., Fraenz, M., and Kunow, H., in *Proc. 16th International Cosmic Ray Conference, Salt Lake City, Utah, USA, August 17-25, 1999*, vol. 7, 1999 p. 117.
11. Keppler, E., Fraenz, M., Korth, A., Reuss, M., Blake, J., and Quenby, J., *Astr. and Astr.* **316** (2), (1996) 464.
12. Christian, E., Binns, W. R., Blake, J. B., and et al., in *Proc. 16th International Cosmic Ray Conference, Salt Lake City*, vol. 7, 1999 p. 519.
13. Heber, B., Sanderson, T. R., and Zhang, M., *Adv. in Spac. Res.* **23** (3), (1999) 567–579.
14. Le Roux, J. A., and Fichtner, H., *Jour. Geophys. Res.* **104**, (1999) 4709 – 4730.
15. Moraal, H., and Steenberg, C. D., in *Proc. 16th International Cosmic Ray Conference, Salt Lake City*, vol. 7, 1999 p. 543.
16. McDonald, F., *AIP* .

Time Variations of the Modulation of Anomalous and Galactic Cosmic Rays

E.R. Christian[1], W.R. Binns[2], C.M.S. Cohen[3], A.C. Cummings[3], J.S. George[3], P.L. Hink[2], J. Klarmann[2], R.A. Leske[3], M. Lijowski[2], R.A. Mewaldt[3], P.L. Slocum[4], E.C. Stone[3], T.T. von Rosenvinge[1], M.E. Wiedenbeck[4], and N. Yanasak[4]

[1]*NASA/Goddard Space Flight Center*
[2]*Washington University, St. Louis*
[3]*California Institute of Technology*
[4]*Jet Propulsion Laboratory*

Abstract. Between the launch of the Advanced Composition Explorer (ACE) in 1997 and the end of 1999, the intensities of galactic cosmic rays at 1 AU have dropped almost a factor of 2, and the anomalous cosmic rays have decreased by an even larger amount. The large collecting power of the Cosmic Ray Isotope Spectrometer (CRIS) and the Solar Isotope Spectrometer (SIS) instruments on ACE allow us to investigate the changing modulation on short time scales and at different rigidities. Using anomalous cosmic ray (ACR) and galactic cosmic ray (GCR) intensities of He, C, O, Ne, Si, S, and Fe, and energies from ~ 6 MeV/nucleon to ~ 460 MeV/nucleon, we examine the differences between the short term and long term effects. We observe the expected correlation of these intensities with neutron monitor data, but see little correlation of GCR and ACR intensities with the locally measured magnetic field.

INTRODUCTION

Although modulation of cosmic rays by the heliospheric magnetic field has been studied and modeled for a long time, the exact roles of drifts, global merged interaction regions (GMIRs), and slow variations in the transport parameters is still unresolved (see, e.g., (1) and references therein). The launch of the Advanced Composition Explorer (ACE) in August 1997 has brought new tools to bear, including large, high-resolution spectrometers such as the Solar Isotope Spectrometer (SIS) (2) and the Cosmic Ray Isotope Spectrometer (CRIS) (3), as well as in-situ heliospheric solar wind and magnetic field measurements.

In this work, we examine data taken with both ACE, situated at the L1 Earth-Sun libration point, and an Earth-based neutron monitor. The data spans approximately two years, from late 1997 to late 1999.

OBSERVATIONS

Using SIS and CRIS data that is available on the web from the ACE Science Center (http://www.srl.caltech.edu/ACE/ASC/level2/index.html), we have begun to study short and long term variations in the modulation of both ACRs and GCRs. Twenty-eight different elemental species and energy bins were used, picked so that the intensities were dominated by either ACRs or GCRs for the entire time period studied here. Three ACR intensities from SIS were used: He 6-10 MeV/nucleon; O 7-13 MeV/nucleon; and O 13-21 MeV/nucleon. SIS measurements of intensities for carbon (33-76 MeV/nucleon), silicon (53-123 MeV/nucleon), and iron (70-168 MeV/nucleon) are included in the GCR data set since ACR contributions for these combinations of charge and energy should be negligible. At CRIS energies, the data is dominated by GCRs and four energy bins each were used from carbon, oxygen, neon, silicon, and

CP528, *Acceleration and Transport of Energetic Particles Observed in the Heliosphere: ACE 2000 Symposium,* edited by Richard A. Mewaldt, et al.
© 2000 American Institute of Physics 1-56396-951-3/00/$17.00

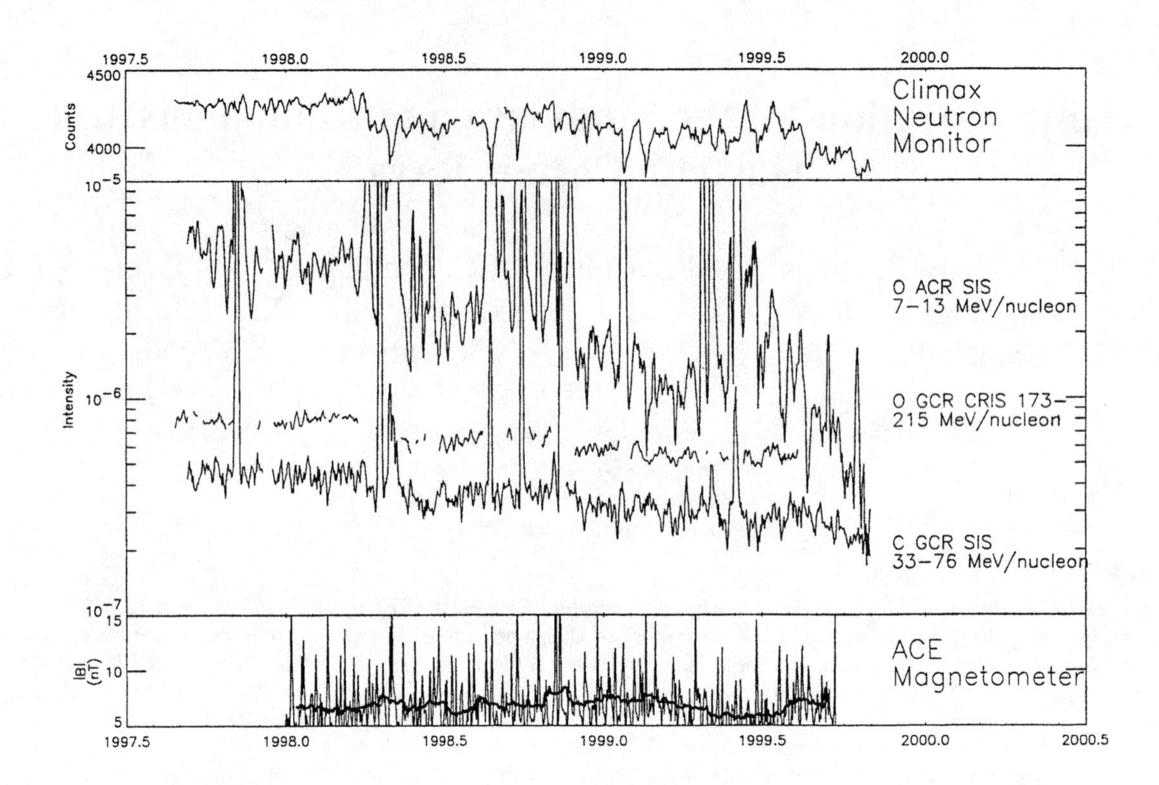

FIGURE 1. Top panel: time profiles for the Climax neutron monitor daily count rate. Middle panel: selected particle intensities (particles/cm^2 sr sec MeV/nucleon); 3-day running average ACR Oxygen between 7 and 13 MeV/nucleon from SIS (+), daily average GCR Oxygen, 173-215 MeV/nucleon from CRIS (solid circle), and 3-day running average GCR Carbon 33-76 MeV/nucleon from SIS (x). Bottom panel: magnetic field amplitude from ACE MAG, shown as both daily averages (thin line) and a 27-day running average (thick line).

iron, and two from sulfur, ranging from 60 MeV/nucleon carbon to 460 MeV/nucleon iron.

Figure 1 shows the daily averages of three of these intensities in the middle panel. Gaps in the CRIS data set are because the CRIS instrument was not designed to work during solar active periods. The top panel shows the Climax neutron monitor daily count rate, and the bottom panel shows the ACE/MAG (4) magnetic field magnitude as both daily average and 27-day running averages. The increase in modulation since the middle of 1997 is obvious in the neutron monitor data and in the ACRs and GCRs. The 27-day magnetic field averages have remained remarkably flat during this same period.

DISCUSSION

The intensity levels early in this time period appear to have decreased in a step-wise fashion, where the steps (see e.g. April 1998) were coincident with increases in solar activity, as has been previously reported (see, e.g., 5,6). However, these apparent steps may be due to medium-term modulation events with recovery ("Forbush decreases") superimposed on the slowly increasing modulation level. Since about the middle of 1999, the increase in modulation has been gradual.

Figure 2 shows the correlation between the high-energy neutron monitor data and ACE intensities for two of the twenty-eight data sets used in this study. In order to improve statistics, each energy bin was averaged for a time period that varied between 1 day and 14 days, depending upon flux. From the straight line fit on the log-log plot, the power law index, α, of the correlation is derived. It has been previously shown ((7) and references therein), that ACR O intensities are well correlated with neutron monitor data taken to the thirtieth power, and that the GCR are correlated with power law index of ~ 6-7 (8). The indices calculated in this work (Figure 3) are consistent with earlier GCR indices and slightly lower for ACRs when compared with (7).

In Figure 3, the power law indices are plotted versus the median rigidity for each data set. The GCR indices are approximately seven at low rigidities, getting somewhat smaller about 1300 MV. This is fully expected because, in the region of the spectra where the intensity is proportional to the kinetic energy (J = AT (9)), the intensities all vary together, but above this region, higher rigidity intensities are less affected by changes in modulation. If we had data that extended up to the rigidity of the Climax neutron monitor (~ 3GV), presumably the power law index of the correlation would be one.

It has been suggested that fundamental processes at the Sun produce heliospheric magnetic field enhancements which are responsible for cosmic ray modulation ((1), (10), (11) and references therein). This is frequently shown as an anticorrelation between the locally-measured average magnetic field strength and the particle intensities. The current observations are not consistent with this picture, as can be seen in Figure 1, which shows no increase in the 1 AU magnetic field that corresponds to the increase in modulation level.

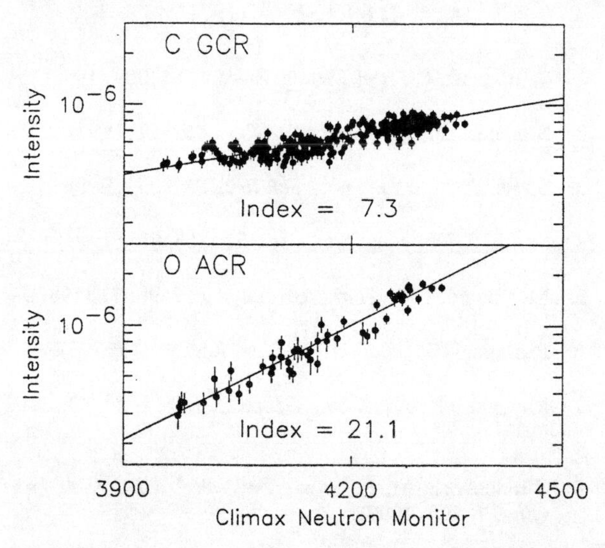

FIGURE 2. Cosmic Ray Intensity vs. Climax neutron monitor counts for (top) GCR Carbon 100-150 MeV/nucleon from CRIS and (bottom) ACR Oxygen 13-21 MeV/nucleon from SIS. The solid lines are the best-fit power laws and the power law indices are shown on each plot. Data in the top panel are daily averages and in the bottom are 7-day averages to improve statistics.

This is shown in a different way in Figure 4 by plotting (for two of the energy bins) the particle intensity versus the average, locally measured magnetic field strength. The data have been averaged over distinct 27-day time periods. The Pearson's correlation coefficient, r, is shown on both parts of Figure 4, and for all 28 data sets varied between 0.2 and –0.3, indicating that there is no strong correlation, which is inconsistent with the observations of (11), however, their data also includes time periods in which the modulation level was increasing without a corresponding increase in the measured magnetic field. This may be partially due to the use of a local magnetic field strength, which does not always correlate with the overall heliospheric magnetic field.

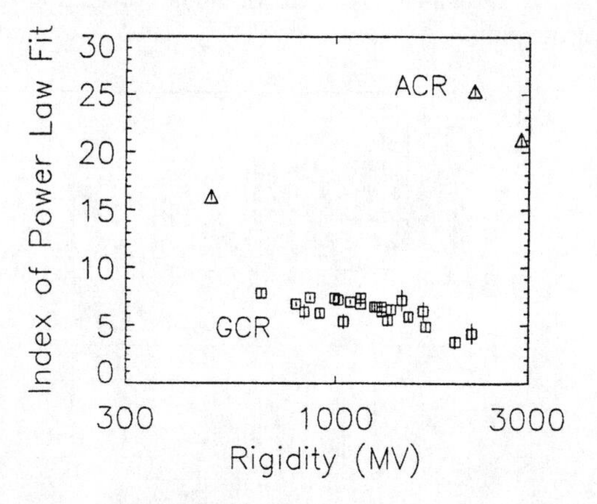

FIGURE 3. Indices of power-law fits between Climax neutron monitor data and ACE particle intensities vs. median rigidity of particle intensities. The triangles are ACR data (charge = +1), and the squares are GCR data (fully stripped).

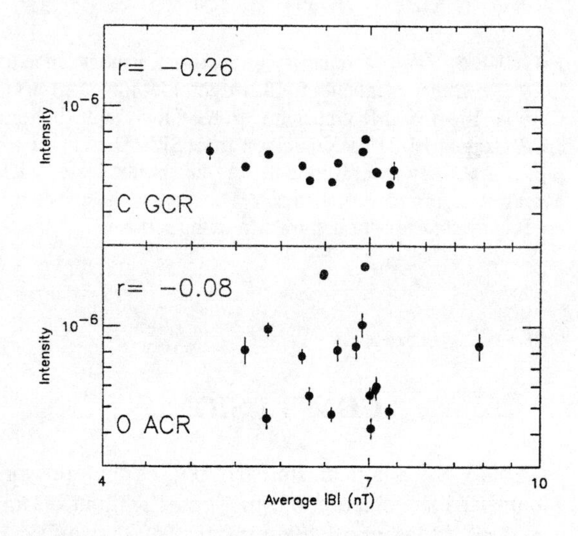

FIGURE 4. Cosmic ray intensity (particles / m^2 sr sec MeV/nucleon) vs. average magnetic field magnitude for (top) GCR Carbon 100-150 MeV/nucleon from CRIS and (bottom) ACR Oxygen 13-21 MeV/nucleon from SIS. The data have been averaged for 27-day periods. The Pearson's correlation coefficient for the data is shown in each section.

One might expect that the lower rigidity intensities measured by SIS and CRIS are more affected by the locally measured magnetic field than the high rigidity intensities (represented by the neutron monitor data). To investigate this, we calculate the ratio of the measured flux to the flux calculated by using the neutron monitor data during the same time period and the power-law fit (i.e. the ratio of each point in Figure 2 and the line). This, in effect, removes from the ACE data any temporal changes in modulation seen by the neutron monitor. In Figure 5 we plot these ratios versus the magnetic field averaged over the same time period. Again, there is no obvious correlation in any of the plots.

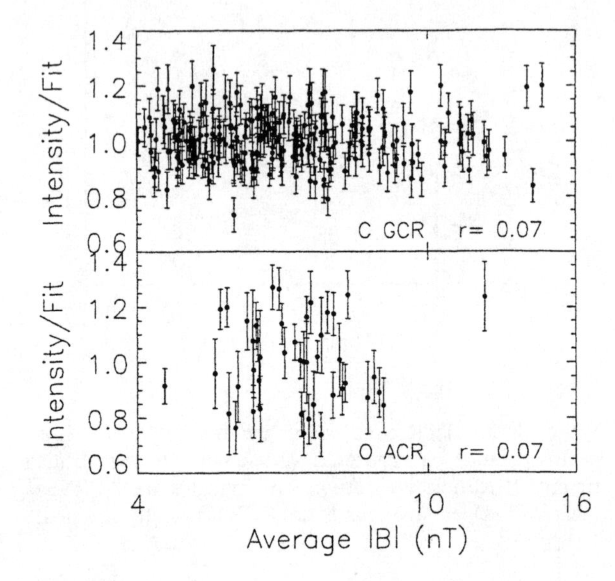

FIGURE 5. Ratio of intensity and Climax neutron monitor fit vs. average magnetic field magnitude for (top) GCR Carbon 100-150 MeV/nucleon from CRIS and (bottom) ACR Oxygen 13-21 MeV/nucleon from SIS. Data in the top panel are daily averages and in the bottom are 7-day averages to improve statistics. The Pearson's correlation coefficient for the data is shown in each section.

CONCLUSION

This work is still in its early phases. Future work includes better selection of quiet times without biasing the data. Measuring the intensity of the ACRs is becoming more and more difficult as the modulation level increases. It is, however, clear that there are steeper power law indices for the ACR component than for the GCR component (which is to be expected), and the power laws derived here are consistent with those calculated by earlier work.

For this time period, we do not see any correlation between the modulation levels and the magnetic field measured at 1 AU. Although this is inconsistent with the observations of (1) and (11), more data is clearly required in order to check this correlation over a larger portion of the solar cycle. Correlation of the particle intensities with the tilt of the neutral current sheet has also been left for the future.

ACKNOWLEDGMENTS

This research was supported by NASA at the Goddard Space Flight Center, the Jet Propulsion Laboratory, and the California Institute of Technology (under grant NAG5-6912). Climax neutron monitor data are courtesy of the University of Chicago and National Science Foundation grant ATM-9613963. ACE MAG data were supplied by the ACE MAG team through the ACE Science Center.

REFERENCES

1. Wibberenz, G., *et al., Space Sci. Rev.*, **83**, 309 (1998).

2. Stone, E. C., *et al., Space Sci. Rev.*, **86**, 357 (1998).

3. Stone, E. C., *et al., Space Sci. Rev.*, **86**, 285 (1998).

4. Smith, C. W., *et al., Space Sci. Rev.*, **86**, 613 (1998).

5. McDonald, F. B., *et al., Astrophys. J.*, **249**, L71 (1981).

6. Burlaga, L. F., *et al., J. Geophys. Res.*, **98**, 1 (1993).

7. Mewaldt, R. A., *et al., Geophys. Res. Lett.*, **20**, 2263 (1993).

8. Wiedenbeck, M. E., *et al., Proc. 26th ICRC (Salt Lake City)*, **7**, 508 (1999).

9. Rygg, T. A., and Earl, J. A., *J. Geophys. Res.*, **76**, 7445 (1967).

10. Burlaga, L. F., and Ness, N. F., *J. Geophys. Res.*, **103**, 29719 (1998).

11. Cane, H.V. *et al., Geophys. Res. Lett.*, **26**, 565 (1999).

Recurrent modulation of galactic cosmic ray electrons and protons: Ulysses COSPIN/KET observations

B. Heber,[1] J. B. Blake,[2] C. Paizis,[3] V. Bothmer,[4] H. Kunow,[4] G. Wibberenz,[4] R.A. Burger[5], and M.S. Potgieter[5]

[1]*Max-Planck-Institut für Aeronomie, Katlenburg-Lindau, Germany* [2]*The Aerospace Corporation, Los Angeles, USA* [3]*Istituto Fisica Cosmica del CNR, Milan, Italy* [4]*Institut für Experimentelle und Angewandte Physik, Universität Kiel, Kiel, Germany* [5]*Potchefstroom University, Potchefstroom, South Africa*

Abstract. Since measurements of space probes in the interplanetary space became available it has been known that associated with the occurrence of recurrent fast and slow solar wind streams, forming Corotating Interaction Regions, recurrent variations in the cosmic ray nuclei flux are observed (1). As pointed out recently by Jokipii and Kota (2) recurrent modulation for positively and negatively charged particles may be different. In the time interval extending from July 1992 to July 1994, Ulysses on its journey to high heliographic latitudes registered ~20 stable and long-lasting Corotating Interaction Regions (CIRs). In this work we use data from the Cosmic Ray and Solar Particle Investigation Kiel Electron Telescope (COSPIN/KET) instrument on board Ulysses to study the recurrent variation of 2.5 GV electrons and protons. We find that 1) electrons are indeed periodically modulated, but that 2) the periodicity of ~29 days is longer than the period of ~26 days for protons, and that 3) the amplitude is larger than the one observed for protons.

INTRODUCTION

Cosmic ray particles respond to the heliospheric magnetic field in the expanding solar wind and its turbulence and so provide a unique probe for conditions in the changing heliosphere. Forbush (3) reported the first conclusive evidence for the existence of recurrent cosmic ray intensity variations. In the 1950's Simpson and coworkers showed that these variations could be attributed to dynamic phenomena in the interplanetary medium (4). Corotating Interaction Regions (CIRs) are the result of the interaction of fast solar wind coming from an extension of the polar coronal hole with the surrounding slow solar wind in interplanetary space. Associated with these CIRs, increases of several MeV nuclei and few keV electrons and decreases in the galactic cosmic and anomalous cosmic ray nuclei flux have been observed (5, 6, 1, and references therein). As pointed out recently by Jokipii and Kota (2) recurrent modulation for positively and negatively charged galactic cosmic rays may be different. From Fig. 1 in (2) one can conclude that the observable amplitudes should be nearly the same.

During Ulysses' slow descent south in 1992 to 1994 from the equatorial plane toward the southern polar regions, the space craft encountered a series of CIRs (Fig. 1), which could be identified by the in situ plasma and magnetic field measurements up to ~35°(7, 8). Intensities of galactic and anomalous cosmic ray nuclei clearly revealed a well-defined variations in connection with the in situ observed CIRs and beyond (9, 10, 11). In this contribution we will investigate the topic of charge sign dependent recurrent modulation when CIRs are observed in situ.

OBSERVATIONS

Particle data analyzed here are obtained with the Cosmic Ray and Solar Particle Investigation Kiel Electron Telescope (COSPIN/KET) on board Ulysses. A detailed description of the instrument is given in Simpson *et al.* (12). Fig. 1 (a) shows the solar wind speed as measured by the *Solar Wind Observations Over the Poles of the Sun* (SWOOPS) experiment on board Ulysses from mid 1992 to mid 1994. At low latitudes (below ~35°) signatures of CIRs are measured in situ by Ulysses (13). The spacecraft is embedded in the uniform fast solar wind stream of the southern coronal hole beyond ~40°. On top of this panel time periods are marked with "c" when Coronal Mass Ejection (CME) could be identified in the plasma and magnetic field measurements (14, and references in there).

Fig. 1 (b) and (c) displays the 9-day and 52-day running mean averaged counting rate of 2.5 GV protons and electrons, respectively. As discussed in (10, 15, 16) protons show clearly a recurrent variation, while the elec-

CP528, *Acceleration and Transport of Energetic Particles Observed in the Heliosphere: ACE 2000 Symposium*, edited by Richard A. Mewaldt, et al.
© 2000 American Institute of Physics 1-56396-951-3/00/$17.00

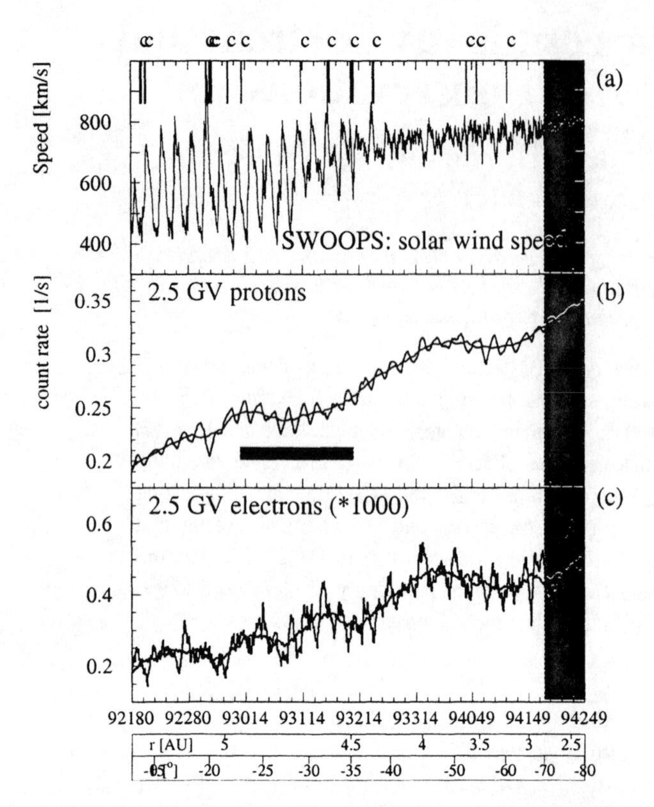

FIGURE 1. Hourly averaged solar wind speed (a), and 9 and 52 day running mean averaged count rate of 2.5 GV protons (b) and electrons (c) from mid 1992 to mid 1994. See text for further details.

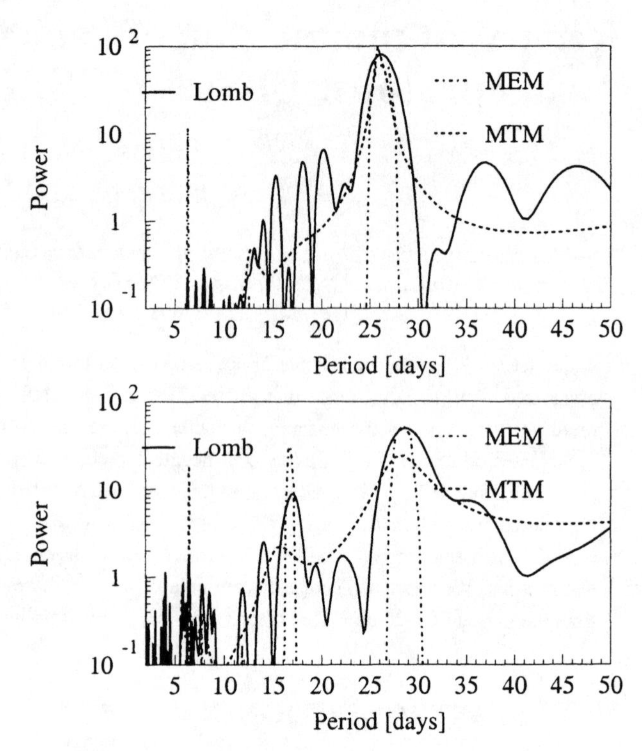

FIGURE 2. Periodicygrams of 9-day running mean averaged detrended proton (upper panel) and electron (lower panel) count rate variation. The different curves display the result of the Lomb algorithm, the MEM and MTM.

tron measurements are more complex. In contrast to the solar wind measurements, recurrent variations in the galactic cosmic ray proton intensity are observed up to highest latitudes with approximately the same periodicity (10, 15, 16). At distances of ~5 AU the influence CMEs on galactic cosmic rays is not clear, since for some of the CMEs Forbush decreases were observed (e.g. time period from day 313 to 323 of year 1992), while for other CMEs no major decrease were measured (e.g. day 10 in 1993).

DATA ANALYSIS

In what follows we analyze the longterm detrended count rate $\Delta I/I$ of 2.5 GV electrons and protons. In the analysis we used 9-day running mean values to reduce statistical uncertainties in the electron measurements: The count rate of 2.5 GV electrons is ~26 per day, and for protons ~25000 entries per day, leading to an statistical uncertainty of ~ 20% and ~0.5% for electrons and protons, respectively. Since the recurrent variation in the electron channel are expected to be of the same magnitude than the one of protons (~6)%, we use 9-day running mean values to obtain a statistical uncertainty of ~6%. The 9-day running mean averages I_9 were detrended by using 52-day running mean averages I_{52}:

$$\Delta I/I = (I_9 - I_{52})/I_{52} \quad (1)$$

Solar activity declined to moderate level at the end of 1992. The last major event, producing an enhanced proton flux in the 38-125 MeV range, was observed in November 1992. To reduce solar influences on the galactic cosmic ray flux we analyzed the time period from day 6 to 206 of year 1993, when Ulysses still observed the in situ signature of CIRs. To investigate the periodicity of the recurrent cosmic ray decreases for protons and electrons we performed different spectral analysis methods:

- As described in Heber (17) we applied the Lomb (18) algorithm on our data set.
- We applied the Maximum Entropy Method (MEM) using the subroutines, described in Faillard *et al.* (19), to the dataset.
- The Multi Taper Method (MTM) of spectral analysis provides a novel means for spectral estimate and signal reconstruction of time series (20) .

Fig. 2 displays the periodicygrams of 2.5 GV protons (upper panel) and electrons (lower panel) from day 6 of year 1993 to day 206 of year 93, for the method used in

Table 1. Frequency and amplitude for 2.5 GV electrons and protons using the Lomb algorithm, the MEM, and MTM. * indicates the result by using the daily averaged data set. Amplitudes and uncertainties are only calculated for the results of the Lomb algorithm; they are simmilar for the other methods.

Method	Period days	Amplitude %
(L) electrons	28.9±2.1	6.3±1.0
(L) protons	26.1±1.7	3.2±0.4
(L) electrons*	28.8±2.1	6.5±1.2
(L) protons*	26.2±1.7	4.1±0.5
	9-day rm	Daily av.
(MEM) electrons	28.4	
(MEM) protons	26.2	
(MTM) electrons	28.5	
(MTM) protons	25.8	

our analysis. The most probable frequencies found are summarized in Table 1.

To estimate the influence of using running mean averages on our results, we applied the different procedures also on the daily averaged detrended data set (not shown here). As expected the running mean filters out low periodicities, so that the power below ~10 days is much smaller in the running mean averaged than in the daily averaged data. Although the power at 26 and 28 days for protons and electrons is reduced for the daily averaged data the shape of the spectra in that frequency range is approximately the same. As can be seen from Fig. 2 and Tab. 1 all three methods give approximately the same frequency for protons as well for electrons, leading us to the conclusion that the most probable frequencies found, especially for the electrons, are real features in our measurement. However, the amplitude of the wave with the most probable frequency is smaller for the running mean averages than for the daily ones.

DISCUSSION

Surprisingly, electrons are not modulated with the same periods as protons. While for protons the 26 day peak is the most significant, we find a periodicity of ~29 days for electrons for all methods used in our analysis.

In Fig. 3 the detrended variation of 9-day running mean averaged protons, electrons, and the hourly averaged solar wind speed and magnetic field strength are

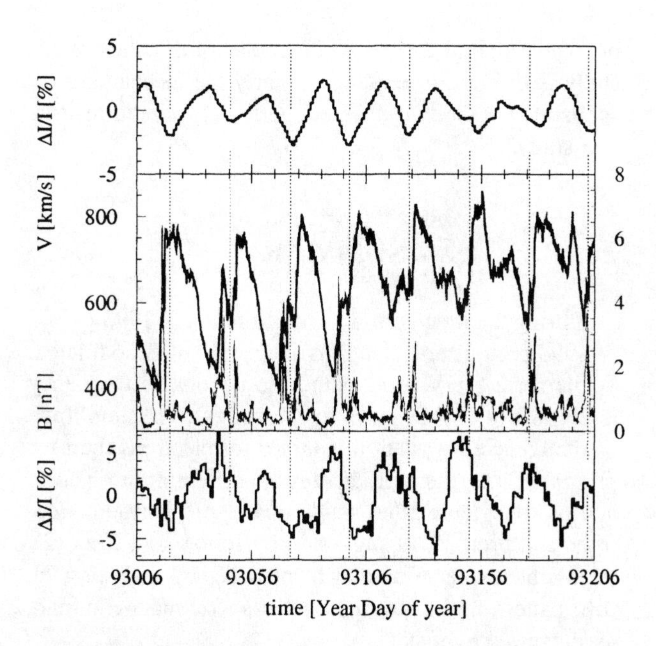

FIGURE 3. Detrended variation of 9 day running mean averaged 2.5 GeV protons (upper panel) and electrons (lower panel). The middle panel displays the hourly averaged solar wind speed and the magnetic field strength. For details see text.

displayed. The shaded area in Fig. 3 indicates the result of the approximation by a sinusoidal law with the most probable frequency to the data. Note, that the observed amplitude is divided by a factor of two and three for protons and electrons, respectively, to emphasize differences between the measurements and the sinusoidal.

Protons: As already mentioned before in the literature (e.g. (21), (10)) recurrent cosmic ray decreases occur regularly over 1 1/2 years. A significant shift to longer periodicities was only observed for a few solar rotations in the beginning of 1994 (15).

Electrons: Show a much complicate variation. For the seven cycles up to mid of 1993, the sinusoidal leads to an acceptable approximation to our measurements. The proton density decreases to its local minimum close to the reverse shock of each CIR during this time period. It seems therefore that the recurrent modulation of protons is caused locally. In contrast, the longer period of the electrons suggests that their recurrent modulation is caused non-locally at higher latitudes, where the rotation period of the sun is between the ~25 days at low latitudes and ~31 days in polar regions. There is also a hint of a 16.5 day period for the electrons shown in Fig. 2. This is rather intriguing, because such a small period is unique to the magnetic field foot point motion theory of Fisk (22, 23) and should occur at intermediate latitudes. The Fisk field allows very efficient latitudinal transport, but why it should apparently favor electrons in the present study is difficult to explain. Clearly the mechanism responsible for the two different periodicities requires further study.

SUMMARY

During the time period from beginning of 1993 to mid of 1993 protons and electrons are recurrently modulated. While for 2.5 GV protons the most probable frequency is 26 days it is ~29 days for electrons. The amplitude of the recurrent variation is larger for electrons than for protons. Since the periodicities are different for electron and protons, there is no single phase shift between electrons and protons, as suggested by Jokipii and Kota (2). A possible explanation may be provided by analyzing the drift pattern in the modified heliospheric magnetic field, as suggested by Fisk (22).

REFERENCES

1. Simpson, J., *Adv. Spac. Res.* **83**, (1998) 7.
2. Jokipii, J., and Kota, J., in *Proc. 26th Int. Cosmic Ray Conf., Salt Lake City, USA, Vol. 6*, vol. 6, Salt Lake City: IUPAP, 1999 p. 504.
3. Forbush, S. E., *Phys. Rev.* **54**, (1938) 975–988.
4. Simpson, J., *Phys. Rev.* **94**, (1954) 425.
5. Heber, B., Sanderson, T. R., and Zhang, M., *Adv. in Spac. Res.* **23** (3), (1999) 567–579.
6. Simnett, G., Kunow, H., Flückiger, E., Heber, B., Horbury, T., Kota, J., Lazarus, A., Roelof, E., Simpson, J., Zhang, M., and Decker, R., *Adv. Spac. Res.* **83**, (1998) 215–258.
7. Phillips, J., Bame, S., Barnes, A., Barraclough, B., Feldman, W., Goldstein, B., Gosling, J., Hoovegreen, G., McComas, D., Neugebauer, M., and Suess, S., *GRL* **22** (23), (1995) 3301–3305.
8. Balogh, A., Gonzales-Esparza, J., Forsyth, R., Burton, M., Goldstein, B., Smith, E., and Bame, S., *Interplanetary Shock Waves: Ulysses Observations In and Out of the Ecliptic Plane*, vol. 39A, 1995 pp. 171–181.
9. Zhang, M., *Astrophys. Jour.* **488**, (1997) 841+.
10. McKibben, R., Connell, J., Lopate, C., Simpson, J., and Zhang, M., in *The High Latitude Heliosphere*, edited by R. Marsden, vol. 254, Kluwer, 1995 pp. 367–379.
11. Kunow, H., Dröge, W., Heber, B., Müller-Mellin, R., Röhrs, K., Sierks, H., Wibberenz, G., Ducros, R., Ferrando, P., Rastoin, C., Raviart, A., and Paizis, C., *Space Sci. Rev.* **72**, (1995) 397–402.
12. Simpson, J., Anglin, J., Barlogh, A., Bercovitch, M., Bouman, J., Budzinski, E., Burrows, J., Carvell, R., Connell, J., Ducros, R., Ferrando, P., Firth, J., Garcia-Munoz, M., Henrion, J., Hynds, R., Iwers, B., Jacquet, R., Kunow, H., Lentz, G., Marsden, R., McKibben, R., Müller-Mellin, R., Page, D., Perkins, M., Raviart, A., Sanderson, T., Sierks, H., Treguer, L., Tuzzolino, A., Wenzel, K.-P., and Wibberenz, G., *Astron. Astrophys. Suppl.* **92** (2), (1992) 365–399.
13. Gosling, J. T., Feldmann, W. C., McComas, D. J., Phillips, J. L., Pizzo, V. J., and Forseyth, R. J., *Geophys. Res. Let.* **22** (23), (1995) 3333–3336.
14. Henke, T., *Ionisationszustände und magnetische Topologie koronaler Massenauswürfe*, Dissertation, TU Braunschweig, 1999.
15. Blake, J. B., Looper, M. D., Keppler, E., Heber, B., Kunow, H., and Quenby, J. J., *Geophys. Res. Let.* **24**, (1997) 671–674.
16. Paizis, C., Heber, B., Ferrando, P., Raviart, A., Falconi, B., Marzolla, S., Potgieter, M. S., Bothmer, V., Kunow, H., Müller-Mellin, R., and Posner, A., *Jour. Geophys. Res.* **104** (A12), (1999) 28241.
17. Heber, B., Dröge, V. B. W., Kunow, H., Müller-Mellin, R., Posner, A., Ferrando, P., Raviart, A., Paizis, C., McComas, D., Forsyth, R., Szabo, A., and Lazarus, A., in *Proc. 31st Eslab symposium*, Noordwijk, 1997 .
18. Lomb, N. R., *Astrophys. and Space Science* **39**, (1976) 447–462.
19. Faillard, D., Labeyrie, L., and Yiou, F., *EOS Trans. AGU* **77**, (1996) 379.
20. Thomson, D., *Proc. IEEE* **70**, (1982) 1055 – 1096.
21. Dröge, W., Kunow, H., Heber, B., Müller-Mellin, R., Sierks, H., Wibberenz, G., Raviart, A., Ducros, R., Ferrando, P., Rastoin, C., Paizis, C., and Gosling, J., in *Proc. SW8*, 1996 .
22. Fisk, L., *Jour. Geophys. Res.* **101** (15547 - 15553).
23. Zurbuchen, T. H., Schwadron, N. A., and Fisk, L. A., *Jour. Geophys. Res.* **102**, (1997) 24175–24181.

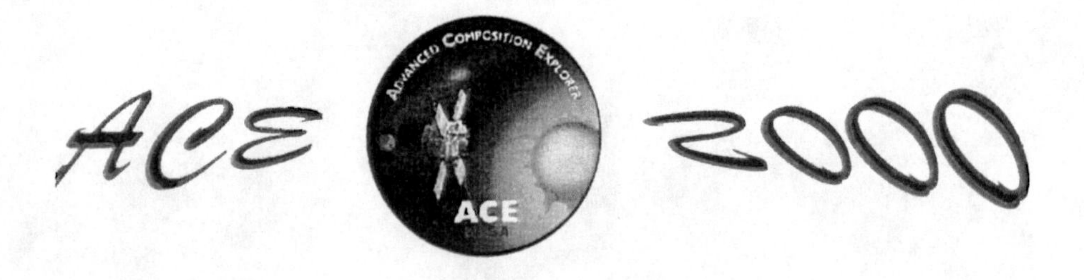

Cosmic Rays in the Galaxy

Constraints on Cosmic-Ray Acceleration and Transport from Isotope Observations

M. E. Wiedenbeck[1], W. R. Binns[2], E. R. Christian[3], A. C. Cummings[4], J. S. George[4], P. L. Hink[2], J. Klarmann[2], R. A. Leske[4], M. Lijowski[2], R. A. Mewaldt[4], E. C. Stone[4], T. T. von Rosenvinge[3], and N. E. Yanasak[1]

[1]*Jet Propulsion Laboratory, Pasadena, CA 91109, USA*
[2]*Washington University, St. Louis, MO 63130, USA*
[3]*NASA/Goddard Space Flight Center, Greenbelt, MD 20771, USA*
[4]*California Institute of Technology, Pasadena, CA 91125, USA*

Abstract. Observations from the Cosmic Ray Isotope Spectrometer (CRIS) on ACE have been used to derive constraints on the locations, physical conditions, and time scales for cosmic-ray acceleration and transport. The isotopic composition of Fe, Co, and Ni is very similar to that of solar system material, indicating that cosmic rays contain contributions from supernovae of both Type II and Type Ia. The electron-capture primary ^{59}Ni produced in supernovae has decayed, demonstrating that a time $\gtrsim 10^5$ yr elapses before acceleration of the bulk of the cosmic rays and showing that most of the accelerated material is derived from old stellar or interstellar material rather than from fresh supernova ejecta.

INTRODUCTION

The Cosmic Ray Isotope Spectrometer (CRIS) instrument (1) on the Advanced Composition Explorer (ACE) has returned more than two years of isotopically-resolved measurements of the composition of low-energy galactic cosmic rays (~ 50 to ~ 500 MeV/nucleon) under conditions of relatively low solar modulation. Because of the instrument's large geometrical acceptance, ~ 250 cm^{2}sr, this data set allows statistically-significant determinations of the abundances of essentially all stable and long-lived cosmic-ray isotopes from He to Cu (atomic numbers $Z = 2$ to 29), with exploratory measurements for elements with $30 \leq Z \lesssim 34$.

These data are being used to investigate a wide range of topics. In this paper we discuss the isotopic composition of Fe, Co, and Ni in the cosmic-ray source and its implications for the environment from which cosmic rays are accelerated. We also examine the abundances of primary electron-capture nuclides and their daughter products and use these results to derive constraints on the time that elapses between nucleosynthesis and acceleration and on the conditions in the acceleration volume.

Additional results based on data from CRIS are discussed in other papers in this volume. The topics covered include: the cosmic-ray confinement time in the Galaxy (Yanasak et al.); the distribution of matter thicknesses traversed during transport (Davis et al.); the effects of in-flight electron attachment and decay on the energy spectra of pure electron-capture nuclides (Niebur et al.); elemental fractionation of the cosmic-ray seed population (George et al.); contributions of special sources (particularly of ^{22}Ne) to the source material (Binns et al.); energy spectra of major cosmic ray elements (Leske et al., Davis et al.); and solar-modulation of galactic cosmic rays on short time scales (Christian et al.).

SOURCE COMPOSITION

The composition of the source material that is accelerated to produce galactic cosmic rays (GCRs) can help in determining the objects in which the nucleosynthesis of this material occurs and the regions of the Galaxy from which particles are accelerated. This information then constrains models for the acceleration and transport of the particles. It is well established that the elemental composition of the cosmic-ray source is very similar to solar system composition, once fractionation effects related to atomic properties such as first ionization potential or volatility are taken into account (2). In addition, it has been shown that the isotopic composition of major cosmic-ray elements including Mg, Si, Fe, and Ni does not differ markedly from solar-system (i.e., terrestrial) composition. Indeed, at present there is only one element, Ne, with a well established isotopic difference between the cosmic-ray source and the solar system (3, 4, 5, 6, 7).

CP528, *Acceleration and Transport of Energetic Particles Observed in the Heliosphere: ACE 2000 Symposium*, edited by Richard A. Mewaldt, et al.
© 2000 American Institute of Physics 1-56396-951-3/00/$17.00

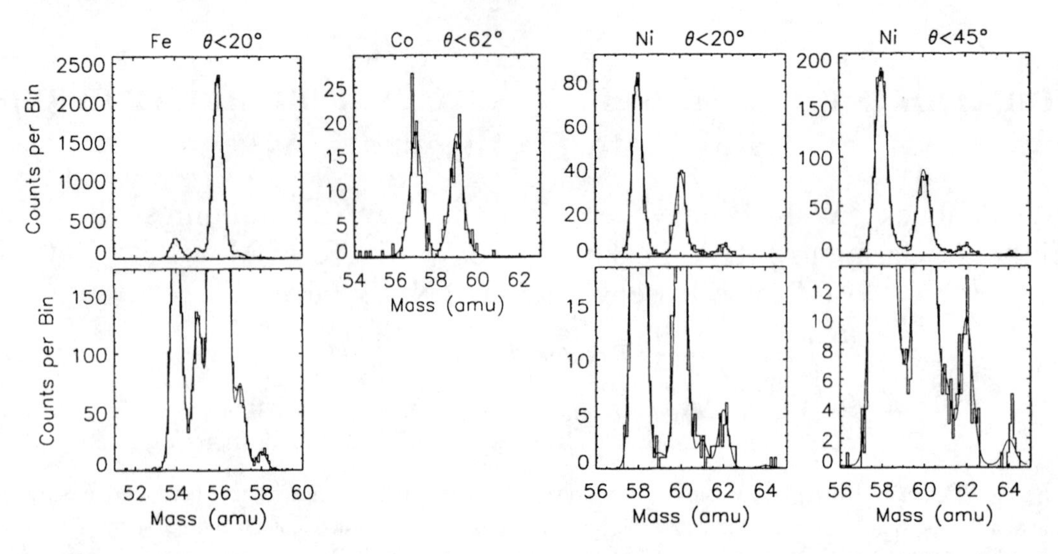

FIGURE 1. Measured mass distributions for Fe, Co, and Ni with selected cuts on angle of incidence, θ, as indicated. The lower panels show the same histograms with vertical scales expanded to better show the rare isotopes.

ACE/CRIS Observations

Using ACE data it is possible to compare these samples of matter in greater detail by improving the statistical accuracy of the measurements and by determining abundances of very rare cosmic-ray isotopes. Of particular interest are the isotopes of Fe, Co, and Ni. Among these three elements there are ten stable isotopes, and among them there is a general trend of decreasing abundances with increasing mass. As a result, there are only relatively minor contributions to the observed abundances of stable cosmic-ray nuclides (other than ^{54}Fe) from fragmentation of heavier species during transport through the interstellar medium (ISM), and source abundances can be derived with reasonable accuracy.

Figure 1 shows mass histograms for Fe, Co, and Ni obtained from CRIS data collected between August 1997 and December 1998. The plots in the upper row of the figure show the complete histograms while those in the lower row show the same data with the vertical scales expanded to better show the rare isotopes. For Ni, two sets of histograms are shown differing in the maximum angle (θ, measured from the detector normal) to which particles were accepted. The rms mass resolution for iron-group isotopes is 0.23 amu at small angles of incidence and has an approximately parabolic dependence on θ, increasing to about twice the $\theta = 0°$ value for $\theta \simeq 45°$. By considering only events with θ less than some appropriately-chosen maximum value, it is possible to optimize the trade-off between resolution and statistics for the isotopes of interest. Thus, in the $\theta < 20°$ histogram, the spill-over into the ^{59}Ni region from the abundant adjacent isotopes is minimized, while in the $\theta < 45°$ histogram the rare isotope ^{64}Ni is present with reasonable statistics. On the histograms, fits to the mass distributions are shown. For each isotope the fitted peak shape is a superposition of Gaussians with a distribution of widths corresponding to the angular distribution of particles in the data set, and it has been assumed that the shape is the same for all isotopes of an element. For Co there is negligible overlap between the two isotope peaks so the abundances were obtained by event counting rather than from the fits.

To obtain isotopic abundances for an element, the measured relative numbers of events were corrected for differences in energy interval and differences in the probability for loss by nuclear interaction in the CRIS instrument. These corrections are relatively small, amounting to 6.6% and 1.5%, respectively, in the most extreme case (^{64}Ni/^{58}Ni).

The elemental abundances are derived by calculating energy spectra for a wide range of elements, fitting that set of spectra with a common shape, and obtaining the relative abundances from the normalization factors for the various spectra. Figure 2 shows the measured spectra for Fe, Co, and Ni, along with the fits.

To obtain the cosmic-ray source composition from the measurements, a leaky-box propagation model was used to estimate the required secondary corrections to the measured abundances. The model parameters were taken from (8). The solar modulation level was modified to correspond to the near-solar-minimum conditions under which the CRIS data were collected (modulation parameter $\phi \simeq 500$ MV). This model adequately accounts for a variety of secondary nuclides in the sub-iron region, Sc through Mn.

For the present discussion we consider isotopes with mass number $A \geq 56$ and do not present a source abundance for ^{54}Fe. With the exception of ^{56}Fe, this nuclide

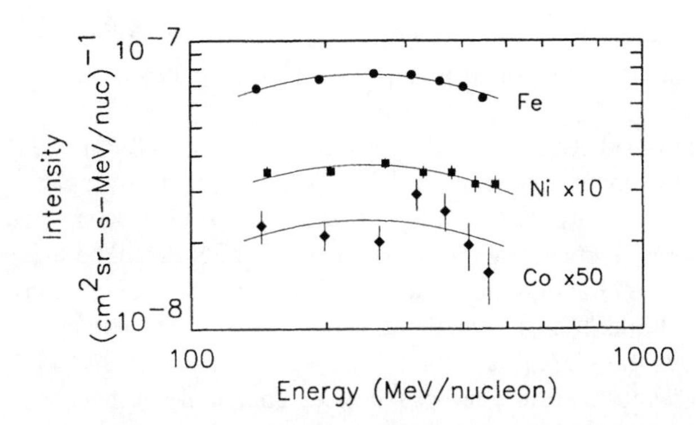

FIGURE 2. Energy spectra of Fe, Co, and Ni measured with CRIS. Error bars indicate statistical uncertainties only and are smaller than the plotted points in some cases.

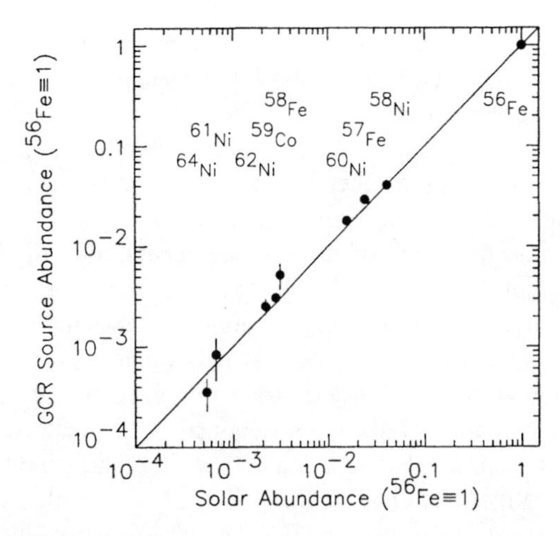

FIGURE 3. Comparison of GCR source abundances with solar system values.

has the largest source abundance among the isotopes of Fe, Co, and Ni and it has been reported (5) that the ^{54}Fe/^{56}Fe source ratio is enhanced by $\sim$ 50% over the solar value. The derivation of the ^{54}Fe source abundance is complicated by a significant secondary contribution from fragmentation of ^{56}Fe (directly and through ^{55}Fe) and by the fact that a fraction of the secondary ^{54}Mn produced during propagation decays by β^- emission to ^{54}Fe (9, 8). CRIS results on the ^{54}Fe source abundance will be reported after a self-consistent analysis of these effects is carried out.

Derived Source Abundances

Figure 3 compares the cosmic-ray source abundances we obtain with those found in solar system material (10). Abundances in each sample have been normalized so that $^{56}\mathrm{Fe} \equiv 1$. As seen in the figure, there is a strong correlation between the abundances of the Fe, Co, and Ni isotopes in the two populations, even though the relative abundances range over a factor of $\sim$ 2000. Abundance differences for individual isotopes are less than a factor of 2, and commonly no more than a few tens of percent. This result is in line with recent reports (3, 4, 6) that the abundances of the isotopes of Mg and Si are in good agreement with solar system values, typically differing by $\lesssim$20%.

The great similarity between GCR source and solar system compositions is all the more striking when compared with abundance distributions expected from various stellar nucleosynthesis models. Attempts to account for solar system abundances as sums of the ejecta from supernovae of various types and initial masses have found (11, 12) that contributions are required both from Type II supernovae (hereafter, SN II), which result from the core collapse of massive stars ($\gtrsim 8M_\odot$), and Type Ia supernovae (SN Ia), which are caused by the accretion of a critical mass of matter onto white dwarf stars in binary systems. Whereas the abundances of intermediate-mass elements such as Mg and Si can be reasonably explained in terms of SN II contributions alone, the abundances of iron-group elements are not adequately explained in this way (13). Rather, one finds that a mixture in which $\sim$ 10% if the mass comes from SN Ia and the other $\sim$ 90% from SN II can account for the iron-group nuclides (11). Because of the preponderance of iron-group material in SN Ia ejecta, these explosions would provide $\gtrsim$50% of the Fe and Ni by contributing only $\sim$ 10% of the total mass.

In addition to comparisons such as Fe+Ni versus Mg+Si, one can also consider the origins predicted for specific nuclides in the various nucleosynthesis models. For example, the SN II yield calculations of Thielemann et al. (14) indicate that very neutron-rich nuclides such as ^{58}Fe and ^{64}Ni should be nearly absent in the ejecta. They attribute the origin of these species to production in SN Ia. Alternative SN II calculations by Woosley and Weaver (15) do not confirm this deficit of very neutron-rich species but do indicate that SN II can only produce sufficient ^{58}Fe in stars more massive than $15M_\odot$, while these massive stars systematically over-produce ^{64}Ni. It is clear that differences among available supernova yield calculations dictate caution in using them for interpreting detailed abundance observations. Indeed, we find that the agreement between the isotopic composition of the cosmic-ray source (with the exception of ^{22}Ne) and that of the solar system is generally better than the agreement among available supernova yield calculations by different research groups.

The similarity between the GCR source and solar system compositions may arise naturally if cosmic rays are accelerated out of the general interstellar medium and if there has been relatively little evolution of the composition of that medium during the 4.5 Gyr since the solar system condensed out of it. Two scenarios of this type have been widely discussed: first, direct acceleration from the gas and dust in the general ISM and second, acceleration of material from the coronae of stars.

In the first class of models, the acceleration mechanism must be capable of injecting atoms from grains with comparable efficiency to injection from the gas phase, since refractory elements are largely depleted from the gas phase of the ISM and reside in grains. In fact, the grain material should be accelerated with greater efficiency than gas-phase material by the factor ~ 5 found for the relative fractionation of high- and low-volatility elements in the GCR source (16). Theoretical work has led to models (17) in which interstellar shocks efficiently accelerate charged grains to relatively high velocities where ions are sputtered from the grains and injected into the shock for acceleration along with the gas-phase ions.

In the second class of models, the initial step in the energization of the material occurs in stellar coronae where flares and CME-driven shocks analogous to those from the Sun produce particles in the MeV energy range (18, 19). This energetic material is subsequently further accelerated by shocks in the ISM. Since the composition of matter at the surfaces of most stars remains essentially unaltered from the composition of the material from which those stars formed, one expects abundances characteristic of the average ISM, including both gas and dust components. This scenario has the attractive feature that it accounts for the observed elemental fractionation of the GCR source as due to the same mechanism, presumably related to first ionization potential or first ionization time, that produces the observed fractionation of the solar wind and solar energetic particles. However, there remains an unresolved problem of how the stellar ejecta can remain energetic long enough to encounter the shocks needed for acceleration to cosmic-ray energies.

The cosmic-ray data provide observational constraints which must be accounted for by any successful model of cosmic-ray origin and acceleration. The need for significant contributions of material from both SN II and SN Ia restricts models which involve acceleration of material from a limited portion of the ISM. Stars form preferentially in associations as giant molecular clouds collapse. The more massive of the stars evolve rapidly and undergo supernovae explosions while still closely grouped. The evolutionary time scales for low-mass stars are much longer and these stars survive long after the original associations have lost their identity and the stars have been spread extensively into the Galaxy. Thus contributions of material from SN Ia should be broadly and rather smoothly distributed through the Galaxy, not clustered like the SN II contributions.

It has been proposed (20) that cosmic rays are preferentially accelerated in super-bubbles produced as winds and explosions of massive stars in associations blow hot, low-density voids in the general ISM. For such a model to account for the observed GCR source composition requires that this hot, tenuous phase of the ISM occupy a large fraction of galactic volume so that SN Ia will be sufficiently abundant (relative to SN II) in the acceleration volume. It is thought that this fraction is $\sim 50\%$ averaged over the Galaxy's volume (see, for example, (20) and references therein), and possibly a significantly larger fraction inside the solar system's galactocentric radius (21).

In addition to their significance for understanding the origin of galactic cosmic rays, the GCR source isotopic abundances being obtained from CRIS are directly relevant to understanding the composition of present-day interstellar matter and the rate of chemical evolution in the Galaxy. Isotopic (as opposed to elemental) abundance determinations in material outside the solar system are presently limited to a relatively few elements, primarily those that form interstellar gas molecules with rotational and/or vibrational transitions that can be used to distinguish the mass of the atoms involved by spectroscopic means (22). For most elements, and particularly for noble gases and refractory elements that are depleted from the gas phase, very little is known about isotopic composition. (See, however, (23) concerning the isotopic composition of Ne in the very local ISM as derived from studies of the anomalous component of cosmic rays.)

It can be argued that the GCR source composition presently provides the best estimate of the isotopic composition of certain important elements in the ISM, including Mg, Si, Fe, Ni, Cu, and Zn. Further analysis of data from CRIS should add several elements to this list. Uncertainties remain concerning the details of the location and physical conditions of the galactic material from which cosmic rays are derived, but these appear no more limiting than those which are encountered in attempting to extract overall galactic isotopic composition from measurements of interstellar molecules. In fact, comparison of isotopic composition results for elements such as C, O, and S which can be investigated by both techniques should be useful for assessing the magnitude of these uncertainties.

ACCELERATION CONSTRAINTS FROM ELECTRON-CAPTURE PRIMARIES

Additional constraints on the location and nature of cosmic-ray acceleration are related to the time that elapses between nucleosynthesis of the cosmic-ray seed material and acceleration to high energies. It is generally accepted that the major contributors to cosmic-ray acceleration in the Galaxy are shocks produced by supernovae. This view is supported both by the fact that supernovae are the only objects that are known to output sufficient power to sustain the observed GCR intensity and by the fact that observations of synchrotron radiation from supernova remnants show that at least electrons have been accelerated to relativistic energies there. However, a long-standing question is whether the bulk of the material accelerated by supernova shocks originates from the ejecta of the same exploding star or from surrounding interstellar matter.

It was pointed out by Soutoul et al. (25) that this elapsed time can be derived from the observed abundances of radioactive cosmic-ray nuclides that should be produced predominantly in supernovae explosions (rather than as secondaries from fragmentation during transport through the Galaxy) and that can decay only by orbital electron capture. After ejection by the explosion, nuclei reside in the thermal matter of the remnant for an unknown time before acceleration occurs. Under these conditions, electron-capture decays proceed with the laboratory halflife. Once acceleration occurs and a small amount of matter has been traversed at high energies (a few tens of mg/cm^2 is sufficient), the nuclei become stripped of their orbital electrons and are effectively stable. Thus the presence (absence) of a significant primary component of an electron-capture nuclide in the cosmic rays that are observed indicates an elapsed time between nucleosynthesis and acceleration which is short (long) compared to the halflife.

Acceleration Time Delay

A sizeable number of pure-electron-capture nuclides are observed with CRIS, but most of these are dominated by secondaries produced during propagation. While of interest for studying processes occurring during transport, these secondary-dominated nuclides are not useful for investigating the acceleration time delay. Two electron-capture nuclides, ^{59}Ni ($T_{1/2} = 7.6 \times 10^4$ yr) and ^{57}Co ($T_{1/2} = 0.74$ yr), have small secondary contributions because they are too massive to be produced by fragmentation of abundant cosmic-ray species such as ^{56}Fe. As seen in Figure 1, ^{59}Ni appears to be absent, or nearly so, while its stable daughter product ^{59}Co is clearly present.

The upper two panels of Figure 4 show how the observed abundances of the mass-59 isotopes would vary depending on the delay time between nucleosynthesis and acceleration. The observed abundances contain contributions from both secondaries (dashed lines) and primaries. The primary $^{59}\mathrm{Ni} + {}^{59}\mathrm{Co}$ is constrained by the observed abundances, but the fraction $f_0 = ({}^{59}\mathrm{Ni}/({}^{59}\mathrm{Ni} + {}^{59}\mathrm{Co}))_{\mathrm{primary}}$ is not. The various solid

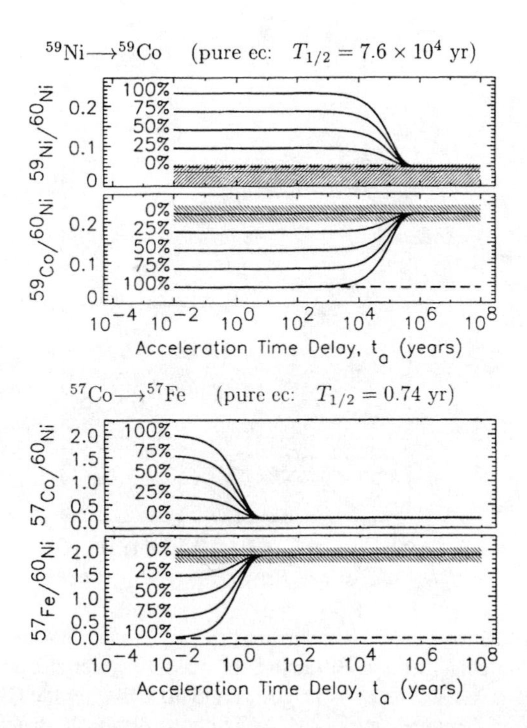

FIGURE 4. Dependence of arriving GCR abundances of electron-capture nuclides and their daughters on the time delay between nucleosynthesis and acceleration. (See description in the text.)

curves are parameterized by the value of f_0, which represents the fraction of the primary mass-59 material synthesized in the form of ^{59}Ni. The hatched regions show the observed abundances from CRIS with their one standard deviation uncertainties. The CRIS observations, which indicate that in the arriving cosmic rays all of the mass-59 primaries are in the form of ^{59}Co, are consistent with either a time delay before acceleration which is long compared to the 7.6×10^4 yr halflife of ^{59}Ni, or a nucleosynthesis process which produces ^{59}Co rather than ^{59}Ni.

To distinguish between these two possibilities one can examine theoretical predictions for the relative yields of the ^{59}Co and ^{59}Ni from supernova explosions. Figure 5 shows the allowed combinations of the delay time, t_a, and the ^{59}Ni fraction, f_0, as a 98% confidence region (hatched area). The dashed lines show the minimum, maximum, and average (weighted by a Salpeter initial mass function [IMF]) values of f_0 found from numerical calculations (15) of ejected yields from SN II resulting from stars of various initial masses. Other calculations (12) indicate that ^{59}Ni should constitute a major fraction of the mass-59 ejecta from SN Ia. Thus there should have been a significant initial fraction of ^{59}Ni from any combination of SN II and SN Ia sources, and one can conclude that a delay time $\gtrsim 10^5$ yr between nucleosynthesis and acceleration is needed to account for the CRIS observations (26).

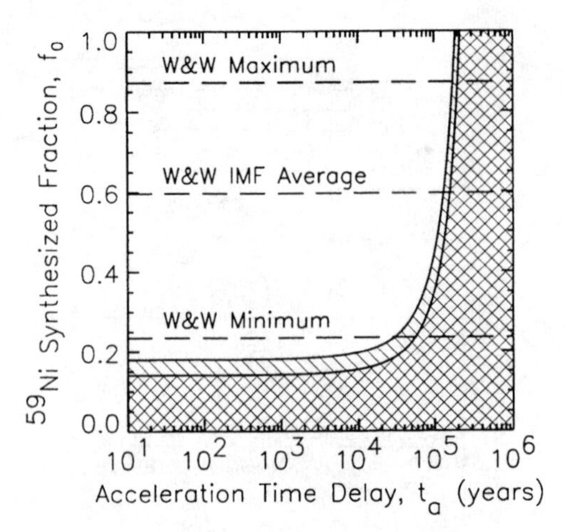

FIGURE 5. Allowed combinations (98% confidence) of acceleration delay times and fractions of mass-59 material synthesized as ^{59}Ni. The cross-hatched region is based on the CRIS measurement uncertainties alone while the diagonally-hatched extension also takes into account estimated cross-section uncertainties.

This time delay before acceleration is significantly longer than the time required for a supernova shock to dissipate its energy (27), so we conclude that the seed population for cosmic-ray acceleration is predominantly old stellar or interstellar material rather than fresh supernova ejecta. This is consistent with the above results (Fig. 3) which show that the isotopic composition of the GCR source is very similar to that of the solar system. To obtain such a result in a model where supernovae promptly accelerate a portion of their own ejecta would require that acceleration efficiencies be comparable over a sizeable range of stellar masses so that the cosmic-ray source would contain a mix of contributions similar to that which produced abundances found in the general ISM.

To be consistent with the mass-59 observations, one expects that all of the observed mass-57 primaries should be in the form of the daughter product, ^{57}Fe, rather than the parent ^{57}Co, since the halflife of the latter nuclide (0.74 yr) is much shorter than the halflife of ^{59}Ni. The lower two panels of Figure 4 show how the observed abundances of the mass-57 nuclides should depend on the time between nucleosynthesis and acceleration, assuming that the ^{57}Co is stable after acceleration. (As will be discussed below, this assumption will be valid only under a limited set of acceleration conditions.) As expected, the ^{57}Co and ^{57}Fe data are consistent with complete decay of primary ^{57}Co.

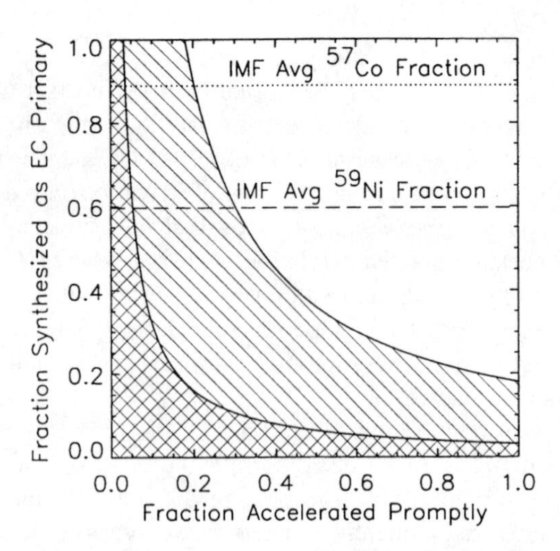

FIGURE 6. Fractions of the cosmic rays that could have been accelerated promptly compared to the halflives of ^{59}Ni ($\sim 10^5$ yr, diagonal hatching) and ^{57}Co ($\sim$ 1yr, cross hatching). The hatched regions correspond to 98% confidence levels and take into account both the CRIS measurement errors and estimated uncertainties on the fragmentation cross sections.

Fraction Accelerated Promptly

In the above discussion it was assumed that all of the cosmic rays are accelerated with a common delay time between nucleosynthesis and acceleration. In reality, one should expect some distribution of delay times, and the observed abundances of electron-capture primaries can be used to provide information about this distribution. For simplicity we assume that the accelerated material can be divided into two populations: a "prompt" population for which the time delay was much less than the halflife of the isotope being considered and a "delayed" population accelerated on times scales much longer than the halflife. From the observational upper limits on the abundances of the electron-capture primaries one can set upper limits on the fraction of material in the prompt category. This upper limit also depends on f_0, the fraction of the material of the particular mass number synthesized as the electron-capture primary (rather than as its daughter product). Figure 6 shows 98% confidence regions indicating the fractions of the material accelerated promptly compared to the 7.6×10^4 yr halflife of ^{59}Ni (diagonal hatching) and compared to the 0.74 yr halflife of ^{57}Co (cross hatching).

Also shown are dashed and dotted lines indicating the values of f_0 predicted for ^{59}Ni and ^{57}Co, respectively, by taking IMF-weighted averages of the calculated yields from the SN II models of Woosley and Weaver (15). From the ^{59}Ni result we conclude that up to $\sim 30\%$ of the cosmic-rays could have been accelerated on time scales $\ll 10^5$ yr without causing a significant conflict with the observed composition. The ^{57}Co results potentially provide a more stringent limit ($\lesssim$5%) on the fraction accelerated on time scales $\ll$ 1 yr. However, the applicability of the ^{57}Co results depends strongly on the physical environment in which the acceleration occurs since this influences the chances of in-flight electron attachment and subsequent decay. This problem is discussed in the next subsection.

Electron Attachment and Loss in Flight

In the simple interpretations presented above of the abundances of primary electron-capture nuclides and their daughters, the assumption was made that cosmic-ray acceleration occurs over a time interval short compared to the $\sim 2 \times 10^7$ yr confinement time in the Galaxy (28) and that during acceleration and propagation the nuclei remain fully stripped of any electrons.

However, the probability for electron attachment in flight is not completely negligible and can affect the interpretation of the observed abundances of electron-capture-decay nuclides. The upper panel in Figure 7 shows, as a function of energy per nucleon, the mean amounts of interstellar matter (assumed to be composed only of H and He in a 10:1 ratio by number) that a Co nucleus must traverse to attach an electron or to strip off an electron that was previously attached (29). In the lower panel the corresponding mean times for attachment and stripping are shown, assuming a medium with a uniform density of 0.25 H atoms/cm^3, as indicated by the surviving fractions of β-unstable secondaries such as ^{10}Be (28).

Energetic ^{57}Co ions passing through interstellar matter will tend to be stripped of their electrons in passing through $\sim$ 1–10 mg/cm^2 of material (Fig. 7). If this matter is traversed on a time scale short compared to the $\sim$ 0.74 yr ^{57}Co halflife, decays will be prevented. This will be the case if the acceleration occurs in a medium having a density of at least several thousand atoms/cm^3.

As shown in Figure 7, the mean free path for electron attachment by a bare Co nucleus at 600 MeV/nucleon (the approximate interstellar energy of the particles measured with CRIS) is $\sim$ 11 g/cm^2. This is long compared to the mean free path for fragmentation of the nuclei in collisions with interstellar matter ($\sim$ 2–3 g/cm^2), so the accelerated population of ^{57}Co will only undergo minor alteration by subsequent electron attachments followed by electron-capture decays. Thus the simple interpretation of the ^{57}Co observations presented in the previous section is applicable only if the cosmic-ray acceleration occurs in a medium dense enough to strip the nuclei before electron capture decays can occur.

One can also examine the effects of in-flight electron attachment and loss on the interpretation of the measured ^{59}Ni and ^{59}Co abundances. The mean free paths and related quantities shown for Co in Figure 7 differ little from the corresponding quantities for Ni since the high-energy attachment and loss cross sections should be smooth functions of the atomic number of the energetic nucleus. Because the halflife of ^{59}Ni is a factor of 10^5 longer than that of ^{57}Co, the required density to prevent decay during acceleration is smaller by this factor, amounting to just a few hundredths of an atom/cm^2. Since this value is small compared to the average density encountered by the cosmic rays during propagation, this constraint should easily be met. Thus the conclusions presented based on the decay of ^{59}Ni to ^{59}Co should be relatively robust.

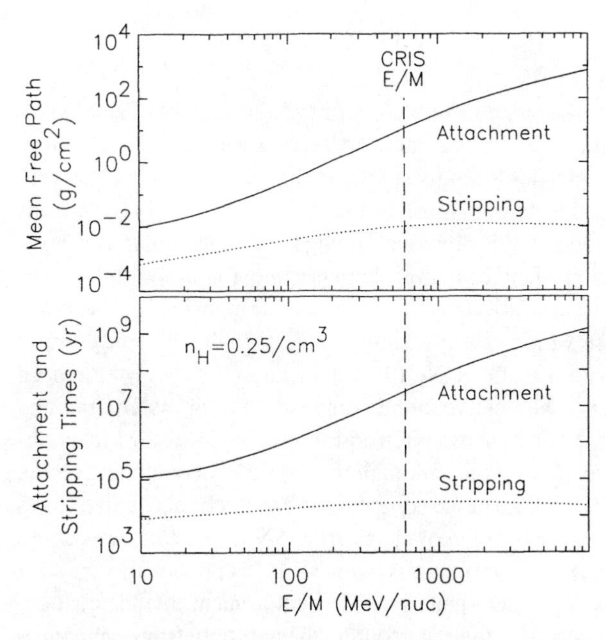

FIGURE 7. Electron attachment and loss mean free paths (29) and related quantities as a function of the energy of cosmic-ray Co nuclei. An interstellar He/H ratio of 0.1 by number has been assumed. Dashed vertical lines indicate approximate interstellar energy of Co nuclei measured by CRIS.

To construct a scenario that would allow the in-flight decay of ^{59}Ni, it is necessary to assume that particle acceleration occurs over a relatively long time span. At energies $\lesssim$150 MeV/nucleon the electron-attachment mean free path is shorter than the mean free paths for nuclear interaction losses and for escape from the Galaxy. If the ^{59}Ni remain below this energy for a time long compared to their halflife, electron attachment and subsequent decay will convert the nuclei to ^{59}Co. Although at present there is no observational indication that a slow acceleration of this sort was involved in producing galactic cosmic rays, it is also difficult to completely rule out.

CONCLUSIONS

Detailed examination of the abundances of stable isotopes of Fe, Co, and Ni reveals a composition pattern which is strikingly similar to that of solar system material, differing by at most several tens of percent. This confirms and extends previous findings that the isotopic composition of other major elements, most notably Mg and Si, differs little from solar system composition. These observations are in striking contrast to the isotopic composition of GCR Ne (3, 4, 6) which is strongly enhanced in ^{22}Ne relative to the composition of the solar wind and the composition inferred for the very local ISM from observation of the anomalous cosmic-ray component of Ne. The GCR source composition can not be accounted for as a superposition of ejecta from SN II alone but appears to require contributions from SN Ia, as previously noted for solar system material. The requirement that the cosmic-ray source material contain the appropriate mix of contributions from SN II, which occur preferentially in associations, and SN Ia, which tend to be rather smoothly distributed in the Galaxy, provides a potentially useful constraint on the sites from which cosmic rays could originate.

Measurements of mass-59 nuclides in cosmic rays show that the pure electron-capture nuclide ^{59}Ni synthesized in supernovae has decayed. The simplest interpretation of this result leads to the conclusion that at least $\sim 10^5$ yr elapses before acceleration of the bulk of the cosmic-ray material. An upper limit $\sim 30\%$ can be set on the fraction of the material which could have been accelerated on shorter time scales.

These observations can be accounted for in a model in which cosmic rays are accelerated by supernova-produced shocks as they propagate through the general interstellar medium accelerating a mix of nuclei previously synthesized and returned to the ISM by stars with a broad distribution of masses and corresponding evolutionary time scales, similar to the distribution that produced the material from which the solar system condensed. The population of material accelerated by supernova shocks contains at most minor contributions promptly accelerated from the ejecta driving the shocks.

ACKNOWLEDGMENT

This research was supported by the National Aeronautics and Space Administration at the California Institute of Technology (under grant NAG5-6912), the Jet Propulsion Laboratory, the Goddard Space Flight Center, and Washington University.

REFERENCES

1. Stone, E. C. et al., *Sp. Sci. Rev.* **86**, 283 (1998).
2. Engelmann, J. J. et al., *Astron. Ap.* **233**, 96 (1990).
3. DuVernois, M. A. et al., *Ap. J.* **466**, 457 (1996).
4. Webber, W. R., Lukasiak, A., and McDonald, F. B., *Ap. J.* **476**, 766 (1997).
5. Connell, J. J. and Simpson, J. A., *Ap. J. Lett.* **475**, L61 (1997).
6. Connell, J. J. and Simpson, J. A., *Proc. 25th Internat. Cosmic Ray Conf.* (Durban) **3**, 381 (1997).
7. Thayer, M. R., *Ap. J.* **482**, 792 (1997).
8. Leske, R.A., *Ap. J.* **405**, 567 (1991).
9. Grove, J. E. et al., *Ap. J.* **377**, 680 (1991).
10. Anders, E. and Grevesse, N., *Geoch. Cosmoch. Acta* **53**, 197 (1989).
11. Tsujimoto, T. et al., *Mon. Not. Royal Astron. Soc.* **277**, 945 (1995).
12. Iwamoto, K. et al., *Ap. J. Suppl.* **125**, 439 (1999).
13. Meyer, J.-P. and Ellison, D. C., in *LiBeB, Cosmic Rays, and Related X- and Gamma-Rays*, edited by R. Ramaty et al., Astro. Soc. Pacific, San Francisco, ASP Conf. Series vol. 171, 1999, pp. 187–206.
14. Thielemann, F.-K., Nomoto, K., and Hashimoto, M., *Ap. J.* **408** (1996).
15. Woosley, S. E. and Weaver, T. A., *Ap. J. Suppl.* **101**, 181 (1995).
16. Meyer, J.-P., Drury, L. O'C., and Ellison, D. C., *Ap. J.* **487**, 182 (1997).
17. Ellison, D. C., Drury, L. O'C., and Meyer, J.-P., *Ap. J.* **487**, 197 (1997).
18. Meyer, J.-P., *Ap. J. Suppl.* **57**, 173 (1985).
19. Shapiro, M. M., in *LiBeB, Cosmic Rays, and Related X- and Gamma-Rays*, edited by R. Ramaty et al., Astro. Soc. Pacific, San Francisco, ASP Conf. Series vol. 171, 1999, pp. 138-145.
20. Higdon, J.C., Lingenfelter, R.E., and Ramaty, R., *Ap. J. Lett.* **509**, L33 (1998).
21. Lingenfelter, R. E., Higdon, J. C., and Ramaty, R. "Cosmic Ray Acceleration in Superbubbles and the Composition of Cosmic Rays", in *Acceleration and Transport of Energetic Particles Observed in the Heliosphere*, edited by R. Mewaldt et al., New York, AIP Conf. Proceedings, 2000 (this volume).
22. Wilson, T. L., *Reports on Prog. in Phys.* **62**, 143 (1999).
23. Leske, R. A. et al., *Sp. Sci. Rev.* **78**, 149 (1996).
24. Woosley, S. E. and Weaver, T. A., *Ap. J.* **243**, 651 (1981).
25. Soutoul, A., Cassé, M., and Juliusson, E., *Ap. J.* **219**, 753 (1978).
26. Wiedenbeck, M. E. at al., *Ap. J. Lett.* **523**, L61 (1999).
27. Longair, M. S., *High Energy Astrophysics*, vol. 2, Cambridge Univ. Press, Cambridge, England, 1994.
28. Connell, J. J., *Ap. J. Lett.* **501**, L59 (1998).
29. Crawford, H. J., Ph.D. thesis, Univ. of Calif. Berkeley, Lawrence Berkeley Laboratory Report LBL-8807 (1979).

The Composition and Energy Spectra of High Energy Cosmic Rays

Simon P. Swordy

Enrico Fermi Institute and Department of Physics University of Chicago, Chicago, Illinois 60637

Abstract. The existing paradigm of the origin of Galactic cosmic rays places strong supernovae shocks as the acceleration site for this material. However, although the EGRET gamma-ray telescope has reported evidence for GeV gamma rays from some supernovae, it is still unclear if the signal is produced by locally intense cosmic rays. Although non-thermal x-ray emissions have been detected from supernova remnants and interpreted as synchrotron emission from locally intense electrons at energies up to $\sim$100 TeV, these results seem inconsistent with the electron source spectrum inferred by direct measurements. It remains the case that simple energetics provide the most convincing argument that supernovae power the bulk of cosmic rays. Two characteristics which can be used to investigate this issue at high energy are the source energy spectra and the source composition derived from direct measurements.

INTRODUCTION

The detailed composition of cosmic rays arriving at Earth has been studied for many years. Other papers at this symposium show the spectacular advances in detailed isotopic composition determinations for cosmic rays which have been made by ACE. In this work we try to examine how elemental composition at the highest energies can be used to investigate a favorite paradigm - that of diffusive shock acceleration by supernovae remnants. To derive the nature of the sources from the measurements made near Earth a crucial component is a detailed understanding of the history of cosmic rays during their passage through our Galaxy. This process is often referred to as propagation, but this is really something of a misnomer since there are reasons to believe that cosmic rays may also be accelerated during this process by the general magnetohydrodynamic turbulence of the Galaxy. The discovery, nearly 30 years ago(1, 2), of an energy dependence to the apparent "propagation" pathlength of cosmic rays prompted the realization that the measured cosmic-ray energy spectra are significantly steeper than would be observed near the source. As a consequence our direct knowledge of the source spectra and history of cosmic rays in our Galaxy is limited in scope to the energy ranges where this energy dependence is known. Unfortunately these measurements only extend to energies of $\sim$100 GeV/n, far lower than the highest energy direct measurements and 10 orders of magnitude below the highest energy cosmic rays observed through air-showers. To make progress, we have to extrapolate what we know about this history at lower energies and introduce some underlying simplifying assumptions.

THE HISTORY OF HIGH ENERGY COSMIC RAYS IN OUR GALAXY

Although there have been many measurements of the secondary spallation produced nuclei in cosmic rays, the best observations were made by the HEAO-3 satellite during the 1980s(3, 4). These data have by far the highest statistics of any measurement and, at least in case of the Cherenkov counter measurements, a well calibrated energy scale. They also have excellent charge resolution which is vitally important for good separation of secondaries from neighbors which are predominantly primary in nature. The data from this satellite for the Boron/Carbon ratio are shown in Figure 1, together with some data from the CRN Space Shuttle experiment, albeit with lower statistics. Also shown on this picture are simple propagation models based on a power law of magnetic rigidity ($\propto R^{-\delta}$) dependent escape from the Galaxy. As can be seen models where $\delta \sim 0.65$ provide a reasonably good fit to these data. A study of this history can also be made to the sub-Fe/Fe ratio where a similar value for δ is obtained. If reacceleration is significant the highest energy particles should have the longest pathlengths - since the energy (rigidity) would be expected to increase with residence time. The absence of this trend from these secondary to primary ratios provides constraints on the amount of reacceleration which could occur over this rigidity range. Even the "weak limit"

CP528, *Acceleration and Transport of Energetic Particles Observed in the Heliosphere: ACE 2000 Symposium*, edited by Richard A. Mewaldt, et al.
© 2000 American Institute of Physics 1-56396-951-3/00/$17.00

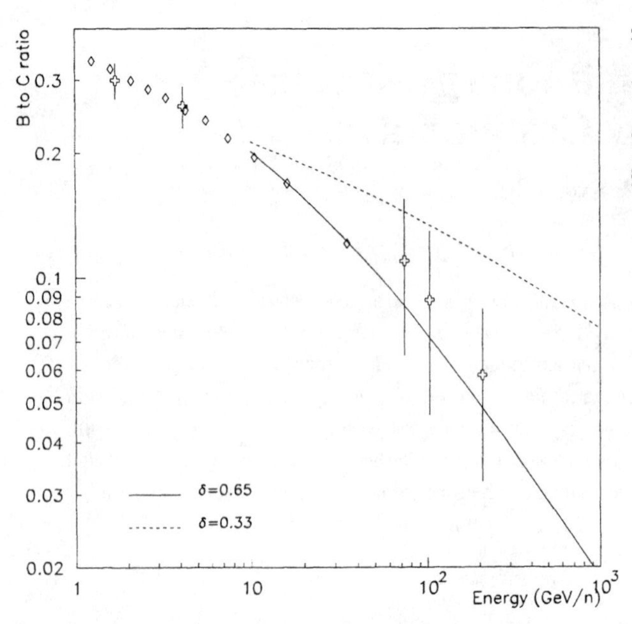

FIGURE 1. The ratio of Boron to Carbon in cosmic rays(5, 4)

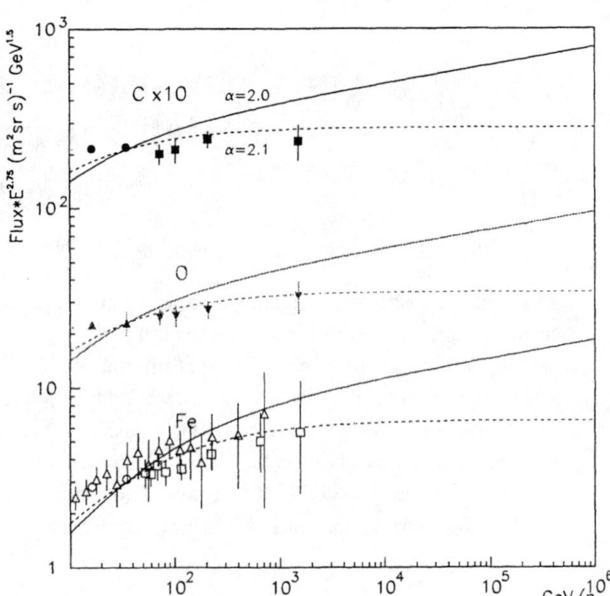

FIGURE 2. Cosmic ray fluxes at high energy for Carbon, Oxygen and Iron(8, 4), the solid lines are models for a source $\alpha = 2.0$ and dashed line $\alpha = 2.1$

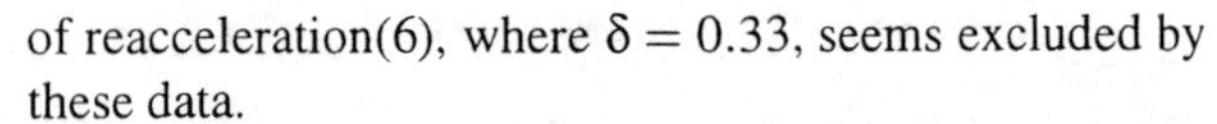

of reacceleration(6), where $\delta = 0.33$, seems excluded by these data.

As discussed in the introduction, to infer the source spectra at higher energies we must make extrapolations of this history to higher energy. Indeed, we must further assign this pathlength energy dependence to other elemental species, including protons and helium nuclei, to provide an overall framework for the interpretation of the observed elemental spectra. As a further simplification we assume that re-acceleration is insignificant at all the energy scales investigated. With the caveat that this is a pretty substantial extrapolation of cosmic-ray history at low energies, we can derive the source properties of cosmic-ray spectra at high energy.

ELEMENTAL SOURCE SPECTRA

For carbon and oxygen nuclei this derivation is reasonably accurate since the energy dependence of the pathlength is directly measured for these nuclei as shown in Figure 1. A source of cosmic rays produced by diffusive strong shock acceleration driven by a supernova explosion is expected to have a population of particle magnetic rigidities which is close to a decreasing power law with increasing rigidity(7). Since at high energy (>10GeV/n) the nucleus rest mass is essentially negligible we can also expect a simple power law in particle energy/nucleon , E, of the form $dN/dE \propto E^{-\alpha}$. Here, α is the spectral index of the source. Using this source model we can calculate an expected form for spectra of various nuclei, under the assumptions of pathlength variation with energy discussed above, with a simple propagation calculation which also ignores particle energy losses, a reasonable approximation above 10GeV/n. Figure 2 shows a comparison between the expected spectral shapes and the measured data for C, O, and Fe nuclei. These curves have an arbitrary flux normalization near 10GeV/n and the ordinate shows the intensity multiplied by a factor of $E^{2.75}$ to make differences between the data and the model curves more apparent. The two different curves correspond to possible source spectral indices of $\alpha = 2.0$ or $\alpha = 2.1$. With this comparison we can examine wether or not all nuclei have similar rigidity spectra at the source and try to discover the best fit value of α. In this energy range theories of supernova diffusive shock acceleration predict that all nuclei should have similar rigidity spectra at the source and for strong shocks α should be $\sim$2(7). These data show a reasonably good fit to these predictions for $\alpha = 2.1$ and therefore lend support to the origin paradigm, at least into the TeV/n range.

Figure 3 shows similar calculations for protons and helium nuclei. These seem to indicate that protons have a source spectral index, α, relatively close to the heavier nuclei shown in Figure 2, but the helium seem signifi-

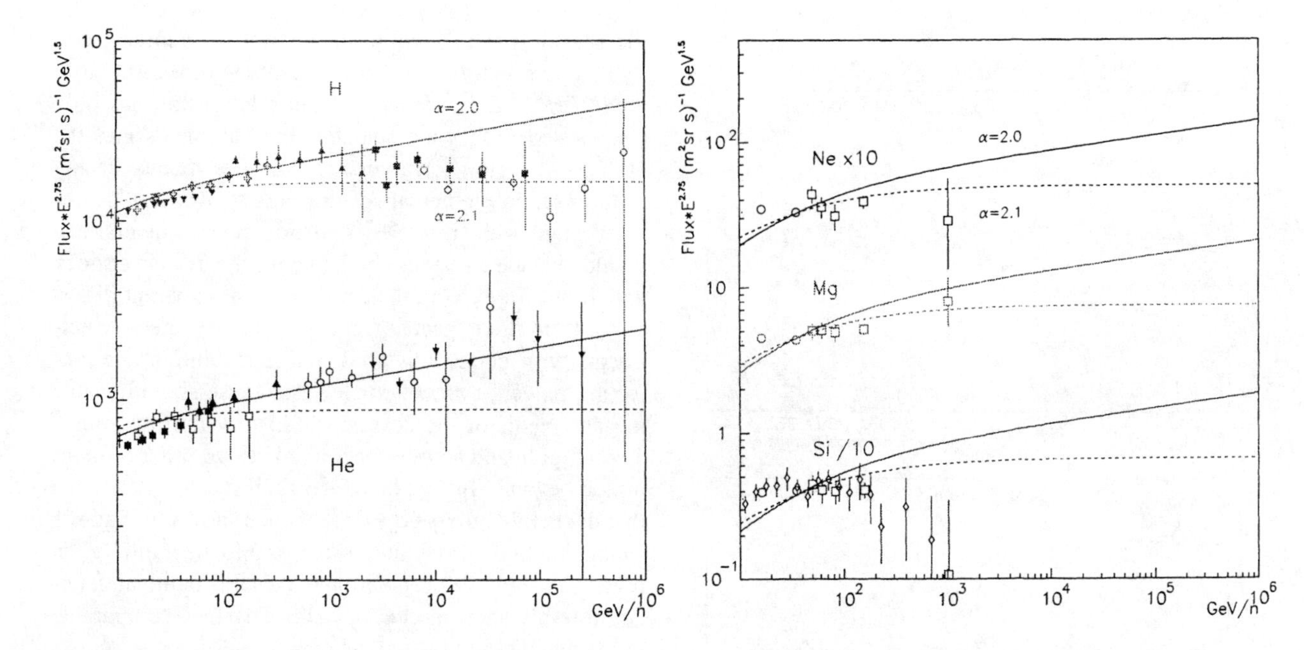

FIGURE 3. Cosmic ray fluxes(12, 10, 11) at high energy for Protons and Helium, the solid lines are models for a source $\alpha = 2.0$ and dashed line $\alpha = 2.1$

FIGURE 4. Cosmic ray fluxes(8, 9) at high energy for Neon, Magnesium and Silicon, the solid lines are models for a source $\alpha = 2.0$ and dashed line $\alpha = 2.1$

cantly flatter. The source spectra of helium seems closer to $\alpha = 2.0$ than 2.1. We can extend this type of analysis to the other major primary nuclei, Ne, Mg and Si as shown in Figure 4. Here the statistics are not as large at the higher energies but there is general agreement with a source value $\alpha = 2.1$, although the Silicon seem significantly steeper at high energy in data sets from two separate experiments.

SOURCE COMPOSITION

Aside from the source spectra we can directly examine the elemental composition of the source at high energies. This can be done for a wider range of elements since elements which have a statistical precision too poor for accurate spectral measurements can still provide useful abundances at high energy. In addition the establishment of source abundances at high energies is less prone to systematic errors in corrections because of the short propagation pathlength shown in Figure 1. The ratio of some cosmic-ray elemental source abundances to the local galactic abundance at 100 GeV/n are plotted versus atomic first ionization potential (FIP) in the upper panel of Figure 5. These are mostly derived from CRN measurements(8). This figure shows the charateristic dependence of source abundance on FIP which has been discussed by several authors for lower energy source abundances. That this pattern persists to high energy is another indication that the source apparently accelerates elements in similar rigidity spectra. This figure also provides graphic confirmation that these particles were initially selected by some kind of process at the eV energy scale before acceleration by a factor of $\sim 10^{11}$. Since other atomic properties, in particular volatility, are strongly correlated with FIP this selection could also be due to these other atomic effects(15).

Another striking feature of Figure 5 is the similarity of the cosmic ray source heavy nuclei distribution to the local galactic values, after correction for the FIP/volatility effect. Since most of the measured particles at 100GeV/n travel on average through a few g/cm^2 of interstellar material they are certainly not local in origin but they have a very similar chemical composition. This implies the cosmic ray source material is not particularly unusual, a feature that points away from ideas of the bulk of the cosmic rays originating in exotic or unusual systems or associations(16).

In fact the general abundance distribution of the source of high energy cosmic rays is very similar to solar energetic particles which also show a similar type of FIP effect. The lower panel of Figure 5 shows the abundances of Solar Energetic Particles (SEP) relative to the local Galactic values at 50MeV/n. This could be a simple coin-

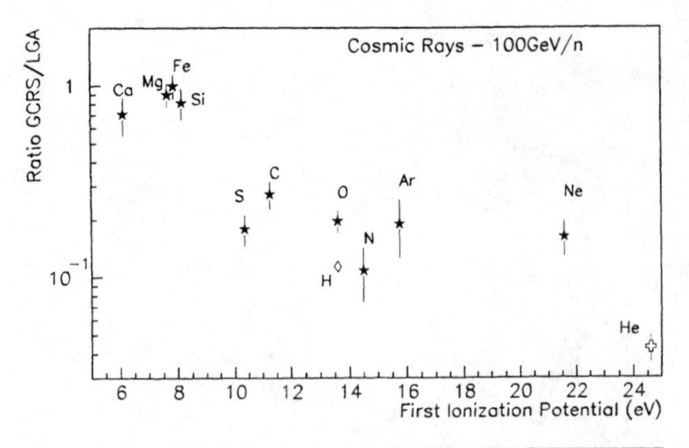

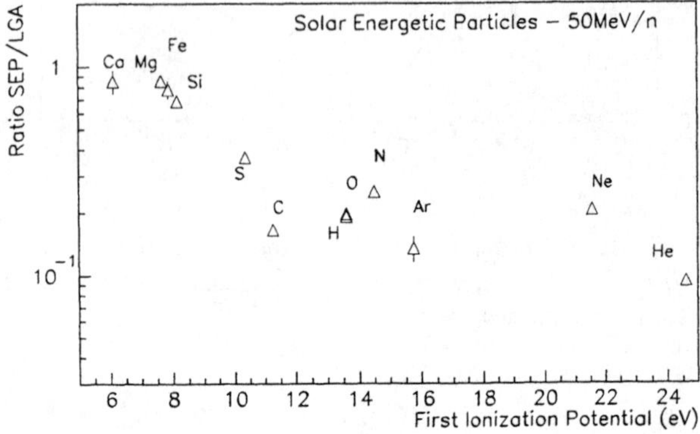

FIGURE 5. Top panel: cosmic ray source abundance at 100 GeV/n relative to local galactic abundances versus atomic first ionization potential(8, 13). Bottom panel: solar energetic particle abundances at 50 MeV/n relative to local galactic abundances(14, 13).

cidence, but we can speculate that the FIP effect in cosmic rays has something to do with the outpouring of energetic particles from ordinary stars. Something similar to this SEP flux is presumably streaming out from all stars and could provide the seed population which is accelerated to become energetic cosmic rays. The cosmic-ray source abundance mechanism might be observable in our own Sun. The transfer process of particles from convective regions in the photosphere to the energetic outbursts which produce the SEP might set the cosmic-ray source abundances.

CONCLUSIONS

With the caveats discussed in the introduction there are features in cosmic rays which directly suggest the supernova diffusive shock paradigm is correct. Apart from a couple of notable exceptions discussed above, the bulk of the energy spectra seem to originate from similar rigidity source spectra across all elements with a form close to $dN/dR \sim R^{-2}$. However this relationship has only been tested with good statistics measurements into the 100GeV/n region for a wide range of elements. If any differences in elemental spectra persist over a wide energy range with improved statistics measurements they would provide a serious challenge for the diffusive shock paradigm. To explore the elements with apparent differences from the expected rigidity spectra we need experiments with good individual charge resolution and collecting power at $m^2 sr\ year$ levels. These should enable measurements of the cosmic ray history into the multi-Tev/n region and source spectra at least an order of magnitude beyond. In fact there are good reasons to believe that the pathlength energy dependence shown in Figure 1 cannot continue to decline so steeply into the multi-TeV/n region. A simple extrapolation makes this pathlength uncomfortably short at energies above 10TeV, comparable to the thickness of the galatic disk.

On the other hand the overall spectrum of arriving cosmic ray particles has almost the same power law shape from energies of 10^{10}eV up to 10^{20}eV. The highest energy particles are many orders of magnitude in energy beyond that thought possible to create under the supernova diffusive shock mechanism. An overall scheme which unifies cosmic ray origin over this enormous energy range remains elusive.

(Work supported by NASA via grant NAG5-5071.)

REFERENCES

1. Juliusson and Meyer, Ap. Lett., 14 (1973) 153.
2. Smith et al., Ap. J., 180 (1973) 987.
3. Binns et al., Ap. J., 324 (1988) 1106.
4. Engelmann et al., A. & A., 233 (1990) 96.
5. Swordy et al., Ap. J., 349 (1990) 625.
6. Seo & Ptuskin, Ap. J., 431 (1994) 705.
7. see e.g. review Axford, proc. 17th ICRC 12 (1981) 155.
8. Muller et al., 374 (1991) 356.
9. Ichimura et al., Phys. Rev. D48 (1993) 1949.
10. references in Swordy et al., Proc. 23rd ICRC, Invited and Rapp. (1993) 243.
11. Seo et al., Ap. J., 378 (1991) 763.
12. Asakimori et al., Ap. J., 502 (1998) 278.
13. Meyer, Ap. J. S., 57 (1985) 173.
14. Reames, Adv. Sp. Res. 15 (1995) 41.
15. Meyer et al., Ap. J., 487 (1997) 182.
16. This is a restatement of a remark made by Prof. M. Israel (Wash. U.) at the symposium.

COSMIC RAY ACCELERATION IN SUPERBUBBLES AND THE COMPOSITION OF COSMIC RAYS

R.E. Lingenfelter[1], J.C. Higdon[2], & R. Ramaty[3]

1) Center for Astrophysics and Space Sciences, University of California, San Diego, La Jolla, CA
2) W. M. Keck Science Center, Claremont Colleges, Claremont, CA
3) Laboratory for High Energy Astrophysics, NASA Goddard Space Flight Center, Greenbelt, MD

Abstract. We review the evidence for cosmic ray acceleration in the superbubble/hot phase of the interstellar medium, and discuss the implications for the composition of cosmic rays and the structure and evolution of the interstellar medium (ISM). We show that the bulk of the galactic supernovae, their expanding remnants, together with their metal-rich grain and gas ejecta, and their cosmic ray accelerating shocks, are all confined within the interiors of hot, low-density superbubbles, generated by the multiple supernova explosions of massive stars formed in giant OB associations. This superbubble/hot phase of the ISM provides throughout the age of the Galaxy a cosmic ray source of essentially constant metallicity for acceleration by the shocks of many supernovae over time scales of a few Myr, consistent with both Be/Fe evolution and ACE observations of ^{59}Ni/^{59}Co. We show that the expected metallicity ($>$ 2 times Solar) and filling factor ($>$ 50%) of the superbubble/hot phase is high enough that the composition of cosmic rays accelerated from fast, supernova grains in these superbubbles is quite consistent with the both Be/Fe and cosmic ray data, while their acceleration from grains in the well-mixed cooler phases of the ISM is *not* consistent with observations. We also show that if the refractory cosmic ray metals come from the sputtering of fast refractory grains then the accompanying scattering of ambient gas by these fast grains can also account for the relative abundance of cosmic ray volatiles.

INTRODUCTION

ACE measurements by Wiedenbeck et al. (53) of the cosmic ray ^{59}Ni/^{59}Co abundance ratio and optical measurements of the Be/Fe abundance ratios in old stars, e.g. Molaro et al. (34) and Boesgaard et al. (4), have shown us that the making of cosmic rays is like the making of fine wine. Both have to be aged and blended. Cosmic rays can not be too young and fresh, because their low ratio of K-capture ^{59}Ni to its daughter ^{59}Co requires an age of $> 10^5$ years before acceleration, and they can not be too old and diluted, because the roughly constant ratio of cosmic ray produced Be to supernova produced Fe requires the metallicity in the matter from which the cosmic rays are accelerated to be roughly constant within a factor of $\sim$2. This can be achieved if the metallicity is dominated by supernova ejecta.

The cellar where all of this aging and blending happens quite naturally is the vast metal-enriched superbubble/hot phase of the interstellar medium (SB/HISM), where most supernovae occur. This can be seen from the extensive studies of Galactic and extragalactic supernovae, their progenitors and their cosmic ray accelerating remnants, and of the chemical enrichment and evolution of the interstellar medium, that we will review here with special emphasis on the metallicity and filling factor of the SB/HISM. We also discuss how ACE and other measurements of the cosmic ray composition can define these two properties in the region of cosmic ray acceleration.

The source of energy for cosmic ray acceleration is thought to be shock waves driven by the expansion energy of supernova ejecta, e.g. Blandford & Ostriker (3) and Axford (1). The power required to maintain the Galactic cosmic rays is about 10^{41} ergs^{-1}, e.g. Lingenfelter (23). The Galactic supernova rate is about 3 supernovae per century, e.g. van den Bergh & McClure (49), most of which (80 to 90%) are core-collapse (Type II and Ib/c) supernovae of relatively young ($<$ few 10^7 yrs) massive O and B stars; the remainder are Type Ia thermonuclear explosions of much older accreting white dwarfs. Thus, the average cosmic ray energy needed per supernova is about 10^{50} ergs, which requires an acceleration efficiency of about 10% from the blast wave shocks of the supernovae, since supernovae all seem to have similar ejecta kinetic energies of about 10^{51} ergs, e.g. Woosley & Weaver (56) and Nomoto et al. (36).

The source of the particles that are accelerated as cosmic rays and the site of their acceleration, however, are still debated. But if the energy comes from supernova shocks, the site and source of the particles clearly must

CP528, *Acceleration and Transport of Energetic Particles Observed in the Heliosphere: ACE 2000 Symposium*,
edited by Richard A. Mewaldt, et al.
© 2000 American Institute of Physics 1-56396-951-3/00/$17.00

be in the material through which the shocks pass. New clues to origin of the particles come from recent measurements, e.g. Molaro et al. (34) and Boesgaard et al. (4), of Be/Fe abundances in old halo stars that show that the ratio of cosmic ray spallation produced Be relative to core-collapse supernova-produced Fe has remained roughly constant throughout the evolution of the Galaxy. This constancy requires (Ramaty, Kozlovsky & Lingenfelter (38); Ramaty, Lingenfelter & Kozlovsky (40)) that the cosmic rays be accelerated out of matter that is only partially diluted by mixing with the interstellar medium (ISM) so that it is still sufficiently enriched in supernova-synthesized metals that its metallicity did not changed by more than $\sim$2 over galactic evolution. Detailed calculations by Ramaty et al. (42) and Ramaty, Lingenfelter & Kozlovsky (41) of the production and evolution of the Galactic Be/Fe ratio clearly show that the bulk of the cosmic rays can not be accelerated from the well-mixed ISM, as has been recently assumed, e.g. Meyer et al. (33) and Ellison et al. (11).

Just such a metal-enriched environment is where most supernovae do in fact occur and through which most of the cosmic ray accelerating, supernova shocks propagate. As we have discussed in Higdon, Lingenfelter & Ramaty (17,18), extensive observations show that the bulk of the core-collapse supernova progenitors are formed in OB associations in giant molecular clouds and that the combined winds and supernova ejecta of these stars form hot, low density superbubbles, that reach dimensions of several hundred pc and last for tens of Myr. During this time the bulk of the supernova ejecta, the supernova shocks, and their cosmic ray acceleration are all confined within the superbubble/hot phase of the ISM.

SUPERNOVAE & THE SUPERBUBBLE/HOT ISM

Core-collapse (Type II and Ib/c) supernovae are highly correlated in space and time, e.g. McCray & Snow (28). Such supernovae thus create giant cavities, or superbubbles, in the interstellar medium rather than many smaller, isolated bubbles, e.g. Mac Low & McCray (26) and Tomisaka (48). This is expected because 1) the massive O and B star supernova progenitors ($>$8 $M_\odot$; e.g. Woosley & Weaver (56); Nomoto et al. (36)) are not distributed uniformly in interstellar space, but tend to form clusters, since the majority of these massive stars are born in the most massive ($> 10^5$ $M_\odot$) molecular clouds in gravitationally unbound OB associations, while less massive clouds are destroyed by the intense UV irradiation with the birth of their first O star (McKee & Williams (30)); and 2) these stars are short-lived and slow moving; the progenitors of core-collapse supernovae have main sequence lifetimes of $\sim$ 3 to 35 Myr and OB stars in associations have dispersion velocities of only $\sim$4 km s^{-1} (Blaauw (2)), so they do not travel too far ($\sim$120 pc in 30 Myr) from their birthplaces before they die in supernova explosions. Consequently, the combined effect of these clustered supernova explosions is to create superbubbles, which expand and merge to form the hot ($>10^6$ K), tenuous ($<10^{-3}$ cm^{-3}) phase of the ISM with an average filling factor of $\sim$ 50%, or more (e.g. Yorke (57); Spitzer (45); McKee (29); Rosen & Bregman (43); Korpi et al. (20)) of the Galactic disk (with a scale height $\sim$100 pc) and essentially all of the corona/halo (with scale height of $\sim$3 kpc). We discuss the spatial distribution of the SB/HISM filling factor in more detail below.

An analysis of the surface brightness distribution of the remnants of historical supernovae in our Galaxy by (Higdon & Lingenfelter (16) has shown that 85$\pm$10%, of the observed Galactic supernovae occured in the superbubble hot phase of the ISM. This is quite consistent with more extensive observations of supernovae in other late type galaxies. As discussed in detail in Higdon et al. (17), the combined observations of van Dyk et al. (51) and Kennicutt, Edgar, & Hodge (19) show that the great majority, $\sim$ 90$\pm$10%, of the core-collapse supernovae in late type galaxies also occur within superbubbles, and because of the large filling factor of the superbubble, hot-phase of the ISM, half, or more, of the Type Ia should also occur within the superbubbles just by chance. Thus, since core-collapse supernovae account for 80 to 90% of all supernovae in our galaxy and Type Ia make up the remainder, roughly 80% of all supernovae occur in the superbubble hot phase of the ISM, and the bulk of the cosmic rays accelerated by their shocks are also produced there.

The observed concentration of supernovae in the SB/HISM and the subsequent cosmic ray acceleration in this hot ($> 10^6$ K) phase also argues strongly against a first-ionization-potential (FIP) injection bias for cosmic ray enrichment which requires warm partially ionized gas, and does not work in the nearly fully ionized gas of the SB/HISM. However, the SB/HISM environment is quite consistent with cosmic ray source particle injection from the sputtering of the high velocity refractory grains formed in supernovae, see Lingenfelter, Ramaty & Kozlovsky (25) and Lingenfelter & Ramaty (24). Such a volatility bias for cosmic ray refractory metal injection from sputtering of supernova grains was first suggested by Cesarsky & Bibring (7) and it was also recently proposed by Meyer et al. (33) for shock-accelerated, ice-stripped, refractory cores of grains in the warm ISM.

The occurrence of most supernovae in the SB/HISM and cosmic ray acceleration in that hot phase also argues strongly against the mass/charge (A/Q)-dependent acceleration model of Ellison et al. (11) for the volatile elements, because this too requires warm partially ionized gas, and Ellison & Meyer (12) argue it does not work in the highly ionized gas of the SB/HISM. As we have shown in Lingenfelter & Ramaty (24) and discuss below, however, a mass dependent injection of volatiles appears to result quite naturally from the scattering of the ambient gas atoms in direct collisions with fast grain atoms that must accompany the sputtering of the grains.

METALLICITY OF THE SUPERBUBBLE/HOT ISM

The bulk of the metals (elements with Z>5) in the Galaxy have been produced by supernovae and ejected into the ISM. The relative abundances of most elements have remained relatively constant (e.g. Timmes, Woosley & Weaver (47)), because they simply reflect the IMF-averaged supernova yields which do not depend strongly on the interstellar metallicity, e.g. Woosley & Weaver (56). The averaged relative abundances of the present ISM, e.g. Savage & Sembach (44), the older (4.5 Gyr) Solar system material, e.g. Grevesse, Noels & Sauval (15), and the IMF-averaged fresh supernova ejecta, e.g. Lingenfelter et al. (25), are all within about $\sim 10\%$ of one another. But their overall abundance in the ISM (i.e. the interstellar metallicity) has grown steadily over time with the accumulation of fresh supernova ejecta continuously injected and mixed into the ISM. The time scale for thorough mixing is generally thought (e.g. McWilliam (31); Thomas, Greggio & Bender (46)) to be on the order of 30 to 100 Myr. This is comparable to the typical mean life of the SB/HISM reservoir into which the bulk of the supernova ejecta with a metallicity 10 times Solar (e.g. Woosley & Weaver (56)) is injected and in which the bulk of the mixing is expected to occur. Thus, we would expect significant variations in the metallicity, but not in the abundances of most elements relative to one another, within the SB/HISM as a function of the age of individual superbubbles and their generating OB associations.

The average, or equillibrium, metallicity of the SB/HISM is not known, but the supernova ejecta appear to be able to provide sufficient metals to produce a metallicity >2 times Solar, as is required (Ramaty et al. 2000b) for the cosmic ray source from the constancy of the Be/Fe abindances in old stars. This can be seen from a simple comparison of the total mass of the SB/HISM and the mass of supernova ejecta produced during the mean life of the SB/HISM. The total mass of the SB/HISM is $\sim 10^8$ $M_\odot$, assuming an average, e.g. Yorke (57) and Spitzer (45), SB/HISM density of $\sim 10^{-3}$ H/cm^3, a SB/HISM scale height of $\sim$ 3 kpc and an effective Galactic radius of $\sim$ 15 kpc. Taking a nominal SB/HISM mean life, or mixing time, $t \sim 100$ Myr, the required SB/HISM input is ~ 1 $M_\odot$/yr(t/100Myr). The present Galactic SNII/Ibc rate of about 1 SN every 40 yr, producing an IMF-averaged ejecta mass of 18 $M_\odot$, gives a Galactic SNII/Ibc ejecta input of ~ 0.45 $M_\odot$/yr with a metallicity, z_{SN} of 10 times Solar. If all of the remaining SB/HISM mass comes from evaporated clouds and swept up gas in the well-mixed ISM with Solar metallicity $z_\odot$ of 1, then the averaged SB/HISM metallicity, $z_{HISM} \sim [10M_{SN} + 1(M_{HISM} - M_{SN})]/M_{HISM}$, or $z_{HISM} \sim 1 + 9M_{SN}/M_{HISM} \sim 1 + 4(t/100\text{Myr})$. Thus the SB/HISM metallicity $z_{HISM} > 2$ times Solar for any SB/HISM mean life, or mixing time, $t > 25$ Myr, consistent with the estimated values, e.g. McWilliam (31) and Thomas, Greggio & Bender (46).

Such a mixing time, or mean age, of metals from supernova ejecta in these superbubbles is more than a couple orders of magnitude longer than the minimum age (<100 kyr) of cosmic ray source metals required by the ACE observations of Wiedenbeck et al. (53), showing that the bulk of the ^{59}Ni had decayed (with a 110 kyr mean life) in the cosmic ray source material prior to acceleration.

Observational evidence of such supernova ejecta enriched superbubble metallicity may be found in the x-ray emission from the interiors of giant HII regions in the Large Magellanic Cloud (LMC), thought to be superbubbles powered by supernovae. The observed x-ray luminosities of these bubbles, which should scale directly with metallicity, are an order of magnitude higher than would be expected (Chu & Mac Low (8)) if they had a typical LMC metallicity of only 1/3 Solar. Thus, we suggest that the x-ray observations do in fact imply an average metallicity of roughly 3 times Solar in these superbubbles.

Such metallicities are larger than that calculated from the simple analytic superbubble model of Mac Low & McCray (26), which assumes conductive heating and evaporation of swept-up ISM as the primary source of superbubble gas and predicts an averaged metallicity of only ~ 1.1 for a ~ 50 Myr old superbubble of ~ 700 pc radius. But this model was based on several assumptions that greatly reduce the superbubble metallicity. First, the model neglected the interstellar magnetic fields that would greatly supress the conductive heating normal to the field lines, overestimating the ISM input, and also provide additional external confining pressure, underestimating the supernova input required to generate the bubble, e.g. Tomisaka (48). In addition, the model assumed the unit density ISM extended to heights much larger than the 700 pc bubble radius, instead of the measured scale height of < 200 pc, which would greatly reduce the assumed ISM input, and moreover would allow the blowout of the superbubble into the Galactic halo, also greatly underestimating the supernova input required to generate the bubble. As a result of these underestimates of the required supernova power, the model was able to generate a $\sim$ 50 Myr old superbubble of $\sim$ 700 pc radius, with a very low effective rate of SNII/Ibc supernovae of only $\sim$ 2 SN/Myr kpc^2 in the Galactic plane. This assumed rate is only 3% of the estimated local SNII/Ibc rate $\sim$ 70 SN/Myr kpc^2 (as we show below), and thus the model underestimates the supernova ejecta input by a factor of 35! When an appropriate supernova rate is used and even minimal effects of the magnetic fields and gas scale height are considered, superbubble metallicities of more that 2 times Solar would be expected.

A large fraction of the C, O and refractory metals in this ejecta may be in graphite and oxide grains, since in the core-collapse supernova 1987A roughly 0.2 $M_\odot$ of this material condensed out of the cooling, expanding ejecta as high velocity ($\sim$ 2500 km/s) grains within 2 years after the explosion, see Kozasa, Hasegawa & Nomoto (21), and as much as 1 $M_\odot$ could be expected, see Dwek (9), to condense before the ejecta is reheated and slowed by the reverse shock and the grains with a much smaller charge to mass ratio begin to move separately from the ejecta plasma. In fact, Dwek (9,10) suggests that supernova ejecta are the major source of refractory grains in the Galaxy and interactions with supernova shocks are the major cause of their destruction.

Thus, supernova ejecta and winds can be expected to dominate the metallicity and grains within the SB/HISM, where the bulk of supernova shock waves are dissipated and the bulk of cosmic rays should be accelerated. These supernova grains should therefore be the major injection source required for the cosmic ray metals, because of their high initial velocity (Lingenfelter et al. (25)) and possible subsequent acceleration (Ellison et al. (11)). Moreover, because the metallicity of the supernova ejecta is essentially independent of progenitor metallicity (Woosley & Weaver (56)), the SB/HISM can provide the essentially constant source of cosmic ray metals required by Be/Fe observations. Therefore, we would expect that throughout the age of the Galaxy, the bulk of the core-collapse supernovae occur in the metal enriched SB/HISM, and the blast wave shocks of their remnants accelerate the bulk of the Galactic cosmic rays out of the enriched gas and dust in the SB/HISM.

FILLING FACTOR OF THE SUPERBUBBLE/HOT ISM

The hot ($\sim 10^6$ K), tenuous ($\sim 10^{-3}$ H/cm^3) phase of the interstellar medium is powered primarily by Galactic supernovae and formed through the merger of superbubbles, generated by the clustered supernovae in OB associations. This can be seen energeticly from a comparison of the power required to maintain the pressure in the SB/HISM and that provided by Galactic supernovae, which suggests that the filling factor, i.e. the fractional volume, of the SB/HISM should be large. The total energy in SB/HISM is $\sim 3 \times 10^{56} f_{HISM}$ ergs, assuming a SB/HISM filling factor, f_{HISM}, a SB/HISM pressure of $\sim 3 \times 10^{-12}$ erg/cm^3, a Galactic radius of 15 kpc and a scale height of 3 kpc. For a SB/HISM mean life t of 100 Myr, the power required to maintain the SB/HISM is $\sim 3 \times 10^{48} f_{HISM}$ ergs/yr(t/100Myr). The Galactic SNII/Ibc rate of $\sim$ 1 SN/40 yr with an average ejecta energy $\sim 10^{51}$ ergs/SN, gives a Galactic SNII/Ibc power of $\sim 2.5 \times 10^{49}$ ergs/yr. Thus even with significant (>50%) energy losses SNII/Ibc could completely fill (f_{HISM} = 1) the Galaxy with the SB/HISM in t > 25 Myr.

The overall Galactic average value of the filling factor of SB/HISM is, in fact, generally taken to be $\sim$ 50%, or more, depending on the assumed Galactic scale height, e.g. Yorke (57), Spitzer (45), McKee (29). Because of the strong dependence of the SB/HISM filling factor on local supernova rates, it is thought to be high $\sim$ 90% in the inner Galaxy (i.e. within the Solar radius of 8.5 kpc) where most Galactic supernovae occur, as well as in the Galactic halo where the superbubbles blow-out, and low < 50% in the outer Galaxy beyond the Solar radius where few supernovae occur.

A more quatitative estimate of the dependence of the SB/HISM filling factor on Galactic radius (see Table 1) can be made from the radial dependence of the Galactic supernova rate and the calculated filling factor versus the supernova rate.

The dependence of the SB/HISM filling factor on local supernova rates has been quantified by recent calculations of 2D hydrodynamics by Rosen & Bregman (43), and 3D magnetohydrodynamics by Korpi et al. (20). These calculations determined the filling factors of all phases of the ISM as a function of height z above the Galactic plane for a range of supernova rates. Generally these calculations suggest that the SB/HISM filling factor is lowest at the Galactic plane where the superbubble expansion is most constrained by the warm and cold phases of the ISM, and increases to $\sim$ 100% in the halo at large distances above the plane. For the purposes of cosmic ray acceleration and composition, what is important is the height averaged filling factor of the SB/HISM for $|z| < 300$ pc, which is the range of heights where most of the supernovae occur. For assumed supernova rates (adjusted to 10^{51} ergs/SN) of 5, 20, 40 and 80 SN/Myr kpc^2 in the Galactic plane these calculations give height averaged ($|z|<300$ pc) SB/HISM filling factors of $\sim$ 0.1, 0.4, 0.6 and 0.9 respectively. This suggests that at low supernova rates (< 40 SN/Myr kpc^2) the SB/HISM filling factors within

Table 1. RADIAL DEPENDENCE OF GALACTIC SUPERNOVA RATE & EXPECTED SB/HISM FILLING FACTOR

Galactic Radius kpc	MoleCloud Density $M_\odot/pc^2$	OBAssoc Density N/kpc^2	Supernova Rate * SN/kpc^2Myr	SB/HISM Filling Factor† $\|z\|<300pc$
1	1	0.3	35	~ 0.4
4	7	2.3	250	~ 0.9
6	6	1.6	210	~ 0.9
8	2.5	0.6	90	~ 0.9
10	1.5	0.4	50	~ 0.5
15	0.4	-	12	~ 0.1

* Galactic SN rate of 3 SN/100yr normalized to surface density distribution of molecular clouds from Williams & McKee (54) and OB associations from McKee & Williams (30)

† Expected filling factor based on the SN rate from the hydrodynamic calculations by Rosen & Bregman (43) and Korpi et al. (20).

$|z|<300$ pc scale roughly linearly with the supernova power, while at higher supernova rates ($\geq$ 80 SN/Myr kpc^2) the SB/HISM filling factors within $|z|<300$ pc reach a maximum value of ~ 90%.

The Galactic radial dependence of the supernova rate can be estimated by normalizing the Galactic SN rate of 3 SN/100yr to the radial dependence of the surface density of either molecular clouds from Williams & McKee (54) or OB associations from McKee & Williams (30), which are proportional to one another, as we see in Table 1. Such a normalization gives a local supernova rate at the Solar distance (8.5 kpc) of about 80 SN/Myr kpc^2 and a peak rate at about 4 kpc of 250 SN/Myr kpc^2. From the calculated dependence of the filling factor on supernova rate, we thus estimate the Galactic radial dependence of the SB/HISM filling factor within $|z|<300$ pc, as shown in Table 1. We see that the SB/HISM is expected to fill most (~ 90%) of the ISM within $|z|<300$ pc from somewhere inside of 4 kpc out to roughly the Solar distance of 8.5 kpc, decreasing thereafter with Galactic radius to ~ 50% at 10 kpc and ~ 10% at 12 kpc where the supernova rate is very low. As we show below, a SB/HISM filling factor of > 50% can provide a cosmic ray injection composition in the SB/HISM that is consistent with current estimates of the required cosmic ray source composition. We note that one recent estimate by Ferriere (14) of the radial dependence of the SB/HISM filling factor gives only 20% locally, but this is for a very low local supernova rate from a very steep assumed radial dependence that is not consistent with the molecular cloud and OB association observations.

A local SB/HISM filling factor of ~ 90% would appear to be quite consistent with observations within the local kpc, see Blaauw (2) Fig. 8, which show that the Sun presently lies inside the ~ 500 pc radius superbubble produced by the ~ 30 Myr Cas-Tau OB association, e.g. Olano (37). This local superbubble is defined in the Galactic plane by a ring of young OB associations know as Gould's Belt which have formed from the ring of cooling gas swept up by the superbubble. The Cas-Tau association inturn is part of a larger (~ 1 kpc radius) ring of OB associations, including Cam-1, Aur-1, Gem-1 and Mon-2, formed by an older, now vanished OB association.

COSMIC RAY ACCELERATION IN SUPERBUBBLE/HOT ISM

These hot, low density superbubbles are the hot phase of the ISM, where shock acceleration of cosmic rays is expected, e.g. Axford (1), to be "most effective", because the energy losses of the accelerated particles are greatly reduced and the supernova shocks do not suffer major radiative losses, as they would in a denser medium. The rapid radiative loss of supernova remnant energy in the average ISM sets in at a radius of ~20 pc, while the undiminished shock energy of nonradiative remnants in the superbubble hot phase expand out to radii of ~200 pc. At full shock energy, supernovae in the low density SB/HISM expand to ~10^3 times the volume of those in the average ISM. Thus, the supernova shocks in low density, but metal enriched SB/HISM process a comparable masses of gas and for $z > 2$ at least twice the metals as those in the average ISM, contrary to the estimate of Ellison & Meyer (12).

Also, since the energy of supernova shocks in the SB/HISM, unlike that of shocks in the denser ISM, is not dissipated by radiation losses before the shocks slow to sound speed, cosmic rays are accelerated in SB/HISM primarily by low Mach number shocks. Such low Mach number (e.g. <4) shocks can produce, e.g. Axford (1), the power-law index of ~ 2.3 required for the cosmic ray source spectrum, while the lower spectral indices (~ 2) produced by high Mach number shocks in the denser ISM are not consistent with the required source value.

The observed concentration of supernovae in the superbubble hot phase and the much higher acceleration efficiency expected there clearly show that the bulk of the cosmic rays must be accelerated in the SB/HISM. Such an acceleration site also argues strongly against a first-ionization-potential (FIP) injection bias, e.g. Meyer (32), which requires warm partially ionized gas, not the highly ionized gas of the hot phase. Acceleration in the SB/HISM further argues against a mass/charge (A/Q) dependent acceleration model for the volatile elements, which Ellison & Meyer (12) argue does not work in highly ionized hot gas. As we have shown in Lingenfelter et al. (25) and Lingenfelter & Ramaty (24) and discuss further below, however, sputtering and scattering of hot gas by high velocity refractory grains from supernovae in the SB/HISM can provide a self-consistent cosmic ray injection source for both refractory and volatile elements.

The transient acceleration of low energy (<100 MeV/nucleon) cosmic rays (LECRs) in superbubbles has also been suggested, e.g. Bykov (5), as an alternative source of Be production in the Galaxy. To account for the measured Be/Fe evolution solely by LECRs, however, would require (Ramaty et al. (41)) that there be as much or more energy in the LECRs as there is the relativistic cosmic rays. Bykov (5) suggests that such LECRs might be accelerated in supernova shocks during the early (<3 Myr) stages of superbubble formation and that these LECRs are later further accelerated to relativistic cosmic ray energies by the ensemble of supernova shocks as the superbubble fully develops, e.g. Bykov & Fleishman (6). But since the energy in such LECRs persists for only a small fraction ($<10\%$) of the age (~ 50 Myr) of the superbubble and then more energy is added as the LECRs become relativistic cosmic rays which persist for most of the age of the superbubble, such a model can not produce a time averaged LECR energy comparable to that of the relativistic cosmic rays. Even if the LECRs were not further accelerated to relativistic energies, comparable total energy densities in LECRs and relativistic cosmic rays would require that roughly half of the supernovae in superbubbles accelerate LECRS, but $<5\%$ of the superbubble supernovae occur during the first few Myr of superbubble growth when condition favorable to LECR acceleration might be expected (Bykov (5)).

EXPECTED ABUNDANCES OF REFRACTORY COSMIC RAYS

We have shown in Lingenfelter et al. (25), Higdon et al. (17) and Lingenfelter & Ramaty (24) that the observed enrichment of the cosmic ray refractory elements can be produced by the preferential acceleration in the SB/HISM of suprathermal ions sputtered off high velocity (few 1000 km s^{-1}) refractory grains, which formed as condensates in the expanding ejecta of supernovae, e.g. Kozasa et al. (21) and Dwek (9). The measured (Naya et al. (35)) broad width (5.4 ± 1.4 keV) of the Galactic 1.809 MeV line from the decay of long-lived (1.0×10^6 yr mean life) ^{26}Al, most likely produced in Type II supernovae, e.g. Woosley & Weaver (56), clearly suggests that refractory grains, containing most of the live Galactic ^{26}Al, are still moving at velocities of ~ 450 km s^{-1} some 10^6 yrs after their formation, and that the bulk of the grains are in low density superbubbles because the grains would have been stopped much earlier in the much denser average ISM. We also showed that only a very small fraction ($\sim 10^{-4}$) of the grains formed in a typical supernova need be accelerated to account for the average injection of cosmic ray metals.

The similarity of the cosmic ray source and solar abundance ratios of refractory elements, mainly Mg, Al, Si, Ca relative to Fe, simply reflects the fact that supernovae are the primary source of these elements, e.g. Timmes et al. (47), and that the SB/HISM filling factor is large where cosmic rays are accelerated, so that the bulk of the Fe grains from the SNIa also contribute to the high velocity grain population in the SB/HISM. In particular, since the Si, Mg, Al, and other refractory elements are primarily produced in core-collapse SNII/Ibc, while only about half of the Fe is made in them and the other half is made in thermonuclear SNIa, a SB/HISM filling factor of $\sim$ 90% leads to differences of only $\sim 5\%$ between the Si/Fe ratio in SB/HISM and the average Galactic production ratio, which determines that in the well-mixed ISM. This is well within the present uncertainties in the inferred cosmic ray source ratios shown in Table 2, where we see from a much more detailed estimate in Lingenfelter & Ramaty (24) that the injection abundances expected for cosmic ray acceleration predominantly in the SB/HISM is consistent with the present cosmic ray source ratios of Engelmann et al. (13) even for an assumed SB/HISM filling factor of only 50%. Similar small differences $< 10\%$ in relative abundances from a SB/HISM filling factor of $\sim 90\%$ would be expected for those s-process elements which appear to come primarily from the winds of less massive stars.

The estimated mean refractory abundances in supernova grains (Table 2) are based on the calculations by

Table 2. COSMIC RAY INJECTION ABUNDANCE RATIOS IN %

	ISMGrains	ISMCores *	SNGrains†	SBGrains**	CRInject‡	CRSource§	Solar¶
C/Fe	690	-?-	210–510	–	–	422±14	1122±139
O/Fe	1400	400	320–520	460–690	455–665	522±11	2344±414
Mg/Fe	115	110	50–150	90–190	90–185	103±3	120±4
Al/Fe	10	10	5–16	8–20	8–20	7.7±1.5	9.8±0.3
Si/Fe	105	65	110–170	105–185	100–175	99±2	115±4
Ca/Fe	6	6	4–8	5–9	5–9	6.0±0.9	7.1±0.2
Ni/Fe	6	6	6–14	6–9	6–9	5.6±0.2	5.6±0.2

* ISMGrains and ISMCores – HST interstellar depletion determined abundance from Savage & Sembach (44).
† SNGrains – Range of IMF averaged supernova ejecta mixes weighted with relative SNII:SNIb:SNIa rates of 67-75%:13-15%:20-10% from van den Berg & Tammann (50) and van den Berg & McClure (49), except for O; for the SNII and SNIb contributions, refractory O is assumed to be bound in $MgSiO_3$, Fe_3O_4, Al_2O_3, CaO and NiO, and for (the very small) SNIa contribution, all the produced O is assumed bound to Fe.
** SBGrains – Modified SNGrains for 85% of SNII and SNIb and 50% of SNIa in superbubbles plus ISM refractory grain ISMCores for a mean superbubble metallicity range of 2–5 times that of ISM, as discussed in the text.
‡ CRInject – Galactic supernova averaged grain abundances for cosmic ray injection, taking a mix of SBGrain abundances for supernova acceleration in superbubbles and ISMCore grain abundances (without any supernova enrichment) for supernova acceleration outside the superbubbles, weighted by the relative swept-up metal masses and supernova rates, as discussed in the text.
§ CRSource – elemental abundances from Engelmann et al. (13).
¶ Solar system – elemental abundances from Grevesse, Noels & Sauval (15).

Woosley & Weaver (56), Woosley, Langer & Weaver (55) and Nomoto et al. (36) of supernova ejecta abundances for Types II, Ib and Ia, averaged over the initial mass function and supernova rates of van den Berg & Tammann (50) and van den Bergh & McClure (49), except that we assume the grain O abundance is limited to that bound in Al_2O_3, $MgSiO_3$, Fe_3O_4, CaO and NiO. We also show for comparison, the refractory abundances in the typical, older icy interstellar grains (ISMGrains) and their refractory cores (ISMCores) recently determined by HST observations, see Savage & Sembach (44). Here we see that the Si/Fe of 65% in refractory cores of ISM grains, which Meyer et al. (33) proposed as the cosmic ray source, is not consistent with the required cosmic ray source value of 99±2%.

EXPECTED ABUNDANCES OF VOLATILE COSMIC RAYS

In addition to the sputtering of refractory ions, the interactions of the high velocity, supernova grains can also provide a simultaneous, self consistent cosmic ray injection source of H, He and other volatiles. Cesarsky & Bibring (7) suggested that high velocity grains may temporarily pick up by implantation volatile atoms from the gas through which they pass, and their subsequent sputtering could provide a source of less enriched suprathermal volatiles. We suggest a much more direct injection process for the volatiles. Since direct collisions of fast grains with ambient gas atoms and ions are thought to be the primary means of grain momentum loss, e.g. Ellison et al. (11) §2.3, we would expect that the supernova grains should simply scatter ambient H, He and other volatile atoms to the same suprathermal injection velocities as the grains and their sputtered refractory products. Such a process would, in fact, directly account for the measured cosmic ray abundance ratio by number of the refractory (including C and "bound" O) to volatile elements, i.e. (C,O,Mg,Al,Si,Fe,etc)/(H,He,etc) = 0.010 of Engelmann et al. (13), since Ellison et al. (11 §2.4) assume that roughly 0.5%-1% of grain collisions with ambient gas atoms, predominantly scattering volatile atoms, result in the sputtering of a refractory atom from the grain surface, all of which come off with essentially the same injection velocity. Moreover, because the geometric scattering cross section increases with mass to the 2/3 power, such scattering should also lead to a mass-dependent enrichment of heavier volatiles with respect to H, as is observed in the cosmic rays, e.g. Meyer et al. (33), and which Ellison & Meyer (12) argue can not be accounted for by an A/Z dependent acceleration bias in the hot ISM.

The composition of the grain-scattered suprathermal volatiles can be further enriched by the fact that most of the supernova shocks will be interacting with grains and gas in the supernova-ejecta and progenitor-wind enriched superbubbles. Since the $^{22}Ne/^{20}Ne$ ratio in the Wolf Rayet winds of massive, supernova progenitors may exceed the solar system value by more than two orders of magnitude, e.g. Maeder & Meynet (27), grain-scattering of such wind enriched could account for the high ^{22}Ne/^{20}Ne observed in the cosmic rays, e.g. Leske et al. (22). The existence of such a Wolf Rayet signature in the cosmic rays also provides further evidence for the acceleration of cosmic rays in the superbubble hot phase where the bulk of the massive Wolf Rayet, supernova progenitors are also confined.

This work was supported by NASA ATP and ACE/GI Programs.

REFERENCES

1. Axford, W.I., 17th ICRC Papers 12, 155 (1981)
2. Blaauw, A., in The Physics of Star Formation and Early Stellar Evolution, eds. C. Lada, and N. Kylafis, (Dordrecht: Kluwer), 125 (1991)
3. Blandford, R.D., & Ostriker, J.P., ApJ, 237, 793 (1980)
4. Boesgaard, A.M., et al., AJ, 117, 1549 (1999)
5. Bykov, A., ASP Conf. Series, 71, 146 (1999)
6. Bykov, A., & Fleishman, G., MNRAS, 255, 269 (1992)
7. Cesarsky, C.J., & Bibring, J-P., in Origin of Cosmic Rays, G. Setti et al. eds. (Dordrecht: Reidel), 361 (1981)
8. Chu, Y.H., & Mac Low, M-M., ApJ, 365, 510 (1990)
9. Dwek, E., ApJ, 329, 814 (1988)
10. Dwek, E., ApJ, 501, 643 (1998)
11. Ellison, D., Drury, L., & Meyer, J., ApJ, 487, 197 (1997)
12. Ellison, D. & Meyer, J., ASP Conf. Series, 71, 207 (1999)
13. Engelmann, J.J., et al., A&A, 233, 96 (1990)
14. Ferriere. K.M., ApJ, 503, 700 (1998)
15. Grevesse, N., Noels, A., & Sauval, A.J., ASP Conf. Series, 99, 117 (1996)
16. Higdon, J.C., & Lingenfelter, R.E., ApJ, 239, 867 (1980)
17. Higdon, J.C., Lingenfelter, R.E., & Ramaty, R., ApJ, 509, L33 (1998)
18. Higdon, J.C., Lingenfelter, R.E., & Ramaty, R., 26th ICRC Conf. Papers, 4, 144 (1999)
19. Kennicutt, R.C., Edgar, B.K., & Hodge, P.W., ApJ, 337, 761 (1989)
20. Korpi, M.J., et al., ApJ, 514, L99 (1999)
21. Kozasa, T., Hasegawa, H., & Nomoto, K., A&A, 249, 474 (1991)
22. Leske, R.A., et al., Space Sci. Rev., 78, 149 (1996)
23. Lingenfelter, R.E., in Astronomy & Astrophysics Encyclopedia, S. Maran ed. (New York: Van Nostrand), 139 (1992)
24. Lingenfelter, R.E., & Ramaty, R., 26th ICRC Conf. Papers, 4, 148 (1999)
25. Lingenfelter, R.E., Ramaty, R., & Kozlovsky, B., ApJ, 500, L153 (1998)
26. Mac Low, M-M., & McCray, R., ApJ, 324, 776 (1988)
27. Maeder, M., & Meynet, G., A&A, 278, 406 (1993)
28. McCray, R., & Snow, T.P., ARA&A, 17, 213 (1979)
29. McKee, C., ASP Conf. Ser., 80, 292 (1995)
30. McKee, C., & Williams, J., ApJ, 476, 144 (1997)
31. McWiliam, A., ARA&A, 35, 503 (1997)
32. Meyer, J., ApJSupp, 57, 173 (1985)
33. Meyer, J., Drury, L., & Ellison, D., ApJ, 487, 182 (1997)
34. Molaro, P., Bonifacio, P., Castelli, F., & Pasquini, L., A&A, 319, 593 (1997)
35. Naya, J.E., et al., Nature, 384, 44 (1996)
36. Nomoto, K., et al., in Thermonuclear Supernovae, P. Ruiz-Lapuente et al. eds. (Dordrecht: Kluwer), 349 (1997)
37. Olano, C.A., A&A, 112, 195 (1982)
38. Ramaty, R., Kozlovsky, B., & Lingenfelter, R.E., Phys. Today, 51:4, 30 (1998)
39. Ramaty, R., & Lingenfelter, R.E., ASP Conf. Ser. 71, 104 (1999)
40. Ramaty, R., Lingenfelter, R.E., & Kozlovsky, B., 26th ICRC Conf. Papers, 4, 140 (1999)
41. Ramaty, R., Lingenfelter, R.E., & Kozlovsky, B. 2000a. in The Light Elements and Their Evolution, L. da Silva, M. Spite and J. R. de Medeiros, eds., IAU, in press (2000)
42. Ramaty, R., Scully, S.T., Lingenfelter, R.E., & Kozlovsky, B., 2000b. ApJ in press astro-ph/9909021 (2000)
43. Rosen, A., & Bregman, J.N., ApJ, 440, 634 (1995)
44. Savage, B., & Sembach, K., ARA&A, 34, 279 (1996)
45. Spitzer, L., ARA&A, 28, 71 (1990)
46. Thomas, D., Greggio, L., & Bender, R., MNRAS, 296, 119 (1998)
47. Timmes, F.X., Woosley, S.E., & Weaver, T.A., ApJS, 98, 617 (1995)
48. Tomisaka, K., PASJ, 44, 177 (1992)
49. van den Bergh, S., & McClure, R.D., ApJ, 425, 205 (1994)
50. van den Bergh, S., & Tammann, G., ARA&A, 29, 363 (1991)
51. van Dyk, S.D., Hamuy, M., & Filippenko, A.V., AJ, 111, 2017 (1996)
52. Waddington, C.J., ApJ, 470, 1218 (1996)
53. Wiedenbeck, M., et al., ApJ, 523, L61 (1999)
54. Williams, J.P. & McKee, C.F., ApJ, 476, 166 (1997)
55. Woosley, S.E., Langer, N., & Weaver, T.A., ApJ, 448, 315 (1995)
56. Woosley, S.E., & Weaver, T.A., ApJS, 101, 181 (1995)
57. Yorke, H., ARA&A, 24, 49 (1986)

The Cosmic Ray – X-ray Connection: Effects of Nonlinear Shock Acceleration on Photon Production in SNRs

Donald C. Ellison

Department of Physics, North Carolina State University, Box 8202, Raleigh NC 27695, U.S.A.; don_ellison@ncsu.edu

Abstract. Cosmic-ray production in young supernova remnant (SNR) shocks is expected to be efficient and strongly nonlinear. In nonlinear, diffusive shock acceleration, compression ratios will be higher and the shocked temperature lower than test-particle, Rankine-Hugoniot relations predict. Furthermore, the heating of the gas to X-ray emitting temperatures is strongly coupled to the acceleration of cosmic-ray electrons and ions, thus nonlinear processes which modify the shock, influence the emission over the entire band from radio to gamma-rays and may have a strong impact on X-ray line models. Here we apply an algebraic model of nonlinear acceleration, combined with SNR evolution, to model the radio and X-ray continuum of Kepler's SNR.

INTRODUCTION

More than twenty years of spacecraft observations in the heliosphere have proven that collisionless shocks can accelerate particles with high efficiency, i.e., 10-50% of the ram energy can go into superthermal particles (e.g., Eichler [17]; Gosling et al. [21]; Ellison et al. [19]; Terasawa et al. [30]). Energetic particles exist throughout the universe and shocks are commonly associated with them, confirming that shock acceleration is important beyond the heliosphere as well. In fact, shocks in supernova remnants (SNRs) are believed to be the main source of Galactic cosmic rays, and these shocks are expected to be much stronger than those in the heliosphere and can only be more efficient and nonlinear.

The conjecture that collisionless shocks are efficient accelerators is strengthened by results from plasma simulations, which show efficient shock acceleration consistent with spacecraft observations (e.g., Scholer, Trattner, & Kucharek [29]; Giacalone et al. [20]), and other indirect evidence comes from radio emission from SNRs (see Reynolds & Ellison [27]) and equipartition arguments in AGNs and γ-ray bursts (see Blandford & Eichler [6] for an early review). There is also clear evidence that shocks can produce strong self-generated turbulence. This has long been seen in heliospheric shocks (e.g., Lee [24, 25]; Kennel et al. [23]; Baring et al. [3]) and there is evidence that it occurs at SNRs as well (i.e., Achterberg, Blandford, & Reynolds [1]).

While the importance of nonlinear (NL) shock acceleration is evident, NL solutions to diffusive shock acceleration are complicated and results are often unwieldy and difficult to use for astrophysical applications. Therefore, we have developed a simple, algebraic model of diffusive shock acceleration, based on more complete studies, which includes the essential nonlinear effects (Berezhko & Ellison [4]; Ellison, Berezhko, & Baring [18]). This technique is computationally fast and easy-to-use, yet includes (i) the modification of particle spectra when the backpressure from energetic ions smooths the shock structure, and (ii) the influences on the shock dynamics when the magnetic turbulence is strongly amplified by wave-particle interactions.

The complications of NL shock acceleration and the many parameters required to characterize it are offset somewhat by the fact that the entire particle distribution function, from thermal to the highest energies, is interconnected and must be accounted for self-consistently with a nonthermal tail connecting the quasi-thermal population to the energetic one. Because energy is conserved, a change in the production efficiency of the highest energy particles *must* impact the thermal properties of the shock heated gas and vice versa. If more energy goes into relativistic particles, less is available to heat the gas. In contrast, the power laws assumed by test-particle models have no connection with the thermal gas, energy conservation does not constrain the normalization of the power law, and the spectral index can be changed with no feedback on the thermal plasma. Furthermore, there is a direct linkage between protons and electrons (which produce most of the photon emission associated with shocked gas) in nonlinear models, so the entire emission from radio to gamma-rays, plus cosmic-ray observations, can, in principle, be used to constrain the models.

Here we describe some of the nonlinear features expected to occur in young SNRs and investigate some

CP528, *Acceleration and Transport of Energetic Particles Observed in the Heliosphere: ACE 2000 Symposium,* edited by Richard A. Mewaldt, et al.

© 2000 American Institute of Physics 1-56396-951-3/00/$17.00

implications of efficient cosmic-ray production on the broad-band continuum from Kepler's SNR. We refer to Berezhko & Ellison [4] and Ellison, Berezhko, & Baring [18] for details of the NL shock model and its application to particle and photon production in SNRs. Work on the NL X-ray line emission from Kepler is in progress, i.e., Decourchelle, Ellison, & Ballet [12]. Previous test-particle calculations of the X-ray emission from Kepler have been reported by Borkowski, Sarazin, & Blondin [7], who used a two-dimensional hydrodynamic simulation, and by Rothenflug et al. [28], who investigated the emission from the reverse shock. The work of Decourchelle & Ballet [11] is mentioned below.

NONLINEAR SHOCK ACCELERATION

The nonlinear effects in shock acceleration are of two basic kinds: (i) the self-generation of magnetic turbulence by accelerated particles and (ii) the modification (i.e., smoothing) of the shock structure by the backpressure of accelerated particles. Briefly, (i) occurs when counter-streaming accelerated particles produce turbulence in the upstream magnetic field which amplifies as it is convected through the shock. This amplified turbulence results in stronger scattering of the particles, and hence to more acceleration, quickly leading to saturated turbulence levels near $\delta B/B \sim 1$ in strong shocks. The wave-particle interactions produce heating in the shock precursor which may be observable. Effect (ii) results in the overall compression ratio, $r_{\rm tot}$, being an ever increasing function of Mach number (as if the effective ratio of specific heats $\gamma_{\rm eff} \to 1$ and analogous to radiative shocks), i.e.,

$$r_{\rm tot} \simeq 1.3 M_{\rm S0}^{3/4} \quad \text{if} \quad M_{\rm S0}^2 > M_{A0} \,, \tag{1}$$

or by

$$r_{\rm tot} \simeq 1.5 M_{A0}^{3/8} \tag{2}$$

in the opposite case ($M_{\rm S0}$ is the sonic and $M_{\rm A0}$ is the Alfvén Mach number). Simultaneously, shock smoothing causes the viscous subshock to be weak ($r_{\rm sub} \ll r_{\rm tot}$) and the temperature of the shocked gas to drop below the test-particle value.

There are many parameters associated with NL shock acceleration (see Ellison, Berezhko, & Baring [18] for a listing), but the most important ones that determine the solution are the Mach numbers (i.e., the shock speed, pre-shock density, and magnetic field) and the injection efficiency, $\eta_{\rm inj,p}$ (i.e., the fraction of total protons which end up with superthermal energies). As described in Berezhko & Ellison ([4]), we use Alfvén heating in the precursor which reduces the efficiency compared to adiabatic heating. Significantly, parameters typical of young SNRs should result in NL acceleration.

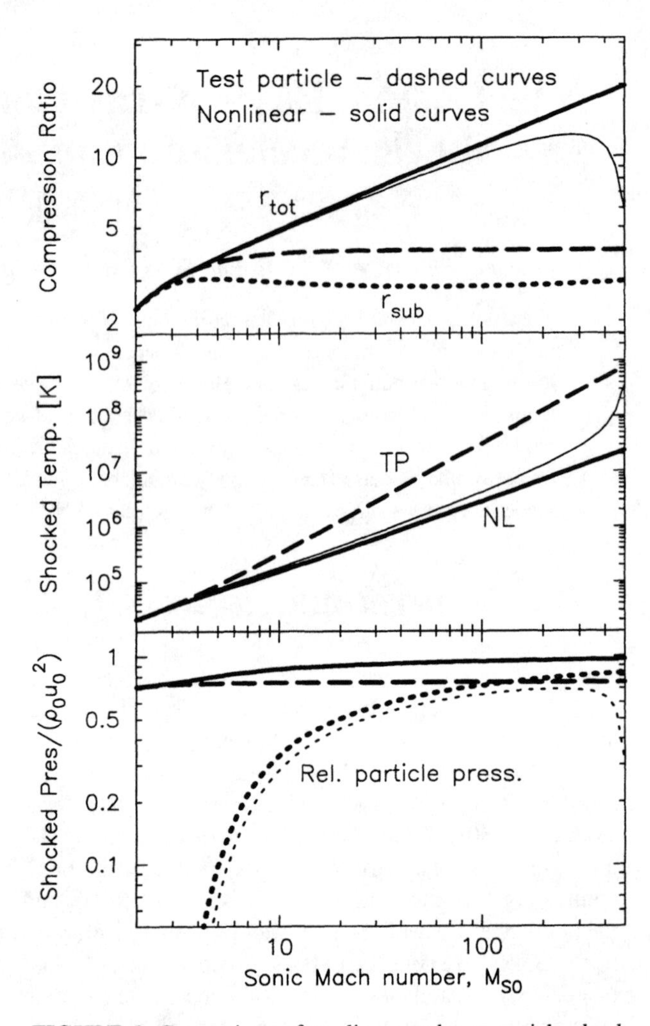

FIGURE 1. Comparison of nonlinear and test-particle shock results as a function of upstream sonic Mach number. The top two panels show that, at $M_{\rm S0} > 100$, the downstream density can be > 5 times *larger* and the shocked temperature > 10 times *smaller* if the acceleration is NL. The bottom panel shows that for $M_{\rm S0} > 100$ and $\eta_{\rm inj,p} = 10^{-3}$, more than 70% of the pressure in the shocked gas ends up in relativistic particles in NL shock acceleration. These results depend on parameters other than $M_{\rm S0}$, and in particular are a strong function of $\eta_{\rm inj,p}$, particularly at high $M_{\rm S0}$. The heavy-weight curves are for $\eta_{\rm inj,p} = 10^{-3}$ and the light-weight curves are for $\eta_{\rm inj,p} = 10^{-4}$.

Figure 1 shows a comparison between test-particle (dashed lines) and nonlinear results as a function of sonic Mach number, $M_{\rm S0}$. As illustrated in the top panel, the overall compression ratio, $r_{\rm tot}$, is generally > 4 and can increase without limit in the NL case for $\eta_{\rm inj,p} = 10^{-3}$. Simultaneously, the subshock compression ratio, $r_{\rm sub}$, is less than 4 (typically $r_{\rm sub} \sim 3$). The overall compression determines the shocked density, while the subshock is mainly responsible for heating the gas. Thus, the shocked gas will be cooler (middle panel) and denser if the acceleration is NL compared to a test-particle (TP) case. This is an important consideration for X-ray line models. The particle spectrum at the highest energies is determined by $r_{\rm tot}$, and thus will be flatter than in a TP shock. However, the spectral shape at lower energies depends on $r_{\rm sub}$, indicating that the particle spectrum will go from being steeper than the TP prediction at low energies to flatter at the highest energies, this is the so-called concave signature of NL shock acceleration. The bottom panel shows that a large fraction of the shocked pressure can end up in relativistic particles (dotted line) for large $M_{\rm S0}$. These results depend on the injection efficiency assumed and the turnover in $r_{\rm tot}$ at large $M_{\rm S0}$ seen in the $\eta_{\rm inj,p} = 10^{-4}$ example is a transition to a strong, *unmodified* shock (see Berezhko & Ellison [4] for a full discussion).

In Figure 2 we show the phase space distribution functions, $f(p)$, for electrons (dashed line), protons (solid line), and helium (dotted line). The parameters used for this forward shock are those discussed below for Kepler's SNR and given in Table 1. These particles produce photons from synchrotron, bremsstrahlung, inverse-Compton, and pion-decay processes (see Baring et al. [2] for details) and these spectra are shown in Figure 3 for both the forward and reverse shocks. As discussed below, we have chosen parameters to cause the reverse shock to contribute substantially to the total X-ray emission, but this choice impacts the emission over the entire range from radio to gamma-rays. While the radio synchrotron comes primarily from the reverse shock, the forward shock dominates emission at TeV energies with a pion-decay component. Any changes in parameters to accommodate the X-ray observations must be consistent with observational constraints in all other bands.

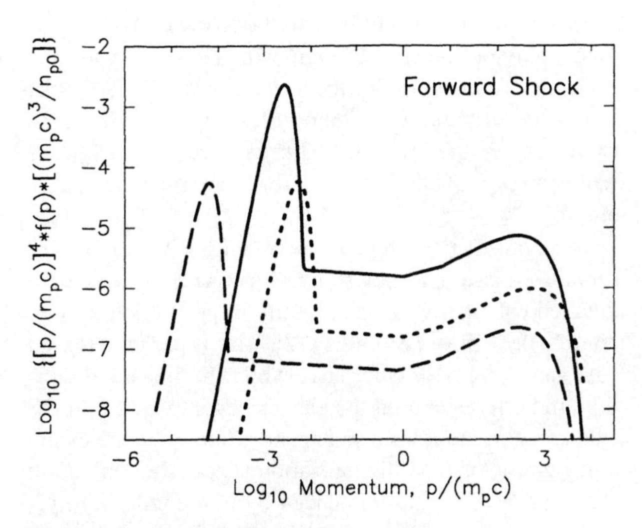

FIGURE 2. Downstream phase space distribution functions, f, versus momentum, p. We have multiplied $f(p)$ by $[p/(m_{\rm p}c)]^4$ to flatten the spectra, and by $[(m_{\rm p}c)^3/n_{p0}]$ to make them dimensionless (n_{p0} is the upstream proton number density). The solid line shows protons, the dotted line shows helium, and the dashed line shows electrons.

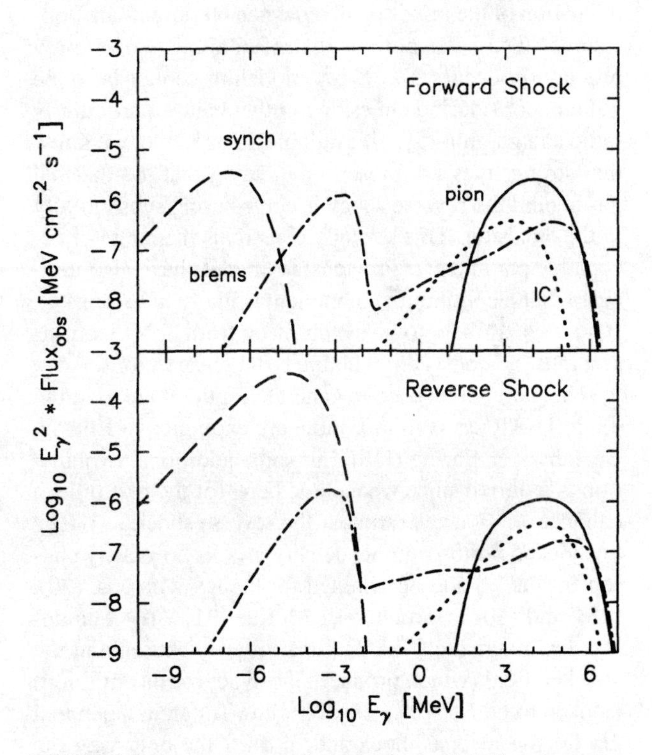

FIGURE 3. Photon spectra from the forward (FS) and reverse shocks (RS) obtained with parameters used to generate the spectra shown in Figure 2. Notice that emission in the X-ray band (1-100 keV) is totally bremsstrahlung for the FS but has a synchrotron component in the RS. Emission at gamma-ray energies is dominated by the FS, while the radio comes totally from the RS.

APPLICATION TO KEPLER'S SNR

We now apply our model to the radio and X-ray emission seen in Kepler's SNR. The X-ray emission shows the presence of lines and these have been interpreted, in models without acceleration (see Decourchelle & Petre [13]; Decourchelle et al.. [14]), as coming from the ejecta material heated by the reverse shock. However, these previous models also required the presence of a power law photon continuum, presumably coming from synchrotron emitting TeV electrons (e.g., Reynolds [26]).

Here, we use the evolutionary model of Truelove & McKee [31] to obtain the forward and reverse shock parameters at the $\sim$ 395 yr age of Kepler's SNR and, using these shock parameters, calculate the NL particle and then photon spectra. Figure 4 shows the synchrotron and bremsstrahlung emission from Figure 3 and indicates that we are able to obtain a good fit to the data within the constraint that the continuum emission from the *reverse shock* contribute substantially to the X-ray emission (our model only produces continuum which lies below any X-ray emission lines). Since the forward and reverse shocks have very different emission profiles (Figure 3), the ability to resolve the individual shocks in radio and X-rays will be exceptionally important for constraining the models.

We note that the relation between the line and continuum emission is not straightforward and requires a detailed calculation as is currently in progress, i.e., Decourchelle, Ellison, & Ballet [12]. This is particularly the case in young SNRs because, while the lines and continuum both depend on the electronic temperature in the shocked gas, the lines also depend strongly on the ejecta composition and on the ionization state. The ionization state can be strongly influenced by non-equilibrium effects and by the history of the shocked plasma. What can be said, however, is that the presence of lines restricts the synchrotron intensity to some level comparable to (with a factor of 10 say) or below the bremsstrahlung continuum.

As mentioned above, NL shocks are extremely complicated with many (often poorly defined) parameters and our purpose in this preliminary work is only to give some indication of the issues involved when NL effects are considered. The basic point is that, if nonlinear cosmic-ray production occurs, the X-ray modeling cannot be done without considering emission in other bands, particularly radio and gamma-rays (if available). For Kepler we know that strong lines are present, indicating that the thermal gas behind the reverse shock is contributing substantially to the emission. This strongly constrains the range of acceptable parameters. Previous TP models have also indicated that a continuum component is likely to be present. If so, this is likely to be synchrotron from TeV electrons and must be consistent with the radio observations.

The parameters used to generate Figure 4 are given in Table 1. All terms in this Table are explained in Ellison, Berezhko, & Baring ([18]) but some additional explanation is required since we include here, for the first time, a calculation of acceleration at the reverse shock. (i) The unshocked proton number density, n_{p0}, is arbitrarily chosen for the FS and determined for the RS with eqs. (20), (28), and (30) in Truelove & McKee [31]. (ii) The unshocked magnetic field, B_0, is arbitrarily chosen and the shocked field (which produces the synchrotron emission) is taken to be $B_2 = r_{tot} B_0$. We allow for an independent B_0 for the reverse shock and, in fact, the only way the RS can contribute substantially to the X-ray emission is for the RS B_0 to be much larger (25 vs. 1.5 μG) than the FS B_0. (iii) The supernova energy, E_{sn}, and ejecta mass, M_{ej}, are standard values. (iv) We assume the same unshocked proton temperature, $T_{p0} = 10^4$ K, for the FS and RS. The solutions are relatively insensitive to this parameter. (v) The shocked electron to proton temperature ratio, T_{e2}/T_{p2}, is a free parameter in the NL shock model and is set to 1 for both shocks. (vi) The injection efficiency, $\eta_{inj,p}$, is an important model parameter determining the overall acceleration efficiency. Values much larger than $\eta_{inj,p} = 2\times10^{-4}$ tend to produce strong nonthermal tails on the bremsstrahlung emission which may be inconsistent with the observations. Values much smaller than 2×10^{-4} yield test-particle solutions with typical temperatures higher than allowed by the X-ray observations. We note that adjustments in $\eta_{inj,p}$ produce extremely large modifications in the TeV results and can be highly constrained with gamma-ray observations. (vii) The electron to proton ratio at relativistic energies, $(e/p)_{rel}$, is an arbitrary parameter in the simple model but is expected to be between 0.01 and 0.05 if $(e/p)_{rel}$ in galactic cosmic rays is typical of that produced by the strong shocks in young SNRs. This factor is very important for determining the pion-decay contribution to TeV gamma-rays. Furthermore, lowering $(e/p)_{rel}$ lowers the overall emission in the radio to X-ray band but increases the relative contribution of thermal to synchrotron photons in the X-ray band. A firm determination of the synchrotron contribution to X-rays, combined with a gamma-ray detection, will help determine $(e/p)_{rel}$, one of the most important unknown parameters in shock acceleration. (viii) The maximum energy cosmic rays obtain in the NL model depends on the scattering mean free path, λ, which is as-

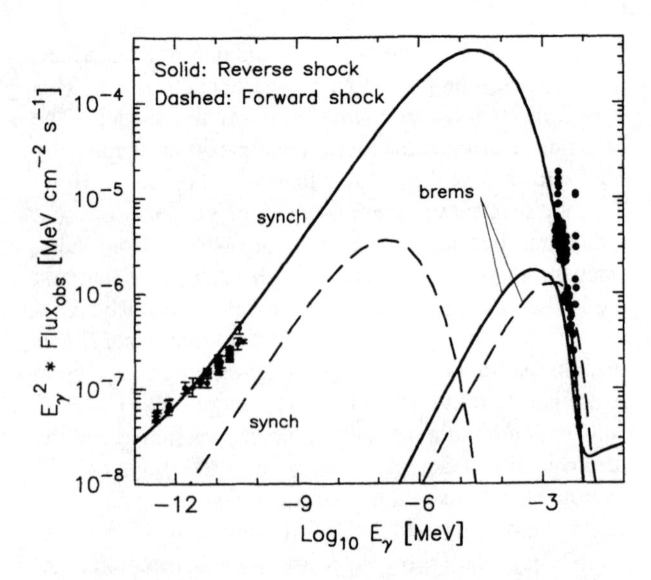

FIGURE 4. Radio and X-ray data for Kepler's SNR compared to the emission from the forward and reverse shocks. The radio data is from Reynolds & Ellison [27] and X-ray data is adapted from Decourchelle & Petre [13] (Decourchelle, private communication). We use a standard νF_ν representation and have suppressed the error bars on the X-ray points.

sumed to be,

$$\lambda = \eta_{\mathrm{mfp}}\, r_{\mathrm{g,max}}\, (r_{\mathrm{g}}/r_{\mathrm{g,max}})^{\alpha} \ , \qquad (3)$$

where η_{mfp} is taken to be independent of particle momentum (Baring et al. [2], use the notation η), $r_{\mathrm{g}} = p/(qB)$ is the gyroradius in SI units, $r_{\mathrm{g,max}}$ is the gyroradius at the maximum momentum, p_{max}, and α is a constant parameter ($\alpha = 1 = \eta_{\mathrm{mfp}}$ is roughly the Bohm limit). Here we only consider $\alpha = 1$. A fairly large value of $\eta_{\mathrm{mfp}} = 15$ (giving a maximum cosmic-ray energy well below that obtained in the Bohm limit) is required to avoid having the X-ray emission totally dominated by synchrotron from TeV electrons. (ix) We assume that the initial density profile in the ejecta has a power law density distribution, $\rho \propto r^{-n}$, with $n = 9$, and that the unshocked ISM is uniform.

The output values are also given in Table 1. Of these, V_{sk} and R_{sk} are taken directly from the Truelove & McKee [31]) solution. Once these are obtained, the Mach numbers are determined (given the relevant input parameters), and then the NL shock model determines the compression ratios, temperatures, and particle spectra. Using the particle spectra, the continuum emission is calculated. It is important to notice that the shocks are highly efficient and NL. They yield total compression ratios > 4, shocked temperatures nearly 10 times lower than the corresponding test-particle shocks, and place the majority of the total energy flux, ϵ_{rel}, into relativistic particles (mainly protons), yielding γ_{eff}'s less than 5/3.

Finally, our model includes a rough estimate of the emission volume

$$V_{\mathrm{emis}} \approx (4\pi/3)\, R_{\mathrm{sk}}^{3}/r_{\mathrm{tot}} \ , \qquad (4)$$

as described in Ellison, Berezhko, & Baring [18]. This volume is considerably less than the total remnant volume. If we assume a distance to Kepler of $D_{\mathrm{snr}} = 5$ kpc (e.g., Decourchelle & Ballet [11]), the normalization given by eq. (4), and shown in the Table, matches the observations.

Table 1. Parameters for Kepler's SNR model

Input parameters	Forward shock	Reverse shock
n_{p0} [cm^{-3}]	0.5	1.25
B_0 [μG]	1.5	25
E_{sn} [10^{51} erg]	1	—
M_{ej} [$M_{\odot}$]	5	—
T_{p0} [K]	10^4	10^4
$T_{\mathrm{e2}}/T_{\mathrm{p2}}$	1	1
$\eta_{\mathrm{inj,p}}$	2×10^{-4}	2×10^{-4}
$(e/p)_{\mathrm{rel}}$	0.03	0.03
η_{mfp}	15	15
n	9	9
Output values		
V_{sk} [km s^{-1}]	4100	1750
R_{sk} [pc]	2.5	2.1
M_{S0}	260	110
M_{A0}	1040	43
r_{tot}	11	6.4
r_{sub}	3.9	3.5
B_2 [μG]	16	160
$E_{\mathrm{max,p}}$ [eV]*	4.1×10^{11}	1.7×10^{12}
$\eta_{\mathrm{inj,e}}$ †	5.1×10^{-4}	6.3×10^{-4}
T_{p2} [K]	2.8×10^{7}	1.3×10^{7}
T_{tp} [K]**	2.0×10^{8}	9.4×10^{7}
ϵ_{rel}	0.79	0.54
γ_{eff}	1.40	1.45
Flux parameters		
D_{snr} [kpc]	5	—
V_{emis} [pc^3]	6.0	6.4

* This is the maximum proton energy, but since synchrotron losses are unimportant here, electrons have the same E_{max}.
† This is the electron injection efficiency.
** This is the temperature the shock gas would have obtained if no acceleration took place.

DISCUSSION AND CONCLUSIONS

X-ray line and continuum emission contains a vast amount of information on supernova (SN) ejecta elemental composition, ISM density and elemental composition, the SN explosion energy, and the mass of ejecta. In addition to heating the plasma, the forward and reverse shocks accelerate some fraction of the shocked material to cosmic-ray energies and this acceleration is believed to be quite efficient, removing energy from the thermal plasma (e.g., Kang & Jones [22]; Dorfi & Böhringer [16]; Berezhko, Ksenofontov, & Petukhov [5]). Despite the expected efficiency of shock acceleration, virtually all current X-ray line models assume that the shocks that heat the gas *do not* place a significant fraction of their energy in cosmic rays (exceptions to this are Chevalier [10] and Dorfi [15]). Here, we investigated the broad-band continuum emission expected in Kepler's SNR from efficient shock acceleration, by coupling self-similar hydrodynamics (Chevalier [10]; Truelove & McKee [31]) with nonlinear diffusive shock acceleration (e.g., Berezhko, Ksenofontov, & Petukhov [5]). We were able to show that the radio and X-ray continuum can be fit with reasonable parameters in a way that allows the reverse shock to contribute substantially to the total X-ray emission. This constraint is required since X-ray line models of Kepler (e.g., Decourchelle & Petre [13]) require emission from the shock-heated, metal-rich ejecta material.

This preliminary calculation is not fully self-consistent for several reasons. Most importantly, we use self-similar results (i.e., Truelove & McKee [31]) to model the SNR evolution. These solutions neglect the effects of energetic particle escape from the FS and assume that the ratio of cosmic-ray pressure to total pressure at the shock front is a constant. They also assume that $\gamma_{\rm eff} = 5/3$. As indicated in the bottom panel of Figure 1, the total shocked pressure doesn't differ much with or without cosmic-ray production and we have demonstrated (i.e., Decourchelle, Ellison, & Ballet [12]) that NL shock results do not change dramatically over most of the age of Kepler for typical values of the injection parameter, $\eta_{\rm inj,p} > 10^{-4}$. However for lower $\eta_{\rm inj,p}$, the nonlinear solutions can have test-particle, unmodified solutions at very high sonic Mach numbers with a rapid transition to the NL solution as the Mach number drops (see Fig. 1 and Berezhko & Ellison [4] for a detailed discussion). The self-similar solutions we use are still approximate, however, because we have not yet modified them for the change in $\gamma_{\rm eff}$ that results when a substantial fraction of the shocked pressure is in cosmic rays. We also neglect cosmic-ray diffusion and assume they are spatially coupled to the gas – an excellent approximation for all but the highest energy particles. However, the highest energy electrons produce the X-ray synchrotron photons so there may be differences that are not modeled in the emission volumes and other important parameters between the radio and X-ray bands. Finally, we have not included absorption in our models which is probably not important for the Kepler radio observations, but will be required to model the low energy X-rays.[1]

Besides providing a more self-consistent model of photon production, predictions from NL shock models provide a test of the fundamental assumption that SNRs are the primary source of galactic cosmic-ray ions. If this is so, the acceleration is almost certainly nonlinear since 5-30% of the total ejecta kinetic energy is required to replenish cosmic rays as they escape from the galaxy. Since shocks put more energy into accelerated ions than electrons, nonlinear effects seen in X-ray emission will be evidence for the efficient shock acceleration of ions as well as electrons. X-ray observations potentially provide *in situ* information on cosmic-ray ion production, complementing observations of pion-decay γ-rays in this regard. Any inference of nonthermal tails on electron distributions in X-ray observations will provide information on electron injection, the least well understood aspect of shock acceleration. Such *in situ* information on high Mach number shocks is available nowhere else.

[1] Note that we have only plotted the X-ray observations above 2 keV in Figure 4 to avoid conflict with the absorbed low energy end of the X-ray distribution.

Our modeling of Kepler's SNR suggests that typical source parameters produce large nonlinear effects in the broad-band spectrum and suggest that the test-particle approximations that are almost universally used are inadequate for SNRs as young as this. Besides the differences discussed above, one might expect that the growth rate of the Rayleigh-Taylor instability will be greater in a cosmic-ray modified shock because of the larger spatial gradients of density, pressure, etc. Furthermore, the high compression ratios result in a considerably thinner region between the forward and reverse shocks than predicted in the TP case (Decourchelle, Ellison, & Ballet [12]). This places the contact discontinuity closer to the shock and may make it easier for the Rayleigh-Taylor "fingers" to distort or overtake the FS, a situation that appears difficult with normal TP parameters (Chevalier, Blondin, & Emmering [9]; Chevalier & Blondin [8]). Another important difference concerns electron heating and equilibration. The higher densities in the NL models mean that electron heating is much more efficient than in TP shocks. Our initial calculations [12] suggest that it may be possible to obtain full equipartition between electrons and ions, at least for high values of $\eta_{\rm inj,p}$, in the shocked ejecta.

ACKNOWLEDGMENTS

I wish to thank the organizers of the ACE-2000 Symposium for putting on a very useful and enjoyable meeting and for providing support. I'm especially grateful to A. Decourchelle for helpful comments and suggestions and to A. Decourchelle and L. Sauvageot for furnishing the Kepler X-ray data.

REFERENCES

1. Achterberg, A., Blandford, R.D., Reynolds, S.P. 1994, A.A., 281, 220
2. Baring, M.G., Ellison, D.C., Reynolds, S.P., Grenier, I.A., & Goret, P., 1999, Ap.J., 513, 311
3. Baring, M.G., Ogilvie, K.W., Ellison, D.C., & Forsyth, R.J., 1997, Ap.J., 476, 889
4. Berezhko, E.G., Ellison, D.C., 1999, Ap.J., 526, 385
5. Berezhko, E. G., Ksenofontov, L., & Petukhov, S. I., 1999, Proc. 26th Int. Cosmic Ray Conf. (Salt Lake City), 4, 431.
6. Blandford, R.D., & Eichler, D., 1987, Phys. Repts., 154, 1
7. Borkowski, K.J., Sarazin, C.L., & Blondin, J.M., 1994, Ap.J., 429, 710
8. Chevalier, R.A., & Blondin, J.M., 1995, Ap.J., 444, 312
9. Chevalier, R.A., Blondin, J.M., & Emmering, R.T., 1992, Ap.J., 392, 118

10. Chevalier, R.A. 1983, Ap.J., 272, 765

11. Decourchelle, A., & Ballet, J., 1994, A.A., 287, 206

12. Decourchelle, A., Ellison, D.C., & Ballet, J., in preparation.

13. Decourchelle, A., & Petre, R., 1999, Astron. Nachr., 320, 203

14. Decourchelle, A., et al., 2000, in preparation

15. Dorfi, E. A. 1994, Ap.J.Suppl., 90, 841

16. Dorfi, E.A., & Böhringer, H. 1993, A.A., 273, 251

17. Eichler, D. 1981, Ap.J., 247, 1089

18. Ellison, D.C., Berezhko, E.G., & Baring, M.G., 2000, Ap. J., in press- **astro-ph/0003188**

19. Ellison, D.C., Möbius, E., & Paschmann, G., 1990, Ap.J., 352, 376

20. Giacalone, J., Burgess, D., Schwartz, S.J., Ellison, D.C., & Bennett, L. 1997, J.G.R., 102, 19,789

21. Gosling, J.T., Asbridge, J.R., Bame, S.J., Feldman, W.C., Zwickl, R.D., Paschmann, G., Sckopke, N., and Hynds, R.J. 1981, J.G.R., 86, 547

22. Kang, H. & Jones, T. W. 1991, M.N.R.A.S., 249, 439

23. Kennel, C.F., Edmiston, J.P., Scarf, F.L., Coroniti, F.V., Russell, C.T., Smith, E.J., Tsurutani, B.T., Scudder, J.D., Feldman, W.C., Anderson, R.R., Moser, F.S., and Temerin, M., 1984, J.G.R., 89, 5436

24. Lee, M.A., 1982, J.G.R., 87, 5063

25. Lee, M.A., 1983, J.G.R., 88, 6109

26. Reynolds, S.P. 1996, Ap.J.(Letts), 459, L13

27. Reynolds, S.P., & Ellison, D.C. 1992, Ap.J.(Letts), 399, L75

28. Rothenflug, R., Magne, B., Chieze, J.P., & Ballet, J., 1994, A.A., 291, 271

29. Scholer, M., Trattner, K.J., & Kucharek, H. 1992, Ap.J., 395, 675

30. Terasawa, T., et al. 1999, Proc. 26th Int. Cosmic Ray Conf. (Salt Lake City), 6, 528.

31. Truelove, J.K., & McKee, C.F.: 1999, Ap.J.Suppl., 120, 299

Cosmic Ray Transport in the Galaxy

Vladimir S. Ptuskin

Institute for Terrestrial Magnetism, Ionosphere and Radio Wave Propagation of the Russian Academy of Sciences (IZMIRAN), Troitsk, Moscow region 142092, Russia

Abstract. A discussion of the propagation of the nuclear component of cosmic rays in the Galaxy is presented. The principal topics are the following: parameters of the basic diffusion model, radioactive secondary isotopes in cosmic rays, effects of distributed reacceleration, dependence of the diffusion coefficient on energy and anisotropy constraints, and cosmic rays in a galactic wind model.

INTRODUCTION

While the basic features of the model of cosmic ray propagation in the Galaxy seem to be well-established [1], some essential issues are still not clear. The continuous flow of data from new space and ground based experiments causes us to continually refine the model. Below we present a few examples in which recent development of the basic model has responded to a challenge of the experimental results.

DIFFUSION COEFFICIENT OF COSMIC RAYS

The modern space experiments, Ulysses, and in particular the CRIS experiment with great collecting power on board the ACE spacecraft, provide highly accurate data on radioactive secondary isotopes ^{10}Be, ^{26}Al, ^{36}Cl, and ^{54}Mn in low-energy cosmic rays [2, 3]. Heavier primary nuclei produce these isotopes in the course of diffusion and nuclear fragmentation in the interstellar gas. The typical decay time of the isotopes under consideration is $t_{dec} \sim 1$ Myr and hence they may come to an observer from distances not much larger than $\sqrt{Dt_{dec}} \approx 300\,\mathrm{pc}$, where D is the diffusion coefficient of cosmic rays, and the estimate was made for D = $3\mathrm{x}10^{28}$ cm^2/s. Knowing the number density of primary nuclei from the observations at the Earth, the production cross sections from the laboratory experiments, and the gas distribution from astronomical observations, one can calculate the production rate of secondary nuclei. The observed abundance of radioactive isotopes determines then the actual value of the diffusion coefficient. The detailed procedure was described in [4]. At an energy of E = 400 MeV/nucleon in interstellar space a value for the diffusion coefficient D = $5\mathrm{x}10^{28}$ cm^2/s, with an uncertainty of ~50%, can be deduced from the cosmic ray data presented in [2, 3].

It is significant that the interpretation of the present-day precise measurements of radioactive isotopes in cosmic rays necessitates rather accurate modeling. For example, the diffusion model with a simplified gas distribution in a form of a single gas layer with constant gas density is commonly used in cosmic ray research. This model would give a diffusion coefficient which is a factor of 2 larger than the value found in the model with a realistic gas distribution. The most essential ingredient, which is not included yet in the calculations, is the tensor character of cosmic ray diffusion. The reason is a poor knowledge of the structure of the galactic magnetic field in the vicinity of a few hundred parsecs around the Sun.

The diffusion character of cosmic ray propagation in the Galaxy is explained by the action of the interstellar magnetic field. The average magnetic field in the Galaxy is predominantly azimuthal and has a strength B_0 = (2-3)x10^{-6} G. The strength of the random component is close to B_1 = $5\mathrm{x}10^{-6}$ G at the principle scale $L \sim 100$ pc. There is an extended spectrum of magnetic field

CP528, *Acceleration and Transport of Energetic Particles Observed in the Heliosphere: ACE 2000 Symposium*, edited by Richard A. Mewaldt, et al.
© 2000 American Institute of Physics 1-56396-951-3/00/$17.00

inhomogeneities down to ~ 10^8 cm that can be approximated by the spectrum of fluctuations $W(k)dk \propto k^{-2+a}dk$, where $a = 0.33 \pm 0.2$ on wave number k. Note that $a = 1/3$ for a Kolmogorov spectrum.

The theory of cosmic ray diffusion in the Galaxy is constructed analogously to the well-studied case of cosmic ray diffusion in the heliosphere. The charged energetic particles are resonantly scattered by random magnetic field so that a particle with Larmor radius r_g is scattered by the inhomogeneities with wave number $k = (r_g \cos\vartheta)^{-1}$, ϑ is the particle pitch-angle. The diffusion coefficient parallel to the magnetic field line can be estimated as

$$D_{par} \sim \frac{vr_g}{3}\left(\frac{B_{1,res}}{B}\right)^{-2} \propto vr_g^a. \qquad (1)$$

Here v is the particle velocity, $B_{1,res}$ is the amplitude of the random magnetic field at $k \sim 1/r_g$, and B is the total magnetic field. The calculation of a perpendicular diffusion coefficient presents a challenge to the theory because it has to take into account the coupled processes of the particle resonant scattering and the stochastic behavior of magnetic field lines. The approximate equation for the perpendicular diffusion coefficient averaged over long-wave wandering of field lines is [5]

$$D_{per} \sim \left(\frac{B_{1,tot}}{B}\right)^4 D_{par} \qquad (2)$$

(see also [6] for results of a numerical simulation).

The calculated diffusion coefficient of cosmic rays in the interstellar medium is approximately equal to $D_c \sim 3\cdot 10^{28}\beta\left(R/1\text{ GV}\right)^{0.33\pm0.2}$ cm^2/s ($\beta = v/c$) . This theoretical formula is in reasonable agreement with the low-energy empirical value given above.

DEPENDENCE OF DIFFUSION ON ENERGY

The dependence of cosmic ray diffusion on energy can be determined from the content of stable secondary nuclei measured over a wide energy range. The leaky box approximation can be used in these calculations instead of a more complicated diffusion model. In the leaky box approximation, the diffusion term $-\nabla(D\nabla I)$ for cosmic ray intensity I is substituted by the term I/T_e, where T_e has the meaning of the cosmic ray exit time from the Galaxy. The characteristic which determines nuclear fragmentation is the escape length $X_e = \bar{\rho} vT_e$, the mean matter thickness traversed by cosmic rays in the course of their exit from the Galaxy ($\bar{\rho}$ is the mean gas density). For an observer at the galactic disk, the relation between the escape length X_e and the parameters of the diffusion model is the following [1]:

$$X_e = \frac{v\mu H}{2D}. \qquad (3)$$

Here $\mu \approx 2.4\text{x}10^{-3}$ g/cm^2 is the surface gas density of the galactic disk; H is the height of the cosmic ray halo. Eq. (3) is valid for not very heavy nuclei with the nuclear attenuation length $X_N >> \frac{h}{H}X_e$, where h is the height of the galactic gas disk. (It should be pointed out that the equivalence between the diffusion and leaky box models for radioactive isotopes could be proved only under the additional condition $\tau >> H^2/D$. The last condition is not fulfilled for the rapidly decaying isotopes discussed in the present paper. The difference between different models for ^{10}Be is illustrated below in Figure 1.)

The calculations [7] carried out with the use of the modified weighted slab method (which gives the exact solution of transport equations) and with the set of new nuclear cross sections give

$$X_e = 11.3\beta \text{ g/cm}^2, \quad R < 5 \text{ GV};$$
$$X_e = 11.3\beta(R/5\text{ GV})^{-0.54} \text{ g/cm}^2, \quad R \geq 5 \text{ GV} \qquad (4)$$

based on a least squares fit to observed B/C and (Sc+Ti+V)/Fe ratios in cosmic rays at energies 0.5 – 200 GeV/n. Here R is the particle magnetic rigidity. The derived source spectrum is $Q \propto R^{-2.35}$. Notice that this shape of the source spectrum holds good for all propagation models discussed below including the model with reacceleration.

The power-law dependence of the escape length on rigidity at $R > 5$ GV (4) is consistent with the theoretical prediction. However, the low-rigidity regime in eq. (4) can not be naturally explained by the resonant scattering on the random magnetic field with a power law spectrum on k. Eq. (4) implies a constant leakage time T_e of cosmic rays from the Galaxy at rigidities $R < 5$ GV since $X_e = \bar{\rho} vT_e$.

In principle, the energy-independent leakage can be explained by the convective transport of cosmic rays by the galactic wind [8] or turbulent diffusion. The attempt to make the models with constant wind velocity and turbulent diffusion fit the peak in the secondary/primary ratio at $R \sim 5$ GV, lead to too strong a rigidity dependence of the escape: $X_e \sim R^{-0.73}$, and $X_e \sim R^{-0.85}$ respectively [7]. Extrapolated to higher energies 10^{12} - 10^{14} eV, such a strong dependence on energy comes into conflict with the observed slope of the energy spectra of primary nuclei (close to $E^{-2.7}$ at high energies) and with the observed high isotropy of cosmic rays, see discussion below.

The distributed reacceleration of cosmic rays in the interstellar medium after their exit from the compact sources (supernovae remnants) changes the shape of the particle spectra and presents an alternative explanation of the decrease of secondary/primary ratios at small energies [9, 10]. In the minimal model the stochastic reacceleration occurs as a result of scattering on randomly moving waves responsible for the spatial diffusion. The efficiency of reacceleration is determined by the dependence of the particle diffusion coefficient on momentum $D_{pp} \sim p^2 V_a^2 / D_{par}$, where V_a is the Alfven velocity. A fit to the observations can be achieved with the diffusion coefficient

$$D = 3.8 \cdot 10^{28} \beta(H/5 \text{ kpc})(R/1 \text{ GV})^{1/3} \text{ cm}^2/s, \quad (5)$$

and an Alfven velocity of V_a = 21 km/s [7]. The dependence of diffusion on rigidity in (5) corresponds to particle scattering on random fields with a Kolmogorov spectrum.

The ratio of the characteristic time for reacceleration to the time for cosmic ray leakage from the Galaxy is approximately equal to $T_a / T_e \approx 1.2\beta^2 (R/1 \text{ GV})^{2/3}$. This reacceleration is essential at low energies and could in principle manifest itself in the observations of the secondary electron-capture isotopes ^{49}V, ^{51}Cr, and others, see [11, 12] for discussion. The efficiency of reacceleration decreases with energy and is insignificant at $E > 20$ GeV/n. The signature of the model with reacceleration is the predicted relatively weak energy dependence of the primary/secondary ratio at high energies (according to the scaling $X_e \sim R^{-1/3}$) compared to the model without reacceleration ($X_e \sim R^{-0.54}$).

The model with reaccelearation is distinguished by its low predicted anisotropy. The anisotropy perpendicular to the galactic disk is caused by the general diffusion of cosmic rays out of the Galaxy and is determined by the equation

$$A = -\frac{3D}{vI}\frac{fI}{fz} = \frac{3D}{vH}\frac{z}{h}, \quad (6)$$

where I is the total cosmic ray intensity, z is the distance of an observer position from the galactic midplane, h is the height of cosmic ray source distribution. Eq. (3) shows that the ratio D/H, which appears in eq. (6), is determined by the escape length characteristic of the given model. Using eq. (6) and assuming that $z/h = 0.1$, one can find the anisotropy $A = 5\text{x}10^{-3}(E/10^{14} \text{ eV})^{0.54}$ in the model without reacceleration, and $A = 0.6\text{x}10^{-3}$ $(E/10^{14} \text{ eV})^{1/3}$ in the model with reacceleration. The observed value $A_{obs} = (0.5 - 1)\text{x}10^{-3}$ at $E = 10^{12} - 10^{14}$ is in favor of the model with reacceleration. It is worthy of note, however, that the anisotropy constraint is not very severe, since the measured anisotropy may differ from the average value (6) in the galactic disk because of specific structure of the local galactic magnetic field. For example, the anisotropy is suppressed when the local magnetic field is directed perpendicular to the gradient of cosmic ray density, whereas it has some nonzero parallel component on average over the galactic disk.

SELFCONSISTENT GALACTIC WIND MODEL

The effect of a hypothetical galactic wind flow on cosmic ray transport was usually studied on a phenomenological basis, see [1, 13, 14] and references therein. In this context the cosmic ray diffusion coefficient $D(r, p)$ and the wind velocity $\boldsymbol{u}(r)$ are considered as free parameters fitted by comparison with cosmic ray, radio-astronomical and gamma-ray observations. A simple example was mentioned above. The wind model might explain the observed decrease of secondary/primary ratio at low energies if there is a wind with constant velocity $u = \frac{3}{2(\gamma+2)}\frac{\mu c}{X_0} \approx 24$ km/s. Here γ is the exponent of the cosmic ray differential spectrum $I \propto p^{-\gamma}$, and X_0 determines the low rigidity asymptotics of the escape length in eq. (4), $X_e \approx \beta X_0$. (There is a significant indication from the analysis of low-energy data that the commonly accepted scaling $X_e \propto \beta^n$, $n = 1$ should be replaced by more strong dependence with $n > 1$ [3, 15, 16]. The proposed interpretation of this dependence [16] assumes the wind model with non-monotonic dependence of cosmic ray convection velocity as a function of distance from the galactic plane.)

A different approach was used in papers [17, 18]. They describe a self-consistent model, which simultaneously includes the magnetohydrodynamic calculation of the galactic wind flow sustained by the cosmic ray pressure, and the consideration of transport of cosmic rays in this flow. The link between relativistic particles and thermal plasma is established through the cosmic-ray stream instability.

It is assumed that galactic cosmic rays are produced in the galactic disk. Cosmic rays largely determine the structure of the galactic wind flow in a rotating galaxy with a frozen-in large-scale magnetic field. The stream instability of cosmic rays, moving away from the galaxy along the spiral magnetic field, creates small-scale Alfvenic turbulence in the system. Wave-particle interaction proceeds through the cyclotron resonance. The equilibrium spectrum of turbulence that is determined by the nonlinear Landau damping of waves on the thermal ions defines the value of the diffusion coefficient.

The disk-halo transition at distance z_0 ~ 1-3 kpc above the galactic midplane is an important part of the picture in which ion-neutral friction damps short-scale magnetohydrodynamic waves below this level. The structure of this internal region adjacent to the galactic disk is not yet well studied. The wind velocity is ~ 30 km/s at 3 kpc. The wind flow goes through slow, Alfvenic, and fast magnetosonic points at about 5 kpc, 7 kpc, and 19 kpc respectively. The asymptotic value of the wind velocity is close to 480 km/s and is formally attained at very large distances, of the order of 1 Mpc. Actually, an external intergalactic pressure can decelerate the galactic wind flow through a termination shock supposedly located at around 300 kpc.

The diffusion coefficient at $z < z_0$ is determined by the turbulence created by "external" sources (not by cosmic rays) and is probably close to the value $D_c \sim 3 \cdot 10^{28} \beta \left(R / 1\,\mathrm{GV} \right)^{1/3}$ cm²/s, which was given above for the Kolmogorov type spectrum of the interstellar turbulence. The diffusion coefficient at $z > z_0$ is created by cosmic rays themselves through the stream instability. According to [16], the value of this diffusion coefficient can be estimated as $D_s \sim 10^{27} \beta (R/1\,\mathrm{GV})^{1.2}$ cm²/s, that is one - two orders of magnitude smaller than D_c at R = 1 GV but more rapidly rises with rigidity. (The exponent a = 1.2 in the expression for D_s is calculated from the relation $a = 2(\gamma - 1)/3$ using the exponent of observed cosmic ray spectrum γ = 2.75. The exponent of cosmic ray source spectrum in this case is $\gamma_s = \dfrac{2\gamma + 1}{3} \approx 2.2$.)

The calculated wind velocity u is approximately linear function of distance z from the galactic disk, $u = wz$, w = const at $z_0 < z \le 20$ kpc. Diffusion in this case is more important than convection for cosmic ray transport at distances $z \le z_m$, $z_m = (D/w)^{1/2}$. Convection dominates at $z \le z_m$. The critical distance z_m depends on particle energy through the dependence of the diffusion coefficient D on energy. (Note that the distance z_m is defined by the condition that the corresponding Peclet number is equal to unity: $u(z_m)z_m/D = 1$.)

The described wind model for an observer at the galactic disk can be roughly approximated by a pure diffusion model with the effective size of the halo equal to z_m. In particular, H should be substituted by z_m in eq. (3). The relations $D_s \sim R^{1.2}$ and $z_m \sim D_s^{1/2}$ lead then to the scaling $X_e \sim R^{-0.6}$ that is in reasonable agreement with the empirical formula (4) at $R > 5$ GV. This picture holds for particles with high rigidities. The low-rigidity particles ($R < 5$ GV) are trapped in the internal region $z < z_0$ by the wave barrier generated by stream instability in the upper halo. Their escape from the internal region goes with a small probability ~ $u(z_0)/v$ determined by the value of the convection velocity at $z = z_0$ that leads to the scaling $X_e \propto \beta$ consistent with eq. (4).

The preceding characteristics can be obtained from the following approximate expression for the escape length in the wind model that is derived on condition that adiabatic losses in the expanding wind are neglected:

$$X_e = \frac{\mu v}{2} \left[\sqrt{\frac{2 z_0}{u(z_0) D_s}} \exp\left(\frac{u(z_0) z_0}{2 D_s} \right) \times \left(1 - \mathrm{erf}\left(\sqrt{\frac{u(z_0) z_0}{2 D_s}} \right) \right) + \frac{z_0}{D_c} \right]. \quad (7)$$

Here erf(x) is Error function. Eq. (7) gives the dependence $X_e \approx \dfrac{\mu v}{u(z_0)}$ at low rigidities (presumably, at $R < 5$ GV), and $X_e \approx \dfrac{\mu v}{2} \left(\sqrt{\dfrac{\pi z_0}{u(z_0) D_s}} + \dfrac{z_0}{D_c} \right)$ at high rigidities. The relative yield of the term with D_c in the last equation can be very roughly estimated as $\left(R / 5 \times 10^7\,\mathrm{GV} \right)^{0.54}$. This yield is insignificant except at very high energies (E > 10 TeV) where it could change the exponent of cosmic ray spectrum by $\Delta\gamma \approx 0.27$ and make the spectrum more flat. Such behavior is probably seen in cosmic ray data [19].

Even when different models give similar predictions for stable nuclei, they can in principle be discriminated by analyzing the radioactive species. Specific is the energy dependence of the abundance of radioactive isotopes. As was discussed earlier, the content of radioactive secondary isotopes in cosmic rays is determined by the gas distribution in a sphere of radius $R_{dec} \sim (Dt_{dec})^{1/2}$. Let us assume that secondaries are produced in the gas disk with a half thickness $h << R_{dec}$. The production rate q, in units particles/cm^{3}s, is proportional to the density of primaries multiplied by the gas density. The number density of radioactive isotopes can be estimated as $N \sim qht_{dec}/R_{dec}$ (with no regard for nuclear fragmentation and energy losses, and under the assumption that $R_{dec} << (H, z_m)$). Thus the content of radioactive isotopes varies with energy and reflects the dependence $D(E)$.

The results of calculations of surviving fraction of the ^{10}Be isotope in cosmic rays in several models that were discussed in the present paper are shown on Figure 1, see also [20]. The surviving fraction of the decaying isotope is defined as $S = I(\tau)/I(\tau = \infty)$. It is the ratio of the flux for a radioactive isotope to the flux for the same isotope calculated on the assumption of infinite decay time. The surviving fractions at interstellar energy 0.4 GeV/n were calculated from the isotope measurements on Ulysses (labeled U) [2], ACE (labeled A) [3], and Voyager (labeled V) [21]. The error bars present the statistical errors alone. The surviving fraction at higher energies was derived in [22] from the elemental ratios measured in the HEAO-3 experiment (labeled H).

The calculations were made for the static diffusion model with diffusion coefficient obtained from eqs. (3), (4) (solid line), the diffusion model with reacceleration (dash-dot line), the wind model (short-dash line), and the leaky box model (dashed lines). The characteristic time of reacceleration in the interstellar medium is large compared to the decay time. This means that reacceleration has no significant direct effect on the radioactive isotope. The difference between diffusion models with and without reacceleration arises from different dependence of the diffusion coefficient on energy in these two cases. The effect of adiabatic losses was not taken into account in the calculations made for the wind model. The curve "wind" in Figure 1 approximately merges with the curve "diffusion + reacceleration" at energies less than few GeV/nucleon.

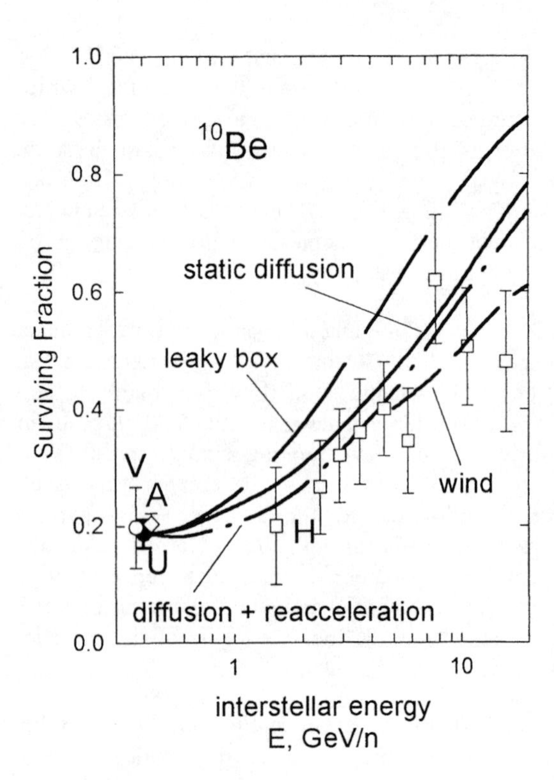

FIGURE 1. The surviving fraction of ^{10}Be as a function of interstellar kinetic energy.

The difference between models is clearly visible on Figure 1. The determination of the energy dependence of the surviving fraction in the low-energy measurements on ACE [3], and the measurements at about 3 GeV/nucleon in the balloon experiment ISOMAX [23] may significantly improve our understanding of cosmic ray transport in the Galaxy. The expected statistical error in the determination of the surviving fraction of ^{10}Be isotope for a 10 days flight is about +/- 0.05, comparable to the splitting between the leaky-box and diffusion curves shown on Figure 1.

CONCLUSION

The data on the nuclear component of cosmic rays can be well explained in the framework of the galactic diffusion model. The role of some processes in cosmic ray transport such as the rate of particle reacceleration in the interstellar medium and the convection velocity of the hypothetical galactic wind remain to be refined.

ACKNOWLEDGMENTS

The author is grateful to the Organizers of the ACE-2000 Symposium for grand hospitality and sponsorship. This work was also supported at the University of Maryland by NASA grant NAG5-7069, and at IZMIRAN by RFBR grant 98-02-16347.

REFERENCES

1. Berezinskii, V. S., Bulanov, S. V., Dogiel, V. A., Ginzburg, V.L., and Ptuskin, V. S., *Astrophysics of Cosmic Rays*, Amsterdam: North-Holland, 1990.
2. Connell, J. J., DuVernois, M. A., and Simpson, J. A., 25th *Intern. Cosmic Ray Conf., Durban* **3**, 397 (1997).
3. Yanasak, N. E., Binns, W. R., Cummings, A. C., Christian, and E. R., George, J. S. *et al.*, 26th *Intern. Cosmic Ray Conf., Salt Lake City* **3**, 9 (1999).
4. Ptuskin, V. S., and Soutoul, A., *Space Sci. Rev.* **337**, 859 (1998).
5. Chuvilgin, L. G., and Ptuskin, V. S., *A&A* **279**, 278 (1993).
6. Giacalone, J., and Jokipii, J. R., *ApJ* **520**, 204 (1999).
7. Jones, F.C., Ptuskin, V. S., Lukasiak, A., and Webber, W. R., 26th *Intern. Cosmic Ray Conf., Salt Lake City* **4**, 215; **4**, 291 (1999).
8. Jones, F. C., *ApJ* **229**, 747 (1979).
9. Simon, M., Heinrich, W., and Mattis, K. D., *ApJ* **300**, 32 (1986).
10. Seo, E. S., and Ptuskin, V. S., *ApJ* **431**, 705 (1994).
11. Connell, J. J., and Simpson, J. A., 26th *Intern. Cosmic Ray Conf., Salt Lake City* **3**, 33 (1999).
12. Mahan, S. E., Binns, W. R., Christian, E. R., Cummings, A. C., George, J.S. *et al.*, 26th *Intern. Cosmic Ray Conf., Salt Lake City* **3**, 17 (1999).
13. Webber, W. R., Lee, M. A., and Gupta, M., *ApJ* **90**, 96 (1992).
14. Bloemen, J. B. G. M., Dogiel, V. A., Dorman, V. L., and Ptuskin, V. S., *A&A* **267**, 372 (1993).
15. Krombel, K. E., and Wiedenbeck, M. E., *ApJ* **328**, 940 (1988).
16. Soutoul., A., and Ptuskin, V. S., 26th*Intern. Cosmic Ray Conf., Salt Lake City* **4**, 184 (1999).
17. Zirakashvili, V. N., Breitschwerdt, D., Ptuskin, V. S., and Volk, H. J., *A&A* **311**, 113 (1996).
18. Ptuskin, V. S., Volk, H. J., Zirakashvili, V. N., and Breitschwerdt, D., *A&A* **321**, 434 (1997).
19. Swordy, S. P., 24th *Intern. Cosmic Ray Conf.,Rome* **2**, 697 (1995).
20. Ptuskin, V.S., Soutoul A., and Streitmatter, R. E., 26th *Intern. Cosmic Ray Conf., Salt Lake City* **4**, 195 (1999).
21. Lukasiak, A., McDonald, F. B., and Webber, W. R., *ApJ* **430**, L72 (1994).
22. Webber, W. R., and Soutoul, A., *ApJ* **506**, 335 (1998).
23. Mitchell, J. W., Barbier, L. M., Bremerich, M., Christian, E. R., Davis, A. J. *et al.*, 26th *Intern. Cosmic Ray Conf., Salt Lake City* **3**, 113 (1999).

The Diffusion of Cosmic Rays in The Galaxy -A Monte Carlo Approach

W.R. Webber

New Mexico State University, Department of Astronomy, Las Cruces, NM 88003

Abstract. We have used a Monte Carlo diffusion model to describe the propagation of cosmic ray particles in the Galaxy. This model uses realistic dependencies for the matter density, the size of the galactic halo, the diffusion coefficient, etc., to describe this motion. The results are similar in many respects to those of Leaky Box models, but interesting differences, subject to measurement by ACE and other high precision instruments are evident. In this paper we focus on these differences as they relate to: 1) The effect of the matter density on the spectra of nuclei from ~1-100 GeV and how it influences the determination of the primary spectra of these nuclei; 2) The effect of the matter density and the halo size on the $^3He/^4He$, B/C and sub Fe/Fe ratios; 3) The effect of all of these propagation parameters on the surviving fraction of the radioactive clock isotopes including ^{14}C and also on the K-capture isotopes.

INTRODUCTION

The traditional Leaky Box Model (LBM) used to describe cosmic ray propagation in the Galaxy provides very precise answers to a wide variety of observations. As such it is useful for an initial comparison of predictions with measurements. However, it provides very little insight into the details of this propagation and the parameters used to describe it. All of these details are basically contained in two parameters, the total path length in g/cm^2 and its energy dependence, and an average matter density, n, for the entire propagating region. Diffusion models can be made physically more realistic but analytical diffusion models are also limited in the description of the many parameters and their spatial and energy dependence necessary to fully describe this propagation in the Galaxy.

We have recently described a Monte Carlo diffusion model (MCM), (1, 2, 3), that overcomes many of these difficulties by introducing detailed spatial and energy dependencies of all of the relevant propagation parameters Here we discuss its precise application to the propagation of cosmic ray nuclei including radioactive ones. For these calculations we take source spectra, s, in the range $P^{-2.25 \text{ to } -2.35}/\beta$ and a diffusion coefficient given by $K=2 \times 10^{28} \times P^{0.50}$ cm^2 sec^{-1} that accurately fits the B/C ratio at all energies above ~1 GeV/nuc. The spectrum at sufficiently high energies after propagation is then $dj/dp \sim P^{-(s+a)} = P^{-2.75\text{-}2.85}$ where a is taken to be fixed at 0.50. The IS matter density n_0 at Z=0 and its Z dependence characterized by a simple exponential $n=n_0 \exp^{-Z/Zm}$ as well as the boundary of the propagating region Z_B (size of the halo) are variables. In these calculations 10^4 particles are injected at each logarithmically spaced energy interval -8 to the decade- from 10 MeV to 10 TeV. These particles are followed for the course of their lifetime (up to > 2 x 10^7 yrs) with E loss by ionization and stochastic energy gain, nuclear interaction, radioactive decay and convection effects included and the particles are summed in 20 intervals from Z=0 to Z_B as well as in the energy intervals. The results reported here are for the first Z interval corresponding to the suns location near Z=0.

CALCULATION OF THE IS PROTON AND HELIUM SPECTRA

From the Monte Carlo calculations we find that the spectra of all nuclei begin to flatten slightly below their high energy asymptotic exponent = (s+a), below ~100 GeV/nuc. This flattening is also noticed in equivalent Leaky Box Model calculations (note: the fact that pure rigidity spectra are plotted in the form of energy/nuc spectra is a cause for part of this flattening, particularly at lower energies). From the Monte Carlo calculation we find that the degree of this flattening depends on the value of the matter integral, $\int n_0 dZ$, in the Z direction at the location of the sun, along with the diffusion properties including the location of the boundary (halo size) (3). These

CP528, *Acceleration and Transport of Energetic Particles Observed in the Heliosphere: ACE 2000 Symposium*, edited by Richard A. Mewaldt, et al.

© 2000 American Institute of Physics 1-56396-951-3/00/$17.00

parameters are discussed in more detail in the following sections, however, a set of values, n_0=1.0 cm^{-3}, Z_m = 0.25 Kpc and Z_B = 3.0 Kpc provide a good fit to a wide variety of data and is used here to calculate the shape of the IS proton spectrum. These parameters give a value of the matter line integral = 7.5 x 10^{20} cm^{-2} in the Z direction.

The IS spectrum is determined from a set of measurements of the proton or helium spectrum made at the Earth at a certain solar modulation level. In the example used here we take the BESS proton spectrum measured from ~0.3 to over 100 GeV in 1997 (4). The absolute value and spectral index of the IS proton spectrum are then determined in an iterative fashion by requiring that the BESS proton spectrum provide a best fit to a particular IS spectrum that is modulated by a rigidity dependent modulation which is ~$P^{-1.0}$ above ~1-2 GV as found from neutron monitor and spacecraft measurements near the Earth (5). This is equivalent to a force field modulation with a given value of ϕ, the modulation parameter in MV (6). The situation for the 1997 BESS measurements is shown in Figure 1, where

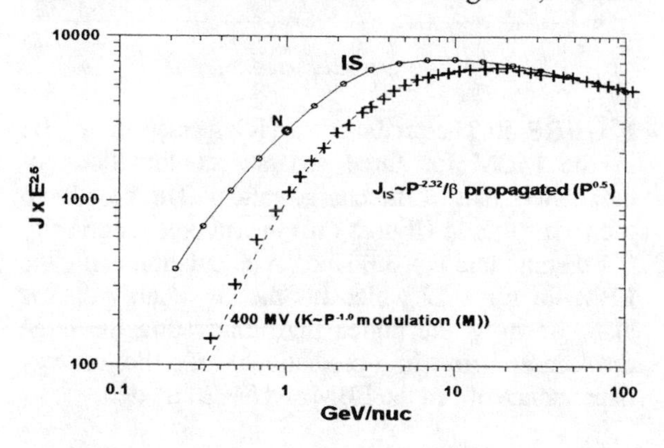

FIGURE 1. IS proton Spectrum derived from BESS 1997 proton measurements (4) shown as crosses, assuming that the solar modulation is ~$P^{1.0}$. The IS spectrum has a value = 2680 at 1 GeV, and a spectral exponent given by (s+a) -2.82. Solar modulation for a value of ϕ = 400 MV is shown as a dashed line.

J_{IS} (1 GeV) is determined to be = 2.68 P/m^2·sr·s·MeV and s+a -2.82. Using these values for the IS spectrum the BESS data is remarkably well fit above ~1 GeV by a value for the modulation of ϕ = 400 MV (Figure 2).

This BESS data is the first available with the overall accuracy: 1) To illustrate the flattening of the IS spectrum due to the local matter density and, 2) To be able to be used to actually determine the value and rigidity dependence of the residual modulation given a propagated IS spectrum.

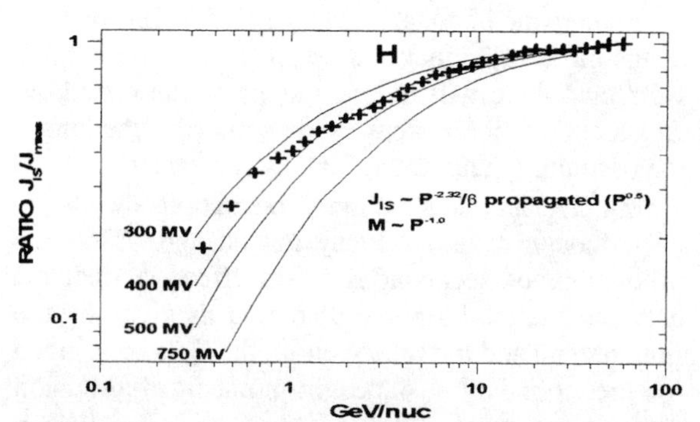

FIGURE 2. Fractional modulation deduced from BESS 1997 proton measurements and deduced IS proton spectrum. The fit of this fraction to a $P^{-1.0}$ modulation with ϕ=400 MV is a confirmation of the intensity and spectral index of the IS spectrum.

THE $^3He/^4He$, B/C AND SUB Fe/Fe RATIOS AND THE DECAY OF THE RADIOACTIVE ISOTOPES

The propagation of primary nuclei and the production of secondary nuclei in Monte Carlo diffusion models is governed mainly by three factors. The 1st is the matter density along the galactic plane (Z=0), n_0. The 2nd is the characteristic height of the matter distribution Z_m. These two quantities combine to form the line integral, $\int n_0 \exp^{-Z/Z_m} dz$, of the matter density perpendicular to the plane as described earlier. This is a quantity that independently can be determined from astrophysical measurements. The best astrophysical estimates of the line integral of the sum of the HI, H_2 and HII components at the location of the sun is ~(6-8) x 10^{20} cm^{-2} where HI contributes ~4 x 10^{20} cm^{-2}, HII ~1 x 10^{20} cm^{-2} and the remainder is due to the very uncertain contribution of H_2. The ionized hydrogen, HII, observed by its absorption of radio waves and its H alpha emission, is sometimes called the Reynolds layer (7) and is believed to have a much flatter Z distribution (Z_m ~1 Kpc) than the neutral gas. Each of these components has a different Z dependence so the resulting Z dependence is more complicated than the simple exponential we have used. We have also made calculations with two exponentials, Z_{m1} = 0.1 kpc and Z_{m2} = 1.0 kpc, with the same total line integral as the single exponential. Any differences in the secondary/primary ratios are not observable to a level of ±5% in these calculations.

The 3rd factor governing the propagation is the distance to the boundary Z_B in Kpc. This is taken to be a fixed distance independent of r. This distance is normally expressed in terms to the number of steps λ, (diffusion MFP) to the boundary, thus bringing in the diffusion coefficient, K. For example, in a typical case where the boundary distance is at 3 Kpc, and the diffusion coefficient = 2 x 10^{28} cm^2 sec^{-1} at 1 GeV/nuc, there will be 5500 steps to the boundary. In general the more steps to the boundary, the longer the particles remain in the Galaxy (characteristic time $\tau = Z_B^2/2K$) and thus the more interactions they have, even though n_0 and Z_m may not change. Thus the production of secondaries is, in effect, a trade-off between the total amount of matter as given by the line integral and the time spent in the Galaxy as given by the boundary distance in number of diffusion steps.

The 1st step in the study of the secondary production is to match the observed $^3He/^4He$, B/C and sub Fe/Fe ratios as a function of energy. This can be achieved at ~1 GeV/nuc for the combination of parameters listed in Table 1. Any of these com-

TABLE 1. Parameters used to fit $^3He/^4He$, B/C and sub Fe/Fe Ratios

Density (n_0-cm^{-3})	Halo Size (z_B=L-kpc)	Line integral (l_m-10^{20} cm^{-2})
1.6	2.25	9.6
1.2	3.00	7.4
0.9	4.00	5.6

binations (or values in between) will fit the observed ratios to an accuracy of ± 5% at 1 GeV/nuc and with a rigidity dependence of $K \sim P^{0.5}$, also will fit the observed ratios of B/C and sub Fe/Fe up to energies ~100 GeV/nuc (Figure 3). These combinations give values of the matter line integral between 5.6

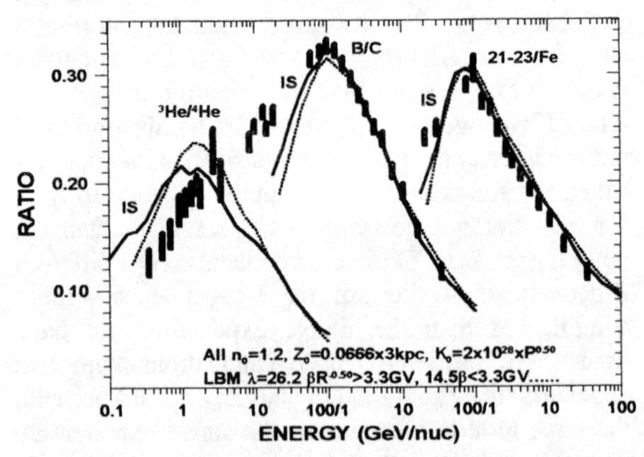

FIGURE 3. The measured and predicted ratios of $^3He/^4He$, B/C and sub Fe/Fe. See text and Table 1 for parameters used in MCM calculation. Data points are from (8).

and 9.6 x 10^{20} cm^{-2} and so are consistent with the direct astrophysical estimates of this quantity mentioned earlier. They also have values of n_0 that are within the measured range 0.5 to 1.5 for the sum of the individual components.

Note that the low energy predictions are for the interstellar ratios, the measurements are after solar modulation, hence the poor fit at low energies.

The combinations which fit the B/C and sub Fe/Fe ratios give, however, greatly different values for the surviving fractions of the radioactive nuclei. This is illustrated in Figure 4 for the surviving fraction of ^{10}Be and the curves for n_0 = 1.6, 1.2 and 0.9.

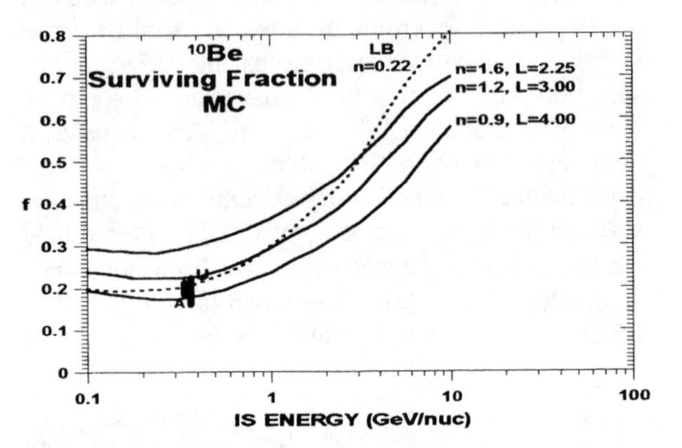

FIGURE 4. The predicted surviving fraction of ^{10}Be in the MCM for three separate combinations of parameters that fit the charge ratios. The best fit to the ACE (9) and Ulysses (10) measurements gives n_0 = 1.03 cm^{-3} and L = 3.6 Kpc. A calculation using the LBM for n = 0.22 which fits the low energy data is also shown as the dotted line, illustrating the large differences in the predictions of the energy dependence of f in the LBM and MCM models.

The observations of ACE and of Ulysses (9, 10) give a value of 0.20 for the surviving fraction, f, of this isotope with an error ~6% at ~100 MeV/nuc which from the MCM calculation implies a value of n_0 = 1.03±0.06, and a corresponding line integral = 6.2± 0.4 x 10^{20} cm^{-2} for the matter density (for Z_m = 0.2 Kpc) and a boundary distance of 3.6 Kpc (6600 diffusion steps). This use of the decaying isotopes to determine n_0 thus gives a value for the line integral consistent with but more precise than even the direct astrophysical measurements!

Figure 4 also illustrates the large differences in the surviving fraction of ^{10}Be as a function of energy as predicted by the LBM and the MCM. This is a result of the time dilation of the ^{10}Be decay lifetime which then allows the decay of these nuclei to sample the matter density further and further from the plane. In the LBM this matter density is independent of position but in the MCM the decrease of the density with Z results in a lower surviving fraction at large distances (high energies). This could be explored using ^{10}Be measurements at high energies but unfortunately this difficult measurement has not yet been made. It is possible, however, to obtain similar information on f using the highly precise Be/B ratio measurements from HEAO (11). The results of this study are shown in Figure 5 along with calculations for a LBM and MCM both of

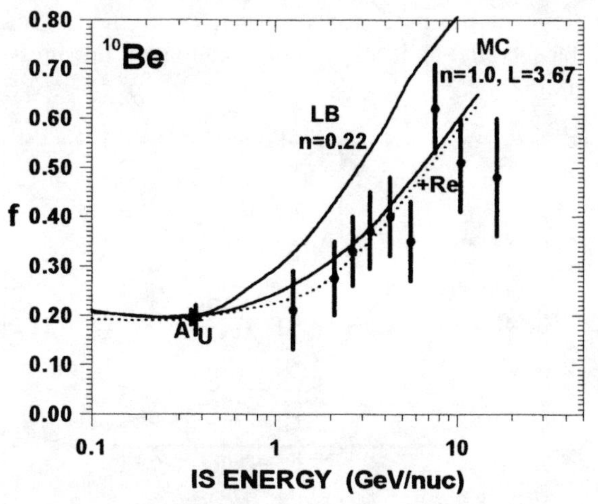

FIGURE 5. The predicted surviving fraction of ^{10}Be as a function of energy for the LBM and MCM that fit the low energy spacecraft data. Values of f inferred from the Be/B ratio measured by HEAO (11) are shown.

which fit the lower energy ^{10}Be surviving fraction. The data clearly demonstrates that diffusion models in which the matter density falls off rapidly with Z are needed to explain the surviving fraction obtained from the higher energy radioactive decay isotopes. Also on this curve we show the effects of reacceleration on the surviving fraction as a dotted line. The differences are small and not distinguishable at the present level of accuracy of the data.

There still is more to this story, however. The data at low energies only can now be used, because of the high precision of the ACE data, to unfold the Z distribution of matter as well, using the different lifetimes of the clock isotopes. Figure 6 shows the Z=0 matter density that is required to fit the observations from ACE and Ulysses for each of the four radioactive clock isotopes from ^{36}Cl to ^{10}Be spanning a range of lifetimes ~ a factor of 5. The calculations are shown for both the LBM and the MCM. For the LBM the density required to fit the data increases from ~0.22 cm^{-3} for ^{10}Be to ~0.31 cm^{-3} for ^{36}Cl (an increase of 43%). This increase, which can be observed clearly for the first time because of the accuracy of the new ACE data, is attributable to the fact that although the matter density is taken to be constant with distance away from the plane in this model, in reality it is decreasing with increasing Z. Thus ^{10}Be with its longer half-life samples the density further from the plane than ^{36}Cl and both sample an average matter density less than the value n_0 at Z=0, which is expected to be ~1 cm^{-3}. Earlier observations of this effect for ^{10}Be only have been used to argue that the cosmic rays must spend much of their time in an extended halo several times the thickness of the matter disk.

If this is the correct argument, then these parameters can be determined more specifically in the MCM. In the example we use a halo thickness that is ~15 x the characteristic matter thickness. The

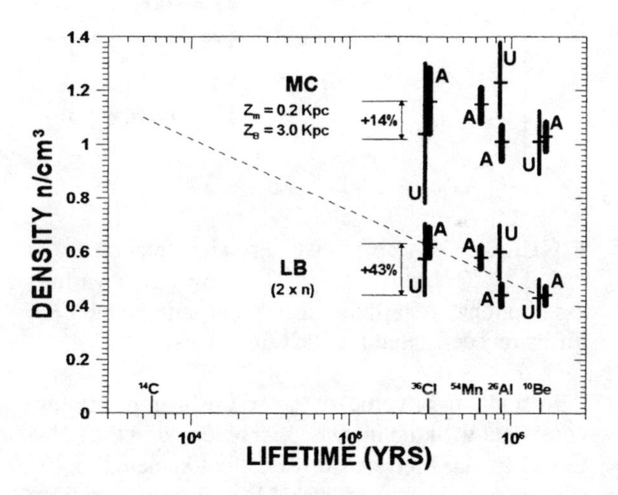

FIGURE 6. The value of n_0 that gives a best fit to the ACE (9) and Ulysses (10) measurements of the surviving fraction of each of the clock isotopes ^{36}Cl, ^{54}Mn, ^{26}Al and ^{10}Be for both the MCM and the LBM.

required density to explain the observed surviving fractions of each of the radioactive decay nuclei increases from 1.03 cm^{-3} for ^{10}Be to 1.16 cm^{-3} for ^{36}Cl, an increase ~13%. This is much less than the 43% in the LBM and just about the overall statistical accuracy of the ACE data itself (or the uncertainties in the decay constants themselves). It is nevertheless worth examining if a set of slightly different parameters can be chosen to make the value of n_0 the same for all of the clock isotopes. We find that this % increase goes through a broad minimum ~10-12% centered on n_0 = 0.9, Z_m = 0.25 Kpc in the MCM. Thus within the present uncertainties in both the measurements (±5%) and the decay constants (±10%) it is not possible in the MCM to make the value of n_0 completely independent of radioactive species.

This behavior of the deduced matter density on the decay constant has important implications when studying the radioactive decay isotope ^{14}C which has a half life ~1/50 of that of ^{36}Cl and therefore should provide a very local sample of the matter density. The situation for ^{14}C is illustrated in Figure 7 which shows the surviving fraction (and also the more directly measured ^{14}C/^{13}C ratio) calculated using recently measured cross sections (12). The predictions are for a LBM with n=0.22 which will fit

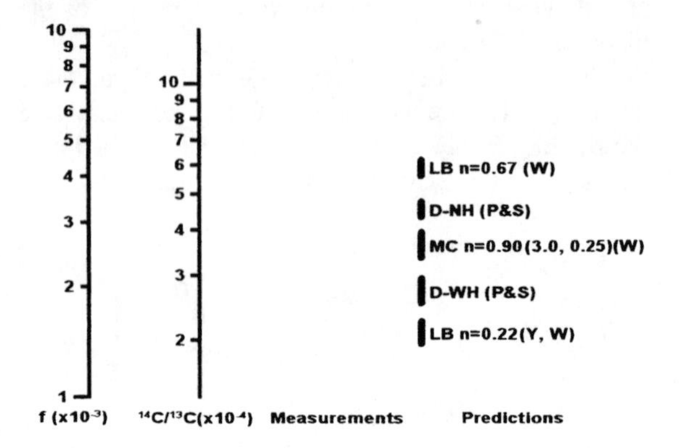

FIGURE 7. The predicted surviving fraction f, (or the ^{14}C/^{13}C ratio) for ^{14}C decay for various assumptions regarding the propagation and the diffusion coefficient (see text for details).

^{10}Be and for a value of n=0.67 which is roughly consistent with the increase in n observed in the LBM model as the lifetime decreases, extrapolated to ^{14}C (Figure 6). The points labeled D in Figure 7 are from a diffusion model calculation (13) with a Z dependence of matter density (NH) and with a local hole in the matter density (WH). The band labeled MC is from the Monte Carlo calculation in this paper which shows the smallest change in n_0 that will fit the data from ^{36}Cl to ^{10}Be as described earlier. At the present time there are no measurements of this ratio so this column is left blank. Any measurement of the ^{14}C/^{13}C ratio with a factor ~2 or less uncertainty from ACE will be of great value and help clarify our understanding of the factors governing the propagation of nuclei and the production of secondaries in the matter disk of the Galaxy.

K-CAPTURE DECAY

Finally in this paper we also discuss the production and decay of K-capture nuclei. This work is still on going but again we first start with calculations using the LBM and then try to understand any differences with respect to the MCM and how they arise. Figure 8 shows the LBM calculations for the ^{51}V/^{49}V ratio (similar to those described in (14)) along with the data from earlier Voyager measurements and the new high resolution data from Ulysses (10) and ACE (15).

The calculation is performed by including the K-capture cross sections directly in the input cross section file of the LBM since when attachment occurs the nuclei basically decay because of their short lifetime. The simplest interpretation of the figure is to assume that an excess of the measured ratio over the predicted one at a fixed energy and solar modulation level is an indication of the pres-

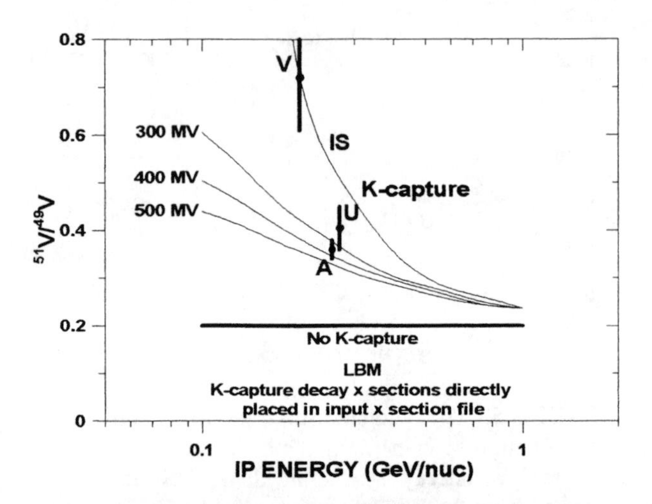

FIGURE 8. The predicted ^{51}V/^{49}V ratio in the LBM with and without K-capture decay. The predicted ratios are shown for various values of solar modulation ϕ, but the data; A=ACE, U=Ulysses and V= Voyager, are put on without reference to the modulation level.

ence of reacceleration in which lower energy particles with a higher K-capture decay cross section end up at higher energies. This re-acceleration would tend to move the data points to the right in the figure.

The actual situation is complicated by several factors, however. Uncertainties in both the K-capture cross sections and the details of the propagation model itself can contribute significant errors. For example the values for the K-capture cross sections are based on theoretical calculations dating back over 50 years (16) and are not substantiated by any measurements that we are aware of. Also the determination of the appropriate modulation level for the measurement is uncertain by perhaps as much as 100 MV.

In the initial measurement of this ratio (14), believed to be made at a modulation level ~300 MV, using Voyager 1 and Voyager 2 data, the high value of 0.72±0.11 for the ^{51}V/^{49}V ratio could be interpreted as a re-acceleration of possibly ~100

MeV/nuc (above the expected 300 MV modulation level) albeit with large errors because of the uncertainties in the measurement. For the ACE measurement, if it is assumed that the modulation level is ~500-550 MV then some re-acceleration might also be indicated in Figure 8, but if the modulation level is only ~400 MV on the basis of the BESS proton measurement as given earlier in this paper, then essentially no reacceleration is apparently needed to explain the ACE data.

The Monte Carlo model calculations tell a somewhat different story, however. First of all, energy gain by reacceleration is always competing with energy loss by ionization. In the energy and charge range of interest here the loss is greater than the gain so the net effect is that the particles are still losing energy and reacceleration does not boost particles from a lower to a higher energy. The overall solution to this problem is more complicated than the original simple picture. It is discussed further in a separate paper on reacceleration (17) at these proceedings.

SUMMARY AND CONCLUSIONS

We have examined the propagation of cosmic rays in the galaxy using a Monte Carlo diffusion model, paying particular attention to the differences in the predictions using this model and the standard LBM. In the case of the calculation of the IS proton and helium spectra the density of matter perpendicular to the disk in the MCM is found to significantly effect the bending of these spectra in the energy range below 100 GeV/nuc. This bending along with the form of the rigidity dependent solar modulation need to be carefully taken into account in order to deduce the IS spectra from those measured at the Earth.

The observed $^3He/^4He$, B/C and Sub Fe/Fe ratios can be fit by a range of combinations of the values of n_0 and Z_B in the MCM. These fits are very close to the ones obtained using the LBM. In other words, the observed details of these ratios can be matched equally well using either propagation model. When radioactive decay nuclei are used the redundancy in n_0 is removed. A value of $n_0 = 1.03\pm0.06$ and a corresponding matter line integral = $6.2\pm0.4 \times 10^{20}$ cm^{-2} are obtained in the MCM. These values are more precise than the current astrophysical measurements thus making a strong argument that cosmic rays are the best probe of the matter distribution perpendicular to the disk of the Galaxy. The MCM predicts a much different energy Dependence of the surviving fraction of the radioactive isotopes that the LBM, however. This difference again arises because of the specific matter distribution perpendicular to the disk. The HEAO results on ^{10}Be decay at higher energies clearly favor the MCM predictions. The MCM also leads to nearly the same value for n_0 for each of the radioactive isotopes with different decay times in contrast to the greatly different values of n needed in the LBM. This leads to different predictions for the surviving fraction of ^{14}C, the shortest lived isotope. Observations of the surviving fraction of this isotope, when they are available, will greatly help our understanding of the more local production and propagation of secondary nuclei in the disk of our galaxy.

The MCM also leads to different predictions on the amount of K-capture decay relative to the LBM. We believe this is again related to the specific distribution of matter used in the MCM. These differences and the effects of re-acceleration are discussed more fully in a separate paper (17).

ACKNOWLEDGMENTS

This work would not have been possible without the computer support of Dr. John Rockstroh and Jason Peterson. This research is supported as part of the ACE Guest Investigator Program by NASA Grant NAG 5-7795.

REFERENCES

1. Webber, W.R. and Rockstroh, J.M., *Adv. Space Res.*, **19**, 817-820, (1997)
2. Webber, W. R., Peterson, J.D. and Rockstroh, J.M., *Proc. 26th ICRC*, **4**, 222-224, (1999)
3. Webber, W.R. *Proc. 26th ICRC*, **4**, 219-221, (1999)
4. Sanuki, T., et al., *Proc. 26th ICRC*, **3**, 93-96, (1999)
5. Lockwood, J.A. and Webber, W.R., *J.G.R.*, **101**, 21, 573-580, (1996)
6. Gleeson, L.J. and Axford, W.I., *Ap.J.*, **154**, 1011-1019, (1968)
7. Reynolds, R.J., *Ap.J.*, **282**, 191-196, (1983)
8. Engelmann, et al., *A & A*, **233**, 96-111, (1990)
9. Binns, W.R., et al., *Proc. 26th ICRC*, **3**, 21-24, (1999)
10. Connell, J.J. and Simpson, J.A., *Proc. 26th ICRC*, **3**, 33-36, (1999)
11. Webber, W.R. and Soutoul, A., *Ap.J.*, **506**, 335-340, (1998)
12. Webber, W.R., et al., *Ap.J.* **508**, 949-958, (1998)
13. Ptuskin, V.S. and Soutoul, A., *A & A*, **337**, 859-865, (1998)
14. Soutoul, A., et al., *A & A*, **336**, L61-65, (1998)
15. Niebur, S.M., *These proceedings*, (2000)
16. Soutoul, A., *Private communication*, (2000)
17. Webber, W.R., *These proceedings*, (2000)

Abundances of the Cosmic Ray β-decay Secondaries and Implications for Cosmic Ray Transport

N. E. Yanasak[1], W. R. Binns[2], E. R. Christian[3], A. C. Cummings[4], A. J. Davis[4], J. S. George[4], P. L. Hink[2], J. Klarmann[2], R. A. Leske[4], M. Lijowski[2], R. A. Mewaldt[4], E. C. Stone[4], T. T. von Rosenvinge[3], M. E. Wiedenbeck[1]

[1]*Jet Propulsion Laboratory*
[2]*Washington University*
[3]*NASA/Goddard Space Flight Center*
[4]*California Institute of Technology*

Abstract. Galactic cosmic rays (GCRs) pass through the interstellar medium (ISM) and undergo nuclear interactions that produce secondary fragments. The abundances of radioactive secondary species can be used to derive a galactic confinement time for cosmic rays using the amount of ISM material traversed by the cosmic rays inferred from stable GCR secondary abundances. Abundance measurements of long-lived species such as ^{10}Be, ^{26}Al, ^{36}Cl, and ^{54}Mn allow a comparison of propagation histories for different parent nuclei. Abundances for these species, measured in the energy range $\sim 50-500$ MeV/nuc using the Cosmic Ray Isotope Spectrometer (CRIS) aboard the Advanced Composition Explorer (ACE) spacecraft, indicate a confinement time $\tau_{esc} = 16.2\pm0.8$ Myr. We have modeled the production and propagation of the radioactive secondaries and discuss the implications for GCR transport.

INTRODUCTION

The abundances of secondary cosmic rays, produced by fragmentation of primary galactic cosmic rays (GCRs) as they pass through the interstellar medium (ISM), are a measure of the amount of material through which the GCRs pass before escaping from the galaxy. The value of the average pathlength before escape at a particular GCR energy λ_{esc}, determined by measurements of the ratio of secondary to primary GCR abundances, corresponds to an ISM density ρ_{ISM} and a propagation time in the galaxy τ_{esc} (confinement time). Within the "leaky box" model (LBM), the relationship between these three quantities and GCR velocity v is assumed to be $\lambda_{esc} = \rho_{ISM} v \tau_{esc}$. The radioactive β-decay secondaries will undergo decay after production during propagation and serve as "clocks", measuring the average rate of escape in comparison to the decay time. Therefore, the surviving abundance of these species will be sensitive to τ_{esc} and the ISM density. Abundances reported from the IMP-7,8 and ISEE-3 spacecraft were consistent with a mean ISM number density $n_{ISM} \sim 0.1-0.3$ atoms/cm^3 and τ_{esc} between 6-20 Myr (for a review, see (1)). Precise measurements of ^{26}Al, ^{36}Cl, and ^{54}Mn which allow for a comparison of propagation histories for different parent nuclei have been made recently by several experiments (2, 3), leading to improved values for τ_{esc} and the average ISM density characteristic of the cosmic ray confinement region.

Interpretation of the secondary radionuclide abundances is specific to the framework of the chosen propagation model. The LBM replaces the spatial diffusion and loss of particles at the galaxy boundary in more complex models by a mean energy-dependent, spatially-independent escape pathlength. Despite the simple nature of the LBM, its predictions can be made to fit a variety of GCR abundance measurements. The LBM is a suitable approximation to more realistic propagation models which may incorporate diffusion, convection, and reacceleration of the GCRs (e.g., (4)). However, differences between these models arise in the predictions for the shorter halflife radioactive species if one assumes inhomogeneities in the local ISM. We report new measurements of the β-decay secondaries and attempt to provide a consistent picture of GCR propagation for all the secondary radionuclides using our steady state LBM and predictions from other models.

OBSERVATIONS

The data reported here were obtained using the Cosmic Ray Isotope Spectrometer (CRIS) instrument aboard

CP528, *Acceleration and Transport of Energetic Particles Observed in the Heliosphere: ACE 2000 Symposium,* edited by Richard A. Mewaldt, et al.
© 2000 American Institute of Physics 1-56396-951-3/00/$17.00

the Advanced Composition Explorer (ACE) spacecraft (5). This instrument uses the dE/dx versus total E technique to measure elemental and isotopic composition of the galactic cosmic rays. It consists of four Si detector stacks to measure the particle energy deposition and a scintillating optical fiber trajectory (SOFT) hodoscope that acts as part of the trigger and determines particle trajectories (6). The period of observation for these data is between 27 August 1997 and 10 April 1999, excluding periods of significant solar energetic particle intensity. Events were selected for this study using charge and mass consistency requirements and a cut to reject particles stopping near the edge or the surfaces of the silicon detectors where there are larger uncertainties in their identification (6).

The ratios of the radioactive isotope abundances to the abundances of stable isotopes of the same elements are shown in Table 1. The isotopic abundances for Al, Cl, and Mn were determined using a maximum likelihood fit of Gaussian peaks to their mass histograms. Isotopes of Be are clearly resolved and their abundances were obtained by counting events in each peak (see (6)). The mass resolution of the Si detectors degrades with increasing angle of incidence because of multiple Coulomb scattering, and zenith angle cuts were applied for improved resolution of the mass peaks (6). Corrections to these abundances (<6% for all isotopes) were made to account for differences in fragmentation, SOFT detection efficiency, and geometry factor for a given energy interval.

Table 1. GCR Radionuclide Abundances from CRIS

Ratio	Energy (MeV/nuc)	Measurement
$^{10}Be/^{9}Be$	70-145	0.120±0.008
$^{26}Al/^{27}Al$	125-300	0.048±0.002
$^{36}Cl/Cl$	150-350	0.062±0.008
$^{54}Mn/Mn$	178-400	0.114±0.006

PROPAGATION MODEL

The observed abundances from CRIS were compared with model abundances calculated using a steady state LBM described in Yanasak, et al.(6). Our model source abundances were based on solar system isotopic observations (7) in conjunction with a source composition from HEAO-3 observations for predominant elemental species (e.g., C, Fe) (8) and Ulysses for rarer elemental species (e.g, P) and for ^{22}Ne (9), adjusted to match the CRIS data. Source spectra J were assumed to be power laws in momentum p: $dJ/dE \propto p^{-2.35}$. Energy loss, nuclear decay, spallation, and attachment and stripping of electrons from the GCR nuclei during propagation were considered in the model. The ISM composition was chosen to be 90% H, 10% He by number, with a 24% ionized H fraction.

The value of the escape length λ_{esc} was adjusted to match the CRIS data for the secondary to primary ratios B/C, F/Ne, P/S, and (Sc+V)/Fe as well as the ratio of elements for charges Z=17-25 to Fe to sample the secondary production for Cl. We assumed this following form of the escape length, motivated by the form given in (10):

$$\lambda_{esc} = \frac{28.0\beta}{(\beta R)^{0.6} + \alpha(\beta R/1.4GV)^{-1.4}} \quad (1)$$

where R is the particle rigidity, α is a normalization factor, and β is the ratio of the particle velocity to the speed of light v/c. Previous low energy experiments which have measured secondary to primary ratios at multiple energies indicate a dependence of $\beta^{3.5-4.5}$ to fit the data over the full range in energy space (e.g., (11)) rather than the $\beta^{1.0}$ dependence commonly used in other studies (10). At lower energies, our form for λ_{esc} gives a dependence $\propto \beta(\beta R)^{1.4}$, and this strong dependence on energy provides a good fit for CRIS GCR secondary to primary ratios (6).

Although a value of $\alpha = 1.0$ provides a good fit for secondary to primary ratios with $Z < 21$, we were unable to account for all of the secondary to primary ratios with a unique value of α. Using α=1.0 tends to underestimate the production of sub-iron secondaries by ~15% at CRIS energies, while model predictions at higher energies are in good agreement with HEAO-3 data. For this study, we adjust λ at low energies for each clock species to match the secondary to primary ratios for parent nuclei which contribute most to that species, setting α=0.15 for ^{54}Mn analysis and α=1.0 for the other species.

Solar modulation of the cosmic ray spectra was simulated using a spherically symmetric model (12), with a diffusion coefficient proportional to βR. The diffusion coefficient used corresponds to a modulation parameter value of ϕ=325 MV (6).

DISCUSSION

Figure 1 shows a comparison between CRIS abundances for the secondary radionuclides measured at Earth and those obtained previously from Ulysses (13, 2, 14, 15), ISEE-3 (16, 17) and Voyager (18, 19, 20). Curves of constant ISM hydrogen density $n_H = n_{ISM} - n_{He}$ from our model calculations are shown overlying the data in Figure 1. Previously reported abundances were measured at different levels of solar modulation, and these have been adjusted to the level of modulation appropriate to CRIS for direct comparison. The largest adjustment required was 25% for the ISEE-3 measurement of ^{10}Be (17). Previous observations were reported as one value per experiment averaged over a wide energy range, owing to the limited counting statistics. Because of the larger number of events that CRIS collects, the data can be divided into several energy intervals, enabling comparisons with the predicted energy dependence. To within statistical uncertainties, the CRIS results are consistent with those reported from previous experiments. Because of the excellent statistical significance of the CRIS data, the effect of uncertainties in solar modulation on model predictions must be considered. A change in the solar modulation level $\Delta\phi \sim 50$ MV corresponds to a variation in the ISM number density of $\sim \pm 7\%$ from the ^{10}Be data and somewhat smaller variations for the other species (6).

From Figure 1, the observations imply $n_H = 0.28 - 0.43$ H atoms/cm^3. Differences in these results for n_H and those presented previously (3) are primarily due to the use of different total and partial fragmentation cross-section formulas (6). Average total ISM density values indicated by the secondary clock abundances are somewhat less than the typical assumption of a nominal ISM density in the galactic disk $n_{ISM} \sim 1$ atom/cm^3. Taken together, these four clock species indicate a hydrogen density $n_H \sim 0.34 \pm 0.02$ H atoms/cm^3, which is consistent to within two standard deviations for all four species. A systematic uncertainty in the halflife of ^{54}Mn must also be considered. From the combined observations of Wuosmaa, et al.(22) and Zaerpoor, et al.(23), the halflife of GCR ^{54}Mn is $6.8 \pm 1.5 \times 10^5$ yr. The size of this systematic uncertainty in determining n_H is comparable in amount to the statistical uncertainty shown for each CRIS ^{54}Mn data point in Figure 1. The abundance for the shorter-lived species ^{36}Cl does not indicate an ISM density which is significantly different than those implied by the other radioactive secondary abundances which sample a larger volume around the solar system. The densities indicated by the GCRs may imply significant propagation in a lower density galactic halo or propagation in a local ISM cavity (4).

The increase of the clock isotope abundances with increasing energy predicted by our LBM propagation model is consistent with the observed abundances of all clock species within the statistical limitations of the data. However, the data are also consistent with calculations from the diffusion model of Strong and Moskalenko (24), which shows a flatter energy dependence at CRIS energies. Variations in the energy dependence can result from the energy dependence in the isotopic fragmentation cross sections, which may contribute as much as 10% uncertainty to the predicted ISM density at the lowest energies probed by the CRIS data and less uncertainty at higher energies where more cross-section measurements have been made (6). Both the energy dependence and the value of the calculated density are dependent on the cho-

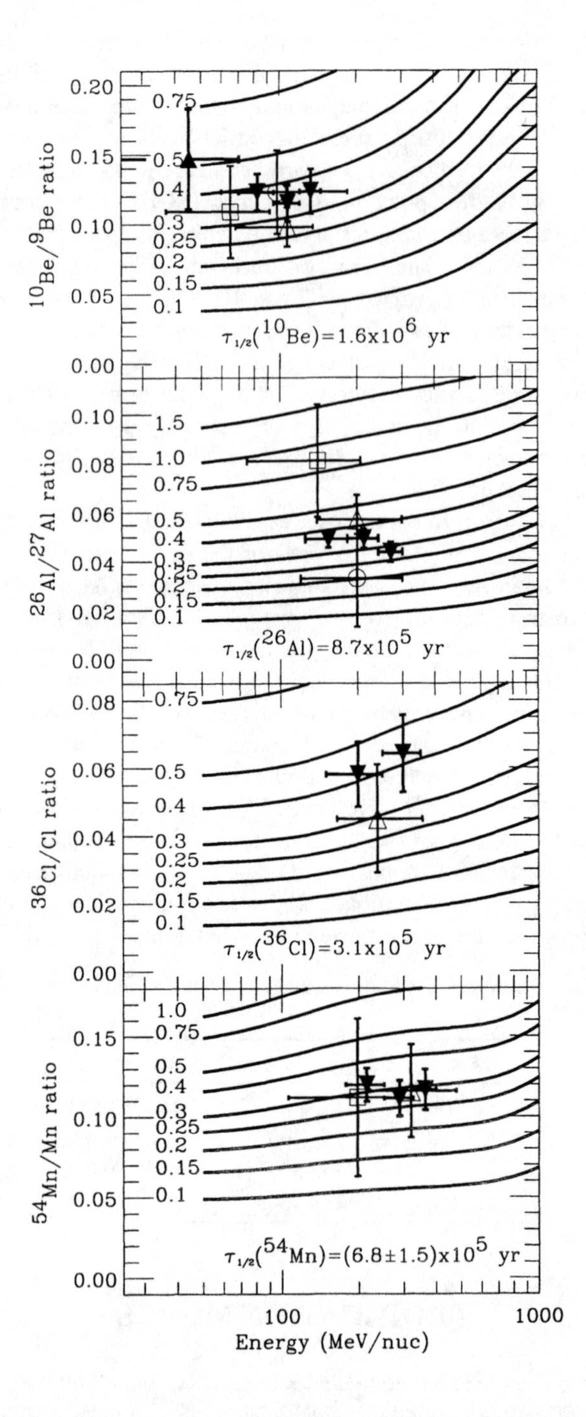

FIGURE 1. Secondary radionuclide abundance ratios from CRIS and previous experiments. Curves show predicted abundance ratios for different n_H in atoms/cm^3. An ISM helium component of n_{He}/n_H=0.11 has been assumed. Ratios from previous experiments have been adjusted to a level of solar modulation ϕ=325 MV for comparison. References for the data are as follows: *Filled triangle* this work; *Open circle* (16, 17); *Open square* (19, 18); *Open triangle* (13, 2, 14, 15)

sen parameterization of λ_{esc} within the formalism of the LBM.

Using the appropriate value of ρ_{ISM} for each species, the confinement times and statistical uncertaintites calculated from the CRIS clock abundances are presented in Table 2 along with previous results for comparison. The CRIS data imply a confinement time of $\tau_{esc} = 16.2 \pm 0.8$ Myr. This value is consistent with results from all four clock abundances to within two standard deviations, considering the systematic uncertainty in the halflife of ^{54}Mn in addition to the statistical uncertainty shown in Table 2. It should be noted that the CRIS abundance for ^{54}Mn is in good agreement with the Ulysses result (15). Although measured abundances from Ulysses and CRIS generally agree for all four clock species (Figure 1), the derived confinement times for ^{10}Be and ^{54}Mn are in slight disagreement. However, the higher value of τ_{esc} from ^{10}Be data measured by Ulysses cannot be explained by a difference in λ_{esc} in the respective models. We find $\lambda_{esc} \sim 6.7$ g/cm^2 from our model, which is higher in value than the pathlength used in the Ulysses analysis ($\sim$6.0 g/cm^2) and which should predict a higher value of τ_{esc} for similar ISM densities. This difference may result in part from a difference in partial fragmentation cross sections in the respective models.

Table 2. GCR Confinement Time

Clock	Experiment	τ_{esc}*(Myr)
^{10}Be	ACE/CRIS	$14.9^{+1.4}_{-1.2}$
	Ulysses (13)	$26.0^{+4.0}_{-5.0}$
	Voyager (19)	$27.0^{+19.0}_{-9.0}$
	ISEE-3 (16)	$8.4^{+4.0}_{-2.4}$
^{26}Al	ACE/CRIS	$18.0^{+1.2}_{-1.1}$
	Ulysses (15)	$19.0^{+3.0}_{-3.0}$
	Voyager (18)	$13.5^{+8.5}_{-4.5}$
	ISEE-3 (17)	$9.0^{+20.0}_{-6.5}$
^{36}Cl	ACE/CRIS	$13.8^{+2.5}_{-2.0}$
	Ulysses (2)	$18.0^{+10.0}_{-6.0}$
^{54}Mn	ACE/CRIS	$33.7^{+3.7}_{-3.2}$
	Ulysses (2)†	$\sim 11.0^{+4.7}_{-3.1}$

* The confinement times from experiments other than CRIS are compiled from the listed references.

† Recalculated from (14) using $\tau_{1/2} = 0.63$ Myr from (22).

CONCLUSION

Average ISM hydrogen number densities between $n_H = 0.28 - 0.43$ H atoms/cm^3 are necessary to account for the secondary radionuclides produced during propagation within the context of the leaky box model. A value of n_H=0.34±0.02 H atoms/cm^3 is consistent with all species to within two standard deviations. The surviving fractions of the secondary radionuclides indicate a mean confinement time for cosmic rays in the galaxy of $\tau_{esc} = 16.2 \pm 0.8$ Myr.

ACKNOWLEDGMENTS

This research was supported by the NASA at the California Institute of Technology (under grant NAG5-6912), the Jet Propulsion Laboratory, NASA/Goddard Space Flight Center, and Washington University.

REFERENCES

1. Simpson, J. A., *Ann. Rev. Nucl. Part. Sci.* **33**, 323 (1983).
2. Connell, J. J., DuVernois, M. A., & Simpson, J. A., *Ap. J. Lett.* **509**, L97 (1998).
3. Yanasak, N. E. et al., *Proc. 26th Int. Cosmic Ray Conf.* (Salt Lake City) **3**, 9 (1999).
4. Ptuskin, V. T., & Soutoul, A., *A&A* **337**, 859 (1998).
5. Stone, E. C. et al., *Space Sci. Rev.* **86**, 285 (1998).
6. Yanasak, N. E. et al., *Ap. J.* (in preparation) (2000).
7. Anders, E. & Grevesse, N., *Geochim. Cosmochim. Acta* **53**, 197 (1989)
8. Englemann, J.J., et al, *A&A*, **233**, 96 (1990)
9. DuVernois, M. A., and Thayer, M. R., *Ap. J.* **465**, 982 (1996).
10. Soutoul, A., and Ptuskin, V. T., *Proc. 26th Int. Cosmic Ray Conf.* (Salt Lake City) **4**, 184 (1999).
11. Krombel, K. E., & Wiedenbeck, M. E., *Ap. J.* **328**, 940 (1988)
12. Fisk, L. A., *JGR*, **76**, 221 (1971)
13. Connell, J. J. *Ap. J.* **501**, L59 (1998).
14. DuVernois, M. A., *Ap. J.* **481**, 241 (1997).
15. Simpson, J. A., & Connell, J. J., *Ap. J.* **497**, L85 (1998).
16. Wiedenbeck, M. E., & Greiner, D. E., *Ap. J. Lett.* **239**, L139 (1980).
17. Wiedenbeck, M. E., *Proc. 18th Int. Cosmic Ray Conf.* (Bangalore) **9**, 147 (1983).
18. Lukasiak, A., McDonald, F. B., & Webber, W. R., *Ap. J. Lett.* **430**, L69 (1994).
19. Lukasiak, A., et al., *Ap. J.* **423**, 426 (1994).
20. Lukasiak, A., McDonald, F. B., & Webber, W. R., *Ap. J.* **448**, 454 (1997b).
21. DuVernois, M. A., Simpson, J. A., and Thayer, M. R., *A&A* **316**, 555 (1996).
22. Wuosmaa, A. H., et al., *Phys. Rev. Lett.* **80**, 2085 (1998)
23. Zaerpoor, et al., *Phys. Rev. Lett.* **82**, 2219 (1999)
24. Strong, A. W., and Moskalenko, I. V. *Proc. 26th Int. Cosmic Ray Conf.* (Salt Lake City) **4**, 255 (1999).

Secondary Electron-Capture-Decay Isotopes and Implications for the Propagation of Galactic Cosmic Rays

S.M. Niebur[1], W.R. Binns[1], E.R. Christian[2], A.C. Cummings[3], J.S. George[3], P.L. Hink[1], M.H. Israel[1], J. Klarmann[1], R.A. Leske[3], M. Lijowski[1], R.A. Mewaldt[3], E.C. Stone[3], T.T. von Rosenvinge[2], M.E. Wiedenbeck[4], and N.E. Yanasak[4]

[1]*Washington University in St. Louis*
[2]*NASA/Goddard Space Flight Center*
[3]*California Institute of Technology*
[4]*Jet Propulsion Laboratory*

Abstract. We report the first observation of an energy dependence in the titanium, vanadium, and chromium isotopic abundances in cosmic rays. The observations were made in the 100 – 500 MeV/nucleon energy interval using data from the Cosmic Ray Isotope Spectrometer on the ACE spacecraft. The ^{51}Cr and ^{49}V isotopes in cosmic rays are produced by fragmentation of heavier cosmic ray nuclides and decay only by electron capture. The observations indicate that electron-capture decay occurred primarily at the lower energies measured and that there is a resulting energy dependence in the abundances of these isotopes and their decay products.

INTRODUCTION

As galactic cosmic rays propagate through the interstellar medium, they may gain and lose energy as they encounter dense areas of the interstellar medium, supernova remnant shocks, and turbulent magnetic fields. Energy gain due to an encounter with one or more remnant supernova shocks capable of accelerating particles to higher energies has been called distributed acceleration (1). Energy gain due to interaction of cosmic rays with turbulent magnetic fields has been termed diffusive acceleration (2). If cosmic rays undergo either distributed or diffusive acceleration, their energy as they enter the heliosphere could be higher than the energy to which they were initially accelerated.

Certain cosmic-ray isotopes can be used to trace whether the cosmic rays spent much of their lifetime at a lower energy than their energy at heliospheric entry. These isotopes, ^{37}Ar, ^{44}Ti, ^{49}V, ^{51}Cr, ^{55}Fe, and ^{57}Co, are produced by fragmentation of heavier cosmic rays and decay only by electron capture, which occurs primarily at lower energies where the probability of electron attachment (and thus subsequent decay) is larger. The laboratory half-lives of these isotopes range from 28 days (^{51}Cr) to 47 years (^{44}Ti); their half-lives with a single attached electron are approximately twice the laboratory values (3). Since any of these short-lived isotopes would have decayed during the $>10^5$ year delay between nucleosynthesis and initial acceleration indicated by Co and Ni isotopic ratios (4), these isotopes must have been produced by fragmentation of heavier cosmic rays during galactic propagation. These secondary cosmic rays are produced as bare nuclei that cannot decay by electron capture until they pick up an electron from the medium through which they propagate. At energies above several hundred MeV/nucleon the probability of electron attachment is low and these isotopes are essentially stable. If evidence of decay were observed at these higher energies, it may be a sign that the nuclei had once propagated at lower energies, where electron attachment and subsequent decay is possible, and thus must have experienced distributed acceleration. The energy dependence of nuclear cross sections, electron-attachment cross sections, ionization energy loss, and solar modulation all contribute to energy dependence and must be taken into account before firm conclusions about reacceleration are drawn.

CP528, *Acceleration and Transport of Energetic Particles Observed in the Heliosphere: ACE 2000 Symposium*, edited by Richard A. Mewaldt, et al.
© 2000 American Institute of Physics 1-56396-951-3/00/$17.00

The energy dependence of electron-capture isotopes and how they can be used to determine the effect of reacceleration has been discussed by Raisbeck *et al.* (5), Silberberg *et al.* (1, 6), Letaw *et al.* (7), and others. These authors describe distributed acceleration and suggest tests, including measurements of secondary electron-capture decay isotopes. In 1998, Soutoul *et al.* (8) analyzed a combined data set of ISEE-3 and Voyager vanadium isotopic measurements (vanadium isotopes were not clearly resolved in either set) to address this problem with a $^{51}V/^{49}V$ data point, thereby combining effects of the $^{51}Cr \rightarrow {}^{51}V$ and $^{49}V \rightarrow {}^{49}Ti$ decays. The CRIS data set is the first large enough (currently 20 times the number of ISEE-3+Voyager vanadium events used by Soutoul *et al.*) to examine the parent/stable and daughter/stable ratios over a range of energies for both the $^{51}Cr \rightarrow {}^{51}V$ and $^{49}V \rightarrow {}^{49}Ti$ decays; CRIS has already collected significant amounts of all isotopes involved. In this paper we will present Ti, V, and Cr isotopic measurements that suggest that electron-capture decay has occurred at the lower cosmic-ray energies.

CRIS DATA ANALYSIS

The Cosmic Ray Isotope Spectrometer (CRIS) on ACE detects cosmic rays with energy ~50 – ~600 MeV/nucleon and charge 2 to 30 (He to Zn) with excellent mass resolution (< 0.25 amu) and a large collecting power (geometrical factor of 250 cm^2-sr)(9). The CRIS instrument consists of a scintillating optical fiber trajectory (SOFT) hodoscope and four silicon solid-state detector telescopes. The SOFT hodoscope, which includes three *xy* scintillating-fiber planes and a trigger fiber plane, is used to determine trajectories of incident nuclei. The silicon detector telescopes, each a stack of fifteen 3-mm-thick silicon wafers, are used to determine the rate of energy loss and total energy of the detected nuclei.

The CRIS data set contains enough Ti, V, and Cr events to be divided into several energy bins over the 100 – 500 MeV/nucleon energy range (corresponding to interstellar energies between 200 – 1000 MeV/nucleon; the solar modulation parameter ϕ = 500 MV for this time period). After simple data cuts were applied to select a high resolution data set and minimize backgrounds, each of the seven bins we used contains 400 – 2500 events, depending on element and energy bin. The energy intervals measured for isotopes of neighboring elements differed by less than 3%.

The resulting histograms each have clear mass separation and an adequate number of particles, as shown in Figure 1. The abundances presented here were obtained by counting the number of particles in each peak, where the peaks are defined as separated at the lowest point of the valley between them. It is estimated that the error in the abundances introduced due to spill-over between adjacent isotopes is approximately 2%, small in comparison to the statistical errors shown in Figures 2, 3, and 4.

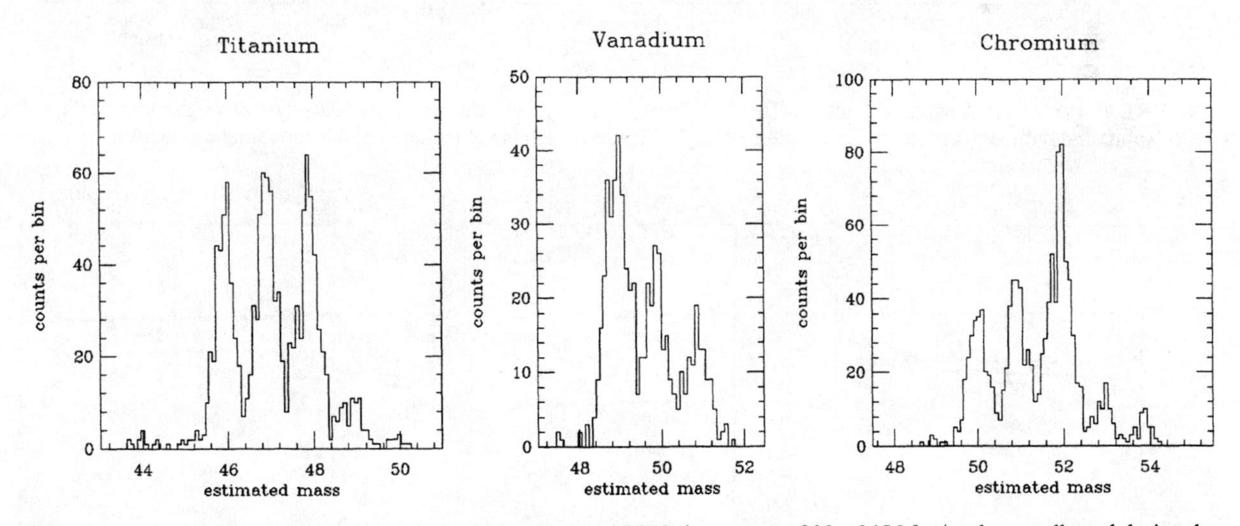

FIGURE 1. Titanium, vanadium, and chromium histograms of CRIS data, energy 300 – 345 Mev/nucleon, collected during the first 750 days of operation (bin size = 0.1 amu). The data was divided into seven energy bins; the other 18 histograms also have a mass resolution of <.25 amu.

RESULTS

The electron-capture-decay parent abundances, daughter abundances, and parent+daughter abundances relative to the abundances of nearby stable reference isotopes are shown in Figures 2 and 3. ^{52}Cr is used as the stable reference isotope for the ^{51}Cr → ^{51}V decay; the sum of the stable isotopes ^{46}Ti, ^{47}Ti, and ^{48}Ti is used for the ^{49}V → ^{49}Ti decay. In each case, the daughter isotope abundances increase with decreasing energy, indicating enhanced electron capture and decay at the lower energies. The abundances of the parent isotopes and the sum of the parent and daughter isotopes are consistent with a corresponding depletion in the parent isotopes at lower energies. Nearby stable isotopes do not show these trends; their abundances are not so energy-dependent. Therefore, the ^{51}Cr → ^{51}V and ^{49}V → ^{49}Ti data (particularly the daughter isotopic abundances) indicate that electron-capture decay has occurred more substantially in the lower energy cosmic rays.

These isotopes are compared in Figures 2 and 3 with results from a leaky-box propagation model (described in (10)) that uses energy-dependent nuclear cross sections (11) and solar modulation corrections (12) to produce predictions for observations at 1 AU. The dashed line shows the modeling results when electron attachment and subsequent decay are not allowed to occur at any energy. The solid line shows results using the same model but with energy-dependent electron attachment cross sections (13), allowing electron-capture decay to occur. The two model results converge at higher energies, as expected.

The effect of electron-capture decay would be more pronounced outside the heliosphere, before solar modulation. Adiabatic deceleration in the solar wind stochastically lowers the energy of the cosmic rays, spreading nuclei of a given energy over a range of lower energies (12, 14). Therefore, the cosmic-ray nuclei observed at a particular energy propagated through the Galaxy at a range of higher energies.

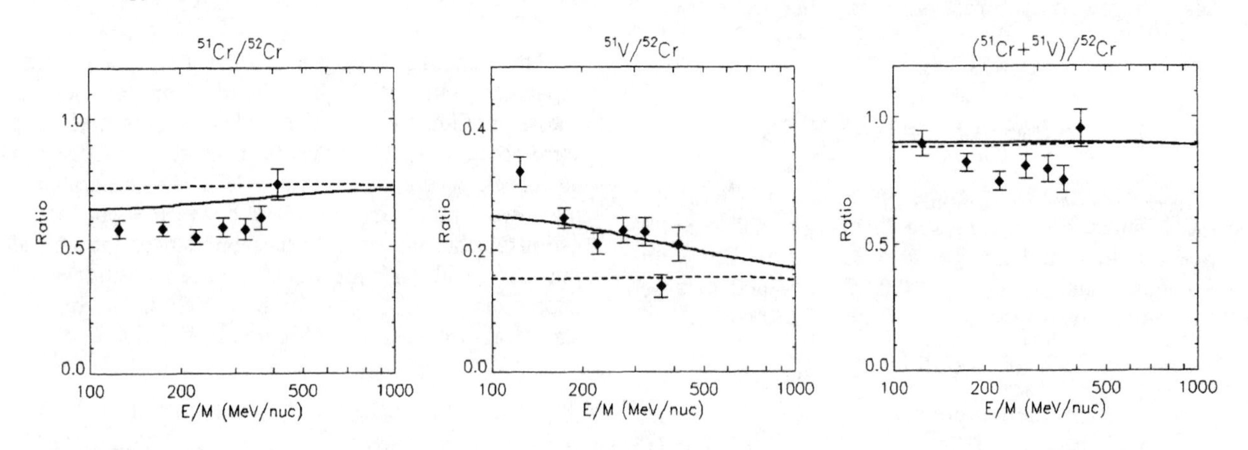

FIGURE 2. Parent and daughter abundances for the ^{51}Cr → ^{51}V decay. Plotted curves are leaky-box propagation model results, with (solid) and without (dashed) electron-capture decay. Reacceleration is not included in this modeling calculation.

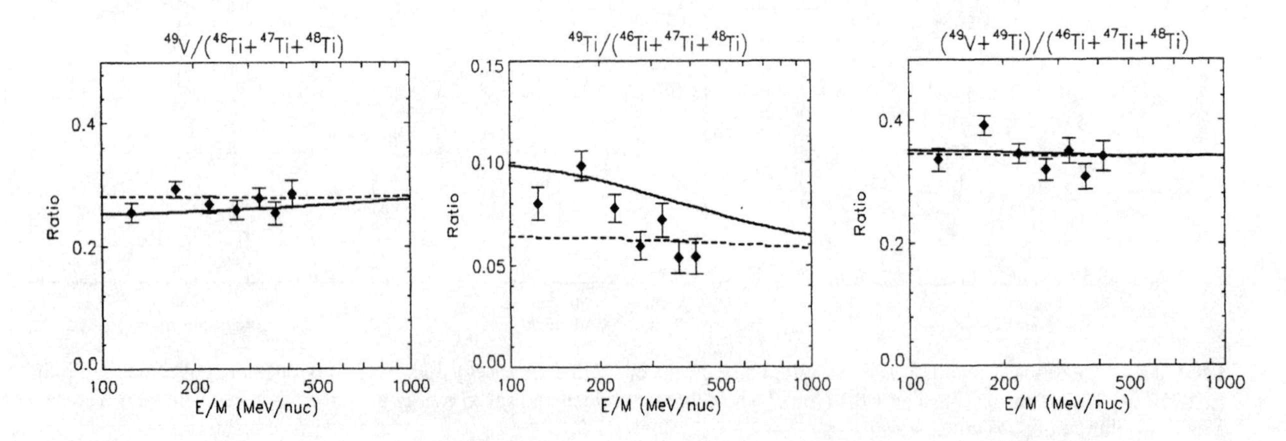

FIGURE 3. Parent and daughter abundances for the ^{49}V → ^{49}Ti decay. Plotted curves are leaky-box propagation model results, with (solid) and without (dashed) electron-capture decay. Reacceleration is not included in this modeling calculation.

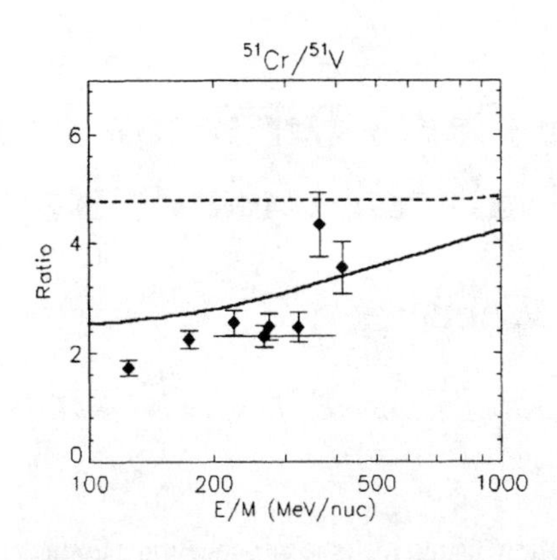

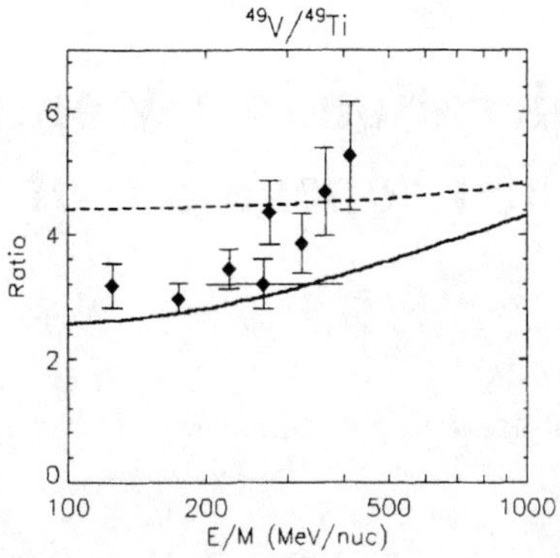

FIGURE 4. Parent/daughter ratios for $^{51}Cr \rightarrow {}^{51}V$ and $^{49}V \rightarrow {}^{49}Ti$. Plotted curves are the same as in figures 2 and 3. The points with horizontal bars show Ulysses results at the energy interval indicated (15).

Figure 4 shows the same data as parent/daughter ratios in order to compare the CRIS data with the recently reported data points of Connell and Simpson (15). The CRIS data in the same energy interval are consistent with the Ulysses data point for each of the two ratios. Although we prefer to examine the parent and daughter energy dependence separately to better isolate effects and check for internal consistency between the observations and the model, we note that these parent/daughter ratios also indicate that electron-capture decay has occurred more often at the lower energies.

DISCUSSION AND CONCLUSION

This preliminary data analysis indicates that electron capture and decay have occurred primarily at the lower cosmic-ray energies. These data are in qualitative agreement with our leaky-box propagation model that includes energy-dependent electron attachment cross sections (solid line in figures 2, 3, and 4), but does not include reacceleration during transport. The offset observed in several ratios may be due to errors in the nuclear cross sections.

We plan to incorporate reacceleration into our model and compare our data on the secondary electron-capture decay isotopes (^{37}Ar, ^{44}Ti, ^{49}V, ^{51}Cr, ^{55}Fe, and ^{57}Co) and their decay products with this and other models in order to quantify the amount of acceleration that could have occurred during cosmic ray propagation through the interstellar medium. As CRIS continues to collect data, the statistical uncertainties will be reduced, allowing more stringent comparisons with various model results. We may also be able to use these data to better evaluate the effects of solar modulation.

ACKNOWLEDGMENT

This research was supported by the National Aeronautics and Space Administration at the California Institute of Technology (under grant NAG5-6912), the Goddard Space Flight Center, the Jet Propulsion Laboratory, and Washington University.

REFERENCES

1. Silberberg, R. et al., *Phys. Rev. Lett.* **51**, 1217-1220 (1983).
2. Heinbach, U., and Simon, M., *Astrophys. J.* **441**, 209-221 (1995).
3. Wilson, L. W. Ph.D. thesis. University of California at Berkeley. Lawrence Berkeley Laboratory Report LBL-7723 (1978).
4. Wiedenbeck, M. E. et al., *Astrophys. J.* **523**, L61-L64 (1999).
5. Raisbeck, G. M. et al., *Proc. 14th Internat. Cosmic Ray Conf. (Munich)* **2**, 560-563 (1975).
6. Silberberg, R., and Tsao, C. H., *Phys. Reports* **191**(6), 351-408 (1990).
7. Letaw, J. R., Silberberg, R., and Tsao, C. H., *Astrophys. J. Supp.* **56**, 369-391 (1984).
8. Soutoul, A. et al., *Astron. Astrophys.* **336**, L61-L64 (1998).
9. Stone, E. C. et al., *Space Sci. Rev.* **88**(1), 285-356 (1998).
10. Leske, R. A., *Astrophys. J.* **405**, 567-583 (1993).
11. Webber, W. R. et al., *Phys. Rev. C.* **58**(6), 3539-3552 (1998).
12. Goldstein, M.L., Fisk, L. A., and Ramaty, R. *Phys. Rev. Lett.* **25**, 832-835 (1970).
13. Crawford, H. J. Ph.D. Thesis. University of California at Berkeley. Lawrence Berkeley Laboratory Report LBL-8807 (1979).
14. Labrador, A. W., and Mewaldt, R. A., *Astrophys. J.* **480**, 371-376 (1997).
15. Connell, J. J., and Simpson, J. A., *Proc. 26th Internat. Cosmic Ray Conf. (Salt Lake City)* **3**, 33-36 (1999).

Reacceleration in a Monte Carlo Diffusion Model for the Propagation of Galactic Cosmic Rays

W.R. Webber

New Mexico State University, Department of Astronomy, Las Cruces, NM 88003

Abstract. We have examined the effects of reacceleration using realistic propagation parameters for the galactic matter density, halo size, diffusion coefficient, etc., in a Monte Carlo diffusion model. Taking a diffusion coefficient, K, $\sim P^{0.33}$ along with reacceleration we can reproduce the observed B/C and sub Fe/Fe ratios using a rigidity dependence of dP/dt similar to that used earlier by Seo and Ptuskin. In this case the ratios at low energies can be explained without a break in K (a plus), but the observed ratios at high energies are steeper than calculated (a minus). The primary/primary ratios such as H/He, He/O, etc., are changed significantly ,thus requiring different assumptions for the source abundance ratios. The amount of reacceleration needed to reproduce the observed B/C ratio is 35% of the particle's energy below 1 GeV/nuc. We have searched for more decisive examples that this level of reacceleration is actually occurring. The behavior of the $^2H/^4He$ ratio at low energies and the K-capture nuclei ratios are two possible tests that are examined.

INTRODUCTION

One of the more elusive questions involving cosmic ray propagation in the Galaxy is the role and magnitude of cosmic ray reacceleration. Here reacceleration is taken to be an additional (slower) acceleration after an assumed initial rapid burst of acceleration wherein the cosmic rays receive most of their energy up to at least 10^{14} ev. This initial burst of acceleration must be very rapid because it completely overpowers the not insignificant energy loss by ionization for $Z \sim 26$ nuclei or the rapid E loss by synchrotron emission for 10^{12} ev and above electrons. After this initial period of acceleration the cosmic ray nuclei will continuously lose energy by ionization loss as they propagate through the IS medium producing certain characteristic spectral features, unless reacceleration plays a significant role. Thus the real key in trying to understand the role of reacceleration is not so much how much energy the particles gain by reacceleration as it is to determine how much the well known ionization energy loss is modified by this reacceleration. Since this reacceleration is presumably a rather slow stochastic process its spatial dependence may be similar to the galactic matter and magnetic field distribution. Ptuskin and colleagues have discussed the effects of reacceleration within the framework of the Leaky Box model (LBM). In this paper we follow the discussion in (1, 2).

CALCULATION OF THE EFFECT OF REACCELERATION ON THE B/C AND SUB Fe/Fe RATIOS

The original arguments supporting the significance of reacceleration were based on its effect on the B/C ratio (3). In our Monte Carlo (MCM) calculation (4), we assume that the diffusion coefficient is $\sim P^{0.33}$ instead of the usual $P^{0.50}$. We also take the momentum (energy) dependence of dP/dt to be $\sim P^{2.0\text{-}0.33}$, and the thickness of the accelerating region to be = 0.6 Kpc all approximating the calculations of (1, 2). The magnitude of the reacceleration term is referenced to the energy loss term for Carbon at 1 GeV/nuc in the calculation. The matter density = 1.0 $\exp^{-0.25/Z}$ cm^{-3}, where Z is the distance in Kpc perpendicular to the galactic plane and the value of the boundary distance, z_B = 3.0 Kpc. These are the same values used in the normal MCM propagation calculation. We can reproduce a best fit to the B/C ratio for an E gain = (2.5 ±0.5) x E loss (^{12}C) at 1 GeV/nuc as shown in Figure 1 (Curve B). Also shown in this figure is the B/C ratio obtained in the reacceleration model of (1) (curve A), as well as two samples of a Leaky Box model calculation using slightly different values for the path length (curves L and W). All four curves provide equally good fits to the HEAO data on the B/C ratio. For the reacceleration models this is accomplished without a break in the diffusion coefficient at low rigidities (a plus) but the observed energy dependence

CP528, *Acceleration and Transport of Energetic Particles Observed in the Heliosphere: ACE 2000 Symposium*, edited by Richard A. Mewaldt, et al.
© 2000 American Institute of Physics 1-56396-951-3/00/$17.00

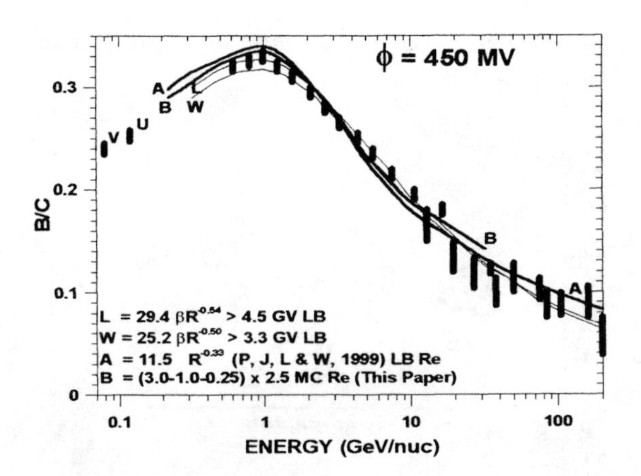

FIGURE 1. The B/C ratio calculated using the simple LB model (curves L and W) and with reacceleration in the LB model (1) curve A and in the MCM (this paper) curve B. Data is from (5, 6, 7)

of the ratios (particularly the sub Fe/Fe ratio not shown here because of space limitations) is steeper than calculated (a minus).

Our calculation provides an estimate of the actual total amount of reacceleration required. In the energy range from ~0.1 – 1 GeV/nuc this is ~35% of the particle's observed IS energy, independent of energy. Thus a particle at a current IS energy of 316 MeV/nuc will have gained ~110 MeV/nuc energy. This requires a source of energy ~1/3 as large as that responsible for the initial acceleration. During this time this particle will have lost ~200 MeV/nuc by ionization loss from its initial energy.

This competition between E-loss and E-gain will significantly distort the energy spectra below ~1 GeV/nuc, a point often overlooked in earlier calculations. Deviations as large as a factor of 2 from a simple $P^{-2.30}/\beta$ normally propagated spectrum (no reacceleration) will occur e.g., for the ^{12}C spectrum. Significant changes in the primary/primary ratios such as H/He, He/O, etc., will also occur. These spectral changes due to reacceleration can be compensated for, to some extent, by different source abundance ratios, and different source spectra but overall the introduction of reacceleration seems to introduce as many (or more) complications than it solves. What is needed are more decisive examples to show that this level of reacceleration is occurring. In what follows we consider two possibilities, the $^2H/^4He$ ratio at low energies and the K-capture nuclei ratios.

EFFECT OF REACCELERATION ON THE $^2H/^4He$ RATIO AND ON THE K-CAPTURE NUCLEI RATIOS

Here we consider two possible sets of ratios that might give a more unambiguous clue as to the magnitude of reacceleration. The 1st is the $^2H/^4He$ (or $^2H/^3He$) ratio. ^2H in cosmic rays is mostly produced as secondaries from cosmic ray ^4He and ^3He reactions with IS H and He. However, the reaction $^1H_{CR} + ^1H_{IS} \rightarrow ^2H + \pi+$ also produces a significant flux of ^2H at low energies (8). This reaction has a maximum cross section of several mb, rather sharply peaked at an energy of H_{CR} ~600 MeV. The resulting ^2H spectrum has a broader but still well defined peak at ~130 MeV/nuc. In the absence of reacceleration this reaction would produce a peak in the IS $^2H/^4He$ ratio after propagation at an energy ~100 MeV/nuc or lower. Interplanetary energy loss would move this peak down to still lower energies and thus it would be unobservable within the heliosphere. The calculated $^2H/^4He$ ratio without reacceleration (curve MC) and with the same reacceleration that reproduces the B/C ratio is shown in Figure 2 (curve MC + Re). These

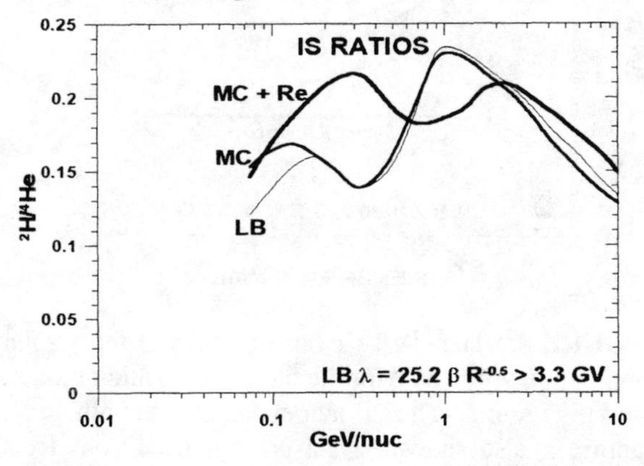

FIGURE 2. The IS $^2H/^4He$ ratio calculated for three propagation models: 1) LBM (λ=25.2 $\beta R^{-0.50}$>3.3 GV) (curve LB); 2) MCM as described in (9) (curve MC); 3) MCM with reacceleration (curve MC + Re)

two ratios are significantly different because of the large variations in the IS ratio caused by relatively large energy dependent cross section variations for the production of ^2H between ~300-1000 MeV/nuc and the aforementioned $^1H + ^1H$ reaction at lower energies. Notice that in the model with reacceleration the H + H bump at low energies is moved up to ~200-300 MeV/nuc by reacceleration. It should thus now be observable at low solar modulation levels. We plan to study this ratio further including the effects of solar modulation to see whether perhaps the Voyager spacecraft measurements could distinguish the differences shown in Figure 2 at energies <100 MeV/nuc at low modulation levels (ϕ ~ 200 MV).

There are several K-capture nuclei ratios that can also be examined for the effects of reacceleration. In the examples here we take the ratios $^{51}V/^{51}Cr$ and $^{49}Ti/^{49}V$. In Figure 3 we show the $^{51}V/^{51}Cr$ ratio calculated in the LBM, with ϕ = 400 MV and with K-capture, and no K-capture. For the MCM we also calculate the ratio with ϕ=400 MV with no reacceleration and then using the same reacceleration necessary to reproduce the B/C ratio. We see that there is a significant difference between the predictions of the LBM and the MCM for the same value of modulation. The effect of K-capture decay is much less in the MCM. We believe that this is due to the fact that in the MCM, this decay only takes place in the immediate disk whereas in the LBM decay occurs throughout the propagating volume because of the uniform density.

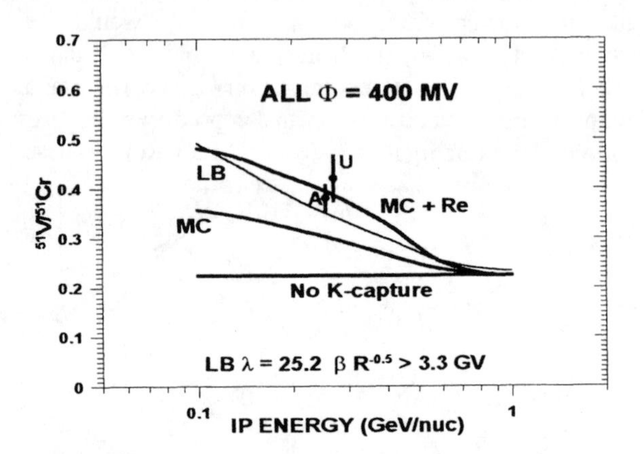

FIGURE 3. The $^{51}V/^{51}Cr$ ratio calculated for a solar modulation = 400 MV for the three propagation models used in Figure 2. The LB model calculation with no K-capture is also shown. Data point labeled A is from ACE (10), point labeled U is from (11)

The $^{51}V/^{51}Cr$ ratio observed by ACE (10) and Ulysses (11) is much larger than the prediction for a modulation of ϕ = 400 MV in the MCM although the ratio agrees more closely with the predictions of a simple LBM. With the addition of reacceleration in the MCM the predicted ratio agrees well with the measurements.

For the $^{49}Ti/^{49}V$ ratio shown in Figure 4 the conclusions are different. Here the measured ratio is less than the predictions of the LBM but agrees better with the predictions of the MCM without reacceleration. Reacceleration in the MCM now makes the calculated ratio larger than the measurements. So the conclusions based on this ratio are almost opposite to those based on the $^{51}V/^{51}Cr$ ratio. Perhaps more detailed measurements of these ratios now being made in several energy intervals by ACE experimenters will help to clarify this difference. Or perhaps cross section uncertainties are contributing to the differences between the predictions and observations of the two ratios examined here. If this is the case uncertainties as large as ± 20% in both the cross sections into ^{49}Ti and ^{51}V are apparently necessary.

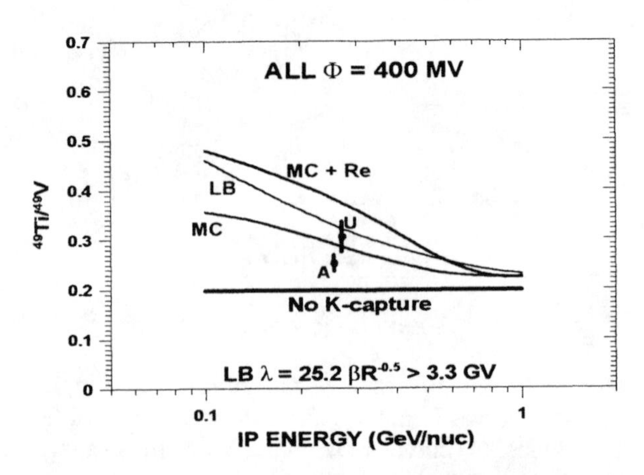

FIGURE 4. THE $^{49}Ti/^{49}V$ ratio calculated for a solar modulation = 400 MV for the three propagation models used in Figure 2. The LB model calculation with no K-capture is also shown. Data point references are the same as Figure 3.

We stress that these results on K-capture nuclei are preliminary. Further calculations with different propagation parameters are underway in the MCM to try and understand the origin of the differences between the LB and MC models.

ACKNOWLEDGEMENTS

This work would not have been possible without the computer support of Dr. John Rockstroh and Jason Peterson. This research is supported as part of the ACE Guest Investigator Program by NASA Grant NAG 5-7795.

REFERENCES

1. Ptuskin, V.S., et al., *Proc. 26th ICRC*, **4**, 291-294, (1999)
2. Seo, E.S. and Ptuskin V.S. , *Ap.J.*, **431**, 705-714, (1994)
3. Heinbach, U. and Simon M, *Ap.J.*, **441**, 209-221, (1993)
4. Webber, W.R. and Rockstroh, J.M., *Adv. Space Res.*, **19**, 817-820, (1997)
5. Engelmann, et al., *A & A*, **233**, 96-111, (1990)
6. Chappell, J.H. and Webber, W.R., *Proc. 17th ICRC*, **2**, 59-64, (1981)
7. Swordy, S.P., et al., *Ap.J.*, **349**, 625-636, (1990)
8. Webber, W.R., *Adv. Space Res.*, **19**, 755-758, (1997)
9. Webber, W.R., *These proceedings*, (2000)
10. Niebur, S.M., et al., *These proceedings*, (2000)
11. Connell, J.J. and Simpson, J.A., *Proc. 26th ICRC*, **3**, 33-36, (1999)

Galactic Cosmic Ray Neon Isotopic Abundances Measured on ACE

W.R. Binns[1], M.E. Wiedenbeck[2], E.R. Christian[3], A.C. Cummings[4], J.S. George[4], P.L. Hink[1], J. Klarmann[1], R.A. Leske[4], M. Lijowski[1], S.M. Niebur[1], R.A. Mewaldt[4], E.C. Stone[4], T.T. von Rosenvinge[3], and N.E. Yanasak[2]

1. Washington University, St. Louis, MO 63130 USA
2. Jet Propulsion Laboratory, Pasadena, CA 91109 USA
3. NASA/Goddard Space Flight Center, Greenbelt, MD 20771 USA
4. California Institute of Technology, Pasadena, CA 91125 USA

Abstract. Measurements of neon isotopic abundances from the ACE-CRIS experiment are presented. Abundances have been obtained in six energy intervals over the energy range of ~100≤E≤280 MeV/nucleon. These measurements are compared with the ACE-SIS data for lower energies extending down to ~8 MeV/nucleon. We find that the CRIS $^{22}Ne/^{20}Ne$ abundance ratio at the source is a factor of 5.0±0.3 greater than for the solar wind. The CRIS measured abundances agree well with previous experiments. The CRIS and SIS measurements are in good agreement for the higher SIS energies. However for the lower energy SIS data, where the anomalous cosmic rays (ACR) are being sampled, the ratio decreases and agrees with solar wind abundances. The implications of these results are discussed.

INTRODUCTION

The $^{22}Ne/^{20}Ne$ ratio in the galactic cosmic rays (GCR) has been shown by several experiments (1-4) to be overabundant when compared to solar wind (SW) abundances (5), which is the sample of solar system (SS) matter taken to best represent the composition of material from which the solar system formed (6). Cassé and Paul (7) suggested that this overabundance could be due to the presence of an admixture of Wolf-Rayet (WR) star material mixed into the general GCR source material. Woosley and Weaver (8) developed a "supermetallicity" model that recognizes that the synthesis of neutron-rich isotopes in massive stars is directly proportional to their initial metallicity (fraction of elements heavier than He). If GCRs originate in a region of the Galaxy with higher metallicity than the SS, this could result in an overabundance of $^{22}Ne/^{20}Ne$. However, an overabundance of the neutron rich isotopes of Mg and Si should also be observed, and that is not the case. Olive and Schramm (9) suggest that instead of the GCR $^{22}Ne/^{20}Ne$ ratio being anomalously high, it is actually the solar system that is anomalously low. This could result if supernovae in the near vicinity at the time of formation of the solar system injected large amounts of ^{20}Ne into the pre-solar nebula. These models are reviewed by Mewaldt (10). Recently Soutoul and Legrain (11) have suggested that in diffusion models GCRs may come preferentially from the inner Galaxy if the cosmic ray density is greater toward the center of the Galaxy. (Gamma-ray observations weakly support a GCR density gradient (12)). In this model the GCRs sample material preferentially from higher metallicity regions closer to the galactic center than the SS.

In this paper we present measurements of the Ne isotopes obtained by the Cosmic Ray Isotope Spectrometer (CRIS) instrument on the ACE spacecraft (13). We have obtained the $^{22}Ne/^{20}Ne$ ratio energy dependence for the energy range 100≤E/M≤281 MeV/nucleon and compare these data with those from the ACE-Solar Isotope Spectrometer (SIS) instrument (14) for lower energies. At the lowest energies (~8-35 MeV/nucleon) SIS measures the anomalous cosmic rays (ACR) that are believed to be a sample of the very-local interstellar-medium (VLISM). Thus we can compare the Ne isotopic composition in the GCR source, which is believed to be a sample of the ISM over a significant part of the galactic disk, with that of the VLISM. The CRIS abundance ratios are compared with results from other experiments. We also derive the $^{22}Ne/^{20}Ne$ source ratio using ^{21}Ne measured abundances as a "tracer" (15) for secondary production of the neon isotopes.

CP528, *Acceleration and Transport of Energetic Particles Observed in the Heliosphere: ACE 2000 Symposium,* edited by Richard A. Mewaldt, et al.
© 2000 American Institute of Physics 1-56396-951-3/00/$17.00

MEASUREMENTS

In Figure 1a-f we show mass histograms of the CRIS neon data for events in 6 energy bins spanning the energy range of 100<E/M<281 MeV/nucleon.

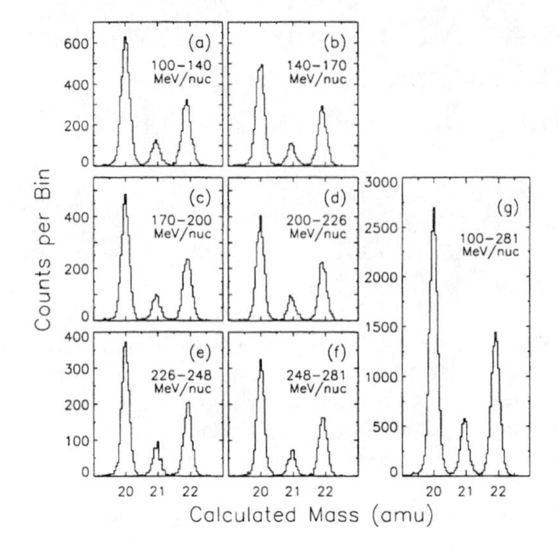

FIGURE 1. Mass histograms of neon events in 6 energy bins (a-f). Figure 1g is the sum of all events in Figs. 1a-f.

Figure 1g shows combined data from the 6 energy intervals. These events are a selected, high-resolution data set that were collected from Jan. 21, 1998 through Sept. 24, 1999. The primary selection was to use only events with angle ≤25° relative to the detector normal. This selected data set corresponds to ~35% of the good events for all angles. The mass resolution is sufficiently good (0.10 amu for events plotted in Figure 1) so that there is only a very small overlap of the mass distributions for adjacent masses. To obtain abundances, we have taken mass cuts at the minimum between peaks and simply counted events. The number of counts in the full data set is 3.0×10^4. The statistical accuracy of the CRIS and SIS data are such that the neon isotopic abundances as a function of energy can be studied in greater detail than was previously possible.

In Figure 2 we plot the $^{22}Ne/^{20}Ne$ ratio measured by CRIS (circles) and by SIS (16) (squares) respectively. (The CRIS measured values are corrected so that the E/M range for each data point is equal for the neon isotopes). We also show the SW abundance ratio as a horizontal dotted line (5,6). The CRIS GCR $^{22}Ne/^{20}Ne$ ratio is approximately constant with a small decrease in the ratio toward lower energies, and is considerably enhanced over the SW ratio. SIS measures a mix of GCRs (dominant at the higher SIS energies) that have a high $^{22}Ne/^{20}Ne$ ratio, and ACRs (dominant at the lower SIS energies) that have a ratio similar to SW (16,17). Thus we see that there is a dramatic difference between the CRIS GCR and the low-energy SIS ACR measurements.

The $^{21}Ne/^{20}Ne$ ratios measured by CRIS are plotted in Figure 3. The data show a roughly constant ratio with a slight decline at lower energies as was the case for $^{22}Ne/^{20}Ne$. The SW $^{21}Ne/^{20}Ne$ ratio is very low (2.4×10^{-3} (5,6)) and the much higher abundance of ^{21}Ne in GCRs is thus believed to be almost entirely secondary (i.e. resulting from nuclear interactions of primary cosmic rays during propagation). This makes it possible to use ^{21}Ne as a nearby "tracer" to estimate the fractions of ^{22}Ne and ^{20}Ne that are secondary GCR nuclei (15).

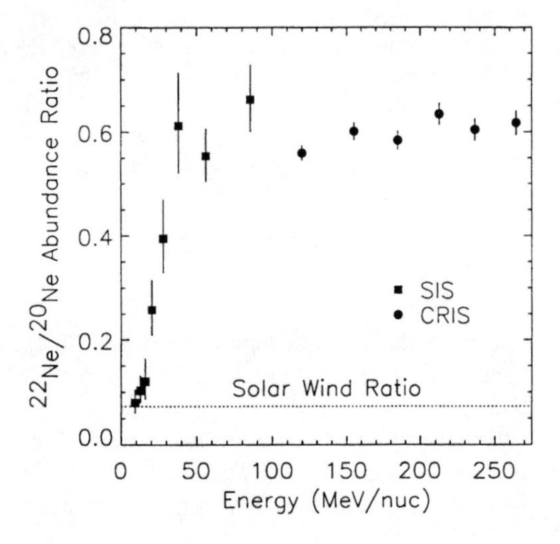

FIGURE 2. The $^{22}Ne/^{20}Ne$ ratio is plotted as a function of energy for CRIS (circles) and SIS (16) (squares).

In Figure 4 we plot the CRIS Ne isotopic ratios averaged over the full energy range (taken from Fig 1g) and compare them with measurements made by other experiments (1-4). We have adjusted the ratios of the other experiments to correspond to the CRIS modulation level (~500 MV). The CRIS measurements have smaller statistical uncertainties than previous experiments (the CRIS error bars are smaller than the data points), and there is reasonably good agreement with those measurements.

We have corrected for secondary production during propagation using the tracer method (15) to obtain the source abundance ratio for $^{22}Ne/^{20}Ne$. In the tracer method the observed abundances of secondary nuclides such as ^{21}Ne, ^{19}F, or ^{17}O are scaled to determine the secondary contribution to the observed ^{22}Ne, and thereby to derive the ^{22}Ne source abundance.

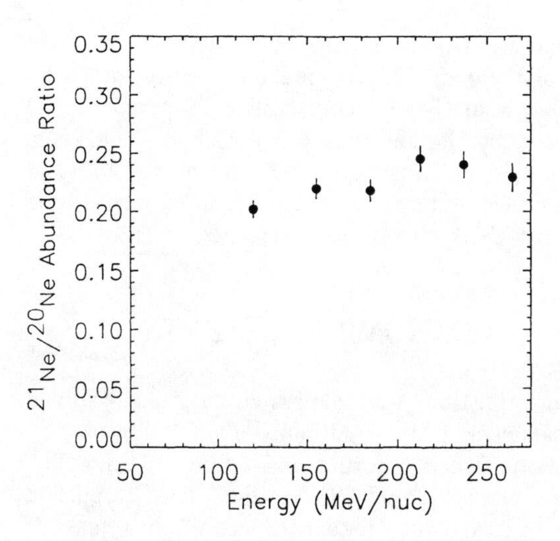

FIGURE 3. The $^{21}Ne/^{20}Ne$ ratio measured by CRIS is plotted as a function of energy.

A series of leaky-box propagation calculations was carried out using various assumed escape mean free paths. Figure 5 shows the calculated relationship between the $^{22}Ne/^{20}Ne$ source ratio and the abundance of the secondary nuclide ^{21}Ne at Earth predicted by the model (plotted as a difference from the observed value). The point on the abscissa corresponding to a 0% difference for ^{21}Ne at Earth gives the best estimate of the ^{22}Ne source abundance using that secondary as a constraint. Source abundances obtained using ^{19}F and ^{17}O secondary tracers were also obtained (not shown) and averaged with the ^{21}Ne value to obtain a best overall ratio. The measurement uncertainty in the ^{21}Ne abundance is represented by the vertical dotted lines in Figure 5. The points at which they intersect the plotted line indicate the uncertainty in the $^{22}Ne/^{20}Ne$ ratio due to the uncertainty in the ^{21}Ne measurement.

In Figure 6 we plot the preliminary CRIS source abundance of $^{22}Ne/^{20}Ne$ that is 0.366±0.006±0.014. The source abundance ratio was calculated by averaging the source ratio estimates obtained from the three tracer isotopes (^{19}F and ^{17}O tracers both give source abundance ratios of 0.375 compared to 0.350 for the ^{21}Ne tracer source estimate in Figure 5). The first uncertainty in the abundance ratio is the quadratic sum of the $^{22}Ne/^{20}Ne$ ratio uncertainty from counting statistics and the uncertainty from counting statistics for ^{21}Ne (Figure 5). The second uncertainty is our estimate of the systematic uncertainty calculated using the "sample standard deviation" obtained from the three tracer isotope estimates. We note that the differing source ratio estimates from the three different tracer isotopes and the derived systematic uncertainty are comparable to that expected from nuclear cross-section uncertainties. The error bar on the CRIS data

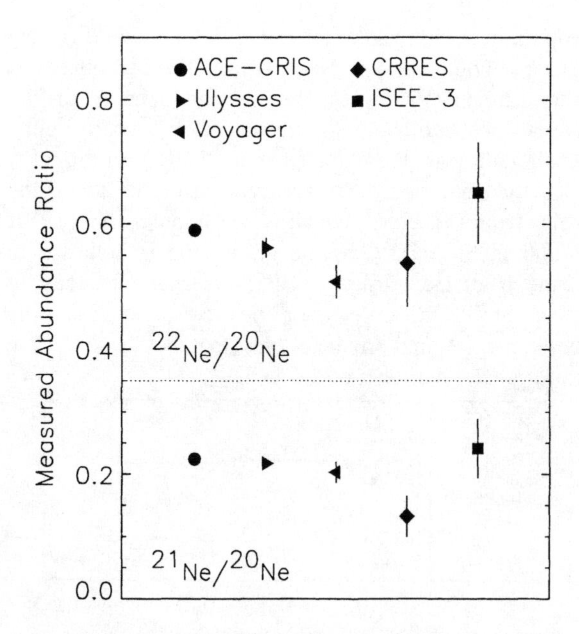

FIGURE 4. Ratios of $^{22}Ne/^{20}Ne$ and $^{21}Ne/^{20}Ne$ measured by ACE-CRIS (circle), Ulysses (2)(right triangle), Voyager (3)(left triangle), CRRES (4)(diamond), and ISEE-3 (1)(square). Ratios have been adjusted for solar modulation (4).

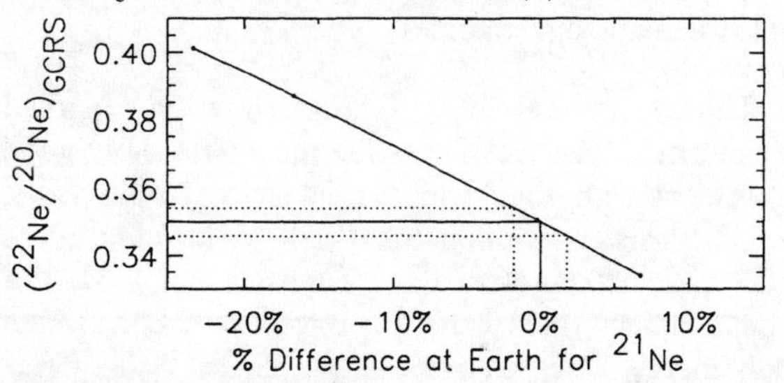

FIGURE 5. The plotted curve shows the value of the $^{22}Ne/^{20}Ne$ source ratio as a function of percent difference between the calculated value of the tracer isotope ^{21}Ne and our measured value. The horizontal dotted lines indicate the uncertainty in the $^{22}Ne/^{20}Ne$ source estimate due to our measurement uncertainty for ^{21}Ne (vertical dotted lines).

point in Figure 6 is the linear sum of the statistical and systematic uncertainties.

DISCUSSION

The values for the SW (5), which are believed to give the best estimate of the solar system neon abundances (6), a range of values for SEPs (18), and meteoritic abundances (19) are plotted in Figure 6. The GCR $^{22}Ne/^{20}Ne$ ratio is 5.0±0.3 times the SW value of 0.073 (5). Leske et al. (17) give similar results obtained by SAMPEX. Neon, unlike most other elements, is observed to have a number of separate populations in the solar system with distinctly different isotopic compositions. Ne-A is found in carbonaceous chondrite meteorites and may be pre-solar in origin. Ne-B is believed to result from solar wind implantation in grains and has a $^{22}Ne/^{20}Ne$ ratio very close to that of contemporary measurements of the solar wind (5). Ne-E, which is nearly pure ^{22}Ne (19) is found in SiC and Graphite grains and is believed to come from He burning AGB stars and the decay of ^{22}Na produced in SN or novae. It is the only component found in meteorites that has a $^{22}Ne/^{20}Ne$ ratio greater than that of the GCRs.

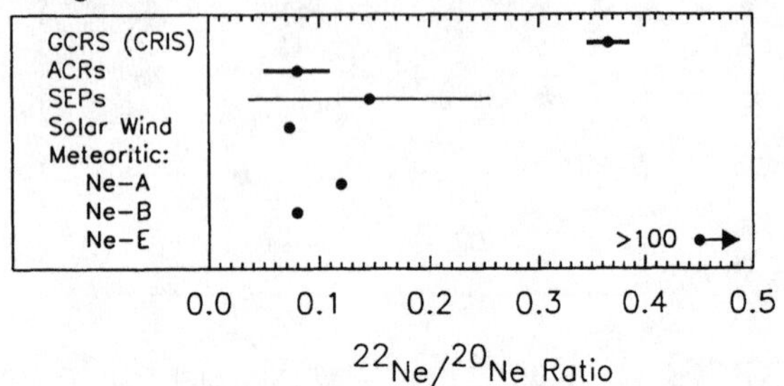

FIGURE 6. $^{22}Ne/^{20}Ne$ source abundance for CRIS (GCR) and SIS (ACR) data compared to SW (5,6), SEPs (18) and meteoritic abundances (19). The error bar plotted for CRIS is the linear sum of the statistical and systematic uncertainties.

The fact that the GCRs show a marked enhancement in $^{22}Ne/^{20}Ne$ while the ACRs show good agreement with solar wind abundances is remarkable and is consistent with the SAMPEX results (17). If the ACR ratio is representative of the general ISM, this would indicate that there is a large overabundance of high-energy ^{22}Ne relative to ^{20}Ne in the Galaxy (i.e. cosmic rays) compared to low energy matter (i.e. dust grains and gas). We note that it is not known at present to what extent the ACRs actually represent the general ISM. It is also important to note that the Solar System may not have formed in the local ISM. Based on the higher metallicity of the sun relative to that of nearby stars, Wielen et al (20) suggest that the sun formed ~2kpc closer to the Galactic center than our present position at Galactocentric radius of ~8.5kpc.

This overabundance in the GCRs could result from the acceleration of WR star material (7) since the high velocity winds which result in mass losses of $\sim 10^{-5}$ solar masses per year expel large quantities of He-burning material rich in ^{22}Ne (21,22). Since these massive stars have short lifetimes on average before they become supernovae ($\sim 10^5$ y) (23) some of this material may be swept up and accelerated by nearby SN, or by the supernova shock from the star that ejected the material in the first place. This would occur without substantial mixing into the ambient ISM since ~90±10% of WR stars are believed to exist in OB associations within superbubbles formed in giant molecular clouds (21) (based upon observations by van Dyk et al (24) on core collapse supernovae and comments by Higdon et al. (23)). Thus it would seem that WR stars in superbubbles might be an ideal astrophysical setting for WR material to provide the enhanced $^{22}Ne/^{20}Ne$ observed in GCRs.

ACKNOWLEDGMENTS

This research was supported by the National Aeronautics and Space Administration at Washington University, the California Institute of Technology (under grant NAG5-6912), the Jet Propulsion Laboratory, and the Goddard Space Flight Center. It was also supported by the McDonnell Center for the Space Sciences at Washington University.

REFERENCES

1. Wiedenbeck, M.E., and Greiner, D.E., *Phys. Rev. Lett.*, **46**, 682, 1981.
2. Connell, J.J., and Simpson, J.A., *Proc. of the 25th International Cosmic Ray Conference*, **3**, 381, 1997.
3. Lukasiak, A. et al, *ApJ*, **426**, 366, 1994.
4. DuVernois, M.A. et al, *ApJ*, **466**, 457,1996.
5. Geiss, J., *Proc. of the 13th International Cosmic Ray Conference*, **5**, 3375, 1973.
6. Anders, E, and Grevesse, N., *Geochim. Et Cosmochim. Acta* **53**, 197, 1989.
7. Cassé, M., and Paul, J.A., *ApJ*, **258**, 860, 1982.
8. Woosley, S.E., and Weaver, T.A., *Ap.J.*, **243**, 651,1981.
9. Olive, K.A., and Schramm, D.N., *Ap.J.*, **257**, 276, 1982.
10. Mewaldt, R.A., in *Cosmic Abundances of Matter*, Ed. By C.J. Waddington, AIP Conference Proc. **183**, 124, 1988.
11. Soutoul, A., and Legrain, R. *Proc. of the 26th International Cosmic Ray Conference*, **4**, 180, 1999.
12. Hunter, S.D., et al, *ApJ* **481**, 205 (1997).
13. Stone, E.C., et al, *Space Science Reviews*, **86**, 285, 1998.
14. Stone, E.C. et al, *Space Science Reviews* **86**, 357, 1998.
15. Stone, E.C., and Wiedenbeck, M.E., *ApJ* **231**, 606, 1979.
16. Leske, R.A. et al, *Proc. of the 26th International Cosmic Ray Conference*, **7**, 539, 1999.
17. Leske, R.A. et al, *Space Sci. Revs.* **78**, 149, 1996.
18. Leske, R.A. et al, *GRL*, **26**, 2693, 1999.
19. Ozima, M., and Podosek, F.A., *Noble Gas Geochemistry*, Cambridge: Cambridge University Press, 1983.
20. Wielen, R., Fuchs, B., and Dettbarn, C, A, *Astron. Astrophys.*, **314**, 438, 1996.
21. Maeder, A. and Meynet, G., *Astron. Astrophys.*, **278**, 406, 1993.
22. Chiosi, C, and Maeder, A, *Ann. Rev. Astron. Astrophys*, **24**, 329, 1986.
23. Higdon, J.C., Lingenfelter, R.E., and Ramaty, R., *ApJ*, **509**, L33, (1998), and Private Communication.
24. Van Dyk, S.D. et al, *AJ*, **111**, 2017, 1996.

^{22}Ne excess in cosmic rays from the inner Galaxy.

A. Soutoul and R. Legrain

DAPNIA/Sap, SPhN CEA/SACLAY 91191 Gif sur Yvette CEDEX, France

Abstract. The abundances of heavy cosmic ray nuclei at the solar system are calculated in a 3 dimension galactic model with cylindrical symmetry and with diffusive confinement. In this model the overabundance of ^{22}Ne is energy dependent. A comparison with HEAO data is suggestive of a radial dependent source of ^{22}Ne in the galaxy steeper than the source of the other heavy cosmic rays.

INTRODUCTION

The isotopic anomalies of heavy ions abundances at the cosmic ray source have received various explanations none of them being widely accepted yet. The overabundance of neutron rich isotopes of neon at the cosmic ray source was first reported by Maehl et al. [1-2]. The ^{22}Ne overabundance at the cosmic ray source is now well established [3-7]. Note that this overabundance is with respect to the solar (either solar wind or meteoritic neon A) abundance. Also note it is obtained with one-dimensional exponential model for cosmic ray propagation.

It is generally accepted that ^{22}Ne which is copiously produced in Wolf Rayet (WR) stars and further expelled in the strong stellar winds of these stars can ultimately be injected in the cosmic ray accelerator provided this accelerator is situated in the vicinity of the star [8]. Both the steady shock from the powerful stellar wind and that from the subsequent supernova can accelerate particles to cosmic ray energies [8]. ^{12}C may also be overabundant in cosmic rays from WR stars. Its overabundance could be biased by elemental selection effects [9]

^{22}Ne synthesis is also expected to be enhanced in regions of the inner galaxy with metallicities higher than solar [10]. Both stars and the interstellar medium have a higher than local metallicity. Woosley and Weaver [10] note that ^{22}Ne synthesis scales with metallicity the radial dependence of which is approximately given by:

$$\mathrm{Log}_{10}(Z/Z_\odot) \approx -.07(r-R_\odot) \qquad (1),$$

where Z is the mass fraction of elements heavier than helium at distance r from the galactic center and $Z_\odot \approx .02$ is its value at the solar system. Here $R_\odot = 8.5$ kpc. The enhancement of ^{22}Ne in the super metallic model is too small (see Table 1 in [10]) to account for the overabundance of ^{22}Ne observed in low energy cosmic rays [3-7].

However the increase of WR stars towards the inner galaxy is steeper than the corresponding increase of O stars [11]. Further WR stars in the WC phase contribute to the local galactic enrichment in ^{12}C and ^{22}Ne [12]. Maeder and Meynet [12] have proposed a model in which cosmic rays are extracted from inner galactic matter originating from WR and supernovae. Their model possesses attractive features from both previous models.

In this contribution we discuss primary to primary ratios with nuclear charges $6 \geq z \geq 14$ in the energy range ~ 1-30 GeV/N in cosmic rays related to the ^{22}Ne anomaly at low energy. We use a simple galactic model for the confinement of cosmic rays. The overabundance of a primary isotope at the solar system results from the steeper than average radial dependence of its abundance at the cosmic ray source.

CP528, *Acceleration and Transport of Energetic Particles Observed in the Heliosphere: ACE 2000 Symposium,* edited by Richard A. Mewaldt, et al.
© 2000 American Institute of Physics 1-56396-951-3/00/$17.00

CALCULATION

Model

Diffusive confinement of cosmic rays takes place in the cylindrical galactic volume with radius R and half-thickness H_h. Interstellar gas and cosmic ray sources are distributed in a disk with half-thickness H_g. Isotropic diffusion of cosmic rays is assumed with uniform values of the diffusion coefficients D_g in the gas disk and D_h in the halo. Cosmic rays freely escape at the boundary of the volume. For simplicity the gas density in the halo is taken equal to zero and $D_g = D_h = D$. Here R=20 kpc and H_h=7.3 kpc. The half column density of gas at the solar system is $\delta = 7.2*10^{20}$ nucleons.cm^{-2}. The gas density depends on the radial distance to the galactic center. $\chi(r)$ the density of cosmic ray sources reads:

$$\chi(r) = f(r)g(r) \qquad (2),$$

where f(r) is the supernovae rate at radial distance r from the galactic center and the trial function g(r) reads:

$$g(r)=\exp(-\mu(r-R_\odot)) \qquad (3),$$

where the radial gradient μ is a free parameter. The supernovae rate and the radial dependent gas density are as in [13] (Fig. 1).

Solution

The solution for the equilibrium density of cosmic ray nuclei as a function of position in the galactic volume was given in this model with no continuous (ionization) losses by Ginzburg et al. [14], see also [15]. It holds for uniform density of the interstellar gas in the disk. Since the gas density is not constant we make use of a perturbation method, where the solution of Ginzburg et al. is the main term. The solution is obtained in the form of a series [13] and is not presented here for lack of space. The interstellar density of a primary isotope is calculated with the source term given by eq. 2-3. This density (more precisely its mean value in the gas disk at each radial distance r) is then used in the source term for calculating the density of the secondary isotope (see [13] for details). This provides the equilibrium density of both one primary nucleus and its secondary product as a function of the radial distance to the galactic center.

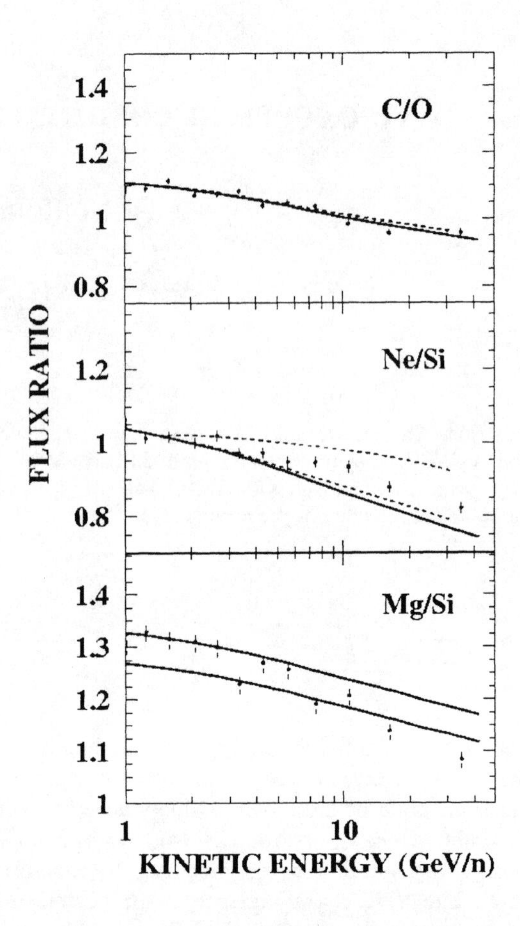

FIGURE 1. Flux ratios as a function of kinetic energy. The solid lines are leaky box calculations and dashed lines are galactic model calculations normalized to the leaky box result at low energy. From top to bottom: a) C/O, dashed curve is for μ_c-μ_o =0.023 where μ_c and μ_o are the radial gradients of the cosmic ray sources of carbon and oxygen. b) Ne/Si dashed curves are for two overabundances of ^{22}Ne at ~1 GeV/N with the galactic model (2.7 lower dashed curve and 8. upper dashed curve). c) Mg/Si. The full curves are for 6 % difference of the Mg/Si ratios at the cosmic ray source. Data points: Ref. [17].

Results and discussion

We examine primary to primary ratios with charges $6 \geq z \geq 14$ in the energy range ~ 1-30 GeV/N. For this we first compare the outcome from a leaky box calculation to the observation (Fig. 1). In this calculation the energy dependencies at the cosmic ray source are identical for all isotopes and elements. We make use of the latest cross sections [16] to calculate the boron to carbon (B/C) ratio which closely agrees to the B/C ratio in the HEAO data [17]. From this one gets the mean pathlength for an exponential distribution, traversed by cosmic rays which reads $\lambda_e = 30\ \beta^{1.5}\ \rho^{-.55}$ g/cm^2 where β is the ratio of the interstellar velocity of particles to the velocity of light and ρ is their rigidity. This value is valid above ~ 1 GeV/N. It is very close to that obtained by Ptuskin et al. [18]. In Fig 1 the full lines are the result of this calculation where appropriate relative abundances at the cosmic ray source are adopted to match the HEAO data towards the low energies. Clearly the agreement between the calculated C/O ratio and the observation is excellent. For Ne/Si and Mg/Si the agreement although good is not as good as for the C/O ratio. The observed Ne/Si ratio is flatter than the calculated one while the observed Mg/Si ratio is steeper than the calculated one. These deviations have been already noted [19-20]. The full curves of Fig 1-c are for a 6 % difference of the Mg/Si ratios at the cosmic ray source.

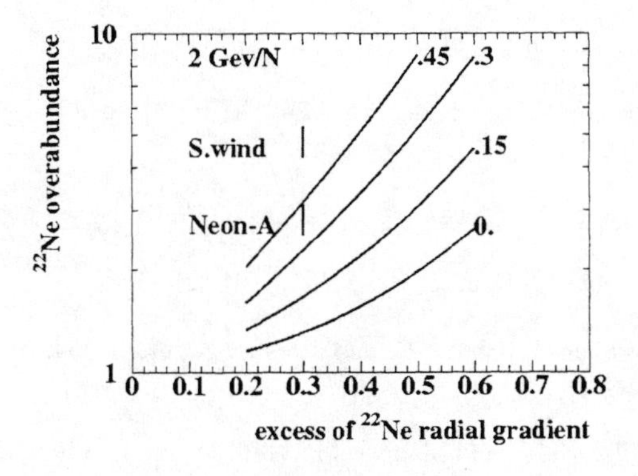

FIGURE 2. ^{22}Ne overabundance as a function of the excess of ^{22}Ne gradient μ_{22} -$\mu_{o,}$ in the galactic model for different values of the gradient μ_0 (0., 0.15, 0.3, 0.45)and at kinetic energy 2 GeV/N. Overabundance in solar wind and in Ne-A adapted from [4] and [7].

Now we consider the galactic model. It is known that the one dimensional diffusion model and the leaky box are formally equivalent provided that the diffusion coefficient D and the escape length λ_e are such that $D\lambda_e = \delta\beta c m_n H_h$, so that D ~ $\rho^{+.55}$, where m_n is the mean mass of a nucleon of the interstellar medium, with 90 % H and 10 % He [18]. In the galactic model the dependence of D with ρ is still close to $\rho^{+.55.}$ As ρ and D increase the mean distance between cosmic ray sources and the solar system is increasing as well up to ~ H_h the halo size. This conclusion is valid in all the range of radial gradients considered here. This results in a splitting of the energy spectra calculated with the galactic model and the leaky box with identical source spectra, unless $\chi(r)$ ~ constant (eq.2). These small spectral deviations although interesting are not discussed here. In the galactic model splitting between spectra of isotopes and elements with distinct $\chi(r)$ can also occur. As observed above there is no indication of a splitting between carbon and oxygen spectra. The agreement of the calculated leaky box energy dependence of the C/O ratio with the data is so good that one may question the need of an inner galactic origin of cosmic rays.

In the inner galaxy the contribution from massive stars to the C/O ratio moderately depends on metallicity (Table 2 and 3 in [12], Table 10 in [21]). Models of galactic evolution yield deviations as small as .01 dex.kpc^{-1} (μ_c-μ_o ~.023) between the gradients of C and O in the inner galaxy [22]. Such small deviations are within the uncertainties of the C/O data in Fig1-a. Here we determine the value of D at which the amount of boron from primary oxygen in the galactic model is equal to its value in the leaky box calculation at the same energy for $\mu = \mu_o$ =.3 (eq. 3). Next the calculation is performed with the same value of D for primary carbon with $\mu = \mu_c$ =.323. The corrected C/O ratio after normalization to the leaky box calculation at low energy is shown in Fig 1-a (dashed curve). It is seen that the same gradient μ_o (the same function $\chi(r)$) holds for the main progenitors of boron (but for ^{22}Ne μ_{22}-$\mu_o \geq 0.$, see below).

^{22}Ne overabundance from WR stars is nearly constant at $Z \geq Z_\odot$ (Table 2 and 3 in [12], Table 10 in [21]). Within 2 kpc of the Sun the ratio of the number of WR stars N_{wr} to the number of OB stars N_{ob} scales as exp(-(μ_{wr} -μ_{ob}) r) with μ_{wr} -$\mu_{ob} \leq .3$ [11-12]. The ^{22}Ne abundance in interstellar matter from the sole WR stars increases more steeply than that of C and O and than metallicity in the inner galaxy.

The ^{22}Ne density in cosmic rays at the solar system with gradients $\mu_{22} \geq \mu_o$ is calculated in the galactic model. The overabundance is defined as the ratio of intensity of ^{22}Ne at μ_{22} to its intensity at $\mu_{22} = \mu_o$ and at the same energy. This ratio is plotted in figure 2 as a function of μ_{22} -μ_o for various values of μ_o (0., 0.15, 0.30,.0.45). For comparison we show the combined determinations of the observation and uncertainties from Voyager and Ulysses [4,7} with respect to solar wind and to Neon A (the overabundance from Ulysses has been converted by us from Neon A to solar wind). To compare the excess gradient in cosmic rays to the excess gradient of ^{22}Ne in galactic matter we assume this last one is ~μ_{wr} -μ_{ob} ~ .3. We see from Fig. 2 that values of μ_o larger than the gradient due to the sole metallicity .16 (.07 dexkpc^{-1}) are demanded. Values of the source gradient as small as μ_o ~0. for carbon and oxygen are excluded. Since the source $\chi(r)$ is unknown we have chosen in this exploratory work the simple dependence in eq 2-3. Further, heavy nuclei are overabundant in cosmic rays. The energy in the charge range $z \geq 6$ is a small fraction of the total. But more energy than locally *per* supernova is required with μ_o

> 0. Here we have discussed gradients from metallicity only, but other dependencies can be present in $\chi(r)$. For example some dependence on the amount of gas may also be present [13]. Alternately it is possible that some (or all) ^{22}Ne from WR stars in cosmic rays is injected in the cosmic ray accelerator prior to dilution in the surrounding interstellar matter [8].

In Fig. 1-b we have plotted the Ne/Si ratio as a function of energy from our leaky box calculation (full curve). In the galactic model the ^{22}Ne overabundance in cosmic rays increases with energy. The Ne/Si ratio is flatter accordingly as a function of energy. This ratio, normalized to the leaky box result at low energy, is shown on Fig. 1-b for two overabundances of ^{22}Ne at this energy (2.7 lower dashed curve and 8. upper dashed curve, both with μ_o =.3). The dashed curves bracket the HEAO data. This is suggestive of a radial dependent source of ^{22}Ne in the galaxy steeper than the source of the other heavy cosmic rays. On Fig. 1-c the Mg/Si ratio from the same leaky box calculations is shown and compared to the HEAO data. (upper full curve). The lower full curve results from a similar calculation with 6 % less magnesium at the cosmic ray source. The HEAO data suggests that the abundance of neutron rich isotopes of magnesium is lower in the inner solar circle than locally. Further, the SN+WR model predicts an excess of neutron rich isotopes of Mg larger than observed at low energies [4,7,12]. New isotopic data are desirable at intermediate energies to clarify the questions discussed in this work.

REFERENCES

1. Maehl R. C., Fisher A. J., Hagen F. A. and Ormes J. F., *Ap.J.* **202**, L. 119, (1975).
2. Fisher A. J. et al., 1976, *Ap. J.* **205**, 938, (1976).
3. Wiedenbeck M. E. and Greiner D. E., *Ap. J.* **247**, L.119, (1981).
4. Connel J. J. and Simpson J. A., Proc. 25rd ICRC 3, pp 381-384, (1997).
5. Lukasiak A. et al., *Ap. J.* **426**, 366, (1994).
6. Duvernois M. A.,et al., *Ap. J.* **466**, 457, (1996).
7. Webber W. R. et al., *Ap. J.* **476**, 766, (1997).
8. Cassé M. and Paul J., 1982, *Ap. J.* **258**, 860, (1982).
9. Meyer J. P., Drury L. O'C. and Ellison D. C., *Ap.J.* **487**, 182, (1997).
10. Woosley S. E. and Weaver T. A., *Ap. J.* **243**, 651, (1981).
11. Van Der Hucht K. A. et al., 1988, *A&A,* **199**, 217, (1988).
12. Maeder A. and Meynet G., *A&A* **278**, 406, (1993).
13. Soutoul A. and Legrain R., 1999, Proc. 26th ICRC 3, pp 180-183.
14. Ginzburg V. L., Khazan Ya. M. and Ptuskin V. S., *Astrophys. And Space Sci.* **68**, 295, (1980).
15. Berezenskii V. S. et al., *Astrophysics of cosmic rays*, North Holland., 1990, p. 48
16. Webber W. R. et al.., *Ap. J.* **508**, 949, (1998).
17. Engelmann J. J. et al., *A&A* **233**, 96, (1990).
18. Ptuskin V. S. et al., Proc. 26th ICRC 3, pp 291-294, (1999).
19. Webber W. R. et al., *Ap. J.* **348**, 611, (1990).
20. Webber W. R. et al., *Ap. J.* **392**,L. 91, (1992).
21. Portinari l., Chiosi C. and Bressan A., *A&A* **334**, 505, (1998).
22. Wilson T. L. and Matteucci F., *A&A review* **4**, (1992).

ERRATUM

In [13], Eq. 2 and Eq. 6 read:

$$N(z,r) = N_m(z,r) + \sum_k N_{k+1}(z,r) \quad (2)$$

$$N_{k+1}(z,r) = \int_0^R - n_1(r_0) v\sigma < N_k(r_0) > \Phi(r,z,r_0) dr_0$$

(6)

On the Low Energy Decrease in Galactic Cosmic Ray Secondary/Primary Ratios

A.J. Davis[1], R.A. Mewaldt[1], W.R. Binns[3], E.R. Christian[2], A.C. Cummings[1], J.S. George[1], P.L. Hink[3], R.A. Leske[1], T.T. von Rosenvinge[2], M.E. Wiedenbeck[4], N.E. Yanasak[4]

[1] *California Institute of Technology, Pasadena, CA 91125*
[2] *Goddard Space Flight Center, Green Belt, MD 20771*
[3] *Washington University in St. Louis, MO 63130*
[4] *Jet Propulsion Laboratory, Pasadena, CA 91109*

Abstract. Galactic cosmic ray (GCR) secondary/primary ratios such as B/C and (Sc+Ti+V)/Fe are commonly used to determine the mean amount of interstellar material through which cosmic rays travel before escaping from the Galaxy (Λ_{esc}). These ratios are observed to be energy-dependent, with a relative maximum at $\sim$ 1 GeV/nucleon, implying a corresponding peak in Λ_{esc}. The decrease in Λ_{esc} at energies above 1 GeV/nucleon is commonly taken to indicate that higher energy cosmic rays escape more easily from the Galaxy. The decrease in Λ_{esc} at energies < 1 GeV/nuc is more controversial; suggested possibilities include the effects of a galactic wind or the effects of distributed acceleration of cosmic rays as they pass through the interstellar medium. We consider two possible explanations for the low energy decrease in Λ_{esc} and attempt to fit the combined, high-resolution measurements of secondary/primary ratios from $\sim$ 0.1 to 35 GeV/nuc made with the CRIS instrument on ACE and the C2 experiment on HEAO-3. The first possibility, which hypothesizes an additional, local component of low-energy cosmic rays that has passed through very little material, is found to have difficulty simultaneously accounting for the abundance of both B and the Fe-secondaries. The second possibility, suggested by Soutoul and Ptuskin, involves a new form for Λ_{esc} motivated by their diffusion-convection model of cosmic rays in the Galaxy. Their suggested form for $\Lambda_{esc}(E)$ is found to provide an excellent fit to the combined ACE and HEAO data sets.

INTRODUCTION

The simplest form of the leaky box model of GCR propagation (1) is characterized by an exponential path-length distribution (PLD) for the GCRs. The mean of this PLD is $\Lambda_{esc}(E)$ (where E denotes energy/nucleon). The energy dependence of $\Lambda_{esc}(E)$ is a free parameter in the model, and GCR secondary/primary ratios and spectra can be used to deduce empirically the form of this energy dependence. At high energies, the form of $\Lambda_{esc}(E)$ is constrained by the high-precision HEAO3-C2 data from a $\sim$ 1 GeV/nucleon up to $\sim$ 35 GeV/nucleon (2). Depending on the shape of the GCR source spectra input to the model, one finds $\Lambda_{esc} \propto R^{\alpha}$, where $\alpha \simeq -0.6$ (see, eg. (3), (4)). This decrease in $\Lambda_{esc}(E)$ at high energies is generally taken to indicate that GCRs escape more easily from the Galaxy at higher energies.

Below $\sim$ 1 GeV/nucleon, the GCR secondary/primary ratios are observed to decrease, thus implying a decrease in $\Lambda_{esc}(E)$ from a peak occurring at $\sim$ 1 GeV/nucleon (see e.g. (3), (4)). Various theoretical explanations exist for a decrease in $\Lambda_{esc}(E)$ towards lower energies, but none have gained widespread acceptance (see (5) and references therein). Many studies have simply introduced an artificial break in $\Lambda_{esc}(E)$ at 1-2 GeV/nucleon without theoretical justification. For example, a commonly used form is $\Lambda_{esc}(E) \propto \beta^{\gamma}$, for $R < R_0$, where R_0 is a constant rigidity typically chosen to be $\sim 5GV$, and $\gamma \sim$ 1 to 4.

The Cosmic Ray Isotope Spectrometer (CRIS) on NASA's Advanced Composition Explorer (ACE) spacecraft has made the most precise measurements to date of the solar minimum energy spectra of GCR nuclei with Z = 4 to 28 in the energy range 50 - 500 MeV/nucleon. The data considered here (taken during the period August 1997 to April 1998) provide significantly tighter constraints on the form of $\Lambda_{esc}(E)$ at low energies than were previously possible. The data can also be used to test new interpretations of the observed low-energy decrease in secondary/primary ratios. In this paper, we attempt to avoid artificial fits to $\Lambda_{esc}(E)$ and investigate whether introducing a second, low-energy source consisting of cosmic rays that have passed through very little matter (i.e. a relatively local source of cosmic rays) can reproduce the observations without invoking a decrease in $\Lambda_{esc}(E)$

CP528, *Acceleration and Transport of Energetic Particles Observed in the Heliosphere: ACE 2000 Symposium,*
edited by Richard A. Mewaldt, et al.
© 2000 American Institute of Physics 1-56396-951-3/00/$17.00

at low energies. We also consider a new form for $\Lambda_{esc}(E)$ suggested by Soutoul and Ptuskin (6) that is motivated by a galactic diffusion-convection model of GCR propagation in the Galaxy.

STANDARD LEAKY BOX MODEL

We first attempt to fit the CRIS and HEAO3-C2 composition and energy spectra using a leaky box model with no local source of cosmic rays. We use a steady-state model based on the formalism of Meneguzzi, Audouze, and Reeves (1), described previously in (7), and (8). The model includes the effects of escape from the Galaxy, energy losses in and nuclear interactions with the ISM, and decay of radioactive species. The increased energy losses due to the ionized fraction of hydrogen in the ISM are accounted for as in (9), and the ISM is assumed to be 90% hydrogen and 10% helium by number. The source spectra for all GCRs are taken to be power laws in momentum, with indices of -2.35 ($dQ/dE \propto P^{-2.35}$), and the source elemental abundances used are within 5% of the source abundances quoted in the HEAO3-C2 analysis of Engelmann et al. (2). Energy-dependent partial cross sections for the GCRs were calculated using the semi-empirical cross section formulae of Silberberg, Tsao and Barghouty (10), scaled to experimental data from (11) and (12) and references therein. Total inelastic cross sections were taken from (13) and references therein. Solar modulation was calculated using the spherically symmetric model of Fisk (14), with a solar wind speed of 400 km/s, a diffusion coefficient $\kappa(R) = \kappa_0\beta R$, where R is the rigidity and β the velocity of the cosmic ray, and a modulating volume extending to 120 AU.

We have investigated a variety of proposed forms for $\Lambda_{esc}(E)$ but find that a parameterization proposed by Soutoul and Ptuskin (6) in the context of a diffusion-convection model of GCR propagation in the Galaxy provides the best fit to the data:

$$\Lambda_{\mathrm{esc}} = \frac{29.5\beta}{(\beta R)^{0.6} + (\beta R/1.3\mathrm{GV})^{-2.0}} \quad (1)$$

Figures 1 and 2 show the results of the model fit to the B/C and (Sc+Ti+V)/Fe secondary/primary ratios, and the energy spectra. Also shown is the energy dependence of $\Lambda_{esc}(E)$ used to obtain the fit. Equation 1 gives the usual $R^{-0.6}$ dependence at high energies. At low energies, $\Lambda_{esc}(E) \propto \beta^3 R^2$, a significantly stronger energy-dependence than used in many previous studies (e.g. (3) or (4)). This behavior at low energies is a consequence of the assumed presence of a galactic wind that convects cosmic rays from the Galaxy with a convection velocity which is a non-monotonic function of distance from the galactic plane. This form provides an excellent fit to the

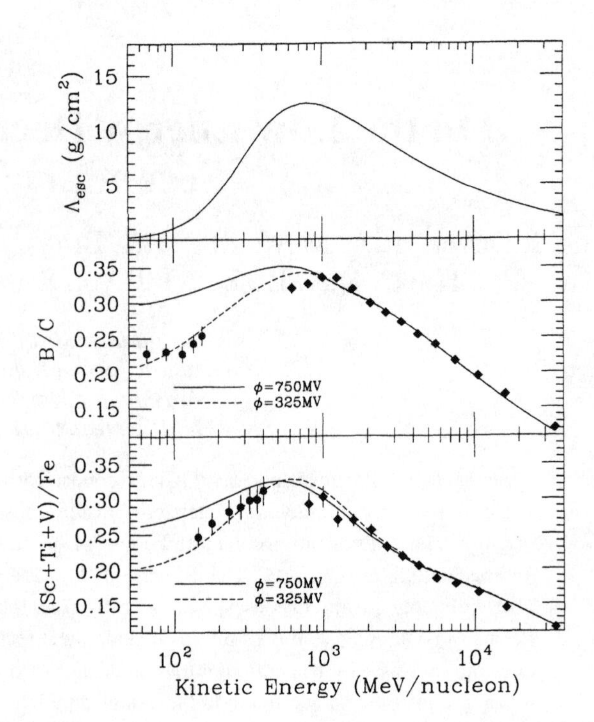

FIGURE 1. Leaky box model fits to CRIS (circles) and HEAO3-C2 (diamonds) B/C and (Sc+Ti+V)/Fe secondary/primary ratios. The top panel shows the energy dependence of $\Lambda_{esc}(E)$ used to obtain the fit (Equation 1).

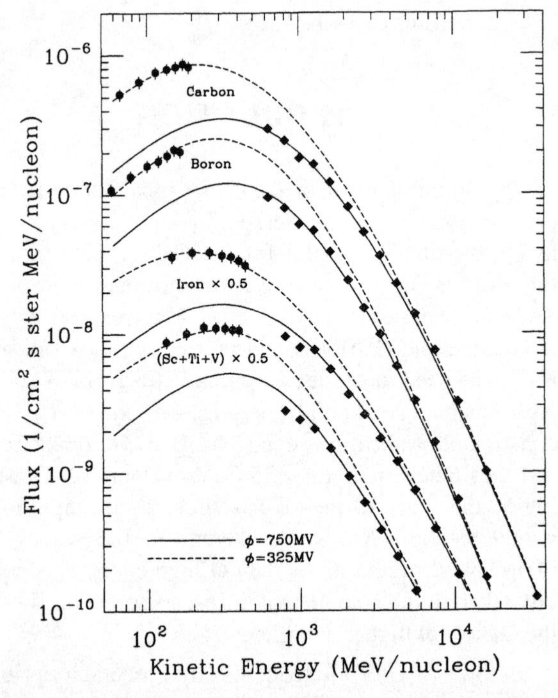

FIGURE 2. Leaky box model fits to CRIS (circles) and HEAO3-C2 (diamonds) B, C, Fe and (Sc + Ti + V) spectra, using $\Lambda_{esc}(E)$ as in Equation 1.

new, high-precision CRIS data, and also data from previous space experiments, such as (7) and (15). Note that a strong energy-dependence of $\Lambda_{esc}(E)$ at low energies was also used in the analysis of (7).

The CRIS data were obtained during solar minimum, while the HEAO3-C2 data were obtained during intermediate solar modulation conditions in 1980, so the two data sets cannot be fit using the same modulation parameters. Using our assumed source spectra, we find that a modulation parameter $\phi = 325$ MV best fits the CRIS data, while $\phi = 750$ MV best fits the HEAO3-C2 data.

LEAKY BOX MODEL WITH A LOCAL SOURCE OF PRIMARY COSMIC RAYS

We now consider a leaky box model without a strong decrease in $\Lambda_{esc}(E)$ at low energies: i.e.

$$\Lambda_{\text{esc}} = 29.5\beta R^{-0.6}, \quad \text{for all R} \tag{2}$$

Figure 3 shows a plot of Equation 2. With this $\Lambda_{esc}(E)$, the leaky box model over-produces secondary GCRs relative to primaries at low energies. However, by introducing a second, low-energy source of cosmic rays that have passed through very little matter, we may be able to again construct a model which fits the observations. This 'local source' of cosmic rays should have energy spectra which are steep relative to GCR interstellar spectra, so that they contribute significantly only below $\sim$ 1 GeV/nucleon. The composition of this local source would be determined by requiring that the combination of local-source CRs plus GCRs fit the observations.

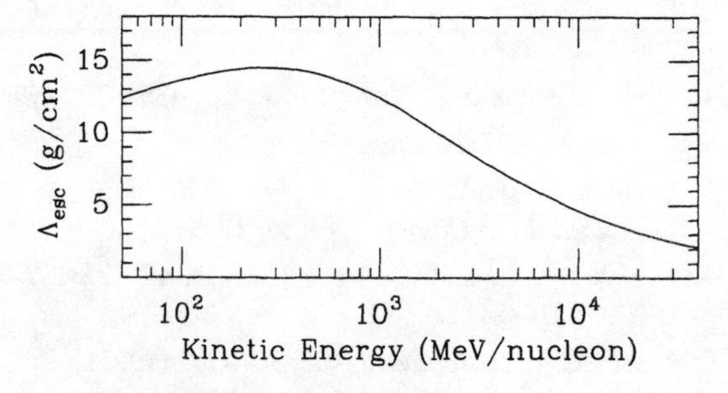

FIGURE 3. $\Lambda_{\text{esc}} = 29.5\beta R^{-0.6}$, for all R

Figure 4 illustrates the idea. A GCR interstellar carbon spectrum is shown, calculated using a leaky box model with $\Lambda_{esc}(E)$ as in Equation 2. Also shown is a possible local-source carbon spectrum, with $Q(E) \propto E^{-1}\exp(-E/E_0)$ where $E_0 = 500$ MeV/nucleon (the 'superbubble' spectrum of Bykov and Fleishman (16)). After modulating these spectra to 400 MV, the local source contributes a significant fraction of the total carbon intensity at $\sim$ 200 MeV/nucleon, while contributing negligibly above 1 GeV/nucleon.

Since CRs from the local source are assumed to travel through little or no material before reaching the solar system, they cannot contribute any secondaries to the mix. Therefore all the secondaries must be produced by the GCR component, and we must try to fit the CRIS and HEAO boron and (Sc+Ti+V) spectra with the leaky box model using $\Lambda_{esc}(E)$ as in Equation 2. The spectra of primary elements such as C and Fe are taken to be an appropriate mix of the GCR and local components.

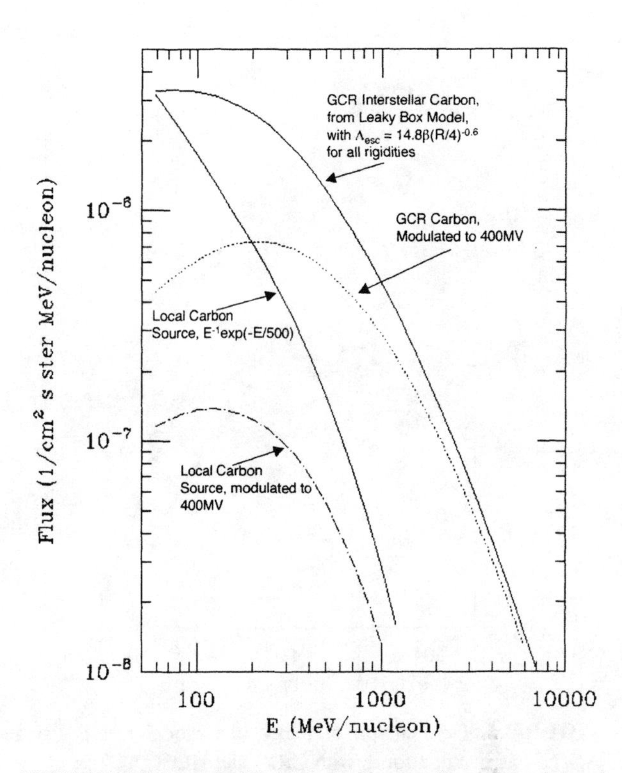

FIGURE 4. Schematic illustration of interstellar and modulated GCR and local source carbon spectra. The local source contributes significantly to the total carbon spectrum only below $\sim$ 1 GeV/nucleon.

Figures 5 and 6 show our attempt to achieve this fit. In Figure 5 the modulation level for the CRIS data is adjusted to 475 MV, which fits the CRIS boron spectrum. In Figure 6 a local source of C and Fe have been added in, which makes it possible to fit the measured CRIS and HEAO spectra for these elements. The required local contributions of C and Fe (evaluated at 200 MeV/nuc at 1 AU with $\phi = 475$ MV) amount to 22% and 35% of the GCR C and Fe at this energy. While the resulting fits to B, C, and Fe are all satisfactory at both CRIS and HEAO modulation levels, note in Figure 5 that this model does not produce sufficient Fe-secondaries. Indeed, using this form for $\Lambda_{esc}(E)$ we were unable to find any combination of parameters that provided satisfactory fits to both the boron and Fe-secondaries at the same level of solar modulation. While this does not rule out the possibility of contributions from a local source, it does indicate that our particular choice of $\Lambda_{esc}(E)$ and local source spectra do not lead to an improved fit to the available GCR data.

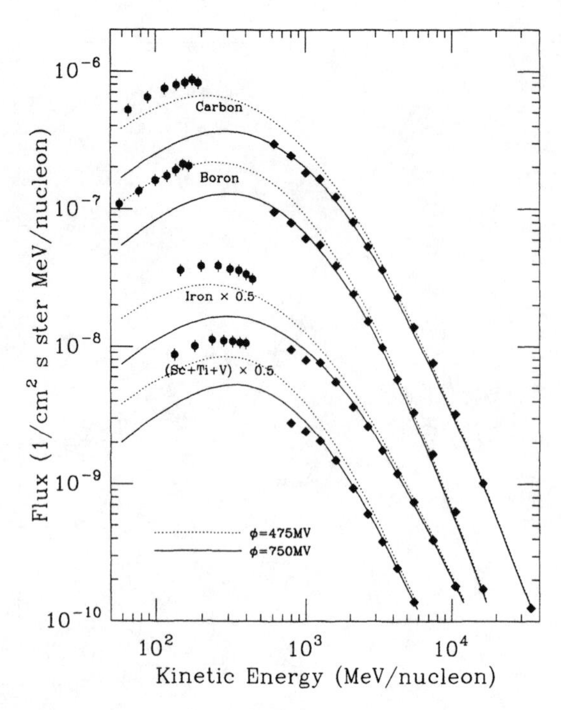

FIGURE 5. Comparison of leaky box model results using $\Lambda_{esc}(E)$ as in Equation 2 with CRIS and HEAO3-C2 B, C, Fe and (Sc+Ti+V) spectra. The boron and sub-Fe secondaries cannot be fit simultaneously.

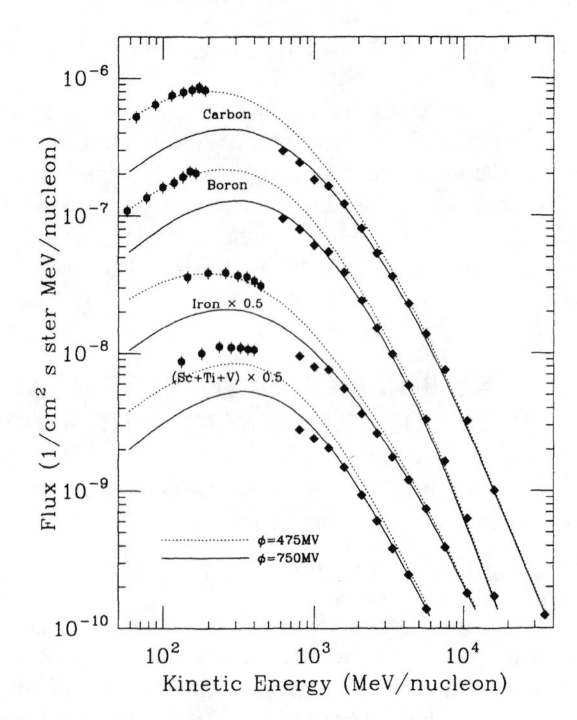

FIGURE 6. As in Figure 5, with C and Fe from the proposed local source added to the GCRs from the leaky box model.

CONCLUSIONS

The leaky box model of GCR propagation successfully reproduces the combined ACE and HEAO observations. However, this fit requires a stronger energy-dependence of $\Lambda_{esc}(E)$ at energies below $\sim$ 1 GeV/nucleon than has been used in many previous studies. We find that the functional form proposed by Soutoul and Ptuskin (6) in the context of a galactic diffusion-convection model provides an excellent fit to the data.

We investigated whether introducing a second, low-energy, local source of cosmic rays might lead to a model which also reproduces the observations, without invoking a decrease in $\Lambda_{esc}(E)$ at low energies. This approach does not lead to an improved fit to the data, since it does not reproduce the GCR secondaries as well as the standard leaky box model. However, a local source of cosmic rays in combination with other forms of $\Lambda_{esc}(E)$ is not ruled out. CRIS data spanning a complete solar cycle will allow for more precise tests of this idea during the coming years, since the relative contributions of the local and GCR components arriving at 1 AU should change as a function of the level of solar modulation.

ACKNOWLEDGEMENTS

This research supported by NASA at Caltech (grant NAG5-6912), JPL, NASA/GSFC, and Washington U.

REFERENCES

1. Meneguzzi, M., Audouze, J., and Reeves, H., *Astron. Astrophys.* **15**, 337 (1971).
2. Englemann, et al., *Astron. Astrophys.* **233**, 96–111 (1990).
3. Stephens, S. A., and Streitmatter, R., *Astrophys. J.* **505**, 266–277 (1998).
4. Webber, W. R., et al., *Astrophys. J.* **457**, 435–439 (1996).
5. Ptuskin, V. S., *This Volume.*
6. Soutoul, A., and Ptuskin, V. S., *Proc. 26th Int. Cosmic Ray Conf. (Salt Lake City)* **4**, 184–186 (1999).
7. Krombel, K. E., and Wiedenbeck, M. E., *Astrophys. J.* **328**, 940–953 (1988).
8. Leske, R. A., *Astrophys. J.* **405**, 567–583 (1993).
9. Soutoul, A., Ferrando, P., and Webber, W. R., *Proc. 21st Int. Cosmic Ray Conf. (Adelaide)* **3**, 337–341 (1990).
10. Silberberg, R., Tsao, C. H., and Barghouty, A. F., *Astrophys. J.* **501**, 911–919 (1998).
11. Webber, W. R., et al., *Astrophys. J.* **508**, 940–948 (1998).
12. Webber, W. R., et al., *Astrophys. J.* **508**, 949–958 (1998).
13. Tripathi, R. K., Cucinotta, F. A., and Wilson, J. W., *NASA Technical Paper 3621* (1997).
14. Fisk, L. A., *J. Geophys. Res.* **76**, 221 (1971).
15. Garcia-Munoz, M., and Simpson, J. A., *Proc. 16th Int. Cosmic Ray Conf. (Kyoto)* **1**, 270–274 (1979).
16. Bykov, A. M., and Fleishman, G. D., *MNRAS* **255**, 269 (1992).

A Measurement of Cosmic Ray Deuterium from 0.5-2.9 GeV/nucleon

G.A. de Nolfo[1], L.M. Barbier[2], E.R. Christian[2], A.J. Davis[1], R.L. Golden[5], M. Hof[4], K.E. Krombel[2], A.W. Labrador[6], W. Menn[4], R.A. Mewaldt[1], J.W. Mitchell[2], J.F. Ormes[2], I.L. Rasmussen[7], O. Reimer[3], S.M. Schindler[1], M. Simon[4], S.J. Stochaj[5], R.E. Streitmatter[2], W.R. Webber[5]

[1]*California Institute of Technology, Pasadena, California 91125*
[2]*NASA/Goddard Space Flight Center, Greenbelt, MD 20771*
[3]*Max-Planck-Institut fuer extraterrestrische Physik, Garching, Germany*
[4]*Universitat Siegen, Siegen, Germany*
[5]*New Mexico State University, Las Cruces, NM 88003*
[6]*University of Chicago, Chicago, IL 60637*
[7]*Danish Space Research Institute, Copenhagen, Denmark*

Abstract. The rare isotopes ^{2}H and ^{3}He in cosmic rays are believed to originate mainly from the interaction of high energy protons and helium with the galactic interstellar medium. The unique propagation history of these rare isotopes provides important constraints on galactic cosmic ray source spectra and on models for their propagation within the Galaxy. Hydrogen and helium isotopes were measured with the balloon-borne experiment, IMAX, which flew from Lynn Lake, Manitoba in 1992. The energy spectrum of deuterium between 0.5 and 3.2 GeV/nucleon measured by the IMAX experiment as well as previously published results of ^{3}He from the same instrument will be compared with predictions of cosmic ray galactic propagation models. The observed composition of the light isotopes is found to be generally consistent with the predictions of the standard Leaky Box Model derived to fit observations of heavier nuclei.

INTRODUCTION

Extensive observations of cosmic ray abundances over a wide range in energy help to form a comprehensive picture of cosmic ray origin and propagation. In the simplest picture such as the standard Leaky Box Model, cosmic rays propagate within the Galaxy influenced by the competing processes of nuclear interactions and escape from the Galaxy. The light isotopes such as Li, Be, B are significantly enhanced over solar system abundances indicating that these elements are produced as secondary or spallation products of primary C, N, and O elements. Thus, the determination of the secondary/primary ratio provides a measure of the amount of material traversed by primary cosmic rays during propagation. At 1 GeV/nucleon, the mean free pathlength for escape from the Galaxy is found to be $\lambda \sim 10$ g/cm^2.

The isotopes ^{2}H and ^{3}He are of particular interest as these isotopes are considered to be interaction products of the more abundant hydrogen and helium nuclei. In addition, the abundance of ^{2}H and ^{3}He can, in principle, provide a more sensitive determination of the escape pathlength than heavier cosmic ray nuclei, since H and He are affected by fewer nuclear destruction processes during propagation. These isotopes may also provide a test of whether cosmic rays undergo continuous acceleration or "reacceleration" during their passage through the Galaxy (8). Hydrogen and helium isotopes have been measured by the Isotope Matter-Antimatter Experiment (IMAX) instrument over a wide range in energies extending to 2.9 GeV/nucleon (6). In this paper, we present IMAX measurements of ^{2}H as well as the previously published results of ^{3}He (6) and compare these observations with predictions from current propagation models.

INSTRUMENT AND FLIGHT

IMAX was designed to measure antiprotons and the light isotopes over a wide energy range. IMAX employed a combination of detectors including a superconducting magnetic spectrometer (2), a time-of-flight (TOF) system, scintillation counters (S1,S2), and large-area aerogel Cherenkov detectors. Particle identification is accomplished by measuring the particle velocity β, charge Z, and rigidity R (momentum/charge). For further details on the performance of the IMAX instrument see (1).

CP528, *Acceleration and Transport of Energetic Particles Observed in the Heliosphere: ACE 2000 Symposium*, edited by Richard A. Mewaldt, et al.
© 2000 American Institute of Physics 1-56396-951-3/00/$17.00

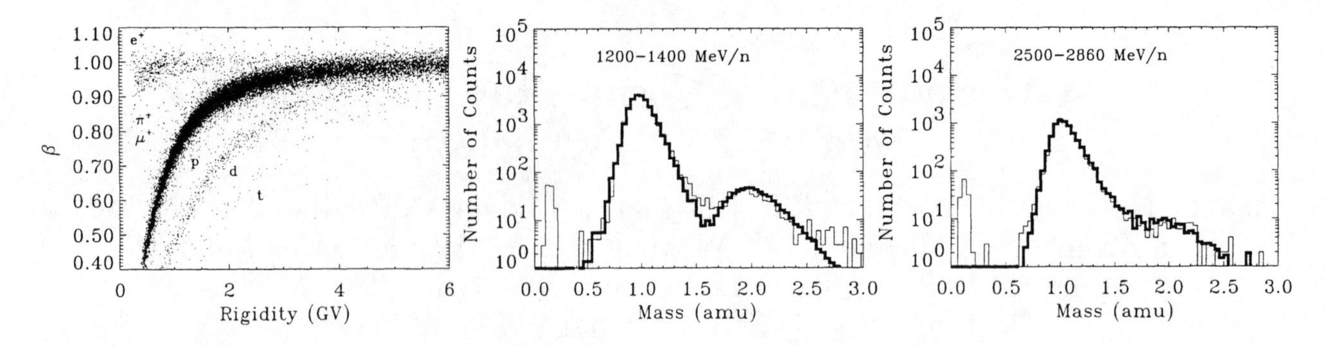

FIGURE 1. IMAX mass separation for Z=1, β=1 particles along with the corresponding mass histograms in two separate energy intervals. The mass is determined between 2.5-2.860 GeV/nucleon using the velocity obtained from the Cherenkov counters. The thick solid lines represent an instrument simulation, while the thin solid lines refer to measured distributions.

IMAX flew in July, 1992 from Lynn Lake, Manitoba, Canada. The flight lasted 16 hours at float including a long ascent of ~7 hours. IMAX reached a maximum float altitude of 36 km (5 g/cm^2 of residual atmosphere). The geomagnetic cutoff varied between 0.35 GV at Lynn Lake and 0.63 GV at Peace River, Alberta.

DATA ANALYSIS

IMAX events are accepted based on a four-fold coincidence between the photomultiplier signals from the opposite sides of the top and bottom TOF scintillators. The selection criteria employed to obtain a clean sample of charge one particles are discussed in (1).

Figure 1 shows the isotopic separation for charge one particles using the β-rigidity technique. The corresponding mass histograms in two representative energy intervals are shown in Figure 1, where the velocity in the higher energy interval is obtained from the aerogel Cherenkov counter. Due to the non-gaussian behavior of the distributions, a simulation was developed to accurately model the instrument response (6),(7). The simulation takes into account, on an event by event basis, the actual spectral shape of the incoming particles, the TOF timing resolution, the photoelectron statistical fluctuations and δ-ray contributions to the Cherenkov light yield. It also takes into account the spatial resolution and rigidity resolution of the tracking system, and the effects of multiple coulomb scattering. The simulation results are shown in Figure 1 as the thick solid lines. ;

RESULTS

In order to determine the flux of deuterium at the top of the atmosphere, it is necessary to account for nuclear interaction losses within in the instrument and atmosphere as well as for the secondary population of ^{2}H produced from the interaction of protons and helium in the 5 g/cm^2 residual atmosphere above the instrument. The attenuation of ^{2}H within the instrument and atmosphere is determined using a universal parametrization for the total reaction cross section given by Tripathi et al. (10). This model is in good agreement with current measurements for the inelastic cross sections of ^{2}H+p and ^{3}He+p reactions (10), (11). IMAX has a mean grammage in the instrument of 13.8 g/cm^2. The different materials encountered during the particle's traversal through the IMAX instrument are accounted for in this calculation.

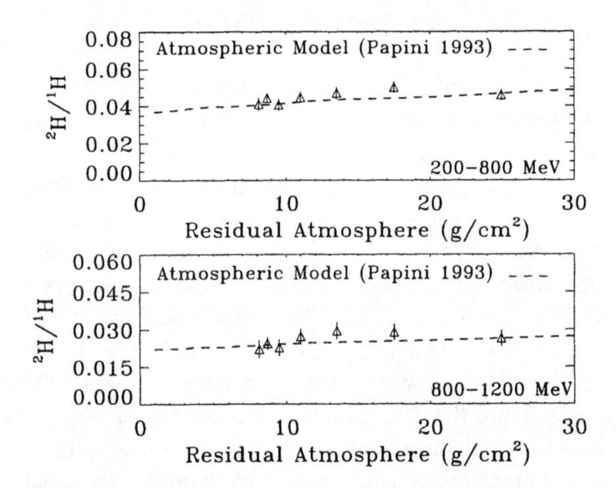

FIGURE 2. IMAX ^{2}H/^{1}H as a function of atmospheric depth in two separate energy intervals. The model of Papini is normalized to the IMAX data as discussed in the text.

The contribution from atmospheric secondary ^{2}H is determined from calculations by Papini et al. (4),(5). Several other calculations have been performed including those of Webber (12) and Lijowski et al. (3). Current calculations differ by as much as a factor of 2-3 where the differences are most likely due to the assumed primary proton and helium spectra and the assumed interaction cross sections. In order to better constrain our estimate of the contribution of secondary ^{2}H, we obtain the ^{2}H/^{1}H ratio as a function of depth in the atmosphere during IMAX's long ~7 hour ascent to float altitudes. Figure 2 shows the ^{2}H/^{1}H ratio at seven separate depths in the atmosphere. The dashed curve is the calculation by Papini (4) for the production of secondary deuterium in 5 g/cm^2 residual atmosphere during solar minimum conditions. The Papini calculations are fit to the IMAX ^{2}H/^{1}H ratio as a function of depth resulting in a top of the atmosphere ratio of 0.036 ±.004 at 600 MeV/nucleon and 0.022 ±.003 at 1 GeV/nucleon, as shown in Figure 2. The contribution of secondary deuterium predicted by Papini's model is consistent with the rate of growth of ^{2}H/^{1}H as a function of depth in the atmosphere measured by IMAX. Secondary deuterium produced within the instrument is vetoed by the instrument trigger and event selection criteria.

DISCUSSION AND CONCLUSION

Figure 3 shows the ^{2}H spectrum at the top of the atmosphere along with recent measurements from the BESS experiment during three separate flights from 1993 to 1995 (11). The solar modulation during the IMAX 1992 flight is consistent with a modulation parameter in the spherically symmetric force-field model of ϕ = 750 MV (1). At low energies, where the effects of atmospheric

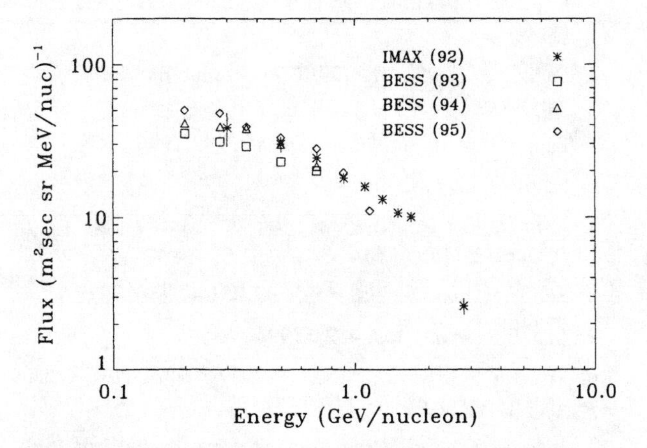

FIGURE 3. IMAX top of the atmosphere ^{2}H flux compared with measurements from the BESS experiment (11).

secondaries are significant, IMAX observes a higher deuterium flux than BESS observations during 1993, contrary to what one might expect from solar modulation effects alone. The discrepancy appears to be in the choice of the absolute value for the calculated contribution of atmospheric secondary deuterium. We find the IMAX growth curves are consistent with Papini's atmospheric secondary calculation at solar minimum conditions. On the other hand, the BESS secondary calculations result in fluxes a factor of ~2 lower, closer to Papini's solar maximum calculation (11). Table 1 lists the IMAX measurements for the ^{2}H flux at the top of the instrument and atmosphere.

Table 1. ^{2}H Flux at the top of the instrument (TOI) and atmosphere(TOA).

Energy (GeV/n)	Mean Energy (GeV/n)	$^2H_{TOI}$	$^2H_{TOA}$
0.4-0.6	0.5	35.8 ± 2.2	30.0 ± 3.5
0.6-0.8	0.7	25.5 ± 1.5	24.3 ± 1.9
0.8-1.0	0.9	18.3 ± 1.2	18.0 ± 1.5
1.0-1.2	1.0	15.5 ± 1.0	15.8 ± 1.1
1.2-1.4	1.2	12.8 ± 0.8	13.1 ± 0.9
1.4-1.6	1.5	10.4 ± 0.7	10.7 ± 0.7
1.6-1.8	1.7	9.7 ± 0.7	10.1 ± 0.7
2.5-2.9	2.7	2.6 ± 0.3	2.6 ± 0.3

The ratios of ^{2}H/^{1}H and ^{2}H/^{3}He are shown in Figures 4[a] and 4[b] along with the predictions of propagation models based on a standard Leaky Box calculation by Seo et al. (9) and a reacceleration model by Seo & Ptuskin (8). The ^{2}H/^{1}H ratio is in excess of the model predictions at low energies where the atmospheric secondary contribution is largest. The IMAX results for the ^{2}H/^{3}He ratio are in better agreement at higher energies, not unexpectedly, since the ^{2}H/^{3}He ratio is essentially independent of pathlength in the interstellar medium.

A measure of the ^{2}H/^{4}He ratio over a wide energy range may help to distinguish between existing propagation models, especially as we expect the ^{2}H/^{4}He ratio to exhibit a strong energy dependence resulting from the ^{2}H production cross sections. Reacceleration, on the other hand, would smear out this energy dependence. The ^{2}H/^{4}He and ^{3}He/^{4}He ratios are shown in Figures 4[c] and 4[d] and are compared with predictions from Seo & Ptuskin (8), Webber (14) and Reimer et al. (6). Webber's calculation is a standard Leaky Box Model with a pathlength $\lambda=31.6\beta R^{-0.6}$ for R>4.7 GV and $\lambda=12.5\beta$ below 4.7 GV that is based on B/C measurements (15). The solid curve in Figure 4[c] is a calculation by Seo & Ptuskin that includes the affects of reacceleration (8). Finally, the solid curve in Figure 4[d] is a standard Leaky Box calculation by Reimer et al. (6) that assumes similar input parameters to the Webber calculations (14), though with slightly different cross sections.

The IMAX observations of the ^{2}H are generally consistent with predictions of the standard Leaky Box Model in which protons and helium have the same propagation history as the heavier component of cosmic rays. How-

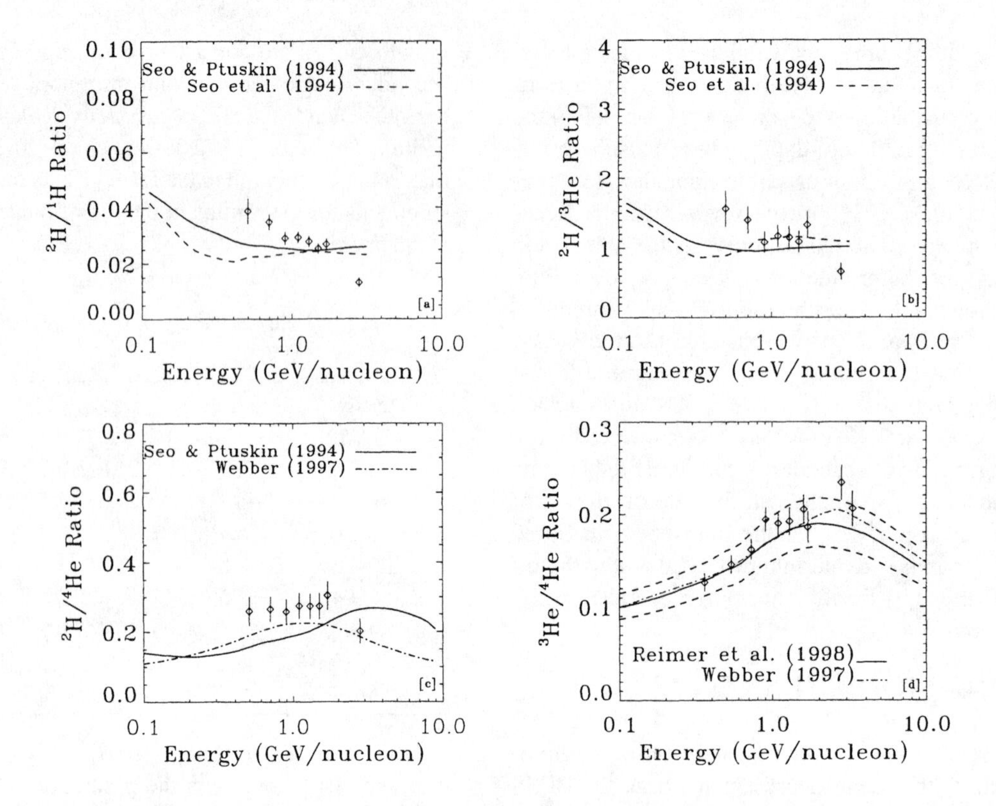

FIGURE 4. [a]-[b]: IMAX ^{2}H/^{1}H and ^{2}H/^{3}He ratios compared with the propagation model with reacceleration of Seo & Ptuskin (8) (solid curve) and the standard Leaky Box calculation of Seo et al. (9) (dashed curve), [c]-[d]: ^{2}H/^{4}He, and ^{3}He/^{4}He ratios compared with propagation models of Webber (14) (dot-dashed curve), and Reimer et al. (6) (solid curve). The dashed curves in [d] represent the uncertainties in the Reimer et al. calculation.

ever, there is a tendency for the ^{2}H/^{4}He and ^{2}H/^{1}H observations to lie somewhat above the model predictions. Understanding this possible excess will require a thorough investigation of the dependence of these ratios on the relevant production cross sections for ^{2}H and on the assumed pathlength distribution.

ACKNOWLEDGMENTS

The IMAX project was supported in the United States by NASA under NAG5-5227 (Caltech) and under RTOP 353-87-02 (GSFC) and grants NAGW-1418 (NMSU/BBMF) and in Germany by the Deutsche Forschungsgemeinschaft (DFG) and the Bundesministerium für Bildung, Wissenschaft, Forschung und Techologie (BMBF).

REFERENCES

1. Menn, W., *et al.*, accepted for publication in *ApJ*, **533**, No. 1, (2000).
2. Hof, M., *et al.*, *Nucl. Inst. & Meth.*, **A345**, (1994), 561.
3. Lijowski, M., (1994), Ph.D. Dissertation, Louisiana State University.
4. Papini, P., *et al.*, *Proc. 23rd Int. Cosmic Ray Conf.* (Calgary), (1993), 503.
5. Papini, P., *et al.*, *Il Nuovo Cimento*, **19**, (1996), 367.
6. Reimer, O., *et al.*, *ApJ*, **496**, (1998) 490.
7. Reimer, O., *et al.*, *Proc. 24th Int. Cosmic Ray Conf.* (Rome), **2**, (1995), 614.
8. Seo, E.S., and Ptuskin, V.S., *ApJ*, **431**, (1994), 705.
9. Seo, E.S., *et al.*, *ApJ*, **432**, (1994), 656.
10. Tripathi, R.K., *et al.*, *NASA/TP-1999-209726*, NASA Langley Research Center, (1999).
11. Wang, J.Z., *et al.*, *Proc. 26th Int. Cosmic Ray Conf.* (Salt Lake City) **3**, (1999), 37.
12. Webber, W.R., *et al.*, *ApJ*, **380**, (1991), 230.
13. Webber, W.R., *et al.*, *ApJ*, **392**, (1992), L91.
14. Webber, W.R., *Adv. Space Res.*, **19**, No. 5, (1997), 755.
15. Webber, W.R., *et al.*, *ApJ*, **457**, (1996), 435.

COSMIC RAY PATH LENGTH DISTRIBUTIONS FROM SUPERBUBBLE/GIANT HII REGIONS

J.C. Higdon[1], & R.E. Lingenfelter[2]

1) W. M. Keck Science Center, Claremont Colleges, Claremont, CA
2) Center for Astrophysics and Space Sciences, University of California, San Diego, La Jolla, CA

Abstract. We have constructed a Monte Carlo Simulation program for the space and time dependent distribution of Galactic supernovae, as a source of both shock energy and particles for cosmic ray acceleration, based on recent astronomical observations and theory. From these simulated spatial and temporal distributions of discrete sources, we then determine the form and variance of the expected local cosmic ray path length distributions, together with anisotropies and local intensity histories, for a wide range of effective mean free paths.

INTRODUCTION

Assuming that galactic cosmic rays are accelerated by supernova shocks (e.g. Axford (1)), we calculated from Monte Carlo simulations the cosmic ray energy density in the Galaxy for times $\geq$ 150 Myr from clustered type II supernova explosions of massive stars born in OB associations. Such type II supernovae are the dominant ($\sim$85%) supernova population in our Galaxy, e.g. van den Bergh & McClure (7). This simulation is based on energy-independent, diffusive cosmic-ray propagation, and calculations are presented for diffusion mean free paths, λ, in the range, 0.1 pc $\leq \lambda \leq$ 10 pc. The distribution and properties of the parent OB associations are based on the works of McKee & Williams (4) and Williams & McKee (8), which investigated multigenerational massive star formation in our Galaxy. For a wide range of diffusion mean free paths we determined probability distributions for the cosmic ray energy density at Earth, the cosmic ray directional anisotropy, the cosmic ray age, the number of contributing supernovae, the distances of the contributing supernovae, and the modeled path lengths. Finally, we simulated the cosmic-ray intensity histories at Earth over the last 50 Myr.

DISTRIBUTION & OCCURRENCE OF TYPE II SUPERNOVAE

We based our galactic distribution of type II supernovae on the work of McKee & Williams (4) and Willams & McKee (8). We distributed OB associations randomly in the Galaxy, uniformly in azimuthal angle, but nonuniformly in galactic radius, employing a source density as a function of galactic radius, R, $exp[-R(kpc)/3.5]$ over 3 kpc $\leq R \leq$ 11 kpc. We used their association luminosity distribution, i.e., the probability distribution of the number of massive stars born in a single generation, N, as $(N_{max}/N-1)$, where N_{max} is 1500. We employed this probability distribution to determine the number of first generation massive stars in a given association. Following McKee & Williams (4) we assumed that five generations of star formation typically occur in each OB association. They also suggested that subsequent generations of massive stars in typical associations must have the same total ionizing luminosities. Thus we assumed that an identical number of massive stars were created in each subsequent generation of star formation. The number of supernova precursors created in such multigenerational OB associations ranged from 5 to 7500 with a mean number of $\sim$40. Following McKee & Williams (4) we further assumed that in each generation of star formation the distribution of initial stellar masses was $[(M_U/M)^{3/2}-1]$, where M is stellar mass and the upper limiting mass M_U = 120 $M_\odot$. Finally, to determine when each massive ($> 8M_\odot$) star collapsed in a type II supernova, we compared their age to the stellar main sequence lifetimes derived by Schaller et al. (5).

In our simulation we terminated the creation of OB associations in a test galaxy when the total ionizing flux of all the galactic OB stars reached $1.9\times10^{53}s^{-1}$, the value found by McKee & Williams (4) from analyses of thermal radio emission and *COBE* observations of far infrared NII line emission. To calculate the total galactic ionizing flux, we used the relation between stellar ionizing flux and initial stellar mass from McKee & Williams (4), and taking the threshold mass for a core collapse supernovae of 8

CP528, *Acceleration and Transport of Energetic Particles Observed in the Heliosphere: ACE 2000 Symposium*, edited by Richard A. Mewaldt, et al.
© 2000 American Institute of Physics 1-56396-951-3/00/$17.00

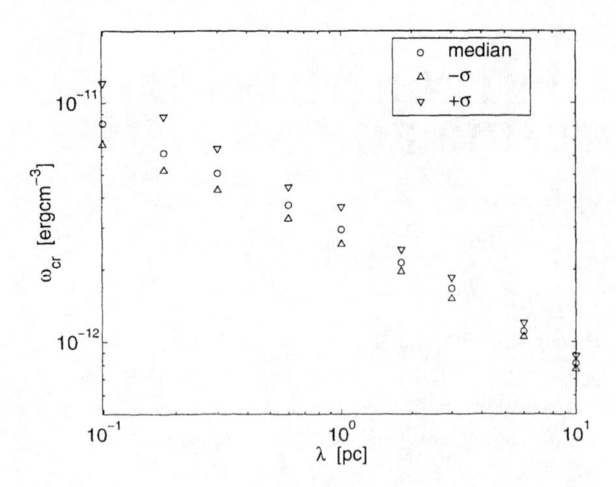

FIGURE 1. Cosmic ray energy density at Earth as function of the diffusion mean free path λ for $W_{sn} = 10^{50}$ ergs and $\tau = 30$ Myr, with the $\pm 1\sigma$ variance.

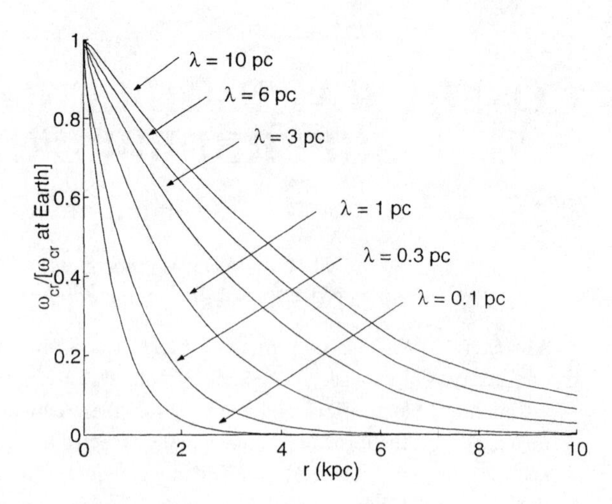

FIGURE 2. Relative cosmic ray energy density at Earth as function of the source distance r from the Earth for a range of the diffusion mean free path, λ.

$M_{\odot}$, we found that the mean time between galactic type II supernovae was ~25 years.

COSMIC RAY PROPAGATION

We employed the simplest, physically realistic model (Ramaty, Reames, & Lingenfelter (6) of cosmic ray propagation, three-dimensional isotropic diffusion with an energy independent scattering mean free path, λ. The probability, μ, that a cosmic ray propagates a distance, r, from a single point source of age, t, located at the origin is the solution of

$$\frac{\partial \mu}{\partial t} = \frac{1}{r^2}\frac{\partial}{\partial r}\frac{\lambda c}{3}r^2\frac{\partial \mu}{\partial r} - \frac{\mu}{\tau},$$

where λ is the diffusion mean free path, c is the light speed, and τ is the effective mean lifetime, including both escape and nuclear destruction. Thus the cosmic ray energy density from a single supernova at a distance, r, and age, t, is $\omega_{cr}(r,t) = W_{sn}\mu(r,t)$, where W_{sn}, taken to be 10^{50} ergs, is the total energy of cosmic rays produced in the supernova, and

$$\omega_{cr}(r,t) = W_{sn}(\frac{4\pi\lambda ct}{3})^{-3/2}exp(-\frac{3r^2}{4\lambda ct} - \frac{t}{\tau}).$$

MONTE CARLO SIMULATION

In the Monte Carlo simulations we used this diffusive cosmic ray propagation model for each supernova source, and summed the contributions of all galactic supernovae, with $\tau = 30$ Myr for a range of λ from 0.1 pc to 10 pc. Since the acceleration lifetime of supernova shocks in the superbubble/hot phase of the interstellar medium is of the order of 10^5 yrs (e.g. Higdon, Lingenfelter & Ramaty (2,3), we took that time as the temporal step size in our simulations. We evolved each test galaxy for 1.5×10^8 years using the above prescription for locating OB associations. In each test galaxy we created $\sim 10^7$ supernovae, each with a distinct age and galactic location. We treated each OB association as the spatial but not temporal point source of all of the supernovae occurring in the association since the effect of the < 100 pc dispersion of supernova progenitors is negligible except for the time dependent intensity, if the Sun happened to be inside such an association. For each choice of a diffusion mean free path, λ, we created an ensemble of 500 test galaxies.

We kept statistics on a variety of variables: ω_{cr} at Earth, cosmic ray anisotropy at Earth, age of cosmic rays, number of contributing supernovae, and distances of the contributing supernovae. Figure 1 shows the dependence of the cosmic ray energy density at Earth as a function of diffusion mean free path, λ. Figure 2 illustrates the contributions of supernovae as functions of distance and mean free path. This plot demonstrates that a significant ($\geq 25\%$) fraction of the cosmic rays, detected at Earth, can be accelerated by distant (≥ 3 kpc) supernovae, residing in the inner Galaxy where interstellar properties, e.g., metallicity and the volume filling factor of tenuous medium, can be very different from those in the solar vicinity. Figure 3 shows for λ = 0.3 and 3 pc the probability distributions for: cosmic ray energy density at Earth, mean cosmic ray age, anisotropy, number of supernovae

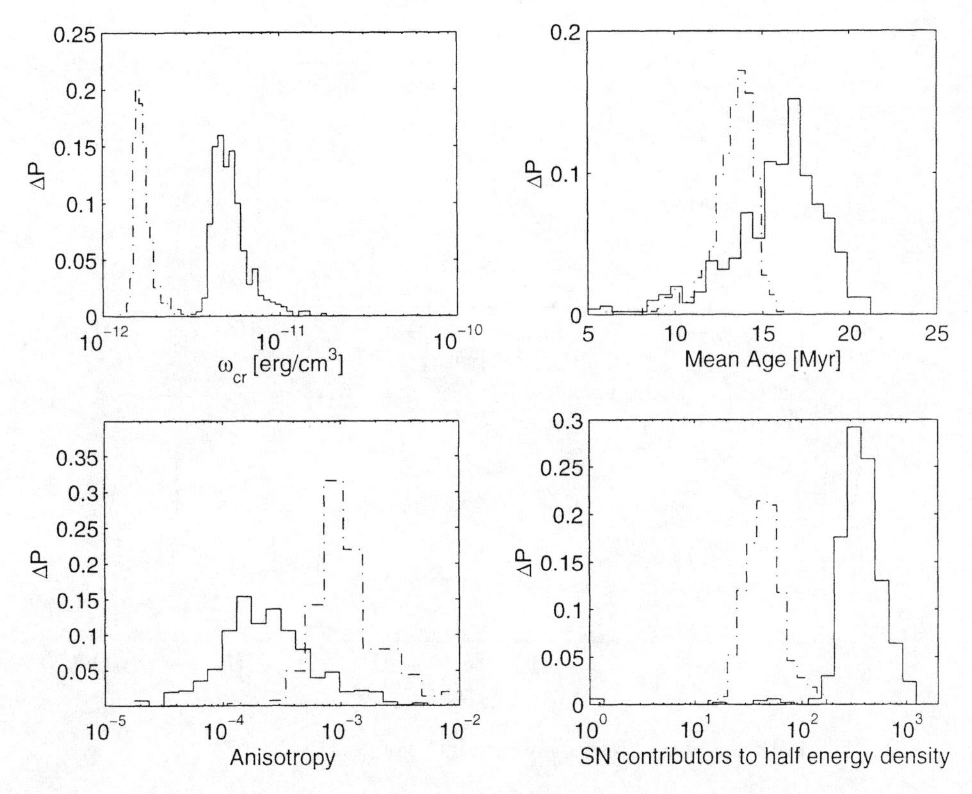

FIGURE 3. Probability distributions of the local cosmic ray energy density, mean age, anisotropy, and number of supernovae contributing half of the local energy density for $\lambda = 0.3$ pc (solid) and 3 pc (dashed).

contributing to half the local cosmic ray energy density. We found that the local cosmic ray energy density is proportional to $\lambda^{-0.5}$, the mean anisotropy, as defined in equation (4) of Ramaty et al. (6), increases as $\lambda^{0.5}$, and number of supernovae contributing to half the local energy density decreases as λ^{-1}. Also we find that the mean cosmic age changed by only ~30% when λ changed by an order of magnitude.

PATH LENGTH DISTRIBUTIONS & COSMIC RAY HISTORIES

A determination of the cosmic ray path length distribution is complicated by our lack of knowledge of the interstellar medium properties. However, on kpc scales, we expect some differences in interstellar properties, such as cloud and intercloud densities, to average out. As a first order approximation we assume that the interstellar medium density is uniform, and, consequently, the path length distribution scales with the cosmic ray age distribution. Clearly, this approximation breaks down at small λ, < 0.1 pc, where the contribution of supernovae residing in the local, tenuous superbubble dominate, and at

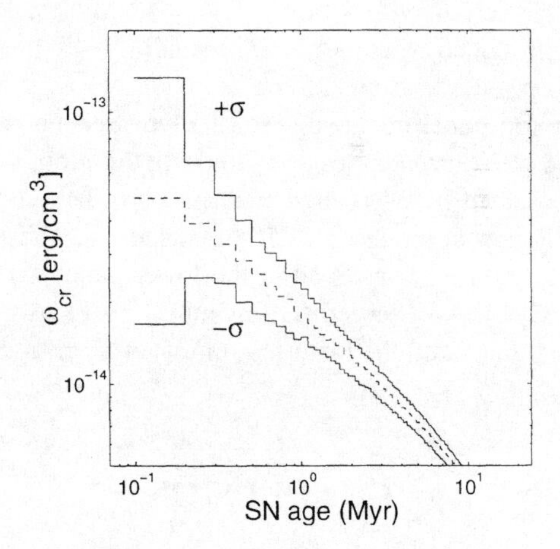

FIGURE 4. Modeled path length versus supernova age t, or path length ct, for λ of 3 pc, with the $\pm 1\sigma$ variance.

large λ, > 10 pc, where the contribution from supernovae in the inner Galaxy dominate. Figure 4 shows the mean age dependence of the local cosmic rays for $\lambda = 3$ pc with the $\pm 1\sigma$ variance.

We modified our Monte Carlo calculation to simulate the history of cosmic ray intensity at Earth. For each test

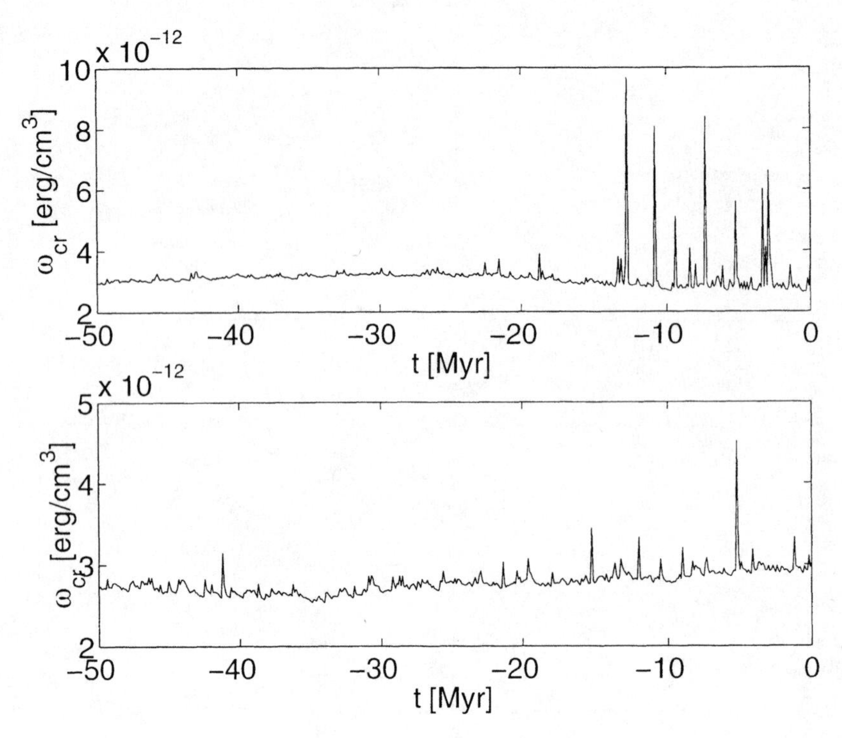

FIGURE 5. Sample cosmic ray intensity histories at Earth for λ = 1 pc

galaxy we calculated ω_{cr} at Earth every 0.1 Myr. Figure 5 shows sample realizations, displaying two distinct effects. First, *correlated* narrow ($\Delta t \leq 0.1$ Myr) spikes are seen, produced by supernovae in small ($N \leq 20$), nearby (≤ 50 pc) associations. Such common behavior is seen in the top panel at times ≥ -13 Myr. Second, we observed weak, broad rises and falls with durations of ~ 50 Myr, created by correlated supernovae in large (≥ 500) associations at modest (~ 250 pc) distances. The start of such a feature is observed in the lower panel starting at -35 Myr. We observed significant ($\geq 50\%$) cosmic ray fluctuations in all the histories simulated.

CONCLUSION

We created a Monte Carlo simulation of correlated type II supernovae distributed throughout the Galaxy. For a range of diffusion mean free paths we investigated ensembles of 500 test galaxies; in each test galaxy we created a cosmic-ray distribution from $\sim 6 \times 10^6$ supernovae. We determined probability distributions for: cosmic ray energy density at Earth, anisotropy, number of contributing supernovae, and cosmic ray ages. We derived the dependence of these variables on the diffusion mean free path, λ. Finally, we calculated histories of local cosmic ray intensity over the last 50 Myr.

From our investigation we suggest that any meaningful determination of the diffusion mean free path must be based on the measured values *all* these properties. Further, we find that models of cosmic ray source properties based solely on the properties of the local interstellar medium and local supernovae may be inappropriate in view of the possibility that a significant fraction of local cosmic rays might be accelerated by distant supernovae, where properties can be very different.

REFERENCES

1. Axford, W.I., 17th ICRC Papers 12, 155 (1981)
2. Higdon, J.C., Lingenfelter, R.E., & Ramaty, R., ApJ, 509, L33 (1998)
3. Higdon, J.C., Lingenfelter, R.E., & Ramaty, R., 26th ICRC Conf. Papers, 4, 144 (1999)
4. McKee, C., & Williams, J., ApJ, 476, 144 (1997)
5. Schaller, G., Schaerer, D., Meynet, G., and Maeder, A., A&AS, 96, 269 (1992)
6. Ramaty, R., Reames, D.V., and Lingenfelter, R.E., PRL, 24, 913 (1970)
7. van den Bergh, S., & McClure, R.D., ApJ, 425, 205 (1994)
8. Williams, J.P. & McKee, C.F., ApJ, 476, 166 (1997)

Co/Ni Element Ratio in the Galactic Cosmic Rays between 0.8 and 4.3 GeV/nucleon

S.H. Sposato[1], L.M. Barbier[2], W.R. Binns[1], E.R. Christian[2], J.R. Cummings[1], G.A. deNolfo[3], P.L. Hink[1], M.H. Israel[1], R.A. Mewaldt[3], J.W. Mitchell[2], S.M. Schindler[3], R.E. Streitmatter[2], C.J. Waddington[4]

[1]*Department of Physics & McDonnell Center for the Space Sciences, Washington University, St. Louis, MO 63130*
[2]*Laboratory for High Energy Astrophysics, Goddard Space Flight Center, Greenbelt, MD 20771*
[3]*Space Radiation Laboratory, California Institute of Technology, Pasadena, CA 91125*
[4]*School of Physics and Astronomy, University of Minnesota, Minneapolis, MN 55455*

Abstract. In a one-day balloon flight of the Trans-Iron Galactic Element Recorder (TIGER) in 1997, the instrument achieved excellent charge resolution for elements near the Fe peak, permitting a new measurement of the element ratio Co/Ni. The best fit to the data, extrapolated to the top of the atmosphere, gives an upper limit for this ratio of 0.093 ± 0.037 over the energy interval 0.8 to 4.3 GeV/nucleon; because a Co peak is not seen in the data, this result is given as an upper limit. Comparing this upper limit with calculations by Webber & Gupta [14] suggests that at the source of these cosmic rays a substantial amount of the electron-capture isotope ^{59}Ni survived. This conclusion is in conflict with the clear evidence from ACE/CRIS below 0.5 GeV/nucleon that there is negligible ^{59}Ni surviving at the source. Possible explanations for this apparent discrepancy are discussed.

INTRODUCTION

The Trans-Iron Galactic Element Recorder (TIGER) is a balloon-borne cosmic-ray experiment that utilizes plastic scintillation counters, plastic and aerogel Cherenkov counters, and scintillating optical fiber hodoscopes. With this combination of detectors we have achieved charge resolution of 0.23 charge units (cu) for elements near the Fe peak for energies < 4.3 GeV/nucleon. This resolution has enabled a new measure of the Co/Ni abundance ratio using data from the one-day balloon flight from Fort Sumner, NM in September 1997.

The Co/Ni measurement is interesting because it can put limits on the time between the nucleosynthesis and acceleration of cosmic rays [8]. ACE/CRIS has measured isotopes of Co and Ni in the energy interval 150 MeV/nucleon – 500 MeV/nucleon. That measurement found that ^{59}Ni, which decays only by electron capture (half-life 76,000 yr), has all decayed to ^{59}Co, implying that acceleration occurred more than 10^5 years after nucleosynthesis [15]. At higher energies isotope data are not available, but elemental abundances can put constraints on the decay of ^{59}Ni. Engelmann *et al.* [2] measured a Co/Ni ratio of ≈ 0.12 near 1 GeV/nucleon. Previous balloon measurements of this ratio, extrapolated to the top of the atmosphere, have been 0.17 ± 0.05 [1] and 0.31 ± 0.15 [3].

THE TIGER INSTRUMENT

The TIGER instrument contains three scintillation counters with wave-length-shifter-bar readout and two Cherenkov detectors: an acrylic radiator (n = 1.5) in a light box and an aerogel radiator (n = 1.04) in a light box. (See Figure 1.) There are two scintillators on the top of the stack and one at the bottom. The scintillation counters and the Cherenkov detectors provide charge and velocity measurements of the cosmic rays, provided that pathlength through the instrument can be determined.

CP528, *Acceleration and Transport of Energetic Particles Observed in the Heliosphere: ACE 2000 Symposium*,
edited by Richard A. Mewaldt, et al.
© 2000 American Institute of Physics 1-56396-951-3/00/$17.00

The trajectory of each cosmic ray through the instrument is measured to within a few millimeters using a Coarse Hodoscope and a coded Fine Hodoscope. Each consists of two planes of 1.5mm x 1.5mm square scintillating optical fibers, one at the top of the stack and one at the bottom. Each plane has two layers of fibers, one each in the X and Y directions. The Coarse Hodoscope determines position to within 8cm, and the Fine Hodoscope determines position to within 6mm [5,9].

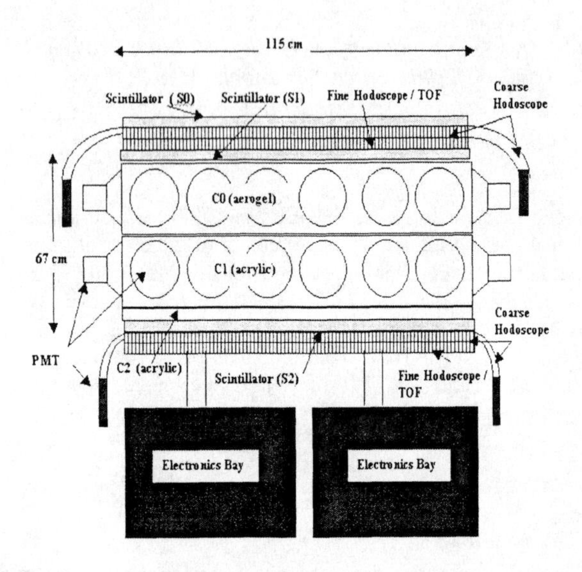

FIGURE 1. Cross-section of the TIGER instrument

DATA ANALYSIS

There are several steps in the data analysis. The first is a sec(θ) correction for variations in pathlength made using the hodoscopes. Mapping corrections are made to the data from each detector to remove areal response variations. Corrections are also made to remove a residual θ dependence in the data. This dependence is probably due to the change in energy deposition of knock-on electrons with depth in the detector stack. The Z-dependence of the signals from the scintillators and Cherenkov detectors is also determined empirically. Interaction cuts are made to remove particles that changed charge in the detector stack. The difference in charge measured in the top two scintillators is required to be < 0.6cu. Also the difference in charge between the average of the top two scintillators and the bottom scintillator is required to be < 0.6cu.

After these corrections are made, the data are divided into energy intervals. Interval 1 contains E < 2.7 GeV/nucleon. Interval 2 contains 2.7 < E < 4.3 GeV/nucleon. The data have an inherent energy dependence in these regions. Empirical fits are made to the energy variations in each region.

A histogram of the data from TIGER in Energy Intervals 1 and 2 (E < 4.3 GeV/n) is shown in Figure 2. The charge is determined using the sum of the signals from the three scintillators ($Z_{(S0+S1+S2)}$) and C1 for the energy fits in Interval 1, and the sum of the signals from the scintillators and C1 ($Z_{(S0+S1+S2)} + Z_{C1}$) and C0 for the energy fits in Interval 2.

To calculate the abundance of each element, the histogram is fit with multiple Gaussians, one for each elemental peak. The results from the Gaussian fit are shown superimposed over the scintillator histogram in Figure 2 and Figure 3.

Inefficiencies in the hodoscopes presented the only problem with TIGER during the 1997 Balloon Flight. Multiple hits in the Course Hodoscope produced by knock-on electrons reduced the instrument's ability to detect high-Z, high-energy cosmic rays. The hodoscope bias introduced a charge dependent bias in the TIGER data. To determine this bias, the TIGER data are compared to HEAO-3-C2 data [2] and corrected empirically [10].

The Co/Ni ratio grows with depth in the atmosphere as the Ni and Co interact and break up into lighter nuclei. We extrapolated the observed Co/Ni to the top of the atmosphere using growth curves for Mn/Fe [4]. This method is valid because the values of Mn/Fe and Co/Ni are both on the order of 0.11, assuming that the partial cross-section of Ni to Co is approximately the same as that of Fe to Mn.

The Co/Ni ratio measured by TIGER corrected to the top of the atmosphere is 0.093 ± 0.037 (Table 1). In the calculation of this number it is assumed that the elemental abundance peaks are Gaussian. However, as seen in Figure 3 there is no resolved Co peak, and the Fe peak appears to have a slight non-Gaussian tail, which could account for the amount of Co inferred from the Gaussian fits. Therefore, the above value of the Co/Ni ratio should be taken as an upper limit.

Comparison of Co/Ni Measurements

Figure 4 shows the TIGER Co/Ni ratio compared with other measurements. The lines in Figure 4 are calculations [14] that assumed either that ^{59}Ni had not decayed (solid line) or that it had fully decayed

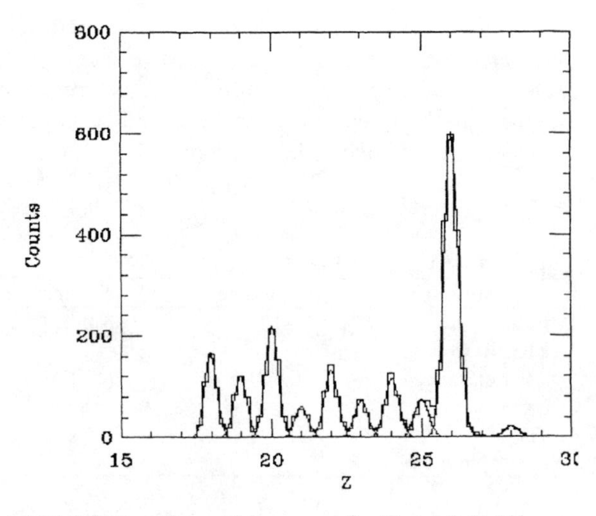

FIGURE 2. Charge histogram for E < 4.3 GeV/n

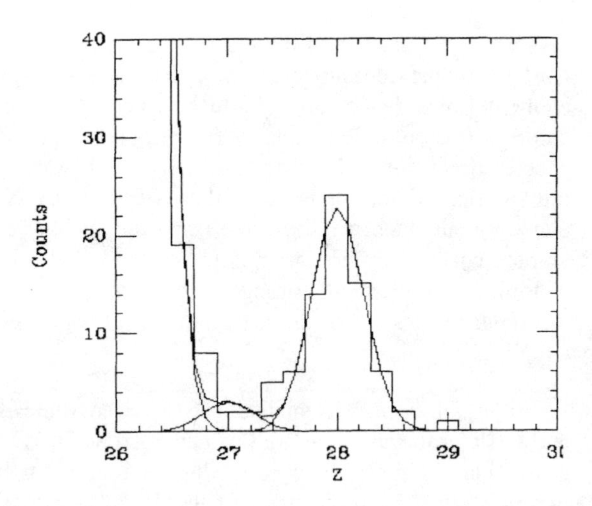

FIGURE 3. Charge histogram for E < 4.3 GeV/n in Co-Ni region.

(dotted line). The TIGER Co/Ni ratio (upper limit) of 0.093 ± 0.037, which is lower than previous balloon measurements, is consistent with the HEAO-3-C2 Co/Ni ratio for energies between 0.8 and 4.3 GeV/nucleon. This value is also consistent with the solid Webber & Gupta [14] line suggesting that the ^{59}Ni at the source has not had time to decay to ^{59}Co. At face value, the TIGER measurement and the general trend of the HEAO data suggest that a substantial amount of ^{59}Ni has not decayed, implying a short time delay of <7.6x10^4 years between nucleosynthesis and acceleration, although the "complete-decay" line is only 2σ above the TIGER measurement.

On the other hand, the ACE/CRIS Co/Ni elemental measurement of 0.137 ± 0.13 [7] at energies between 150 - 500 MeV/nucleon is 2σ away from the "no-decay" line and 2σ away from the "complete-decay" line. We note this interpretation of the elemental Co/Ni ratios is not consistent with the isotopic measurement made by ACE/CRIS at < 500 MeV/nucleon. The ACE-CRIS mass histograms of Ni and Co show that at earth there is a substantial amount of ^{59}Co in the GCRs and almost no ^{59}Ni. That result indicates that decay of ^{59}Ni to ^{59}Co has occurred at the source, and a time delay of longer than 7.6x10^4 years has elapsed before the particles were accelerated.

Discussion

Several possible explanations may be suggested for this discrepancy between the indication from our element ratio of Co/Ni that ^{59}Ni has not decayed, and the clear ACE/CRIS evidence that it has decayed. The problem most likely lies with the interpretation of the elemental measurements rather than the highly accurate ACE/CRIS isotopic measurement.

Interpretation of the elemental abundance of Co/Ni is very model dependent. Different models concerning the amount of Ni produced in SN nucleosynthesis and the interstellar propagation of Ni and Co would change the interpretation of the Co/Ni ratio. For example, a shorter propagation grammage would mean less Ni could interact in the ISM, producing less Co, and thereby yielding a lower predicted Co/Ni ratio. Therefore, a low Co/Ni ratio at earth might not mean that there is much Ni back at the source. Because of this, one possible explanation is that the models used in Webber & Gupta [14] are flawed, using source abundances of Co and Ni different from those of the Solar System or incorrect propagation grammages through the ISM.

Another possible explanation is that the total cross-sections of Ni and Co and the partial cross-section of Ni to Co used to produce the curves in Figure 3 are incorrect. The cross-sections used in Webber & Gupta [14] were measured at 600 MeV/nucleon using a CH_2 - C target subtraction technique to determine the cross-section in Hydrogen. More recently, Webber *et al.* [13] have made a measure of the cross-section of ^{58}Ni to Co using a H target with thickness approximating the amount of H traversed by GCR in the ISM. The new ^{58}Ni to Co partial cross-section of 116.8 ± 5.94 mbarn is consistent with the 1990 cross-section of 119.1 ± 7.5 mbarn. This new measure of the partial cross-section was again made at energies between 400 and 650 MeV/n. It is possible that the cross-sections of Ni and Co need to be measured in the GeV/n energy range to account for an energy dependence.

Lastly, the answer to the discrepancy between the Co/Ni isotopic measurements and the higher-energy elemental abundance ratios could be astrophysical in nature. It is possible that the phenomena behind GCR acceleration differ at low and high energies. However, the fact that the interpretation of the ACE/CRIS Co/Ni elemental abundance ratio, made using the Webber & Gupta curves, differs from the clear ACE/CRIS isotopic measurement strongly suggests that the discrepancy lies with the calculations of expected Co/Ni ratio.

All in all, improved models of SN nucleosynthesis and GCR propagation, as well as new total and partial cross sections for Ni and Co at higher energies will permit a cleaner interpretation of the Co/Ni elemental ratio. Those improved models, combined with measurements with further improved resolution and statistical accuracy will permit a better determination of the time between nucleosynthesis and acceleration of cosmic rays above 1 GeV/nucleon.

TABLE 1. TIGER Co/Ni Ratios

Measurement	Abundance	Error
From fit at TIGER	0.128	+/- 0.051
After efficiency correction	0.114	+/- 0.045
Top of Atmosphere	0.093	+/- 0.037

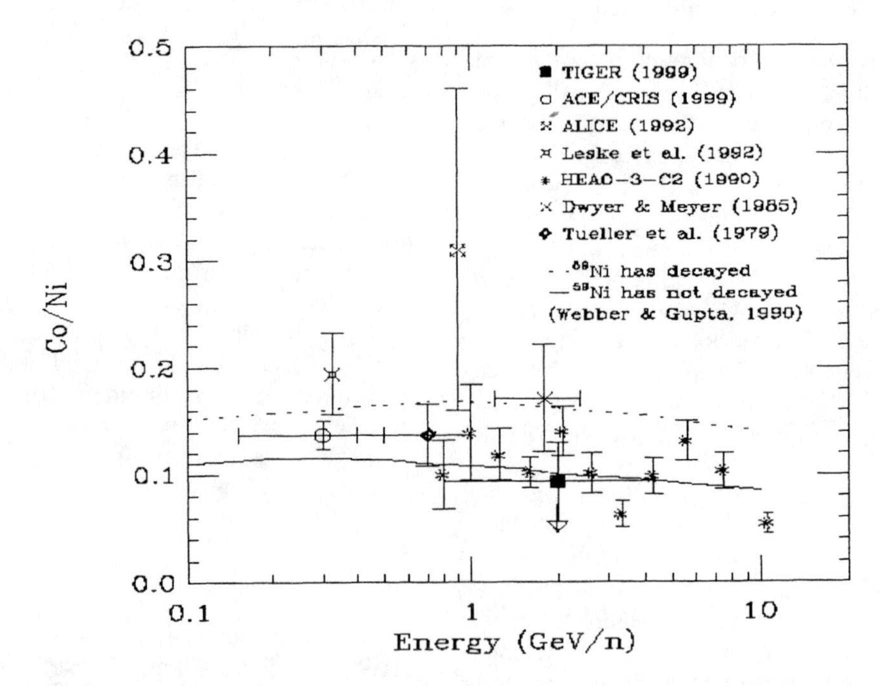

FIGURE 4. Recent measurements of the elemental abundance of Co/Ni. The solid and dotted lines are from Webber & Gupta [14]. Square – TIGER measurement from this work; open circle – ACE/CRIS [7]; asterisks – HEAO-3-C2 [2]; fancy cross – ALICE [3]; fancy square – Leske *et al.* [6]; cross – Dwyer & Meyer [1]; diamond – Tueller *et al.* [11].

This work was supported in part by NASA under grant NAG5-5110.

REFERENCES

1. Dwyer, R., and Meyer, P., *ApJ*, **294**, 441 (1985).
2. Engelmann, J.J, *et al.*, *A & A*, **233**, 96 (1990).
3. Esposito, J.A., *et al.*, *APH*, **1**, 33 (1992).
4. Israel, M.H., *et al.*, *ICRC*, **1**, 323 (1979).
5. Lawrence, D.J., *et al.*, *NIM*, **420**, 402 (1999).
6. Leske, R.A., *et al.*, *ApJL*, **390**, 99 (1992).
7. Lijowski, M., *et al.*, *ICRC*, **3**, 5 (1999).
8. Soutoul, A., *et al.*, *ApJ*, **219**, 753 (1978).
9. Sposato, S.H., *et al.*, SCIFI *97*, p. 527 (1997).
10. Sposato, S.H., *Washington University Ph.D. Thesis*, St. Louis, MO (1999)
11. Tueller, J., *et al.*, *ApJ*, **228**, 582 (1979).
12. Webber, W.R., *et al.*, *ICRC*, **1**, 325 (1987).
13. Webber, W.R., *et al.*, *ApJ*, **508**, 940 (1998).
14. Webber, W.R., & Gupta, M., *ApJ*, **384**, 608 (1990).
15. Wiedenbeck, M.E., *et al.*, *ApJL*, **523**, 61 (1999).

Cosmic Ray Source Abundances and the Acceleration of Cosmic Rays

J. S. George[1], M. E. Wiedenbeck[4], A. F. Barghouty[1], W. R. Binns[2], E. R. Christian[3], A. C. Cummings[1], P. L. Hink[2], J. Klarmann[2], R. A. Leske[1], M. Lijowski[2], R. A. Mewaldt[1], E. C. Stone[1], T. T. von Rosenvinge[3], N. E. Yanasak[4]

[1]*California Institute of Technology*, [2]*Washington University*, [3]*NASA / Goddard Space Flight Center*, [4]*Jet Propulsion Laboratory*

Abstract. The galactic cosmic ray elemental source abundances display a fractionation that is possibly based on first ionization potential (FIP) or volatility. A few elements break the general correlation of FIP and volatility and the abundances of these may help to distinguish between models for the origin of the cosmic ray source material. Data from the Cosmic Ray Isotope Spectrometer instrument on NASA's Advanced Composition Explorer spacecraft were used to derive source abundances for several of these elements (Na, Cu, Zn, Ga, Ge). Three (Na, Cu, Ge) show depletions which could be consistent with a volatility-based source fractionation model.

INTRODUCTION

The elemental abundances of the galactic cosmic rays (GCRs) reflect the abundances at their source and their evolution during propagation through the Galaxy. Careful modeling of the transport of cosmic rays through the intervening matter can provide an estimate of the source composition. The GCR source material is found to be similar in composition to the pool of material from which the solar system was formed, but with an observed elemental fractionation based on chemical properties such as the first ionization potential (FIP) or volatility (as indicated, for example, by the condensation temperature). The Cosmic Ray Isotope Spectrometer (1) (CRIS) on NASA's Advanced Composition Explorer (ACE) spacecraft has been measuring cosmic ray abundances for elements between helium (Z=2) and selenium (Z=34) at energies between 50 and 600 MeV/nucleon for over two years. These new data may help to determine what physical parameter controls the fractionation and provide a vital clue for understanding the source of galactic cosmic rays and the acceleration mechanisms that power them.

There are several explanations for the observed depletion of certain elements in the galactic cosmic ray source. One comes from a similar observation in the abundances of solar energetic particles, coronal material, and the solar wind. In those cases the elemental fractionation is ordered by the first ionization potential (FIP), or possibly first ionization time (FIT) (see e.g. (2)). Apparently, elements with a lower FIP (or FIT) are more easily ionized, enabling them to be more easily transported from the photosphere into the corona (3). The similarity of this depletion pattern to that seen in the GCRs led Cassé et al. (4), Meyer (5), and others to speculate that the GCR seed population might be stellar coronal material. An alternative explanation for the GCR fractionation is that the cosmic rays are grain destruction products whose relative abundances are controlled by their condensation temperatures, or volatility (6, 7). The possibility that the GCR source material comes from dust grains in the interstellar medium has recently been explored again in detail (8).

The two explanations for the GCR fractionation lead to very different views of the cosmic ray source. Fortunately, the general correlation between FIP and volatility is not complete; the abundances of several elements might be used to distinguish between them as an ordering parameter. Four such elements, copper, zinc, gallium, and germanium, lie just beyond the iron-nickel peak where abundances fall rapidly with increasing nuclear charge and secondary contributions from fragmentation of heavier nuclei are minimal. A fifth key element, sodium, is many times more abundant than zinc, but contains an estimated 70% contribution from secondary fragments, notably from magnesium and silicon. The sodium source abundance is much more sensitive to errors in the fragmentation cross-sections and the total amount of interstellar material traversed than the trans-nickel elements for which source abundances can be more reliably determined. For this reason, different methods were used

CP528, *Acceleration and Transport of Energetic Particles Observed in the Heliosphere: ACE 2000 Symposium*, edited by Richard A. Mewaldt, et al.
© 2000 American Institute of Physics 1-56396-951-3/00/$17.00

here to estimate the source abundances of sodium and the heavier elements.

TRANS-NICKEL ELEMENTS

Figure 1 shows a charge histogram of selected heavy element data recorded by CRIS over a two year period. The sub-peaks in copper and zinc correspond to different isotopes which were resolved for the first time in the CRIS instrument. The relative abundances of these isotopes, with relatively large statistical uncertainties, are consistent with the corresponding solar system values.

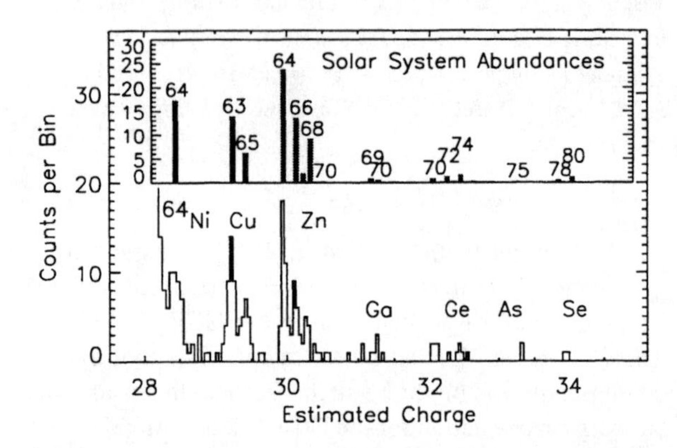

FIGURE 1. CRIS charge histogram for trans-nickel elements compared with solar system abundances (9).

A leaky box propagation model developed at the Naval Research Laboratory for other types of propagation studies (10) was used to model the effects of cosmic ray transport through the Galaxy. All stable isotopes (and those unstable on cosmic ray time scales) from Z=21 to Z=41 were propagated by solving the leaky box equations in the "weighted-slab" approximation in which nuclei are followed as they pass through successive thin slabs of interstellar material. The interstellar medium was taken to be 90% hydrogen and 10% helium by number. Losses due to fragmentation, radioactive decay, and escape from the Galaxy were taken into account, as well as gains from spallation and radioactive decay of heavier nuclides. The fragmentation of elements above zirconium (Z=40) make negligible contributions to the copper-germanium (Z=29-32) region.

The model used the partial cross-sections of Silberberg & Tsao (11, 12). The escape mean free path, Λ_{esc}, was taken to be a function of rigidity, R, and particle velocity, β, and gave results consistent with CRIS observations for the sub-iron/iron ratio.

$$\begin{aligned} \Lambda_{esc} &= 17.2\beta \quad \mathrm{g/cm^2}, & R < 4.0\mathrm{GV} \\ \Lambda_{esc} &= 17.2\beta(R/4.0\mathrm{GV})^{-0.6}, & R > 4.0\mathrm{GV} \end{aligned} \qquad (1)$$

The source abundances were taken initially to be those of the solar system (9) with elemental abundances adjusted as necessary to fit the observed values. Figure 2 shows the relative abundances of 20 elements near Earth. The calculated abundances include the effects of solar modulation with a modulation parameter ϕ=500MV.

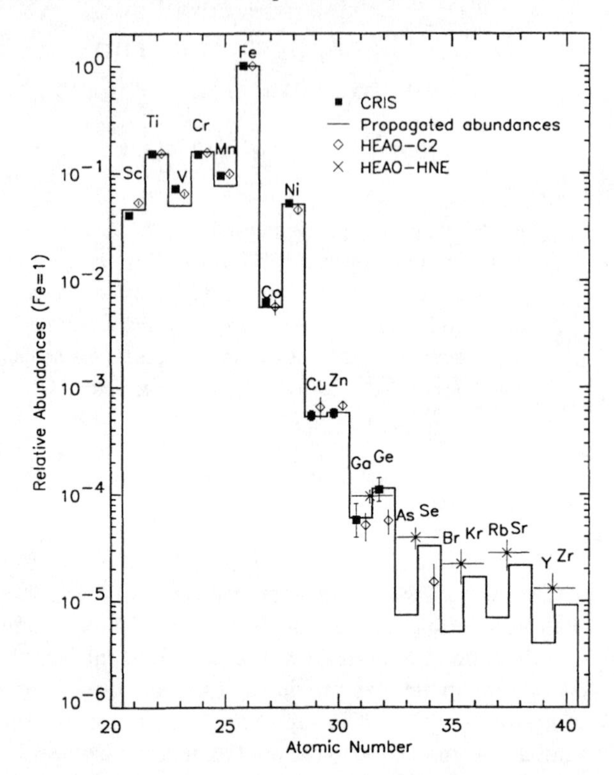

FIGURE 2. Relative near-Earth elemental abundances (Fe≡1). The histogram shows the results of the propagation calculation compared with HEAO-C2 data (13) and HEAO-HNE "low-energy" charge pair measurements (14).

The propagation model accounts reasonably well for the sub-iron elements (21≤Z≤25). These elements have minimal source abundances and are almost purely secondary. The discrepancies for vanadium and manganese are comparable to the uncertainties in the fragmentation cross-sections. Source abundances of elements heavier than iron were adjusted so that the propagated abundances matched CRIS data where available. Above germanium, source values were adjusted in pairs so that the sum of the propagated odd and even charge elements matched the measurements by Binns et al. (14). With the calculated abundances in agreement with observations above and below the elements of interest, secondary contributions to the copper-germanium region should be well represented.

The ratios of the galactic cosmic ray source (GCRS) abundances to the corresponding solar system (SS) values (9) are summarized in Table 1 for the trans-nickel elements. The uncertainties in the GCRS/SS ratios reflect the 1σ statistical uncertainties in the calculated GCR sources and solar system abundances, as well as an estimated uncertainty in the fragmentation cross-sections. The secondary contributions for copper, zinc, gallium, and germanium were calculated to be 16%, 6%, 23%, and 11%, respectively, by repeating the propagation calculation with one element at a time removed from the source. A 20% overall uncertainty in the cross-sections contributed at most a 5% uncertainty to the calculated source abundance.

Table 1. GCR Source relative to solar system (Fe$\equiv 10^6$).

Element	GCRS*	SS†	GCRS/SS
Copper	463±49	580±64	0.80±0.12
Zinc	573±58	1400±62	0.41±0.05
Gallium	45±14	42±2.9	1.07±0.34
Germanium	90±21	132±13	0.68±0.17

* This work.
† Anders & Grevesse (9).

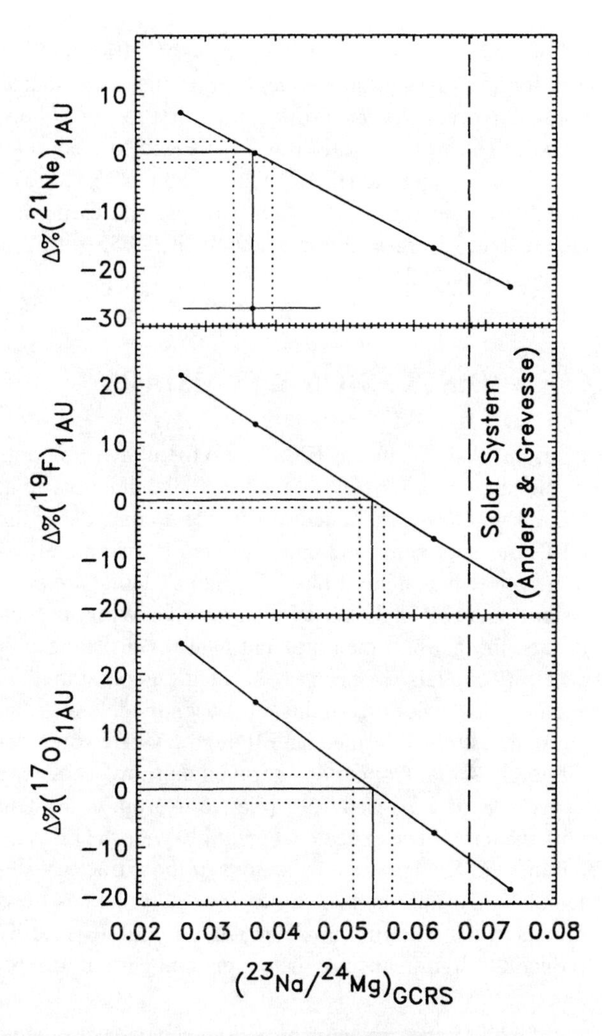

FIGURE 3. Constraints on the $^{23}Na/^{24}Mg$ source ratio. The vertical axis represents the percent difference ($\Delta\%$) between the calculated and observed abundances of nearby secondary isotopes near Earth. The point in the top panel illustrates the statistical uncertainty in the calculated $^{23}Na/^{24}Mg$ source ratio due to uncertainty in the local measurements. The solar system value (9) is indicated by the vertical dashed line.

SODIUM SOURCE ABUNDANCE

To determine the source abundance of sodium, several "purely secondary" nuclides with similar masses were used to constrain the calculated secondary corrections. In particular, the observed abundances of ^{21}Ne, ^{19}F, and ^{17}O should contain fragmentation products from the same parents that produce sodium. A series of steady-state leaky box propagation calculations (model described in (15)) were carried out while varying the escape mean free path, Λ_{esc}. For each Λ_{esc} the source ratios of ^{20}Ne, ^{22}Ne, ^{23}Na, ^{25}Mg, ^{26}Mg, ^{27}Al, ^{28}Si, ^{29}Si, and ^{30}Si relative to the stable primary isotope ^{24}Mg were adjusted to reproduce the locally observed values. Source abundances of heavier nuclides were held fixed at solar system values modified by a FIP fractionation. The source abundances of ^{17}O, ^{19}F, and ^{21}Ne were also based on solar system values, although these are negligible compared to the secondaries produced during propagation.

For each value of Λ_{esc}, the sodium source abundance needed to account for the observed sodium is determined, as well as the relative difference between the calculated and observed values of the three secondary nuclides being used to constrain the model. Figure 3 shows the correlation between the calculated sodium source abundance and these relative differences. When each secondary isotope is taken individually, our best estimate of the sodium source abundance corresponds to the point on the curve where the calculated and observed abundances of the tracer isotope agree ("0% difference"). These values are indicated by the light solid lines in the figure, and the 1σ statistical uncertainty limits are indicated as dotted lines. The differences between the three determinations of the sodium source abundance could be due, for example, to errors in the cross sections for producing the various secondaries.

A weighted average of the three determinations of the sodium source abundance was calculated and the spread among the different values used as an indication of the uncertainty arising from the calculation of the correction for secondary ^{23}Na. The statistical uncertainty in the measurement of sodium was included to obtain a best estimate for the source ratio $^{23}Na/^{24}Mg$ of 0.048 ± 0.015.

Compared to the solar system (9) ratio of 0.067, sodium was found to be marginally depleted in the GCR source relative to other elements with similar first ionization potential. The isotopic ratio was normalized to elemental iron (Fe≡1) using the HEAO (13) Mg/Fe GCRS value of 1.03±0.03 and the solar system isotopic and elemental abundances (9), resulting in a source GCRS/SS ratio for Na/Fe of 0.62±0.19.

DISCUSSION & SUMMARY

Figure 4 shows the GCRS/SS ratio for all five elements studied here as a function of the first ionization potential. HEAO data (13) are added to provide a context for the CRIS measurements and one possible FIP parameterization is drawn as a solid line. Gallium and zinc are consistent with FIP as the ordering parameter. With only ten events, the gallium measurement is also consistent with volatility models. Zinc, in spite of the good statistical accuracy, does not discriminate between models well because it is in the intermediate FIP region where some depletion is likely. Copper and germanium show respective depletions of 1.7σ and 1.9σ relative to iron, consistent with what might be expected if volatility were the relevant parameter. The apparent depletions of the refractory elements magnesium and silicon can be interpreted as mass effects in the volatility model (8) and do not necessarily reduce the significance of the copper and germanium results.

The CRIS instrument is providing new measurements of key elemental source abundances in the galactic cosmic rays. Two of these elements, zinc and gallium, are consistent with either a FIP or volatility fractionation of the source material. Three other elements; sodium, copper, and germanium, show depletions relative to elements of similar first ionization potential which could be consistent with a volatility dependent fractionation model of the GCR source material.

The uncertainty in the sodium source abundance is dominated by the systematic uncertainty in the correction made for the secondary contribution to the observed abundance. The source values for the rare heavy elements beyond nickel are all limited by uncertainties in the measurements and in the solar system abundances. Even with the large CRIS geometry factor, collection rates for the trans-nickel nuclei are low, especially as solar maximum approaches. Nevertheless, a five-year total mission could allow a 50% increase in the available data set. This, along with improvements in propagation models for elements from carbon to nickel, suggests that over the next few years, CRIS measurements do have the potential to discriminate between FIP and volatility as the controlling parameter for the elemental fractionation of the galactic cosmic ray source material.

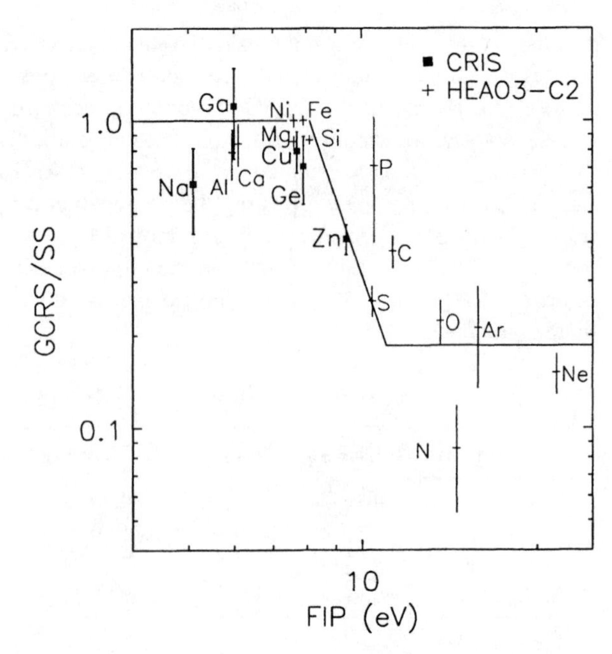

FIGURE 4. GCR Source / Solar System as a function of FIP (Fe≡1). HEAO data (13) added for context.

This work was supported by NASA at the California Institute of Technology (under grant NAG5-6912), the Jet Propulsion Laboratory, the Goddard Space Flight Center, and Washington University.

REFERENCES

1. Stone, E.C., et al., *Space Sci. Rev.* **96**, 285-356 (1998).
2. Geiss, J., *Space Sci. Rev.* **85**, 241-252, (1998).
3. Hénoux, J.-C., *Space Sci. Rev.* **85**, 215-226, (1998).
4. Cassé, M. & Goret, P. *ApJ* **221**, 703 (1978).
5. Meyer, J.-P., *ApJS* **57**, 173-204 (1985).
6. Epstein, R.I., *MNRAS*, **193**, 723 (1980).
7. Bibring, J.-P., and Cesarsky,C.J. *Proc. 17th ICRC (Paris)*, **2**, 289 (1981).
8. Meyer, J.P., Drury, L., & Ellison, D.C., *ApJ* **487**, 182-196 (1997).
9. Anders, E. & Grevesse, N., *Geochim. Cosmochim. Acta* **53**, 197-214 (1989).
10. Tsao, C.H., Silberberg, R., Barghouty, A.F., Sihver, L., *ApJ* **451**, 275 (1995).
11. Silberberg, R., Tsao, C.H., and Barghouty, A.F., *ApJ* **501**, 911 (1998).
12. Tsao, C.H., Silberberg, R., and Barghouty, A.F., *ApJ* **501**, 920 (1998).
13. Engelmann, J.J., et al., *Astron. Astrophys.* **233**, 96-111 (1990).
14. Binns, W.R., et al., *Proc. 18th ICRC (Bangalore)* **9**, 106 (1983).
15. Leske, R.A., *ApJ* **405**, 567-583 (1993).

Propagation of elements just heavier than Nickel

C. Jake Waddington

School of Physics and Astronomy, University of Minnesota
116 Church St. S. E., Minneapolis, MN 55455

Abstract: A study has been made of the sensitivity of observations of the elemental abundances of those nuclei just heavier than $_{28}$Ni to assumptions concerning propagation and source abundances. Calculations of the propagation of elements with $42 \geq Z \geq 27$ have been based on various models for the source abundances. These calculations have compared the predicted elemental ratios at a typical energy in order to look for significant discriminators between the various sources.

INTRODUCTION

Although the abundances of those elements just heavier than Ni decline rapidly with increasing charge, they are still the most abundant ultra heavy, UH, nuclei in the cosmic radiation. While preliminary observations of their abundances have been available for many years, Binns et al. (1989), new and better values should be obtained in the near future. The TIGER detector, Lawrence et al. (1999), will be flown on a 100 day high altitude balloon flight by the end of the next year. The results from this flight will determine the abundances of the individual elements with $Z \leq 40$ for the first time. To understand these results and to use them to predict the source abundances requires calculations of the effects of propagation through the interstellar medium. This paper describes such calculations assuming a wide range of possible source abundances.

PROPAGATION

These calculations use the Tsao et al. (1999) predicted isotopic cross sections for energies between 5.0 and 0.6 A GeV, in a weighted slab leaky box model. For the purposes of direct comparisons all these calculations have been based on the same assumptions. They assume no truncation and no re-acceleration. In addition, the escape length has been set to be independent of energy and = 8.0 g.cm^{-2}. The total charge changing cross sections have been assumed to be independent of energy. All isotopes with half lives < 10^7 y have been allowed to decay to a stable isotope, but K-capture decays have been suppressed unless there is an energetically possible alternate mode. Propagation has been assumed to be in a medium of pure atomic hydrogen, Ginzburg and Syrovatskii (1964)

The cross sections are energy dependent and since the nuclei lose energy due to ionization losses as they propagate it is necessary to use the appropriate cross sections for each element as they move from slab to slab. Fig. 1 shows the energy spectra of source and some fragment elements for $_{40}$Zr source nuclei with an initial energy of 2.0A GeV.

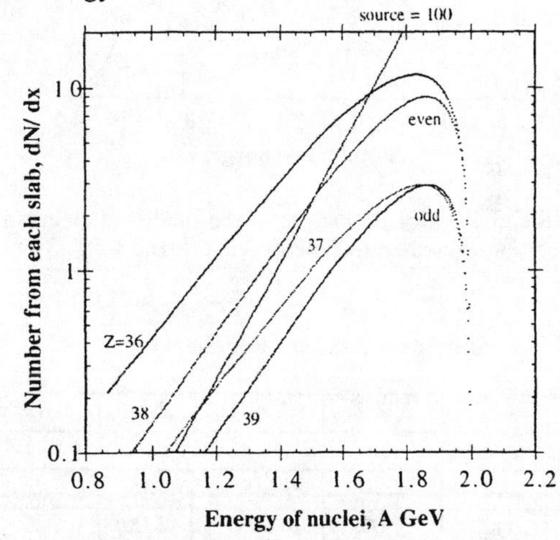

FIGURE 1. The energy spectra of primary and fragment nuclei. The numbers of nuclei in each 0.05 g.cm^{-2} slab are shown for 2.0A GeV $_{40}$Zr source nuclei and several of the heavier fragments produced during propagation.

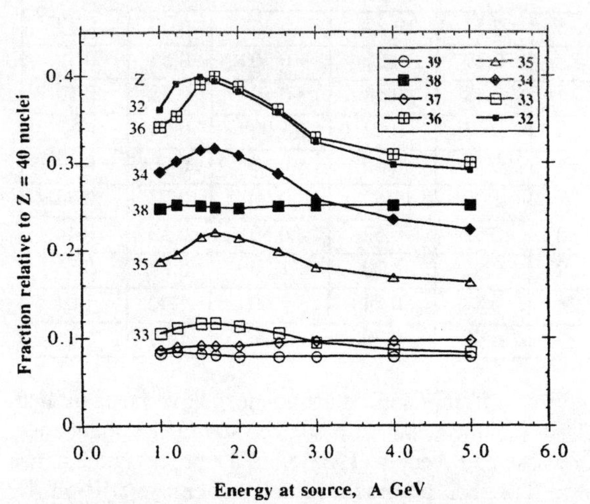

FIGURE 2 .The yields after propagation of various fragments from $_{40}$Zr as a function of source energy.

CP528, *Acceleration and Transport of Energetic Particles Observed in the Heliosphere: ACE 2000 Symposium*, edited by Richard A. Mewaldt, et al.
© 2000 American Institute of Physics 1-56396-951-3/00/$17.00

It can be seen that there will be significant numbers of nuclei detected with energies appreciably less than at the source. The result is that the yields of the various fragments do vary with energy. Fig. 2 shows the yields for a number of fragments resulting from the propagation of $_{40}$Zr at the source as a function of the source energy. However, overall these energy dependent effects are quite small. In Fig. 3 fractional changes in the propagated abundances are shown between two source energies of 5.0 and 1.5 A GeV. At most, in this case, there is a 5% effect between neighboring elements due to the energy difference.

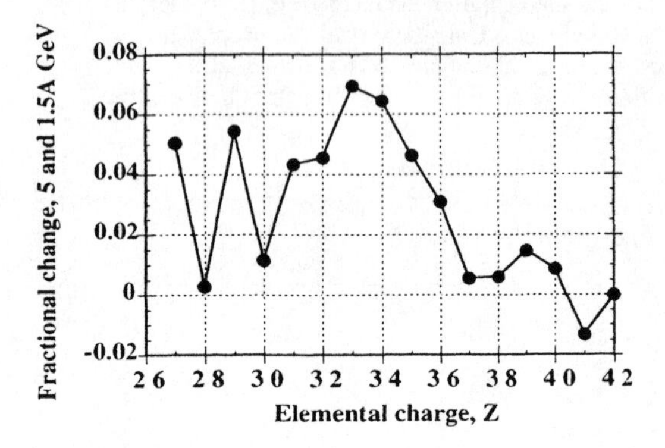

FIGURE 3 Fractional changes in the predicted abundances for an SS source with initial energies of 5.0 and 1.5 A GeV.

TABLE 1 Source correction factors.

Z	FIP	Vol 1	Vol 2	M1	M2
42	0.97	2.50	1.00	1.00	1.13
41	1.00	1.00	1.00	1.00	1.12
40	1.00	1.10	1.00	1.00	1.11
39	1.00	1.00	1.00	1.00	1.11
38	1.00	1.30	0.50	1.00	1.11
37	1.00	1.00	0.50	0.86	0.90
36	0.17	0.25	0.25	0.39	0.39
35	0.27	0.27	0.50	0.55	0.51
34	0.48	0.48	0.50	0.55	0.50
33	0.47	0.47	1.00	0.94	0.88
32	0.78	0.60	0.50	0.66	0.47
31	1.00	1.50	0.50	0.73	0.86
30	0.52	0.52	0.50	0.53	0.44
29	0.82	1.20	1.00	0.84	0.85
28	0.84	0.84	1.00	1.00	1.01
27	0.79	0.79	1.00	1.00	1.01

Eight different source abundances have been studied. These include a standard solar system (SS) abundance, Grevasse and Anders (1988), and a r-process abundance (r-). The SS abundances have been modified by correction factors based on the first ionization potentials (FIP); volatility, (Vol1 and Vol2); and melting point (M1 and M2). These correction factors are shown in Table 1. The r-process abundance has been modified by FIP. Fig. 4 shows these abundances. It can be seen that for many of the elements there are abundance differences of factors of two or greater.

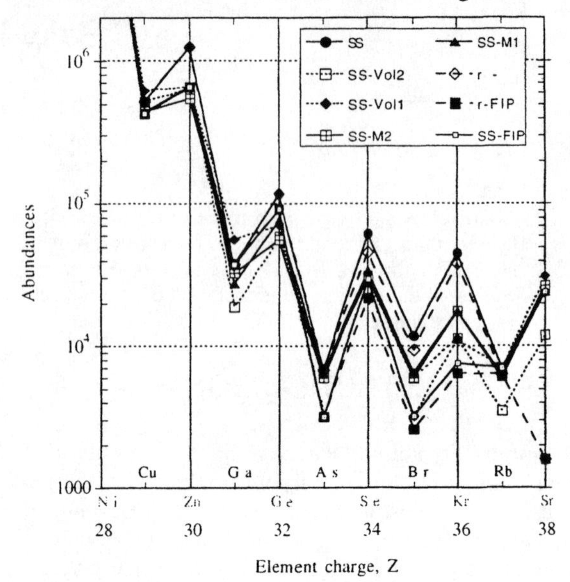

FIGURE 4 Various source abundances, see text.

The propagation calculations show that for many of the odd charged elements in this charge range more secondary fragments will be observed than primary nuclei. Fig. 5 shows the ratios of secondary to primary nuclei in the case of a SS abundance with 1.5A GeV at the source. $_{27}$Co, $_{31}$Ga, $_{33}$As and $_{35}$Br all have more secondary fragments than primary nuclei. Clearly determining the source abundances of these elements will be very sensitive to the assumptions of the propagation calculations.

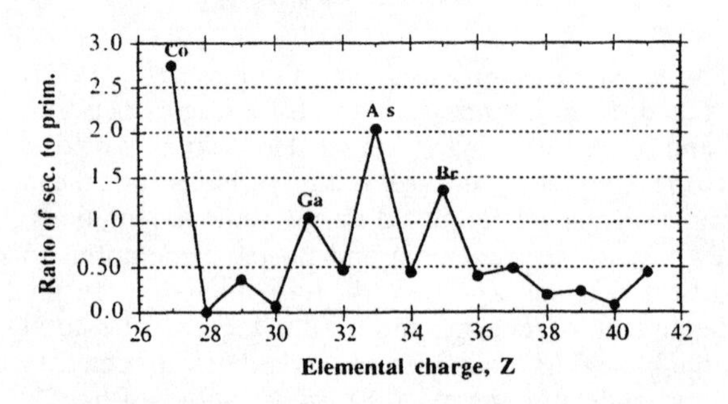

FIGURE 5. The ratio of secondary fragments to primary nuclei after propagation of a SS source abundance with an initial energy of 1.5 A GeV.

COMPARISON WITH DATA

At the present time our best data on the abundances of cosmic ray nuclei in this charge range still rests on the old data from HEAO and Ariel. These observations lacked to resolution needed to separate the individual elements but instead gave the abundances for each even charged element combined with that of the neighboring lighter odd charge element. The "UHGCR" values reported by Binns et al. (1989) are compared with those predicted from these propagation calculations in Fig. 6. In this figure the abundances of each pair have been normalized to that of the ($_{34}$Se + $_{33}$As) pair. Examination of this figure does not show a good match with any of the assumed source compositions, although the best match would appear to be to a SS-FIP source.

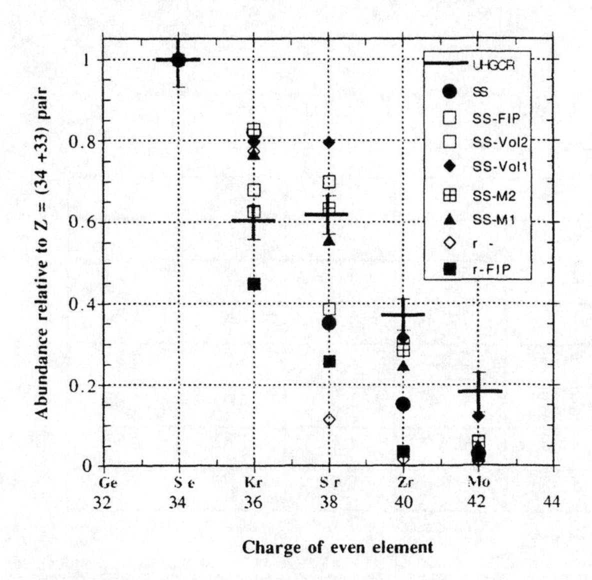

FIGURE 6 The abundances of elemental pairs, normalized to the value for the Se+As pair, as reported by Binns et al. (UHGCR) and as predicted for the various sources at an initial energy of 5.0A GeV. The UHGCR values show the reported uncertainties.

The next generation of such observations should be initiated by the forthcoming TIGER flight. This detector will be able to resolve the individual elements and hence study the ratios of the even to odd charge abundances . These ratios are much more sensitive to the nature of the source composition than are those from the earlier observations. The predicted values for these ratios are shown in Fig. 7 and are listed in Table 1. Also shown in Fig. 7 are the values of these ratios if the production of fragments were suppressed, with all interactions resulting in destruction of the incident nuclei. It can be seen that the effects of fragment production significantly change the predicted ratios. In Table 1 those ratios that are within 10% of the SS values are highlighted. It can be seen that for every assumed source there are ratios which differ from the values for a SS source by more than 10%, in many cases by much more.

CONCLUSIONS

Using the ratios of even to odd elements allows a sensitive comparison between the various assumed source compositions. While estimates of the fluxes of individual elements are strongly dependent on the assumptions in the propagation calculations, the ratios reduce this dependence. Hence, while the actual values of the ratios may still be somewhat uncertain, the comparisons with those for other source compositions should be significant. It can be concluded that an observation that determines these ratios with 10% or better accuracy will be able to distinguish between these different source scenarios. The report at this meeting from ACE of a preliminary value for the Zn/Cu ratio of ≈ 1.0, George et al.(2000) matches nicely with some of the predictions listed in Table 2.

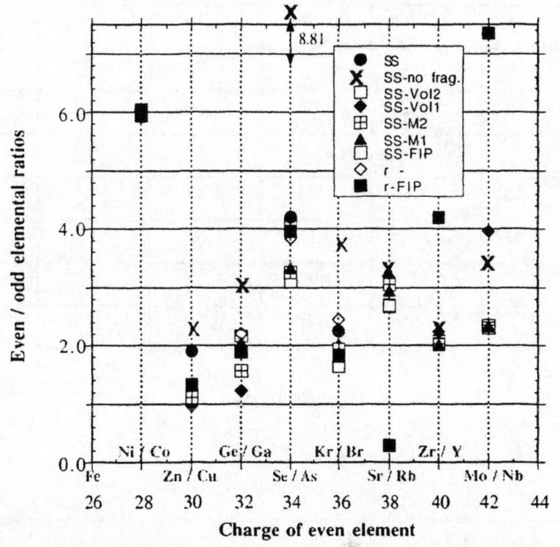

FIGURE 7 Even - odd ratios of elemental abundances as predicted for the various source models. Also shown for a solar system source are the ratios if it is assumed that interactions during propagation produce no secondary fragments.

There are some caveats to these predictions. Since only nuclei with $Z \leq 42$ were considered here, the influence of fragments from the low abundance heavier nuclei have not been included. This would have the most significant effect on the Mo/Nb ratios in Table 2. Also the Ni/Co ratios were calculated with just one set of assumptions regarding the decay or stability of various isotopes. Changing these assumptions can materially affect the values of the ratios, which should not be taken too seriously at this time. Finally, it is clear that these results are critically dependent on the assumed cross sections. Better values, resting on a sound experimental and theoretical foundation, will be needed before the true source composition can be determined with confidence.

ACKNOWLEDGEMENTS

This work partially supported by NASA under contract No.NASA/NAS5-5113.

REFERENCES

C. H. Tsao, R. Silberberg and A. F. Barghouty, 1999, ICRC. Salt Lake City, **1**, 13, and references therein.

W. R. Binns et al., Ap. J.1989, **346**, 997

Grevasse, N. and Anders, E, 1988 AIP Conf. Proc. 183, p.1 "Cosmic Abundances of Matter, ed. C.J. Waddington

Ginzburg, V. L. and Syrovatskii, S. I. 1964 "The Origin of Cosmic Rays", (NY Pergamon).

D. J. Lawrence et al. 1999, Nuc. Ins.& Meth. **A420**, 402

J.S. George et al. 2000, these proceedings

TABLE 2. Elemental ratios

5 A GeV		SS	SS-FIP	r-	r-FIP	SS-Vol2	SS-Vol1	SS-M2	SS-M1	SS-no frag.
42 / 41	**Mo / Nb**	2.34	2.30	7.36	7.36	2.34	3.97	2.36	2.34	3.44
40 / 39	**Zr / Y**	2.02	2.02	4.20	4.20	2.02	2.25	2.03	2.02	2.31
38 / 37	**Sr / Rb**	2.68	2.68	0.30	0.30	2.71	3.18	3.06	2.95	3.34
36 / 35	**Kr / Br**	2.26	1.82	2.46	1.84	1.65	1.99	1.94	1.89	3.76
34 / 33	**Se / As**	4.21	4.12	3.85	3.97	3.11	3.98	3.27	3.34	8.81
32 / 31	**Ge / Ga**	2.18	1.89	2.21	1.89	2.16	1.24	1.57	2.05	3.05
30 / 29	**Zn / Cu**	1.91	1.34	1.91	1.33	1.08	0.97	1.12	1.32	2.35
28 / 27	**Ni / Co**	5.93	6.04	5.93	6.04	5.95	6.04	5.95	5.95	20.94

Ratios shaded are within 10% of SS values

The Origin of Cosmic Rays: What Can GLAST Say?

Jonathan F. Ormes, Seth Digel, Igor V. Moskalenko[1], and Alexander Moiseev,

Goddard Space Flight Center, Greenbelt, Maryland

and Roger Williamson

Stanford University

on behalf of the GLAST collaboration

Abstract. Gamma rays in the band from 30 MeV to 300 GeV, used in combination with direct measurements and with data from radio and X-ray bands, provide a powerful tool for studying the origin of Galactic cosmic rays. Gamma-ray Large Area Space Telescope (GLAST) with its fine 10-20 arcmin angular resolution will be able to map the sites of acceleration of cosmic rays and their interactions with interstellar matter. It will provide information that is necessary to study the acceleration of energetic particles in supernova shocks, their transport in the interstellar medium and penetration into molecular clouds.

INTRODUCTION

The identification of the sites of cosmic-ray acceleration is one of the main unsolved problems in Galactic cosmic-ray astrophysics. We know from radio [1] and recent X-ray observations [2] of synchrotron radiation from supernova remnants that electrons are accelerated to TeV energies. Direct evidence for the acceleration of protons is not yet in hand. There is a considerable theoretical literature that quantifies how this acceleration takes place in the turbulent magnetic fields associated with the shock waves generated by supernova explosions propagating into the interstellar medium [3]. With the Gamma-ray Large Area Space Telescope (GLAST), proposed for launch in 2005, we have the possibility of detecting γ rays from the freshly accelerated cosmic-ray nuclei at their acceleration site. The optimal candidate sites will be those where the shock waves collide with either swept up interstellar matter or nearby clouds. The most commonly discussed site is individual supernova remnants [4] however multiple supernovas inside superbubbles have also been considered [5].

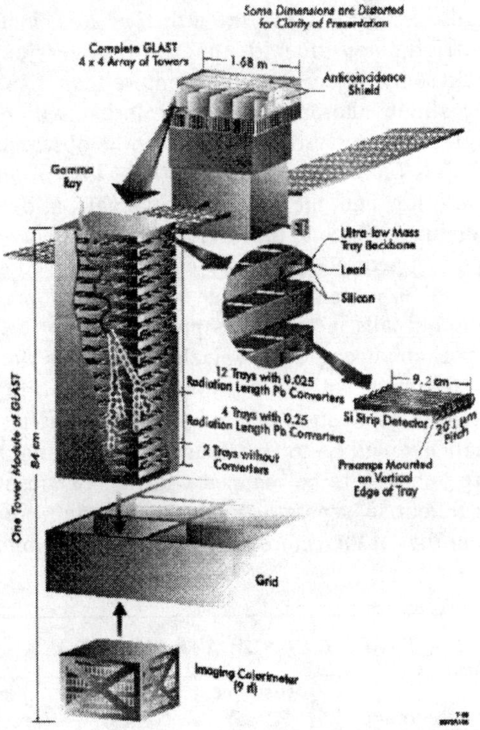

Fig. 1: The GLAST large area telescope

[1]NRC Resident Research Associate visiting GSFC from Inst. for Nucl. Phys., Moscow State U.

CP528, *Acceleration and Transport of Energetic Particles Observed in the Heliosphere: ACE 2000 Symposium*, edited by Richard A. Mewaldt, et al.

© 2000 American Institute of Physics 1-56396-951-3/00/$17.00

While the great majority of the cosmic rays are nuclei, not electrons, most of the electromagnetic signatures of cosmic rays reflect electrons as seen in synchrotron radiation or inverse Compton (IC) interactions of photons with electrons. This includes even the highest energy band, photons around 1 TeV. To identify the site where the bulk of the energy is pumped into relativistic particles we need to find a signature of accelerated nucleons. Such a signature will be the spectrum reflective of neutral-pion decay from collisions of freshly-accelerated nuclei with nearby gas and dust[2]. Depending of the conditions in the medium the neutral-pion decay γ rays will be evident as a spectral feature in the range 50 MeV to a few GeV, while above this range their spectral power law index will match the spectral index of the cosmic rays at the site. This range is ideal for studies with GLAST.

GLAST CAPABILIES & SNR

The GLAST telescope we have proposed is shown in Figure 1. Its capabilities are summarized in Figure 2. At 1 GeV, the effective area, including inefficiencies due to photon conversion and background rejection, will be more than 1 m^2, and our single photon angular resolution will be 0.4 degrees. Supernova remnants can be observed well off axis (up to 60 degrees) without loss of angular resolution and the 2.4 sr solid angle allows any individual source to be observed at a 20% duty cycle. Thus, supernova remnants can be observed in the normal sky-scanning mode of GLAST operation. Minimal tails in the point-spread function (the 95% containment angle is typically 2.5 times the 68% containment angle) optimize the capability for mapping the structure of γ-ray emission. The challenge will be to find sources with angular scales large enough to be mapped with angular resolution sufficient to separate the extended shell emission from that of the central compact source, nominally a pulsar.

TABLE 1. CANDIDATE SUPERNOVA

Remnant	Distance (kpc)	Age (years)	Size (arcmin)
RXJ1713.7-3946	6	??	70
γ Cygni	1.8	7000	60
IC443	1.5	5000	45
SN1006	~3	1000	30
CasA	2.8	300	5
Kepler	4.4	400	3

[2] $P+P \rightarrow \pi^o+X$ and $\pi^o \Rightarrow 2\gamma$

The obvious candidates for this search are those SNR for which EGRET reported a finite flux [6]. We also include RXJ1713.7-3946 for which TeV emission has recently been reported [7]. They are listed in Table 1 along with relevant parameters. To predict the signal GLAST might detect for a typical case, we have modeled the γrays from γ Cygni taking the total flux as measured by EGRET and dividing it into two separate components with 60% of the flux from the pulsar [8].

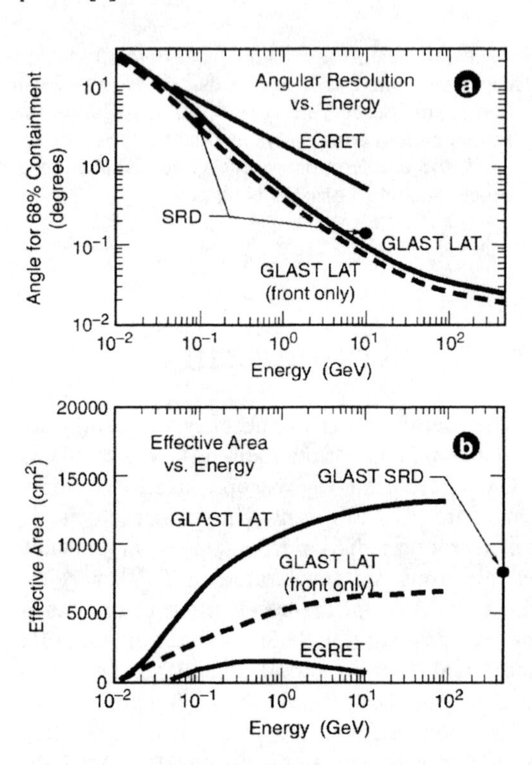

Fig. 2: Performance characteristics of GLAST

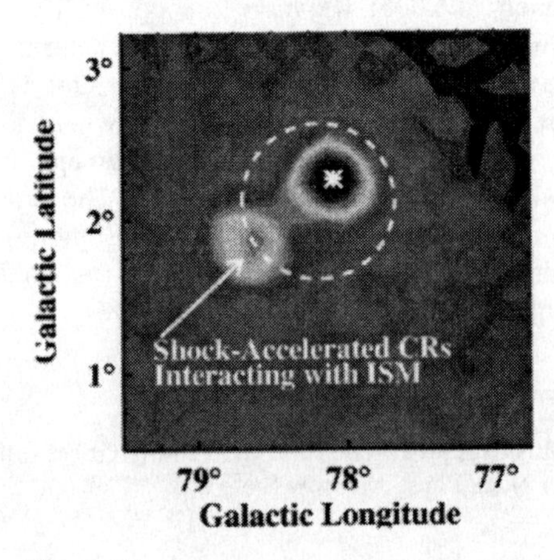

Fig. 3: Simulated observation of γ Cygni

The result of this simulation is shown in Figure 3. Note that the shell component is clearly resolved from the central point source. The spectrum of the γ Cygni region is taken from Ref. 6.

Assuming this source (γ Cygni) is as modeled we expect to spatially resolve the shell source and to be able to measure the spectrum [9] shown in Figure 4. The locally observed cosmic-ray nuclei have spectra proportional to $E^{-2.7}$. This spectrum is the natural result of shock models which predict harder source spectra, $E^{-\delta}$ with spectral exponents δ in the range 2.0 - 2.3 combined with the energy dependent losses by diffusion. The IC and bremsstrahlung components can be deconvolved and subtracted to determine the π^0 contribution. The asymptotic spectral exponent at energies above 10 GeV should be indicative of the source spectrum of freshly-accelerated cosmic rays.

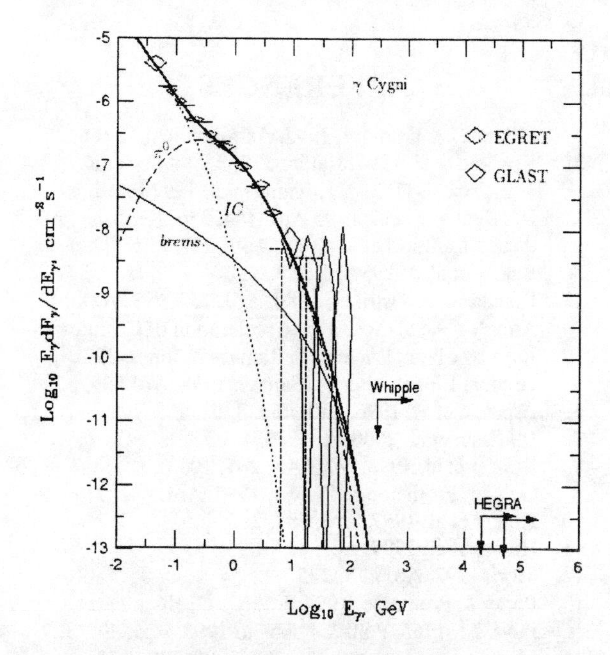

Fig. 4: Expected data from GLAST compared to one of the models of Gaisser et al. [9].

DIFFUSE CONTINUUM GAMMA-RAYS

Once accelerated, the cosmic rays quickly diffuse away from their sources, mix with those from other sources and diffuse out of the Galaxy. Most theoretical treatments of cosmic-ray transport from their sources to the Earth assume the result of this process is a steady flux of cosmic rays throughout the Galaxy with sources in equilibrium with losses. With GLAST we have a chance to probe the spatial structure on a scale that can begin to test this hypothesis.

Of the many possible observables of cosmic-ray transport through the interstellar medium to earth (e.g. Be, B isotopes, positrons, antiprotons) the least exploited to date has been γ rays. The diffuse γ-ray emission in any direction is a measure of the point by point product of the cosmic-ray intensity and the matter density in the interstellar medium. The secondary antiprotons and positrons are produced in the same collisional processes. The EGRET team has modeled this diffuse γ-ray emission [10] on 0.5 degree scale. Their models distribute the cosmic rays proportionally to the matter in the Galaxy on scales of order 1 kpc.

One of the most puzzling findings from EGRET is the excess in the Galactic diffuse γ-ray emission above 1 GeV [11]. Independent confirmation and extension toward the higher energies would help to understand this excess. Is the spectrum of cosmic-ray nuclei [12] or electrons [13] elsewhere in the Galaxy different from what we measure locally? If so, what are the corresponding consequences for all cosmic-ray physics? The range of such possibilities have been recently discussed in the literature [14].

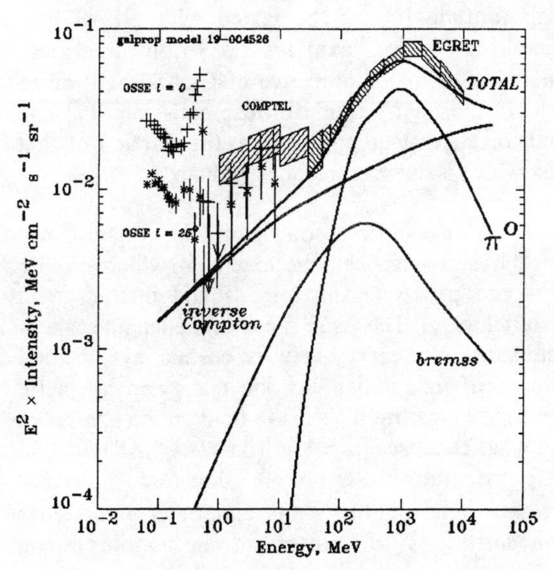

Fig. 5: A model [14] of the Galactic diffuse emission that fits the high energy EGRET data.

The self-consistent approach to this problem developed by Strong et al. [14] models the diffusive propagation of nucleons and electrons. Figure 5 shows their predictions of the various contributing components. Their conclusion, reflected in the figure, is that the harder electron spectrum in the Galaxy is more likely. But their solution is not unique and requires confirmation with the additional data that GLAST will provide. Measurements of high-energy antiproton and positron spectra provide a useful constraint on the nucleon spectrum, while the electron spectrum can be different from place to place due to the large electron energy losses (though radio measurements of synchrotron emission provide a constraint for some electron energies). These hypotheses can be tested with GLAST by measuring the spectrum of Galactic diffuse γ-ray emission above few GeV. First, one can distinguish between the inverse Compton and neutral-pion decay components, and, second, the latitude and longitude γ-ray profiles are different for gas-related nucleon and broader electron components.

Another result of the studies [15] is that the inverse Compton (IC) scattering, especially at high latitudes, plays a more important role than previously thought. The anisotropic distribution of photons above the Galactic plane will result in greater (up to 40%) flux from IC in the halo. Evidence for a large γ-ray halo has been found in an analysis of the EGRET data [16]. This affects our estimates of the extragalactic component, the intensity of cosmic rays in the Galaxy at large and the size of the Galactic cosmic-ray containment halo. The importance of high latitude IC can be tested with GLAST by measuring γ-ray profiles at high energies. Therefore, a major objective of GLAST will be to resolve the high latitude diffuse emission [17] and find, or at least greatly reduce, the fraction of that flux which is due to unresolved point sources.

A longstanding question in cosmic-ray astrophysics has been the extent to which cosmic rays can penetrate into clouds and interact with material there. The issue is one of the length scale of equipartition of energy between cosmic rays, thermal motion of interstellar gas and energy in magnetic fields. Equipartition seems to hold on the kpc scale, but what about scales below 1 kpc? GLAST will be able to study the γ-ray emission from molecular clouds, super-bubble walls and other interstellar concentrations of matter on an angular scale heretofore impossible reaching the tens of arc minute range. At a distance of 1 kpc, spatial scales of 3-10 parsec will be resolved.

CONCLUSION

Gamma rays are a powerful tool to explore the sources and distribution of cosmic rays in our Galaxy. In turn, having improved understanding of the role of cosmic rays is essential for the study of many topics in γ-ray astronomy to be addressed by GLAST. The results of the studies discussed here will have important consequences for understanding of the distribution of cosmic rays in the Galaxy and their dynamic effects thereupon. Besides, it is worth noting that complete understanding of both the Galactic cosmic rays and the γ-ray emissions are essential for the study of the dark matter in the Galaxy.

REFERENCES

1. Reynolds & Gilmore, 1993, Astronomical J., **106**, 272
2. Koyama et al. 1995, Nature **378**, 255; Allen et al. 1997, ApJ, **487**, L97; Koyama et al. 1997 PASJ, **49**, L7; Keohane et al. 1997 ApJ, **484,** 356; Keohane PhD thesis; Tanimori et al. 1998, Ap.J, **497**, L25
3. Baring et al. 1999, ApJ, **513**, 311
4. Blandford, & Ostriker, 1980, ApJ, **237**, 793; Axford, Ann. NY Acad. Sci., **375**, Acceleration of Cosmic Rays by Shock Waves, ed. Ramaty & Jones, 297
5. Higdon, Lingenfelter & Ramaty, 1998, ApJ **509**, L33
6. Esposito et al. 1996, ApJ, **461**, 820
7. Muraishi et al. 2000, A&A, **354**, L57
8. Brazier et al. 1996, MNRAS, **281**, 1033
9. Gaisser, Protheroe & Stanev, 1998, ApJ, **492**, 219
10. Hunter et al. 1997, ApJ, **481**, 205
11. Hunter et al. 1997, ApJ, **467**, L33
12. Mori, 1997, ApJ **478**, 225
13. Porter & Protheroe, 1997, J. Phys. G: Nucl. Part. Phys. **23**, 1765; Pohl & Esposito 1998, ApJ, **507**, 327
14. Moskalenko, Strong & Reimer, 1998, **338**, L75; Strong, Moskalenko & Reimer, 2000 ApJ 537, in press (astro-ph/9811296); Moskalenko and Strong, 1998, ApJ, **493**, 694
15. Moskalenko & Strong, 2000, ApJ, **528**, 357
16. Dixon et al. 1998, New Astron. **3(7)**, 539
17. Sreekumar et al., ApJ, **494**, 52

What Can GLAST Say About the Origin of Cosmic Rays in Other Galaxies?

Seth W. Digel[a,b], Igor V. Moskalenko[a,c], Jonathan F. Ormes[a],
P. Sreekumar[d], and P. Roger Williamson[e],
on behalf of the GLAST collaboration

[a]*NASA/Goddard Space Flight Center, Greenbelt, Maryland,* [b]*Universities Space Research Association,*
[c]*National Research Council and Institute for Nuclear Physics, Moscow State University,*
[d]*Indian Space Research Organization,* [e]*Stanford University*

Abstract. Gamma rays in the band from 20 MeV to 300 GeV, used in combination with data from radio and X-ray bands, provide a powerful tool for studying the origin of cosmic rays in our sister galaxies Andromeda and the Magellanic Clouds. Gamma-ray Large Area Space Telescope (GLAST) will spatially resolve these galaxies and measure the spectrum and intensity of diffuse gamma radiation from the collisions of cosmic rays with gas and dust in them. Observations of Andromeda will give an external perspective on a spiral galaxy like the Milky Way. Observations of the Magellanic Clouds will permit a study of cosmic rays in dwarf irregular galaxies, where the confinement is certainly different and the massive star formation rate is much greater.

INTRODUCTION

High-energy gamma rays are produced in interactions of high-energy cosmic rays with interstellar matter and photons. From the resulting diffuse emission of gamma rays, the properties of the cosmic rays can be inferred (e.g., (5)). Gamma rays have proven to be a useful probe of cosmic rays in the Milky Way, but gamma-ray telescopes to date have lacked the sensitivity and angular resolution to permit the same kind of detailed study of cosmic rays in external galaxies.

The Gamma-ray Large Area Space Telescope (GLAST) is the next generation high-energy (20 MeV–300 GeV) gamma-ray astronomy mission. It is part of the strategic plan of NASA's Office of Space Science and is currently planned for launch in 2005. GLAST will have a factor of 30 greater sensitivity than the Energetic Gamma-Ray Experiment Telescope (EGRET), launched in 1991 on the Compton Gamma-Ray Observatory. Derived performance parameters for our proposed design for the GLAST instrument, which was selected by NASA in February 2000, are presented in Table 1 and Figure 1. See the companion paper by Ormes *et al.* for information about the design and instrumental response of GLAST, and the Web site http://glast.gsfc.nasa.gov for information about the mission.

Table 1. Selected Parameters for GLAST and EGRET

	EGRET	GLAST
Energy Range	0.02–30 GeV	0.02–300 GeV
Field of View	0.5 sr	2.4 sr
Peak Eff. Area	1500 cm^2	13,000 cm^2
Point Source Sensitivity*	5	0.16
Source Location†	$5' - 90'$	$0.2' - 1'$
Mission Life		5 years (10-year goal)

* Sensitivity at high latitude after a 2-year survey for a 5-σ detection, units 10^{-8} cm^{-2} s^{-1}, for $E > 100$ MeV.
† Diameter of 95% confidence region; range: bright sources to sources of flux 10^{-8} cm^{-2} s^{-1} ($E > 100$ MeV).

ADVANCES WITH GLAST

The only external galaxy that EGRET detected in the light of its interstellar gamma-ray emission was the Large Magellanic Cloud (LMC), which was not spatially resolved (10). GLAST will be able to map the diffuse gamma-ray emision of the LMC, as well as the fainter Small Magellanic Cloud (SMC) and Andromeda (M31) galaxies.

CP528, *Acceleration and Transport of Energetic Particles Observed in the Heliosphere: ACE 2000 Symposium,* edited by Richard A. Mewaldt, et al.
© 2000 American Institute of Physics 1-56396-951-3/00/$17.00

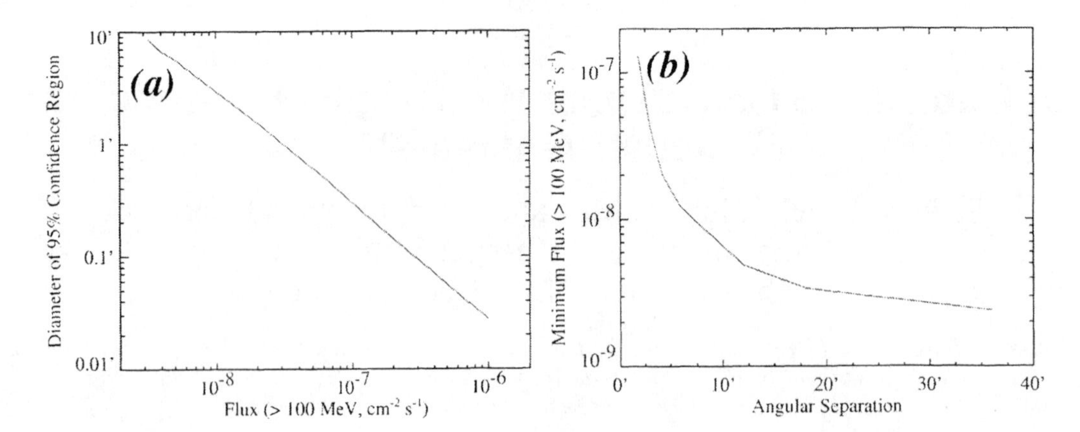

FIGURE 1. Expected performance of GLAST for localizing and resolving point sources. (*a*) Source localization at high latitudes. The position uncertainties for the brightest sources likely will be limited to 10–20″ by uncertainty in spacecraft pointing and instrument alignment. (*b*) Minimum flux required to resolve two closely-spaced sources of equal flux. For both figures, the sources are assumed to have E^{-2} photon spectra and to be observed at high latitudes in a one-year sky survey.

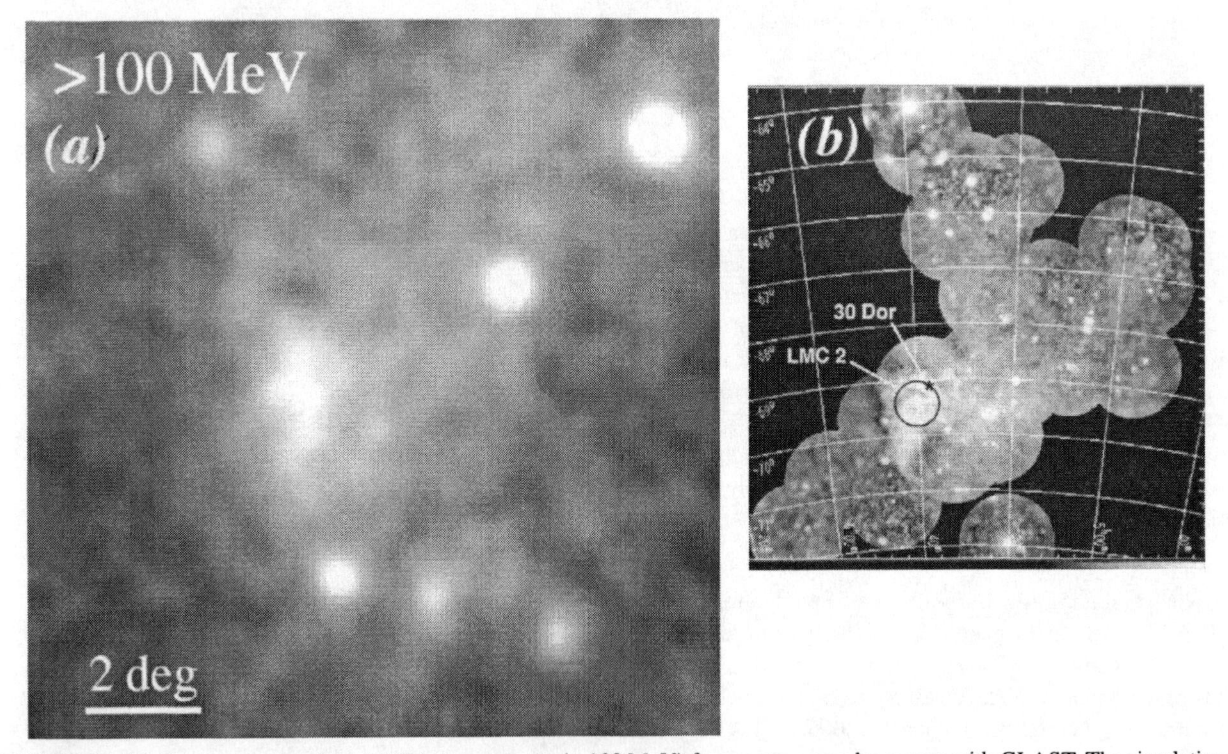

FIGURE 2. (*a*) Simulated map of the LMC in gamma rays (> 100 MeV) from a two-year sky survey with GLAST. The simulation is based on a model of the LMC by Sreekumar (12) and also includes foreground diffuse emission from the Milky Way and an isotropic background consisting of a distribution of faint point sources. (*b*) The LMC in 3/4-keV X-rays, from a mosaic of pointed observations with ROSAT (8). The intense emission regions of 30 Doradus and LMC superbubble 2 are indicated.

LMC

The LMC will be well-resolved by GLAST (Fig. 2). The cosmic-ray distribution can be studied in detail by analyzing the gamma-ray data together with 21-cm H I and 2.6-mm CO surveys of the interstellar medium of the galaxy (see, e.g., (2)). GLAST data should reveal the degree of enhancement of cosmic-ray density in the vicinity of the massive star-forming region 30 Dor and the associated superbubble LMC 2 (6); see the X-ray image in Fig. 2. The LMC 2 superbubble is among the largest of several prominent superbubbles in the LMC,

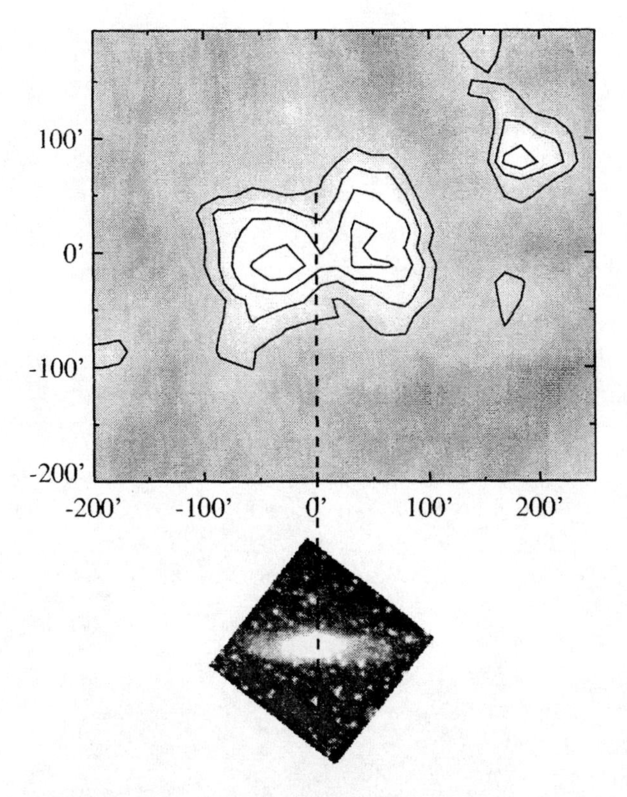

FIGURE 3. Simulated map of M31 from a five-year sky survey with GLAST (approximately equivalent to 6 months of observations with M31 within 30° of the center of GLAST's field of view). The image shows gamma-rays with energies >1 GeV, and has been smoothed to reduce statistical fluctuations. The simulated point source in the upper right indicates the angular resolution of the image, and the inset shows the location and extent of the optical disk of the galaxy. The diffuse emission was modelled based on the distribution of gas in M31, which extends much further than the optical disk, and the EGRET upper limit for the galaxy (1). Contours are spaced by 2×10^{-7} cm^{-2} s^{-1} sr^{-1} from 2.2×10^{-6} cm^{-2} s^{-2} sr^{-1}.

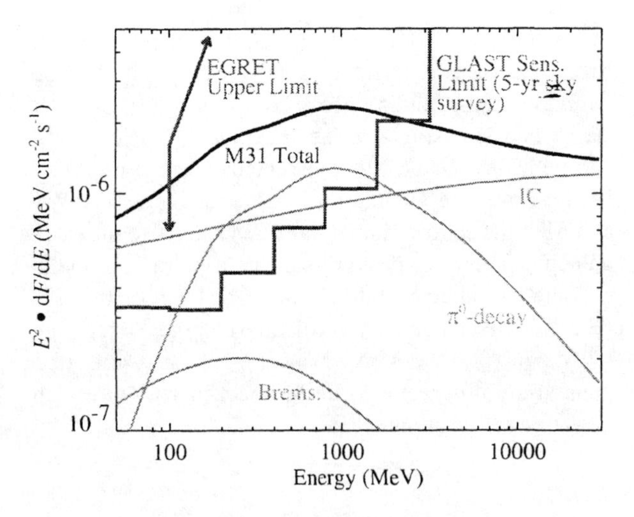

FIGURE 4. Simulated spectrum of M31, obtained by scaling the luminosity spectrum of the whole Milky Way derived by Strong *et al.* (13) (4-kpc halo model) to match the upper limit flux of Blom *et al.* (1) for M31. The differential flux sensitivity of GLAST for a five-year sky survey and the EGRET upper limit of Blom *et al.* are also shown, along with the individual components of the overall spectrum: inverse Compton, Bremsstrahlung, and π^0-decay.

most of which represent regions like 30 Dor that are further evolved. If superbubbles are sources of cosmic rays distinct from individual supernovas (e.g., (4)), superbubble 2, which subtends 1°, may be marginally spatially resolved by GLAST. The extent to which the diffuse emission of the LMC can be attributed to unresolved gamma-ray pulsars has been considered by Hartmann *et al.* (3) and Zhang & Cheng (14). The expectation is that the pulsar contribution could be significant, up to 35%, for energies > 1 GeV. GLAST is unlikely to detect individual pulsars in the LMC, but may be able to address the question of the pulsar fraction with a sensitive measurement of the high-energy spectrum.

SMC

Detection of the diffuse gamma-ray flux of the SMC by GLAST will be useful to verify conclusions about the galactic origin of cosmic rays (e.g., Sreekumar *et al.* (11) based on EGRET data). The non-detection of the SMC, with a 2-σ upper limit of 5×10^{-8} cm^{-2} s^{-1} (>100 MeV) led Sreekumar *et al.* to conclude that the most likely model for the distribution of cosmic rays in the SMC is one for which the galaxy is disintegrating and cosmic rays are only very poorly confined. In this circumstance, the predicted flux is $(2 \pm 3) \times 10^{-8}$ cm^{-2} s^{-1} (>100 MeV) (9), well within the reach of GLAST.

M31

The EGRET 2-σ upper limit for the gamma-ray flux of M31 is 1.6×10^{-8} cm^{-2} s^{-1} (>100 MeV), which is much less than the flux of the Milky Way at M31 (1). The cosmic-ray densities in M31 are certainly lower than in the Milky Way, and it has less ongoing massive star formation. At a flux level of 1×10^{-8} cm^{-2} s^{-1} (>100 MeV), GLAST will resolve the diffuse gamma-ray emission along the major axis of M31, to provide information about the relationship between cosmic rays, star formation rate, and interstellar gas on a large scale (Fig. 3). GLAST may also measure the distribution of cosmic rays in the halo of M31. Spectral measurements may allow a global assessment of inverse-Compton, electron-Bremsstrahlung, and π^0 decay contributions to the interstellar emission (Fig. 4). The unexplained "GeV excess" for the Milky Way (5) will also be detected if present in M31. From the gamma-ray spectra of the Milky Way and M31 the contribution of normal galaxies to the extragalactic gamma-ray background can begin to be assessed.

Detailed studies of M31 that will be possible with GLAST will benefit from further development of cosmic-ray models for the Milky Way, for which results from gamma-ray observations can be checked with direct observations of cosmic rays.

CONCLUSIONS

For the first time, GLAST will enable spatial and spectral studies of diffuse gamma rays from external galaxies. Diffuse, high-energy gamma rays are diagnostic of cosmic-ray densities, which especially for the proton component are difficult to determine from observations at other wavelengths. When considered together with X-ray and radio observations, GLAST data promise a fairly complete understanding of the production, propagation, and confinement of cosmic rays in Local Group galaxies.

IVM acknowledges support from an NAS/NRC Senior Associateship.

REFERENCES

1. Blom, J. J., *et al.*, *Astrophys. J.* **516**, 44, (1999).
2. Cohen, R. S., *et al.*, *Astrophys. J.* **331**, L95, (1988).
3. Hartmann, D. H., Brown, L. E., and Schnepf, N., *Astrophys. J.* **408**, L13, (1993).
4. Higdon, J. C., *et al.*, *Astrophys. J.*, **509**, 33, (1998).
5. Hunter, S. D., *et al.*, *Astrophys. J.* **481**, 205, (1997).
6. Meaburn, J., *MNRAS* **192**, 365, (1980).
7. Ormes, J. F., *et al.*, these proceedings, (2000).
8. Snowden, S. L., and Petre, R., *Astrophys. J.* **436**, L123 (1994).
9. Sreekumar, P., and Fichtel, C. E., *Astron. and Astrophys.* **251**, 447 , (1991).
10. Sreekumar, P., *et al.*, *Astrophys. J.* **400**, L67, (1992).
11. Sreekumar, P., *et al.*, *Phys. Rev. Lett.* **70**, 127, (1993).
12. Sreekumar, P., priv. comm., 1999.
13. Strong, A. W., Moskalenko, I., and Reimer, O., *Astrophys. J.*, **537**, in press (2000), astro-ph/9811296.
14. Zhang, L., and Cheng, K. S., *MNRAS*, **294**, 729, (1998).

Other Contributions to ACE-2000

The following papers were presented at the ACE-2000 Symposium but were not submitted for publication in these proceedings.

Solar Energetic Particles

Particle Acceleration in Impulsive Solar Flares
J. A. Miller
A Simple Model for Ion Acceleration and Transport at an Evolving Coronal/Interplanetary Shock
M. A. Lee
Are Solar Energetic Electron Injection Times Non-Simultaneous in the Corona?
E.C. Roelof et al.
Seed Population for ^{3}He Enhancements in Large Solar Energetic Particle Events
G. M. Mason et al.
Energy Spectra of ^{3}He, ^{4}He and Heavy Ions in Impulsive Solar Energetic Particle Events
J. E. Mazur et al.
The Forecast by On Line Neutron Monitor Data of Great Solar Energetic Particle Events Dangerous for Satellite Electronics and People Health in Spacecraft and Aeroplanes
L. I. Dorman et al.

Interplanetary Acceleration

Isotope Fractionation of Solar Wind Particles
P. Bochsler et al.
Ion Composition and Spectra in Corotating Interaction Regions Observed by ACE/ULEIS
J. R. Dwyer et al.
Energy Spectra and Composition of >30 keV/nucleon ions at Interplanetary Shocks Observed at ACE and WIND
M. I. Desai et al.
Power-law Particle Distributions Generated by Whistler-Mode Turbulence
D. Summers
Energetic Particle Cross-Field Diffusion: Interaction with Magnetic Decreases (MDs)
B. T. Tsurutani et al.
Characteristics of Energetic Ions Observed by the WIND/STEP Instrument Upstream of the Earth's Bow Shock
M. I. Desai et al.

Anomalous and Galactic Cosmic Rays in the Heliosphere

Low-Energy Anomalous Cosmic Rays in the Ecliptic Plane: 1-5 AU
L. J. Lanzerotti et al.
The Onset of Cosmic Ray Modulation (Cycle 23) at 1 AU Coupled with a Transient Low Energy Cosmic Ray Increase in the Outer Heliosphere
F. B. McDonald et al.

ACE-2000 Symposium Participants

A. F. Barghouty
Caltech
MC 220-47
Pasadena, CA 91125
barghouty@srl.caltech.edu

Walter Robert Binns
Washington Univ.
1 Brookings Drive
St. Louis, MO 63130
wrb@howdy.wustl.edu

Paul R. Boberg
Consultant
27338 Brighton Drive
Valencia, CA 91354
pboberg@earthlink.net

Peter Bochsler
Physikalisches Institut Univ. of Bern
Sidlerstrasse 5
CH-3012 Bern
Switzerland
peter.bochsler@phim.unibe.ch

Eric Christian
NASA/GSFC/USRA
Code 661
Greenbelt, MD 20771
erc@cosmicra.gsfc.nasa.gov

Edward W. Cliver
Air Force Research Lab
AFRL/VSBS
Hanscom AFB, MA 01731-3010
cliver@plh.af.mil

Christina M. S. Cohen
Caltech
MC 220-47
Pasadena, CA 91125
cohen@srl.caltech.edu

Alan C. Cummings
Caltech
MC 220-47
Pasadena, CA 91125
ace@srl.caltech.edu

Andrew J. Davis
ACE Science Center at Caltech
MC 220-22
Pasadena, CA 91125
ad@srl.caltech.edu

Georgia Adair de Nolfo
Caltech
MC 220-47
Pasadena, CA 91125
georgia@srl.caltech.edu

Robert B. Decker
JHU/Applied Physics Lab
11100 Johns Hopkins Rd
Laurel, MD 20723
robert.decker@jhuapl.edu

Mihir Desai
Univ. of Maryland
Dept. of Physics
College Park, MD 20742
desai@uleis.umd.edu

Lev Dorman
Cosmic Ray Ctr of Tel Aviv Univ.
Head of Israel Cosmic Ray Center
Qazrin, P.O.B. 2217, 12900
Israel
lid@physics.technion.ac.il,

Joseph Richard Dwyer
Univ. of Maryland
Dept. of Physics
College Park, MD 20742
dwyer@uleis.umd.edu

Donald C. Ellison
North Carolina State Univ.
Physics Dept.
Raleigh, NC 27695-8202
don_ellison@ncsu.edu

Abraham Falcone
Univ. of New Hampshire
Morse Hall
Durham, NH 03824
afalcone@comptel.sr.unh.edu

Joan Feynman
Jet Propulsion Lab
4800 Oak Grove Dr.
Pasadena, CA 91109
joan.feynman@jpl.nasa.gov

Horst Fichtner
Ruhr-Universitaet Bochum
Institut fuer Theoretische Physik IV:
Weltraum- und Astrophysik
Bochum, 44780
Germany
hf@tp4.ruhr-uni-bochum.de

L. A. Fisk
Univ. of Michigan
Space Research Building
Ann Arbor, MI 48109
lafisk@umich.edu

Miriam A. Forman
SUNY/Stony Brook
Dept. of Physics and Astronomy
Z=3800
Stony Brook, NY 11794-3800
miriam.forman@sunysb.edu

Theodore J. Freeman
Space Sciences Lab
UC Berkeley
Berkeley, CA 94720
freeman@ssl.berkeley.edu

Antoinette B. Galvin
Univ. of New Hampshire
EOS SSC Morse Hall Rm 318
Durham, NH 03824
Toni.Galvin@unh.edu

Jeffrey S. George
Caltech
MC 220-47
Pasadena, CA 91125
george@srl.caltech.edu

Joe Giacalone
Univ. of Arizona
Lunar and Planetary
Tucson, AZ 85721
giacalon@lpl.arizona.edu

George Gloeckler
Univ. of Maryland
Department of Physics
College Park, MD 20742
gg10@umail.umd.edu

Robert E. Gold
JHU/APL
11100 Johns Hopkins Rd
Bldg. 23, Room 306
Laurel, MD 20723
robert.gold@jhuapl.edu

John T. Gosling
Los Alamos National Lab
MS D466
Los Alamos, NM 87545
jgosling@lanl.gov

Tohru Hada
ESST, Kyushu Univ.
6-1 Kasuga Koen, Kasuga 816-8580
Japan
hada@esst.kyushu-u.ac.jp

Dennis Haggerty
Applied Physics Lab
11100 Johns Hopkins Rd
Laurel, MD 20723
dennis.haggerty@jhuapl.edu

S. Edward Hawkins III
JHU/APL
11100 Johns Hopkins Road
MS 4-234
Laurel, MD 20707
ed.hawkins@jhuapl.edu

Bernd Heber
Max-Planck-Institut fuer Aeronomie
Max-Planck-Str. 2
Katlenburg-Lindau, 37191
Germany
bheber@linmpi.mpg.de

James C. Higdon
W. M. Keck Center
925 North Mills Ave.
Claremont, CA 91711
jimh@lobach.claremont.edu

Paul L. Hink
Washington Univ.
Department of Physics - 1105
St. Louis, MO 63130
hink@howdy.wustl.edu

George C. Ho
JHU/Applied Physics Lab
11100 Johns Hopkins Rd
Laurel, MD 20723
george.ho@jhuapl.edu

Ke Chiang Hsieh
Univ. of Arizona
Dept. of Physics
Tucson, AZ 85721
hsieh@physics.arizona.edu

Devrie Intriligator
Carmel Research Center
P.O. Box 1732
Santa Monica, CA 90406
devriei@aol.com

Martin H. Israel
Washington Univ.
1 Brookings Drive
Campus Box 1105
St. Louis, MO 63130
mhi@wuphys.wustl.edu

J. R. Jokipii
Univ. of Arizona
Dept. of Planetary Sciences
Tucson, AZ 85721
jokipii@lpl.arizona.edu

Frank C. Jones
NASA/Goddard Space Flight Center
Code 660
Greenbelt, MD 20771
frank.c.jones@gsfc.nasa.gov

Erhard Keppler
MPAE LINDAU
Max Planck Institut f.
Aeronomie Lindau
Katlenburg-Lindau, 37191
Germany
keppler@linmpi.mpg.de

Berndt Klecker
Max-Planck-Institut fuer
extraterrestrische Physik
Giessenbachstr.
Garching, 85740
Germany
bek@mpe.mpg.de

Stamatios M. Krimigis
Applied Physics Lab
11100 Johns Hopkins Road
Laurel, MD 20723
tom.krimigis@jhuapl.edu

Sam Krucker
Space Sciences Lab
UC Berkeley
Berkeley, CA 94720-7450
krucker@ssl.berkeley.edu

Jakobus A. le Roux
Bartol Research Institute
University of Delaware
Newark, DE 19716
leroux@bartol.udel.edu

Martin A. Lee
Univ. New Hampshire
Space Science Center / Morse Hall
Durham, NH 03824
marty.lee@unh.edu

Craig Leff
Jet Propulsion Lab
4800 Oak Grove Drive
Mail Stop 169-315
Pasadena, CA 91109
cleff@jpl.nasa.gov

Richard A. Leske
Caltech
MC 220-47
Pasadena, CA 91125
ral@srl.caltech.edu

Paulett C. Liewer
JPL/Caltech
4800 Oak Grove Drive
Mail Stop 169-506
Pasadena, CA 91109
paulett.liewer@jpl.nasa.gov

Michal Lijowski
Washington Univ.
1 Brookings Drive
Campus Box 1105
St. Louis, MO 63130
lijowski@cosray2.wustl.edu

Robert P. Lin
Space Sciences Lab
UC Berkeley
Berkeley, CA 94720
rlin@ssl.berkeley.edu

James C. Ling
NASA Headquarters
Code SR, NASA Headquarters
300 E. Street S.W.
Washington, DC 20546
jling@hq.nasa.gov

Richard E. Lingenfelter
Univ. of Calilfornia San Diego
CASS 0424, 9500 Gilman Drive
La Jolla, CA 92093-0424
rlingenfelter@ucsd.edu

Yuri Litvinenko
Univ. of New Hampshire
Space Science Center / Morse Hall
Durham, NH 03820
yuri.litvinenko@unh.edu

Andrew J. Lukasiak
Univ. of Maryland
Inst. for Physical Science and
Technology
College Park, MD 20742
al36@umail.umd.edu

Carol G. Maclennan
Bell Labs, Lucent Technologies
Rm 1E436, Bell Laboratories
600 Mountain Ave
Murray Hill, NJ 07974
cgm@bell-labs.com

Dalmiro Jorge Filipe Maia
Observatoire Paris and Lisbon
DASOP, 5 Place Jules Janssen
Meudon, 92190
France
dalmiro@milkyway.oal.ul.pt

Glenn M. Mason
Univ. of Maryland
Department of Physics
College Park, MD 20742
Glenn.Mason@umail.umd.edu

Joseph E. Mazur
The Aerospace Corporation
M2/259 2350 E. El Segundo Blvd.
El Segundo , CA 90245
joseph.mazur@aero.org

Frank B. McDonald
Univ. of Maryland
I.P.S.T., Rm. 3245 CSS Building
College Park, MD 20705-2431
fm27@umail.umd.edu

Richard Mewaldt
Caltech
MC 220-47
Pasadena, CA 91125
rmewaldt@srl.caltect.edu

Karim Meziane
Space Sciences Lab
UC Berkeley
Berkeley, CA 94720
karim@ssl.berkeley.edu

James A. Miller
Univ. of Alabama in Huntsville
Department of Physics
Huntsville, AL 35899
millerj@cspar.uah.edu

Eberhard Moebius
Univ. Of New Hampshire
39 College Road
Durham, NH 03824
eberhard.moebius@unh.edu

Daniel J. Morris
Univ. of New Hampshire
Space Science Center / Morse Hall
Durham, NH 03824-3525
dmorris@comptel.sr.unh.edu

Susan Mahan Niebur
Washington Univ.
1 Brookings Drive
Campus Box 1105
St. Louis, MO 63130
smahan@artsci.wustl.edu

Jonathan F. Ormes
NASA/Goddard Space Flight Center
Code 660
Greenbelt, MD 20771
ormes@lheamail.gsfc.nasa.gov

Mark Popecki
Univ. of New Hampshire
Morse Hall 39 College Rd.
Durham, NH 03824
mark.popecki@unh.edu

Vladimir Ptuskin
IZMIRAN
Institute of Terrestrial Magnetism
Troitsk, Moscow region, 142092
Russia
vptuskin@izmiran.troitsk.ru

Reuven Ramaty
NASA/Goddard Space Flight Center
Code 661
Greenbelt, MD 20771
ramaty@gsfc.nasa.gov

Donald V. Reames
NASA/Goddard Space Flight Center
Code 661
Greenbelt, MD 20771
reames@lheavx.gsfc.nasa.gov

Edmond C. Roelof
JHU/Applied Physics Lab
11100 Johns Hopkins Road
Laurel, MD 20723-6099
edmond.roelof@jhuapl.edu

Ilan Roth
Space Sciences Lab
UC Berkeley
Berkeley, CA 94720
ilan@ssl.berkeley.edu

Alexander Ruzmaikin
Jet Propulsion Lab
4800 Oak Grove Drive
Mail Stop 169-506
Pasadena, CA 91109
alexander.ruzmaikin@jpl.nasa.gov

James Micheal Ryan
Univ. of New Hampshire
Space Science Center / Morse Hall
Durham, NH 03824-3525
james.ryan@unh.edu

Manfred Scholer
Max-Planck-Inst. extrat. Physik
P.O. Box 1603
Garching, 85740
Germany
mbs@mpe.mpg.de

Steve Sears
ACE Science Center at Caltech
MC 220-22
Pasadena, CA 91125
steves@srl.caltech.edu

Gerald Share
Naval Research Lab
Code 7652
Washington, DC 20375
share@gamma.nrl.navy.mil

Penny L. Slocum
Jet Propulsion Lab
4800 Oak Grove Drive
Mail Stop 169-327
Pasadena, CA 91109
penny@heag1.jpl.nasa.gov

Luke Sollitt
Caltech
MC 220-47
Pasadena, CA 91125
sollitt@srl.caltech.edu

Conrad Steenberg
Caltech
MC 220-47
Pasadena, CA 91125
conrad@srl.catech.edu

Edward C. Stone
Jet Propulsion Lab
4800 Oak Grove Drive
Mail Stop 180-904
Pasadena, CA 91109
edward.c.stone@jpl.nasa.gov

Danny Summers
Memorial Univ. of Newfoundland
Dept. of Mathematics and Statistics
St. John's, A1C5S7
Canada
dsummers@math.mun.ca

Simon Swordy
Univ. of Chicago
LASR 933 E. 56th St.
Chicago, IL 60304
swordy@odysseus.uchicago.edu

Bruce T. Tsurutani
Jet Propulsion Lab 169-5
4800 Oak Grove Drive
Mail Stop 169-506
Pasadena, CA 91109
btsurutani@jplsp.jpl.nasa.gov

Allan J. Tylka
Naval Research Lab
Code 7654
4555 Overlook Ave, SW
Washington, DC 20375
tylka@gamma.nrl.navy.mil

Tycho von Rosenvinge
NASA/Goddard Space Flight Center
Code 661
Greenbelt, MD 20771
tycho@lheamail.gsfc.nasa.gov

C. J. Waddington
Univ. of Minnesota
116 Church St. S. E.
Minneapolis, MN 55455
waddington@physics.spa.umn.edu

William R. Webber
New Mexico St. Univ.
Dept of Astronomy MSC 4500
Las Cruces, NM 88003
bwebber@nmsu.edu

Mark E. Wiedenbeck
Jet Propulsion lab
4800 Oak Grove Drive
Mail Stop 169-327
Pasadena, CA 91109
Mark.E.Wiedenbeck@jpl.nasa.gov

Roger Williamson
Stanford Univ.
HEPL MC 4085
Stanford, CA 94305
Roger.Williamson@Stanford.edu

Nathan E Yanasak
Jet Propulsion Lab
4800 Oak Grove Drive
Mail Stop 169-327
Pasadena, CA 91109
yanasak@heag1.jpl.nasa.gov

Masato Yoshimori
Rikkyo Univ.
3-34-1 Nishi-Ikebukuro, Toshima-ku
Tokyo, 171-8501
Japan
yosimori@rikkyo.ne.jp

Gary P. Zank
Bartol Research Institute
University of Delaware
Newark, DE 19716
zank@bartol.udel.edu

Liwei Dennis Zhang
Jet Propulsion Lab
4800 Oak Grove Drive
Mail Stop 169-506
Pasadena, CA 91109
lzhang@jplsp2.jpl.nasa.gov

Thomas Zurbuchen
Univ. of Michigan
Space Research Building
Ann Arbor, MI 48109
thomasz@umich.edu

Ron Zwickl
NOAA/Space Environment Center
325 Broadway
Boulder, CO 80303
rzwickl@sec.noaa.gov

Author Index

A

B

C

D

E

F

G

H

I

J

K

L

M

N

O

P

R

S

SUBJECT INDEX

D

E

F

G

H

I

K

T

X

Related Titles from the AIP Conference Proceedings Subseries on Astronomy and Astrophysics

526 Gamma-Ray Bursts: 5th Huntsville Symposium
Edited by R. Marc Kippen, Robert S. Mallozzi, and Gerald J. Fishman, June 2000, 1-56396-947-5

523 Gravitational Waves: Third Edoardo Amaldi Conference
Edited by Sydney Meshkov, June 2000, 1-56396-944-0

522 Cosmic Explosions: Tenth Astrophysics Conference
Edited by Stephen S. Holt and William W. Zhang, June 2000, 1-56396-943-2

516 26th International Cosmic Ray Conference: ICRC XXVI, Invited, Rapporteur, and Highlight Papers
Edited by Brenda L. Dingus, David B. Kieda, and Michael H. Salamon
May 2000, 1-56396-939-4

515 GeV-TeV Gamma Ray Astrophysics Workshop: Towards a Major Atmospheric Cherenkov Detector VI
Edited by Brenda L. Dingus, Michael H. Salamon, and David B. Kieda,
May 2000, 1-56396-938-6

510 The Fifth Compton Symposium
Edited by Mark L. McConnell and James M. Ryan, March 2000, 1-56396-932-7

499 Small Missions for Energetic Astrophysics: Ultraviolet to Gamma-Ray
Edited by Steven P. Brumby, December 1999, 1-56396-912-2

471 Solar Wind Nine: Proceedings of the Ninth International Solar Wind Conference
Edited by Shadia Rifai Habbal, Ruth Esser, Joseph V. Hollweg, Philip A. Isenberg,
May 1999, 1-56396-865-7

433 Workshop on Observing Giant Cosmic Ray Air Showers from $>10^{20}$ eV Particles from Space
Edited by John F. Krizmanic, Jonathan F. Ormes, and Robert E. Streitmatter,
June 1998, 1-56396-788-X

To learn more about these titles, or the AIP Conference Proceedings Series, please visit the webpage **http://www.aip.org/catalog/aboutconf.html**